D1226329

Wastewater Engineering

collection treatment disposal

McGraw-Hill Series in Water Resources and Environmental Engineering

Ven T. Chow, Rolf Eliassen, and Ray K. Linsley   Consulting Editors

Graf   *Hydraulics of Sediment Transport*

Hall and Dracup   *Water Resources Systems Engineering*

James and Lee   *Economics of Water Resources Planning*

Linsley and Franzini   *Water-Resources Engineering*

Metcalf & Eddy, Inc.   *Wastewater Engineering: Collection, Treatment, Disposal*

Walton   *Groundwater Resource Evaluation*

Wiener   *The Role of Water in Development: An Analysis of Principles of Comprehensive Planning*

Wastewater Engineering

collection treatment disposal

Metcalf & Eddy, Inc.

McGraw-Hill Book Company

New York San Francisco St. Louis Düsseldorf Johannesburg Kuala Lumpur London
Mexico Montreal New Delhi Panama Rio de Janeiro Singapore Sydney Toronto

This book was set in Modern 8A, printed, and bound by The Maple Press Company. The designer was Michael A. Rogondino; the drawings were done by Ayxa Art. The editors were B. J. Clark and Michael A. Ungersma. Charles A. Goehring supervised production.

Wastewater Engineering: Collection, Treatment, Disposal

Printed in the United States of America.

Library of Congress catalog card number: 75-172262

7890 MAMM 798765

07-041675-3

contents

examples

preface

Following the publication of the three volumes of *American Sewerage Practice* in 1914–1915 by Leonard Metcalf and Harrison P. Eddy, the authors were urged to prepare a single-volume abridgment for class use in engineering schools. Their *Sewerage and Sewage Disposal*, a textbook, was published in 1922, with a second edition in 1930.

That classic textbook was used by many of the students who are now leaders of the environmental engineering profession. In the past four decades there have been many developments in wastewater engineering and the basic principles underlying the unit operations and processes are now more clearly understood. Consequently, there has been a great demand from the academic profession and from practicing engineers for a revision of the book. At long last we have taken time to bring out this new volume.

The objective of this book is to bring together a wide body of knowledge from the rapidly changing and expanding field of wastewater engineering and to present it in a format that will make the book useful as a text for students and as a reference work for practicing engineers. The contents have been prepared and coordinated by university professors and by engineers engaged in the design, construction, and operation of wastewater treatment facilities.

The material is arranged in the logical sequence of collection, treatment, and disposal. The chapters on collection systems present those fundamentals of fluid mechanics and hydraulic engineering that the engineer needs to conceive and design wastewater collection and pumping facilities. Emphasis has been placed on the estimation of flowrates from municipalities and industries. No attempt has been made to cover hydrology or construction materials and methods, as information on these subjects is available in other excellent texts.

The extensive section on treatment of wastewaters has a twofold approach. First, a comprehensive presentation is made of the physical, chemical, and biological principles that the student and the practicing engineer must comprehend before tackling complex engineering problems. Then these principles are applied to the design of wastewater treatment facilities. The important subject of disposal is covered in two chapters. One deals with the processing and disposal of the materials removed by treatment, and the other deals with the disposal of the wastewater. Because of the problems encountered in the treatment of industrial wastes, the last chapter is devoted to the conduct of wastewater studies.

The principles and practices are presented in such a manner that the student will be able to apply the fundamentals to the solution of problems. This text is designed to be useful to teachers and students in undergraduate as well as graduate courses. An undergraduate student would concentrate on the engineering aspects in the first six chapters and would obtain an overview of the principles of treatment and disposal of wastewater. A graduate student in environmental engineering and sanitary science would master Chaps. 7 through 16, which present the theoretical concepts and practical solutions to the design and operation of wastewater treatment and disposal facilities. Sample course outlines are presented in Apps. F and G as a guide to instructors. This book permits a flexible approach to undergraduate and graduate courses based on the individual interests of the instructors and the relationship of these courses to the entire curriculum at any particular institution. Numerous examples are included to give students a better working knowledge of the material presented. To test the students' understanding further, problems are included at the end of each chapter.

A practicing engineer will find that this book can serve as the basis for a refresher course for those subjects he might have studied long ago in Metcalf and Eddy's *Sewerage and Sewage Disposal* and in other books. He will also find explanations of some of the newer principles and their application to the concept and design of wastewater treatment systems. The text is cross-referenced and has a detailed index so that subjects of interest can be located readily. In addition, key references are given on each subject so that further study can be carried out. There are numerous tables and summaries of data that will be useful for the student and engineer in seeking limits of applicability of certain unit operations and processes.

The services of many staff members and consultants of Metcalf & Eddy, Inc. are acknowledged as this text is truly a composite effort. Harrison P. Eddy, Jr., as President and more recently as Chairman of the Board, was the motivating force that led to the preparation of this new text. It was his desire to produce a worthy successor to the previous Metcalf and Eddy textbook that served many engineers and students so well and for so long. He made this possible by providing for personnel of the Palo Alto and Boston offices of the firm to contribute their time and effort. I was the Partner-in-Charge of the project, and was responsible for developing the format and scope of the book and for supervising the staff in carrying out the work.

The principal writer and coordinator of this textbook was George Tchobanoglous. He was assisted by Allen J. Burdoin and John W. Raymond, Jr., of the Boston office and Olivia L. Chen, Roger T. Haug, Robert G. Smith, Vernon L. Snoeyink, and James F. Stahl of the Palo Alto office. Ronald W. Crites was responsible for the problem sets and editorial and production assistance. Marcella S. Tennant served as the general technical editor.

Special acknowledgment is made of the contributions of Perry L. McCarty of Stanford University and Clair N. Sawyer, Vice President and Director of Research of Metcalf & Eddy, Inc. and former Professor of Sanitary Science at the Massachusetts Institute of Technology, whose published articles and class materials were used in the preparation of several chapters.

Rolf Eliassen
Senior Vice President
Metcalf & Eddy, Inc.

developments and trends in wastewater engineering

1

Currently the field of wastewater engineering is in a dynamic period of development. Old ideas are being reevaluated and new concepts are being formulated. To play an active role in the development of this field, the engineer must clearly understand the fundamentals on which it is based. Therefore, the purpose of this book is to delineate the fundamental engineering principles involved in the collection, treatment, and disposal of wastewater and to illustrate their application in design. As a general introduction to these subjects, some historical background and some of the recent developments and trends in wastewater engineering are briefly reviewed in this chapter.

COLLECTION

The development of sanitary water supplies and the collection of domestic wastewater are two of the most important factors responsible for the general level of good health enjoyed by the population of the United States. Typically, the planning and design of wastewater collection facilities involves the determination of wastewater flowrates; the hydraulic design of sewers, large conduits, and junction and diversion structures; the selection of appropriate sewer appurtenances; and the design of pump stations. These topics are discussed in Chaps. 2 through 6.

Background

Of the many early sewers that have been described in the literature, most is known about the great underground drains of ancient Rome [1]. On the basis of early writings, it is known that direct connections from the houses to these channels and conduits were not used to any substantial extent because the requirements of public health were little recognized and compulsory sanitation would have been considered an invasion of the rights of the individual. Following Roman practice, early sewers, both here and abroad, were constructed originally for the removal of storm water. All human excreta were excluded from the sewers of London until 1815, from those of Boston until 1833, and from those of Paris until 1880 [3].

It is astonishing to note that, although sewers have been built since the days of the Roman empire, there was little if any progress in the design and construction of wastewater collection facilities until the 1840s. The renaissance began in Hamburg, Germany, in 1842 as a result of a severe conflagration that destroyed part of the city. For the first time, a complete new system was designed according to the modern theories of the day concerning the conveyance of wastewater, taking into account topographic conditions and recognized community needs [3]. This was a spectacular advance when it is considered that the fundamental principles on which the design was based are in use today but had not been widely used before that time.

London Sewers In London as late as 1845, there was no survey of the metropolis adequate as a basis for the planning of sewerage systems. The sewers in adjoining parishes were at different elevations so that junctions with them were impracticable. Some of the sewers were higher than the cesspools that they were supposed to drain, while others had been so constructed that for them to be of any use the wastewater would have had to flow uphill. Large sewers were made to discharge into small sewers.

Following the great epidemic of Asiatic cholera in 1832, cholera again erupted in London in 1848 and claimed over 25,000 victims during the next 6 years. Although the connection between a contaminated water supply and the rapid spread of disease was clearly shown, the filthy living conditions in most houses, due to the absence of domestic sewers, were a great hindrance in combatting the epidemic.

It was not until 1855 that Parliament provided for the Metropolitan Board of Works, which soon after undertook the development of an adequate wastewater collection system.

American Sewers Little is known about the early wastewater collection works in the United States. Often they were constructed by individuals or the inhabitants of small districts, at their own expense and with little or no public supervision. There was a tendency in this country, as elsewhere, to construct the early sewers of needlessly large dimensions. One of the oldest sewers in Brooklyn, which drained less than 20 acres and was on a grade of 1 in 36, was 4 ft high and 5 ft wide. In some cases, the sewers were very large not only at their outlets but also all the way to their heads. It was impossible to obtain adequate velocity in such sewers unless they were laid on steep grades, and consequently some of them became offensive when the accumulated sewage solids underwent decomposition. There were even some instances in which the slopes were laid in the wrong direction.

Although, as noted previously, the fundamental principles governing the flow of wastewater were known from the early 1840s [8], their application to the design of sewers has been evolutionary rather than marked by clear, progressive steps. Many of the same equations are used today, but their fundamental basis and limits of applicability are now better understood.

Recent Developments and Trends

The applications of developments in other fields are responsible for many recent changes in wastewater collection. Three important ones are the application of photogrammetric and computer techniques to the design of both sanitary and combined sewers, the improvement of construction materials, and the application of computers in the control of storm sewers. The problem of whether or not combined or sanitary sewers should be used remains to be resolved in the future.

Design In the past, one of the most time-consuming aspects of sewer design has been the preparation of maps and surface profiles. Today, most of the tedium involved in their preparation has been eliminated through the use of modern photogrammetric techniques.

Several firms and governmental agencies have developed computer programs that can be used to design various parts of wastewater collection systems. In the future, it is anticipated that complete computer design facilities will be developed so that the design of sewer systems can be automated completely.

Materials Recent developments in both the manufacture and use of existing and synthetic materials have had significant applications in wastewater collection. Along with the use of plastics and asbestos-cement for the construction of sewer pipes, the use of protective coatings and epoxy linings is also increasing. The use of joint materials, such as plasticized polyvinyl chloride (PVC), and some of the recently developed synthetic-rubber gasket materials, is ushering in a new era in the construction of sewers.

Operation Increased attention is being given to means of controlling or mitigating the adverse effects caused by the discharge of untreated and unregulated storm water overflows [6]. Coupled to remote rainfall and receiving-water sensing devices, computers operating on a real-time basis will be used in the future to control the various overflow, pumping, and storage facilities so as to minimize the effects of discharging storm water to the environment.

TREATMENT

Ultimately, wastewater collected from cities and towns must be returned to the land or waters of the earth. The complex question of which contaminants in wastewater must be removed to protect the environment—and to what extent—must be answered separately in each case in light of an analysis of local conditions, scientific knowledge, past experience, and engineering judgment.

Background

Although the collection of wastewater dates from ancient times, the treatment of wastewater is a comparatively recent development dating from the late 1800s and early 1900s. Development of the germ theory in the latter half of the nineteenth century by Koch and Pasteur marked the beginning of a new era in sanitation [7, 9]. Before this time, the relationship of pollution to disease had been only faintly understood, and the science of bacteriology, then in its infancy, had not been applied to the subject of wastewater treatment. The early development of wastewater treatment is traced in Table 1·1.

English Practice The subject of wastewater treatment and disposal received only occasional local attention in England until the

TABLE 1·1 HISTORICAL DEVELOPMENTS IN WASTEWATER TREATMENT

Date	Development
B.C.	Irrigation with wastewater in Athens
1550s	Wastewater farming in Germany
1700	Wastewater farming in England
1762	Chemical precipitation of wastewater in England
1860s	Mouras' unit to treat wastewater solids anaerobically
1865	Early experiments on microbiology of sludge digestion in England
1868	Early experiments on intermittent filtration of wastewater in England
1870	Early experiments on intermittent sand filtration in England
1876	First septic tanks in United States
1882	First experiments on aeration of sewage in England
1884	First bar racks in United States
1887	Lawrence Experiment Station established by the Massachusetts State Board of Health for the study of water and wastewater
1887	First chemical precipitation treatment plant in United States
1889	Filtration in contact beds at the Lawrence Experiment Station, Massachusetts
1891	Sludge digestion in lagoons in Germany
1895	Collection of methane from septic tanks and its use for plant lighting in England
1898	Rotary sprinklers for trickling filters
1904	First grit chambers in United States
1904	Travis two-story septic (hydrolytic) tank in England
1904	Imhoff tank patented in Germany
1906	Chlorination of wastewater for disinfection demonstrated by Phelps in United States
1908	First municipal installation of a trickling filter in United States
1908	Formulation of laws of disinfection by Chick in United States
1911	First Imhoff tanks in United States
1911	Separate sludge digestion in United States
1912–13	Aeration of wastewater in tanks containing slate at Lawrence Experiment Station
1914	Experiments by Ardern and Lockett that led to development of the activated sludge process
1916	First municipal activated-sludge treatment plant built in United States
1925	Contact aerator developed by Buswell in United States

construction of sewerage systems after the cholera epidemics of the mid-1800s. Because of the small size of British streams, their pollution by untreated wastewater discharged into them soon became a nuisance. At first, interference with agricultural and manufacturing uses of water was apparently given more attention than any possible danger to health.

Because of the adverse conditions caused by the discharge of wastewater to the environment and the fact that the amount of land suited for wastewater disposal by irrigation was limited, intensive methods of treatment were developed. The objective was to accelerate the forces of nature under controlled conditions in treatment facilities of comparatively small size. Coupled to sedimentation, chemical precipitation was one of the first processes used for the treatment of wastewater (see Table 1·1).

Activated sludge, the most important biological treatment process in use today, can be traced back to England as early as 1882 when the aeration of sewage in tanks, to hasten oxidation of the organic matter, was investigated. The activated-sludge process as we know it today was developed by Ardern and Lockett in 1914 (see Ref. [1] in Chap. 10).

American Practice The treatment and disposal of wastewater in the United States in the late 1800s did not receive as much attention as in England, because the extent of the nuisance caused by discharge of wastewater into the relatively large bodies of water was not as marked and because of the greater areas of land suitable for land disposal.

About 1887, the Lawrence Experiment Station was established by the Massachusetts State Board of Health for the study of both water and wastewater treatment. The influence of the research done there has been profound and far-reaching [4, 9, 10]. In the early 1900s, the increasing public demand for wastewater treatment and the impracticability of procuring sufficient areas for land treatment, particularly for larger cities, led to the rather wide adoption of more intensive methods of treatment, many of which had been developed at Lawrence and are still in use today (see Chaps. 10 to 14).

Recent Trends and Developments

The changing nature of the wastewater to be treated, increased knowledge concerning the fundamental principles involved, and an increased understanding of the environmental effects caused by the discharge of many of the contaminants in wastewater are the principal factors responsible for the many changes that are taking place today in the field of wastewater treatment. The importance of these as well as other factors and their relationship to the engineer is described in the following discussion.

Changing Wastewater Characteristics The number of organic compounds that have been synthesized since the turn of the century now exceeds well over half a million, and some 10,000 new compounds are added each year. As a result, many of these compounds are now found in the wastewater from most cities. While most of them can be treated readily, the number of such compounds that are not, or are only slightly, amenable to treatment is increasing. Moreover, in most cases, little or no information is available on the long-term environmental effects caused by their discharge.

Treatment Methods and Concepts Methods of treatment in which the application of physical forces predominate are known as unit operations (see Chap. 8). Methods of treatment in which the removal of contaminants is brought about by chemical or biological activity are known as unit processes (see Chaps. 9 and 10). Currently most of these methods are undergoing intensive investigation from the standpoint of implementation and application. As a result, many modifications have been developed and implemented; more need to be made.

In the past, unit operations and processes were grouped together to provide what was known as primary and secondary treatment. In primary treatment, physical operations, such as screening and sedimentation, were used to remove the floating and settleable solids found in wastewater. In secondary treatment, biological processes were used to remove most of the organic matter. Recently, the term "tertiary treatment" or "advanced treatment" has been applied to the operations and processes used to remove contaminants not removed in primary and secondary treatment. It should be noted, however, that the terms "primary" and "secondary" are arbitrary and are of little value. A more rational approach is first to establish the degree of contaminant removal (treatment) required before the wastewater can be reused or discharged to the environment and then to group together, on the basis of fundamental considerations, the required operations and processes necessary to achieve that required degree of treatment. For this reason, separate chapters dealing with fundamentals have been devoted to the physical unit operations, the chemical unit processes, and the biological unit processes used for wastewater treatment.

As suburban areas continue to grow, as the variety of industrial wastes discharged to sewers increases, and as the available space for the locations of large treatment facilities decreases, the concept of satellite treatment is again being revived. This is not a new concept; it was proposed by Metcalf & Eddy for the treatment of wastewater in the Los Angeles area 30 years ago [5]. In practice, small treatment plants would be located throughout a system and would be designed to treat principally domestic wastes. The biological solids produced during treatment would be returned to the sewer for central processing. The effluent would be applied to some reuse or would also be returned to the sewer. It is interesting to note that the McQueens plant in Golden Gate Park in San Francisco has been operating on this basis since 1932. Effluent from this plant is used to water the park and to fill some small lakes located within the park.

Advanced Treatment of Wastewater Because of the changing nature of the wastes to be treated, many of the contaminants now found in wastewater are, as previously noted, not affected by conventional treatment operations and processes. Because of the need to remove contaminants, such as nitrogen and phosphorus, that may promote the growth of aquatic plants and algae, additional treatment must be provided. Advanced wastewater treatment methods and means are applied to the removal of these resistant contaminants. In most cases, the means of treatment have been adapted from other fields, such as water treatment and chemical engineering. As the effects of the different contaminants discharged to the environment become more clearly understood, it is anticipated that more emphasis will be placed on the removal of specific contaminants.

The Problem of Industrial Wastes The number of industries that now discharge wastes to domestic sewers has increased significantly during the past 20 to 30 years. However, on the basis of the toxic effects often caused by the presence of these wastes, the general practice of combining industrial and domestic wastes is being reevaluated. For example, in the new wastewater treatment plant designed for the city of Palo Alto, Calif., a separate batch-type industrial-waste treatment plant has been included as part of the overall facility. Toxic or potentially toxic industrial wastes are collected and brought to the treatment plant in a city-owned tank truck and are treated separately. After treatment, the effluent is blended with the incoming domestic wastewater [2].

In the future, it is anticipated that, rather than combining industrial and domestic wastes (a luxury that can no longer be afforded), many municipalities will provide separate treatment facilities or will require that these wastes be treated at the point of discharge to render them harmless before allowing their discharge to domestic sewers.

Wastewater Treatment Studies Because of the problems associated with the discharge of industrial wastes to domestic sewers, the conduct of wastewater treatment studies is increasing. It is anticipated that such studies will be made routinely in the future. Therefore, the engineer must understand the general approach and methodology involved in assessing the treatability of a wastewater (domestic and industrial), the conduct of laboratory and pilot plant studies, and the translation of experimental data into design parameters.

DISPOSAL

The ultimate disposal of treated wastewater and the sludge and concentrated contaminants removed by treatment has been and continues to be one of the most difficult problems in the field of wastewater engineering.

Background

In the past, the disposal of wastewater in most cities was carried out by the easiest method possible, without much regard to unpleasant conditions produced at the place of disposal. Irrigation, practiced in ancient Athens, was probably the first method of wastewater disposal, although dilution was the earliest method adopted by most municipalities. Problems arose when household wastes were admitted to storm sewers because the purification capacity of the watercourse into which they drained was often exceeded. As a result, separate sewers were built and wastewater treatment was instituted. The disposal of sludge became a problem with the application of the more intensive methods of treatment.

Recent Developments and Trends

The most important recent trend in the field of disposal is the establishment of increasingly stringent discharge requirements to protect the environment. As a result, effluent disposal and the disposal of the sludge and concentrated contaminants removed from wastewater are undergoing considerable study.

Effluent Disposal Dilution, land application, and reuse are the methods used currently for effluent disposal. Of these, dilution remains today the most common method of wastewater disposal. To protect the aquatic environment, however, the individual states in conjunction with the federal government have developed (completed in 1970) receiving-water standards for the streams, rivers, and estuarine and coastal waters of the United States (see Chap. 15). It is anticipated that more stringent requirements will be adopted in the future.

As the available assimilative capacity of the environment becomes limited, as discharge requirements become more stringent and a larger number of contaminants must be removed, the reuse potential of the wastewater is increased. In many localities, treatment plants have been designed and located so that a portion of the treated effluent can be disposed of by land application or in conjunction with a variety of reuse applications, such as golf-course irrigation, industrial cooling water, and ground water recharge. This trend will probably continue in the future.

In many locations where the available supply of fresh water has become inadequate to meet water needs, it is clear that the once-used water collected from towns and cities must not be viewed as a waste to be disposed of but rather as a resource. It is anticipated that this concept will become more widely adopted as other parts of the country experience water shortages.

Disposal of Sludge and Concentrated Contaminants

In small treatment plants, sludge and concentrated contaminants are often disposed of in lagoons, sludge drying beds, or sanitary landfills. In large treatment plants serving metropolitan areas, however, the volume of wastes requiring ultimate disposal has become so large that vacuum filtration followed by heat drying and incineration must be used in most cases. Furthermore, the tonnages are also increasing as more and different contaminants are being removed to meet more stringent discharge requirements. Therefore, the continuing search for better methods and means of disposing of concentrated wastewater contaminants will remain high on the list of priorities in the future.

REFERENCES

1. Herschel, C. (trans.): Sextus Julius Frontinus, *The Two Books on the Water Supply of the City of Rome*, Dana Estes & Co., Boston, 1899.
2. Jenks and Adamson, Consulting Engineers: *Regional Waste Water Treatment Works*, City of Palo Alto, Palo Alto, Calif., 1969.
3. Metcalf, L., and H. P. Eddy: *American Sewerage Practice—Volume I Design of Sewers*, 2d ed., McGraw-Hill, New York, 1928.
4. Metcalf, L., and H. P. Eddy: *American Sewerage Practice—Volume III Disposal of Sewage*, 3d ed., McGraw-Hill, New York, 1935.

5. Metcalf & Eddy, Inc.: *Sewage Disposal Problem of Los Angeles, California and Adjacent Communities*, Boston, 1944.

6. Metcalf & Eddy, Inc., Water Resources Engineers, Inc., and University of Florida: *Storm Water Management Model*, vols. 1–4, Environmental Protection Agency, Water Pollution Control Research Series, Rept. Nos. 11024 DOC 07/71, 08/71, 09/71, 10/71, Washington, D.C., 1971.

7. Stanier, R. Y., M. Doudoroff, and E. A. Adelberg: *The Microbial World*, 3d ed., Prentice-Hall, Englewood Cliffs, N.J., 1970.

8. Rouse, H., and S. Ince: *History of Hydraulics*, Dover, New York, 1957.

9. Sedgwick, W. T.: *Principles of Sanitary Science and the Public Health*, Macmillan, New York, 1903.

10. Wagenhals, H. H., E. J. Theriault, and H. G. Hommon: *Sewage Treatment in the United States*, Public Health Bulletin 132, Government Printing Office, Washington, D.C., 1925.

determination of sewage flowrates

2

Determination of the quantity of sewage to be removed from a community is fundamental to the design of collection, pumping, treatment, and disposal facilities. Further, with the recent trend toward regional treatment and disposal, reliable data on current and projected quantities must be available if these facilities are to be designed properly and the associated costs are to be shared equitably. To delineate the methodology involved in establishing sewage flowrates, the material in this chapter has been divided into six major sections dealing with (1) the preparation of comprehensive sewerage plans, (2) population studies, (3) water consumption, (4) sewage flowrates, (5) storm water runoff, and (6) ground water infiltration.

COMPREHENSIVE SEWERAGE PLANS

A comprehensive or master sewerage plan is prepared for the purpose of allowing an orderly development of collection, treatment, and disposal facilities to meet the needs of a community for many years into the future. Although the preparation of a comprehensive plan is a subject not strictly in the purview of this chapter, this brief discussion is included to familiarize the student with the contents of such a plan and to illustrate the functional relationship of data on sewage flow to the other parts of the plan.

Typically, a comprehensive sewerage plan for a community includes five major sections dealing with (1) background data on past, current, and projected community growth; (2) a summary description and evaluation of existing sewerage facilities; (3) the development of project design criteria; (4) the proposed or recommended sewerage facilities; and (5) an implementation program.

Community Growth

Population projections, land-use patterns, and economic trends, both local and regional, must be evaluated as part of a comprehensive plan. Data derived from an analysis of these factors will serve as a basis for the selection of design criteria. Because the subject of estimating population is so basic to the determination of sewage flowrates, a separate section, following this one, has been devoted to it (see "Population Studies").

Land is one of a community's basic resources. In the past, the value of land was related to its ability to support crops and to provide for the essentials of life. Thus, the physical characteristics of the land were most important. By contrast, in an urban setting, the economic value of land is derived partly from its physical characteristics but more importantly from its location for certain productive activities, such as manufacturing and commerce, or for homesites, schools, or recreation areas. Therefore, when making studies of sewerage needs, it is important to note the existing land-use patterns and to predict new and changing patterns.

Trends in the economic development of the community and in the growth of the tax base relative to tax rates and construction cost levels must be studied carefully. These trends will affect the population and growth of an area, the design period to be selected for proposed sewerage facilities, and the financing of the facilities.

Regional trends, such as the decentralization of economic activities, the rapid growth of suburban areas, and the use of regional utility systems, will affect any local plan. The impact of these trends must be evaluated and incorporated into the comprehensive plan.

Existing Sewerage Facilities

A survey (or inventory) of existing facilities usually is conducted to establish their condition and utility and to determine how they can best be integrated into the comprehensive plan.

Design Criteria

The most important design criteria to be selected (or established) are (1) those related to the functional performance and design of the sewerage facilities and (2) the design period. For example, design criteria for treatment and disposal facilities are established in response to regional, state, and federal water-quality guidelines and standards and with regard to the quality of the receiving waters. Unit sewage-flow contributions from various parts of a community are established in this section of a comprehensive report. Total flow quantities are computed after taking into account the background information on past, current, and future growth.

The number of years from the date of design to the estimated date when the conditions of the design will be reached is called the design period or the economic period of design. This period will vary with the type of sewerage facility. Sewer laterals and submain sewers are designed for conditions estimated to be reached far in the future, as the slight additional cost for increased sizes of small pipes is usually immaterial compared with the cost of construction. Large sewers, pumping stations, and treatment works are usually designed for comparatively short periods of time, since the structures can generally be relieved or supplemented by parallel sewers or additional units.

Recommended Sewerage Facilities

It is the responsibility of the sanitary engineer to recommend sewerage facilities on the basis of an analysis of future trends, a review of existing facilities, and the selected design criteria. In most comprehensive plans, facilities are recommended for both short- and long-range needs. Of necessity, long-range recommendations must be flexible and adaptable to changing conditions. For this reason, usually more than one or two alternative plans are proposed. Cost estimates for each of the various alternative plans are included in this section of the comprehensive plan.

Implementation Program

One of the most difficult aspects of comprehensive planning is the development of an implementation program. To develop a successful program, the engineer must recognize the needs and goals of the community as well as the social and environmental constraints within which the system must operate.

POPULATION STUDIES

The quantity of sewage to be removed from any community depends on the population and the per capita contribution of sewage. Therefore, if the quantity of sewage is to be predicted accurately, it will usually be necessary to conduct detailed population studies. Because of the increased life expectancy and mobility of our society, population prediction has become more and more complex. For this reason, it is very important that every sanitary engineer be familiar with the conduct of population studies and the types of data that can be derived from them.

Sources of Information

Although population data are available from a number of sources, the utility of the data obtained from them will vary widely. For example, population data obtained from the decennial census conducted by the Bureau of the Census are quite useful but tend to become less reliable as the length of time from the last census increases. For the interim periods, reliable data can usually be obtained from local sources, such as city and county planning commissions, the chamber of commerce, voter registration lists, the post office, newspapers, and public utilities. Although often overlooked, banking houses located both within the community and in neighboring communities are among the most important sources of population data.

Population Distribution and Density

The distribution of population in any given community depends on a number of factors, including the educational, occupational, and income characteristics of the population; the land-use and zoning patterns within the community; and the influence of national socioeconomic trends. For example, in many of the older communities of the East, the trend in residential districts has been toward smaller households, primarily because younger couples with children are leaving while older persons tend to stay. In other parts of the country, owing to the tremendous population influx, residential areas that were once composed of single-family dwelling units have given way to the development of multiple-family dwelling units, apartments, and even housing projects.

The density of population per square mile for the 40 largest American cities is given in Table 2·1. The density within a city varies greatly, and it is difficult to estimate future changes. For example, a

TABLE 2·1 DENSITY OF POPULATION FOR THE 40 LARGEST CITIES IN THE UNITED STATES IN 1960 [3]

City	Density, persons/sq mi	Population 1960	Area of land, sq mi
New York, N.Y.	25,966	7,781,984	300
Chicago, Ill.	16,014	3,550,404	222
Los Angeles, Calif.	5,447	2,479,015	455
Philadelphia, Pa.	15,584	2,002,512	129
Detroit, Mich.	12,103	1,670,144	138
Baltimore, Md.	11,993	939,024	78
Houston, Tex.	2,923	938,219	321
Cleveland, Ohio	11,542	876,050	76
Washington, D.C.	12,442	763,956	61
St. Louis, Mo.	12,255	750,026	61
Milwaukee, Wis.	8,255	741,321	90
San Francisco, Calif.	16,307	740,316	45
Boston, Mass.	15,157	697,197	46
Dallas, Tex.	2,676	679,684	254
New Orleans, La.	3,057	627,525	205
Pittsburgh, Pa.	10,968	604,332	55
San Antonio, Tex.	3,966	587,718	148
San Diego, Calif.	2,944	573,224	195
Seattle, Wash.	6,810	557,087	82
Buffalo, N.Y.	12,869	532,759	41
Cincinnati, Ohio	6,569	502,550	77
Memphis, Tenn.	3,851	497,524	129
Denver, Colo.	7,295	493,887	68
Atlanta, Ga.	3,587	487,455	136
Minneapolis, Minn.	9,043	482,872	53
Indianapolis, Ind.	6,794	476,258	70
Kansas City, Mo.	3,650	475,539	130
Columbus, Ohio	5,430	471,316	87
Phoenix, Ariz.	2,344	439,170	187
Newark, N.J.	16,814	405,220	24
Louisville, Ky.	6,599	390,639	59
Portland, Oreg.	5,630	372,676	66
Oakland, Calif.	7,041	367,548	52
Fort Worth, Tex.	2,578	356,268	138
Long Beach, Calif.	7,564	344,168	46
Birmingham, Ala.	5,420	340,887	63
Oklahoma City, Okla.	1,086	324,253	299
Rochester, N.Y.	8,682	318,611	37
Toledo, Ohio	6,517	318,003	49
St. Paul, Minn.	6,016	313,411	52

residential section at the present time may become a commercial or manufacturing district in the next decade. Because estimates of the future population in the different parts of the city are part of the basic data upon which sewerage works are designed, the importance of studying present tendencies toward growth or stagnation, and their causes, is evident.

The differences in policies of cities with respect to the extension of their boundaries has been marked. In eastern communities, the city limits have usually been more restricted than in western communities. This tendency has led to rather dense populations in the eastern cities and to somewhat sparser populations in western cities. A scattered population increases the cost of public utilities and services and presents a difficulty in accurately forecasting the direction and degree of development and the increase in population.

Population Projections

The selection of the population projection method depends on the amount and types of data available. The most commonly used methods may be classified as

1. Graphical
2. Decreasing-rate-of-growth (increase)
3. Mathematical or logistic
4. Ratio and correlation
5. Component
6. Employment forecast

Although a detailed discussion of these methods is beyond the scope of this chapter, the elements involved in each one will be briefly reviewed. The basic mathematics involved are presented in Table 2·2. It is recommended that the reader consult current texts or references, particularly McJunkin [12], for further information on these as well as other methods.

Graphical Methods In these methods, graphical projections of the past population-growth curves are used to estimate future population growth. Methods used in the past that are included in this category are arithmetic projection; geometric progression; projections based on linear least-squares regression lines; and comparison methods, using the past growth curves of similar but larger cities. The chief advantages of the graphical methods are their simplicity and the ease with which they may be applied. It should be noted, however, that the results obtained tend to fluctuate over a wide range.

TABLE 2·2 POPULATION PROJECTION METHODS AND EQUATIONS

Method	Basic equation	Definition of terms	Evaluation of constants
Arithmetic	$\frac{dP}{dt} = k_a$	P = population t = time k_a = arithmetic growth constant	$k_a = \frac{P_2 - P_1}{t_2 - t_1}$
Geometric	$\frac{dP}{dt} = k_g P$	k_g = geometric growth constant	$k_g = \frac{\ln P_2 - \ln P_1}{t_2 - t_1}$
Decreasing-rate-of-increase	$\frac{dP}{dt} = k_d(S - P)$	S = saturation population k_d = decreasing-rate-of-increase constant	$k_d = \frac{-\ln \frac{S - P_2}{S - P_1}}{t_2 - t_1}$
Logistic S	$P = \frac{S}{1 + me^{bt}}$	S = saturation population m, b = constants P_0, P_1, P_2 = populations at times t_0, t_1, t_2 n = interval between t_0, t_1, t_2	$S = \frac{2P_0P_1P_2 - P_1^2(P_0 + P_2)}{P_0P_2 - P_1^2}$ $m = \frac{S - P_0}{P_0}$ $b = \frac{1}{n} \ln \frac{P_0(S - P_1)}{P_1(S - P_0)}$
Ratio and correlation	$\frac{P_2}{P_{2R}} = \frac{P_1}{P_{1R}} = k_r$	P_2 = projected population P_{2R} = projected population of a larger region P_1 = population at last census P_{1R} = population of larger region at last census k_r = ratio constant	$k_r = \frac{P_1}{P_{1R}}$

Decreasing-Rate-of-Growth Method As a rule it is found that the larger the city becomes, the smaller will be the rate of growth from year to year. This general reduction in rate of growth as the city increases in size is distinctly marked. There is a similar general decrease in the rate of growth of the entire population of the country (although it varies with geographic location).

As shown in Table 2·2, a saturation population is estimated and the rate constant k_d is computed. Generally, the assumption of a decreasing rate of growth is one of the more reliable methods of estimating future populations, particularly if it is checked by basing the prediction on the experience of comparable cities that have already passed the present population of the city under consideration.

Mathematical Methods The mathematical or so-called logistic methods involve an assumption that population growth follows some logical mathematical relationships in which population growth is a function of time. On this basis, a wide variety of mathematical formulations have been proposed for population forecasting [12, 13].

Briefly, the procedures involved in using these methods are as follows: (1) the ultimate or saturation population is estimated for the area in question; (2) a population versus time plot is prepared showing past population data as well as the computed saturation value; and (3) a curve is fitted through the past data and the saturation value. One of the most commonly used curves is the so-called S-shaped or logistic S curve. An example of a population-growth curve derived by this technique is shown in Fig. 2·1.

The most critical step in using this method is the determination

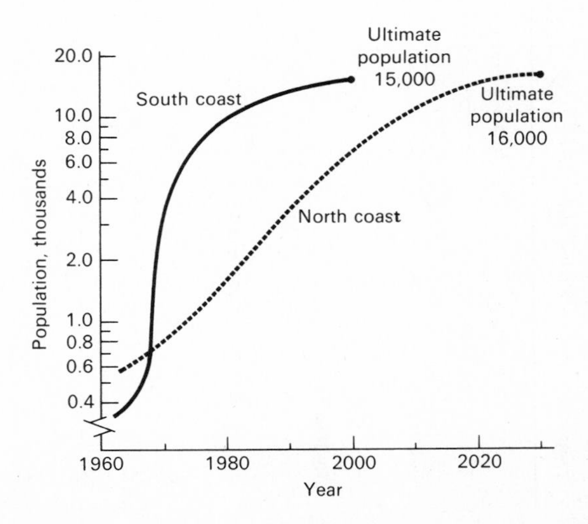

FIG. 2·1 S-shaped population-growth curve for two coastal areas of Ventura County, Calif.

of the saturation population. For example, forecasts using this method, made during the 1930s, have generally proved to be in error due to the use of overly conservative estimates of the saturation population. As with the graphical methods, the major shortcoming of the mathematical methods is that they do not adequately reflect current trends and changes.

Ratio and Correlation Method In this method, it is assumed that the population-growth rate of any given community can be related to that of a larger region, such as the county or state. Thus, by using an appropriate scale factor, population estimates prepared for larger areas can be used to estimate the future growth of smaller areas. As implied by the name of this method, the scale factors used are usually based on simple ratios as shown in Table 2·2 or are derived from correlation studies.

Component Method Here the population forecast is based on a detailed analysis of the components that make up population growth, namely, natural increase and migration. Natural increase represents the increase resulting from the excess of births over deaths. The rate of yearly increase of births over deaths during the past 65 years reflects a continually improving standard of living.

Migration, on the other hand, represents the movement of people into or out of a community. Almost all the previously discussed factors (see "Comprehensive Sewerage Plans") will affect migration, and for this reason accurate determination of this factor is extremely complex. This complexity limits the use of the component method for routine population estimates.

Employment Forecast Methods In these methods, as their name implies, population growth is estimated on the basis of various employment forecasts. In actual practice, the relationship between population and the number of jobs is derived by using the techniques of the ratio and correlation method previously discussed.

The methods of graphical comparison, arithmetic projection, geometric projection, decreasing-rate-of-increase, and logistic S are compared for a single city in Example 2·1.

EXAMPLE 2·1 *Population Projection*

Estimate the population for the city of South San Francisco, Calif., in 1975 using the following methods: (1) graphical comparison, (2) arithmetic projection, (3) geometric projection, (4) decreasing-rate-of-increase, and (5) logistic S. Population data from 1920 to 1960 are

Year	*Population*
1920	4,411
1930	6,193
1940	6,629
1950	19,351
1960	39,418

Solution

1. Graphical comparison. The 1960 population of the city of South San Francisco is compared with the populations of Colorado Springs, Colo., San Mateo, Calif., and Salem, Oreg., which exceeded 40,000 in 1944, 1947, and 1949, respectively. The plots of population above 40,000 for these three cities are shown extended from 1960 in Fig. 2·2. A population projection is then made for South San Francisco as shown in Fig. 2·2. The rate of increase of population is less than that for San Mateo or Colorado Springs but more than that for Salem.

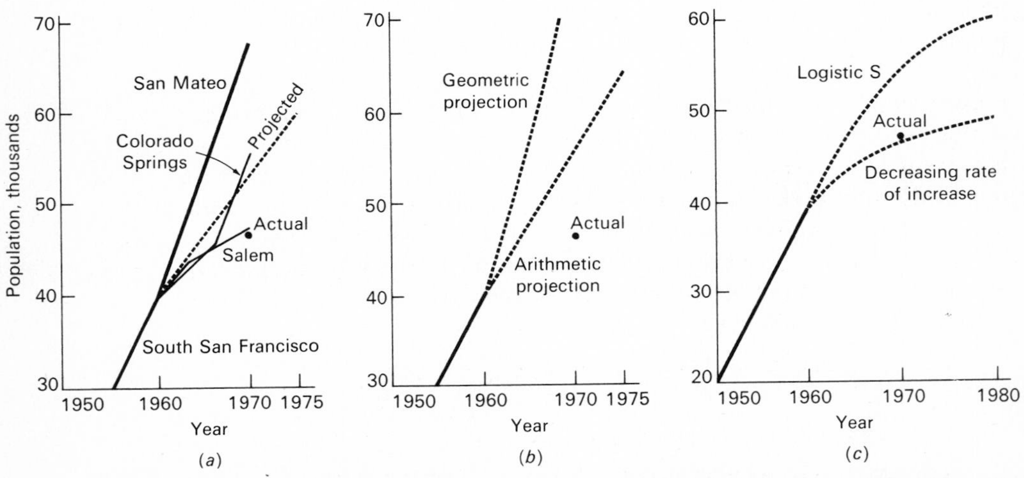

FIG. 2·2 Comparison of population projection methods as applied to South San Francisco, Calif. (*a*) Graphical comparison, (*b*) arithmetic and geometric projection, (*c*) logistic S and decreasing rate of increase.

2. Arithmetic projection. For an arithmetic projection, k_a is determined for two time intervals, 1950–1960 and 1940–1950.

$$k_{a1} = \frac{39{,}418 - 19{,}351}{10} = 2{,}007$$

$$k_{a2} = \frac{19{,}351 - 6{,}629}{10} = 1{,}272$$

An average value of k_a is then computed.

$$k_a = 1{,}640$$

Determine the 1975 population by arithmetic projection.

$$\begin{aligned} P &= P_{1960} + k_a(1975 - 1960) \\ &= 39{,}418 + 1{,}640(15) = 64{,}018 \end{aligned}$$

3. Geometric projection. Determine the geometric-growth constant for 1950–1960:

$$k_g = \frac{\ln 39{,}418 - \ln 19{,}351}{10} = 0.072$$

and the 1975 population by geometric projection.

$$\begin{aligned} \ln P &= \ln 39{,}418 + k_g(1975 - 1960) \\ &= 10.58 + 0.072(15) = 11.66 \\ P &= 116{,}000 \end{aligned}$$

4. Decreasing rate of increase. Assume a saturation population of 50,000. Determine k_d using the 1960 and 1950 populations.

$$k_d = \frac{-\ln \dfrac{50{,}000 - 39{,}418}{50{,}000 - 19{,}351}}{10} = 0.106$$

Determine the 1975 population by the decreasing-rate-of-increase method.

$$\begin{aligned} P &= 50{,}000 - (50{,}000 - 39{,}418)e^{-0.106(15)} \\ &= 50{,}000 - 2{,}152 = 47{,}848 \end{aligned}$$

5. Logistic S. Determine S using $P_2 = 39{,}400$; $P_1 = 19{,}300$; and $P_0 = 6{,}600$ for the years 1960, 1950, and 1940, respectively.

$$\begin{aligned} S &= \frac{2(6{,}600)(19{,}300)(39{,}400) - (19{,}300)^2(6{,}600 + 39{,}400)}{6{,}600(39{,}400) - (19{,}300)^2} \\ &= 63{,}000 \end{aligned}$$

Determine m and b.

$$m = \frac{63{,}000 - 6{,}600}{6{,}600} = 8.55$$

$$b = \frac{1}{10} \ln \frac{6{,}600(63{,}000 - 19{,}300)}{19{,}300(63{,}000 - 39{,}400)} = -0.133$$

Determine the 1975 population using the logistic S method.

$$P = \frac{63{,}000}{1 + 8.55e^{-0.133(\Delta t)}}$$

$$\Delta t = 1975 - 1940 = 35$$

$$P = 58{,}000$$

By including the actual 1970 population on Fig. 2·2 as an indication of the 1975 population, the accuracy of these projection methods in this example is evident.

6. A summary of the population estimates for this example is shown in Table 2·3.

TABLE 2·3 SUMMARY OF POPULATION ESTIMATES FOR SOUTH SAN FRANCISCO, CALIF.

Method	Estimated 1975 population
Graphical comparison	60,000
Arithmetic projection	64,018
Geometric projection	116,000
Decreasing-rate-of-increase	47,848
Logistic S	58,000

Accuracy of Population Projections As a rule, curves of actual population growth are not smooth but fluctuate rather widely. Curves of projected population are always smooth. For this reason, actual populations for any particular place should not be expected to correspond closely to the estimated population for each of several census dates. The deviation is likely to be more marked with smaller cities than with larger cities, since the former are more susceptible to various conditions and factors affecting the rate of growth. Therefore, it is possible that an estimate that may ultimately prove to be a good representation of average growth may depart widely from each of several census figures. Differences amounting to 20 percent or more do not necessarily indicate that the general trend is diverging significantly from the estimate.

In any case, it should be remembered that the accuracy of population estimates decreases as (1) the time period of the forecast increases, (2) the population of the area decreases, and (3) the population rate of change increases. McJunkin [12] gives an excellent account of the relative accuracy of the aforementioned population forecasting methods.

WATER CONSUMPTION

Water Consumption in the United States

Based on detailed studies conducted by the U.S. Public Health Service (USPHS) for the Select Committee on Natural Water Resources of the United States Senate [18], the average water consumption on a national basis was found to be about 147 gallons per capita per day (gpcd). The 1954 and predicted 1980 and 2000 water-consumption rates for the individual states are shown in Table 2·4. In analyzing the data collected for this study, it was estimated that domestic use accounted for 40 percent of the overall daily per capita requirement and that the commercial, industrial, and public uses were 18, 24, and 17 percent, respectively. Thus domestic use on a national scale averaged about 60 gpcd; commercial and industrial use, about 62 gpcd; and public use, about 25 gpcd. In specific instances, however, a wide variation will be observed in these rates, depending upon geographic

TABLE 2·4 PROJECTED PER CAPITA WATER USE FROM MUNICIPAL SYSTEMS IN CONTINENTAL UNITED STATES, BY STATES [18]
(Gallons per capita per day)

State	1954	1980	2000
Total United States	147	148	152
Alabama	105	100	98
Arizona	170	143	141
Arkansas	81	88	90
California	147	135	133
Colorado	174	152	151
Connecticut	137	134	135
Delaware	142	126	125
District of Columbia	133	154	161
Florida	122	99	97

TABLE 2·4 *(Continued)*

State	1954	1980	2000
Georgia	127	119	117
Idaho	170	163	161
Illinois	171	169	170
Indiana	155	151	151
Iowa	110	113	114
Kansas	121	118	118
Kentucky	125	131	136
Louisiana	117	117	118
Maine	142	139	139
Maryland	131	117	116
Massachusetts	121	134	140
Michigan	213	193	187
Minnesota	108	111	113
Mississippi	107	110	113
Missouri	135	165	167
Montana	191	185	185
Nebraska	173	173	173
Nevada	270	230	226
New Hampshire	124	131	133
New Jersey	113	114	116
New Mexico	150	134	132
New York	138	148	153
North Carolina	108	106	106
North Dakota	101	98	100
Ohio	149	147	147
Oklahoma	130	134	138
Oregon	134	132	130
Pennsylvania	147	154	154
Rhode Island	111	110	114
South Carolina	108	98	96
South Dakota	107	101	99
Tennessee	118	119	118
Texas	119	112	112
Utah	223	200	194
Vermont	75	83	83
Virginia	100	90	89
Washington	196	187	185
West Virginia	112	127	132
Wisconsin	158	153	149
Wyoming	196	186	185

Note: Averages of state figures are not the same as the national averages because of different bases used in computing these two sets of figures.

location, climate, size of the community, degree of industrialization, and other influencing factors. Each area should be studied separately if reliable estimates are to be prepared.

Water Consumption in Cities

A direct comparison of water-consumption records from different cities is likely to be misleading because in some cities large quantities of water used for industrial purposes are obtained from privately owned supplies, while in other cities the industries mainly use the municipal supply. Furthermore, the care taken to reduce the waste of water through leaks in mains, service pipes, and plumbing has a decided effect on the per capita consumption. Metering the individual consumer's supply and billing at established meter rates indirectly prevents waste of water by users and tends to keep the consumption reasonably low. The waste and unaccounted-for water in metered systems ranges from 10 to 20 percent of the total water entering the supply-line system. The corresponding range in unmetered systems is much higher. Typical data on water consumption for various cities are shown in Table 2·5.

From an analysis of the data in this table and corresponding population data, it appears that there is a tendency toward a gradual increase in the quantity of water used per capita in those cities where the population has not stabilized and is increasing. In those cities where the population has become somewhat stabilized, the per capita consumption has also become more or less stabilized, and in some large cities (e.g., Chicago) the per capita rate has even decreased.

Water Consumption in Various Establishments

Data on the rates of water consumption for many different types of establishments as well as activities are presented in Tables 2·6 and 2·7. Although these data show wide variations, they are useful in estimating total water consumption for individual establishments and areas.

Industrial water use varies greatly, according to the nature of the manufacturing process. In practical design work, it is therefore desirable to make an actual inspection of the plant concerned and a careful estimate of the quantities of both the water used from all sources and the wastes produced. The same is true of consumption in business districts. The data shown in Table 2·8 may be used as an indication of the magnitude of water usage to be expected from various industrial operations.

TABLE 2·5 AVERAGE WATER CONSUMPTION, PAST AND ESTIMATED FUTURE, BY WATER WORKS SUPERINTENDENTS [14]
(Gallons per capita per day)

City and state	1936	1946	1956	1966	1976
New York, N.Y.	134	146	138	153	162
Baltimore, Md.	131	150	159	174	188
Philadelphia, Pa.	...	163	171	180	188
Springfield, Mass.	100	127	175	230	278
Hartford, Conn.	80	100	111	122	134
Charlotte, N.C.	79	96	109	144	178
Lynchburg, Va.	83	95	114	134	156
Raleigh, N.C.	...	82	102	120	136
Baton Rouge, La.	88	89	85	93	102
Atlanta, Ga.	99	115	122	130	138
Buffalo, N.Y.	210	214	242		
Toledo, Ohio	123	127	149	169	171
Akron, Ohio	91	120	138	150	162
Cedar Rapids, Iowa	...	91	136	170	200
Madison, Wis.	110	135	150	160	175
Des Moines, Iowa	100	100	114	117	125
Omaha, Nebr.	140	156	181	190	207
Wichita, Kans.	93	117	146	148	160
Oklahoma City, Okla.	79	82	114	154	183
Dallas, Tex.	100	116	143	158	175
Austin, Tex.	99	122	140	140	138
Sacramento, Calif.	210	258	236	240	250
Oakland, Calif.	72	102	130	159	177
Portland, Oreg. (Water Auth.)	97	114	104	115	120
San Diego Co., Calif.	...	162	180	190	190
San Diego, Calif.	114	139	126	140	145
Salem, Oreg.	...	139	156	207	224
Chicago, Ill.*	...	244	234	231	
Los Angeles, Calif.*	...	168	177	185	

* Added to table.

TABLE 2·6 ESTIMATED WATER CONSUMPTION AT DIFFERENT TYPES OF ESTABLISHMENTS [16]

Type of establishment	Flow, gpd/ person or unit
Dwelling units, residential:	
Private dwellings on individual wells or metered supply	50–75
Apartment houses on individual wells	75–100
Private dwellings on public water supply, unmetered	100–200
Apartment houses on public water supply, unmetered	100–200
Subdivision dwelling on individual well, or metered supply, per bedroom	150
Subdivision dwelling on public water supply, unmetered, per bedroom	200
Dwelling units, treatment:	
Hotels	50–100
Boarding houses	50
Lodging houses and tourist homes	40
Motels, without kitchens, per unit	100–150
Camps:	
Pioneer type	25
Children's, central toilet and bath	40–50
Day, no meals	15
Luxury, private bath	75–100
Labor	35–50
Trailer with private toilet and bath, per unit ($2\frac{1}{2}$ persons)*	125–150
Restaurants (Including toilet):	
Average	7–10
Kitchen wastes only	$2\frac{1}{2}$–3
Short order	4
Short order, paper service	1–2
Bars and cocktail lounges	2
Average type, per seat	35
Average type, 24-hr, per seat	50
Tavern, per seat	20
Service area, per counter seat (toll road)	350
Service area, per table seat (toll road)	150
Institutions:	
Average type	75–125
Hospitals	150–250
Schools:	
Day, with cafeteria or lunch room	10–15
Day, with cafeteria and showers	15–20
Boarding	75
Theatres:	
Indoor, per seat, two showings per day	3
Outdoor, including food stand, per car ($3\frac{1}{3}$ persons)	3–5

TABLE 2·6 *(Continued)*

Type of establishment	Flow, gpd/ person or unit
Automobile service stations:	
Per vehicle served	10
Per set of pumps	500
Stores:	
First 25-ft frontage	450
Each additional 25-ft frontage	400
Country clubs:	
Resident type	100
Transient type, serving meals	17– 25
Offices	10–15
Factories, sanitary wastes, per shift	15–35
Self-service laundry, per machine	250–500
Bowling alleys, per alley	200
Swimming pools and beaches, toilet and shower	10–15
Picnic parks, with flush toilets	5–10
Fairgrounds (based on daily attendance)	1
Assembly halls, per seat	2
Airport, per passenger	$2\frac{1}{2}$

* Add 125 gallons per trailer space for lawn sprinkling, car washing, leakage, etc.
Note: Water under pressure, flush toilets, and wash basins are assumed provided unless otherwise indicated. These figures are offered as a guide; they should not be used blindly. Add for any continuous flows and industrial usages. Figures are flows per capita per day, unless otherwise stated.

Fluctuations in Water Consumption

While it is important to know the average quantity of water consumption, it is of still greater value to have data on fluctuations in consumption. Representative data on the variations in average water consumption are reported in Table 2·9. The maximum rate of water consumption usually occurs (1) during summer months when water is in demand for street and lawn sprinkling and (2) in the winter when large quantities are allowed to run to prevent freezing of pipes and fixtures.

In addition to the seasonal fluctuations in flow, hourly variations in the rate of water consumption must be considered because of their effect on the rate of sewage flow. In general, the sewage-discharge curve closely parallels the water-consumption curve, but with a lag of several hours. In some places, large quantities of water used by industrial establishments and obtained from sources other than the

TABLE 2·7 MISCELLANEOUS WATER USAGE ESTIMATES [16]

Unit	Normal water consumption
Water closet, tank	4–6 gal/use
Water closet, flush valve, 25 psi	30 gpm
Wash basin	$1\frac{1}{2}$ gal/use
Bathtub	30 gal/use
Shower head	25–30 gal/use
Garden hose, $\frac{5}{8}$ in., 25-ft head	200 gph
Garden hose, $\frac{3}{4}$ in., $\frac{1}{4}$-in. nozzle, 25-ft head	300 gph
Fire hose, $1\frac{1}{2}$ in., $\frac{1}{2}$-in. nozzle, 70-ft head	2,400 gph
Continuous flowing drinking fountain	75 gph
Lawn sprinkler	120 gph
Automatic home laundry machine	30–50 gal/load
Dishwashing machine, home type	6 gal/load
Dishwashing machine,* commercial:	
Stationary rack type, at 15 psi	6–9 gpm
Conveyor type, at 15 psi	4–6 gpm
Garbage grinder, home type	1–2 gpd/person

Water use†	gpm	Total gal	gpcd
Automatic home-type washing machine	3–7	36–50 per load	6.5–9
Automatic home-type dishwasher	2.5–5	4–8 per load	6
Garbage disposal unit, home-type	1.5–2.5	...	3–4
Lawn sprinkler, 3,000-sq-ft lawn, 1-in./week	...	1,850 per week	75
Air conditioner, home-type, water-cooled, 3-ton unit, 8 hr./day, 2 gpm/ton	6	2,880 per day	825

* Does not include water to fill wash tank.
† Adapted from "Land Uses and Water Consumption Requirements," *Public Works*, 90, 120; April, 1959. (Abstract and condensation of thesis by Rodolfo Silva.)

public supply are discharged into the sewers during the working hours of the day. This tends to increase the peak flow beyond the amount resulting from the normal fluctuation in the draft on the municipal water supply.

In the absence of more authoritative information, an additional allowance of 50 percent over the average 24-hr rate of water consumed may be made for the excess of the hourly peak over the average 24-hr

TABLE 2·8 WATER CONSUMPTION IN REPRESENTATIVE INDUSTRIES [Adapted from 11]

Process	Consumption
Cannery:	
Green beans, gal/ton	20,000
Peaches and pears, gal/ton	5,300
Other fruits and vegetables, gal/ton	2,000–10,000
Chemical industries:	
Ammonia, gal/ton	37,500
Carbon dioxide, gal/ton	24,500
Gasoline, gal/1,000 gal	7,000–34,000
Lactose, gal/ton	235,000
Sulfur, gal/ton	3,000
Food and beverage industries:	
Beer, gal/1,000 gal	15,000
Bread, gal/ton	600–1,200
Meat packing, gal/ton live weight	5,000
Milk products, gal/ton	4,000–5,000
Whiskey, gal/1,000 gal	80,000
Pulp and paper:	
Pulp, gal/ton	82,000–230,000
Paper, gal/ton	47,000
Textiles:	
Bleaching, gal/ton cotton	72,000–96,000
Dyeing, gal/ton cotton	9,500–19,000

rate. The figure, however, will vary with the locality and should be applied only when local conditions are found to warrant it. If this peak hourly consumption is applied to the maximum draft for a single day, or 180 percent of the yearly average, and if it is assumed that the portion of the water supply that finds its way into the sewers averages

TABLE 2·9 VARIATIONS IN WATER CONSUMPTION*

Item	Percentage of average for year
Daily average in maximum month	120
Daily average in maximum week	140
Maximum consumption in 1 day	180

* Based on an average water consumption of 147 gpcd.
Note: There are instances in which the maximum rates greatly exceed these averages.

100 gpcd, there will be a maximum flow to the sewers in 1 hr from the public and private water supplies of about 270 gpcd ($100 \times 1.8 \times 1.5 = 270$). Some engineers estimate the maximum rate of contribution originating from water supplies on the basis of their experience and on observed flows in existing sewers (200 gpcd is a typical value).

Proportion of Municipal Water Supply Reaching Sewers

Because sewage consists primarily of used water, the portion of the water supplied that reaches the sewers must be estimated. A considerable part of the water used by commercial and manufacturing establishments and power plants for street and lawn sprinkling and for extinguishing fires and that used by consumers not connected with sewers, does not reach the sewers. There is also some leakage from the water mains and service pipes that does not reach the sewers.

In general, neglecting infiltration of ground water, about 60 to 80 percent of the per capita consumption of water will become sewage. The lower ratios are applicable to the semiarid regions of the southwestern United States. Often, however, excessive infiltration, roof water, and water used in industries that is obtained from privately owned sources make the quantity of sewage larger than that of the public water supply.

With well-built sewers and with roof water excluded, the variation from year to year in the ratio of sewage to water supply in a city is not great, unless there is a substantial change in the industrial uses of water. For this reason, the consumption of water in a city affords basic data for sewerage as well as water-supply engineering.

SEWAGE FLOWRATES

Sewage flowrates are established by considering the source, corresponding water-usage rates, and the type and condition of sewers. Expected variations in the sewage flowrates must be established before the sewers and treatment facilities are designed.

Sources of Sewage

Residential Districts For small residential areas it is common to determine average sewage flows from population density and average per capita contribution of sewage. For large residential districts, it is often advisable to develop flowrates based on land area [8]. Where possible, these rates should be based on actual flow data from selected typical residential areas. In the absence of such data, an estimate of 70 percent of the domestic water consumption is used.

Institutional Facilities Flows from institutional facilities are essentially domestic sewage in nature. Some typical flow values are shown in Table 2·10. Again it is stressed that flow values vary with the region, climate, and type of facility. The actual records of institutions are the best sources of flow data for design.

TABLE 2·10 AVERAGE SEWAGE FLOWS FROM INSTITUTIONAL FACILITIES [1, 6]

Institution	Average flow, gpcd
Medical hospital	175
Mental hospital	125
Prisons	175
High schools	20
Elementary schools	10

Recreational Facilities Flows from recreational facilities are highly seasonal. A complete listing of facilities and design flows for Yellowstone National Park is shown in Table 2·11.

TABLE 2·11 DESIGN UNIT SEWAGE FLOWS FOR RECREATIONAL FACILITIES
Yellowstone National Park

Establishment	Unit	Flow, gpd/unit
Campground (developed)	Person	25
Lodge or cabins	Person	50
Hotel	Person	75
Trailer village	Person	35
Dormitory, bunkhouse	Person	50
Residence homes, apartments	Person	75
Mess hall	Person	15
Offices and stores	Employee	25
Visitor centers	Visitor	5
Cafeteria	Table seat	150
Dining room	Table seat	150
Coffee shop	Counter seat	250
Cocktail lounge	Seat	20
Laundromat	Washing machine	500
Hospital	Bed	200
Gas station	Station	2,000–5,000
Fish-cleaning station	Station	7,500

Commercial Districts Commercial sewage flows are generally expressed in gallons per acre per day (gpad) and are based on existing development or comparative data. Unit flows may vary from 4,500 to more than 160,000 gpad. Flows may also be estimated for certain commercial establishments using the data reported in Tables 2·6 and 2·12.

TABLE 2·12 SEWAGE FLOWS FROM COMMERCIAL DISTRICTS [6]

Establishment	Unit	Average flow, gpd/unit
Shopping center	Employee	60
Small business	Employee	20
Restaurant	Meal	7
Airport	Passenger	5
Theater	Seat	5
Motel	Person	50
Hotel	Person	100

Industrial Districts Industrial sewage quantities vary with the industry type, size, and supervision, and with the waste treatment method. Peak flows are often encountered and may be reduced by use of detention tanks and equalization basins. Large industries tend to reuse wastewaters for process cooling and lawn irrigation. This practice reduces sewage flows. A typical design value for industrial areas is about 5,000 gallons (gal) per acre [19]. Average domestic sanitary sewage contributed from industrial activities may vary from 8 to 25 gpcd [9].

Air-Conditioning and Industrial Cooling Wastes Air-conditioning cooling water requirements vary from 1.5 to 2.0 gallons per minute (gpm) per ton on a once-through basis. A 5-ton unit would produce approximately 14,000 gallons per day (gpd) of wastewater [9]. These flows should be prohibited by law from being discharged to sanitary sewers not only because they overcharge the sewers, but also because they could be collected and reused.

Variations in Sewage Flow

Daily and Weekly As mentioned previously, the sewage-discharge curve closely parallels the water-consumption curve, but with a lag of several hours when extraneous flows are minimal. In the absence of a typical day when home laundering is done, the variation in week-

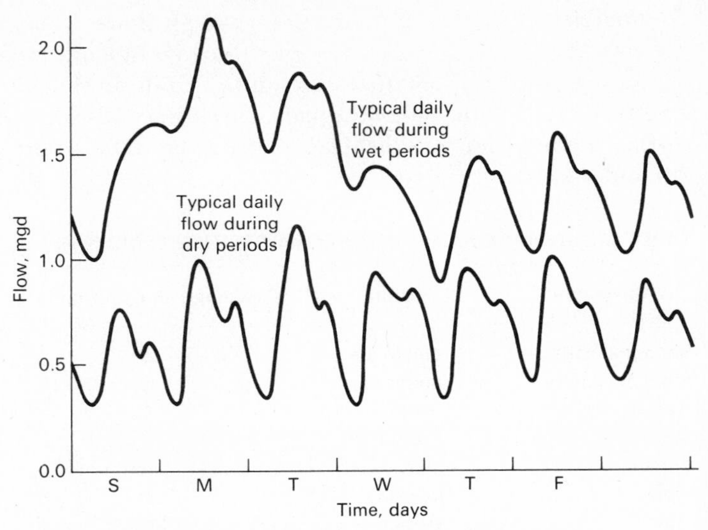

FIG. 2·3 Typical daily and weekly variations in sewage flow.

day flows is negligible. A plot of typical weekly flows for both wet and dry periods is shown in Fig. 2·3.

Hourly The hourly flow of sewage reflects the variation in water consumption during the day. Extreme low flows occur between 2 and 6 A.M., and the maximum flow occurs at about noon. A secondary peak occurs generally at 8 or 9 P.M.; however, this varies with the size of city and the length of sewers. A typical dry-weather-flow curve is shown in Fig. 2·4. (For a more comprehensive treatment of sewage-flow variations see Geyer and Lentz [4].)

Maximum and Minimum For facilities such as hospitals, hotels, schools, and apartment buildings, the fixture-unit method may

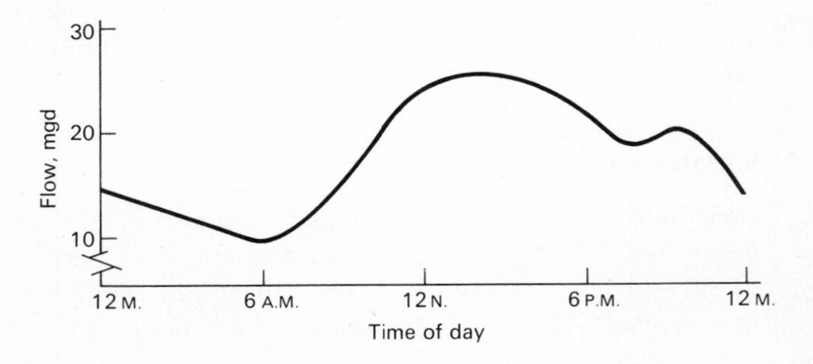

FIG. 2·4 Hourly variation of sewage flow at Allentown, Pa.

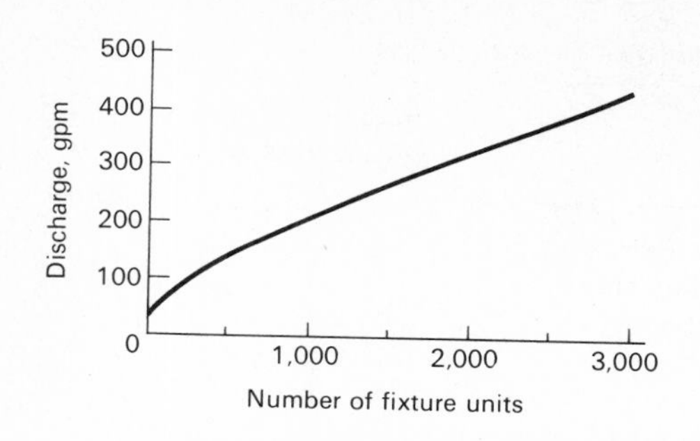

FIG. 2·5 Discharge as a function of the number of fixture units [7].

be used to estimate peak flows. One fixture unit is approximately equal to 1 cubic foot per minute (cfm) of flow. Typical fixture-unit values are shown in Table 2·13. More detailed information on fixtures may be found in Manas [10]. Fixture units have been correlated with peak flows as shown in Fig. 2·5 [7]. The peak flow from a small area as determined by the fixture-unit method is illustrated in Example 2·2.

EXAMPLE 2·2 *Determination of Peak Flow from a School by the Fixture-Unit Method*

Determine the peak flow in gpm from a school serving 500 persons based on three fixture units per person.

Solution

1. Determine the total number of fixture units. 500 persons × 3 fixture units/person = 1,500 fixture units.
2. Using Fig. 2·5, the peak flow will be 270 gpm.

The ratio of the peak flow to the average flow, called the peaking factor, for any one day will range from less than 1.3 for some large sewers to more than 2.0 for some laterals [9].

Since records of existing sewerage systems are usually inadequate to determine peaking factors, such factors for maximum and minimum flows may be estimated from the curves in Fig. 2·6 and 2·7. The curves in Fig. 2·7 are modifications of design curves for the Merrimack River Valley Sewerage District corrected for a number of other cities not included in the district.

Many state agencies have set minimum design flowrates of 400 gpcd for laterals and 240 gpcd for trunk sewers, when no actual measurements are available [9]. These figures assume no extraneous flows other than normal infiltration. The federal government has set minimum average design flows at 75 gpcd unless otherwise justified

TABLE 2·13 FIXTURE UNITS PER FIXTURE OR GROUP [20]

Fixture type	Fixture unit value as load factors
1 bathroom group consisting of tank-operated water closet, lavatory, and bathtub or shower stall	6
Bathtub* (with or without overhead shower)	2
Bidet	3
Combination sink-and-tray	3
Combination sink-and-tray with food-disposal unit	4
Dental unit or cuspidor	1
Dental lavatory	1
Drinking fountain	$\frac{1}{2}$
Dishwasher, domestic	2
Floor drains	1
Kitchen sink, domestic	2
Kitchen sink, domestic, with food waste grinder	3
Lavatory	1
Lavatory	2
Lavatory, barber, beauty parlor	2
Lavatory, surgeon's	2
Laundry tray (1 or 2 compartments)	2
Shower stall, domestic	2
Showers (group) per head	3
Sinks:	
Surgeon's	3
Flushing rim (with valve)	8
Service (trap standard)	3
Service (P trap)	2
Pot, scullery, etc.	4
Urinal, pedestal, syphon jet, blowout	8
Urinal, wall lip	4
Urinal stall, washout	4
Urinal trough (each 2-ft section)	2
Wash sink (circular or multiple), each set of faucets,	2
Water closet, tank-operated	4
Water closet, valve-operated	8

* A shower head over a bathtub does not increase the fixture value.
Note: For a continuous or semicontinuous flow into a drainage system, such as from a pump, pump ejector, air-conditioning equipment, or similar device, two fixture units shall be allowed for each gpm of flow.

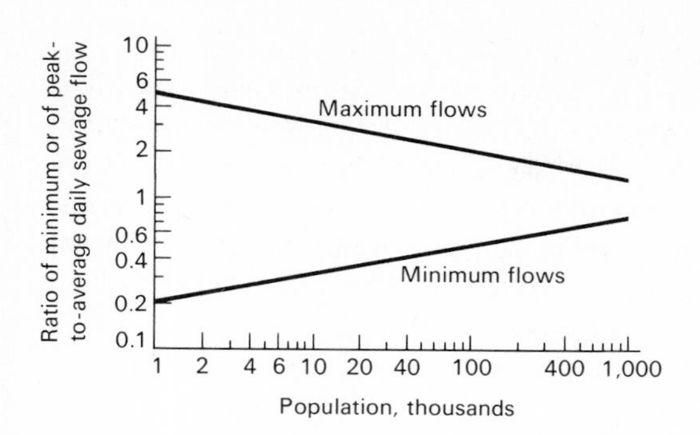

FIG. 2·6 Ratio of minimum and peak flows to average daily sewage flow.

by sound engineering data. On this basis, laterals are designed for a peaking factor of 4.0, and outfall sewers are then designed for a peaking factor of 2.5 [21].

In some cases, maximum flowrates may result almost entirely from extraneous flows. These flows may come from surreptitious roof, yard, or foundation drains, from leaking manhole covers, or from infiltration [9]. Every effort should be made to prohibit illicit connections to sanitary sewers and to require construction and yard-grading techniques that will prevent surface-water entry into basements, manholes or sewer connections.

Some design flows are determined by integrating residential, commercial, and industrial peak flows and obtaining one peak flow for an area of sewerage development [9]. Recommended design practice, however, is to evaluate each component separately and to combine them step by step to determine the maximum flow.

STORM WATER RUNOFF

Although computation of storm-water-runoff flows will not be discussed, as indicated in the preface to this text, it should be noted that

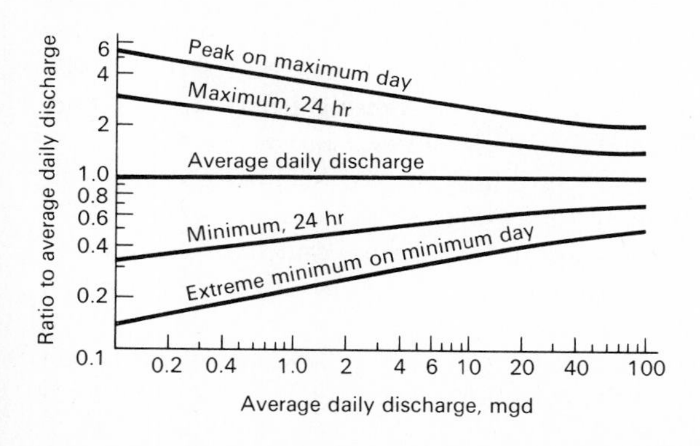

FIG. 2·7 Relation of extreme discharges on maximum and minimum days to the average daily discharge of domestic sewage in Massachusetts [9].

storm water runoff may increase the quantity of sewage flow. This may occur by leakage through manhole covers and other accessible openings or by infiltration caused by the elevation of the ground water table, and it should be considered in the design of interceptor sewers. It has been estimated that leakage through a manhole cover submerged under 1 inch (in.) of water will vary from 20 to 70 gpm, depending on the size and number of openings in the cover [15]. The infiltration resulting from an elevated water table is discussed below.

GROUND WATER INFILTRATION

One portion of the rainfall in a given area ordinarily runs quickly into the storm sewers or other drainage channels. Another portion is evaporated or absorbed by vegetation; and the remainder percolates into the ground, becoming ground water. The proportion that thus percolates into the ground depends on the character of the surface and soil formation and on the rate and distribution of the precipitation according to seasons. Any reduction in permeability, such as that due to buildings, pavements, or frost, decreases the opportunity for precipitation to become ground water and increases the surface runoff correspondingly.

The amount of ground water flowing from a given area may vary from a negligible amount for a highly impervious district or a district with a dense subsoil, to 25 or 30 percent of the rainfall for a semipervious district with a sandy subsoil permitting rapid passage of water into it. The percolation of water through the ground from rivers or other bodies of water sometimes has considerable effect on the ground water table, which rises and falls continually.

Infiltration into Sewers

The presence of high ground water results in leakage into the sewers and in an increase in the quantity of sewage and the expense of disposing of it. This infiltration from ground water may range from 1,000 to 40,000 or more gpd per mile of sewer. During heavy rains, when there may be leakage through manhole covers, as well as infiltration into the sewers themselves, the rate may exceed 100,000 gpd per mile of sewer. It is a variable part of the sewage, depending upon the quality of the materials and workmanship in the sewers and building connections, the character of the maintenance, and the elevation of the ground water compared to that of the sewers.

The sewers first built in a district usually follow the watercourses in the bottoms of valleys, close to and occasionally below the beds of brooks. As a result, these old sewers may receive comparatively large quantities of ground water, while sewers built later at higher elevations will receive relatively smaller quantities of ground water. With an increase in the percentage of area in a district that is paved or built over comes an increase in the percentage of storm water that is conducted rapidly to the storm sewers and watercourses and a decrease in the percentage of the storm water that can percolate into the earth and tend to leak into the sanitary sewers. A sharp distinction is to be made between maximum and average rates of leakage into sewer systems. The former are necessary to determine required sewer capacities; the latter are necessary to estimate such factors as annual costs of pumping sewage.

The rate and quantity of infiltration depends on the length of sewers, the area served, the soil and topographic conditions, and, to a certain extent, the population, which affects the number and total length of house connections. The leakage through defective joints, porous concrete, and cracks is large enough, in many cases, to lower the ground water table to the crown of the sewer and frequently well down toward the invert, although the elevation of the water table varies with the quantity of rain and snow water percolating into the ground.

The results of tests conducted by Santry [17] to determine the minimum achievable infiltration rate for small sewer pipes are shown in Table 2·14. The superiority of the new-sytle PVC joint material for clay pipe and of the rubber joint material for concrete pipe is indicated clearly. The use of the type-3 PVC joint is recommended.

Infiltration Design Allowances

Many old sewer systems and some of recent construction experience excessively high rates of infiltration that may result from extraneous water in addition to infiltration of ground water. Infiltration as used in sewer design may be considered to include extraneous water plus ground water leakage. In some cases, infiltration may be very high because of defective conditions or inferior workmanship in the construction of building connections in areas subject to high ground water.

In reviewing various specification allowances, Velzy and Sprague [22] found that the greatest percentage of specified allowable infiltration rates were within the range of the values shown in Table 2·15. On the other hand, some designers use an infiltration allowance of about 20,000 to 30,000 gpd per mile when computing the required capacity of sewers, usually adding this value to the peak sewage flow.

In many cases, since infiltration can be reduced to a negligible amount in properly constructed sewers, the infiltration allowance is

TABLE 2·14 SUBMERGENCE TEST RESULTS [17]

6-in.-diameter clay pipe			6-in.-diameter concrete pipe*		
Joint material	Head above flow line, in.	Average infiltration rate, gpd/in./mile	Joint material	Head above flow line, in.	Average infiltration rate, gpd/in./mile
Jute only	3	8,270	Jute only	3	6,710
	9	71,050		9	52,800
	15	155,250		15	118,000
	21	258,000		21	205,500
	27	356,000		27	278,000
Cement	3	3,360	Cement	3	680
	9	15,000		9	4,950
	15	28,700		21	16,500
	21	41,200		15	10,450
	27	53,200		27	22,000
Hot pour	3	1,330	Cold mastic	3	990
	9	1,660		9	1,450
	15	3,410		15	3,210
	21	4,720		21	5,130
	27	5,520		27	7,810
Old-style PVC	3	0		3	0
	9	645	Hot pour	9	107
	15	1,450		15	235
	21	1,850		21	419
	27	2,400		27	513
New-style PVC	0–27	Negligible	Rubber gasket	0–27	Negligible

* Not recommended.

determined on the basis of the size of the area, as shown in the three different curves in Fig. 2·8. Curves *A* and *B* may be used for infiltration allowances for areas having old sewers, and curve *C* may be used for the infiltration allowances for areas with new sewers where the ground water will never be above the sewer inverts and there will never be significant rates of flow of extraneous water into the sewers. The choice between curves *A* and *B* will depend upon the condition

TABLE 2·15 INFILTRATION ALLOWANCES FOR SANITARY SEWERS [22]

Pipe size, in.	Infiltration rate range, (gpd/mile)
8	3,500–5,000
12	4,500–6,000
24	10,000–12,000

of the sewers, the elevation of the ground water table, and the method of joint construction. For example, where sewer joints were formed using jute or cement and the presence of a high ground water table is known, curve *A* or higher rates should be used. Curve *B* may be used for new sewers if records or measurements of high flows in recently constructed sewers indicate excessive extraneous water, such as a sudden increase in the rate of flow within a fraction of an hour after the beginning of intense rainfall.

Infiltration and Exfiltration Measurements

When designing a new sewer, the construction specifications should include the performance of both infiltration and exfiltration tests. These tests should be conducted so that infiltration will be kept within the specified design limits and to ensure that good construction practices are followed. Details on the performance and interpretation of the results of infiltration and exfiltration tests may be found in Refs. [2, 5, and 23]. As the name implies, exfiltration is the efflux of

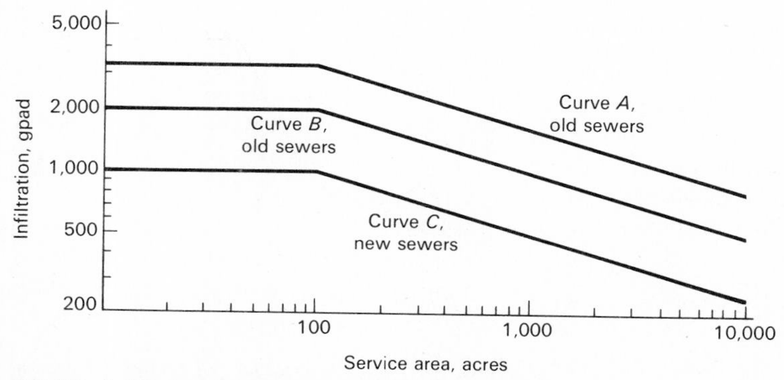

FIG. 2·8 Infiltration rate curves for old and new sewers.

water from the interior of the pipe. The rate of exfiltration from a sewer is directly related to the type of material used for the pipe joints and the character of workmanship pertaining to their construction.

PROBLEMS

2·1 Population data for nine different cities that were approximately the same size in 1960 are presented in the table at the end of this section. (*a*) Plot these data on a single sheet of graph paper. (*b*) List the population projection method or methods that would be applicable to each of these cities. (*c*) Pick a city and estimate the 1970 population. Compare your estimate to that of the 1970 census.

2·2 Predict the population of Ventura County, Calif., in 1970, using arithmetic and geometric projections. Population data from 1951 to 1964 are presented below.

Year	*Population*	*Year*	*Population*
1951	126,700	1958	182,000
1952	132,400	1959	195,200
1953	139,200	1960	208,700
1954	146,000	1961	223,600
1955	152,600	1962	245,900
1956	160,700	1963	281,400
1957	170,600	1964	314,000

2·3 If the saturation population of Boulder, Colo., is taken to be 180,000, estimate the population in 1975 by the decreasing-rate-of-increase method using the data given in Prob. 2·1.

2·4 If the sewage from city *A* flows into an aeration tank having a detention time (volume/flowrate) under average flow conditions of 6 hr, what would be the average detention time from 8 A.M. to 2 P.M.? From 8 P.M. to 2 A.M.? Use the flow variation shown in Fig. 2·9.

2·5 A sewer replacement program including building connections is proposed for city *A*. The existing sewer joints were formed using jute, while the proposed sewers and connections would have PVC joints. If the water table is relatively high, compute the percent change in infiltration for a service area of 1,000 acres.

2·6 Sewers are to be installed in a recreational camping area that contains a development campground for 200 persons, lodges and cabins for 100 persons, and residence homes for 50 persons. Assume that persons staying in lodges use the mess hall and that a 50-seat cafeteria has been constructed. Daily attendance at visitor centers is expected to be 50 percent of the campground capacity. Other facilities include a 10-machine laundromat, a 20-seat cocktail lounge, and three gas stations (2,000 gpd per station). Determine the average waste flow in gpd, using the design unit flows in Table 2·11.

2·7 Determine the peak flow from a hospital with a population of 500, by the fixture-unit method. Assume an average of five fixture units per person.

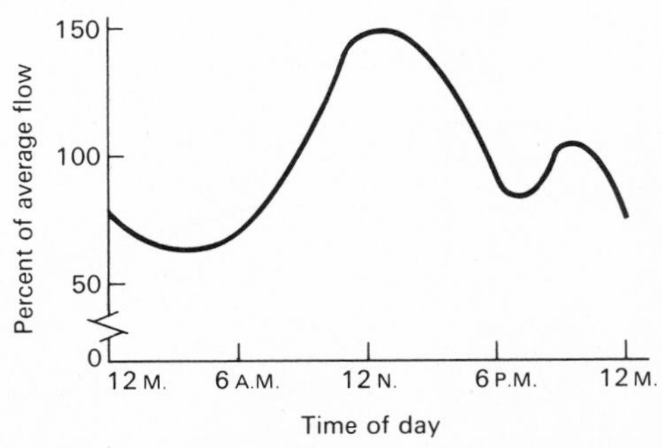

FIG. 2·9 Hourly variation of sewage flow for Prob. 2·4.

2·8 Maximum flowrates are to be determined for sewers in Philadelphia in 1976. For an area with a population of 30,000 and approximately 20 miles of sewers, determine the peak daily flow. Assume 70 percent of the water usage reaches the sewers. Use an infiltration allowance of 10,000 gpd per mile.

City	Population						
	1900	1910	1920	1930	1940	1950	1960
Everett, Mass.	24,336	33,484	40,120	48,424	46,784	45,982	43,544
Beverly, Mass.	13,884	18,650	22,561	25,086	25,537	28,884	36,108
Modesto, Calif.	2,024	4,034	9,241	13,842	16,379	17,389	36,585
Boulder, Colo.	6,150	9,539	11,006	11,223	12,958	19,999	37,718
Daytona Beach, Fla.	1,690	3,721	6,841	16,598	22,584	30,187	37,395
Butte, Mont.	30,470	39,165	41,611	39,532	37,081	33,251	27,877
Provo, Utah	6,185	8,925	10,303	14,766	18,071	28,937	36,047
Bangor, Me.	21,850	24,803	25,978	28,749	29,822	31,558	38,912
Auburn, N.Y.	30,345	34,668	36,192	36,652	35,753	36,722	35,249

REFERENCES

1. Babbitt, H. E., and E. R. Baumann: *Sewerage and Sewage Treatment*, 8th ed., Wiley, New York, 1958.
2. Borland, S.: New Data on Sewer Infiltration-Exfiltration Ratio, *Public Works*, vol. 87, no. 10, 1956.
3. Bureau of the Census, U.S. Department of Commerce: *Statistical Abstract of the United States*, 90th annual ed., Washington, D.C., 1969.
4. Geyer, J. C., and J. L. Lentz: *An Evaluation of the Problems of Sanitary Sewer System Design*, Final Report, Johns Hopkins University, Baltimore, Md., 1964.
5. Horne, R. W.: Control of Infiltration and Storm Flows in the Operation of Sewerage Systems, *Sewage Works Journal*, vol. 17, p. 209, 1945.
6. Hubbell, J. W.: Commercial and Institutional Wastewater Loadings, *J. WPCF*, vol. 34, no. 9, 1962.
7. Hunter, R. B.: *Methods of Estimating Loads on Plumbing Systems*, Report BMS65, National Bureau of Standards, Washington, D.C., 1940.
8. Johnson, R. E. L.: Development of Sanitary Sewer Design Criteria for the City of Houston, Texas, *J. WPCF*, vol. 37, no. 11, 1965.
9. Joint Committee of the American Society of Civil Engineers and the Water Pollution Control Federation: *Design and Construction of Sanitary and Storm Sewers*, ASCE Manual and Report 37, New York, 1969.
10. Manas, V. T. (ed.): *National Plumbing Code Handbook*, McGraw-Hill, New York, 1957.
11. McGauhey, P. H.: *Engineering Management of Water Quality*, McGraw-Hill, New York, 1968.

12. McJunkin, F. E.: Population Forecasting by Sanitary Engineers, *Journal of the Sanitary Division*, ASCE, vol. 90, no. SA4, 1964.

13. McLean, J. E.: More Accurate Population Estimate by Means of Logistic Curves, *Civil Engineering*, vol. 22, no. 2, 1952.

14. Present and Future Estimates of Water Consumption, *Public Works*, vol. 87, no. 12, 1956.

15. Rawn, A. W.: What Cost Leaking Manhole?, *Waterworks and Sewerage*, vol. 84, no. 12, 1937.

16. Salvato, J. A.: The Design of Small Water Systems, *Public Works*, vol. 91, no. 5, 1960.

17. Santry, I. W.: Infiltration in Sanitary Sewers, *J. WPCF*, vol. 36, no. 10, 1964.

18. Select Committee on National Water Resources, United States Senate: *Water Resources Activities in the United States*, Government Printing Office, Washington, D.C., 1960.

19. Studley, E. G., and A. Aarons: Current Sewer Design Practices in Los Angeles City, *J. WPCF*, vol. 38, no. 10, 1966.

20. United States of America Standards Institute, *National Plumbing Code*, USASI A40.8, 1955.

21. U.S. Department of Housing and Urban Development: *Minimum Design Standards for Community Sewerage Systems*, FHA 720, Washington, D.C., 1963.

22. Velzy, C. R., and J. M. Sprague: Infiltration Specification and Tests, *Sewage and Industrial Wastes*, vol. 27, no. 3, 1955.

23. Weller, L. W., and M. K. Nelson: A Study of Stormwater Infiltration into Sanitary Sewers, *J. WPCF*, vol. 35, no. 6, 1963.

hydraulics of sewers

3

The principal factors that affect the flow of wastewater and sewage in sewers are (1) slope, (2) cross-sectional area, (3) roughness of interior pipe surface, (4) conditions of flow, i.e., full, partly full, steady or varied flow, (5) presence or absence of obstructions, bends, etc., and (6) character, specific gravity, and viscosity of the liquid. The purpose of this chapter is to discuss the relationships of these factors and the fundamentals of fluid mechanics as applied to (1) the design of sewers and to those elements in sewage treatment plants involving pipe and open channel flow and (2) the measurement of flow.

FUNDAMENTALS OF FLUID MECHANICS

The analysis of flow in both closed conduits and open channels is based on an adaptation of three basic equations of fluid mechanics: the equation of continuity, the energy equation, and the momentum equation. Before discussing these as well as other concepts, some of the terms commonly used in the field of hydraulics must be defined.

Definition of Terms

The following terms are basic to an understanding of both pipe and open channel flow. A more complete listing of hydraulic terms may be found in Vennard [16].

Pipe and Open Channel Flow The flow of liquid in a pipe may be considered to be open or closed, depending on whether or not the pipe is flowing full. For open channel flow in a pipe, a free liquid surface that is subject to atmospheric pressure must exist. A comparison of pipe flow and open channel flow is shown schematically in Fig. 3·1.

Hydraulic Grade Line The hydraulic grade line, also shown in Fig. 3·1, is a line connecting the points to which the liquid would rise at various places along any pipe or conduit, if piezometer tubes were inserted in the liquid. It is a measure of the pressure head available at these various points. In the case of water flowing in a canal or open channel, as opposed to flow in a pipe under pressure, the hydraulic grade line corresponds with the profile of the water surface.

Energy Grade Line The total energy of flow in any section with reference to some datum is the sum of the elevation head z, the pressure head y, and the velocity head $V^2/2g$. The energy from section to section is usually represented by a line called the energy line or energy gradient (see Fig. 3·1). The form h_L represents the head loss between sections 1 and 2.

Specific Energy The specific energy E, sometimes called the specific head, is the sum of the pressure head y and the velocity head $V^2/2g$

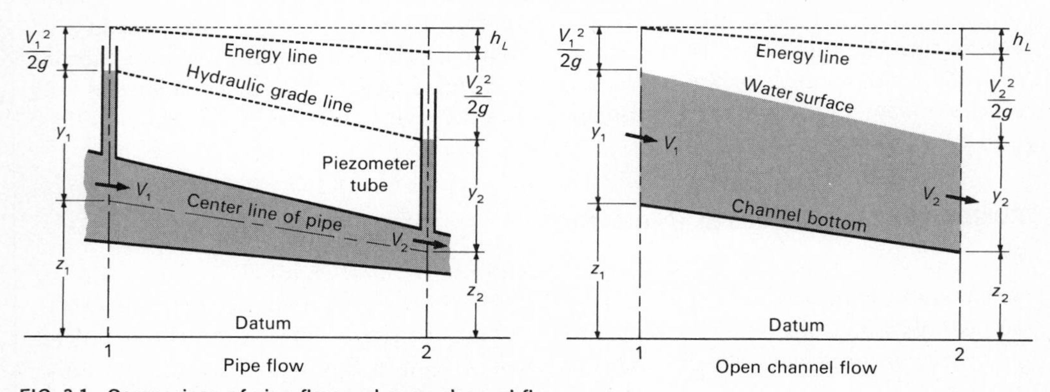

FIG. 3·1 Comparison of pipe flow and open channel flow.

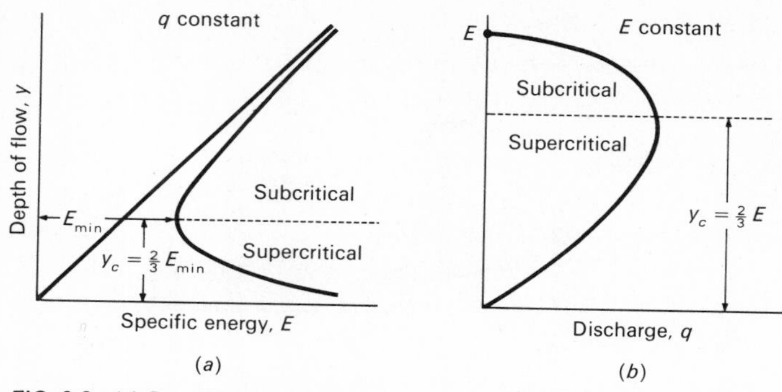

FIG. 3·2 (*a*) Specific-energy diagram and (*b*) q curve [16].

measured with respect to the channel bottom. The specific-energy concept is especially useful in the analysis of flow in open channels.

The relationship between specific energy and depth of flow for a constant rate of flow is illustrated in Fig. 3·2*a*, which usually is called a specific-energy diagram. The relationship between the depth of flow and discharge for constant specific energy is called a q curve and is illustrated in Fig. 3·2*b*.

Steady Flow Steady flow occurs when the discharge or rate of flow at any cross section is constant.

Uniform and Nonuniform Flow Uniform flow exists when the depth, cross-sectional area, and other elements of flow are substantially constant from section to section. Nonuniform flow exists when the slope, cross-sectional area, and velocity are changing from section to section. An example of steady nonuniform flow is the flow through a Venturi section used for pressure measurements.

Varied Flow The flow in a channel is considered varied if the depth of flow changes along the length of the channel. In general, the flow may be gradually varied (GVF) or rapidly varied (RVF) as shown in Fig. 3·3. Rapidly varied flow occurs where the depth of flow changes abruptly.

Equation of Continuity

The equation of continuity expresses the continuity of flow from section to section in a streamtube as shown in Fig. 3·4. According to the principle of conservation of mass, mass can be neither created nor

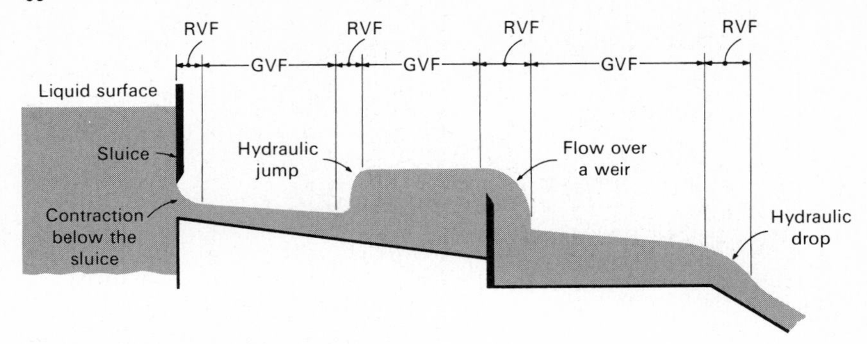

FIG. 3·3 Definition sketch of varied flow.

destroyed between sections A_1 and A_2. Thus, the equation of continuity becomes

$$\rho_1 A_1 V_1 = \rho_2 A_2 V_2 \tag{3·1}$$

where ρ = density, slug/ft
A = cross-sectional area, sq ft
V = velocity, fps

If the fluid is incompressible, then $\rho_1 = \rho_2$, and

$$A_1 V_1 = A_2 V_2 \tag{3·2}$$

The Energy Equation

A flowing fluid may possess four types of energy: displacement or pressure energy E_p, velocity energy E_v, potential energy E_q, and thermal or internal energy E_i. If E_m represents the mechanical energy transferred to or from the fluid (i.e., in a pump, fan, or turbine), and E_h represents

FIG. 3·4 Flow through a streamtube control volume.

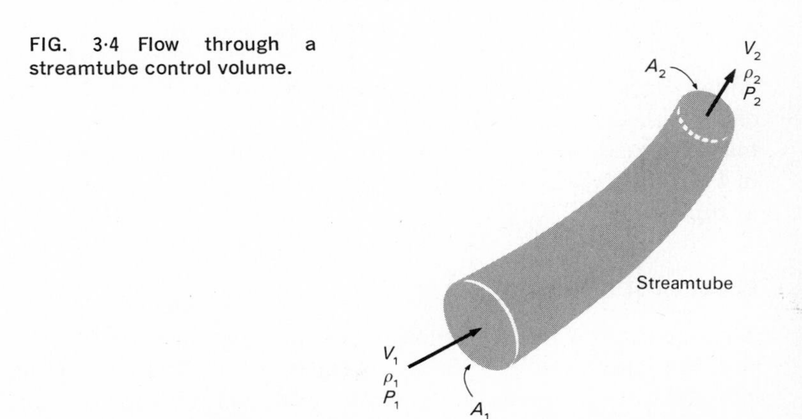

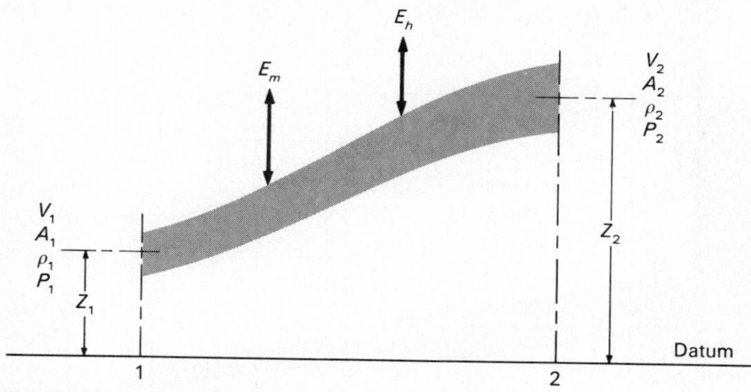

FIG. 3·5 Definition sketch for energy equation.

the heat energy transferred to or from the fluid (i.e., in a heat exchanger), then application of the law of conservation of energy between points 1 and 2 in Fig. 3·5 yields the following equation:

$$(E_p + E_v + E_i)_1 \pm E_m \pm E_h = (E_p + E_v + E_q + E_i)_2 \tag{3·3}$$

The general expression for a fluid in Eq. 3·3 may be rewritten as

$$\frac{p_1}{\gamma} + \alpha_1 \frac{V_1^2}{2g} + z_1 \pm E_m \pm E_h = \frac{p_2}{\gamma} + \alpha_2 \frac{V_2^2}{2g} + z_2 + h_L \tag{3·4}$$

In Eq. 3·4, α_1 and α_2 represent the kinetic-energy correction factors. For turbulent flow in pipes, the value of α usually ranges between 1.01 and 1.10. In practice, α is usually taken to be equal to unity. The head loss h_L between points 1 and 2 in Fig. 3·5 is equal to $E_{i,1} - E_{i,2}$. If the fluid in question is ideal (frictionless) and no mechanical or heat energy is transferred, then Eq. 3·4 reduces to

$$\frac{p_1}{\gamma} + \frac{V_1^2}{2g} + z_1 = \frac{p_2}{\gamma} + \frac{V_2^2}{2g} + z_2 \tag{3·5}$$

which is the familiar form of the Bernoulli equation for incompressible flow. Application of the energy or Bernoulli's equation to flow in pipelines is shown in Fig. 3·6. The energy equation written between points 1 and 2 would be

$$H = \frac{p_1}{\gamma} + \frac{V_1^2}{2g} + z_1 = \frac{p_2}{\gamma} + \frac{V_2^2}{2g} + z_2 + h_{L_{1-2}} \tag{3·6}$$

Although the foregoing analysis is always applicable, it is generally not used for the solution of sewage-flow problems because of the laborious nature of the calculations involved. Solutions of sewer-flow problems are discussed later in this chapter.

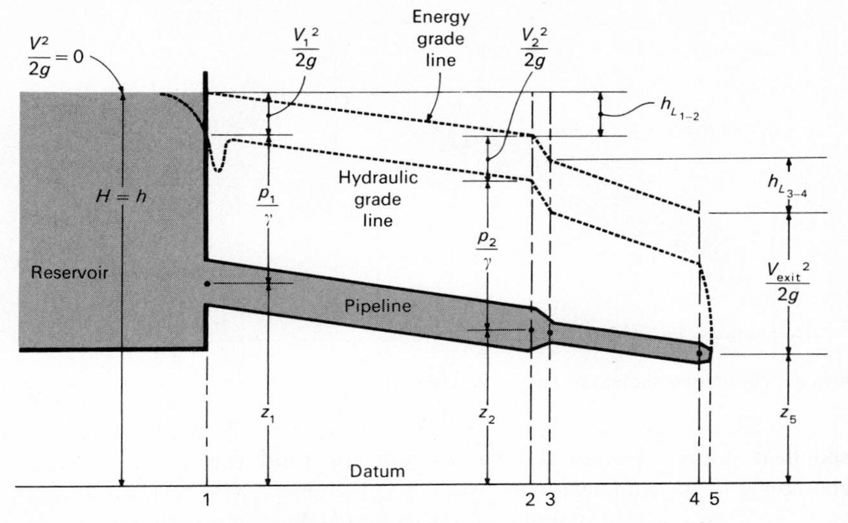

FIG. 3·6 Definition sketch for application of energy equation to a pipeline.

Pumps offer another example of the application of the energy equation, as shown in Fig. 3·7. In this case, the energy equation written between points 1 and 2 is

$$\frac{p_1}{\gamma} + \frac{V_1^2}{2g} + z_1 + E_p = \frac{p_2}{\gamma} + \frac{V_2^2}{2g} + z_2 \tag{3·7}$$

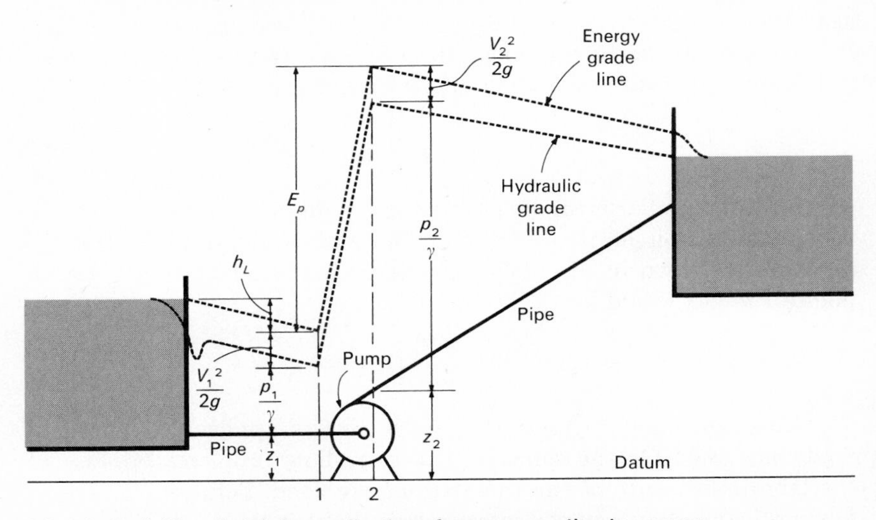

FIG. 3·7 Definition sketch for application of energy equation to a pump.

In Eq. 3·7, E_p is the energy imparted to the water by the pump; friction through the pump passages is neglected.

The Momentum Equation

The momentum carried across area A_1 (see Fig. 3·4) in time dt is equal to

$$(\text{Mass transferred across } A_1)V_1 = (\rho_1 V_1 A_1\, dt)V_1 \tag{3·8}$$

and is in the direction of V_1. In a similar manner, the momentum across A_2 is equal to

$$M_2 = (\rho_2 V_2 A_2\, dt)V_2 \tag{3·9}$$

Thus, the change in momentum between A_1 and A_2 in the streamtube is equal to

$$\Delta M = (\rho_2 V_2 A_2\, dt)V_2 - (\rho_1 V_1 A_1\, dt)V_1 \tag{3·10}$$

The law of conservation of momentum may be stated as

The net force applied = the rate of change of momentum with time

or

$$\Sigma F = \frac{(\rho_2 V_2 A_2\, dt)V_2 - (\rho_1 V_1 A_1\, dt)V_1}{dt} \tag{3·11}$$

In Eq. 3·11, ΣF is the net force on fluid body between points 1 and 2 (see Fig. 3·4).

Analysis of the flow in pipe bends (see Fig. 3·8) affords an excellent

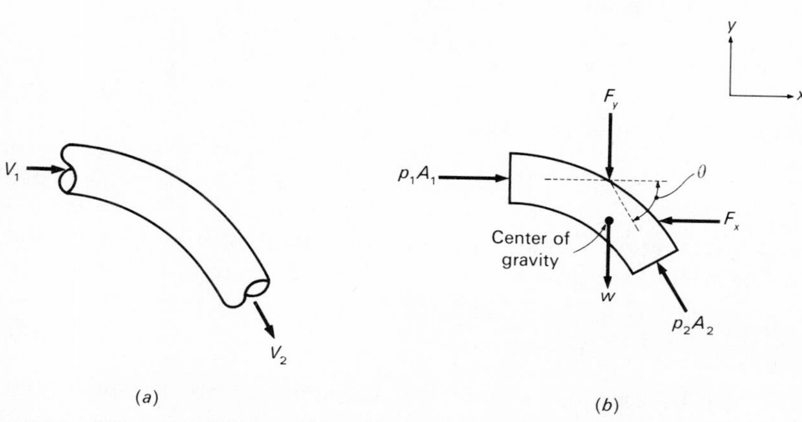

FIG. 3·8 Definition sketch for application of momentum equation.

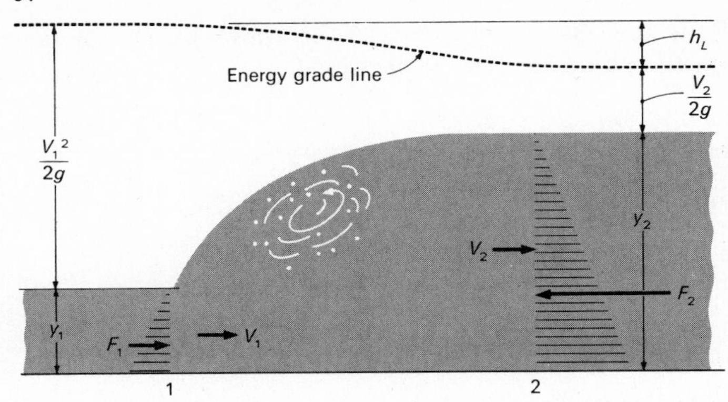

FIG. 3·9 Definition sketch for application of momentum equation to a hydraulic jump.

example of the application of the momentum equation (Eq. 3·11). Application of the momentum equation to the flow in the pipe bend shown in Fig. 3·8 results in the following equations:

$$p_1A_1 - p_2A_2 \cos\theta - F_x = Q\rho(V_2 \cos\theta - V_1) \tag{3·12}$$

$$p_2A_2 \sin\theta - w - F_y = Q\rho V_2 \sin\theta \tag{3·13}$$

Equations 3·12 and 3·13 are used to compute the forces that must be resisted in a pipe bend that is flowing full of water. A typical situation in which the foregoing analysis may be applied is in the design of sewer force mains.

Another application of the momentum and continuity principles to the flow in open channels is in the development of the equation for a hydraulic jump. If two-dimensional flow is considered and friction is neglected, application of the momentum equation (Eq. 3·11) to the flow shown in Fig. 3·9 results in Eq. 3·14.

$$\Sigma F_x = F_1 - F_2 = \frac{\gamma y_1^2}{2} - \frac{\gamma y_2^2}{2} = \frac{qy(V_2 - V_1)}{2} \tag{3·14}$$

From the equation of continuity,

$$V_2 = \frac{q}{y_2} \qquad \text{and} \qquad V_1 = \frac{q}{y_1} \tag{3·15}$$

and substituting for V_1 and V_2 in Eq. 3·14 and rearranging terms, the following result is obtained:

$$\frac{y_1}{y_2} = \frac{1}{2}\left(-1 + \sqrt{1 + \frac{8q^2}{gy_2^3}}\right) \tag{3·16}$$

which may be used to compute the alternate depths of flow shown in Fig. 3·9.

FLOW EQUATIONS

As pointed out earlier, to overcome the difficulties associated with application of the energy principle to the solution of pipe-flow problems, a wide variety of equations have been developed. Some have a rational basis. For the most part, however, the various equations for flow in pipes and channels are essentially empirical. They apply to steady uniform flow and consider only the losses due to friction.

The Chezy and Darcy-Weisbach Equations

One of the earliest expressions for frictional losses in pipes was developed by Chezy in 1775. The Chezy equation was intended to be applicable either to open channels or to pipes under pressure and is perhaps as satisfactory for one case as the other. The other equations that will be discussed were derived and originally applied to either open channels or pressure pipes, but their application has been extended, and they may all be used with a fair degree of satisfaction for open and closed channels. The Chezy equation is

$$V = C\sqrt{RS} \tag{3·17}$$

where V = mean velocity, fps
C = Chezy coefficient
R = hydraulic radius, ft
S = slope

The value of C to be used in Eq. 3·17 for pipes of various construction, size, and shape is difficult to ascertain. Indeed, several of the most widely used equations are, in effect, the Chezy equation with additional terms for determining the value of C more readily. The hydraulic radius is defined as the cross-sectional area of flow divided by the wetted or "frictional" perimeter.

The Darcy-Weisbach equation, which is generally written

$$h_L = f\frac{LV^2}{D2g} \tag{3·18}$$

where h_L = friction loss, ft
f = coefficient of friction
L = length of pipe, ft
D = diameter of pipe, ft
V = mean velocity, fps
g = acceleration due to gravity

is a rearrangement of the Chezy equation with the roughness coefficients related as follows:

$$f = \frac{8g}{C^2} \tag{3·19}$$

The value of f has been found to vary with Reynolds number N_R, pipe roughness, pipe size, and other factors. The relationships among these variables are presented graphically in Figs. 3·10 and 3·11, commonly known as Moody diagrams. The effect of size and roughness is expressed by the ratio of absolute roughness of the pipe ϵ, expressed in feet, to the pipe diameter D, also expressed in feet. By properly estimating or measuring the value of ϵ, the appropriate friction factor f may be determined from Fig. 3·11 or may be computed using the following equation if the flow regime is wholly turbulent:

$$\frac{1}{\sqrt{f}} = 2 \log \frac{D}{\epsilon} + 1.14 \tag{3·20}$$

The Manning Equation

Based on work conducted during the latter part of the nineteenth century, Robert Manning published his now well-known equation for flow in open channels:

$$V = \frac{1.486}{n} R^{2/3} S^{1/2} \tag{3·21}$$

where n = coefficient of roughness
R = hydraulic radius, ft
S = slope, as for Chezy equation (Eq. 3·17)

Although this equation was originally intended for use in the design of open channels, it is now used for both open and closed channels. Because of its simplicity and because the n value is essentially the same as the n value used in Kutter's equation (discussed in the next section), Manning's equation is now the one most commonly used in the design of sewers.

Even though this equation is used extensively, considerable misunderstanding exists concerning the selection of the proper n values. Typical n values for various types of pipes are presented in Table 3·1. In general, n values varying from 0.013 to 0.015 are used in sewer design. Further, these n values are usually assumed to be valid for all depths of flow. Experiments conducted to determine the effect of variation in depth of flow on the friction factor n have conclusively

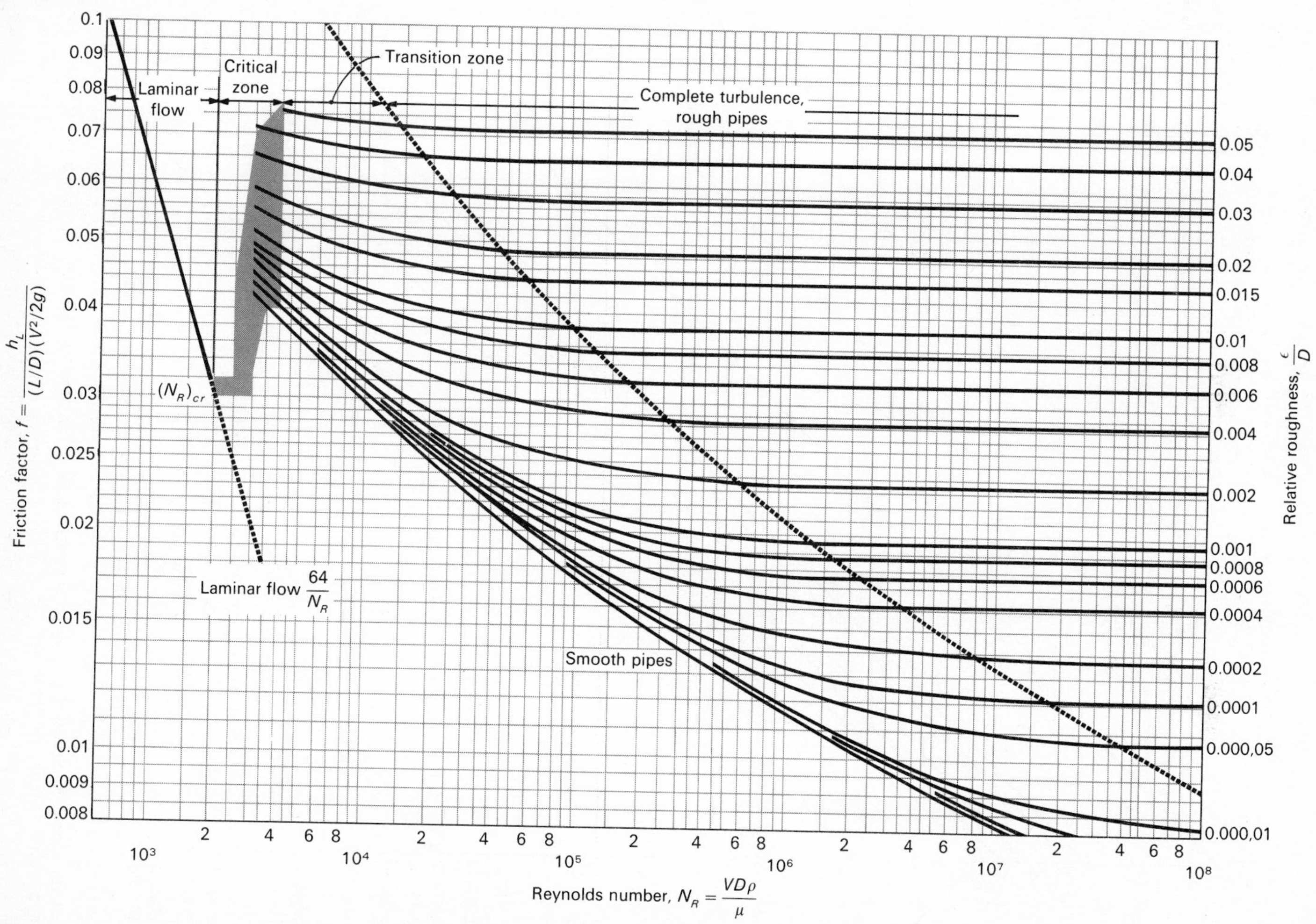

FIG. 3·10 Moody diagram for friction factor in pipes versus Reynolds number and relative roughness [12].

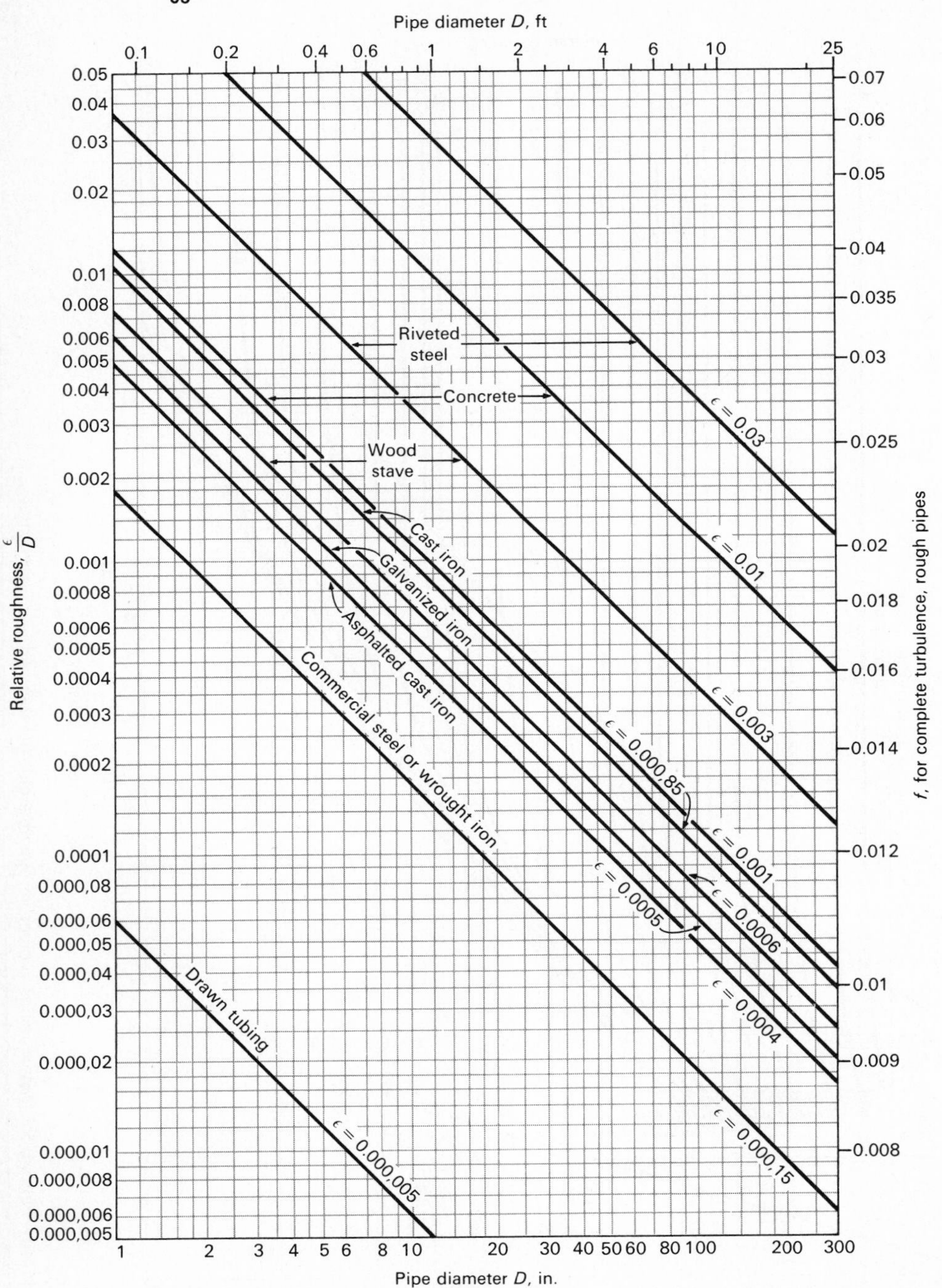

FIG. 3·11 Moody diagram for relative roughness as a function of diameter for pipes constructed of various materials [adapted from 12].

TABLE 3·1 VALUES OF n TO BE USED WITH THE MANNING EQUATION [10]

Surface	Best	Good	Fair	Bad
Uncoated cast-iron pipe	0.012	0.013	0.014	0.015
Coated cast-iron pipe	0.011	0.012*	0.013*	
Commercial wrought-iron pipe, black	0.012	0.013	0.014	0.015
Commercial wrought-iron pipe, galvanized	0.013	0.014	0.015	0.017
Smooth brass and glass pipe	0.009	0.010	0.011	0.013
Smooth lockbar and welded "OD" pipe	0.010	0.011*	0.013*	
Riveted and spiral steel pipe	0.013	0.015*	0.017*	
Vitrified sewer pipe	{0.010, 0.011}	0.013*	0.015	0.017
Common clay drainage tile	0.011	0.012*	0.014*	0.017
Glazed brickwork	0.011	0.012	0.013*	0.015
Brick in cement mortar; brick sewers	0.012	0.013	0.015*	0.017
Neat cement surfaces	0.010	0.011	0.012	0.013
Cement mortar surfaces	0.011	0.012	0.013*	0.015
Concrete pipe	0.012	0.013	0.015*	0.016
Wood stave pipe	0.010	0.011	0.012	0.013
Plank flumes:				
Planed	0.010	0.012*	0.013	0.014
Unplaned	0.011	0.013*	0.014	0.015
With battens	0.012	0.015*	0.016	
Concrete-lined channels	0.012	0.014*	0.016*	0.018
Cement-rubble surface	0.017	0.020	0.025	0.030
Dry-rubble surface	0.025	0.030	0.033	0.035
Dressed-ashlar surface	0.013	0.014	0.015	0.017
Semicircular metal flumes, smooth	0.011	0.012	0.013	0.015
Semicircular metal flumes, corrugated	0.0225	0.025	0.0275	0.030
Canals and ditches:				
Earth, straight and uniform	0.017	0.020	0.0225*	0.025
Rock cuts, smooth and uniform	0.025	0.030	0.033*	0.035
Rock cuts, jagged and irregular	0.035	0.040	0.045	
Winding sluggish canals	0.0225	0.025*	0.0275	0.030
Dredged earth channels	0.025	0.0275*	0.030	0.033
Canals with rough stony beds, weeds on earth banks	0.025	0.030	0.035*	0.040
Earth bottom, rubble sides	0.028	0.030*	0.033*	0.035
Natural stream channels:				
(1) Clean, straight bank, full stage, no rifts or deep pools	0.025	0.0275	0.030	0.033
(2) Same as (1), but some weeds and stones	0.030	0.033	0.035	0.040
(3) Winding, some pools and shoals, clean	0.033	0.035	0.040	0.045
(4) Same as (3), lower stages, more ineffective slope and sections	0.040	0.045	0.050	0.055
(5) Same as (3), some weeds and stones	0.035	0.040	0.045	0.050
(6) Same as (4), stony sections	0.045	0.050	0.055	0.060
(7) Sluggish river reaches, rather weedy or with very deep pools	0.050	0.060	0.070	0.080
(8) Very weedy reaches	0.075	0.100	0.125	0.150

* Values commonly used in designing.

shown n to be greater in partially filled pipes than in pipes flowing full [9]. The variation in n with depth of flow in a circular pipe is shown in Fig. 3·16. Thus, when designing sewers that will usually flow partly full, the n value used should be adjusted accordingly. Selection of n values for use in sewer design is discussed more fully in Chap. 4.

Another area in which considerable misunderstanding exists is in the variation of n with pipe size. If Manning's equation is converted by substituting $D/4$ for R and noting that S may be written as h_L/L, then equating this expression for velocity to the expression for velocity obtained by rewriting the Darcy-Weisbach equation results in the following expression:

$$n = 0.0934R^{1/6}\sqrt{f} = 0.074D^{1/6}\sqrt{f} \tag{3·22}$$

Equation 3·22 shows the relationship between the n in Manning's equation and the f in the Darcy-Weisbach equation. Consideration of Eq. 3·22 shows that if the absolute wall roughness ϵ remains constant with the pipe size, the value of n will increase with pipe size, although the increase will be slight since f decreases with pipe size.

The Kutter Equation

In its general form, Kutter's equation is

$$V = \frac{41.66 + \dfrac{1.811}{n} + \dfrac{0.00281}{S}}{1 + \left(41.66 + \dfrac{0.00281}{S}\right)\dfrac{n}{\sqrt{R}}}\sqrt{RS} \tag{3·23}$$

This equation was derived for open channels, such as canals and natural streams. It was therefore inferred that it was equally applicable to conduits and sewers that are open channels (except when flowing full). However, the cross sections of such conduits when more than half full are radically different from those of ordinary open channels; and, unless the n value is varied, the equation does not give correct values of V for all depths in circular pipes.

Typical n values used in Eq. 3·23 vary between 0.010 and 0.150, depending on the type of channel under consideration. Although the use of this equation has been extensive, its popularity is declining because of its empirical nature and contingent uncertainties.

The Hazen-Williams Equation

Of the numerous exponential types of equations for the flow of water in pipes, the Hazen-Williams equation developed in 1902 has been most commonly used. The Hazen-Williams equation is

$$V = 1.318CR^{0.63}S^{0.54} \tag{3·24}$$

where C may be regarded as a coefficient of roughness. This equation is generally used for pipes discharging under pressure. Typical C values are shown in Table 3·2. When this equation is used in sewer-flow

TABLE 3·2 HAZEN-WILLIAMS COEFFICIENTS [16]

Type of pipe	C
Pipes extremely straight and smooth	140
Pipes very smooth	130
Smooth wood, smooth masonry	120
New riveted steel, vitrified clay	110
Old cast iron, ordinary brick	100
Old riveted steel	95
Old iron in bad condition	60–80

problems, such as the analysis of force mains, the C factor must be adjusted to reflect the composition of the sewage.

Laminar-Flow Equations

In laminar flow, the forces of viscosity predominate in relation to other forces, such as inertia. An example of laminar flow is the pumping of sludge at a sewage treatment plant. Under laminar conditions, the flow rate Q may be expressed as

$$Q = \frac{\pi D^4 \gamma h_L}{128\mu L} \tag{3·25}$$

where π = constant = 3.1416
D = diameter, ft
γ = specific weight, lb/cu ft
h_L = head loss, ft
μ = viscosity, lb-sec/sq ft
L = length, ft

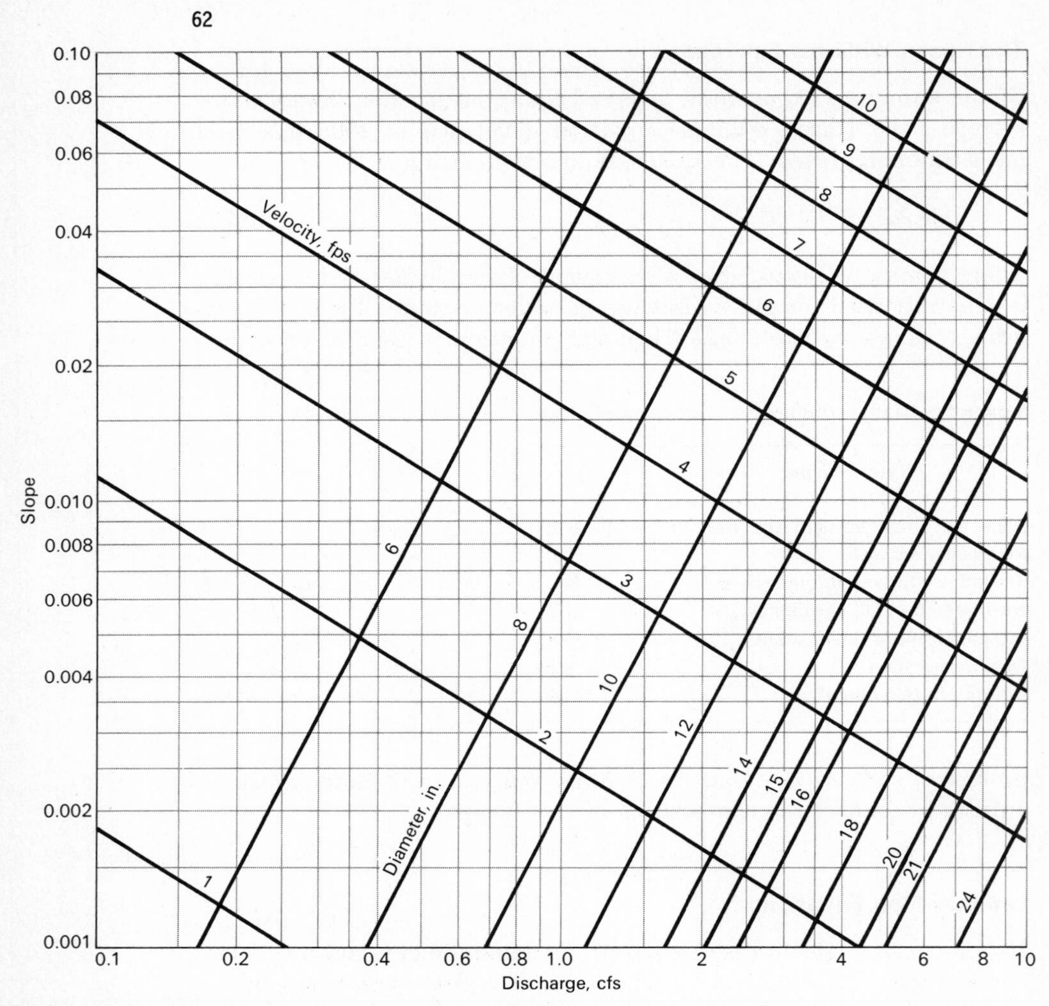

FIG. 3·12 Nomograph for Manning's equation ($n = 0.013$) for flows between 0.1 and 10.0 cfs and slopes between 0.001 and 0.1.

The head-loss term h_L may be expressed as

$$h_L = \frac{32\mu LV}{\gamma D^2} \tag{3·26}$$

By equating Eq. 3·26 to the Darcy-Weisbach equation (Eq. 3·18), it can be shown that the friction factor f in the latter equation is equal to

$$f = \frac{64\mu}{D\rho} = \frac{64}{N_R} \tag{3·27}$$

PIPE AND OPEN CHANNEL FLOW

In general, sewers are designed with the expectation that they will flow full only under maximum flow conditions. It is therefore important to note that the ordinary condition of flow in sewers is that of an open channel in which there is a free water surface in contact with air. When full, sewers are usually under no significant pressure, except in the case of force mains and inverted siphons.

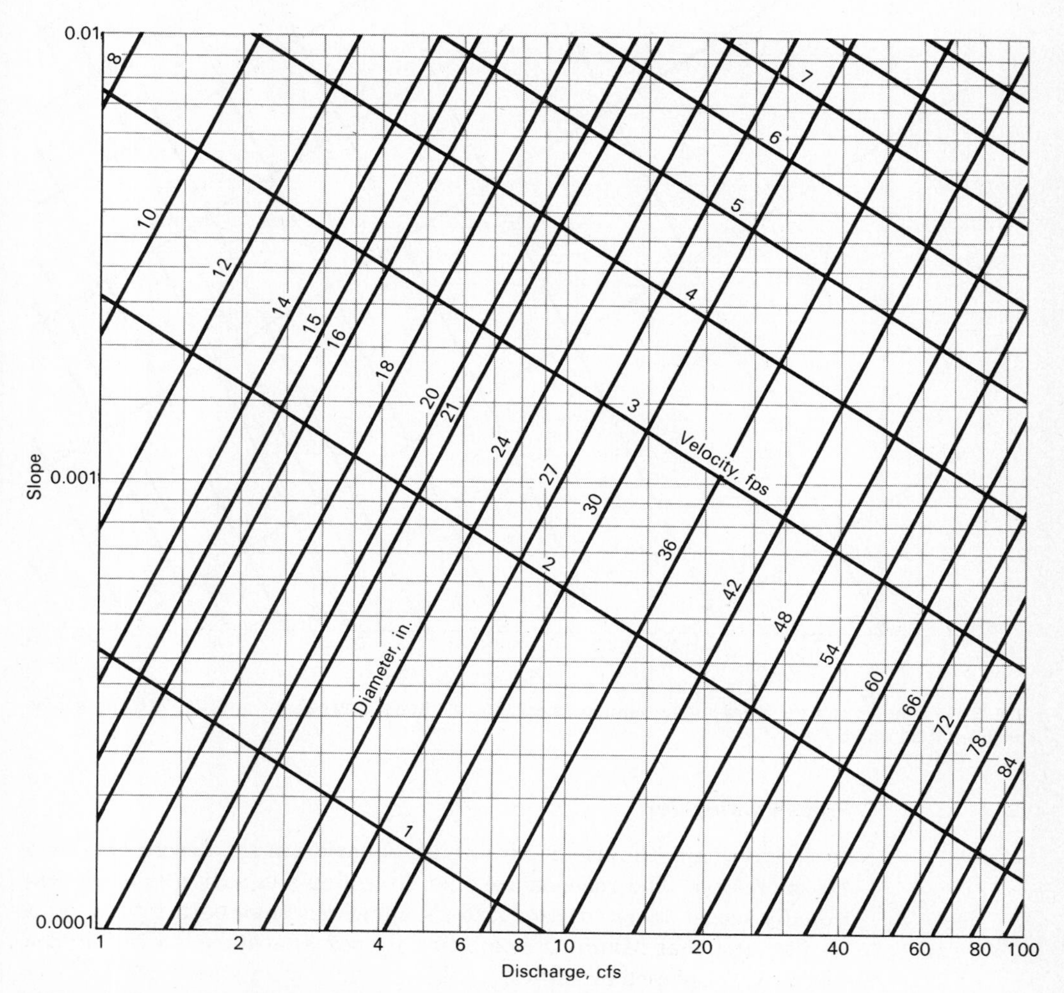

FIG. 3·13 Nomograph for Manning's equation ($n = 0.013$) for flows between 1 and 100 cfs and slopes between 0.0001 and 0.01.

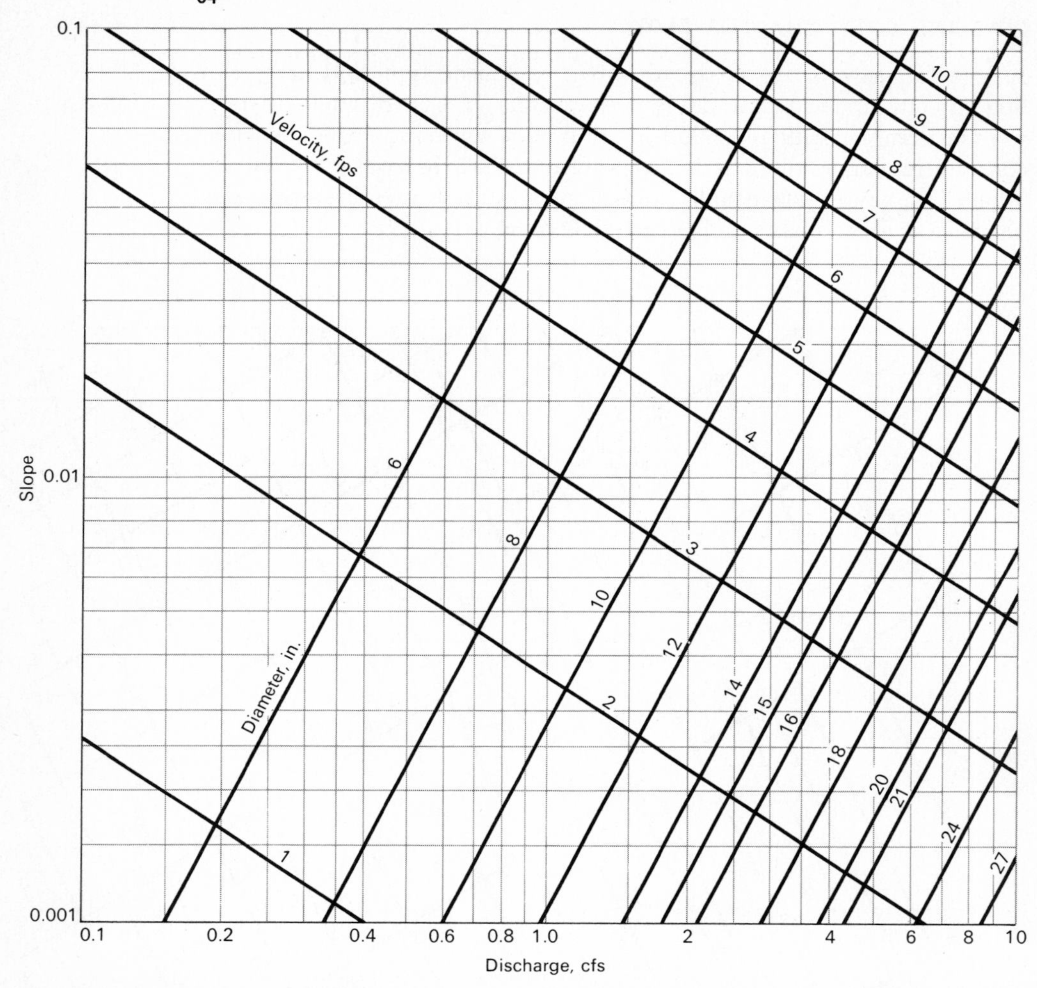

FIG. 3·14 Nomograph for Manning's equation ($n = 0.015$) for flows between 0.1 and 10.0 cfs and slopes between 0.001 and 0.1.

Pipes Flowing Full

In recent years, the use of Manning's equation in the design of sewers has increased. The reasons are that Manning's equation is computationally easier to use than Kutter's equation, that both give similar results, and that Manning's equation is used almost exclusively in the solution of open-channel-flow problems. The most direct method for using Manning's equation for the solution of pipe-flow problems is through the use of diagrams, such as those shown in Figs. 3·12 to 3·15, which can easily be prepared for any encountered design conditions. Values of 0.013 and 0.015 for Manning's n were selected for these diagrams because, as mentioned earlier, these are the most frequently used values in sewer design practice.

Flow in Partly Filled Sections

In many of the problems arising in sewer design, it is necessary to estimate the velocity and discharge when a sewer is partly filled. The

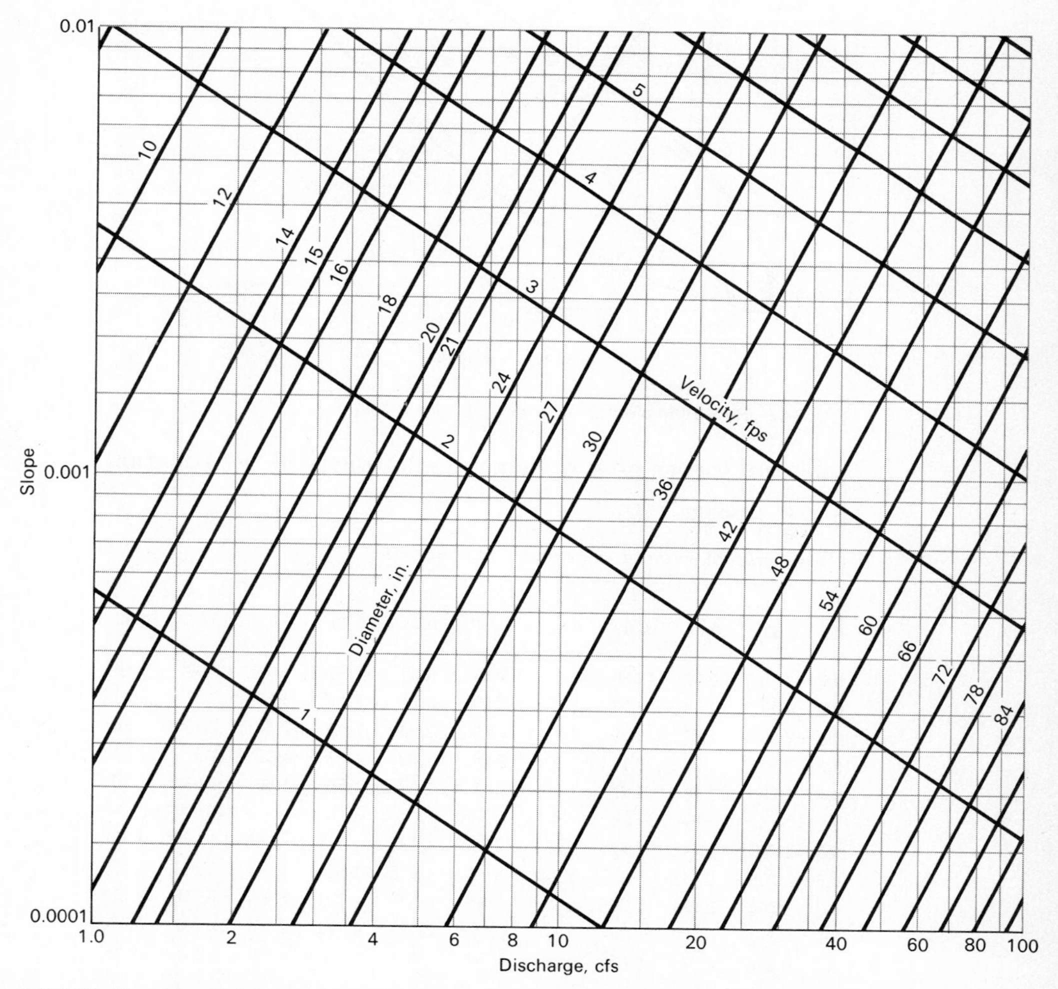

FIG. 3·15 Nomograph for Manning's equation ($n = 0.015$) for flows between 1 and 100 cfs and slopes between 0.0001 and 0.01.

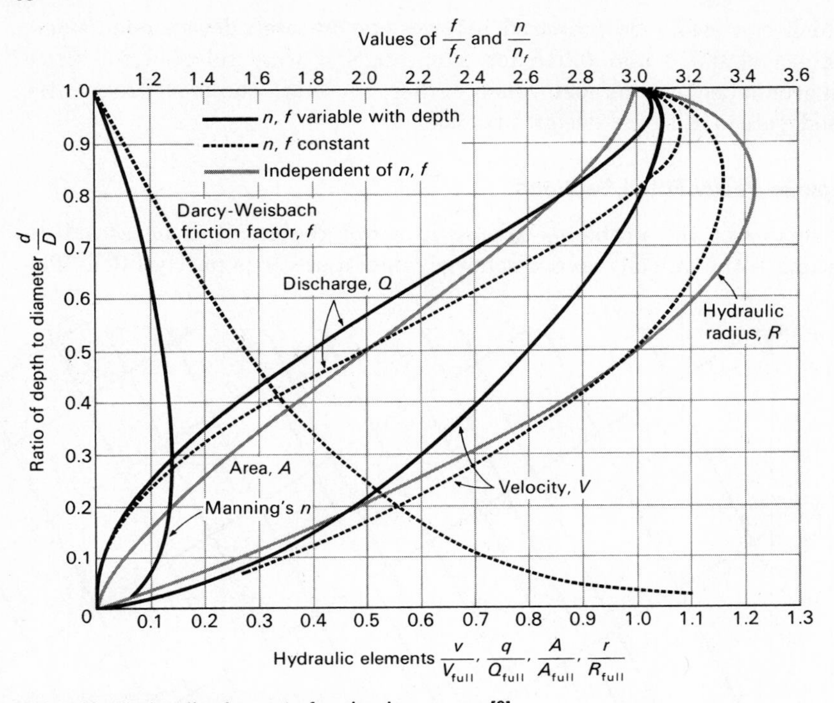

FIG. 3·16 Hydraulic elements for circular sewers [9].

TABLE 3·3 VALUES OF K FOR CIRCULAR CHANNELS IN THE EQUATION $Q = \frac{K}{n} D^{\frac{8}{3}} S^{\frac{1}{2}}$ [10]

(D = depth of water; d = diameter of channel)

$\frac{D}{d}$	0.00	0.01	0.02	0.03	0.04	0.05	0.06	0.07	0.08	0.09
0.0		15.02	10.56	8.57	7.38	6.55	5.95	5.47	5.08	4.76
0.1	4.49	4.25	4.04	3.86	3.69	3.54	3.41	3.28	3.17	3.06
0.2	2.96	2.87	2.79	2.71	2.63	2.56	2.49	2.42	2.36	2.30
0.3	2.25	2.20	2.14	2.09	2.05	2.00	1.96	1.92	1.87	1.84
0.4	1.80	1.76	1.72	1.69	1.66	1.62	1.59	1.56	1.53	1.50
0.5	1.470	1.442	1.415	1.388	1.362	1.336	1.311	1.286	1.262	1.238
0.6	1.215	1.192	1.170	1.148	1.126	1.105	1.084	1.064	1.043	1.023
0.7	1.004	0.984	0.965	0.947	0.928	0.910	0.891	0.874	0.856	0.838
0.8	0.821	0.804	0.787	0.770	0.753	0.736	0.720	0.703	0.687	0.670
0.9	0.654	0.637	0.621	0.604	0.588	0.571	0.553	0.535	0.516	0.496
1.0	0.463									

TABLE 3·4 VALUES OF K' FOR CIRCULAR CHANNELS IN THE EQUATION $Q = \frac{K'}{n} d^{\frac{8}{3}} S^{\frac{1}{2}}$ [10]

(D = depth of water; d = diameter of channel)

$\frac{D}{d}$	0.00	0.01	0.02	0.03	0.04	0.05	0.06	0.07	0.08	0.09
0.0		0.00007	0.00031	0.00074	0.00138	0.00222	0.00328	0.00455	0.00604	0.00775
0.1	0.00967	0.0118	0.0142	0.0167	0.0195	0.0225	0.0257	0.0291	0.0327	0.0366
0.2	0.0406	0.0448	0.0492	0.0537	0.0585	0.0634	0.0686	0.0738	0.0793	0.0849
0.3	0.0907	0.0966	0.1027	0.1089	0.1153	0.1218	0.1284	0.1352	0.1420	0.1490
0.4	0.1561	0.1633	0.1705	0.1779	0.1854	0.1929	0.2005	0.2082	0.2160	0.2238
0.5	0.232	0.239	0.247	0.255	0.263	0.271	0.279	0.287	0.295	0.303
0.6	0.311	0.319	0.327	0.335	0.343	0.350	0.358	0.366	0.373	0.380
0.7	0.388	0.395	0.402	0.409	0.416	0.422	0.429	0.435	0.441	0.447
0.8	0.453	0.458	0.463	0.468	0.473	0.477	0.481	0.485	0.488	0.491
0.9	0.494	0.496	0.497	0.498	0.498	0.498	0.496	0.494	0.489	0.483
1.0	0.463									

relations between hydraulic elements for flow at full depth and at other depths, computed according to Manning's equation, may be found from diagrams such as the one shown in Fig. 3·16.

The hydraulic elements for a circular pipe, as shown in Fig. 3·16, are the hydraulic radius R, the cross-sectional area of the flowing stream A, the average velocity V, and the rate of discharge Q. The curves for R and A are the same for all diameters of pipes; however, the curves for V and Q will depend on the values of slope S and the selected friction coefficient n. To avoid the necessity of preparing a number of diagrams for various S and n values, calculations involving flow in partly filled pipes can more easily be handled using the data in Tables 3·3 and 3·4, taken from King and Brater's *Handbook of Hydraulics* [10]. The following two examples illustrate the use of Fig. 3·16 and Tables 3.3 and 3.4.

EXAMPLE 3·1 *Depth and Velocity in Partly Filled Sections*

Determine the depth of flow and velocity in a 12-in.-diameter sewer laid on a slope of 0.005 with an n value of 0.015 when discharging 0.30 cfs.

Solution

1. Compute K' in the equation $Q = (K'/n)d^{\frac{8}{3}}S^{\frac{1}{2}}$.

$$K' = \frac{nQ}{d^{\frac{8}{3}}S^{\frac{1}{2}}} = \frac{0.015(0.30)}{1(0.07)} = 0.0643$$

2. Determine depth of flow. Entering Table 3·4 with the computed K' value gives a D/d value of 0.25 and the

   Depth of flow = 12(0.25) = 3 in.

3. Compute velocity. From Fig. 3·16 the area of flow is found to be 0.19 times the area of the pipe and is equal to 0.19(0.7854) = 0.149.

$$\text{Velocity} = \frac{0.30}{0.149} = 2.01 \text{ fps}$$

EXAMPLE 3·2 *Diameter of Sewer Flowing Partly Full*

Determine the diameter of a sewer required to handle a flow of 5 cfs when flowing half full. The sewer is to be laid at a slope of 0.001, and the n value is assumed to be 0.013.

Solution

1. Determine K' in the equation $Q = (K'/n)d^{\frac{8}{3}}S^{\frac{1}{2}}$. Entering Table 3-4 with a D/d value of 0.5, the corresponding value of K' is 0.232.

2. Compute diameter d.

$$d = \left[\frac{Qn}{K'S^{\frac{1}{2}}}\right]^{\frac{3}{8}} = \left[\frac{5.0(0.013)}{0.232(0.001)^{\frac{1}{2}}}\right]^{\frac{3}{8}} = 2.27 \text{ ft}$$

Hydraulic Elements of Various Sewer Sections

The majority of sewers presently constructed in this country are of circular cross section. In the past, however, a wide variety of noncircular sewer sections were employed, including egg-shaped, semielliptical, horseshoe, baskethandle, oval, catenary, gothic, parabolic, and elliptical. The first four of these were the more popular shapes. Typical examples of these sections, along with data on their hydraulic elements, are shown in Figs. 3·17 and 3·18 and in Table 3.5.

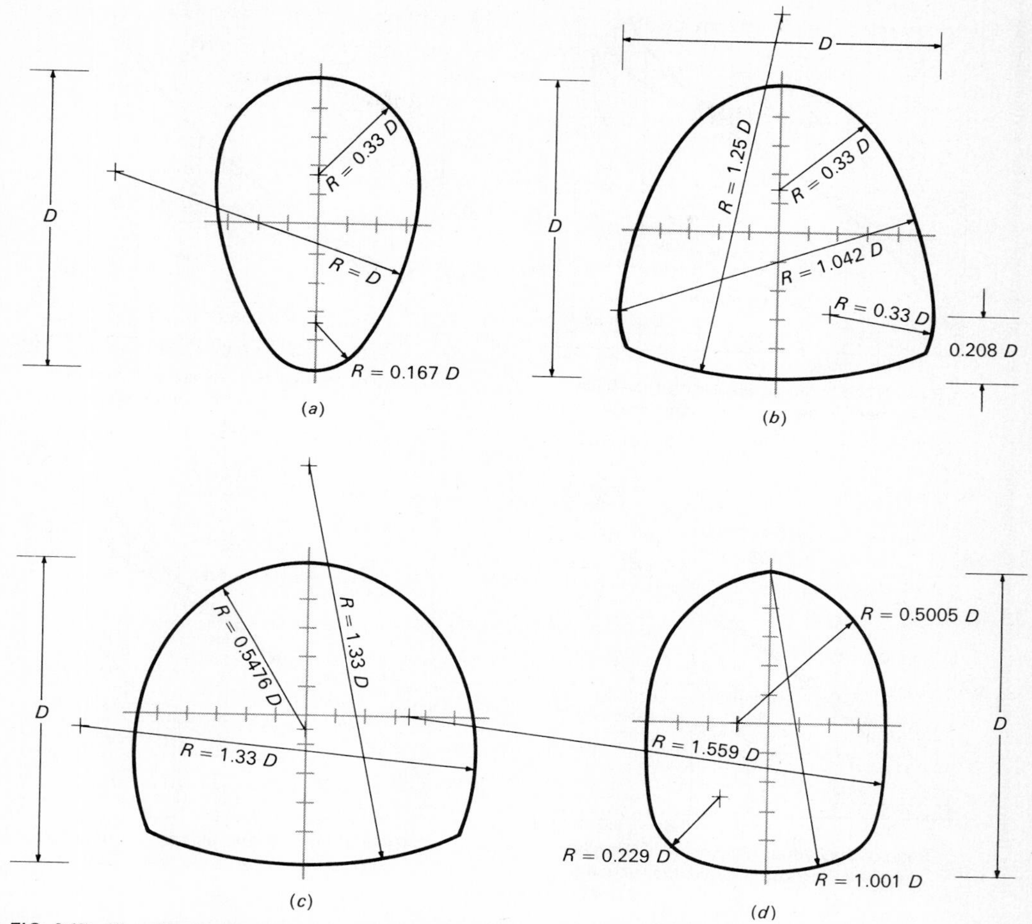

FIG. 3·17 Noncircular sewer sections [11]. (*a*) Egg-shaped section, (*b*) semielliptical section, (*c*) horseshoe section, (*d*) baskethandle section.

Detailed data on these and other sections may be found in Refs. [1 and 11].

Flow in Open Channels

Flow problems in open channels are usually solved using either the Chezy, Manning, or Kutter equations. Here again, the Manning equation is the one most widely used. In the application of Manning's equation to open-channel-flow problems, the most difficult problem is

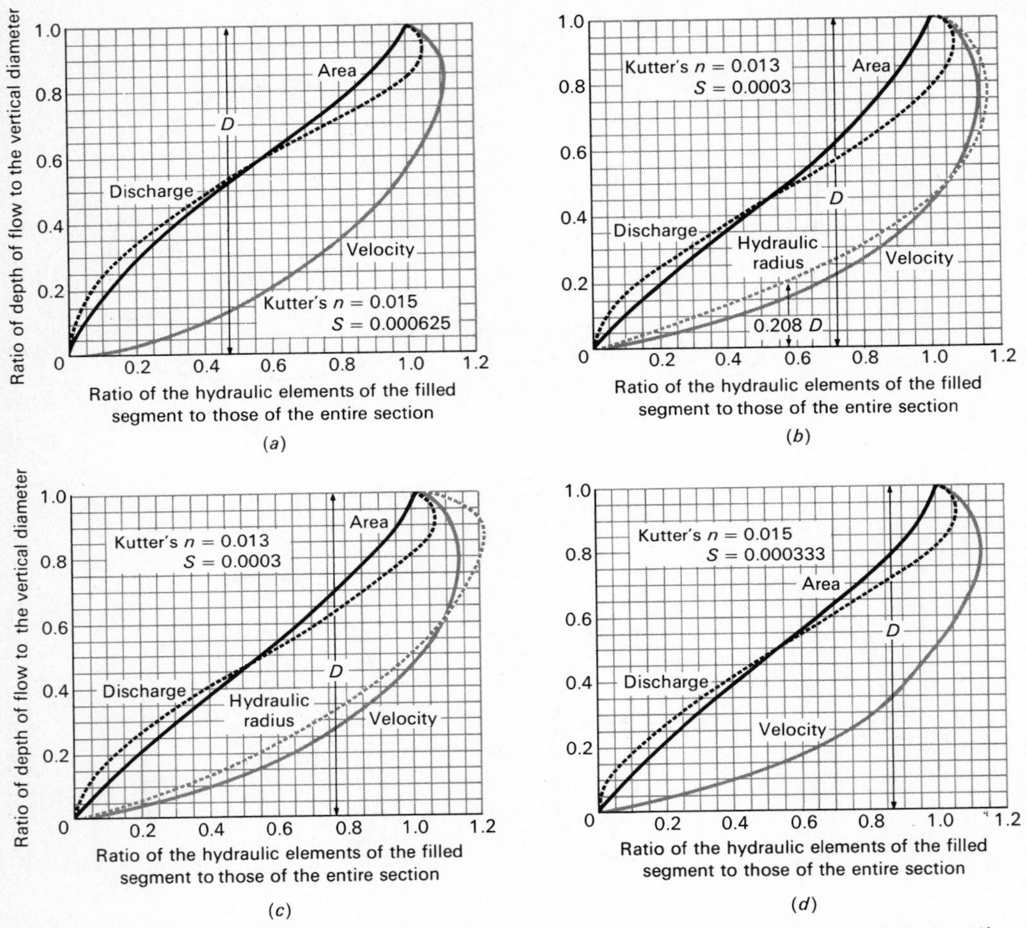

FIG. 3·18 Hydraulic elements of noncircular sewers [11]. (*a*) Egg-shaped section, (*b*) semielliptical section (*c*) horseshoe section, (*d*) baskethandle section.

selection of the friction coefficient n. When dealing with problems involving open channel flow, Refs. [4 and 10] are highly recommended.

NONUNIFORM FLOW

Conditions of steady nonuniform flow exist when a constant quantity of water flows with variable cross sections, slopes, and velocities. The surface of the water, therefore, is not parallel to the invert of the conduit. This condition always exists at points of changing equilibrium,

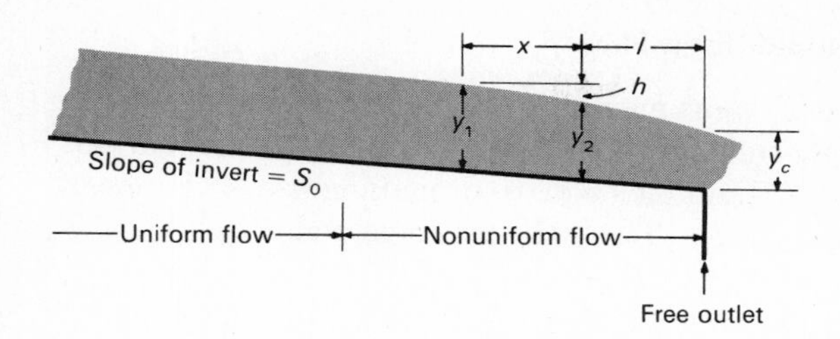

FIG. 3·19 Drawdown curve with free discharge.

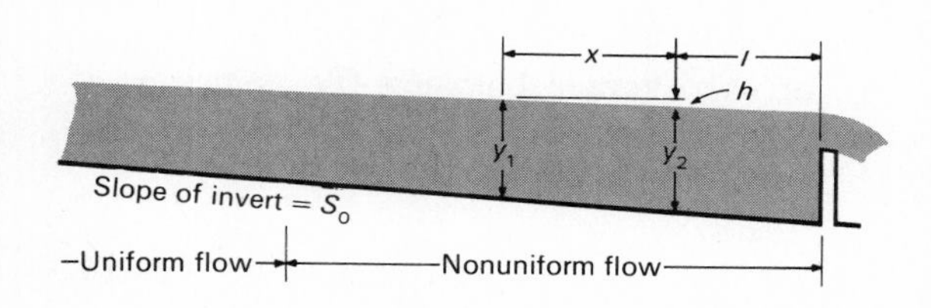

FIG. 3·20 Typical backwater curve.

TABLE 3·5 HYDRAULIC ELEMENTS OF VARIOUS SEWER SECTIONS [11]
(See Figs. 3·17 and 3·18)

	Hydraulic elements of full sections		
Type of sewer	Area	Wetted perimeter	Hydraulic radius
Egg-shaped	$0.510D^2$	$2.643D$	$0.193D$
Semielliptical	$0.783D^2$	$3.258D$	$0.240D$
Horseshoe	$0.913D^2$	$3.466D$	$0.263D$
Baskethandle	$0.786D^2$	$3.193D$	$0.246D$

such as at and near changes in grade and in cross section, and above obstructions of free outlets. Typical examples of nonuniform flow are shown in Figs. 3·19 to 3·21.

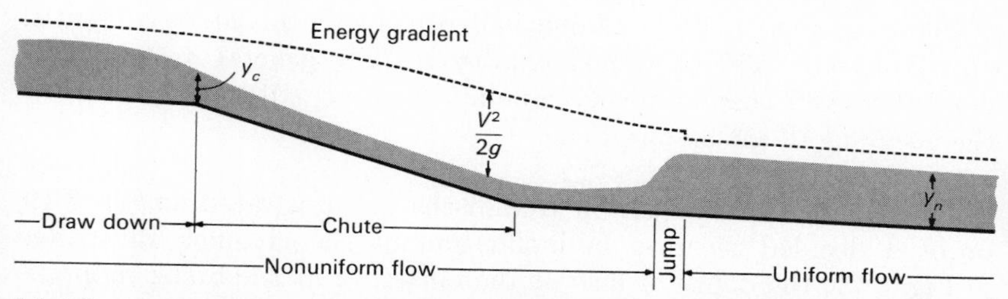

FIG. 3·21 Drawdown curve, chute flow, and hydraulic jump in a prismatic channel.

General Equation for Nonuniform Flow

Assume a reach of conduit short enough so that the loss due to friction may be computed with sufficient accuracy by one of the equations for uniform flow, such as Manning's equation, making use of the mean hydraulic radii and velocities. Then the average velocity (V_{av}) in the reach would be

$$V_{av} = \frac{1.486}{n} R_{av}^{\frac{2}{3}} S^{\frac{1}{2}} \tag{3·28}$$

Taking $V_{av} = \frac{1}{2}(V_1 + V_2)$, or the average between the velocities at the ends of the reach, and $R_{av} = \frac{1}{2}(R_1 + R_2)$, the slope S is then given by Eq. 3·29:

$$S = \frac{n^2(V_1 + V_2)^2}{8.83 R_{av}^{\frac{4}{3}}} \tag{3·29}$$

Since the total fall in the water surface h is equal to the sum of the frictional loss and the difference in the velocity heads,

$$h = xS + \frac{V_2^2}{2g} - \frac{V_1^2}{2g} \tag{3·30}$$

The same expression for the drop in water surface is applicable on the basis of any equation for frictional loss.

If y_1 and y_2 represent the depths of water at the two ends of the reach, and S_0 represents the inclination of the bottom, then

$$h = xS_0 + (y_1 - y_2) \tag{3·31}$$

Equating the values of h,

$$x = \frac{\left(y_1 + \frac{V_1^2}{2g}\right) - \left(y_2 + \frac{V_2^2}{2g}\right)}{S - S_0} \tag{3·32}$$

From this expression the distance x between any two sections of the stream, in which the change in depth is $y_1 - y_2$, can be computed approximately. The foregoing expressions are general and may be applied to any case of steady nonuniform flow, within the limits of the accuracy of the assumptions.

Critical Depth In the case of free discharge, illustrated in Fig. 3·19, or of a decided increase in inclination of the channel, illustrated in Fig. 3·21, the depth of flow at the outlet, or at the break in grade, will be definitely fixed by the rate of discharge for any given conduit. The case is analogous to the discharge of a weir. This depth is called the critical depth, designated as y_c, and is the depth for which the specific energy (that is, $y + V^2/2g$) is a minimum. Differentiating the expression for specific energy for a rectangular channel with respect to depth, and setting it equal to zero, yields

$$\frac{dH}{dy} = 1 - \frac{q^2}{g y_c^3} = 0 \tag{3·33}$$

or

$$y_c = \sqrt[3]{\frac{q^2}{g}} \tag{3·34}$$

Similar computations of the critical depths corresponding to various rates of discharge for other forms of cross section can be made, but they are likely to be very complex.

Within the specified limits, the following equation may be used for computing critical depth in a circular section flowing partly full:

$$y_c = 0.325 \left(\frac{Q}{d}\right)^{\frac{2}{3}} + 0.083d \qquad \left(0.3 < \frac{y_c}{d} < 0.9\right) \tag{3·35}$$

where y_c = critical depth, ft
Q = discharge, cfs
d = pipe diameter, ft

An alignment chart (see Fig. 3·22) may be used to make critical-depth determinations. To determine the critical depth, knowing the flow and the channel geometry, enter the chart from the left. The chart may also be used to determine the flow, knowing the diameter and the critical depth.

Drawdown The transition from a condition of uniform flow in the conduit to the discharge at a free outlet or the drop at a chute is accomplished by a gradual decrease in depth of flow with the stream surface similar to the surface curve above a weir. If the conduit is on a flat slope and the normal velocity is low, the difference between the normal depth and the critical depth will be substantial, the drawdown will be consequential, and the drawdown curve will extend upstream for a considerable distance. In such a case, it may sometimes be possible to make a saving by reducing the size of the conduit and eliminating the drawdown or by lowering the roof while leaving the width unchanged. On the other hand, if the conduit is on such a steep slope that the velocity is high and the critical depth is only slightly lower than the normal depth, no substantial reduction in size of conduit can be made, and the drawdown will not be significant. In those cases, too,

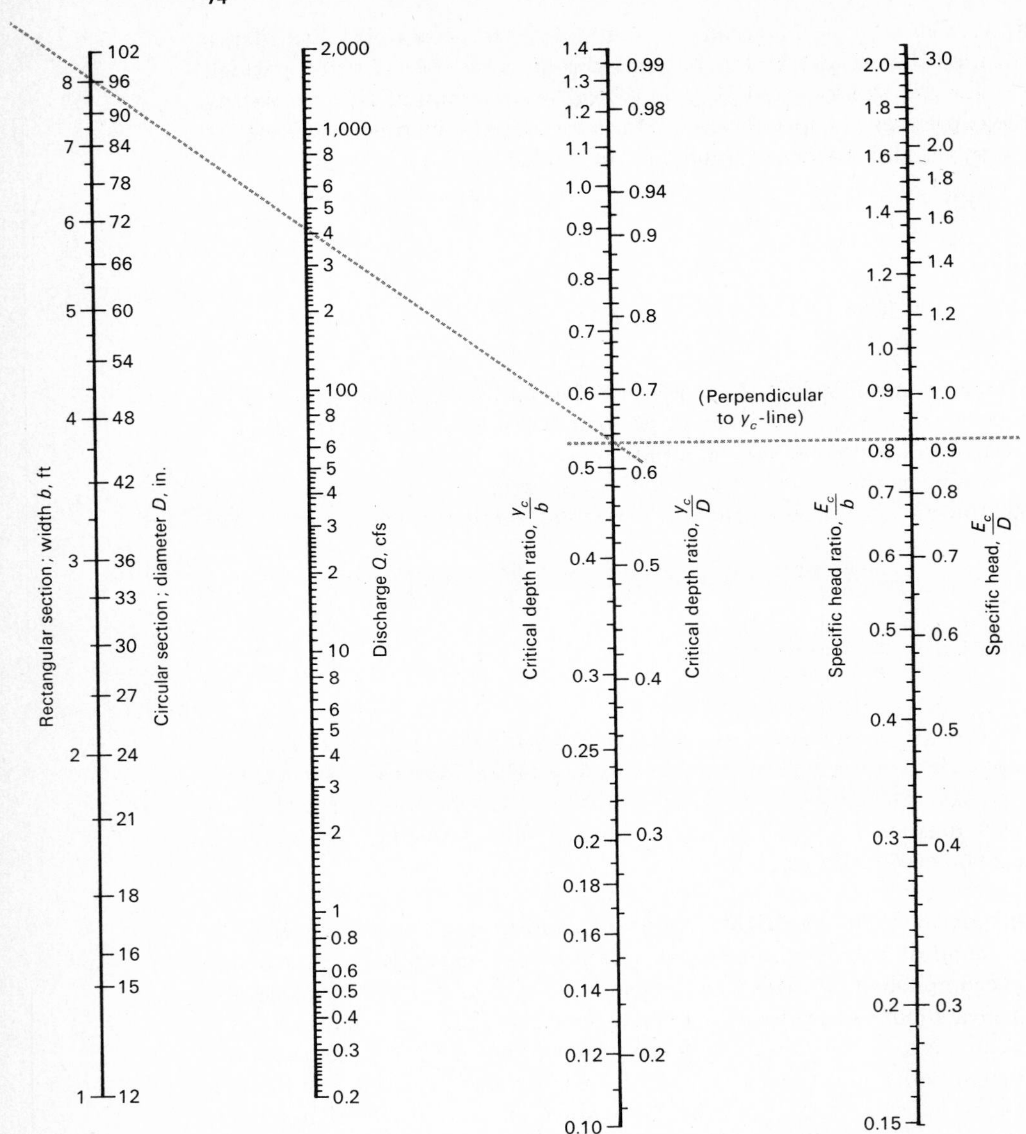

FIG. 3·22 Critical depth of flow and specific head in rectangular and circular conduits [9].

where there is possibility that the conduit may flow full under pressure at times, reduction in section for drawdown would be undesirable.

The length of the drawdown from the control section (the point of critical depth) can be computed approximately by successive steps utilizing the general equation for uniform flow. The computational procedure is illustrated in the following example.

EXAMPLE 3·3 *Length of the Drawdown Curve in Rectangular Section*

Determine the length of the drawdown curve for a rectangular channel 10 ft wide, with a slope of 0.001, with $n = 0.015$, and $Q = 250$ cfs, that discharges freely, as shown in Fig. 3·19.

Solution

1. Using Manning's equation, determine the normal depth of flow. For $S = 0.001$ when $y = 4.0$ ft, $Q = 214$ cfs and when $y = 5.0$ ft, $Q = 288$ cfs. By interpolating when $Q = 250$ cfs, $y = 4.5$ ft, which will be the depth of flow when the effect of the drawdown is no longer felt.
2. Determine the depth at the control section. The critical depth y_c will be about 2.7 ft.
3. Determine the length of the water surface between the depths of 4.5 and 2.7 ft. The stepwise computations are summarized in Table 3·6.

Backwater The surface curve of a stream of water when backed up by a dam or other obstruction is called the backwater curve (see Fig. 3·20). It may be necessary to determine the amount by which the depth is increased at specified points or the distance upstream to which the effect of backwater can be detected. The computational procedure is illustrated in Example 3·4.

EXAMPLE 3·4 *Length of Backwater Curve in Circular Section*

Determine the length of the backwater curve for a 6-ft-diameter reinforced-concrete pipe that discharges 90 cfs to a wet well. The slope is 0.001 and $n = 0.013$. Determine the length of the curve to the point where the depth of flow is 4 ft. Evaluate the effect of assuming n to be constant by also computing the distance assuming n to vary with depth.

Solution

Computations similar to those summarized in Table 3·6 have been made and are summarized in Table 3·7 for n constant and in Table 3·8 for n varying with depth. As can be seen by comparing the tabulated values in these tables, the overall length of the backwater curve is approximately doubled by varying n with depth.

Chute A chute is a channel with so steep a grade that uniform flow can take place at a depth less than the critical depth (see Fig. 3·21). The computation of flow in a chute, in so far as such flow is uniform, is accomplished by the use of the ordinary equations, such as Manning's or Kutter's, although the applicability of these equations at very high velocities is doubtful. Chutes are usually short, however, and the portion of their length in which uniform flow conditions exist is often insignificant. Nonuniform flow occurs at the upper end of the chute, and the hydraulic jump may occur at the lower end if conditions beyond the chute are such as to produce it.

Hydraulic Jump When water moving at a high velocity in a comparatively shallow stream strikes water having a substantial depth, there is likely to be a rise in the water surface of the stream, forming what is called a hydraulic jump (see Fig. 3·21). The jump cannot occur unless the primary depth of the water is less than the critical depth; and, when it does occur, the water surface rises from a point below to a related point above that representing the critical depth.

EXAMPLE 3·5 *Hydraulic Jump*

A rectangular channel 10 ft wide with $n = 0.015$ (see Example 3·3) is carrying a flow of 250 cfs. If the inclination of the upper section (chute) is 0.05, determine the length below the end of the chute where the hydraulic jump will occur.

Solution

1. Determine normal depth in upper section (chute). Using Manning's equation (assumed applicable for very high velocities), the depth of flow in the upper section is found to be 1.17 ft, which is less than the critical depth of 2.7 ft.
2. Compute the depth above the hydraulic jump. The normal depth of flow in the lower section, with its slope of 0.001, is

TABLE 3·6 COMPUTATION OF DRAWDOWN CURVE [11]

y	A	p	R	V	$\frac{V^2}{2g}$	$y + \frac{V^2}{2g}$	Average R	Average V	nV	S	$S - S_0$*	$\Delta\left(y + \frac{V^2}{2g}\right)$	Δx
2.7	27	15.4	1.75	9.28	1.34	4.04							
							1.81	8.81	0.132	0.0035	0.0025	0.04	16
3.0	30	16.0	1.87	8.33	1.08	4.08							
							1.95	7.84	0.118	0.0025	0.0015	0.16	107
3.4	34	16.8	2.02	7.35	0.84	4.24							
							2.09	6.98	0.105	0.0019	0.0009	0.24	266
3.8	38	17.6	2.16	6.61	0.68	4.48							
							2.23	6.32	0.0095	0.0014	0.0004	0.27	675
4.2	42	18.4	2.28	5.97	0.55	4.75							
							2.33	5.76	0.00865	0.0011	0.0001	0.23	2,300
4.5	45	19.0	2.37	5.56	0.48	4.98							
												Total length of drawdown =	3,364

* S_0 = invert slope.

TABLE 3·7 COMPUTATION OF BACKWATER CURVE WITH CONSTANT n

d	$\frac{d}{D}$	K'	$\frac{K'}{n}$	A	V	$\frac{V^2}{2g}$	$d + \frac{V^2}{2g}$	$\frac{K'}{n} d^{\frac{8}{3}}$	$S^{\frac{1}{2}}$	S	S_a	$S_a - S_0$*	$\Delta\left(d + \frac{V^2}{2g}\right)$	Δx
6.0	1.00	.463	35.62	28.72	3.18	.16	6.16	4.24 × 10³	.0212	.000449				
											.000421	− .000579	− .49	846
5.5	0.92	.497	38.23	27.22	3.30	.17	5.67	4.55 × 10³	.0198	.000392				
											.000417	− .000583	− .47	806
5.0	0.83	.468	36.00	25.09	3.58	.20	5.20	4.28 × 10³	.0210	.000441				
											.000492	− .000508	− .46	906
4.5	0.75	.422	32.46	22.75	3.96	.24	4.74	3.86 × 10³	.0233	.000543				
											.000634	− .000366	− .43	1,175
4.0	0.67	.366	28.15	20.12	4.47	.31	4.31	3.35 × 10³	.0269	.000724				
											Total length of backwater =			3,733

* S_0 = invert slope.

TABLE 3·8 COMPUTATION OF BACKWATER CURVE WITH VARIABLE n

d	$\frac{d}{D}$	$\frac{q}{Q_f}$	Q_{full}	V_{full}	$\frac{V}{V_{full}}$	V	$\frac{V^2}{2g}$	$d + \frac{V^2}{2g}$	S	S_a	$S_a - S_0$*	$\Delta\left(d + \frac{V^2}{2g}\right)$	Δx
6.0	1.00	1.00	90	3.18	1.00	3.18	.16	6.16	.00045				
										.00045	−.00055	−.49	891
5.5	0.92	1.01	89	3.15	1.04	3.28	.17	5.67	.00044				
										.00049	−.00051	−.47	922
5.0	0.83	0.91	99	3.50	1.02	3.57	.20	5.20	.00054				
										.00065	−.00035	−.46	1,314
4.5	0.75	0.79	114	4.03	0.98	3.95	.24	4.74	.00075				
										.00087	−.00013	−.44	3,385
4.0	0.67	0.67	134	4.74	0.93	4.41	.30	4.30	.0010				
										Total length of backwater =			6,512

* S_0 = invert slope.

4.5 ft. Assuming that this is the depth below the jump, the depth above the jump may be computed from the equation previously derived (see Eq. 3·16) as follows:

$$\frac{y_1}{y_2} = \frac{1}{2}\left[-1 + \sqrt{1 + \frac{8q^2}{gy_2^3}}\right]$$

$$y_1 = \frac{4.5}{2}\left[-1 + \sqrt{1 + \frac{8(250/10)^2}{32.2(4.5)^3}}\right] = 1.45 \text{ ft}$$

3. Determine the distance below the end of the chute to where the hydraulic jump will occur. When the flow reaches the bottom of the chute, the flat slope retards the flow; and the depth increases. Assuming that the chute has sufficient length for the depth of flow to decrease from the critical depth at the top of the chute to approximately the normal depth of 1.17 ft at the foot of the chute, the flow must continue in the lower section until the retardation has caused an increase in the depth of flow from 1.17 to 1.45 ft. The distance over which the flow must continue before the jump will occur is shown by the computations summarized in Table 3·9.

   The jump will be located about 62 ft below the end of the chute. Under some conditions of channel cross section and slope, the length of the lower section may be insufficient for the water surface to rise to the required depth. It should also be noted that if the cross section of the lower section differs from that of the chute so that the critical depth is below the depth of flow after it reaches the lower section, no jump will occur.

Transitions and Junctions

Transitions in sewerage systems are required between sewers of different sizes, between sewers laid on different slopes, and at sewer junctions. In general, the function of the transition is to change either the shape or the cross section of the flow. The hydraulic-head losses associated with transitions are friction losses and conversion losses. In the design of transition structures, frictional losses can usually be neglected as a first approximation. Conversion losses, such as momentum losses that depend on the geometry of the structure, must be determined or estimated individually for each structure. For this reason, there is no fixed approach to the design of transition structures. When faced with the task of designing a transition structure, it is recommended that the reader refer to Camp [2, 3], Hinds et al. [6], and Jaeger [8].

TABLE 3·9 COMPUTATION OF HYDRAULIC JUMP

y	A	V	$\frac{V^2}{2g}$	$y + \frac{V^2}{2g}$	$\frac{y}{D}$	K'	$\frac{K'}{n}$	$\frac{K'}{n} b^{8/3}$	S	$S^{1/2}$	S_a	$S_a - S$	$\Delta\left(y + \frac{V^2}{2g}\right)$	Δx
1.17	11.7	21.35	7.08	8.25	.117	.0362	2.41	1118	.224	.0502				
											.0480	.0470	.31	6.6
1.2	12.0	20.83	6.74	7.94	.12	.0376	2.51	1165	.214	.0458				
											.0409	.0399	.89	22.3
1.3	13.0	19.23	5.75	7.05	.13	.0425	2.83	1313	.190	.0361				
											.0325	.0315	.70	22.2
1.4	14.0	17.85	4.95	6.35	.14	.0476	3.17	1471	.170	.0289				
											.0274	.0264	.29	11.0
1.45	14.5	17.25	4.61	6.06	.145	.0502	3.35	1554	.161	.0259				
											Total distance below chute		=	62.1

Junctions in sewerage systems usually occur where a branch sewer enters a main sewer (see Chap. 5). At the junction of two or more sewers there will exist a number of flow paths, and the lowest path is common to all the upstream paths. Hence, the design of a junction is similar to the design of a number of transitions, one for each flow path. Additional head-loss allowance must also be made for impact losses resulting from the convergence of the sewage streams and for curvature losses resulting from channel configuration.

In designing junction chambers, it is desirable to have the streams of sewage join in such a way that there is a minimum reduction of the current in either sewer, because this might cause deposit of grit and other suspended matter. To have a minimum reduction of current, the streams of sewage should have the same surface elevation and velocity as they approach each other. This is difficult to accomplish when one sewer is large and the other much smaller. If the invert of the smaller sewer is placed at an elevation that will make the surface elevation of both streams identical during dry weather, the two streams will probably have substantially different elevations when considerable storm water is flowing. Consequently, a study of conditions at the junction when the two sewers carry various quantities of sewage may show that, to prevent sewage being backed up in the smaller sewer during periods of high discharge, it will be desirable to increase the grade of the smaller sewer somewhat for a short distance from the junction.

Velocities of flow in sewers usually are not high enough to require special consideration of the hydraulic features of transition structures at junctions. If reasonable care is taken to avoid sudden enlargements and contractions and to make changes of direction by smooth and reasonably flat curves, no difficulty is likely to develop.

The condition is radically different, however, when high velocities are involved, say, over 6 or 8 fps. In such cases, the theoretical position of the flow line from point to point should be computed by the application of Bernoulli's equation; and the design of the structure should be changed, if necessary, to obtain a smooth curve for this line, tangent to the water surfaces at each end of the transition.

Banking on Curves

In some cases, the banking or superelevation of water surface along the outer wall of a curved channel may be of considerable importance. It is especially important with stream channels or open flumes that should not overflow or with flat-topped conduits in which the water should not touch the roof of the conduit.

The excess in elevation of the water at the outer bank may be computed approximately by the equation

$$E = \frac{V^2 b}{gr} \tag{3·36}$$

where E = difference in elevation of water surface at the two banks
V = average velocity in cross section
b = breadth of channel or stream
r = radius of curvature on centerline of channel

It has been found in some cases that the actual difference in elevation is slightly greater than would be given by this equation.

HEAD LOSSES OTHER THAN THOSE CAUSED BY FRICTION

Head losses that normally occur in sewerage systems, other than those caused by friction, usually consist of velocity head losses, entrance and exit losses, and losses resulting from the geometric configuration of the sewers and their appurtenances.

Velocity Head

Strictly speaking, the head required to produce velocity, $V^2/2g$, is not lost, as it merely corresponds to the transformation of potential energy into kinetic energy; and theoretically it should be reversible when the velocity is checked. This is the case, to a considerable extent, with closed pipes under pressure; but, in open channel conditions, it is rarely possible to recover a significant part of the velocity head, except in the hydraulic jump.

It is usually advisable in sewerage work to consider the head used in producing velocity as lost head; and it is necessary to provide, in every case, sufficient head to develop velocity, in addition to that required to overcome friction and other resistances. At velocities normally found in sewers in flat areas, the velocity head may be insignificant, but this hydraulic element should be considered in design and rejected only if it is relatively unimportant in the problem under analysis. Failure to provide for it has sometimes led to serious loss of capacity, unexpected choking and overflow of sewers, and other problems.

Some cases that require special attention in this regard may be cited: (1) where the flow at low velocity in a large sewer is to discharge into a smaller sewer in which the velocity will be high; (2) in grit chambers where sewage moving slowly through the enlarged chambers must undergo an increase in velocity to flow away through an outlet passage or conduit of smaller dimensions; (3) in effluent conduits or channels leading away from sedimentation tanks; and (4) at restrictions in conduits, such as gates or contractions necessary to pass other structures. In short, situations in which velocity must be built up to initiate flow in a channel or conduit involve the most important head losses encountered in the design of the hydraulic features of treatment plants and require careful consideration.

When velocity is reduced by enlargement of a channel or conduit, a portion of the change in kinetic energy may sometimes be recovered as potential energy, which is manifested by a rise in water surface, provided the change in section is smooth and gradual. The principle of this recovery of velocity head is well illustrated in an analogous problem, that of water flowing under pressure in the Venturi meter. Here the outlet cone is gradually expanded so that a large part of the difference in velocity head between the throat and outlet is recovered.

Entry Head

In the case of a pipe leading from the side of a tank, where the corners are square and sharp, the entry head loss has been found to be $0.505V^2/2g$. This loss is in addition to the velocity head loss. By suppressing or reducing the contraction resulting from the orifice, e.g., by rounding the corners, the entry loss may be substantially reduced, perhaps to $0.1V^2/2g$.

In the case of an open channel or conduit leading from a reservoir, the conditions are similar, but experimental data are lacking. It is probably safe to estimate the loss on the same basis as in a closed pipe.

Sudden Reduction of Cross Section

In addition to the difference in velocity heads $(V_2^2 - V_1^2)/2g$, there is a further loss due to the resistance resulting from the disturbed flow conditions. Experiments on pipes have shown this additional loss to range from zero to approximately $0.5V^2/2g$ (V being the velocity in the smaller pipe). Depending on the ratio of the diameters, the coefficient becomes larger as the difference in diameters increases. There are no experimental data for open channels, but it seems reasonable to assume that similar relationships exist.

Sudden Enlargement of Cross Section

Experimental data [Archer in 11] on losses due to sudden enlargements of closed pipes may be expressed by the equation

$$h_L = 1.098 \frac{V_1^{1.919}}{2g} \left(1 - \frac{A_1}{A_2}\right)^{1.919} \tag{3·37}$$

In the case of an open channel, very little of the velocity head can be recovered. Other disturbances to flow are caused by bends and partly closed valves in pipes and by changes in direction, piers, side inlets, bulkheads, and the effect of wind upon the free surface of the liquid in open channels.

Valves

On the basis of the results of experiments to determine the head loss resulting from partly closed valves in pipes, it has been found that the loss ranges from $0.2V^2/2g$ to $13.5V^2/2g$ as the ratio of the area of the opening to that of the pipe decreases from 0.9 to 0.1. When the area of the opening is half that of the pipe, the loss is about $2.7V^2/2g$.

Curves

It has been observed that even a slight change in direction produces a condition of disturbed flow that increases the frictional resistance and that the length in which such disturbed conditions exist is of greater significance than the sharpness of the deflection; in other words, a curve of short radius and correspondingly short length of curve is likely to result in a smaller loss of head as compared to a curve of long radius with the same change of direction. There are limits, however, beyond which the opposite effect is probable.

Based on such fragmentary data as are available, it appears that the effect of curvature has generally been equivalent to an increase in the value of n by an amount varying from 0.003 to 0.005 in the sections containing much curvature.

METHODS OF MEASURING FLOWING WATER

Direct-discharge and velocity-area are the two major methods used for the measurement of flowing fluids.

Under the category of direct-discharge, the following methods and apparatuses have been used, the choice depending on the flow conditions encountered:

1. Weighing the discharge
2. Volumetric discharge measurement
3. Orifices
4. Standard weirs
5. Venturi meters
6. Flow nozzles
7. Parshall flume (Venturi flume)
8. Palmer-Bowlus flumes (critical-depth meters)
9. Contracted opening
10. California-pipe
11. Computation (based on depth and slope measurements)
12. Chemical tracers
13. Radioactive tracers
14. Magnetic flow meters

The most commonly employed methods and apparatuses for flow measurement in the velocity-area category are as follows:

1. Current meters
2. Pitot tubes
3. Float measurements
4. Dye tracers
5. Salt-velocity
6. Chemical tracers
7. Radioactive tracers
8. Electrical methods

In the following paragraphs a brief discussion is presented of some of these methods, along with the pertinent equations required in their use. When making flow measurements, the selection of the method will depend upon the facilities available, the degree of precision required, and the conditions under which the sewer was built and is operating.

Orifices

The standard orifice, as generally defined, is one in which the edge of the orifice that determines the jet is such that the jet, upon leaving it, does not again touch the wall of the orifice. Practically, this result is obtained by having the outside of the orifice beveled and its throat cylindrical or prismatic in shape, with an axial length between $\frac{1}{16}$ and $\frac{1}{8}$ in., depending upon the thickness of the plate.

In accordance with Torricelli's theorem that the velocity of flow through an orifice is equal to the velocity acquired by a freely falling body in a space corresponding to the head over the orifice, the discharge through an orifice is

$$Q = cAV = cA\sqrt{2gH} \tag{3·38}$$

where Q = discharge, cfs
c = constant
A = orifice area, sq ft
V = velocity, fps
g = acceleration due to gravity, ft/sec^2
H = head, ft

One reason for the need of the coefficient c is that the cross section of the jet, at the point a short distance outside the orifice, has generally a somewhat smaller area than that of the orifice itself, the reduction in area depending upon the character of the orifice. When the edge is sharp so that the water does not adhere to the orifice, the coefficient is at a minimum or the reduction in area is at a maximum. On the other hand, when the orifice is shaped to a bell mouth, the coefficient is at a maximum, and the cross section of the jet may be nearly equal to that of the orifice itself. Typical examples of orifices and their respective coefficients are shown in Fig. 3·23.

Weirs

One of the most accurate methods of measuring water is by the use of a weir, provided the conditions under which the coefficients of dis-

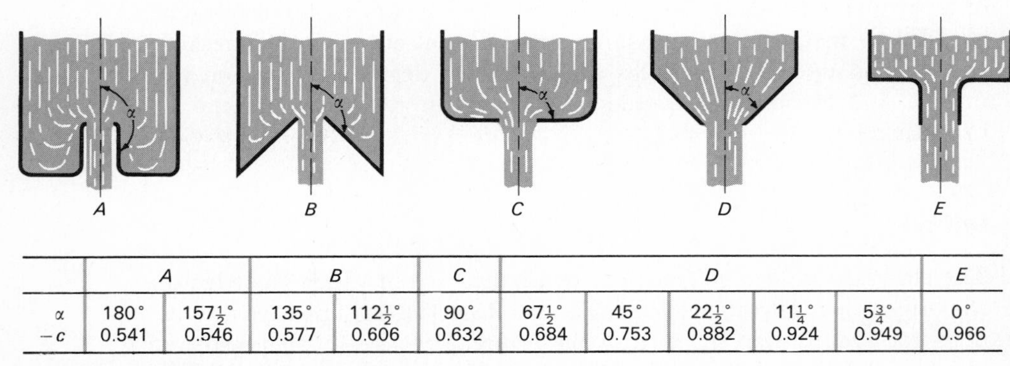

	A		B		C	D					E
α	180°	157½°	135°	112½°	90°	67½°	45°	22½°	11¼°	5¾°	0°
−c	0.541	0.546	0.577	0.606	0.632	0.684	0.753	0.882	0.924	0.949	0.966

FIG. 3·23 Orifice shapes and coefficients [11].

charge of given types of weirs were determined are approximately reproduced in the gagings. The three most common types of weirs—rectangular, triangular, and trapezoidal—and submerged weirs will be discussed here. For the determination of the discharge over broad-crested weirs and dams having different types of crests, the reader may consult such references as Horton [7], King and Brater [10], and Williams and Hazen [18].

Rectangular Weirs The general weir formula for a rectangular weir may be expressed by the equation $Q = CLH^{\frac{3}{2}}$. All the equations now in use may be reduced to this form; but it is better practice, in view of the several methods of correcting for the velocity of approach that are followed by the various experimenters, to use their form of equation.

The Francis equation, developed in 1823, is the most commonly used equation for estimating the flow over rectangular weirs. For contracted weirs, neglecting the approach velocity, the Francis equation is

$$Q = 3.33(L - 0.1nh)h^{\frac{3}{2}} \tag{3·39}$$

where Q = discharge, cfs
L = length of crest of weir, ft
n = number of end contractions
H = observed head corrected to include effect of velocity of approach, ft
h = observed head upon crest of weir, which is difference in elevation (ft) between top of crest and surface of water in channel, at point upstream taken, if possible, just beyond beginning of surface curve

The use of h instead of H in the factor $L - 0.1nH$ used in correcting for end contractions is as precise as ordinary practice warrants. For contracted weirs with the head corrected for velocity of approach, the equation is

$$Q = 3.33(L - 0.1nH)[(h + h_v)^{\frac{3}{2}} - h_v^{\frac{3}{2}}] \tag{3·40}$$

In Eq. 3·40, h_v is equal to head due to the mean velocity of approach, $V^2/2g$. For suppressed weirs, neglecting velocity of approach, the equation is

$$Q = 3.33Lh^{\frac{3}{2}} \tag{3·41}$$

For suppressed weirs, with the head corrected for velocity of approach, the equation is

$$Q = 3.33L[(h + h_v)^{\frac{3}{2}} - h_v^{\frac{3}{2}}] \tag{3·42}$$

These equations are strictly applicable only to vertical sharp-crested rectangular weirs with complete contractions and free overfall when

1. The head h is not greater than one-third the length L
2. The head is not less than 0.5 ft nor more than 2 ft
3. The velocity of approach is 1 fps or less
4. The height of the weir is at least three times the head

These equations are probably usable with heads higher than 2 ft but not much lower than 0.5 ft.

The Rehbock equation, which was first published in Germany in 1911 and revised in 1912, is

$$Q = \tfrac{2}{3}\mu \sqrt{2g}\, Lh^{\frac{3}{2}} \tag{3·43}$$

where $\mu = 0.605 + 1/(320h - 3) + 0.08h/z$
z = distance from bottom of channel to crest of weir, ft

In addition to the foregoing, numerous other equations for rectangular weirs have been developed [10].

Triangular Weirs The basic expression for discharge through a triangular weir is

$$Q = C\left(\tan\frac{\theta}{2}\right)h^{\frac{5}{2}} \tag{3·44}$$

where Q = discharge, cfs
C = constant (experimentally determined)
h = head over angle of weir notch, ft
θ = angle of notch

For a right-angled notched weir in which θ is equal to 90° [that is, $\tan(\theta/2) = 1$], it has been found experimentally that the discharge may be computed by the following equation:

$$Q = 2.5h^{\frac{5}{2}} \tag{3·45}$$

Trapezoidal Weirs The trapezoidal weir differs from the rectangular type in that the sides are inclined rather than vertical. Usually the sides are given an inclination of 1 horizontal to 4 vertical, for the reason that at this angle the slope is just about sufficient to offset the effect of end contractions. When this is done the weir is known as the Cippoletti weir. The general equation of the trapezoidal weir is

$$Q = \tfrac{2}{3}(2g)^{\frac{1}{2}}LH^{\frac{3}{2}} + \tfrac{4}{15}2Z(2g)^{\frac{1}{2}}H^{\frac{5}{2}} \tag{3·46}$$

where Z is the ratio of the vertical projection to the horizontal projection of the side.

For the Cippoletti weir where $Z = 1{:}4$, the equation reduces to

$$Q = 3.367LH^{\frac{3}{2}} \tag{3·47}$$

Submerged Weirs When the water surface in the channel below the weir is higher than the crest, the weir is said to be submerged or drowned. Measurements by submerged weirs are much less certain than by weirs with free discharge, but their use is sometimes unavoidable.

On the basis of a series of experiments conducted on rectangular, triangular, parabolic, cusped, and proportional weirs, Villemonte [17] found that for all types of weirs the discharge could be computed using the following equation:

$$\frac{Q}{Q_1} = \left[1 - \left(\frac{H_2}{H_1}\right)^n\right]^{0.385} \tag{3·48}$$

where Q = submerged discharge, cgs
Q_1 = free discharge, cfs
H_1 = upstream head, ft
H_2 = downstream head, ft
n = exponent in free-discharge equation $Q_1 = CH_1^n$ for each particular weir

Venturi Meters

The principle of this apparatus, based upon Bernoulli's theorem, was discovered about 1791 by the Italian engineer J. B. Venturi. This principle was first practically applied by Clemens Herschel in 1887 in the so-called Venturi meter. As shown in Fig. 3·24, the meter tube, which is the portion of the apparatus to which Venturi's discovery applies, is inserted in a line of pipe and consists of three parts: (1) the inlet cone, in which the diameter of the pipe is gradually reduced; (2) the throat or constricted section; and (3) the outlet cone, in which the diameter increases gradually to that of the pipe in which the meter is inserted.

The throat is lined with bronze. Its diameter, in standard meter tubes, is from one-third to one-half the diameter of the pipe; and its length is but a few inches, sufficient to allow a suitable pressure chamber or piezometer ring to be inserted in the pipe at this point. The length of the upper or inlet cone is approximately one-fourth that of the lower cone. A piezometer ring is inserted at the upper or large end of the inlet cone, and the determination of the quantity of water flowing is based upon the difference in pressures observed or indicated at this point and at the throat of the meter.

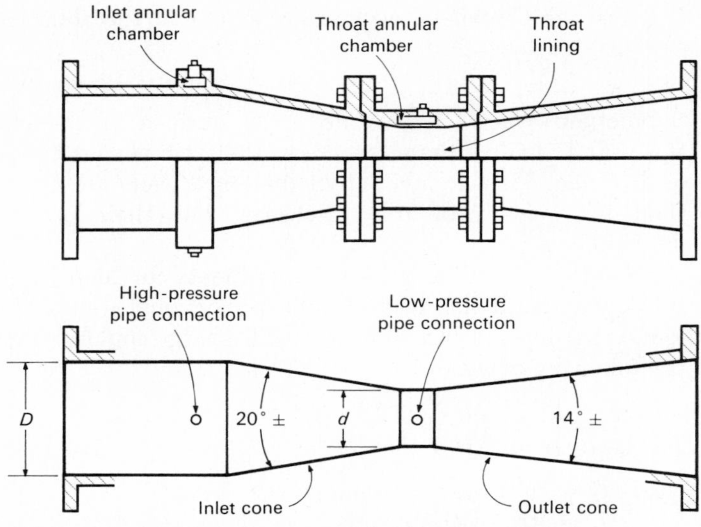

FIG. 3·24 Typical Venturi meter used for flow measurement.

The equation used for computing the discharge through a Venturi meter tube is based on Bernoulli's equation:

$$Q = \frac{A_1 A_2 \sqrt{2g(h_1 - h_2)}}{\sqrt{A_1^2 - A_2^2}}$$

$$= \frac{A_1 A_2 \sqrt{2gH}}{\sqrt{A_1^2 - A_2^2}} \qquad (3\cdot49)$$

where A_1 = area at upstream end, sq ft
A_2 = area at throat of meter, sq ft
h_1, h_2 = pressure heads, ft
$H = h_1 - h_2$

Under actual operating conditions and for standard meter tubes, including allowance for friction, this formula reduces to the form

$$Q = (1.00 \pm 0.02) A_2 \sqrt{2gH} \qquad (3\cdot50)$$

The coefficient written 1.00 ± 0.02 is made up of two parts, or $c = c_1 c_2$ where $c_1 = A_1/\sqrt{A_1^2 - A_2^2}$
c_2 = coefficient of friction

For standard meter tubes in which the diameter of the throat is between one-third and one-half that of the pipe, the values of c_1 range between 1.0062 and 1.0328, while the friction coefficient c_2 varies from

0.97 to 0.99. Thus the range of values of c is from 0.98 to 1.02; accordingly, c has been written above as 1.00 ± 0.02.

In Venturi meters used for measuring sewage, at each annular chamber or piezometer ring there should be valves by which the pressure openings can be closed. These valves may be so designed that, in closing, a rod is forced through the opening to clean out any matter that may have clogged it. When all these valves have been closed, the plates covering the hand holes in the pressure chamber may be removed, and the chamber may be cleaned by flushing with hose or otherwise. Such flushing at short intervals is usually necessary if Venturi meters for sewage are to be maintained in good operating condition. A continuous flushing system is shown in Fig. 3·25.

To prevent the interference with the operation of the register by clogging, an oil seal may be inserted in the pressure pipe, between the meter tube and the register. The pressure is transmitted as far as the seal, through water in the pressure pipes, and from the seal to the register through oil. Thus it is impossible for any sewage to get into the register and interfere with its proper operation.

The Dall flow tube is similar to the Venturi meter, but shorter (see

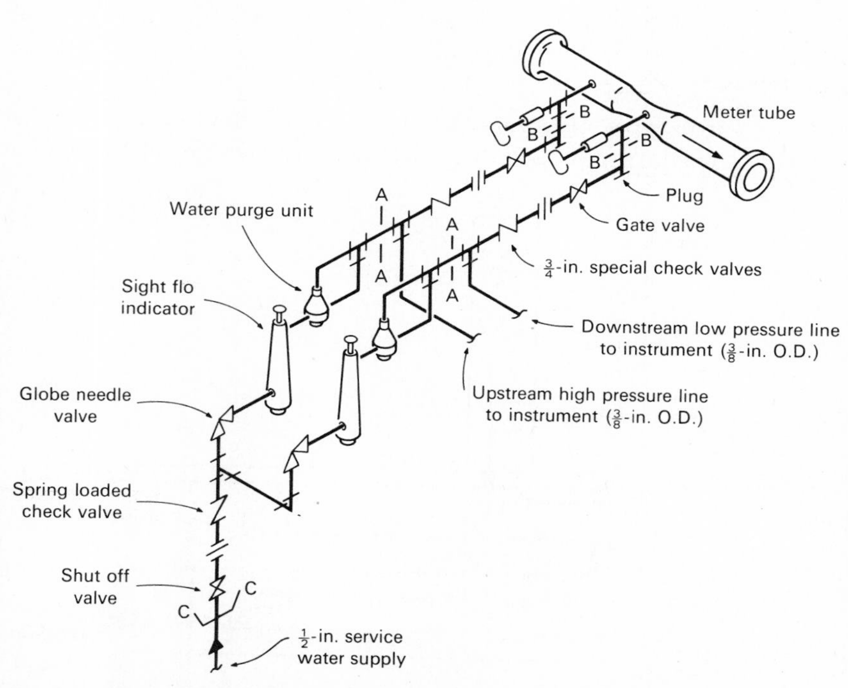

FIG. 3·25 Continuous flushing system for Venturi meter installation.

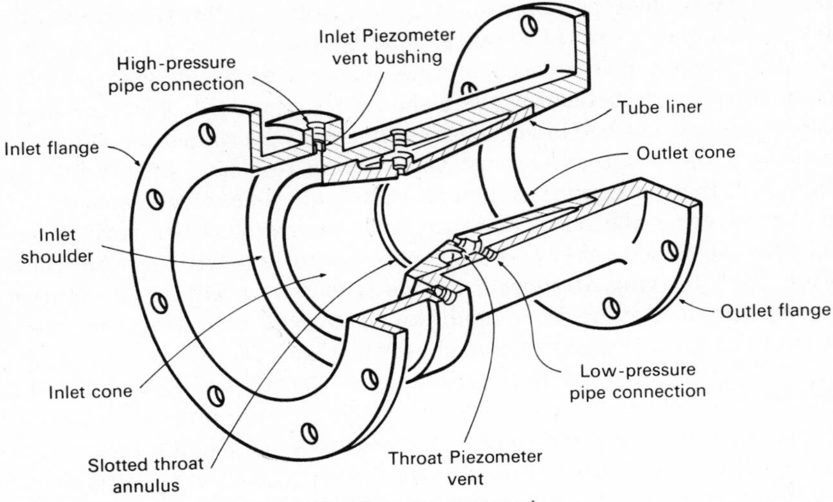

FIG. 3·26 Typical Dall tube used for flow measurement.

Fig. 3·26). It is lightweight and also has less head loss than the Venturi meter.

Flow Nozzles

Two principal types of nozzles are used for the flow measurement: those inserted in a pipe and those attached to the end of the pipe.

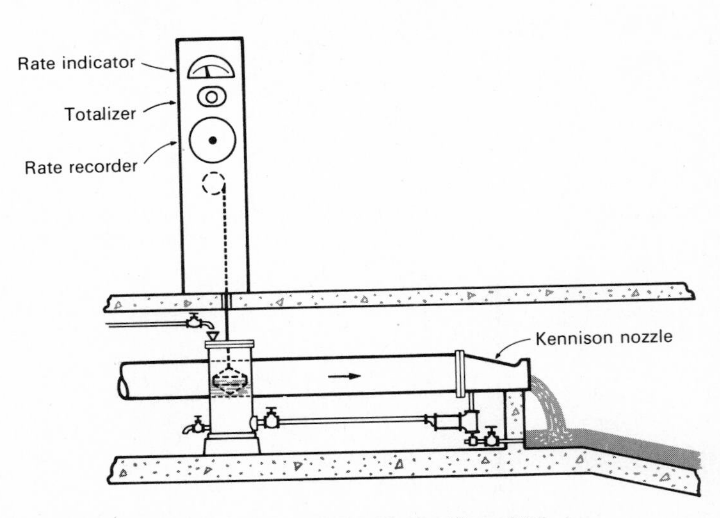

FIG. 3·27 Kennison flow nozzle installation [from BIF].

Nozzle flow meters make use of the Venturi principle but ordinarily employ a nozzle inserted in the pipe instead of the Venturi tube. The form of the nozzle, the method of inserting it in a pipe, and the method of measuring the difference in pressure will vary with the manufacturer. The throat of the nozzle is considerably larger in diameter than the throat of the Venturi tube for the same rate of flow, and the resulting total loss of head is approximately the same.

Flow nozzles attached to the ends of pipes are usually of the Kennison type shown in Fig. 3·27. These nozzles can be used on pipes varying in size from 6 to 36 in. in diameter and can be used for metering flows up to 18 million gallons per day (mgd).

Parshall Flume

Since the Venturi type of meter is applicable only to closed pipes under pressure, it can be used for measuring sewage only in force mains or inverted siphons. The Venturi principle is utilized, however, for the measurement of water flow in open channels by means of the Parshall flume [14], as shown in Fig. 3·28.

Because the throat width is constant, the discharge can be ob-

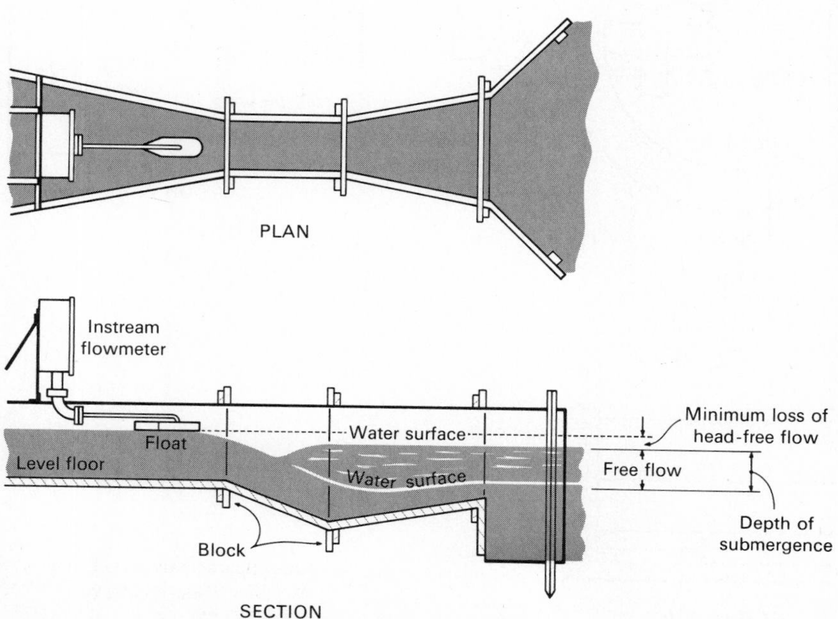

FIG. 3·28 Schematic of Parshall flume metering installation [from Fischer & Porter].

tained from a single upstream measurement of depth. If the flume is operating under submerged conditions, the downstream head must also be measured to determine the discharge. The Parshall flume can also be used as a control device for a grit chamber [1].

Palmer-Bowlus Flumes

The Palmer-Bowlus flume, shown in Fig. 3·29, was developed for the measurement of flow in a variety of open channels. The principle of its operation is the same as that of the Venturi meter or flume. The meter is usually placed in the sewer at a manhole. It backs up the water in the sewer above the meter control section in such a manner that the rate of discharge is related to this upstream depth. Thus, by measuring the upstream depth, the discharge can be read from a calibration curve

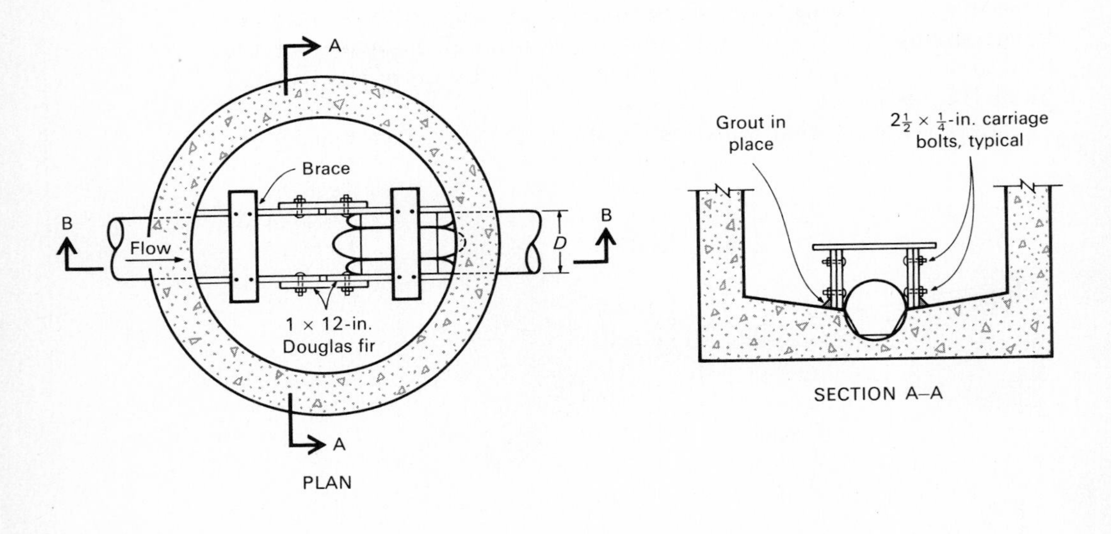

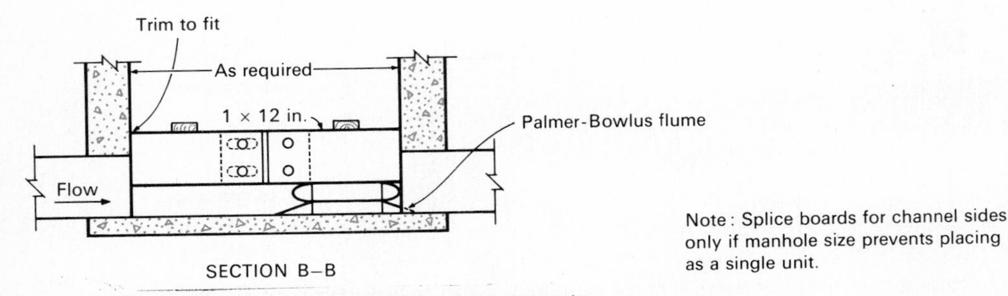

FIG. 3·29 Installation of a Palmer-Bowlus flume in a manhole.

which is usually supplied with each unit. The advantages of the Palmer-Bowlus flume are that it can be installed in existing systems, head loss is insignificant, and it is self-cleansing. Care must be taken to avoid leakage under the flume and conditions in which the flume will be "drowned out." A detailed discussion of the equations required and procedures involved may be found in Palmer and Bowlus [13].

Contracted Opening

As the name indicates, this method of flow measurement requires that the flow pass through a contraction, so it is especially useful for measuring flood flows. In general, it can be used when, due to the channel constriction, there is a drop of a foot or more in the water surface. Discharge through a contracted opening can be computed using an equation similar to the one developed for the Venturi flume.

California-Pipe

This method, developed in 1922 by Van Leer [15], is used for the measurement of the rate of flow from an open end of a partially filled horizontal pipe that is discharging freely to the atmosphere. The discharge pipe should be horizontal and should have a length of at least six pipe diameters. When the pipe is flowing almost full, an air vent should be installed back up the pipe to ensure free circulation of air in the unfilled portion of the discharge pipe. In addition the approach velocity should be kept to a minimum.

The discharge can be computed using the following equation:

$$Q = 8.69\left(1 - \frac{a}{d}\right)^{1.88} d^{2.48} \tag{3·51}$$

where Q = discharge, cfs
a = distance from crown of pipe to top of water surface measured at point of discharge, ft
d = diameter of pipe, ft

Equation 3·51 was derived from the results of experiments conducted on pipes ranging in size from 3 to 10 in. in diameter; however, the method probably could also be used for larger pipes.

Computation

The determination of the quantity of flow by computational methods requires that field measurements be made of the depth of flow and slope of the sewer. In addition, a value for the coefficient of roughness must also be selected. The application of the flow formulas, previously discussed, to the estimation of the quantity of sewage flowing in a sewer requires no explanation. The method, at best, is an approximation dependent upon the steadiness of the flow at the time of observation and the precision with which the coefficient of roughness is assumed for the existing conditions. Nevertheless, this method is the only one that is frequently used in ordinary sewerage work.

Chemical and Radioactive Tracers

Chemical and radioactive tracers, while of great value in measuring clean water, must be used with care in the measurement of sewage flows, because of the relatively large amounts of foreign matter contained in the sewage. Chemical and radioactive tracers are used in two ways when making flow measurements. The first is sometimes called chemical gaging; in the second, velocity is measured.

In chemical or radioactive gagings, a known concentration of a chemical or radioactive substance is continuously added, at a constant rate, to the stream in which the discharge is to be determined. At a distance downstream sufficient to ensure complete mixing of the tracer and stream, the stream is sampled and the concentration of the chemical or radioactive substance is determined. The flow in the stream can then be determined using the following equation:

$$Q_s = \frac{Q_t(C_t - C)}{C - C_s} \tag{3·52}$$

where Q_s = stream discharge, cfs
Q_t = tracer discharge, cfs
C_t = concentration of tracer in tracer-discharge stream
C_s = concentration of tracer in stream before injection
C = concentration of tracer in stream after injection

In cases in which the tracer is nonconservative (i.e., concentration changes with time), suitable corrections must be made in Eq. 3·52.

Where velocity measurements are to be made, the chemical or radioactive tracers are usually injected into the stream upstream of two control points. The time of passage of the prism of water containing the tracer is noted at these control points, and the velocity is then computed by dividing the distance between the control points by the travel time.

When salt (NaCl) is used as the tracer, the time of passage between control points is measured using electrodes connected to an ammeter or recorder. This method of measurement is possible since the injected salt increases the conductivity of the water. When radioactive tracers

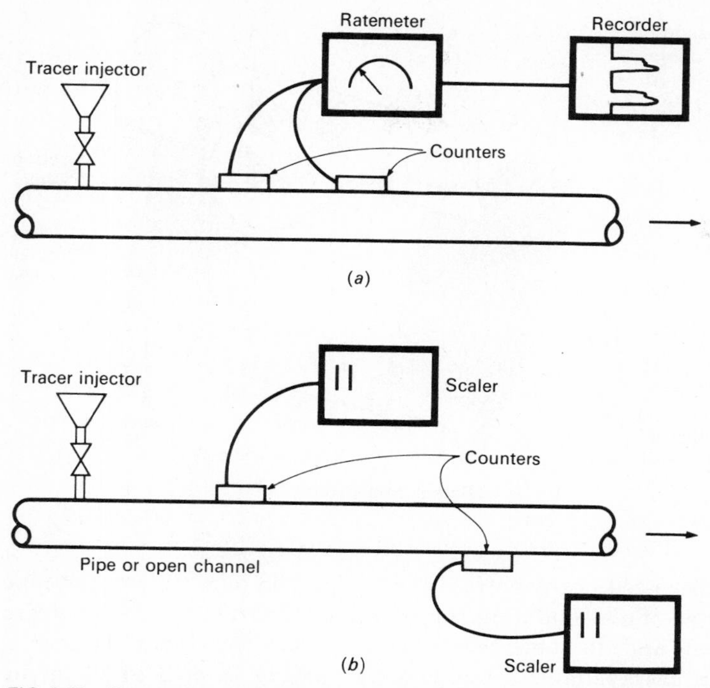

FIG. 3·30 Flow measurement using radioactive tracers. (*a*) Peak timing, (*b*) total count.

are used, the time of passage is noted by radioactive counters attached to the outside of the pipe as shown in Fig. 3·30. The time of passage is the difference between the times when the peak counts were recorded at each counting station.

Magnetic Flow Meters

When an electrical conductor passes through an electromagnetic field, an electromotive force or voltage is induced in the conductor that is proportional to the velocity of the conductor. The voltage thus generated is mutually perpendicular to both the velocity of the conductor and the magnetic field. This statement of Faraday's law serves as the basis of design for electromagnetic flow meters, as shown in Fig. 3·31.

In actual operation, the liquid in the pipe (usually water or sewage) serves as the conductor. The electromagnetic field is generated by placing coils around the pipe. The induced voltage is then measured by electrodes placed on either side of the pipe. If the pipe is a conductor,

FIG. 3·31 Schematic showing magnetic flow meter components [from Fischer & Porter].

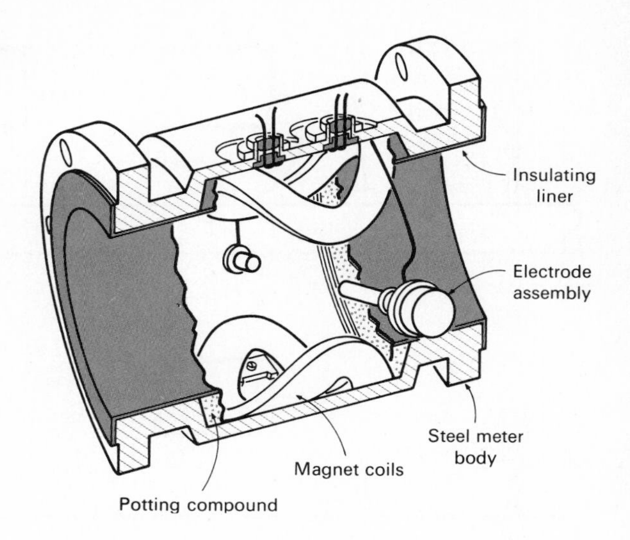

the electrodes need not penetrate the wall of the pipe. Where the pipe is constructed of nonconductive material, the electrodes must penetrate the pipe wall and, in some cases, protrude into the liquid. Magnetic flow meters are available for pipe sizes varying from 2 to 24 in. in diameter without special order. Larger sizes are available on special order.

Current Meters

Current meter measurements may be employed for the accurate determination of the velocity of flow in large sewers or in open channels, provided there is not too much paper or other suspended matter present to clog the meter. Gagings of flow may be made by several methods: the one-point method, the two-point method, the multiple-point method, the method of integrating in sections, and the method of integrating in one operation.

In the one-point method, the meter is held at 0.6 of the depth and in the center of the stream, and the result is assumed to indicate the mean velocity of the stream. This is but a rough approximation, suitable only for hasty observations with no pretense of accuracy. In the two-point method, the velocity is observed at 0.2 and 0.8 of the depth, and the average of these two figures is taken to represent the average velocity in the vertical section. The stream can be divided into a number of vertical sections, and the average velocity in each section is approximately determined by this method. Discussion of the other methods of making gagings along with the methods used for calibrating current meters may be found in King and Brater [10] and in Davis and Sorensen [5].

Pitot Tubes

The Pitot tube, which has proved so useful in water-pipe gagings, is impractical in sewer gagings because the suspended matter contained in the sewage tends to clog the tube.

Float Measurements

Float measurements of the flow in sewers are rarely made, except in rectangular channels or for the approximate determination of the velocity of flow between two manholes; but, in studies of tidal currents or of sewage currents in bodies of water into which sewage may be discharged, floats are universally employed. Occasionally, however, the use of floats to measure the velocity of flow in comparatively small sewers is practicable. Three types of floats may be used: surface, subsurface, and rod or spar.

Only surface velocities can be obtained by the use of surface floats. Owing to the modifying effects of the wind, the results can be considered only as approximations.

Subsurface floats consist of relatively large bodies slightly heavier than water, connected by fine wires to surface floats of sufficient size to furnish the necessary flotation, and carrying markers by which their courses may be traced. The resistance of the upper float and connecting wire is generally so slight that the combination may be assumed to move with the velocity of the water at the position of the submerged float.

Rod floats have been used for measuring flow in open flumes with a high degree of accuracy. They generally consist of metal cylinders so loaded as to float vertically. The velocity of the rod has been found to correspond very closely with the mean velocity of the water in the course followed by the float.

Dye Tracers

The use of dyes for measuring the velocity of flow in sewers, particularly in small pipe sewers, is one of the simplest and most successful methods that has been used. Having selected a section of sewer in which the flow is practically steady and uniform, the dye is thrown in at the upper end, and the time of its arrival at the lower end is determined. If a bright-colored dye, such as eosin, is used and a bright plate is suspended horizontally in the sewer at the lower end, the time of appearance and disappearance of the dye at the lower end can be noted with considerable precision, and the mean between these two observed times may be taken as representative of the average time of flow.

Other dyes which have been successfully used in tracer studies include fluorescein, congo red, potassium permanganate, rhodamine B and Pontacyl Brilliant Pink B. Pontacyl Brilliant Pink B is especially useful in the conduct of ocean outfall dispersion studies.

Electrical Methods

Electrical methods used for measuring the quantity of water flowing in a stream involve the use of equipment such as conductivity cells, hot-wire anemometers, and warm-film anemometers. Although some of these methods have been used in the field, they are not ideally suited for making sewage-flow measurements because the floating and suspended material commonly found in sewage interferes with their operation.

PROBLEMS

3·1 Determine the Chezy C, the Manning n, the Kutter n, and the Hazen-Williams C for a sewer flowing half full. The diameter is 4 ft, $S = 0.0009$, $f = 0.02$.

3·2 A 60-in. circular sewer is on a slope of 0.0025. If $n = 0.013$ at all depths of flow, find

- (*a*) Q and V when flowing full
- (*b*) Q and V when sewage is flowing 15 in. deep
- (*c*) Q and V when sewer is carrying 0.6 of its capacity
- (*d*) V and depth of flow when $Q = 50$ cfs

3·3 Compare the flow in an egg-shaped sewer to that in a semielliptical sewer. The flow depth is 6 in. and the slope is 0.0025. The vertical diameter is 6 ft. and $n = 0.013$.

3·4 The compartments of a grit chamber in which the velocity is 0.5 fps terminate in an outlet channel in which the velocity is 7 fps. The flow from the chamber must pass through gates that reduce the cross section of flow at the point of passage to 60 percent of that available in the grit compartments. Estimate the total head loss, including the head loss through the gates and that necessary to produce an increase in velocity in the outlet channel, but neglecting the friction loss.

3·5 Compute the drawdown curve for a rectangular conduit 6 ft wide with a slope of 0.006, when $n = 0.014$ and $Q = 120$ cfs discharging freely.

3·6 Determine the critical depth for a circular sewer section with a diameter of 3 ft carrying a flow of 10,000 gpm.

3·7 Consider a concrete influent flume of rectangular section to discharge 3 cfs into a series of three Imhoff tanks through six equal-diameter submerged circular orifices spaced 5 ft on centers and with two orifices feeding each tank. The discharge through each orifice is to be 0.5 cfs, and the velocity in the flume is 3 fps throughout its length. Determine the width of the section of flume between each pair of orifices and find the diameters ($c = 0.60$, initial channel width is 1.0 ft).

3·8 For velocities of approach of 5 fps or less, the Francis weir coefficient may be written as

$$c = 3.33\left(1 + 0.259\frac{H^2}{d^2}\right)$$

where H = observed head
$d = P + H$
P = height of weir

Determine the flow rate for a 3-ft-long weir with a 1-ft head. The height of the weir is three times the head.

3·9 A rectangular weir is operating under free flow conditions. The crest of the weir is 0.8 ft above the channel bottom and the length of the weir crest is 1.5 ft. Determine the discharge when the head on the weir is 3 in. and the effect of the approach velocity is considered.

3·10 An accurate measurement of the flow in a pipeline must be made repeatedly. At one point in the system, water is discharged from a horizontal pipe with a 6-in. diameter and the pipe is half full. Design a Cippoletti weir downstream of the discharge where flow conditions have stabilized. The head on the weir must not be more than 1 ft, nor less than 0.2 ft.

REFERENCES

1. Babbitt, H. E., and E. R. Baumann: *Sewerage and Sewage Treatment,* 8th ed., Wiley, New York, 1958.
2. Camp, T. R.: Hydraulics of Sewers, *Public Works,* vol. 83, no. 6, 1952.
3. Camp, T. R.: Hydraulics of Sewer Transitions, *Journal, Boston Society of Civil Engineers,* vol. 19, no. 6, 1932.
4. Chow, V. T.: *Open-Channel Hydraulics,* McGraw-Hill, New York, 1959.
5. Davis, C. V., and K. E. Sorensen: *Handbook of Applied Hydraulics,* 3d ed., McGraw-Hill, New York, 1969.
6. Hinds, J., W. P. Creager, and J. D. Justin: *Engineering for Dams,* vols. 1–3, Wiley, New York 1945.
7. Horton, R. E.: Weir Experiments, Coefficients and Formulas, U.S. Geological Survey Water Supply and Irrigation Paper 200, 1907.
8. Jaeger, C.: *Engineering Fluid Mechanics,* Blackie, London, 1956.
9. Joint Committee of the American Society of Civil Engineers and the Water Pollution Control Federation: *Design and Construction of Sanitary and Storm Sewers,* ASCE Manual and Report 37, New York, 1969.
10. King, H. W., and E. F. Brater: *Handbook of Hydraulics,* McGraw-Hill, New York, 1963.
11. Metcalf, L., and H. P. Eddy: *American Sewage Practice,* 2d ed., vol. 1, McGraw-Hill, New York, 1928.
12. Moody, L. F.: Friction Factors for Pipe Flow, *Transactions ASME,* vol. 66, p. 671, 1944.
13. Palmer, H. K., and F. D. Bowlus: Adaption of Venturi Flumes to Flow Measurements in Conduits, *Transactions ASCE,* vol. 101, p. 1195, 1936.

14. Parshall, R. L.: The Improved Venturi Flume, *Transactions ASCE*, vol. 89, p. 841, 1926.

15. Van Leer, B. R.: The California-Pipe Method of Water Measurement, *Engineering News-Record*, Aug. 3, 1922, Aug. 21, 1924.

16. Vennard, J. K.: *Elementary Fluid Mechanics*, Wiley, New York, 1961.

17. Villemonte, J. R.: Submerged-Weir Discharge Studies, *Engineering News-Record*, p. 866, Dec. 25, 1947.

18. Williams, G. S., and A. Hazen: *Hydraulic Tables*, 3d ed., Wiley, New York, 1920.

design of sewers

4

The facilities for the conveyance of sewage and storm water runoff may be classified as (1) sanitary sewers designed to receive domestic sewage and industrial wastes excluding storm water; (2) combined sewers designed to receive domestic sewage, industrial wastes, and storm water; and (3) storm sewers designed to carry off storm water and ground water, but excluding domestic sewage and industrial wastes.

SANITARY SEWERS

During the past 50 years the trend has been toward the provision of two separate systems, one of sanitary sewers for the collection of sewage and industrial wastes and the other of storm sewers for the collection of storm water. One reason for this trend has been the increased necessity of constructing sewage treatment facilities as a means of controlling pollution in streams and waterways.

In designing a sanitary sewer system, the designer must usually go through the following steps:

1. Preliminary investigations
2. Design considerations
3. Selection of basic design data and criteria
4. Design of the sewers
5. Preparation of contract drawings and specifications

Each of these steps requires a considerable amount of experience and knowledge on the part of the designer. The purpose of this section is to provide the student with some of the necessary knowledge and to illustrate its application in the design of a sanitary sewer system.

Preliminary Investigations

It is desirable to make comprehensive preliminary investigations of the district to be sewered, not only to obtain the data needed by the designing engineer and the construction contractors, but also to place on record pertinent, authentic information as to the local conditions before the construction of the system. Such information may be helpful in connection with future claims for damages. Contractors are justified in making lower bids when they are supplied with complete information about the conditions they will encounter than when they must estimate many of them. All facts ascertained, whether advantageous or not, should be made public, because the judgment of the engineer must be entirely unbiased, and the contractor should in all fairness have all the information available to assist in formulating his bid. Also, if any facts are suppressed, contractors' claims for extra costs based upon such facts may be allowed by the courts.

At the outset of the engineering work, an attempt should be made to obtain all maps and other drawings that furnish information about the area. Municipal and county engineers and surveyors, assessment boards, land title and insurance companies, and public utility officials often have such maps and will permit copies of them to be made. In large sewerage projects, the maps prepared by the U.S. Geological Survey and various state agencies may be useful, and occasionally the Bureau of Soils of the U.S. Department of Agriculture is able to supply helpful maps.

Field Work If satisfactory maps are not available, surveys must be made. The degree of precision required will depend on the conditions of the problem. For large, important projects, it may be advisable to establish a preliminary triangulation system, although usually this is not necessary. The surveys should show the locations of streets, alleys, street railways, public parks and buildings, ponds, streams, drainage ditches, and other features and structures that may influence or be influenced by the sewerage system. In some cases, it is necessary to show property lines.

An accurate, permanent, and complete system of bench levels should be established throughout the area to be covered by the proposed sewerage system. A bench mark should be established on each block of every street in which a sewer is to be laid and where topographic details are to be obtained subsequently. Profiles should then be made of all existing streets and alleys, and if the existing and "established" grades are different, information about the latter should be obtained.

This work should be extended to cover the district within which sewers may be needed during the next 30 to 50 years. In some cases, topographic notes should be obtained for the plotting of a map with contours at intervals of 1, 2, 5, or 10 ft, according to the configuration of the ground. Generally, the surface elevations of streets and alleys are sufficient at street intersections, at all high and low points, and at changes in surface slope; thus surface contours may not be required. The elevations of the beds of streams, ditches, canals, and culverts should be ascertained, and the maximum expected and ordinary water surface elevations should be determined.

Notes pertaining to the condition of all existing structures should be obtained. The elevations of the sills of buildings and the depths of their basements should be determined; and the character, age, and condition of the pavements of streets in which sewers will be laid should be recorded. Available information regarding the location of water and gas mains, electric conduits, and other underground structures should be obtained; and, where such information is lacking, it may be advisable in critical areas to have pits excavated in the streets to obtain the required data.

Local rainfall and runoff data should be collected; or, where these are inadequate, measurements in the field should be undertaken if practicable. Information that builders and contractors can supply regarding ground water should be recorded; and, in the case of low-lying land, it may be desirable to excavate pits or make borings to indicate the ground water conditions.

The character of the soil in which the sewers must be constructed should be ascertained so that the cost can be estimated with fair accuracy. A post auger is often used to obtain samples of earth from shallow depths. For greater depths, subsurface exploration with boring equipment usually is employed. Core drills are sometimes employed at the sites of important structures where it is desirable to obtain true samples of all underlying strata.

Complete information should be obtained concerning the local wages of unskilled and skilled labor, the cost of construction materials and supplies, and the cost of construction of similar work previously done. Freight rates and rental charges for trucks and equipment should be determined. This information is useful in the preparation of reliable estimates of cost.

Preparation of Maps and Profiles Work on preliminary maps and profiles should begin as soon as practicable during the progress of the field work, so that studies preliminary to design may be started before the field work is finished. As a rule, maps on a scale of 200 ft to 1 in. are large enough to permit the data to be shown in adequate detail; but, where there are many subsurface structures, a scale of 50 ft or less to 1 in. may be necessary for clarity.

The maps usually require more than one sheet, and in such cases a key map should be drawn showing the way the detailed maps fit together. The maps should show surface contours, where needed, or street elevations; all streets, railroads, buildings, pipes, conduits, manholes, gate boxes, and catch basins; and the names of streets, parks, public buildings, and watercourses. The magnetic or true north, or both, should be indicated.

Where contours are plotted they should be at sufficiently small intervals to allow the designer to prepare profiles of streets with reasonable accuracy; i.e., where the surface slope is 6 percent or less, the map should show contours at 2-ft intervals; where the surface slope is much greater, 5-ft intervals will usually suffice. Summits in streets should be marked and the elevations given to tenths of a foot, as should also points of depression or "pockets." An example of such a map showing elevations but not pipes or manholes for a residential area is shown on Fig. 4·1.

Profile sheets showing the ground surface along the proposed sewer routes should also be prepared in advance of the computations.

Design Considerations

The major design considerations for a sanitary sewer include estimation of the maximum rates of sewage flow, evaluation of the local factors that may affect the hydraulic operation of the system, and evaluation of alternative designs. In most cases, final determination of these factors of design will be based on the data and information obtained from the preliminary investigations.

Design Sewage Flows In most situations, the total sewage flow consists of three components: domestic sewage, ground water that

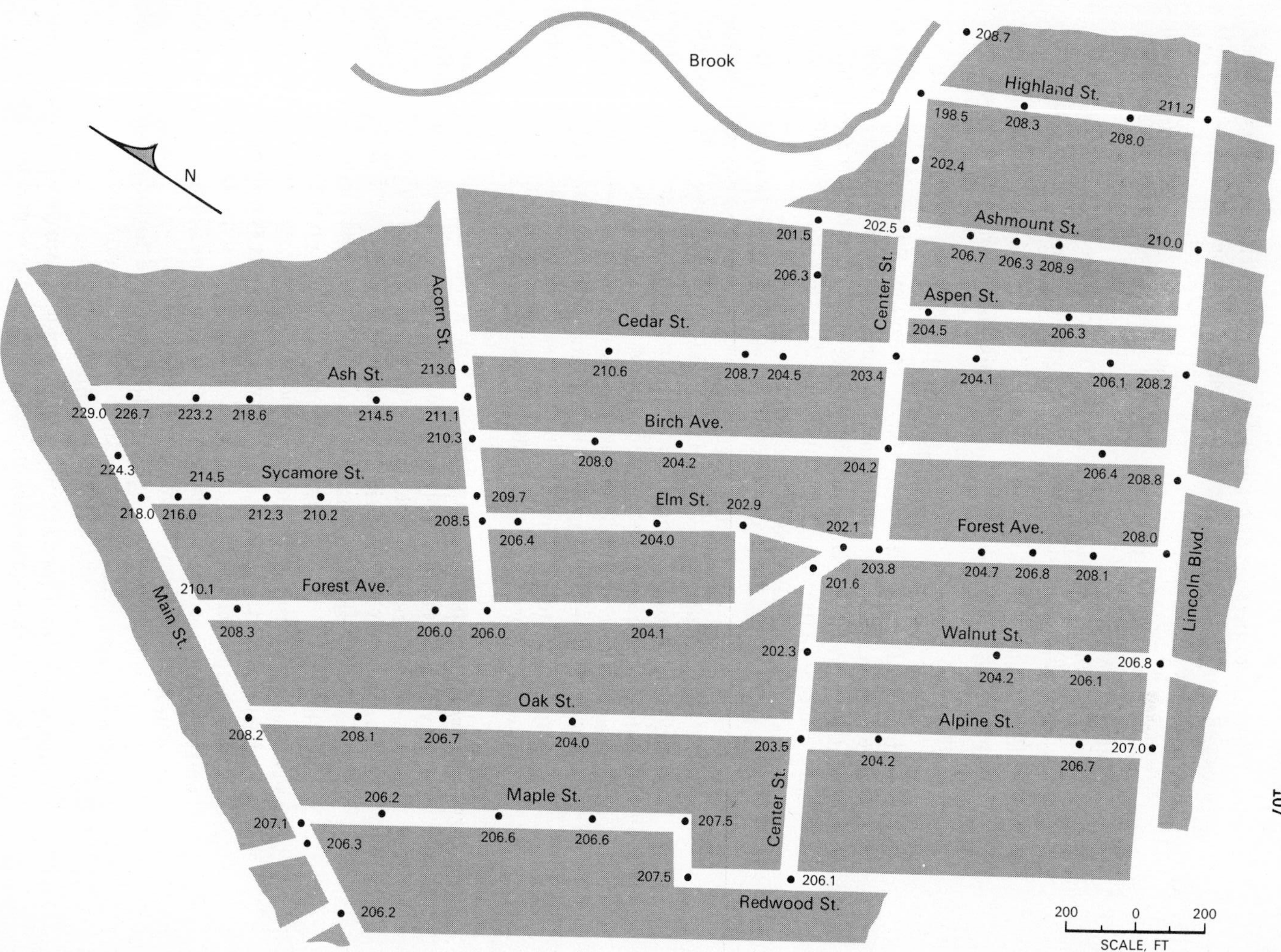

FIG. 4·1 Typical map used for the design of sanitary sewers.

has infiltrated into the sewer, and commercial and industrial contributions. Thus, new sanitary sewers are designed for the following expected future flows:

1. Maximum rate of flow of domestic sewage for the entire service area for a specified time period
2. Infiltration allowance for the entire service area
3. Additional maximum flow rates from commercial and industrial areas

The details of making sewage-flow calculations have been discussed in Chap. 2.

Selection of Design Equation As discussed in Chap. 3, Manning's and Kutter's equations are most commonly used for sewer design. Manning's equation is in wider usage than Kutter's equation, in part because of its simplicity and the fact that both equations give substantially the same results.

It is recommended that Manning's or Kutter's n of 0.013 be used for all proposed and future sewers and 0.015 be used for all existing sewers. Higher values of n should be used for existing sewers if available data indicate deterioration, departures from line and grade, variations of inside dimensions, deposits, or inferior workmanship.

The n value of 0.013 for proposed and future sewers is based on the use of pipe units having not less than 5-ft laying lengths, with true and smooth inside surfaces, and on the assumption that only first-class construction procedures will be followed.

Minimum Velocities Sewage should flow at all times with sufficient velocity to prevent the settlement of solid matter in the sewer. Usual practice is to design the slopes for sanitary sewers to ensure a minimum velocity of 2.0 fps with flow at one-half full or full depth. The velocity at less than one-half full depth will be less than 2.0 fps.

While it is the velocity near the bottom of the sewer that is significant in respect to the transporting power of flowing water, it has been found that a mean velocity of 1.0 fps is sufficient to prevent serious deposition of sewage solids. To prevent deposition of mineral matter, such as sand and gravel, a mean velocity of 2.5 fps will generally be adequate in sanitary sewers.

These are minimum figures. It is desirable to have a velocity of 3.0 fps or more whenever practicable. A minimum velocity of 3.0 fps should be obtained in inverted siphons, to which access for cleaning is difficult. Slopes corresponding with mean velocities as low as 1.5 fps have been used successfully in some special cases, but sewers at such slopes must be built with great care, and their interior surface must also be finished with great care, to achieve successful working conditions.

Repeated removal of sludge and hard materials from sewers is expensive, and if such deposits are not cleaned out they may cause increasingly troublesome conditions. It is desirable, therefore, to use slopes that will give self-cleansing velocities in all cases, even where the resulting increase in cost of construction due to steeper slopes will involve fixed charges greater than the added cost of maintaining the sewers if they are laid on flatter slopes. This is recommended because such maintenance work, if neglected, can result in a substantial deposit. Then the sewer cannot perform its functions properly, and it may fail to carry the sewage at the design rate, resulting in damage to property.

Minimum Slopes Sewers with flat slopes are often required to avoid excessive excavation, to provide a minimum depth of cover, or to satisfy local conditions, such as flat surface slopes or small total fall available.

Where slopes are comparatively flat, the sewer sections and slopes should be designed so that the velocity of flow will increase progressively, or at least will be maintained steady, in passing from the inlets to the outlet of the sewer. This is done so that solids washed into the sewer and transported by the flowing stream may be carried through the sewer and not deposited at some point because of a decrease in velocity. It is seldom possible to attain this condition fully, however, due to topographical characteristics. In general, the minimum slopes given in Table 4·1 for small pipe sewers in a sanitary sewer system have been found to be suitable.

TABLE 4·1 MINIMUM SLOPES FOR SANITARY SEWERS

Size, in.	Slope, ft/ft
8	0.004
10	0.003
12	0.0022
15	0.0015
18	0.0012
21	0.0010
24	0.0009
27 and larger	0.0008

In addition, it may be advisable, after designing a sewer, particularly a trunk or intercepting sewer, for a given service in the future, to consider the actual conditions of operation likely to arise at times of minimum flow during the first few years after its construction. It should be made certain that the velocities will not be so low, for significant periods of time, so as to result in objectionable deposits in the sewer, because the removal of these deposits would involve excessive cost. The construction of a sewer to serve for a long period might not be justified if the cost of cleaning should exceed the cost of building a smaller sewer in the first instance, to serve for a shorter period of time. Later, when the service area has developed to a further degree, a second sewer could be built for the additional expected sewage flow resulting from the increased development. While the latter plan would involve greater cost of construction, enough might be saved in fixed charges and in the cost of operations in the early years of the use of the sewer to more than offset this increased cost.

Minimum Size of Sewers The adoption of a minimum size of sewer is necessary because experience has shown that some comparatively large objects, such as scrub brushes, sometimes get into sewers and that stoppage resulting from them is less likely if sewers are not smaller than 8 in. Obviously, the smallest sewer should be larger than the building sewer connections in general use, so that articles that pass through the building connections may as readily pass through the sewer. A minimum size of 8 in. is recommended for sanitary sewers.

Engineers are not in entire agreement upon the most advantageous size of building connections. The most common size is probably 6 in., but there are many 5- and 4-in. connections.

Maximum Velocities The erosive action of suspended matter depends not only on the velocity with which it is carried along the invert of a sewer but also on its nature. Since this erosive action is the most important factor in determining the safe maximum velocities of sewage, the character of the suspended matter must be considered.

One objection to high velocities in small pipe sewers is that, with reduction in the rate of sewage flow and consequent decrease in depth, large floating objects, which at times enter all sanitary sewerage systems, are likely to be left stranded on the inverts, where they may become so firmly lodged that the next rush of sewage will not detach them. Rags, old brushes, pieces of wood, corncobs, and such things, should not be allowed to enter sanitary sewers, but they are sometimes thrown into the house fixtures. They may be left on the invert of a small pipe sewer where the sewage flows intermittently in swift flushes, as is likely to be the case near the upper ends of sewage systems where the slopes are steep.

Sewer Appurtenances The principal appurtenances associated with sanitary sewers are manholes, drop manholes, building connections, and junction chambers. In addition, a wide variety of special appurtenant structures may be required, depending upon local topographic conditions. Since detailed descriptive material on appurtenances and special structures is presented in Chap. 5, the following discussion will be limited to a few specific design considerations concerning manholes [4].

On the smaller sewers (48 in. and smaller), manholes should be located at changes in size, slope, or direction. On larger sewers, these changes may be made without the requirement of a manhole.

If possible, vertical drops in the flowing sewage should be avoided to minimize splashing. When such drops are necessary, drop inlets or other means of conveying the sewage to a lower elevation should be provided. At such points, a vitrified-clay brick lining may be provided in concrete structures to prevent erosion of the concrete.

In sewers 24 in. and smaller, a turn of 90° or less may be made in one manhole. In sewers ranging from 27 to 48 in., a 90° turn should be made in two manholes, each located about six diameters from the point of intersection, with a straight alignment from manhole to manhole, or possibly in one large manhole that is specially designed. Curves may be used for sewers larger than 48 in. The radius along the centerline of curves for pipe sewers should be the minimum practicable, using pipe units cast with a bevel on one end. The minimum radius for monolithic concrete sewers should be equal to four times the diameter of the sewer.

Where a turn is made in a manhole, compensation for the curve loss is desirable. Where curves in a sewer alignment are used, compensation for the required additional loss of head should be made. One method of compensating for curve loss is to compute the hydraulic grade line along each tangent (to the point of intersection of the forward and back lines) and to apply a steeper slope to the sewer along the short chord of the curve, thus using the additional available head. In other words, the total change of elevation around the curve would be the amount that it would be if it were computed along the tangents.

Manholes on the smaller sewers (24 in. and less) should be placed at intervals not greater than 350 ft. For sewers 27 to 48 in. inclusive, the maximum interval should be 400 ft. On sewers larger than 48 in., manholes may be placed at somewhat greater intervals, depending on circumstances.

Alternative Designs Two or more alternative arrangements may sometimes be found practical for some projects, in certain cases involving locations across private land. Generally, the relative desirability of such alternative designs can be determined by inspection, particularly if no change in size of sewer is probable. In some cases, it may be necessary to prepare design details for each alternative and to make comparative estimates of cost before a decision can be reached. Unless there is a significant advantage in cost or other condition resulting from a location through private property, it is generally inadvisable to build sewers outside of public ways.

Design of a Sanitary Sewer System

The steps and basic data used in the design of a sanitary system are illustrated by the following example.

EXAMPLE 4·1 *Design of a Sanitary Sewer System*

Design a sanitary sewer system for the residential district shown in Fig. 4·1. The district is two-thirds developed; therefore the probable future population density can be estimated without making a detailed population study. It is estimated that the future average saturation population density will be 65 persons per acre. The maximum hourly rate of flow of sewage is estimated at 250 gpcd. The maximum rate of ground water infiltration to the sewers, to be provided for, is 2,000 gpad.

The minimum size of sewer is to be 8 in. The minimum velocity of flow in the sewer when full is to be 2.0 fps. (The more desirable limit of 2.5 fps is impracticable because of the excessive expense that would result from its adoption.) The capacity of the sewers will be determined using Manning's equation with a recommended n value of 0.013.

Since the homes in this area have basements, the minimum depth below the street surface to the top of the sewers will be 7.0 ft. (In areas where basements are not normally constructed, the depth of cover to the top of the sewer may be as little as 3.0 ft.)

Solution

1. Draw a line to represent the proposed sewer in each street or alley to be served. Near the line indicate by an arrow the direction in which the sewage is to flow. Except in special cases, the sewer should slope with the surface of the street. It is usually more economical to plan the system so that the sewage

from any street will flow to the point of disposal by the most direct (and consequently the shortest) route.

The lines representing the system will often resemble a tree and its branches. In general, the laterals connect with the submains; and these, in turn, connect with the main or trunk sewer, which leads to the point of discharge.

2. Locate the manholes, giving each an identification number.
3. Sketch the limits of the service areas for each lateral, unless a single lateral will be required to accommodate an area larger than can be served by the minimum size of sewer with the minimum slope, in which case a further subdivision may be required. Where the streets are laid out, the limits may be assumed as being midway between them. If the street layout is not shown on the plan, the limits of the different service areas cannot be determined as closely and the topography may serve as a guide.
4. Measure the acreages of the several service areas. For this, a planimeter will give results with sufficient accuracy. At this point, the design may be represented as shown in the plan view in Fig. 4·2.
5. Prepare a tabulation, such as that shown in Table 4·2, with columns for the different steps in the computation and a line for each section of sewer between manholes. This tabulation is a concise, timesaving method and shows both the data and the results in orderly sequence for subsequent use.

   Use col. 1 for numbering the lines of the table, for ready reference. Determine by inspection the manhole that is farthest from the point of discharge and enter its identification number in the first line of col. 2, and the number corresponding to the manhole next on the line toward the trunk sewer in col. 3. Enter the name of the street or alley in col. 4, the length between manholes in col. 5, and the area in acres to be served by the sewer at a point just above the lower manhole in col. 6.

   On the next line enter the corresponding data for the next stretch of sewer, and in col. 7 enter the sum of the areas listed in col. 6. The area in col. 7 is the basis for computing the required capacity of the sewer. Enter the data for each section of sewer in the above manner, following the line to the point of discharge, including the trunk or main sewer.

   Enter in col. 8 the rate of flow in the sewer, which is equal to the maximum per capita rate of sewage flow multiplied by the assumed future density multiplied by the area shown in col. 7.

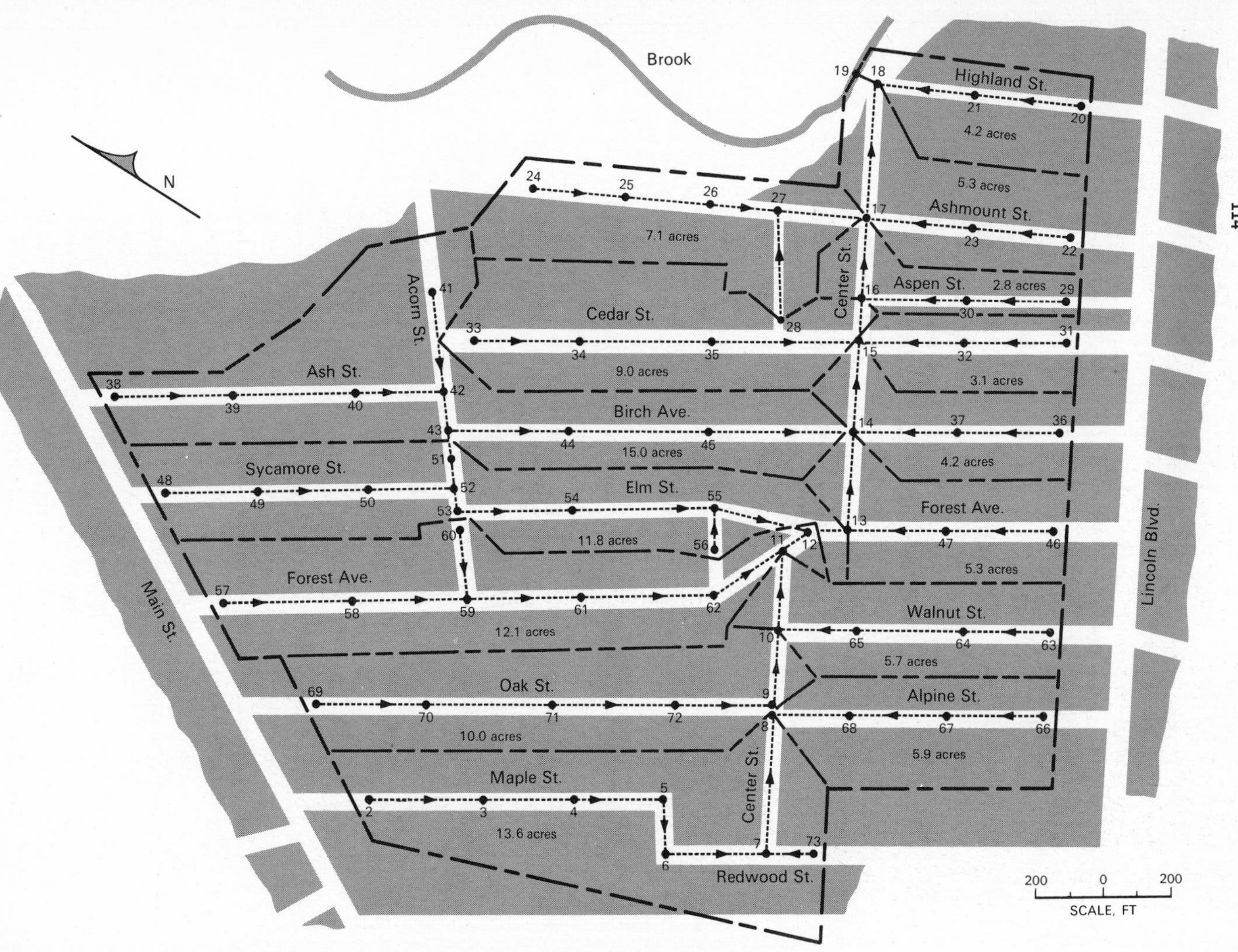

FIG. 4·2 Map showing manholes, sewer lines, and subareas for sanitary sewer design example.

TABLE 4·2 COMPUTATIONS FOR A SANITARY SEWER

					Area, acres				Total maximum flow, sewage and ground water							Invert elevation	
Line (1)	From man-hole no. (2)	To man-hole no. (3)	Location (4)	Length, ft (5)	Increment (6)	Total (7)	Sewage, mgd* (8)	Ground water at 2,000 gpad (9)	mgd (10)	cfs (11)	Size of sewer, in. (12)	Slope, ft/ft (13)	Velocity, fps (14)	Capacity, cfs (15)	Surface elevation, upper end (16)	Upper end (17)	Lower end (18)
1	57	58	Forest Ave.	380	....	.....	.....	.....	.....	....	8	0.004	2.2	0.77	208.2	200.40	198.89
2	58	59	Forest Ave.	370	....	.....	.....	.....	.....	....	8	0.004	2.2	0.77	206.4	198.89	197.40
3	59	61	Forest Ave.	365	....	.....	.....	.....	.....	....	8	0.004	2.2	0.77	205.2	197.40	195.94
4	61	62	Forest Ave.	370	....	.....	.....	.....	.....	....	8	0.004	2.2	0.77	204.3	195.94	194.46
5	62	11	Forest Ave.	240	....	12.1	0.196	0.024	0.220	0.34	8	0.004	2.2	0.77	202.0	194.46	193.50
6	11	12	Forest Ave.	130	35.2	47.3	0.767	0.095	0.862	1.34	12	0.0023	2.2	1.70	201.6	189.30	189.00
7	12	13	Forest Ave.	82	11.8	59.1	0.960	0.118	1.078	1.67	12	0.0023	2.2	1.70	202.1	189.00	188.81
8	13	14	Center St.	280	5.3	64.4	1.046	0.129	1.175	1.82	15	0.0017	2.3	2.70	202.8	188.56	188.08
9	14	15	Center St.	275	19.2	83.6	1.360	0.167	1.527	2.37	15	0.0017	2.3	2.70	203.2	188.08	187.61
10	15	16	Center St.	113	12.1	95.7	1.555	0.191	1.746	2.70	15	0.0020	2.4	2.90	203.6	187.61	187.38
11	16	17	Center St.	245	2.8	98.5	1.600	0.197	1.797	2.78	15	0.0024	2.6	3.20	203.7	187.38	186.79
12	17	18	Center St.	375	12.4	110.9	1.800	0.222	2.022	3.13	15	0.0024	2.6	3.20	202.3	186.79	185.89
13	18	19	Right of way	130	4.2	115.1	1.871	0.230	2.101	3.26	15	0.0025	2.7	3.27	196.0	185.89	185.56

* Based upon a maximum rate of 250 gpcd and 65 persons per acre. Since the capacity of the minimum size of sewer with a minimum velocity of 2.0 fps is 0.7 cfs, equivalent to the maximum rate of discharge from 24.6 acres, and since $\frac{24.6(65 \times 250 + 2{,}000)1.55}{1{,}000{,}000} = 0.7$, all laterals will be 8 in. in diameter (the minimum), as no lateral is to serve an area exceeding 24.6 acres.

Enter in col. 9 the rate of allowance for ground water infiltration, which is equal to the rate per acre to be provided for, multiplied by the area in col. 7.

Column 10 contains the sums of the figures in cols. 8 and 9, in mgd. In col. 11 this rate is converted to cfs, which is the more convenient way of expressing the capacity of sewers, since most diagrams and tables indicate the capacity of circular pipes in cfs.

Column 12 contains the required sewer sizes; col. 13, the slope; col. 14, the velocity when the sewer is full; and col. 15, the capacity. Column 16 contains the elevations of the street surface at the manhole corresponding to the identification number in col. 2. Columns 17 and 18 contain the invert elevations of the upper and lower ends, respectively, of each reach of sewer.

In selecting the sewer sizes and slopes, the designer makes use of profiles, such as the one shown in Fig. 4·3. This allows the designer to select a minimum sewer size and slope that will carry the computed flow and that also will meet minimum depth criteria.

A plan and profile should be prepared for construction purposes for each sewer from data obtained by field observations and surveys. It should show the ground surface, the depth and location of existing basements, the proposed sewer, its slope and size, and the invert elevation at each manhole, as well as the size and elevation of the sewer into which the flow of the sewer under consideration is to discharge.

The scale to be used in preparing the profiles will depend upon the number of obstacles to construction, and hence the amount of detail required. In city work, scales of 20, 40, or 50 ft/in. horizontally and 4, 5, or 10 ft/in. vertically are commonly used. The profile should be drawn either directly above or below the location plan. The plan should be of the same scale as the horizontal scale of the profile and should show all structures, both above and below the ground surface, that may influence the choice of location for the sewer or that may affect the construction operations.

The elevations in Table 4·2 may be used to determine depths of cut and quantities from which estimates of cost may be prepared.

Preparation of Contract Drawings and Specifications

Detailed contract drawings should be completed before bids are requested, so that all of the data will be made available to prospective

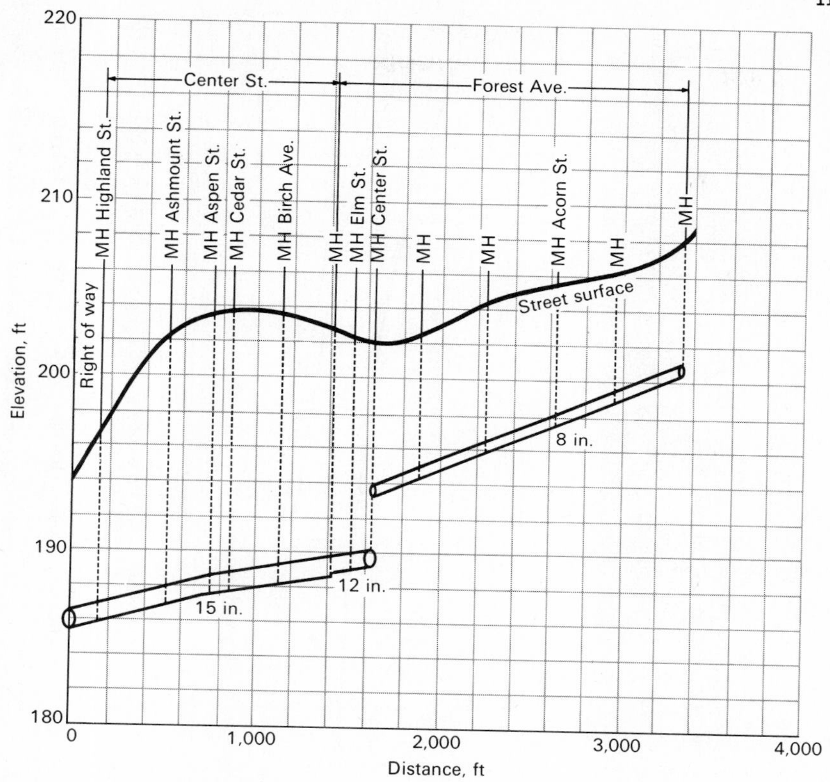

FIG. 4·3 Typical profile for sanitary sewer design example.

bidders. Such drawings should show, as far as practicable, all available information bearing upon surface features, the character of materials to be excavated, and the location, size, and character of structures likely to be encountered in the excavation, together with details of the works to be constructed.

An example of a typical contract drawing is shown on Fig. 4·4. The data usually required for preparing contract drawings are illustrated on this figure, and the method of showing the data, with scales, may be seen. The greater amount of detail and the difference in scale and arrangement that this necessitates are apparent when Fig. 4·4 is compared with the profile on Fig. 4·3. Note that both a plan and a profile are required for indicating the work to be done.

Practice in showing data from borings varies. In some projects, the data are shown on the contract drawings; in others, only the locations are so shown, and details of the test borings are contained in the

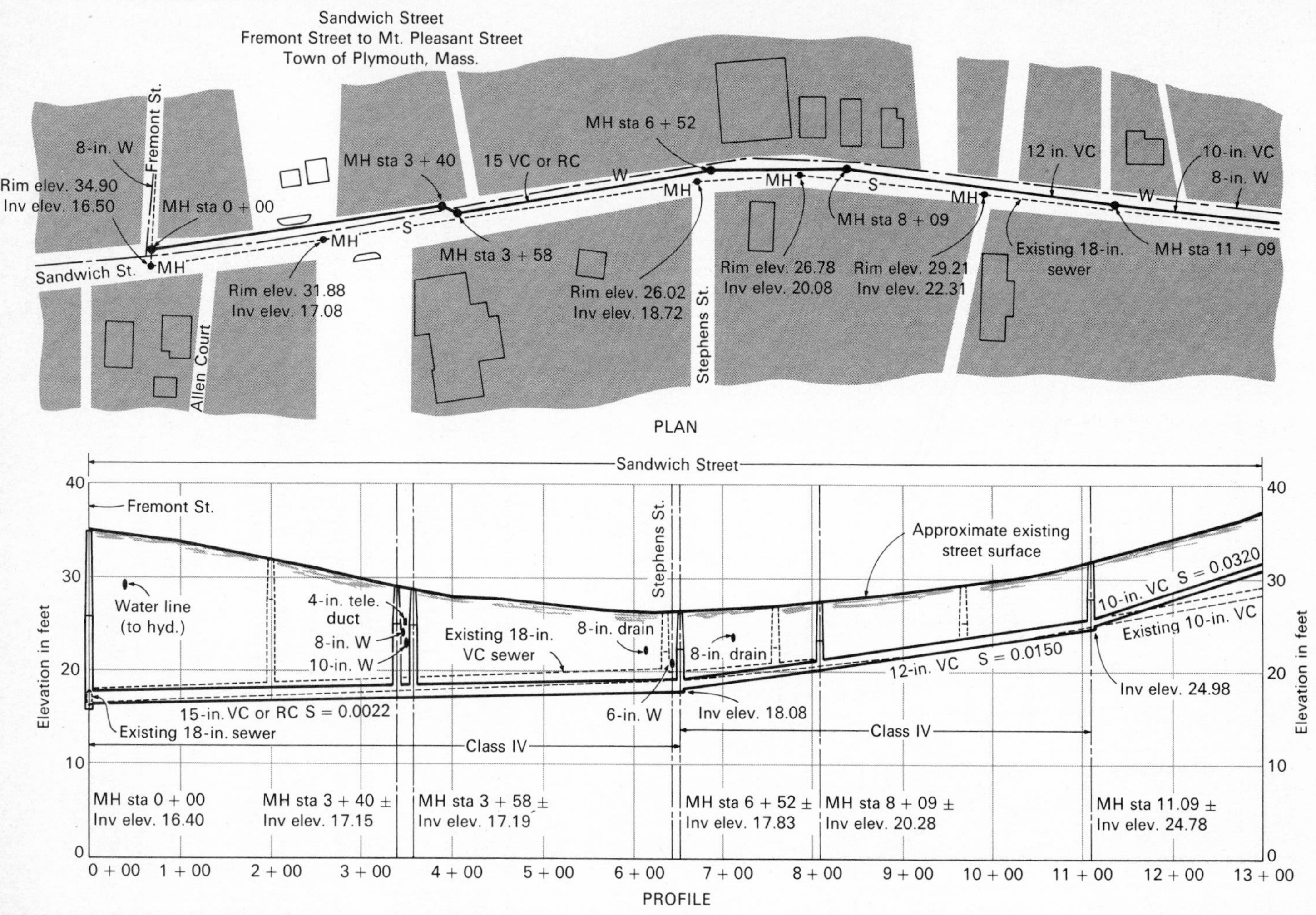

FIG. 4·4 Typical sewer contract drawing.

bound contract documents. Such data consist of copies of the boring contractor's detail logs or reports.

The contract and specifications should be prepared so as to set forth clearly, and as completely as possible, all work, requirements, and conditions included in or affecting the contract. While these details increase the cost of the engineering work, the total cost of the project will usually be less than when the drawings merely show in a general way what is to be done, or where the specifications are incomplete or obscure.

COMBINED SEWERS OR STORM SEWERS

The construction of combined-sewer systems at the present time appears to be almost nil; however, many cities have extensive, old combined-sewer systems. Combined-sewer design generally has been the same as storm-sewer design, with no allowance for sanitary sewage since its volume is so small compared to the expected volume of storm water that it can be neglected. Combined sewers are normally located at greater depth than storm sewers in order to serve basements.

The design of combined sewers or storm sewers requires (1) the preparation of maps and profiles, (2) the determination of the storm water runoff, and (3) the selection of the proper dimensions for the sewer, taking into account the available slope and other topographical and physical conditions.

To illustrate these design procedures, a storm sewer will be designed for the area shown on Fig. 4·1. This area was also used to illustrate the design of a sanitary sewer.

Maps and Profiles

To design a storm sewer properly, it will be necessary to prepare a map and profile of the area to be serviced. The procedure for doing this has been previously discussed in the section dealing with the design of sanitary sewers. An example of such a map, which will also be used for the illustrative design example, is shown on Fig. 4·5.

Storm Water Flow

The problem of determining the rates of storm water runoff to be carried by storm sewers or combined sewers is difficult and somewhat indeterminate, although much progress has been made during recent years in the newer design methods. Such progress has come about with

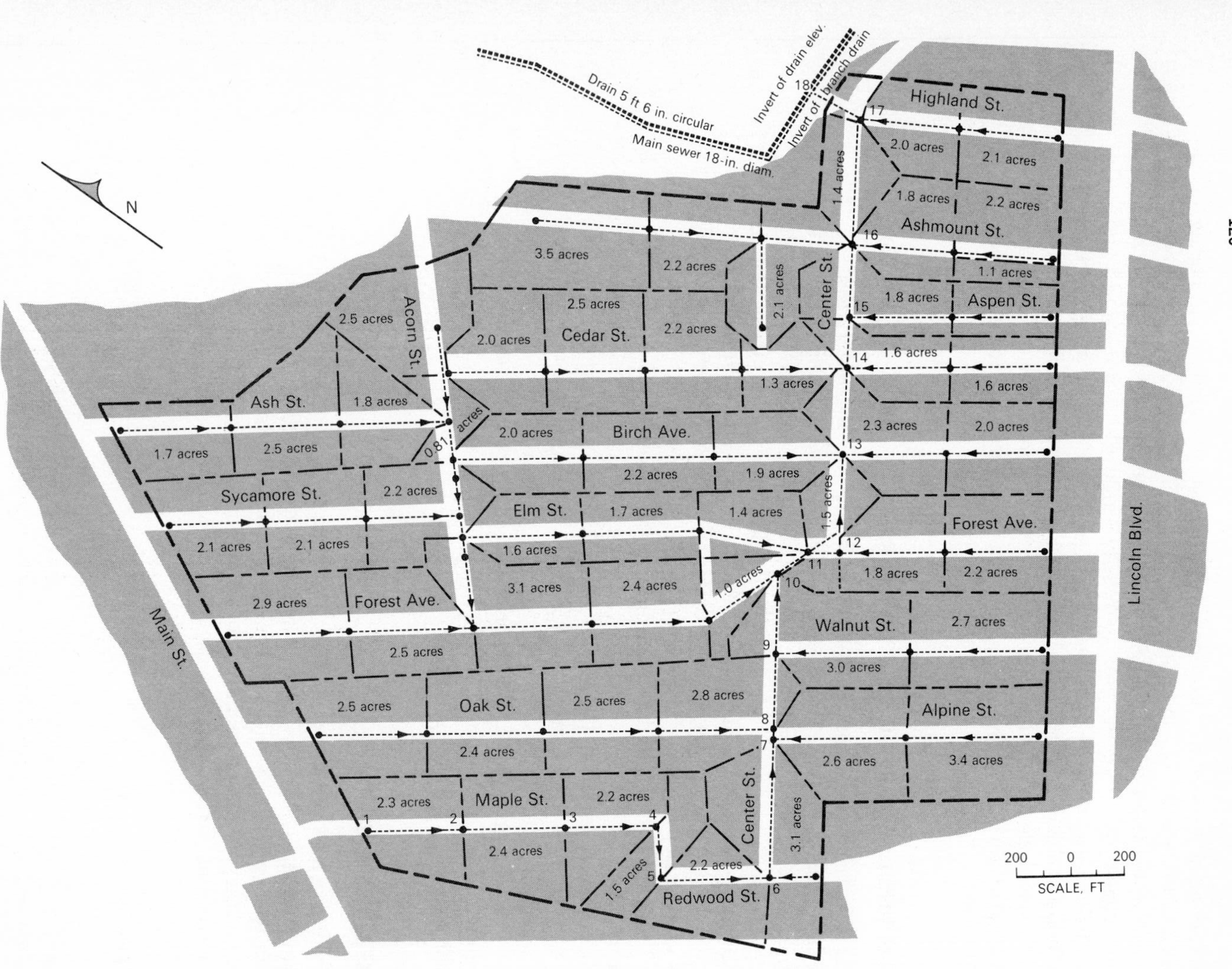

FIG. 4·5 Map for storm-sewer design example.

increased technical knowledge of hydrological events and characteristics in relation to rainfall and runoff, particularly the large amount of data available for study and correlation of rainfall, runoff, topography, and development. The extensive application and use of flow hydrograph data, operation of flood-control reservoirs, pumping of storm runoff, and the requirements for substantially flood-free highways and airfields, have resulted in various methods of estimating rainfall runoff rates and amounts. To assist the engineer in estimating inlet times or in preparing hydrographs, more information is needed.

The determination of storm water flow for the purposes of storm-sewer design may be made by using any of the following techniques: (1) empirical formulas, (2) the rational method, (3) rainfall-runoff correlation studies, (4) hydrograph methods, (5) the inlet method, and (6) digital computer models. The selection of the method will depend on the local geographic and hydrologic conditions, the availability of past rainfall and runoff data, and the degree of protection to be provided.

Detailed data on all these methods have not been included here because they would require a disproportionate amount of space in a book dealing specifically with sewerage and sewage treatment and also because they are already available elsewhere [1 to 3, 5]. The rational method is probably the one most widely used.

To illustrate the procedures involved in the design of a combined sewer or storm sewer, the rational method [1, 5] will be used for estimating the rates of storm water runoff. In this case, it is assumed that the student is already familiar with this method. If not, he should consult one of the references previously mentioned.

Briefly, the computation of storm-water-runoff rates, using the rational method, requires determination of the following basic data:

1. The time-intensity rainfall relation to be used as a basis of design.
2. The probable future condition of the drainage area, i.e., the percentage of impervious surface that may be expected when the district is developed to the extent assumed.
3. The runoff coefficient, i.e., the fraction of the rainfall that will run off the drainage area.
4. The probable time required for water to flow over the surface of the ground to the first inlet, called the inlet time or time of entrance.
5. The area tributary to the sewer at the point at which the size is to be determined.
6. The time required for water to flow in the sewer from the first inlet to the abovementioned point, which, added to the inlet time, gives the time of concentration.

Design of Combined or Storm Sewers

The following example applies equally well to combined sewers or storm sewers. The area shown in Fig. 4·5 is a small portion of the total drainage area of a large district.

EXAMPLE 4·2 *Design of a Storm-Sewer System*

Design a storm-sewer system for the area shown in Fig. 4·5. The location of the proposed main storm sewer that is to receive the storm water from the district is shown on the map, and the invert elevation is known at the point where the proposed branch storm sewer is to be connected and for which provision has been made in the design of the main storm sewer. The required lowest elevation of the invert of the branch storm sewer is therefore known at the proposed point of discharge into the main storm sewer.

A careful study of local conditions, including the present and probable future development of the district, indicates that 70 percent of the surfaces in the district are expected to be impervious. The inlet time has been assumed to be 20 min.

The rate of rainfall is to be taken from the assumed curve of intensity of precipitation represented by the formula $i = 20.4/t^{0.61}$, in which i is rainfall intensity in in./hr and t is rainfall duration in min. This formula represents the average rate of rainfall for a duration of t min which may be expected to be equalled or exceeded on the average once in a 5-year period. The rainfall and runoff curves are shown on Fig. 4·6.

While it was recognized that storm sewers designed on this basis might be overtaxed on the average of once in about 6 years, it was not considered reasonable to provide for storms of greater intensity, because of the greater cost. During the earlier years of the life of the storm sewers, they will be able to carry the runoff from higher rates of rainfall than they will be able to carry later, because the assumed coefficients of runoff are based upon future rather than present conditions. A progressive increase in impervious surface and runoff will be caused by the gradual substitution of roofs and paved areas for presently unimproved areas. In the future, when the district is more densely built up and funds are available, relief sewers can be constructed to provide for higher runoff rates, if flooding has become sufficiently serious to warrant the expenditure.

Figure 4·5 shows the drainage area. Street elevations are

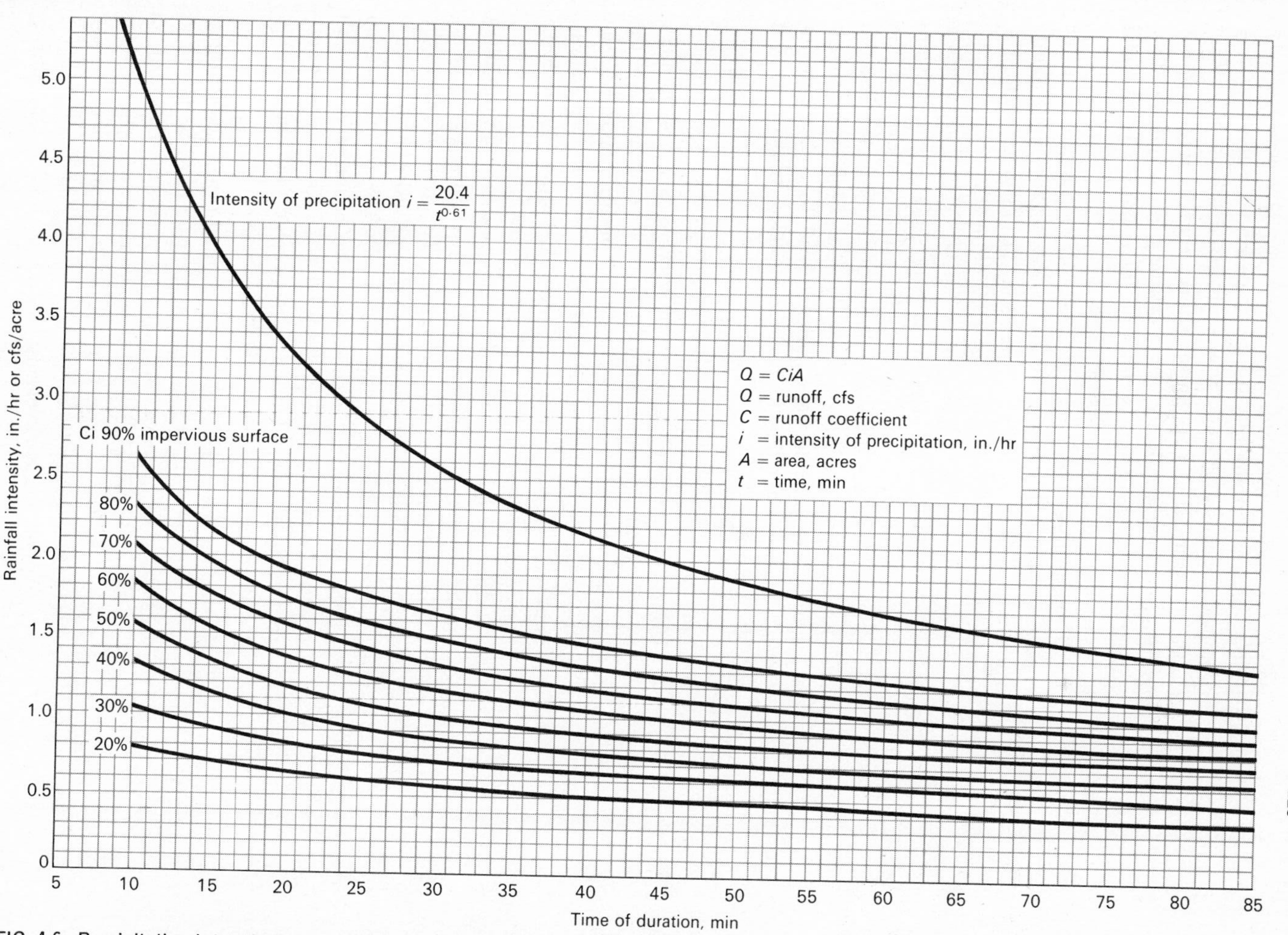

FIG. 4·6 Precipitation intensity curves for various degrees of imperviousness.

shown in Fig. 4·1. The limits of this area are influenced not only by the surface contours but also by the service areas of existing combined sewers and storm sewers. In a district where the surface slopes are moderate and generally uniform, contour maps may not be required. Instead, surface elevations may be adequate if they are obtained for street intersections, for high and low points, and at locations of change of surface slope.

The storm sewers are to be designed, in general, with the crown at a depth of at least 5 ft below the surface of the street. The minimum size of drain is to be 12 in. The assumed minimum mean velocity is 3.0 fps when flow is at full depth.

The capacities of the sewers are to be determined using a value of $n = 0.013$. Velocities for flow at design conditions are much higher for storm sewers than for sanitary sewers for the reason that the design storm flow is many times greater than the peak sewage flow in sanitary sewers. Because of high velocities, it is important to provide for additional head to compensate for losses, such as those due to bends, manholes, transitions, and velocity changes.

Solution

1. Draw a line to represent the storm sewer in each street or alley to be served. Place an arrow near each sewer to show the direction of flow. The sewers should, in general, slope with the street surface. It will usually prove to be more economical, however, to lay out the system so that the water will reach the main storm sewer by the most direct route.

   In some localities the roof water is allowed to discharge on the ground and flow over the surface to the gutter inlets. Under such circumstances, sewers may be provided only to the last gutter inlet, rather than to a point opposite the last house lot, thereby effecting some saving in cost. This practice is open to the criticism that it does not give equal service to all property and is, therefore, inequitable. In the example under consideration, the intention is to provide drainage facilities for all property within the district.

2. Locate the manholes tentatively, giving to each an identification number. In this example, a manhole is to be placed at each bend or angle, at all junctions of storm sewers, at all points of change in size or slope, and at intermediate points where the distance exceeds 350 ft on 12- to 24-in. sections and 400 ft for larger sections.

   Where a good velocity would be maintained during

TABLE 4·3 COMPUTATIONS FOR COMBINED SEWERS OR STORM SEWERS*

	From (1)	To (2)	Location (3)	Length, ft (4)	Increment, acres (5)	Total acres (6)	To upper end, min (7)	In section, min (8)	Rate of run-off, cfs per acre (9)	Required capacity, cfs (10)	Pipe size, in. (11)	Slope, ft/ft (12)	Velocity, fps (13)	Capacity, cfs (14)	Surface elevation (15)	Fall, ft (16)	Upper end (17)	Lower end (18)
1	1	2	Maple St.	300	2.3	2.3	20.0	1.7	1.56	3.6	15	0.003	2.9	3.6	206.2	0.90	199.90	199.00
2	2	3	Maple St.	300	2.4	4.7	21.7	1.7	1.51	7.1	21	0.002	3.0	7.2	206.5	0.60	198.50	197.90
3	3	4	Maple St.	300	2.2	6.9	23.4	1.6	1.47	10.1	24	0.002	3.2	10.2	206.6	0.60	197.65	197.05
4	4	5	Maple St.	165	1.5	8.4	25.0	0.7	1.42	11.9	24	0.0027	3.8	11.9	207.1	0.45	197.05	196.60
5	5	6	Redwood St.	325	2.2	10.6	25.7	1.4	1.40	14.9	27	0.0023	3.8	15.0	207.4	0.75	196.35	195.60
6	6	7	Center St.	400	3.1	13.7	27.1	1.7	1.38	18.9	30	0.0021	3.9	18.8	206.1	0.84	195.35	194.51
7	7	8	Center St.	35	6.0	19.7	28.8	0.2	1.34	26.4	30	0.004	5.2	26.0	203.2	0.14	194.51	194.37
8	8	9	Center St.	230	10.2	29.9	29.0	1.0	1.34	40.0	42	0.0016	4.1	41.0	203.2	0.37	193.37	193.00
9	9	10	Center St.	240	5.7	35.6	30.0	0.9	1.32	47.0	42	0.0022	4.9	47.0	201.9	0.53	193.00	192.47
10	10	11	Forest Ave.	110	11.9	47.5	30.9	0.4	1.31	62.3	48	0.0018	4.9	62.0	201.6	0.20	191.97	191.77
11	11	12	Forest Ave.	95	11.1	58.6	31.3	0.3	1.30	76.3	54	0.0015	4.8	76.0	202.1	0.14	191.27	191.13
12	12	13	Center St.	295	5.6	64.2	31.6	0.9	1.30	83.3	54	0.0018	5.2	84.0	202.8	0.53	191.13	190.60
13	13	14	Center St.	260	17.2	81.4	32.5	0.8	1.28	104.0	60	0.0015	5.3	103.0	203.2	0.39	190.10	189.71
14	14	15	Center St.	145	13.7	95.1	33.3	0.5	1.27	121.0	66	0.0013	5.1	121.0	203.6	0.19	189.21	189.02
15	15	16	Center St.	225	2.9	98.0	33.8	0.7	1.26	124.0	66	0.0014	5.2	126.0	203.7	0.32	189.02	188.70
16	16	17	Center St.	380	13.2	111.2	34.5	1.1	1.25	139.0	66	0.0017	5.8	129.0	202.5	0.65	188.70	188.05†
17	17	18	Private land	165	4.1	115.3	35.6	—	1.24	143.0	48	0.0096	11.4	143.0	196.0	1.59	187.32	185.73

* Figures in col. 8 are obtained by dividing those in col. 4 by 60 and by the figures in col. 13. Figures in col. 10 are obtained by multiplying those in col. 6 by the figures in col. 9. Figures in col. 16 are obtained by multiplying those in col. 4 by the figures in col. 12.

† The difference in the elevations of the 66-in. sewer at point 17 and the 48-in. outlet sewer would allow for velocity head increase and bend losses.

practically all conditions of flow, and the sewer is large enough for workmen to walk without stooping, intervals between manholes up to 600 ft may be used. Sufficient manholes should be built to allow access for inspection and cleaning. Later, when the profiles are drawn and the final slopes are fixed, it may be desirable to change the locations for some manholes so that the sewers would be at the most advantageous depth, particularly where the slope of the street surface is not substantially uniform. Other considerations, such as obstacles underground, may require the installation of additional manholes, due to change in alignment or special forms of construction involved in junctions or connections with other sewers.

3. Sketch the limits of the drainage areas tributary at each manhole. The assumed character of future development and the topography will determine the proper limits.
4. Measure each individual area by planimeter or other methods that will give equally satisfactory results.
5. Prepare a tabulation to record the data and steps in the computations of each section of sewer between manholes.

   The computations for a selected line of this section are shown in Table 4·3. Each lateral is then designed in a similar way. If necessary, the first design of the submain is subsequently modified so as to serve the laterals properly. It is possible in some cases to omit some manholes on lateral storm sewers, using the inlet substructures at which junctions, changes in size, direction, or slope may be made.

Low-Flow Conditions in Combined Sewers

If combined sewers are designed for conditions to be expected many years in the future, the design should be studied for flow conditions to be expected during the early years, particularly where the system is to include relief overflows and the dry weather peak flow is to be collected by an intercepting system and conveyed to a location for treatment. Such a study should include estimates of the depth and velocity of flow and a consideration of the operating cost of the sewer resulting from the necessity of cleaning and flushing to remove deposits. In protracted dry spells, flushing of lateral combined sewers is advisable to remove the solids; flushing of large combined sewers is not likely to be effective, and it may be found advisable to provide a special shape of invert, such as a cunette, to confine the flow and to obtain

TABLE 4·4 ANALYSIS OF COMBINED-SEWER DESIGN UNDER DRY WEATHER FLOW CONDITIONS

		Average rate of sewage flow					Corresponding proportion to		Condition for average dry weather flow	
Line no. (1)	Total tributary area, acres (2)	Sewage, mgd (3)	Ground water at 500 gpad (4)	Total, mgd (5)	Total, cfs (6)	Capacity, % of full capacity (7)	Depth, % of full depth (8)	Velocity, % of full velocity (9)	Depth, in. (10)	Velocity, fps (11)
8	29.9	0.132	0.015	0.147	0.228	0.57	5.0	22.	2.1	0.90
10	47.5	0.209	0.024	0.233	0.361	0.59	6.0	26.	2.9	1.27
11	58.6	0.258	0.029	0.287	0.445	0.59	6.0	26.	3.2	1.25
13	81.4	0.359	0.041	0.400	0.620	0.60	6.0	26.	3.6	1.38
14	95.1	0.420	0.048	0.468	0.725	0.60	6.0	26.	4.0	1.33
16	111.2	0.490	0.055	0.545	0.845	0.59	6.0	26.	2.9	3.00

higher velocities and depths. Low-flow analysis is illustrated in the following example.

EXAMPLE 4·3 *Low-Flow Analysis for Combined Sewers*

Using the design of storm sewers shown in Example 4·2, determine the low-flow conditions if the system were designed as a combined-sewer system. The basic data are

Dry weather flow of domestic sewage = 110 gpcd

Dry weather infiltration of ground water = 500 gpad

Present density of population = 40 persons per acre

Solution

1. Determine the total dry weather flow for each tributary area.
2. Determine the depth and velocity of flow. The ratio of the average dry weather flow to the capacity of the sewer when full shows the proportion of the capacity utilized; the corresponding proportion of depth and of velocity may readily be obtained from a diagram similar to Fig. 3·16 but prepared for various values of D and S approximating those appearing in the design. A diagram similar to the above was used in this example but of larger scale enabling proportions to be read with sufficient accuracy for smaller flows. The results of calculations for this example are shown in Table 4·4.

The velocities in Example 4·3 would be very low for the ordinary dry weather flow, with the likelihood that deposits will be formed. If this sewer were to be part of a combined system, including relief overflows and an intercepting sewer for diverting the dry weather flows to a location for treatment, it would be desirable to design the main sewer with a cunette-shaped invert. Thus, solids would be carried along with the ordinary flows and not deposited above overflow structures where they would be likely to be diverted from the sewer by the first overflow of combined storm water and sewage.

PROBLEMS

4·1 Outline the steps to be followed in designing a separate sanitary sewer and discuss their relative importance.

4·2 As shown in Fig. 4·3, draw a profile for a separate sewer in Maple, Redwood, and Center Sts. to discharge into Forest Ave.

at manhole no. 11. Assume a future population density of 50 persons per acre, with a maximum rate of domestic sewage flow of 275 gpcd and 2,500 gpad of infiltration. Use other basic data as given in the chapter. $V_{min} = 2.0$ fps. The minimum depth of sewer below the surface is 7 ft.

4·3 Compute the yardage of excavation and length of pipe of various diameters required for the sewer shown in the profile in Fig. 4·3. Assume the width of trench to be 1.4 times the inside diameter plus 1 ft, but with a minimum width of 3 ft, and that the depth of excavation will be 0.2 ft below grade of invert.

4·4 List six methods that have been used to determine storm water quantities and briefly discuss their distinguishing characteristics.

REFERENCES

1. American Society of Civil Engineers, *Hydrology Handbook*, Manual 28, 1949.
2. Camp, T. R.: Design of Sewers to Facilitate Flow, *Sewage Works Journal*, vol. 18, no. 1, 1946.
3. Joint Committee of the American Society of Civil Engineers and the Water Pollution Control Federation: *Design and Construction of Sanitary and Storm Sewers*, ASCE Manual and Report 37, New York, 1969.
4. Linsley, R. K., Jr., M. A. Kohler, and J. L. H. Paulhus: *Hydrology for Engineers*, McGraw-Hill, New York, 1958.
5. Metcalf, L., and H. P. Eddy: *American Sewerage Practice*, vol. 1, McGraw-Hill, New York, 1928.

sewer appurtenances and special structures

5

The appurtenant structures that are built on sewerage systems are, as a rule, important in the operation of such works. To clean and inspect sewers, manholes giving access to them are provided. Drop manholes are used in order that sewage may fall vertically from one elevation to another with a minimum amount of disturbance. Where storm water is removed underground, street inlets are provided to permit it to discharge directly into the sewers. Catch basins are provided where the surface runoff contains so much refuse and such a variety of refuse that it should be allowed to settle in a sump, which can be cleaned easily, rather than to flow without check into the sewers.

Where large sewers join, junction chambers are necessary; and these chambers sometimes require structures of considerable complexity. Inverted siphons are used in crossing valleys or below subways and other obstructions. On rare occasions, a true siphon may be used to transmit the flow over a small ridge, although it is usually preferable to incur a considerable expense if necessary to avoid such a detail. Since prestressed reinforced concrete has come into general use, specially designed hollow girders or beams have been employed in some places to cross rivers or deep gulches where inverted siphons or steel bridges would have been used previously. If a combined-sewerage system includes intercepting and relief sewers, some form of regulating device must be used at each location where sewage is discharged from a collecting sewer into an intercepting or relief sewer. Numerous forms of automatic regulators, storm-overflow chambers, and leaping weirs are used for such situations.

Where sewage or treatment plant effluent is discharged into a river, lake, estuary, or ocean, an outlet of some kind is needed. Another type of special structure is the tide gate, which is a large check valve used to prevent the entrance of other water or backflow of sewage into a sewer when its surface elevation reaches such a height that the flow tends to be into rather than out of the sewer.

In the early days of sewerage works, sewer ventilation received much attention and a variety of theories existed concerning the best way to provide for it. It was at one time thought that sewers should be ventilated at manholes to prevent sewer gases from accumulating and finding their way into dwellings, thus affecting the health of the residents. The real needs for ventilation, however, arise from the danger of asphyxiation of workmen in sewers and manholes and from the explosion hazard. Sewer gases have caused fatalities on some occasions, usually where agitation of the sewage caused the release of excessive quantities of gas. The removal of volatile gases by ventilation is an aid in reducing the danger from explosion.

Although some of these special structures offer little opportunity for standardization, due to differences in local conditions, there are certain features that experience has indicated are important in any locality.

MANHOLES

Although manholes are the most familiar feature of a sewerage system, they were not used extensively until some time after many large sewers had been constructed (around 1880). They were introduced to facilitate the removal of grit and silt that had collected on the inverts of sewers having a low velocity of flow. Before that time, when a sewer became so badly clogged that it had to be cleaned, it was customary to excavate down to the sewer, break through its walls, remove the obstruction, and then close in the sewer again. It was not until later, however, that the value of manholes on small sewers was recognized; and the principle became established that there should be no change of grade or alignment in a sewer between points of access to it, unless the sewer is large enough to enable a man to pass through it readily. After the general acceptance of the principle that manholes should be placed at changes in line and grade in small sewers, there was a tendency for a time to go to the opposite extreme and put them in at too frequent intervals. This practice is objectionable because of the unnecessary cost and the inevitable injury to pavements caused by the impact of traffic where there are manhole frames in the roadway.

Manholes are usually placed from 200 to 400 ft apart; but on large sewers, where the diameter is in excess of 48 in., spacings of 1,200 and 1,500 ft are used to some extent, since inspections can be made more easily or work can be carried on under less trying conditions than in sewers of smaller diameters. The majority of manholes are now constructed of precast reinforced concrete, although under some conditions brick masonry is still used.

Manholes should be constructed large enough to provide easy access to the sewer. The clearance opposite the steps should be sufficient for a man to pass up or down without difficulty. On small sewers, there should be room to permit the handling and jointing of the 4-ft rods used for removing obstructions and cleaning the sewer. There should be room to handle a shovel, and the bottom should afford footing for a laborer working in the manhole, but should drain to the sewer. Access structures on large sewers are sometimes constructed so that a boat or a scraper can be lowered into the sewer. The boat may be used for inspection purposes, and the scraper may be used to loosen material that forms a coating on the inside surface of a sewer.

The manholes on small sewers are usually made about 4 ft in diameter when the sewers are of circular cross section. The same size is usually employed for all sewers except when special conditions may require manholes of larger size, such as when a gaging device must be used at the bottom of the manhole. An example of a precast reinforced-concrete manhole is shown in Fig. 5·1.

Care must be taken that the unit pressures on the soil beneath the manhole and beneath the sewer are approximately uniform; otherwise there is danger of settlement of the manhole, which may cause a break in the sewer pipe. If the pressures are not normally the same, a spread foundation may be built to reduce the unit load imposed by the manhole. A pipe joint should be provided at, or very near, the outside surface of the manhole to permit some slight differential movement.

Where the sewer is much larger than the diameter of the manhole, the outside of the latter is usually tangent to one side of the sewer; otherwise it is difficult to enter the sewer, and a special ladder is required to reach the invert. Occasionally, on very large sewers, manholes are built entirely apart from the sewer and shafts are built to lead into it.

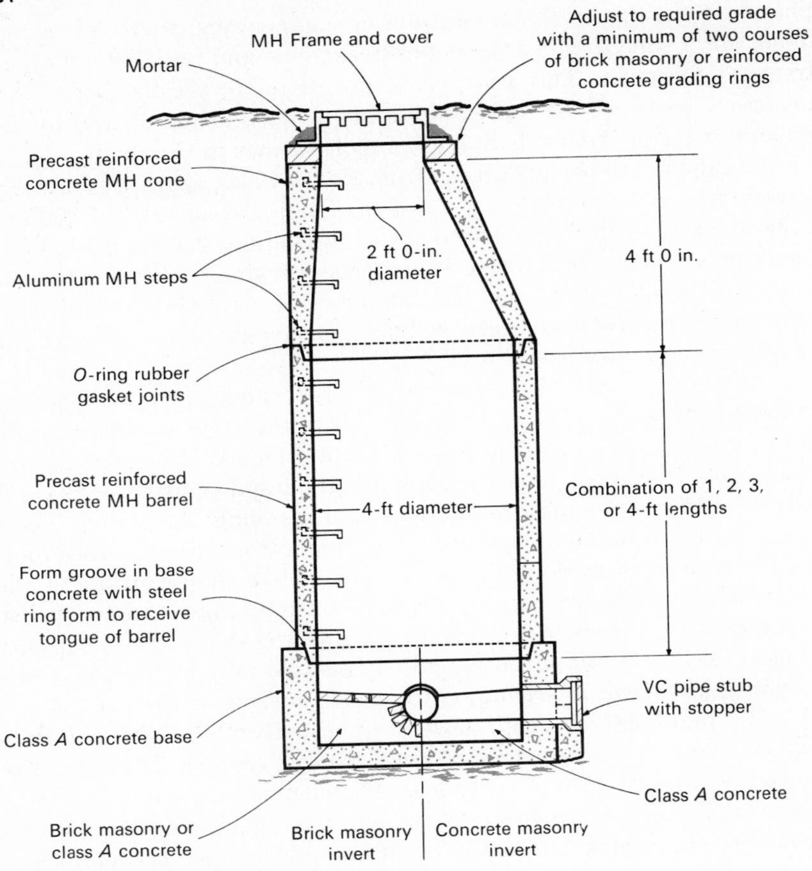

FIG. 5·1 Typical precast concrete manhole.

Changes in size or shape of cross section of the sewer at a manhole produce disturbances in flow with accompanying loss of head. Changes of section, if made by gradual transitions, assist in reducing these head losses. The sidewalls of the channels in manholes, if carried up nearly to the crown, give greater uniformity of section at high flows with lower head losses. There is no uniform practice in this detail, however, as the location of the berm at the top of the sidewall varies from a point mid-depth of the sewer to one level with the crown.

Drop Inlets

Where the difference in elevation between inflow and outflow sewers is to exceed about 1.5 ft, flow from the inflow may be dropped to the

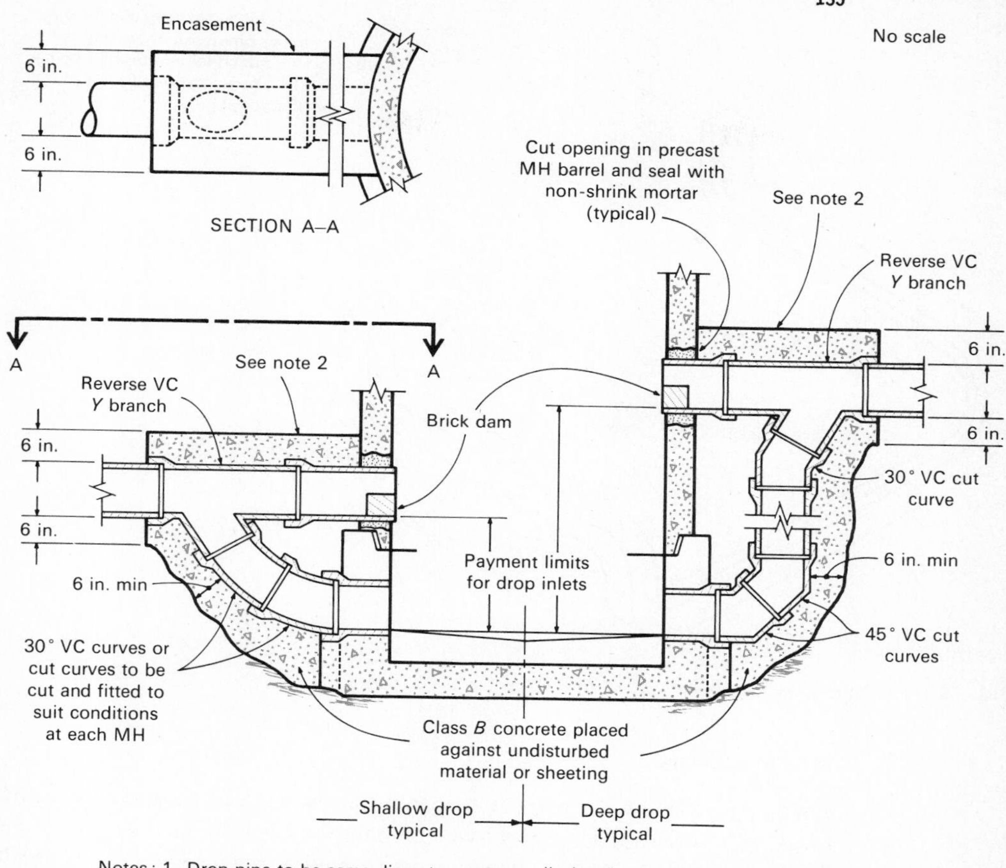

Notes: 1. Drop pipe to be same diameter as sewer discharging into MH for sewers up to and including 12-in. size.
2. Extend encasement to first joint beyond excavation for drop connection.

FIG. 5·2 Typical drop inlet.

elevation of the outflow sewer by a drop inlet or drop manhole, as shown in Fig. 5·2. The purpose of drop manholes is to protect personnel who may at some time have to enter the manhole and to avoid splashing of the sewage.

The drop may be accomplished by an outside connection, as shown in Fig. 5·2, using vitrified-clay pipe. The dimensions of the pipe fittings establish the minimum vertical drop that can be made at a manhole. This minimum distance is about 21 in. for an 8-in. sewer. Because of unequal earth pressures, resulting from backfilling operations in the vicinity of the manhole, the entire outside drop connection is usually encased in concrete, as shown in Fig. 5·2.

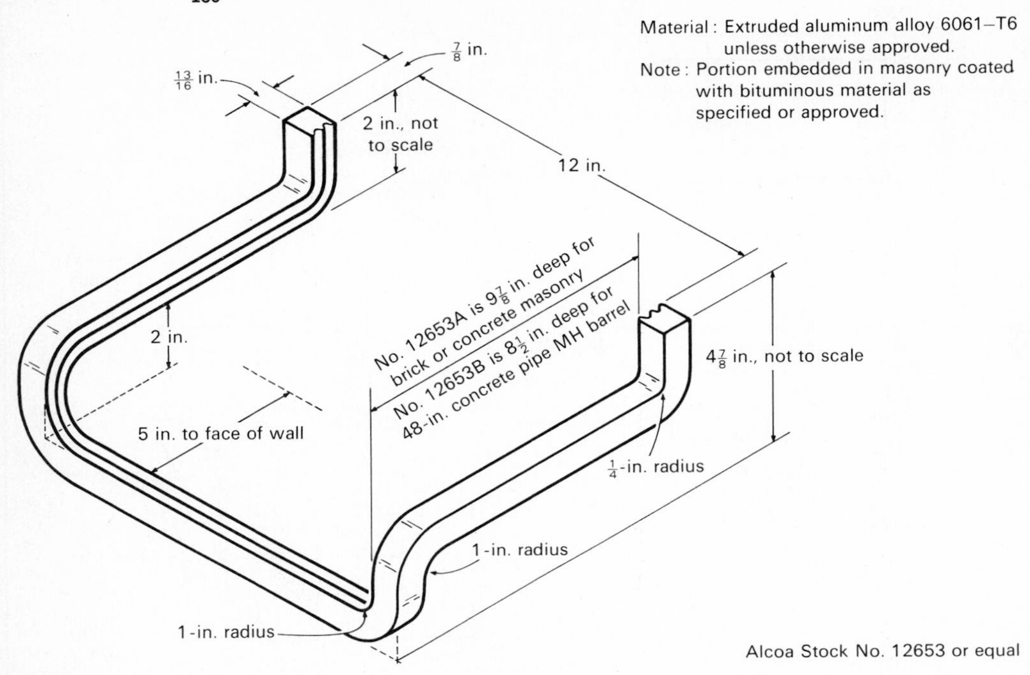

FIG. 5·3 Aluminum manhole step.

Manhole Steps

Former practice with respect to providing steps was to make them of forgings or cast iron and have them embedded in the concrete or brickwork. More recent practice is to make the steps of extruded aluminum alloy and to coat the portion to be embedded in masonry with bituminous material. Examples of steps are shown in Figs. 5·3 and 5·4. The steps are usually placed from 12 to 18 in. apart vertically and somewhat staggered. In a number of projects a vertical spacing of 15 in. has been used. Figure 5·5 shows a cast-iron box-type step for use where the shaft must be kept free from any projections from the wall.

Manhole Frames and Covers

Factors that should be considered in the selection of manhole frames and covers include safety, so that covers will not slip off; convenience of repair and replacement, necessitated by the wear from traffic; strength, sufficient to stand up under the impact from wheel loads; freedom from rattle and noise; cost; possibility of adjustment with the wearing down of pavements to correct for unevenness; sightliness; ventilation to remove gases or admit air to increase safety of workmen in manholes and sewers; protection against entrance of lighted cigars and cigarettes; and protection by locking devices against removal for dumping of refuse into the opening.

The cover should be flat and should lie in the plane of the pave-

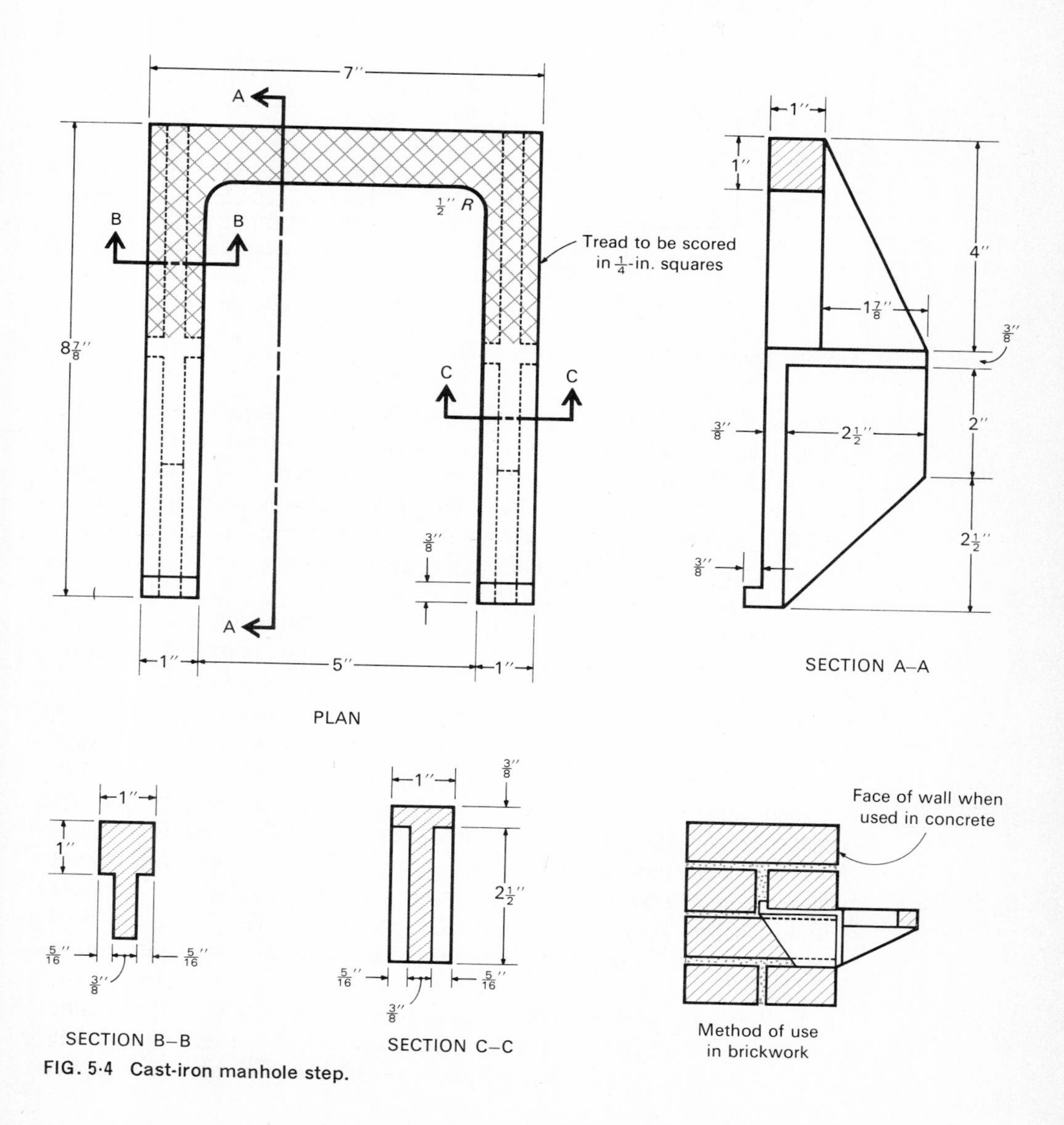

FIG. 5·4 Cast-iron manhole step.

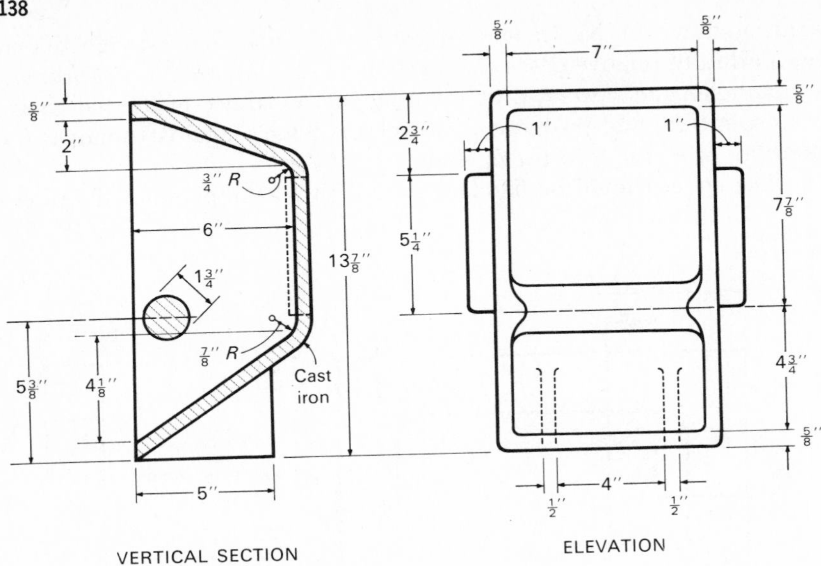

FIG. 5·5 Standard manhole step (box-type).

ment so that it will neither interfere with traffic nor cause excessive wear of the pavement. Covers should be standardized so that those lost by theft or breakage may be readily replaced. The cover should be corrugated or provided with bosses to have a nonskid surface. Circular tops are in almost universal use for sewer manholes. They are inherently stronger than rectangular ones and have the advantage that the cover cannot drop into the manhole.

Frames are usually from 6 to 12 in. in height, depending partly upon the type of pavement in which they are to be installed. Practice as to clear opening varies widely; a 24-in. cover, allowing a 22-in. clear opening, is generally satisfactory. It is more convenient to enter through a larger opening, but the cost and the likelihood of breakage increase substantially with the size, and comparatively little reduction in opening is practicable. In some cases, frames and covers may be used with the opening 30 in. in diameter. This allows suitable space for use of a portable ladder for access by workmen. In general, frames weigh from 250 to 500 lb, and covers from 100 to 150 lb. Manhole frames and covers are usually of cast iron, but semisteel or cast steel may be used. A standard cast-iron frame and cover are shown in Fig. 5·6.

It has become rather common practice to use perforated manhole covers to provide for ventilation. In some designs, the attempt has been made to provide as many holes as possible. Sometimes the per-

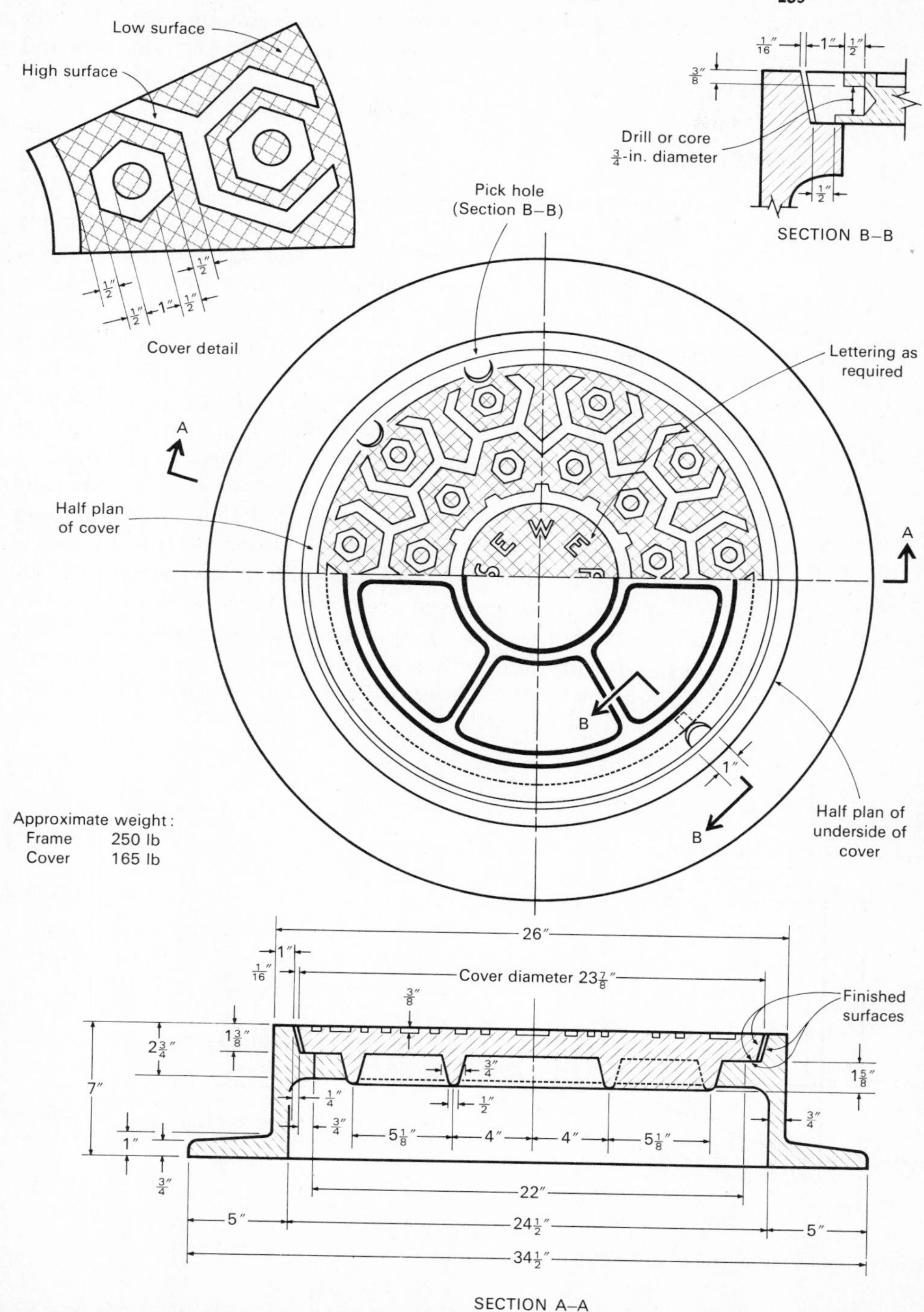

FIG. 5·6 Standard manhole cast-iron frame and cover.

forations are in the bosses, with the object of excluding water. In general, the holes are larger on the inside than on the outside, to avoid plugging with sticks and dirt.

STREET INLETS

Storm water that is not lost by evaporation or percolation drains into the street gutters and is removed at suitable intervals through inlets connecting with the underground storm sewer system. These inlets (see example shown in Fig. 5·7) discharge either directly into the storm sewers or into catch basins that are intended to intercept the refuse and sediment flushed from the street surfaces. Their location is a matter in which the street department is also interested, as it is important that both pedestrian and vehicular traffic be interfered with as little as possible. While it is not practicable to provide instructions that will cover all cases, certain general rules are usually applicable.

Where cross streets are not too far apart, inlets are generally located at street intersections in such a way that they intercept water flowing in the gutters before it reaches the crossings used by pedestrians. Where the distance between intersecting streets exceeds 300 to 500 ft, or where roofs or paved areas outside the street lines drain to the street gutter, the water may accumulate to a sufficient depth in the gutters to interfere with passing vehicles. In such cases, it is better to construct inlets at intermediate points to prevent this con-

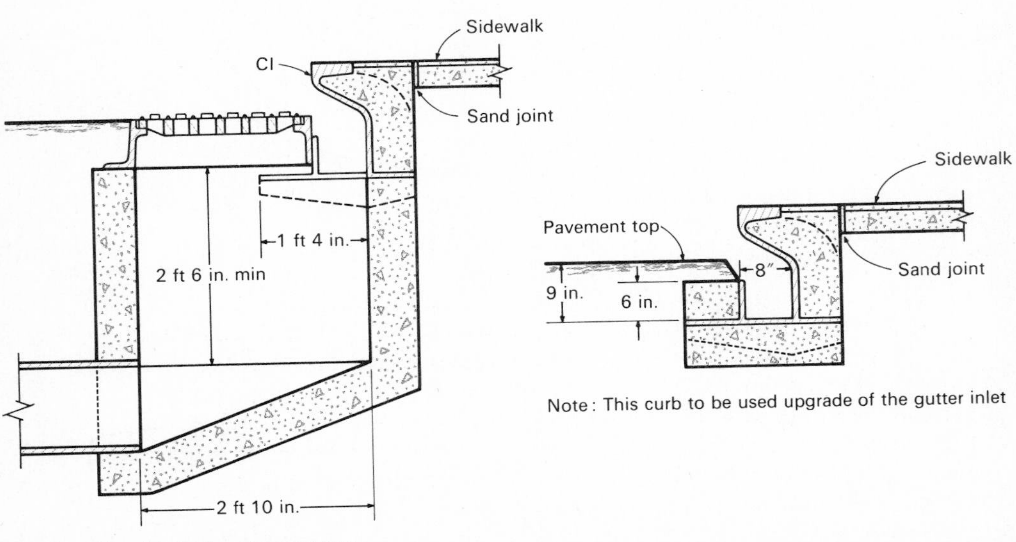

FIG. 5·7 Street inlet for storm water.

dition from occurring. Also, where the slope of the gutter is steep, intermediate inlets should be placed at suitable points to assure a rapid removal of the storm water. In such cases, it may be necessary to provide a small depression of the gutter grade at the inlet, to intercept the water that otherwise would have sufficient velocity to pass by the opening. Such depressions are not favored by street departments, but are sometimes desirable from the standpoint of drainage. On hillside streets, it may be necessary to increase the length of opening in the curb to provide sufficient capacity.

An inlet frequently is located near the angle of intersection of gutters of two streets, so as to serve each street, but this is not considered the best practice. At this location, it is near the line of heaviest travel, which may damage both the inlet frame and grating and the adjacent pavement. Such a location is also unfavorable for the rapid removal of the storm water. The resultant direction of flow of the intersecting streams at this point is away from the inlet rather than toward it. The better arrangement is to place inlets just above each pedestrian crossing.

In case of doubt, it is well to remember that the convenience of the public is better served by having too many rather than too few inlets. The foregoing general practices apply equally well to street inlets and to catch basins, although there is considerable difference between these two classes of structures.

Since a street inlet affords a direct connection between the gutter and the sewer, its design is very important. It should be so designed that as little opportunity as possible exists for its stoppage. The obstruction may arise through the clogging of the opening mouth or gully by which the water enters. It may also occur in the trap or in the pipe connection to the sewer. The objects that cause the most trouble at the openings of the inlet are sticks, waste paper, and leaves. If sticks become lodged against the opening, the leaves and waste paper drawn to it by the next flush of storm water are likely to cause a stoppage. Openings presenting the least possible obstacle to the entrance of these three classes of refuse have been widely used.

Where there is much accumulation of waste paper or leaves, a grating placed in the gutter may be of little value in removing storm water as it may become either partially or completely obstructed at the first runoff from the storm. It would appear advisable in most cases, therefore, to provide an opening in the curb adequate to remove the total flow accumulating at the inlet. Cleanliness of streets and adjacent sidewalks is essential to the successful functioning of inlets.

While a grating, as shown in Fig. 5·8, is ordinarily used in the gutter at inlets, in some cases the curb opening is depended upon en-

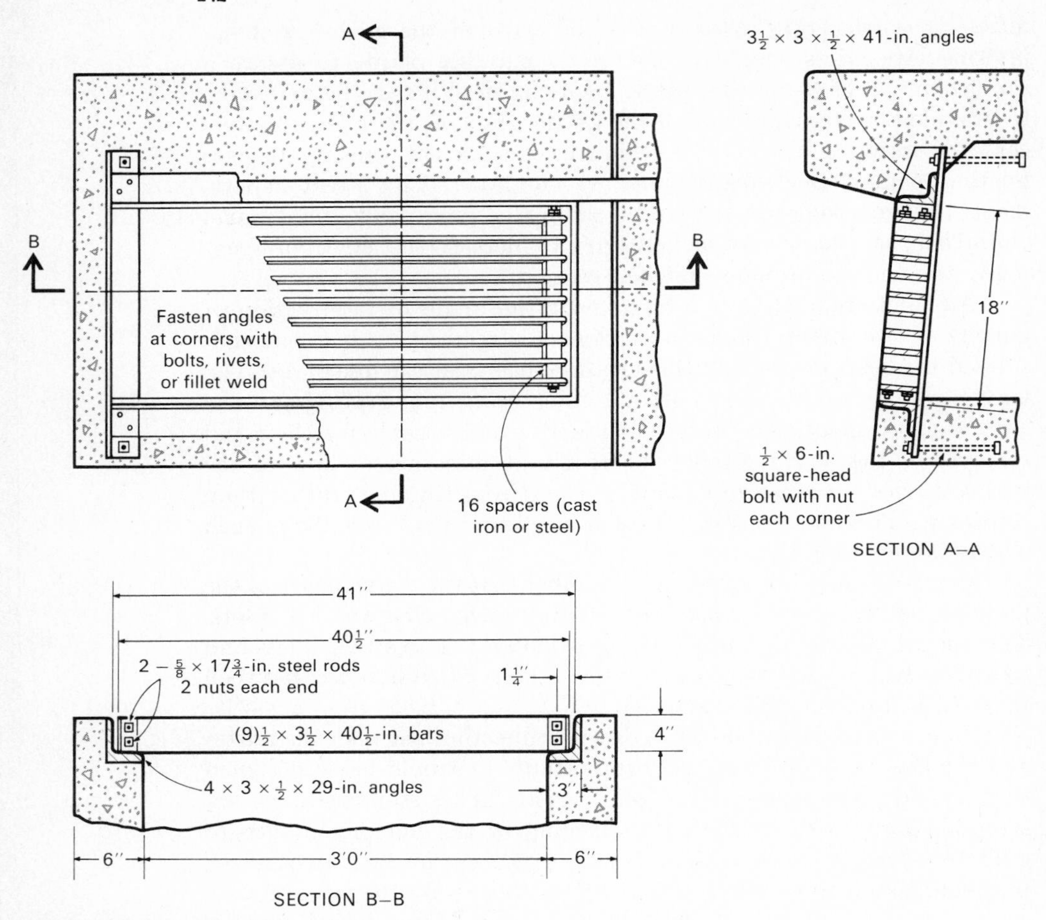

FIG. 5·8 Typical gutter inlet grating.

tirely for receiving the storm water. Since all materials entering these inlets reach the sewer, self-cleansing grades must be used.

Inlet Castings

As with manhole frames and covers, the configuration of inlet castings varies considerably in different localities. The tendency is, however, toward the adoption of a few standard types designed to meet the conditions usually encountered in practice. Typical examples of inlet castings are shown in Fig. 5·9.

Where a larger inlet opening may be required either because of the rate of runoff to be handled or because of a steeper gutter grade, assembled gratings (see Fig. 5·8) are used rather than typical cast grates.

Types of Inlets

There are three general types of storm water inlets: (1) curb inlets, (2) gutter inlets, and (3) combination inlets. A curb inlet consists of a vertical opening in the curb through which the flow passes. A gutter inlet consists of an opening in the gutter beneath one or more grates through which the gutter flow falls. A combination inlet includes a curb opening and a gutter opening, with the gutter opening directly in front of the curb opening. The two openings may be overlapping or offset.

The key dimension of a properly designed gutter inlet is its width perpendicular to the flow. For curb inlets the key capacity parameters are its length of opening and its depression in the gutter.

At low points, the clear opening of an inlet is critical. For this reason, the critical design consideration is in minimizing the clogging of an inlet. Curb inlets are therefore more desirable in such locations. Additionally, at low points two hydraulic conditions can occur, namely, inflow due to weir action, or inflow due to orifice action. Rate of inflow varies with depth of gutter flow to the $\frac{3}{2}$ power in the former and to the $\frac{1}{2}$ power in the latter. Consideration should be given

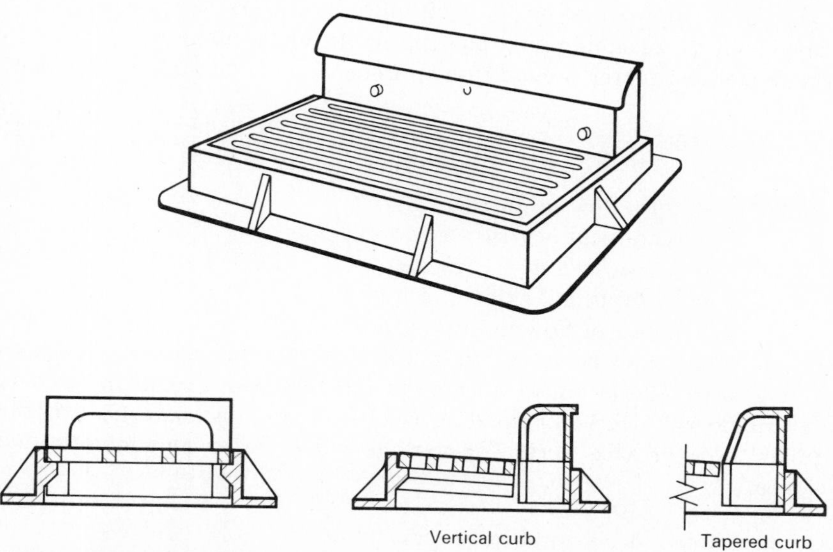

FIG. 5·9 Typical inlet castings.

to this fact when a selection of the type of inlet is made at low points. As a general rule, gutter inlets are used on steep grades, and curb inlets are used on flat grades with steep cross slopes.

Depressed gutter inlets may interfere with vehicular traffic and present a hazard to pedestrians. Moreover, they are not of sufficiently greater capacity over undepressed gutter inlets to justify their use. Curb inlets with depressions close to the curb present only small interference to vehicular traffic and slight hazard to pedestrian traffic, yet are of greater capacity than undepressed curb inlets. Gutter inlets, especially those with depressions, have a greater likelihood of clogging than curb inlets. In some instances, the design capacity of curb and gutter inlets has been reduced by 10 and 30 percent, respectively, to account for clogging [18].

Hydraulics of Inlets

Hydraulic conditions in street gutters, the various types of inlets, and the methods of construction make hydraulic design of such facilities difficult. Field tests are necessary for accurate determinations. Several studies have been conducted to provide design factors for special conditions that may, with due caution, be extrapolated for general design. References [4, 6 to 8, and 18] provide data on inlet design.

Gutter Hydraulics Using the Manning equation and its inherent assumptions (such as uniform flow across the section), Izzard [11] developed an approximation for hydraulics in gutters. Equation 5·1 was developed by assuming that the top width of the channel is equal to the wetted perimeter, a condition approached in wide shallow channels.

$$Q_0 = 0.56 \frac{z}{n} S^{1/2} d^{8/3} \tag{5·1}$$

where Q_0 = total flow in gutter, cfs
z = reciprocal of roadway cross slope
n = Manning's n
S = longitudinal slope of gutter
d = depth of flow in gutter, ft

Equation 5·1 may be solved graphically using the nomograph shown in Fig. 5·10. Based on an n value of 0.020, Fig. 5·11 can be used to determine the degree of flooding (width of flow in gutter) given the gutter flow or vice versa for various gutter cross and longitudinal slopes.

Gutter Inlets A hydraulically efficient gutter inlet will capture all the flow crossing its grate plus a relatively small amount flowing into

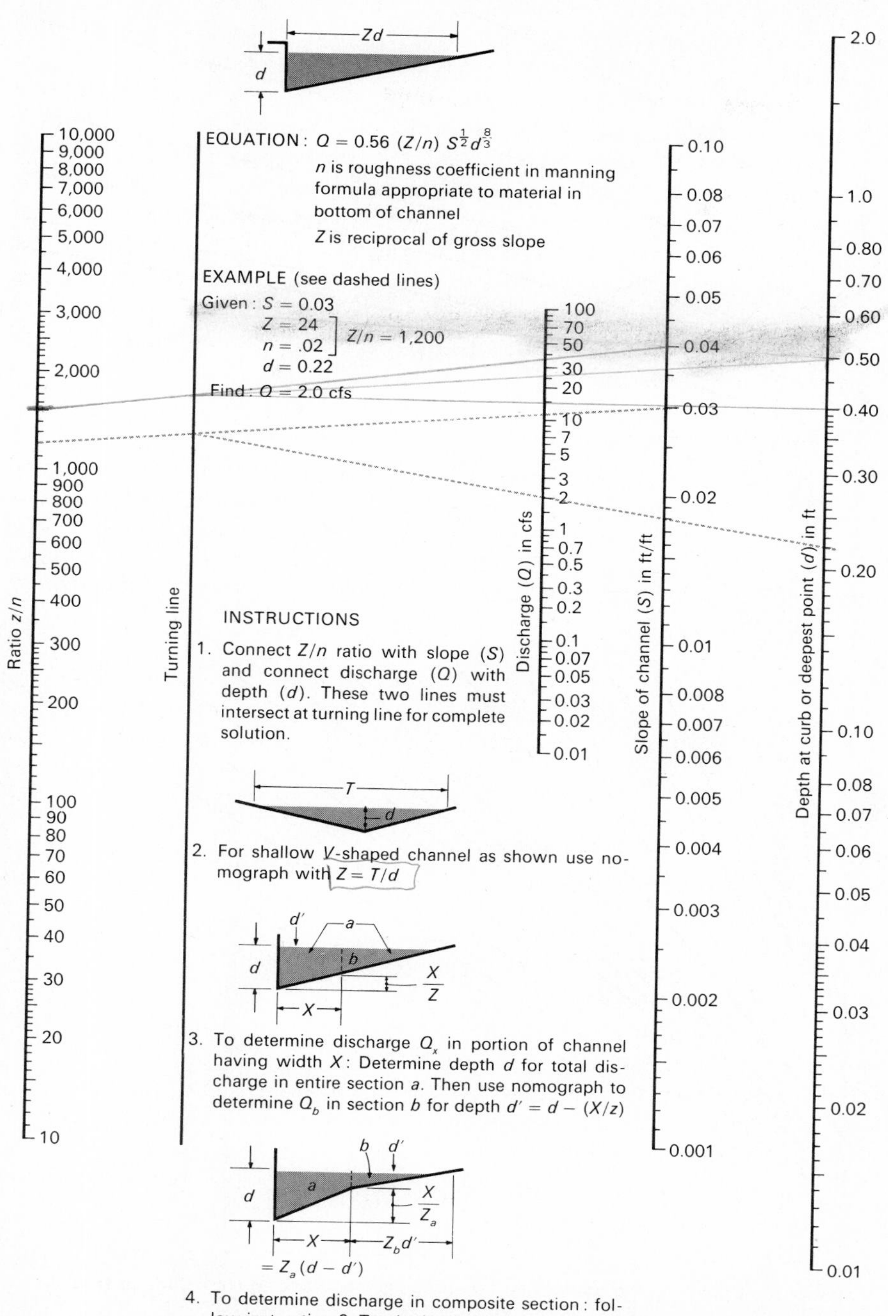

FIG. 5·10 Nomograph for flow in shallow triangular channels [5].

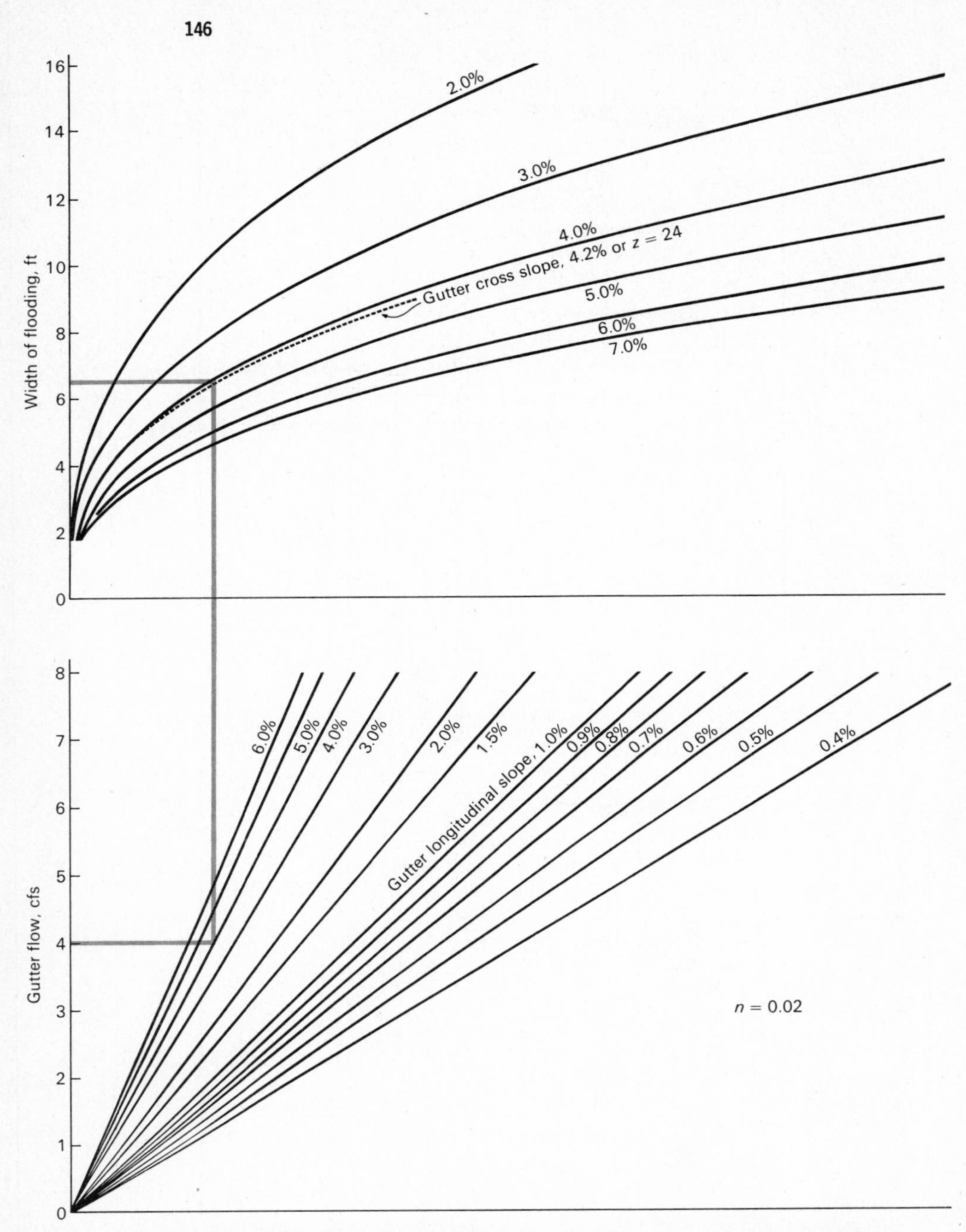

FIG. 5·11 Flow and width of flooding in triangular gutter versus gutter cross and longitudinal slope.

the inlet through weir action along its side. An efficient gutter inlet, in this case, is one having a grate with bars sufficiently long so that water can fall into the opening without striking the downstream grate edge. In addition, bars should be well rounded with a total width of bars less than 50 percent of the width of the inlet [8]. If cross bars are needed for structural reasons, they should be recessed.

Unless details of inlet design are known and correspond to factors used in developing design procedures in Refs. [4, 6 to 8, and 18], the capacity of an inlet should be estimated as equal to the volume of a prism of water in the gutter directly upstream from the inlet and equal to its grate width.

Where the width of the inlet corresponds to the width of flow in the gutter the length of gutter inlet required along the curb so that flow is not carried across the grate can be determined by considering free fall conditions under supercritical flow as follows.

$$L_0 = V_0 t \tag{5·2}$$

where $t = \sqrt{2d/g}$

L_0 = length of inlet along curb required to intercept Q_0, ft

Substituting for t in Eq. 5·2 yields

$$L_0 = V_0 \sqrt{\frac{2d}{g}} \tag{5·3}$$

The value of L_0 found in Eq. 5·3 must be increased by a factor to account for the interference of the grate. Equation 5·4 was developed from test data [18]:

$$L_0 = M V_0 \sqrt{\frac{d}{g}} \tag{5·4}$$

where M varied from 4 for grates with wide bars equal to the width of openings, to 2 for grates with narrow bars.

On the basis of tests at Johns Hopkins University [18], Eq. 5·5 was developed for undepressed gutter inlets to determine the length of inlet required where total inflow along the side of the grate is to be intercepted. This equation is used to develop inlet rating curves as provided in Refs. [18 and 13].

$$L' = 1.2 V_0 z \left(\frac{d - W/z}{g}\right)^{\frac{1}{2}} \tag{5·5}$$

where L' = length of gutter inlet, ft
$V_0 = 2Q_0/(zd^2)$
W = width of gutter inlet, ft
g = acceleration due to gravity

The use of Eq. 5·5 is illustrated in the following example.

EXAMPLE 5·1 *Capacity of a Rectangular Street Inlet*

Determine the capacity of a rectangular street inlet that is 2 ft wide and 4 ft long that is to be located in a street with a crown slope of $\frac{1}{2}$ in./ft and a gutter slope of 4 percent. Use an n value of 0.020.

Solution

1. Set up a coordinate grid as shown in Fig. 5·12 and on it plot the shape of the gutter inlet.
2. Plot flow lines for various discharges. The flow lines shown in Fig. 5·12 are obtained as follows. For a given street slope and crown slope, the maximum width of flow W_0 is determined for selected discharges using either Fig. 5·10 or Fig. 5·11. The coordinates of each flow line are then determined by computing L' (using Eq. 5·5) for a given discharge for various inlet widths W, and plotting the results as indicated (solid curved lines).

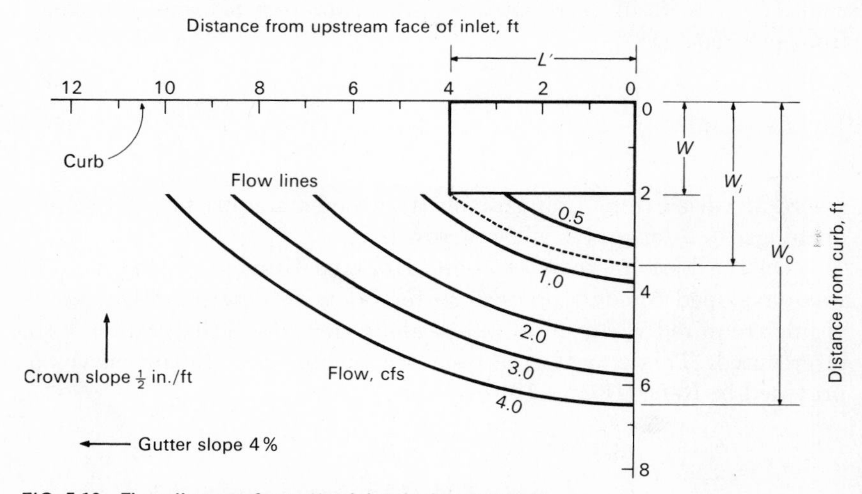

FIG. 5·12 Flow diagram for gutter inlet design example.

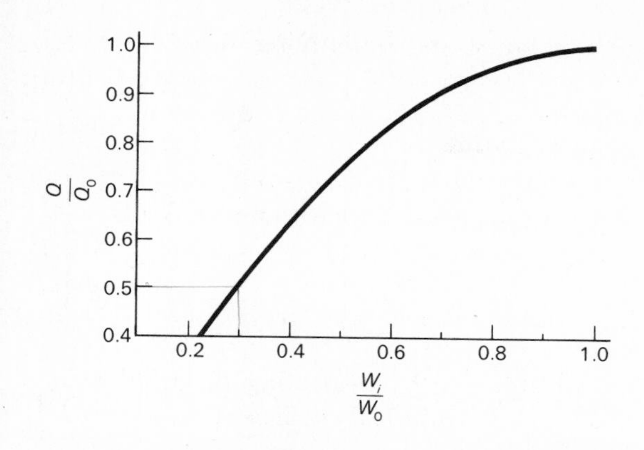

FIG. 5·13 Capacity ratios versus interception ratios for triangular gutter inlets [18].

3. Determine the capture ratio W_i/W_0 where W_i is the distance from the curb to the maximum flow line intercepted by the gutter inlet and W_0 is the distance from the curb to flow line of a larger flow (4 cfs in this example). For a flow of 4 cfs, W_0 is about 6.5 ft and W_i is 3.6 ft. The capture ratio is 3.6/6.5 or 0.56.

4. Determine the capacity of the inlet using Fig. 5·13. Entering Fig. 5·13 with a capture ratio of 0.56 the capacity ratio Q/Q_0 is found to be 0.81. The capacity of the inlet is therefore

   $$Q = 0.81(4) = 3.24 \text{ cfs}$$

   Under the stated conditions, 3.24 cfs will be intercepted and 0.76 cfs will move down the gutter to the next inlet.

Curb Inlets Referring to Fig. 5·14 and neglecting the frictional force offered by the gutter, the acceleration toward the curb opening is $g(\cos\theta_0)$ [18] and the width of flow is $d \tan\theta_0$. Substituting these values into Eq. 5·3, the required length of curb inlet is given by Eq. 5·6.

$$L_0 = V_0\sqrt{\frac{2d\tan\theta_0}{g\cos\theta_0}} \tag{5·6}$$

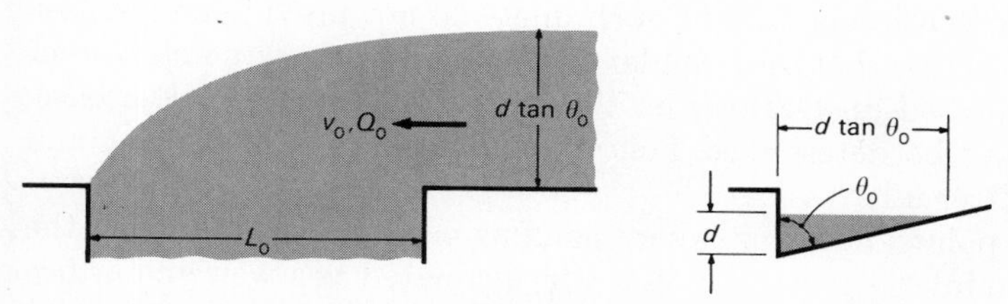

FIG. 5·14 Definition sketch for analysis of undepressed curb inlet.

Using $V_0 = 2Q_0/(2d^2)$ with $z = \tan \theta_0$ and simplifying,

$$\frac{Q_0}{L_0 d \sqrt{gd}} = \sqrt{\frac{\sin \theta_0}{8}} \tag{5·7}$$

For gutters used in practice, $\sin \theta_0$ is nearly equal to unity. Thus

$$\frac{Q_0}{L_0 d \sqrt{gd}} = \frac{1}{\sqrt{8}} = 0.35 \tag{5·8}$$

Since this expression has been obtained by neglecting the friction in the gutter, it is necessary to apply an empirical coefficient to the equation. From test data, it has been found that for cross slopes ranging between $z = 12$ and $z = 48$,

$$\frac{Q_0}{L_0} = 0.2 \sqrt{g}\, d^{\frac{3}{2}} \tag{5·9}$$

On the basis of test results [18] it has been shown that as long as the carry-over flow past the inlet is less than 40 percent of the gutter flow, the capacity of the inlet is practically proportional to the length of inlet. For such a condition, Eq. 5·10 can be used to compute the flow into an inlet:

$$\frac{Q}{L} = 0.2 \sqrt{g}\, d^{\frac{3}{2}} \tag{5·10}$$

where d is the total depth in the gutter flow obtained using Eq. 5·1 or Fig. 5·10.

The flow into depressed gutter inlets is too complex to provide a generalized formula. The reader is referred to Refs. [4 and 18] for design aids.

Combination Inlets Combined curb and gutter inlets are normally used in locations, such as low points, where clogging is expected to be a problem. On continuous slopes, such inlets do not provide an increase in capacity above that of a similar gutter inlet, unless the curb inlet portion is located upstream from the gutter inlet. In the latter case, capacity may be determined first by evaluating the curb inlet and then the gutter inlet.

In low points, however, where ponding occurs, the capacity of the combination inlet is greater and may be evaluated as a weir and orifice if flooded.

CATCH BASINS

The catch basin (see Fig. 5·15) was formerly considered an essential part of any American combined-sewerage or storm-sewer system. Experience had shown that in many sewers the velocity of the sewage was insufficient to prevent the formation of sludge deposits, and it was much more expensive to remove this sludge from the sewers than from catch basins. This experience was gained in days when the pavements

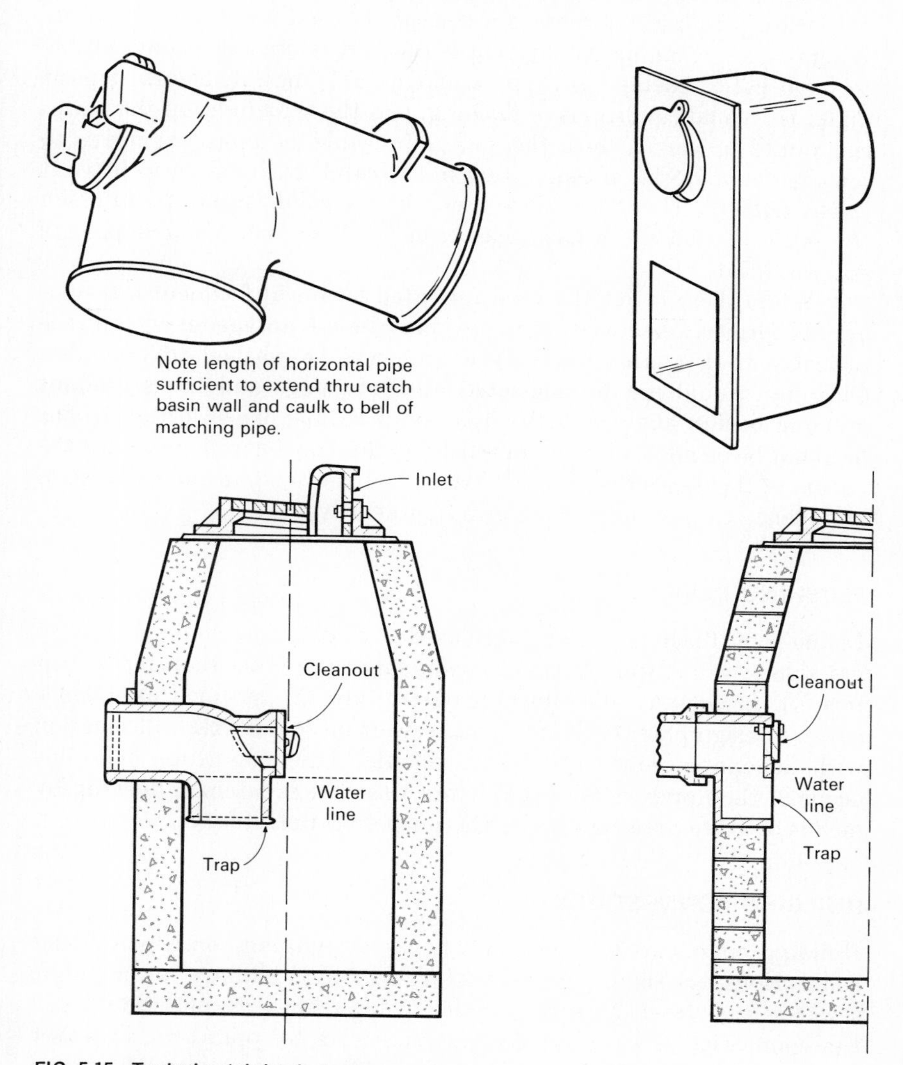

FIG. 5·15 Typical catch basin trap.

of American streets were crude and uneven, and little attention was given to keeping them clean. The sewers themselves were not laid with the present regard for self-cleansing velocities. Under such conditions it was natural that catch basins should find more favor than at the present time. Durable, smooth pavements, more efficient street cleaning, and sewers laid on steeper slopes have reduced the need for such structures to a few special situations.

Where storm-sewer slopes are sufficient to ensure self-cleansing velocities, flushing of streets is a commonly used method of cleaning. Where street cleaning by flushing is done, it is obviously undesirable to have catch basins, as these would quickly be filled with deposit under the usual conditions of flushing, and the remainder of the material would be carried into the sewer. It would be more expensive to remove the material accumulating in the catch basins than to leave it in the gutters. Therefore, since much of the solid material will reach the sewer after a catch basin has been filled, self-cleansing slopes are still essential.

While there is not the need for catch basins at frequent intervals as was formerly believed, their use may be advantageous where it is probable that a large quantity of grit will be washed to the inlet. Cleaning should not be neglected until stoppage and the attendant flooding occurs but should be done on a regular basis. Catch basins need not be cleaned when there is little accumulation in them unless the nature of the deposit is such as to create offensive odors and a source of annoyance to persons passing or living nearby.

FLUSHING DEVICES

In the past, flushing devices were used so that sewers could be laid on flat slopes, which were at times inadequate to produce scouring velocities. At the present time, flushing devices are almost never used, since current design practice is to provide sewers on slopes that will function with adequate scouring velocity. Should some deposition of solids occur in the upper ends of sewers or in the sewer laterals, flushing by means of a fire hose may serve the purpose satisfactorily.

BUILDING CONNECTIONS

Building connections, also called house or building sewers or house connections, are small pipe sewers leading from buildings to the public sewer in the street. In some municipalities, they are constructed and the connection is made to the public sewers by plumbers. In other places, the portion of the work in the public street or even the entire connection from the street sewer to the building is installed by the municipality.

A majority of cities require the use of cast-iron pipe for the house sewer for a distance of several feet outside the wall of the building. Even if cast-iron pipes are used, care must be taken to support them properly so that they will not be damaged by settling. If the pipe passes through the foundation wall, a joint should be provided at the outside face of the wall.

The minimum size of house connection pipes should be 4 in., and 5- or 6-in. sizes are preferable. The minimum slope for a connection is usually fixed by local regulation, and less than 0.25 in./ft is rarely permitted or desirable.

Building sewers should be constructed carefully to avoid stoppages and infiltration of ground water. Sewer lines should have a uniform grade and a straight alignment where possible, as indicated in Fig. 5·16. If a bend must be made in the line, a wye branch and cleanout should be installed.

The building connection enters the sewer at a branch fitting for the smaller-size sewers or at a stub or slant for the larger-size sewers. Where the sewer is in a deep trench, a vertical pipe encased in concrete, called a chimney, is sometimes used. It ends at a specified depth below the surface, and the house sewer is connected to the branch at the top

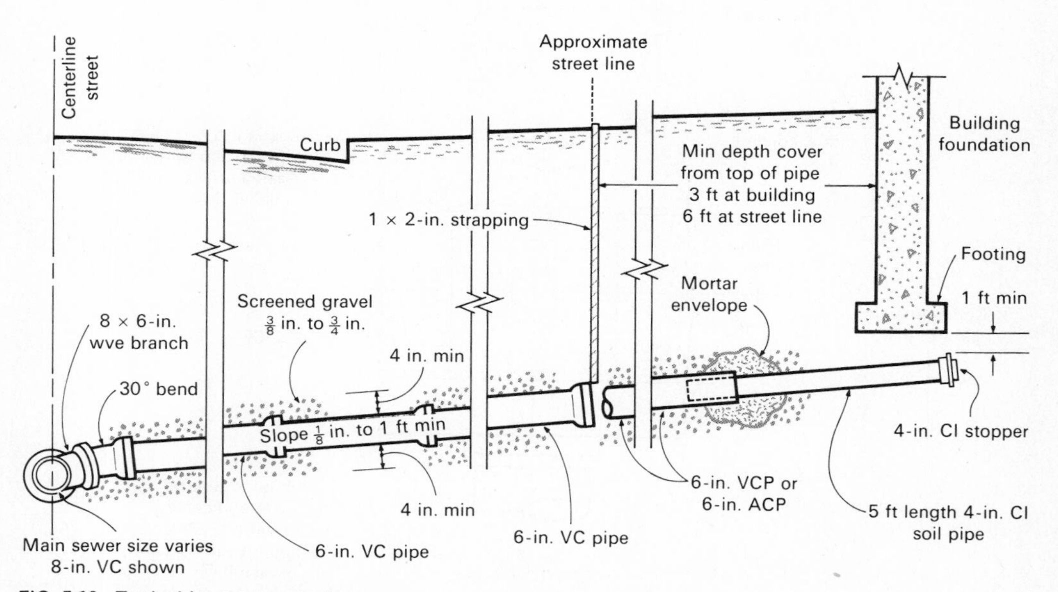

FIG. 5·16 Typical house connection.

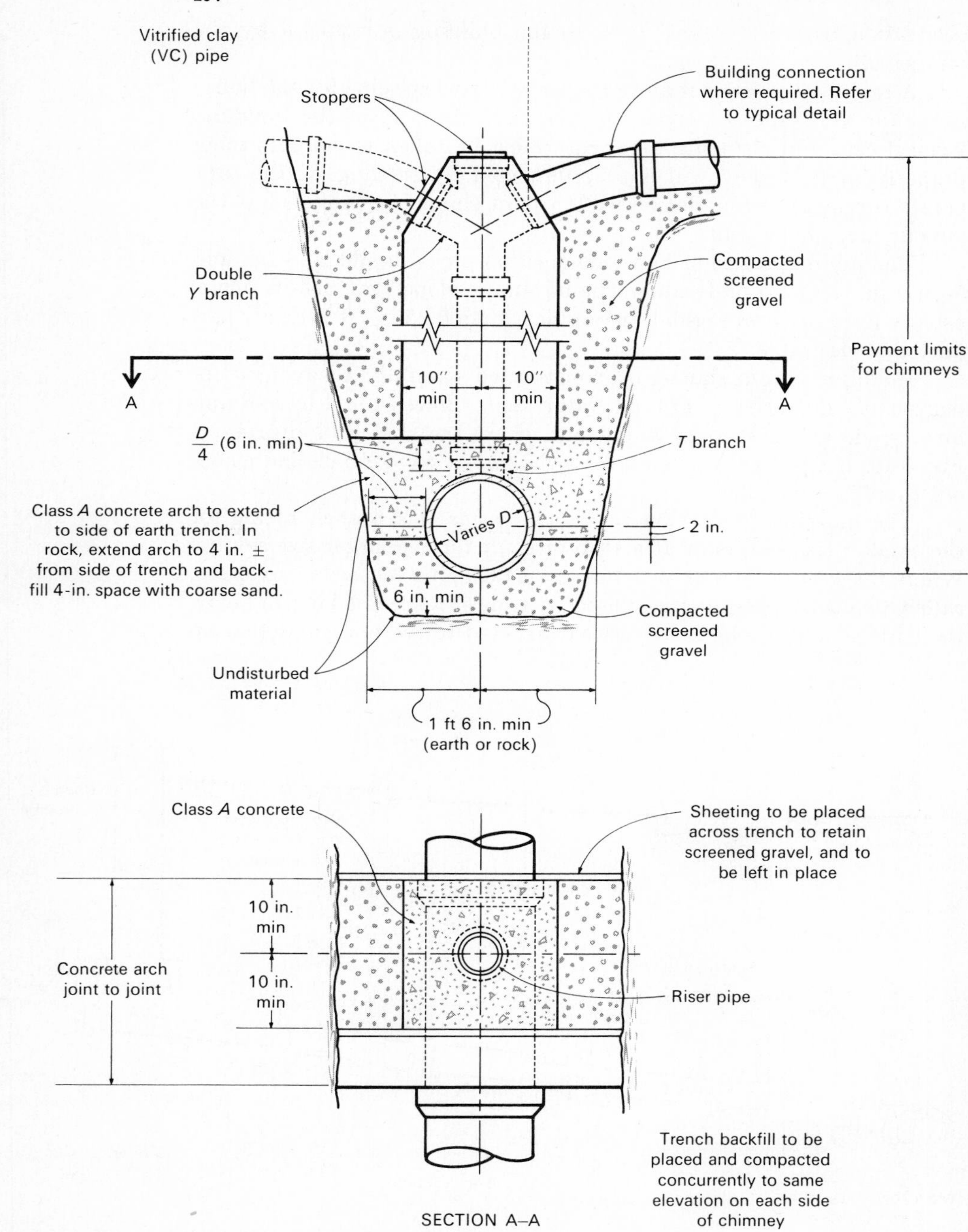

FIG. 5·17 Sewer chimney for house connection.

of the chimney. Typical details for a sewer chimney are shown in Fig. 5·17.

SPECIAL STRUCTURES

Special structures in sewerage systems are usually constructed to overcome conditions imposed by the local topographic characteristics. Typical examples of such structures include junctions, transitions, inverted siphons, energy dissipators, vortex chambers, etc. Such special structures, in each case, must be designed and built to meet the local conditions.

A situation requiring the construction of a special structure is shown in Fig. 5·18. In this case, it was necessary to develop some method for sewage collected near the surface to discharge to a deep tunnel interceptor sewer. The solution of this problem was based on a detailed laboratory study [2] and consisted of using vortex action to spiral the flow down and against the inside wall of the shaft to the tunnel interceptor.

SEWER JUNCTIONS

Sewer junctions are required where one or more branch sewers join or enter a main sewer. With small sewers where it is impractical for a man to enter, the changes in size, direction, and slope should be made at

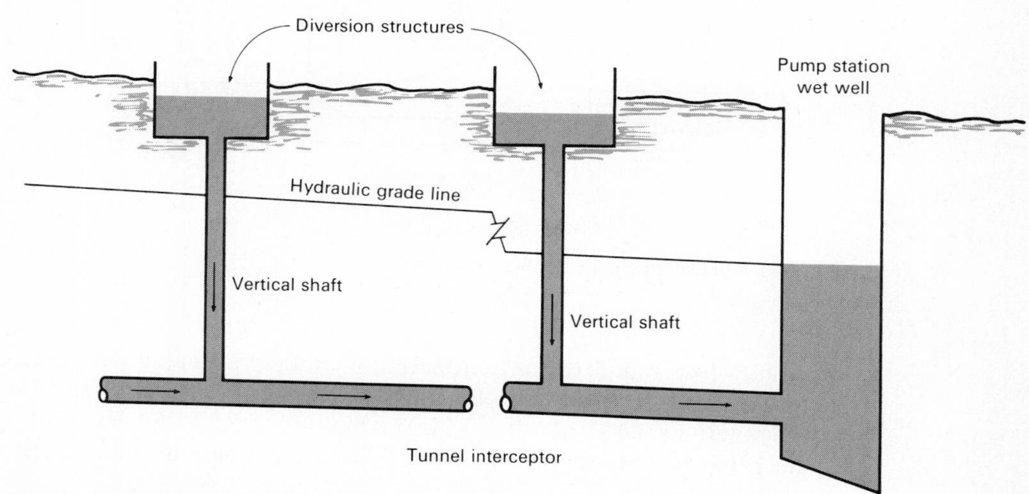

FIG. 5·18 Generalized profile of deep shaft interceptor system.

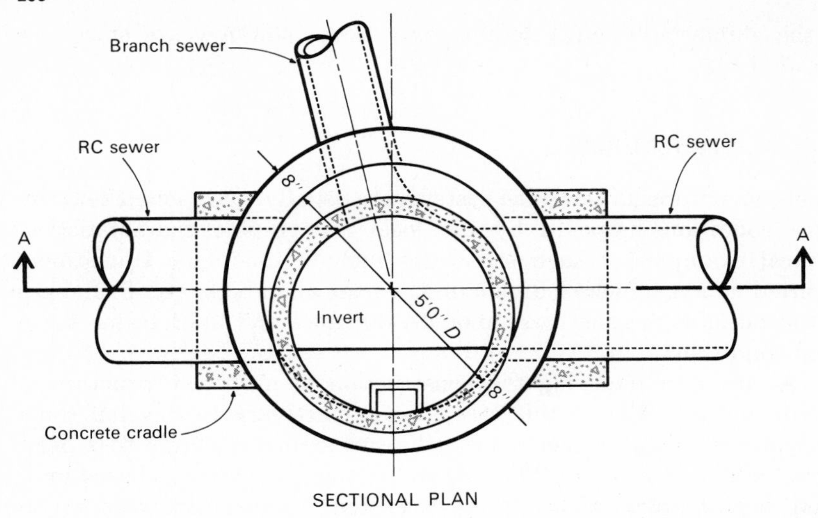

SECTIONAL PLAN

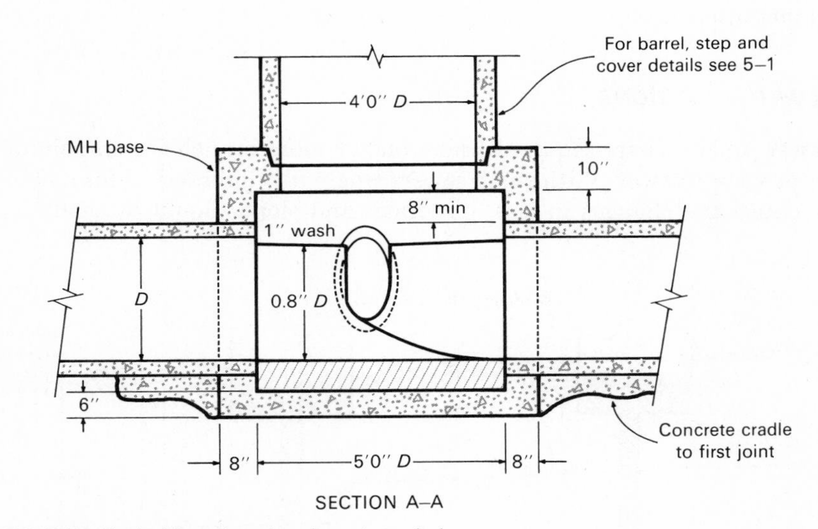

SECTION A–A

FIG. 5·19 Typical junction chamber at manhole.

manholes (see Fig. 5·19) or at junction chambers provided with manholes [12]. Where the sewers are large enough to be entered, so that their junctions need not be made in manholes, and the horizontal angle between them is 45° or less, junction chambers are commonly used.

INVERTED SIPHONS

Any dip or sag introduced into a sewer to pass under structures encountered, such as conduits or subways, or under a stream or across a valley, is termed an inverted siphon. It is misnamed, for it is not a siphon; and the term "depressed sewer" has been suggested as more appropriate. Since the pipe constituting the inverted siphon is below the hydraulic grade line, it is always full of water under pressure although there may be little flow in the sewer.

Practical considerations, such as the increased danger of stoppage in small pipes, tend to fix the minimum diameters for inverted siphons at about the same size as for ordinary sewers: 6 or 8 in. in sanitary systems, and about 12 in. in combined systems. As obstructions are much more difficult to remove from an inverted siphon than from a sewer, special care should be taken to prevent their formation. As high a velocity as practicable should be maintained in the inverted siphon, say 3 fps or more for domestic sewage, and 4 to 5 fps for storm flows.

In some cases, catch basins or grit chambers have been built just above the inverted siphons, but these are troublesome to clean, and the material removed from them usually has an offensive odor. Inverted siphons should be flushed frequently and their operation inspected regularly to assure prompt removal of obstructions.

Experience has shown that the smaller pipes are sometimes clogged as a result of sticks catching in the bends. When such conditions have been experienced or are likely to occur, it is advantageous to provide for racks in front of the inlets to the smaller pipes, preferably so arranged that the material collected on them may be washed off and carried through the larger pipes at times of high flow.

Flushing or cleaning may be accomplished in various ways, depending upon the available facilities and surrounding conditions. It may be done by backing up the sewage and then releasing the backed-up flow; by admitting clean water at the head of the inverted siphon; by providing a permanent scraping mechanism in the siphon; or by hand cleaning, using jointed rods with suitable scrapers or other tools, after draining the siphon.

Manholes or cleanout chambers should be provided at each end of an inverted siphon to give access for rodding, pumping, and, in the case of pipes of larger size, for entrance. The introduction of intermediate manholes on an inverted siphon in such a manner that the sewage will be free to rise in them is objectionable, since grease and other scum tends to fill the shaft. They are advantageous, however, if the sewage is confined within the pipes as it passes through the manhole, affording access or means of ridding the siphon of deposit through a gated connection or similar device.

Since an inverted siphon is subject at all points of its cross section to an internal pressure, the pipe walls will be in tension, although the magnitude of the tension will be affected by external water and soil pressures. Because of these tensile stresses, inverted siphons are usually constructed of steel, cast iron, or reinforced concrete.

The computation of the pipe sizes for inverted siphons is made in the same way as that for sewers and water mains. The diameter depends upon the hydraulic slope and the maximum flow of sewage to be carried. The head loss or drop in the hydraulic gradient actually required at any time for the flow will be the difference in level in the free water surfaces at the two ends. It will equal the sum of the friction-head and other losses. The losses are relatively small for low velocities, other things being equal, but they increase roughly as the square of the velocity. For a clean 12-in. siphon 50 ft long and a velocity of 3 fps, a total loss of 6 in. would probably be a maximum figure; for a velocity of 2 fps, 3 in.; but for a velocity of 6 fps, the total loss would be 24 in. The friction loss alone would be not over one-third of the above values.

Experience in this country has shown the advantage of using several pipes, instead of one pipe, for the inverted siphon, arranged in such manner as to bring additional pipes progressively into action with increase in sewage flow. In this way, reasonable velocities are maintained at all times. Care must be taken that inverted siphons built on or under river beds have sufficient weight or anchorage to prevent their flotation when empty. The design of an inverted siphon is illustrated in Example 5·2. The design of inlet and outlet structures for the inverted siphon in Example 5·2 is illustrated in Example 5·3.

EXAMPLE 5·2 *Inverted Siphon Design*

This problem involves the design of an inverted siphon, consisting of several pipes, to replace an existing single-pipe siphon that has not performed properly because of sedimentation due to low velocities. Its solution is given as an illustration. The basic data are as follows:

Length of inverted siphon = 440 ft

Available fall (invert to invert) = 3.6 ft

Maximum depression of siphon = 9 ft

Gravity sewers connected by inverted siphon: diameter = 30 in.

Material is reinforced concrete

Slope = 0.0022

Typical rates of flow:

Minimum = 4 cfs

Maximum dry weather = 9 cfs

Ultimate maximum is capacity of gravity sewer

Available fall = 3.6 ft

Assumed loss at inlet = 0.4 (see Example 5·3)

Available for friction in inverted siphon = 3.2 ft

Approximate available hydraulic slope = 3.2/440 = 0.0073

Solution

The capacity of a 30-in. concrete gravity sewer laid on slope 0.0022 is 19.3 cfs, as computed using Manning's equation with an n value of 0.013. The velocity at this rate of flow is about 4.0 fps.

Two conditions must, if possible, be satisfied: the velocity in the pipes selected should be sufficient to ensure scouring (3 fps, if possible) and the available hydraulic slope of 0.0073 must not be exceeded.

The pipes of inverted siphons are rather susceptible to lessening of capacity resulting from formation of coatings of grease or other materials. It is considered advisable, therefore, to use a value of 0.015 for Manning's n. Pipe sizes should be selected that will be particularly adapted to the minimum flow, the maximum dry weather flow, and the ultimate maximum flow. Three siphon pipes are found to meet these requirements, with regulation at the inlet so that the minimum flow will be confined to one pipe, the maximum dry weather flow to two pipes, and the combined capacity of the three pipes will be equivalent to that of the influent sewer. The required computations are as follows:

1. For minimum flow of 4.0 cfs. The required capacity of the first pipe is 4.0 cfs. From Fig. 3·15, the capacity of a 14-in. cast-iron pipe ($n = 0.015$) with a hydraulic gradient of 0.0073 is 4.0 cfs. The velocity would be 3.7 fps. The capacity of a 14-in. pipe at $S = 0.0073$ is 4.0 cfs.
2. For maximum dry weather flow of 9.0 cfs. Since the capacity of the 14-in. pipe is 4.0 cfs, the required capacity of the second

pipe is 5.0 cfs. From Fig. 3·15, the capacity of a 16-in. cast-iron pipe ($n = 0.015$) with a hydraulic gradient of 0.0057 is 5.0 cfs. The velocity would be 3.5 fps. The capacity of a 16-in. pipe at $S = 0.0073$ is 5.7 cfs and combined with the capacity of the 14-in. pipe of 4.0 cfs gives a total of 9.7 cfs beyond which a third pipe begins to operate.

3. For ultimate maximum flow of 19.3 cfs. Since the combined capacity of the 14- and 16-in. pipes is 9.7 cfs, the required capacity of the third pipe is 9.6 cfs. From Fig. 3·15, the capacity of a 20-in. cast-iron pipe ($n = 0.015$) and S of 0.0065 is 9.6 cfs. The velocity would be 4.5 fps. The capacity of a 20-in. pipe at $S = 0.0073$ is 10.4 cfs. The combined capacity of 14- and 16-in. pipes is 9.7 cfs and the total available capacity is 20.1 cfs (19.3 required).

EXAMPLE 5·3 *Design of an Inlet and an Outlet Chamber*

Design an inlet chamber and an outlet chamber for the inverted siphon in Example 5·2.

Solution

1. Inlet chamber. The invert of the 14-in. pipe is continuous with that of the 30-in. gravity sewer, with a wall on each side of a central channel extending from the 30-in. sewer to the 14-in. pipe. One wall crest corresponds to the flow depth in the 30-in. sewer with the minimum flow. The other wall crest corresponds to the depth in the 30-in. sewer with the maximum dry weather flow. Thus the minimum flow will be in the 14-in. pipe, and the flow up to the maximum dry weather flow will be in the 14- and 16-in. pipes. Higher flows will cause a portion in excess of the maximum dry weather flow in the 20-in. pipe.

   To determine the elevations of the tops of the walls, which will permit overflow to the 16- and 20-in. pipes, it is necessary to compute the depth of flow in the 30-in. gravity sewer when the rate of flow is the same as the capacity (4.0 cfs) of the 14-in. pipe, and when the combined capacity (9.7 cfs) of the 14- and 16-in. pipes is reached. The results of these calculations, determined with the use of Tables 3·3 and 3·4, indicate the heights to be about 9.3 and 15 in., respectively. The accuracy of these calculations is adequate provided that the velocity in the gravity sewer is not in excess of about 5 fps.

   Under maximum flow conditions the overflow walls will

be submerged. They cannot, therefore, be considered as weirs, but as obstructions causing certain losses of head in passing the design rates of flow. The flow over these walls is approximately at right angles to that in the gravity sewer (conservative assumption), so that this loss may be taken as the head required to produce the necessary velocity across the top of the wall, assuming the energy of velocity in the gravity sewer to be lost in the change of direction.

A maximum of 5.7 cfs must pass across the wall to the 16-in. pipe. The depth on the wall will then be at least 5.7 in., or the difference in elevation between the two walls, the higher wall having been designed to overflow just as the capacity of the first two pipes is exceeded.

The higher wall must pass at least 10.4 cfs, and the depth available is the distance from the top of the wall to the crown of the 30-in. sewer, or 15 in. Assuming the length of the wall is 6 ft, the inlet losses may be approximated as described below.

In the 14-in. pipe, the loss is negligible since the invert is continuous with that of the 30-in. sewer, and transition can be made to take advantage of the velocity of approach. Even at low flows in the gravity sewer, this velocity is equal to or greater than that in the 14-in. pipe. (For the partly filled sewer, the approach velocity may be estimated by the method given in Chap. 3.)

With the 16-in. pipe at maximum dry weather flow, the velocity over the wall is equal to $5.7/(0.50 \times 6)$ or 1.9 fps; the corresponding velocity head is 0.06 ft. After passing over the wall, a velocity of 4.1 fps in the 16-in. pipe must be obtained in a new direction (conservative assumption), which requires a head of 0.27 ft. Thus, the total head loss is 0.33 ft (less than the assumed inlet allowance of 0.4).

With the 20-in. pipe, the velocity over the wall is equal to $10.4/(1.25 \times 6)$ or 1.4 fps; the corresponding head is 0.03 ft. The velocity in the 20-in. pipe is 4.3 fps; the head is 0.29 ft. The total head loss is 0.32 ft. Allowance of 0.4 ft for inlet losses appears ample, and no revision of the computation of available slope is required.

The relative elevations of the pipe inverts are now readily found. When the capacity of the 14-in. pipe is reached, the depth of flow in the 30-in. gravity sewer is 9.3 in. Since there is no velocity head loss to be allowed for, the crown of the 14-in. sewer is 9.3 in. above the 30-in. invert. The crown of the 16-in. pipe will be at an elevation of 15 in. minus the entrance head loss of 0.33 ft (4 in.), or 11 in. above the 30-in. invert. The crown of the 20-in. pipe is 0.32 ft (4 in.) below the 30-in. crown. From these figures, the relative invert elevations may be determined easily.

The above design is based on the use of flexible joint pipe to accomplish the deflections at joints that permit the three pipes to be depressed the required amount. If bends amounting to 10 or 20° are required to accomplish the depression, an additional head loss allowance must be made for each bend.

2. Outlet chamber. At the outlet end of the inverted siphon, the junction of the three pipes with the 30-in. gravity sewer should be designed to reduce the opportunity for eddies to carry sediment back into the pipes that frequently have no flow through them, but are full of standing water. This is especially important in the case of the 20-in. pipe, which will not be in operation except at unusual rates of flow.

   The design may be accomplished by locating the two larger pipes, or the inverts of corresponding channels within the junction chamber, as high as possible, to avoid pooling and reduction of velocity in the chamber. As a further precaution, the outlet of the 20-in. pipe (least frequently required) may be raised so that the invert of its channel has a sharp forward pitch toward the invert of the 30-in. sewer. The crown of the pipe must not, however, be raised above that of the 30-in. sewer, or a portion of it will be above the hydraulic gradient.

REGULATING DEVICES

One function of a sewage-flow regulator is to prevent surcharge of an intercepting sewer by closing an automatic gate upon the branch sewer connection, thus cutting off the sewage and forcing it to flow to another outlet. Another function is to regulate the flow to the interceptor in time of storm so that discharge from one sewer carrying a heavily polluted storm flow may be admitted to the interceptor in greater proportion than that from another sewer carrying a more dilute storm flow.

A storm overflow is designed to allow the excess sewage flow above a definite rate to escape from the sewer in which it is flowing, through an opening into a relief sewer. Whenever an outlet is into a body of water the level of which may rise above the outlet, and it is necessary to prevent this backwater from entering the sewer, a tide gate is employed. The purposes of the regulator and the storm over-

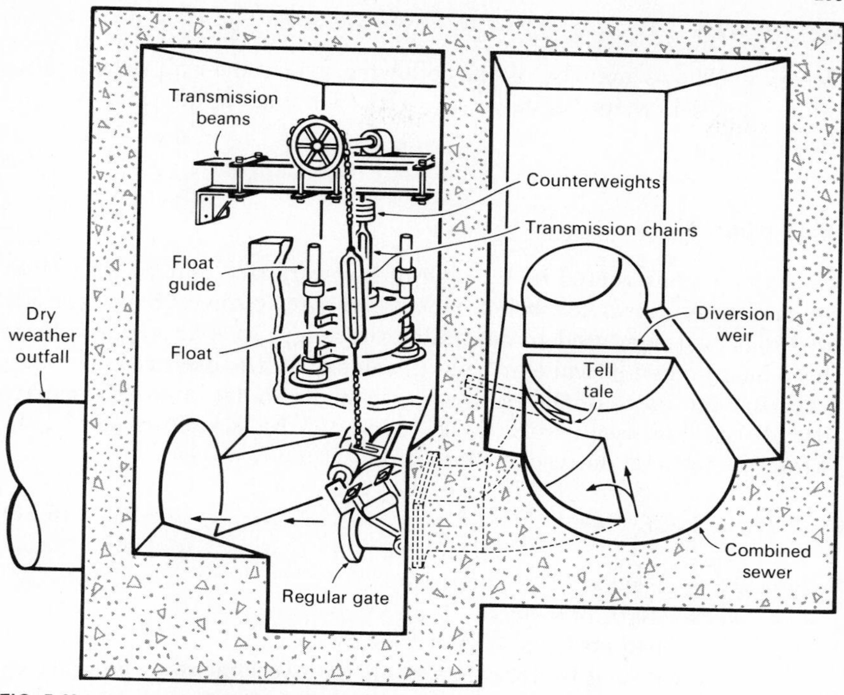

FIG. 5·20 Automatic sewer regulator [Brown and Brown type A].

flow are supplementary, namely, to cause the ordinary flow of sewage to be delivered to a distant point of discharge and, in time of storm, to allow the excess flow to pass into a nearer watercourse.

Recent types of regulators include the Brosius or Krajewski-Pesant regulator, the Milwaukee regulator, and the Brown and Brown regulator, which is shown in Fig. 5·20.

In general, automatic mechanical regulators, either float operated or of the flap-gate type, are impractical for small flows because of the small waterways and the consequent danger of clogging. Because of the size of structure required and the high cost of equipment, mechanical regulators are not considered economically justified for flows under 2 cfs. Furthermore, all such devices require periodic inspection, skillful adjustment, and careful maintenance of all parts by workmen of some mechanical skill. Lack of necessary periodic inspection and maintenance accounts for the extensive history of failure and abandonment of automatic mechanical regulators.

OVERFLOWS

Storm overflows may be of the following types: overfall or side-flow weirs, baffled weirs, transverse weirs, leaping weirs, and siphon spillways.

Side Weirs

Side weirs, constructed in the side of a sewer, are commonly used to bypass excess storm flow in combined-sewerage systems. In general, the methods of design used to determine the length of weir required have been based on empirical equations developed on the basis of laboratory experiments. Some of the earliest experiments to determine the capacity of side-flow weirs were those conducted by Engels in 1917 [14]. From his tests he derived the equation

$$Q_w = 3.32L^{0.83}h^{1.67} \qquad (5{\cdot}11)$$

where Q_w = discharge over weir, cfs
L = length of weir, ft
h = head on lower end of weir, ft

It is interesting to note that in all the experiments conducted by Engels, he observed a water-surface profile of the type shown in Fig. 5·21*b*. The significance of this observation will be discussed later.

In 1923, Smith and Coleman [17] reported the results of experiments conducted on a small flume $4\frac{3}{4}$ in. wide by 6 in. deep. In all cases they observed a falling profile over the weir section, as shown in Fig. 5·21*c*. The weir-discharge equation developed from their model is

$$Q_w = 0.671L^{0.72}E_w^{1.645} \qquad (5{\cdot}12)$$

where E_w is the channel specific energy referenced to the weir. This equation, along with those proposed by Engels, has served as the basis of design for the great majority of side-overflow weirs now in existence.

In addition, other empirical relationships were developed or studied by Parmley in 1905 [16], Babbitt in 1917 [3], and Tyler, Carollo, and Steyskal in 1929 [19].

A theoretical approach to this problem was presented in 1928 by Nimmo [15]. Nimmo approached the problem of determining the discharge over a side weir by considering changes in momentum. This was done by considering upstream and downstream channel cross sections along the weir and thus writing: The change in momentum in the length dL minus the momentum lost in the overflowing water is equal

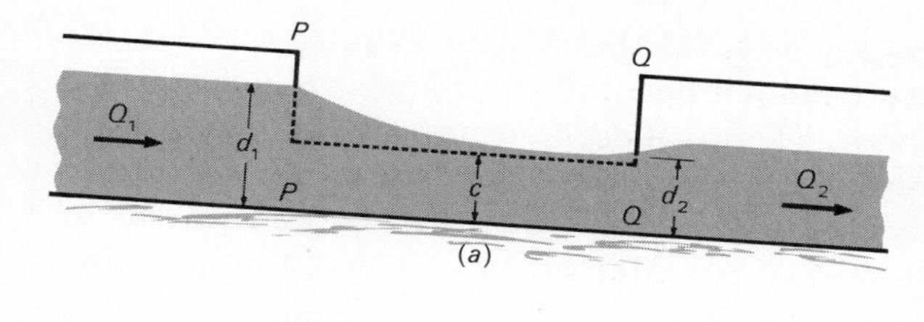

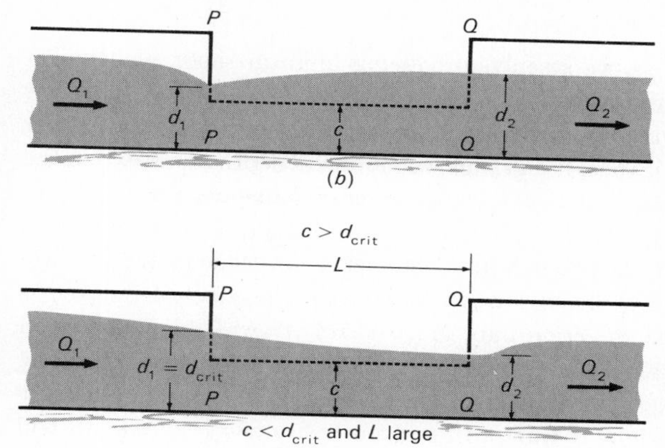

FIG. 5·21 Possible types of water-surface profiles [9]. (a) Steeply sloping bed, (b) bed with low slope, (c) another bed with low slope.

to the summation of the external forces. He applied his theory to a rectangular channel; and, by assuming a constant channel width, horizontal channel bottom, and neglecting friction, he arrived at the following equation for the water-surface profile over a weir:

$$\frac{dd}{dL} = \frac{AQ}{BQ^2 - QA^3} \frac{dQ}{dL} \qquad (5\cdot13)$$

In Eq. 5·13, dQ/dL is the flow per unit length over the weir and is given by the following normal weir equation:

$$\frac{dQ_w}{dL} = C\sqrt{2g}\,(d - c)^{3/2} \qquad (5\cdot14)$$

where dQ_w/dL = discharge per unit length, cfs/ft
C = constant
g = acceleration due to gravity, ft/sec^2
d = depth of flow in channel, ft
c = height of weir, ft

Working from the assumption of constant total energy along the weir, de Marchi in 1934 [9] published the results of a theoretical investigation in which he arrived at an equation identical to the one developed by Nimmo, Eq. 5·13. De Marchi demonstrated that if dd/dL in Eq. 5·13 is positive (tranquil flow), there will be a rising water-surface profile over the weir, whereas if dd/dL is negative (shooting flow), a falling water-surface profile results. This analysis by de Marchi explains the previously mentioned results observed by Engels and by Smith and Coleman. In fact, de Marchi showed that three possible profiles existed [9].

1. If the channel bed slopes steeply, producing uniform shooting flow upstream of the weir, the resultant profile is as shown in Fig. 5·21*a*. The weir has no effect in the upstream direction because the flow is shooting. Along the length of the weir there is a gradual reduction in depth; and beyond the weir the depth increases as the flow is retarded, tending asymptotically to the normal depth.
2. Considering next the condition with a channel of low slope giving uniform tranquil flow some distance upstream of the weir, the effect of the weir is felt in the upstream direction only (Fig. 5·21*b*). Downstream of section QQ, the depth will be normal depth corresponding to the flow Q_2 remaining in the channel. Along the length of the weir there will be a gradual increase in depth (since the flow is tranquil), and upstream of PP the depth will tend asymptotically to the normal depth for the initial flow Q_1.
3. If the weir crest is below the critical depth corresponding to the initial flow Q_1 and the weir is long, then the reduction in depth at sections progressively farther upstream of section QQ will lead to a depth less than the critical depth. This will cause the flow to become shooting, and thus lead to the profile shown in Fig. 5·21*c*.

In some recent work, Frazer [10] noted that two additional flow conditions can exist, although he has not observed or studied them. In these cases, the flow starts out as shown in Fig. 5·21*a* and *c*, and then a hydraulic jump develops within the weir section.

From a review of these and other theoretical studies of weirs conducted within the past 20 years, it can be concluded that in a theoretical analysis of this problem two equations are required: one accounts for either the momentum or energy along the channel and the other relates the discharge over the weir to the weir length.

In general, the flow situation normally encountered in the design of weirs in sewers is of the type shown in Fig. 5·21*c*, where the height of the weir is less than the critical depth in the channel. A rising water-surface profile may occur in a situation where only small amounts of flow are to be diverted. The method used in the design of overflow weirs will depend upon the type of flow. In this section, equations will be presented for computing weir lengths for both falling and rising

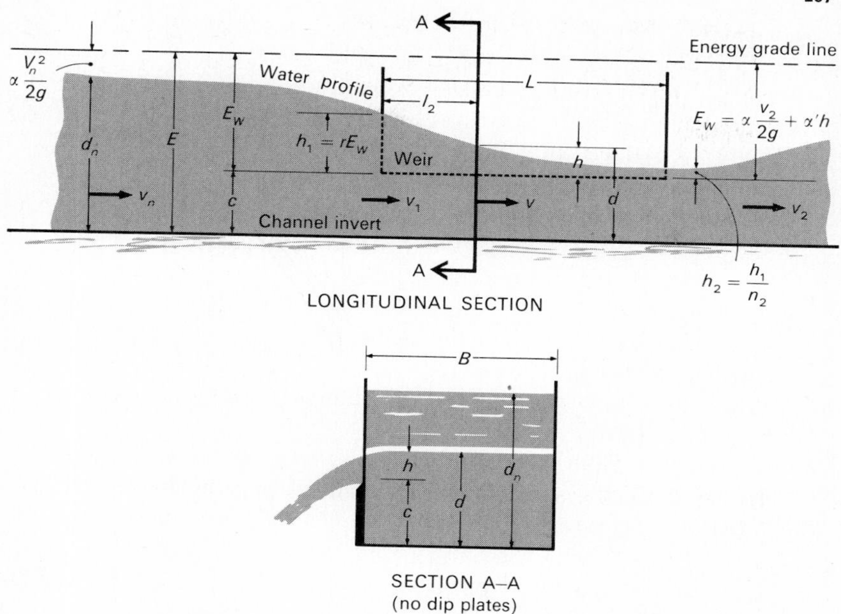

FIG. 5·22 Theoretical profile and nomenclature for side weirs [1].

water-surface profiles. In addition, two illustrative examples are also presented.

Falling Water Surface The equations and procedures for computing weir lengths when the water-surface profile is falling are based on an analysis of this problem presented by Ackers [1] in 1957. A definition sketch for this problem is shown in Fig. 5·22. In his development, by combining Bernoulli's theorem, including an allowance for variation in channel velocity with a weir-discharge formula, Ackers was able to develop simplified design equations for computing required weir lengths. The results of this analysis are shown graphically in Fig. 5·23. This figure shows the relationship between the ratio of the incoming to the outgoing head on the weir n_2, the term c/E_w, which is the height of the weir, divided by the channel specific energy referenced to the top of the weir, and the ratio of the length of weir to the channel width or diameter L_2/B.

For example, for an n_2 value of 10, the resulting equation would be

$$L_2 = 2.03B\left(5.28 - 2.63\frac{c}{E_w}\right) \tag{5·15}$$

FIG. 5·23 Side weir design chart [1].

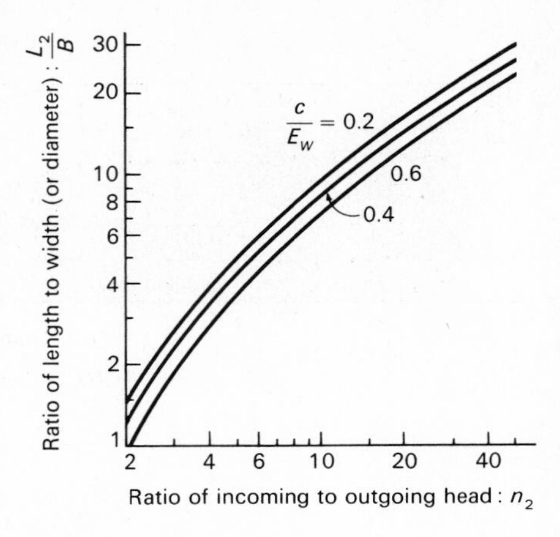

In Eq. 5·15, the specific energy term E_w may be computed using Eq. 5·16:

$$E_w = \alpha \frac{V_n^2}{2g} + \alpha'(d_n - c) \qquad (5\cdot16)$$

where α = velocity coefficient
V_n = normal velocity in approach channel, fps
g = acceleration due to gravity, ft/sec^2
α' = pressure-head correction
d_n = normal depth of flow in approach channel, ft
c = height of weir above channel bottom, ft

When using Eq. 5·16, α and α' may be taken as 1.2 and 1.0, respectively in the approach channel. At the lower end of the weir, α and α' should be taken at about 1.4 and 0.95, respectively.

As shown in Fig. 5·22, the value of h, the head at the upstream end of the weir, was found to be a function of the specific energy E_w. Experimentally, the value of h_1 was found to be equal to $0.5E_w$. It should be noted that the aforementioned equations apply only if there is a falling water-surface profile along the weir; however, it can be shown that a falling profile results if c/E_w is less than about 0.6. The application of these equations will be illustrated by the design examples.

Rising Water Surface The analysis of weirs under the conditions of a rising water-surface profile is based on the solution of Eq. 5·13 developed by de Marchi:

$$L_2 - L_1 = \frac{B}{C}\left[\phi\left(\frac{d_2}{E}\right) - \phi\left(\frac{d_1}{E}\right)\right] \qquad (5\cdot17)$$

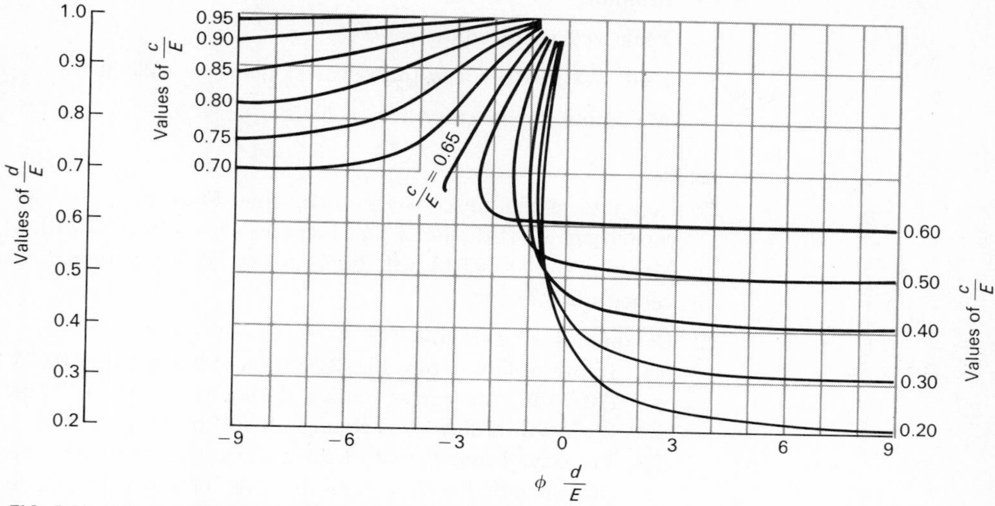

FIG. 5·24 Values of $\phi(d/E)$ for various values of c/E [9].

where $L_2 - L_1$ = length of weir, ft
B = channel width, ft
C = constant (0.35 for a free nappe)
$\phi(d/E)$ = varied flow function
d_1, d_2 = depths in channel, ft
E = specific energy, ft

In Eq. 5·17, if the discharge and the type of flow are known, then d/E can be determined at each end of the weir, and the length of the weir can be determined using Eq. 5·17 and Fig. 5·24. In general, it is recommended that this analytical method be used only in those cases where a rising water surface will occur. As a rule Eq. 5·17 works best when the Froude number is between about 0.3 to 0.92.

EXAMPLE 5·4 *Side Weir in Existing Pipe*

Determine the length of weir that must be placed in an existing 48-in. pipe, which is to be now used as a combined sewer, if the maximum wet weather flow is 70 cfs and the maximum allowable wet weather flow to the treatment plant is not to exceed 25 cfs. The maximum dry weather flow is 5 cfs. All this flow is to be treated. The data and assumptions are as follows:

Pipe diameter = 48 in.

Slope = 1:400 or 0.0025

Manning's n = 0.013

Peak wet weather flow = 70 cfs

Peak wet weather flow to treatment plant = <25 cfs

Maximum dry weather flow = 5 cfs

Solution

1. Compute maximum capacity of pipeline. From Fig. 3·13 the capacity of a 4-ft-diameter pipe laid on a slope of 0.0025 with a Manning's n of 0.013 will be 71.9 cfs. The corresponding velocity is 5.72 fps.

2. Compute flow characteristics of 70 cfs.
   (*a*) Depth of flow. From Fig. 3·16 for a q/Q_{full} value of 0.975 (70/71.9), the corresponding d/D_{full} value is 0.8. Thus the depth of flow d_n is equal to 0.8(4) or 3.2 ft.
   (*b*) Velocity. From Fig. 3·16 for a d/D_{full} value of 0.8, the corresponding v/V_{full} value is 1.14. The velocity V_n is equal to 5.72(1.14) or 6.52 fps.
   (*c*) Type of flow (tranquil or shooting). From Fig. 3·22 for a flow of 70 cfs the critical depth in the 4-ft pipe is equal to 0.629(4) or 2.51 ft. Therefore, the flow is tranquil since $d_n > d_c$.

3. Compute flow characteristic at 5 cfs.
   (*a*) Depth of flow. From Fig. 3·16 for a q/Q_{full} value of 0.0696 (5/71.9), the corresponding d/D_{full} value is 0.18. Thus, the depth of flow d_n is equal to 0.18(4) or 0.72 ft. Since all the dry weather flow must be retained, the weir height c will be 0.72 ft above the bottom of the pipe.
   (*b*) Velocity. From Fig. 3·16 for a d/D_{full} value of 0.18, the corresponding v/V_{full} value is 0.55. The velocity V_n is equal to 5.72(0.55) or 3.11 fps.

4. Weir calculations.
   (*a*) Compute the specific energy E_w.

$$E_w = \frac{1.2\,V_n^2}{2g} + (d_n - c)$$

$$= 1.2\,\frac{(6.52)^2}{2g} + (3.2 - 0.72) = 3.27$$

   (*b*) Compute c/E_w.

$$\frac{c}{E_w} = \frac{0.72}{3.27} = 0.220$$

   Since $c/E_w < 0.6$, a falling head is possible.

(*c*) Compute required weir length. For an assumed n_2 value of 10, the required length is given by

$$L_2 = 2.03B\left(5.28 - 2.63\frac{c}{E_w}\right)$$
$$= 2.03(4)[5.28 - 2.63(0.22)] = 38 \text{ ft}$$

5. Estimate peak wet weather discharge to treatment plant.
   (*a*) Compute velocity at lower end of weir.

$$1.4\frac{V_2^2}{2g} + 0.95h_2 = 3.27$$

but $h_2 = 0.1(0.5 \times 3.27) = 0.16$ ft, so

$$V_2^2 = 143 \qquad \text{and} \qquad V = 11.95 \text{ fps}$$

   (*b*) Outflow depth.
   Outflow depth $d_2 = 0.72 + 0.16 = 0.88$ ft.
   (*c*) Outflow area. From Fig. 3·16 for a d/D_{full} value of 0.22 (0.88/4), the corresponding a/A_{full} value is 0.16.
   Outflow area $= 0.16\pi(2)^2 = 2.01$ sq ft.
   (*d*) Discharge to treatment plant.
   Discharge $= 2.01(11.95) = 24$ cfs
   Since 24 cfs < 25 cfs, the design is satisfactory.

EXAMPLE 5·5 *Side Weir with Junction Structure*

Determine the length of weir required if a smaller diameter pipe with a capacity of 5 cfs is to be used as a control. In this case, some sort of junction structure will be required. The data and assumptions are as follows:

Large pipe diameter = 48 in.

Slope of large sewer = 0.0025

Manning's $n = 0.013$

Capacity of large sewer = 71.9 cfs

Small pipe diameter = 15 in.

Capacity of small sewer = 5 cfs

Peak wet weather flow = 70 cfs

Maximum dry weather flow = 5 cfs

Solution

In this example, the weir height will be set at the same elevation as the top of the pipe. Thus, when the flow exceeds 5 cfs, the smaller pipe will be slightly surcharged by the incoming flow.

1. Weir calculations.

(a) Compute the specific energy E_w.

$$E_w = 1.2\frac{V_n^2}{2g} + (d_n - c)$$

$$= 1.2\frac{(6.52)^2}{2g} + (3.2 - 1.25) = 2.74$$

(b) Compute c/E_w.

$$\frac{c}{E_w} = \frac{1.25}{2.74} = 0.45$$

Since $c/E_e < 0.6$, a falling head is possible.

(c) Compute required weir length. For an assumed n_2 value of 10, the required weir length is given by

$$L_2 = 2.03B\left(5.28 - 2.63\frac{c}{E_w}\right)$$

$$= 2.03(4)[5.28 - 2.63(0.45)] = 33.3 \text{ ft}$$

Baffled Side Weirs

As the name indicates, a baffled side weir is one in which a baffle has been placed across the main channel for the purpose of increasing the capacity of the side weir.

In considering the hydraulic problems involved in their design, the analysis must be largely theoretical, as there are few experimental data available upon which to base the design. If the area below the deflecting baffle is considered as an orifice, contraction will occur on the side in contact with the baffle. This contraction may appreciably depress the surface of the sewage below and thus determine the elevation at which the baffle or skimmer should be set. The increase in velocity due to pressure head should also be taken into account, together with the rise in level further downstream where the conditions of gravity flow in the sewer will have been reestablished. The elevation of the overflow weir and of the lower edge of the baffle may then be determined.

In experiments Tyler, Carollo, and Steyskal [19] found that a horizontal skimming plate had no advantage and also that a deflecting bulkhead placed squarely across the conduit, with its bottom edge at the elevation of the weir crest, was more efficient in deflecting excess flow over a side weir than one set at an angle. They suggested, as a result of their experiments, that by making the elevation of the lower edge of this bulkhead adjustable, it could be set to pass the desired amount of sewage down the main conduit. This would force much larger discharges over a side weir of given length of crest than could be discharged by a similar weir without the bulkhead. This increase in discharge is due to an increase in head on the weir, which is caused by the change of velocity head to static head above the bulkhead. A condition of nonuniform flow exists at and below the bulkhead, and a jump may be produced. A bend or a flat gradient downstream may force the jump back against the bulkhead, which will also increase the discharge over the side weir.

Transverse Weirs

One method of avoiding the uncertainty involved in the use of side weirs is to place the weir directly across the line of flow and deflect the sewer to one side, thus reversing the arrangement used with side weirs. It should be possible to determine the discharge of a weir in this position by the usual weir equation with greater certainty than when the weir is in the side wall of the conduit, as the variation in head along the crest will presumably be smaller.

Another scheme for diverting storm water flow using a transverse weir is shown in Fig. 5·25. Under peak wet-weather-flow conditions, the sewer leading to the sewage treatment plant will be surcharged and should be designed accordingly. In addition, a deflector should be installed to avoid the formation of a vortex whenever the sewer leading to the treatment plant becomes surcharged.

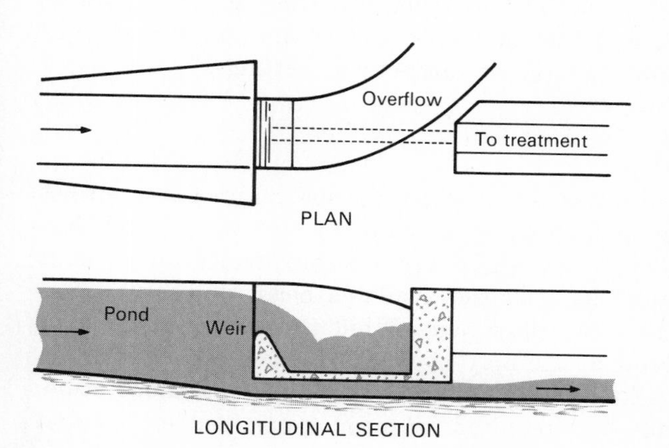

FIG. 5·25 Weir overflow structure.

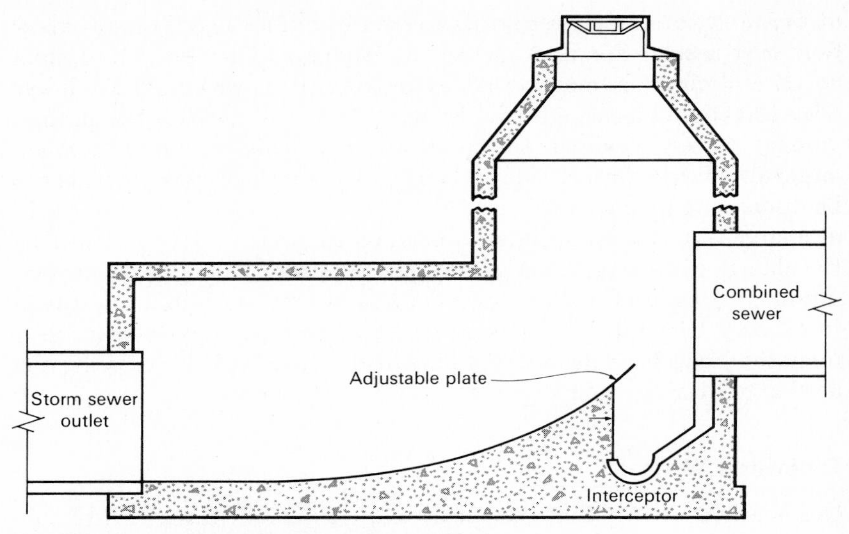

FIG. 5·26 Adjustable leaping weir at Seattle, Wash.

Leaping Weirs

A leaping weir is formed by a gap in the invert of a sewer so that the ordinary flow of sewage falls through the opening and passes to the interceptor. At times of storm, the increased velocity of flow causes most of the flow to leap the opening and pass on to the storm outlet. An example of a leaping weir is shown in Fig. 5·26. The steel weir plate is inclined to give a spouting effect and is designed so that it can be adjusted for various flow conditions.

The design of leaping weirs is usually a trial-and-error process involving computation of the trajectory of flow and determination of the percentage of flow captured at various weir settings.

Siphon Spillways

The siphon affords a means of regulating the maximum water-surface elevation in a sewer with smaller variations in high-water level than can be secured with other devices. It works automatically and without mechanism. Since it utilizes all the available head, it discharges at higher velocities than do overflow weirs. While this device has obvious advantages, its infrequent use may be due to inadequate information concerning certain matters pertaining both to its design and operation, such as the minimum head required for priming or possible noise and vibration from sudden starting and stopping of the siphon, especially with high heads. Figure 5·27 shows a typical siphon relief structure.

The approximate cross section for the siphon throat can be determined by Eq. 5·18:

$$Q = ca\sqrt{2gh} \tag{5·18}$$

where Q = discharge, cfs
c = coefficient of discharge (0.6 to 0.8)
a = area of cross section of throat, sq ft
h = head, ft = difference in water level above and below siphon

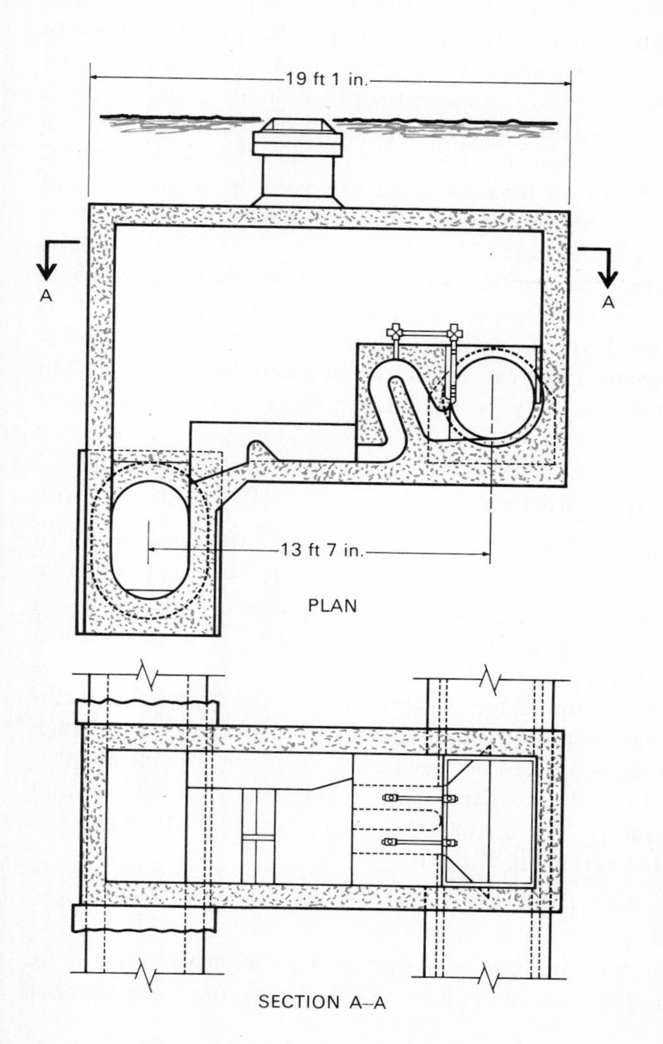

FIG. 5·27 Siphon spillway.

A trial section may then be drawn; the losses may be determined for this section; and the corrected values of Q or a may be computed. The design then involves working out such details as the method of venting, the shape of section to meet the particular requirements, and the fixing of elevations of inlet, spillway, outlet, and air vent. An air vent with area equal to $a/24$ has been found to be ample. The siphon inlet should be large so that the velocity at entrance will be small, thus preventing large head losses. The section tapers gradually to the throat; and the lower leg, which may be either vertical or inclined, is of uniform section or slightly flared.

When the water rises to the elevation of the crest in the siphon, it will flow over the crest in a thin sheet, falling against the opposite wall and carrying with it some air. This condition will continue until the water has risen sufficiently to cause the air remaining in the siphon to be exhausted quickly and siphonic action will begin. If the capacity of the siphon is greater than the amount of excess flow in the sewer, the sewage surface will be lowered until the vent is uncovered and siphonic action will cease. Generally, it is desirable to have the elevation of the vent lower than that of the siphon crest, so that there will be only a little overflow before the siphonic action begins or after it ceases. Theoretically, the maximum operating head for a siphon is about 33.9 ft at sea level. It decreases at the rate of about 1 ft for every 850 ft of increase in elevation above sea level. This is more than ample for such uses as may be found for siphon reliefs on sewer construction.

Fixed-Orifice Diversion Structure

A fixed-orifice diversion structure employs a fixed orifice, as shown in Fig. 5·28, for the purposes of (1) diverting relatively small rates of dry weather sewage flow from an existing combined sewer into an interceptor and (2) during storm flows limiting the discharge of storm water to the interceptor as much as possible without the use of mechanical regulators. In operation, during storms, the hydraulic gradients in the diversion chambers will rise, and the total discharge of each combined sewer will be divided between the stream and the interceptor in proportion to the relative capacities of the diversion orifices and the storm water outlet. The latter is eventually controlled by the elevation of the water surface in the stream.

OUTLETS

Strictly speaking, the outlet of a sewerage system is the end of an outfall sewer where the sewage is discharged. There may be a number

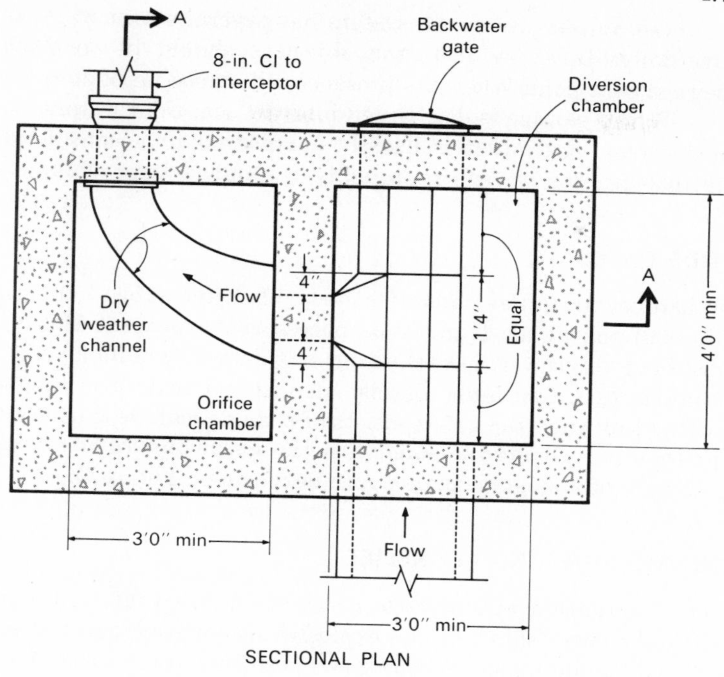

SECTIONAL PLAN

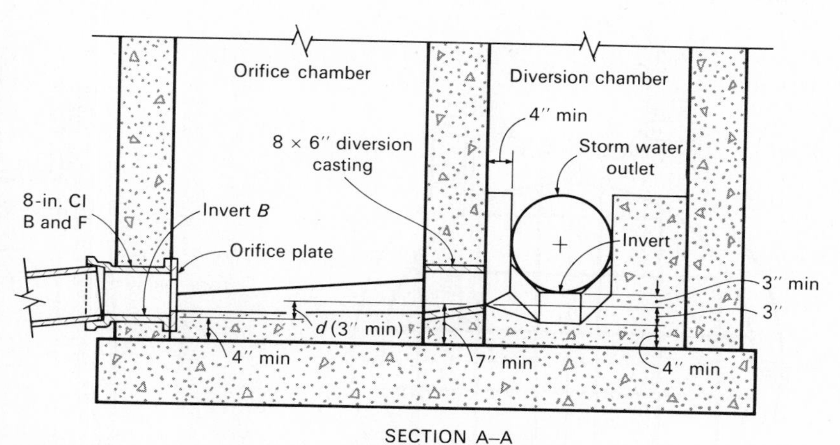

SECTION A–A

FIG. 5·28 Fixed-orifice diversion and control structure.

of these outlets where the system has several storm water outfalls or overflows. In every case, the objective should be to discharge the sewage at a point where its presence will cause no offense.

Where sewage is discharged into the sea, tidal waters, a lake, or a deep river, the outlet is usually submerged to a considerable depth to disperse the sewage thoroughly through a large volume of water.

TIDE GATES

Wherever an outlet ends at a body of water subject to considerable fluctuations in level and it is necessary to prevent this water from entering the sewer, a backwater or tide gate is employed. This gate consists of a flap hung against an inclined seat. The hinges may be at the top when the gate consists of a single leaf, as is usually the case, or they may be at the side when the gate consists of two leaves. An example of a single-leaf tide gate is shown in Fig. 5·29.

VENTILATION AND AIR RELIEF

The ventilation of sewers is necessary to prevent the accumulation of sewer gases that may be explosive or corrosive; to prevent the occasional buildup of odorous gases resulting from the putrefaction of

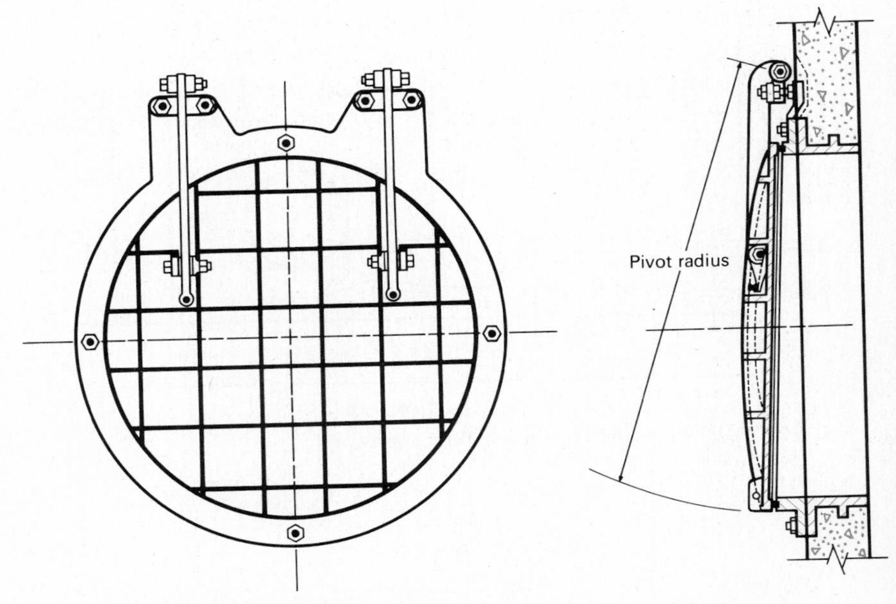

FIG. 5·29 A single-leaf tide gate [from Armco].

sewage; to reduce the accumulation of hydrogen sulfide which corrodes exposed concrete and metal; and to prevent the creation of pressures above or below atmospheric which may occasionally break household plumbing water seals. The actual composition of gases found in sewers will vary with the character of the sewage, but may include air, natural or manufactured gas (illuminating), gasoline vapor, carbon monoxide, carbon dioxide, hydrogen sulfide, and methane, in varying amounts.

The movement of air in sewers is caused by a number of factors, such as the difference in unit weight between the outer air and that in the sewers, the difference in elevation of the various openings into the sewers, the flow of the sewage which tends to move the sewer air, and the effect of the wind blowing in through openings, particularly through the outlets of large sewers.

The most effective openings for ventilation should be those in the manhole covers. The connections through which sewage and rainwater are delivered to the sewers from houses are likely to be filled from time to time by the discharges from those properties, while, for combined systems, the traps used on street inlets and catch basins will be sealed by the water within them so that no air can enter or escape there.

The effect of the temperature inside and outside the sewers upon ventilation depends upon the unit weights of air due to these temperature differences, which may be large in winter. The movement of the sewer air is theoretically toward the end of the laterals, since they are at higher elevations than the trunk sewers, and the colder outer air will enter at the lower openings of the sewers. Practically, however, the wind and the drag on the sewer air due to sewage flowing down grade have some effect in checking movement of the air up the sewers. The result, at times, is downstream movement of the sewer air.

Normally, house sewers connected to vent stacks, perforated manhole covers, and unobstructed outlets provide all the ventilation that is necessary in sanitary or combined sewers. Forced draft is required occasionally. Storm sewers may be ventilated through untrapped inlet connections, unobstructed outlets, and perforated manhole covers. It is undesirable to ventilate combined sewers through untrapped inlet connections because of the probability of the escape of objectionable odors to the street surface. There is some objection to the use of perforated manhole covers for sewer ventilation because of the entrance of surface water through the perforations, and the escape of odors and visible vapors from manholes.

If the outlet of a sewer is submerged and there are few ventilated house connections, some special provision for ventilation is usually necessary. A ventilating shaft with a cross-sectional area at least one-half that of the sewer and tall enough to extend above nearby roofs should suffice. Perforations in manhole covers or other special means of ventilation should always be provided where main traps are installed in house sewers.

A change of sewer section at depressions should be vented from upstream to downstream of the depression, so that sewer air will not be compressed and may move either upstream or downstream. On large sewers having one or more inverted siphons, the siphon inlet chambers should be vented, possibly with exhaust fans, and should be equipped with facilities to deodorize the exhausted sewer air.

PROBLEMS

5·1 A curb opening is to be designed for a street with a 1.5 percent grade. If the depth of flow is 0.18 ft and the street has a 1:16 crown, determine the flow in the inlet and the length of opening when $n = 0.018$, $k = 0.22$.

5·2 A storm sewer of 48-in. diameter has a slope of 0.0015. The maximum dry weather flow is 10 cfs and an average storm will produce 25 cfs. Design a three-pipe inverted siphon for maximum dry weather flow, average storm flow, and maximum storm flow (capacity of the 48-in. sewer). The available hydraulic slope is 0.005 and $n = 0.013$. The minimum scouring velocity is 3 fps.

5·3 A side weir is designed to have a falling water surface profile. The height of the weir is such that a flow of 10 cfs is contained in a 5-ft-diameter pipe with a slope of 0.0016 and an n of 0.013. Find the length of the weir if the ratio of incoming to outgoing head on the weir is 5 and $Q = 65$ cfs.

5·4 Discuss the importance of ventilation and air relief in sewers.

REFERENCES

1. Ackers, P.: A Theoretical Consideration of Side Weirs as Stormwater Overflows, *Proceedings, Institution of Civil Engineers, London,* vol. 6, no. 2, 1957.
2. Allegheny County Sanitary Authority, *Carnegie Technical Report,* Pittsburgh, 1952.
3. Babbitt, H. E., and E. R. Baumann: *Sewerage and Sewage Treatment,* 8th ed., Wiley, New York, 1958.
4. Bauer, W. J., and D. C. Woo: Hydraulic Design of Depressed Curb-Opening Inlets, *Highway Research Record,* Highway Research Board, no. 58, p. 61, 1964.
5. Bureau of Public Roads: *Drainage of Highway Pavements,* Hydraulic Engineering Circular 12, Washington, D.C., 1969.
6. Bureau of Public Roads: *Urban Storm Drainage,* Hydraulic Information Circular 2, Washington, D.C., 1951.

7. Cassidy, J. J.: Generalized Hydraulic Characteristics of Grate Inlets, *Highway Research Record*, Highway Research Board, no. 123, p. 36, 1966.

8. Chow, V. T.: *Handbook of Applied Hydrology*, McGraw-Hill, New York, 1964.

9. Collinge, V. D.: The Discharge Capacity of Side Weirs, *Proceedings, Institution of Civil Engineers, London*, vol. 6, no. 2, 1957.

10. Frazer, W.: The Behavior of Side Weirs in Prismatic Rectangular Channels, *Proceedings, Institution of Civil Engineers, London*, vol. 6, no. 2, 1957.

11. Izzard, C. F.: Hydraulics of Runoff from Developed Surfaces, *Proceedings, Highway Research Board*, vol. 26, 1946.

12. Jaeger, C.: *Engineering Fluid Mechanics*, Blackie, London, 1956.

13. Joint Committee of the American Society of Civil Engineers and the Water Pollution Control Federation: *Design and Construction of Sanitary and Storm Sewers*, ASCE Manual and Report 37, Washington, D.C., 1969.

14. Metcalf, L., and H. P. Eddy: *Sewerage and Sewage Disposal*, 2d ed., McGraw-Hill, New York, 1930.

15. Nimmo, W. H. R.: Side Spillways for Regulating Diversion Canals, *Transactions, ASCE*, vol. 92, 1928.

16. Parmley, W. C.: The Walworth Sewer, Cleveland, Ohio, *Transactions, ASCE*, vol. 55, 1905.

17. Smith, D., and G. S. Coleman: The Discharging Capacities of Side Weirs, *Proceedings, Institution of Civil Engineers, London*, 1923.

18. Storm Drainage Research Committee, *The Design of Storm Water Inlets*, The Johns Hopkins University Storm Drainage Research Project, 1956.

19. Tyler, R. G., J. A. Carollo, and N. A. Steyskal: Discharge over Side Weirs with and without Baffle, *Journal, Boston Society of Civil Engineers*, vol. 16, 1929.

pumps and pumping stations

6

This chapter is intended to serve as an introduction to the study of pumps and pumping stations. Because the subject matter in this area is so broad in scope, this chapter is limited to that material with which the sanitary engineer should be most familiar. Therefore, stress is placed on the types and operating characteristics of pumps and pump drive units, on the fundamentals of pump selection, and on the design of sewage pumping stations. The theoretical basis and engineering reasons for pumping station design decisions, especially in the simpler cases, are discussed in the latter section.

Reference books such as [1 to 3, 5, and 6], journal articles such as [9 and 14] and other current publications should be used to complement the material presented in this chapter. In practice, the engineer must also be familiar with the catalogs and data of leading pump manufacturers, because engineering design problems cannot be solved without reference to such information. Inspection of existing pumping stations and the study of plans and specifications for existing or proposed stations will provide additional information and experience that cannot be obtained in any other way.

Descriptions of the most commonly encountered pumps in the wastewater field, their operating characteristics, and their construction are discussed in this section. Some special pumps that have found application are also mentioned.

Centrifugal Pumps

Centrifugal pumps are classified as (1) radial-flow, (2) mixed-flow, and (3) axial-flow [11]. The relationship of these pumps to the many other types of pumps that are used in the wastewater field and other fields is shown schematically in Fig. 6·1. In general, radial-flow and mixed-flow pumps are used for the pumping of sewage and storm water. Axial-flow pumps may be used for pumping storm drainage unmixed with sewage or treatment plant effluent.

Radial-Flow Pumps In these pumps, centrifugal force is used to impart energy to the fluid. It should be noted that in the eleventh edition of *Hydraulic Institute Standards* (1965) radial-flow pumps were called centrifugal pumps. The principal components of a radial-flow centrifugal pump are indicated in Fig. 6·2. As shown in this figure, water enters the impeller axially and is discharged at right angles to the shaft. It is collected by a channel of gradually increasing area, called a volute, which extends to the pump discharge nozzle. This type of pump is known as a single-end suction, volute type. If water enters the impeller axially from both sides, the impeller is called a double-suction impeller. Because rags and trash in sewage, even though screened, would quickly clog the small passages in typical clear-water radial-flow pumps, those used for raw sewage are generally of the single-end suction, volute type fitted with nonclog impellers, such as the one shown in Fig. 6·2.

Shafts may be horizontal but usually are vertical; however, horizontal-shaft, double-suction pumps have been used in some of the largest stations in the United States, notably those of the Metropolitan Sanitary District of Greater Chicago. It is noted that on the pump shown in Fig. 6·2 the shaft does not extend into the suction passage. In the double-suction type of pump, the shaft extends completely through both suction passages, and any rags contained in the sewage will have a tendency to wrap around the shafts.

Nonclog pumps have wide-open passages and a minimum number of vanes (not exceeding two in the smaller sizes and limited to three, or at the most four, in the larger sizes). Impellers are almost univer-

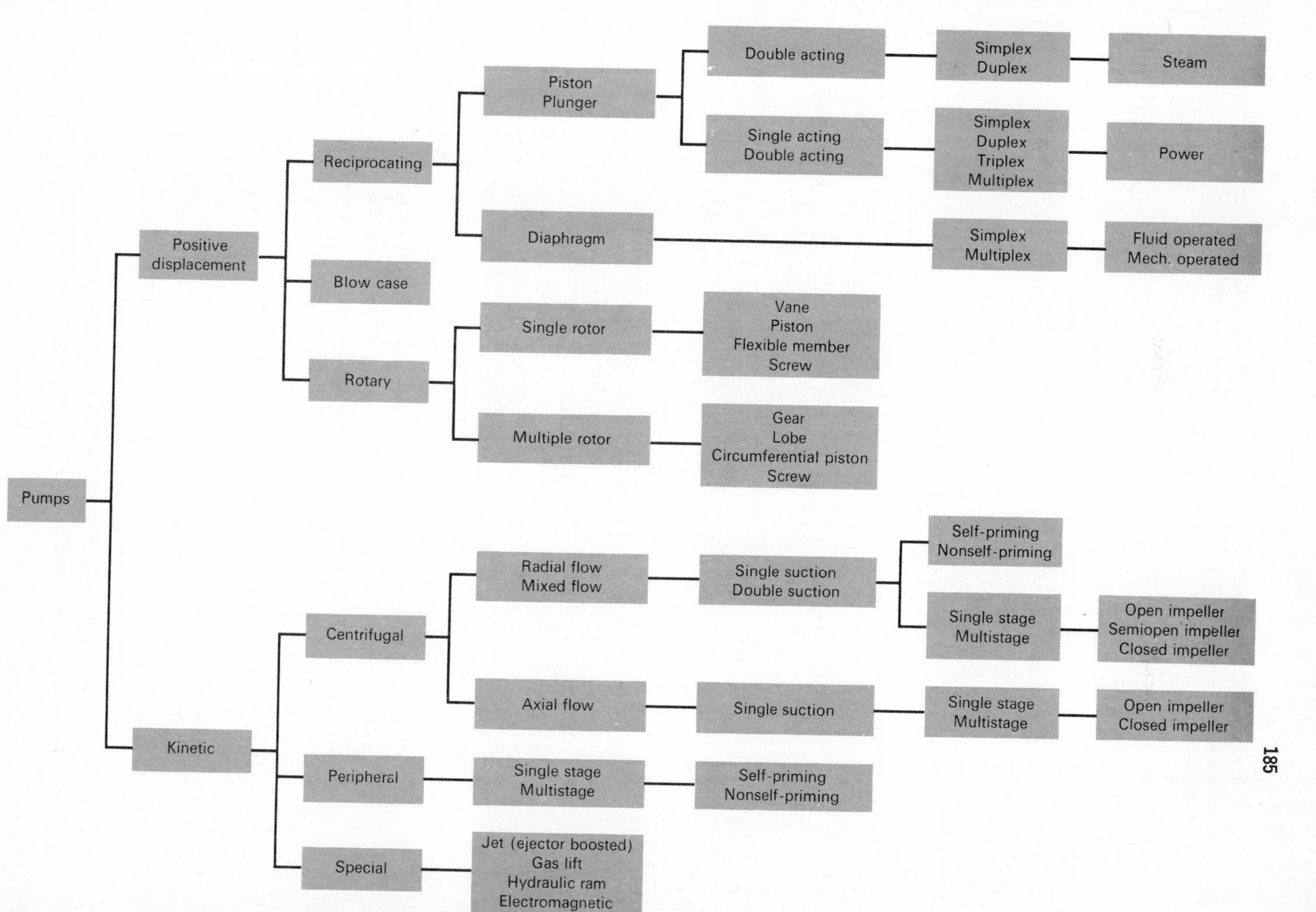

FIG. 6·1 Classification of pumps [11].

FIG. 6·2 Typical vertical radial-flow sewage pump [from Fairbanks Morse].

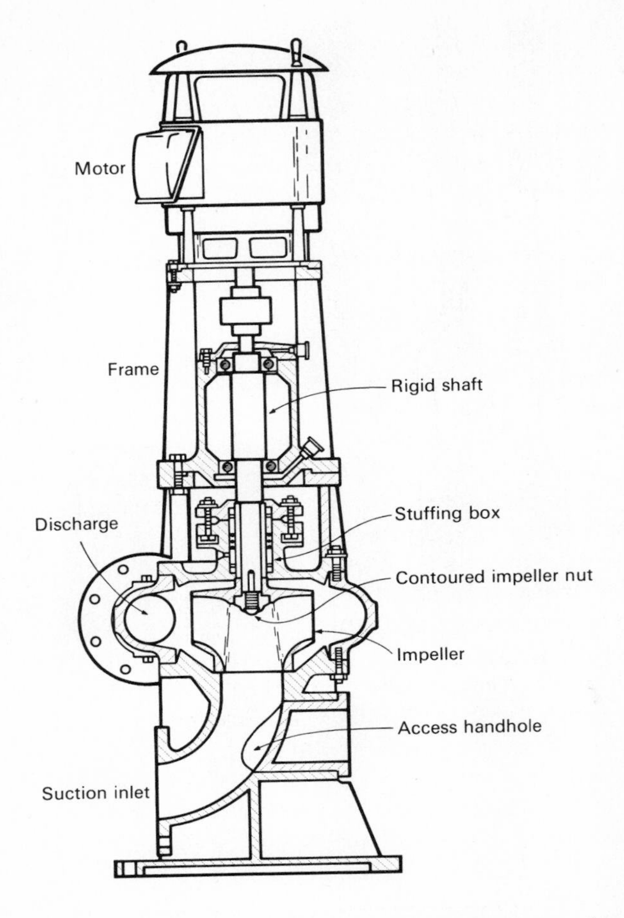

sally of the enclosed type. For raw-sewage service, 4-in. (diameter of discharge opening) pumps normally should be able to pass 3-in.-diameter spheres; 8-in. pumps, 4-in.-diameter spheres; etc. The sphere size becomes correspondingly larger as the pump size increases (up to 8 in. or larger spheres for 36-in. pumps, depending on the design). Nonclog pumps smaller than 4 in. in size should not be used in municipal pumping stations for handling raw sewage.

Mixed-Flow Pumps Pumps with mixed-flow impellers occupy an intermediate range between radial-flow pumps in which a forced vortex is superimposed on radial outward flow and axial-flow pumps in which a forced vortex is superimposed on axial flow. The specific speed (a type number used to characterize pumps—see "Pump Operating Characteristics") for these pumps varies from about 4,000 to 10,000.

As the specific speed increases from 4,000 to 6,500 to 10,000, the pump characteristics become more like those of an axial-flow pump. The shape of the impeller and the design of the pump also vary. Mixed-flow impellers may be installed either in volute-type casings, in which case they are designated as mixed-flow volute pumps, or in diffuser-type casings similar to propeller pumps, in which case they are designated as mixed-flow propeller pumps.

Mixed-flow volute pumps are suitable for pumping raw sewage and storm water, especially in the specific speed range between 4,000 and 6,000. They are available in 8-in. and larger sizes and for heads up to 50 to 60 ft. They operate at higher speeds than the nonclog pumps, are usually of lighter construction, and, where applicable, cost less than corresponding nonclog pumps. The size of spheres that the mixed-flow volute pump will pass is much smaller than can be handled by a nonclog pump of the same size, but the 8-in. pump will pass a 3-in.-diameter sphere. Impellers may be of either the open or enclosed type, but the enclosed type is preferred.

Axial-Flow Pumps For pumping storm water, particularly in those cases where a large quantity has to be pumped against a low head, this type of pump is less expensive than those already described. The pump itself consists of a multiple-bladed screw rotor or propeller placed in a casing, with fixed guide vanes before and after the propeller. The action is then similar to that of a ship's propeller as it draws the water through the inlet guide vanes and discharges it through the outlet guide vanes. This type of pump should not be used for raw sewage or sludges, as rags are apt to lodge and build up on the guide vanes.

Displacement Pumps

Plunger-type reciprocating-power pumps are commonly used for transferring sludge from primary settling tanks to digestion tanks and from one digester to another in sewage treatment plants. The Moyno (trade name) progressing cavity rotary pump has been used for handling particularly heavy concentrated sludge. Rotary gear-type pumps are used in the lubrication systems of sewage plant equipment, such as engines and blowers.

Pump Operating Characteristics

Operating characteristics depend on the size, speed, and design of the pump. Characteristic curves, commonly called pump curves, show the

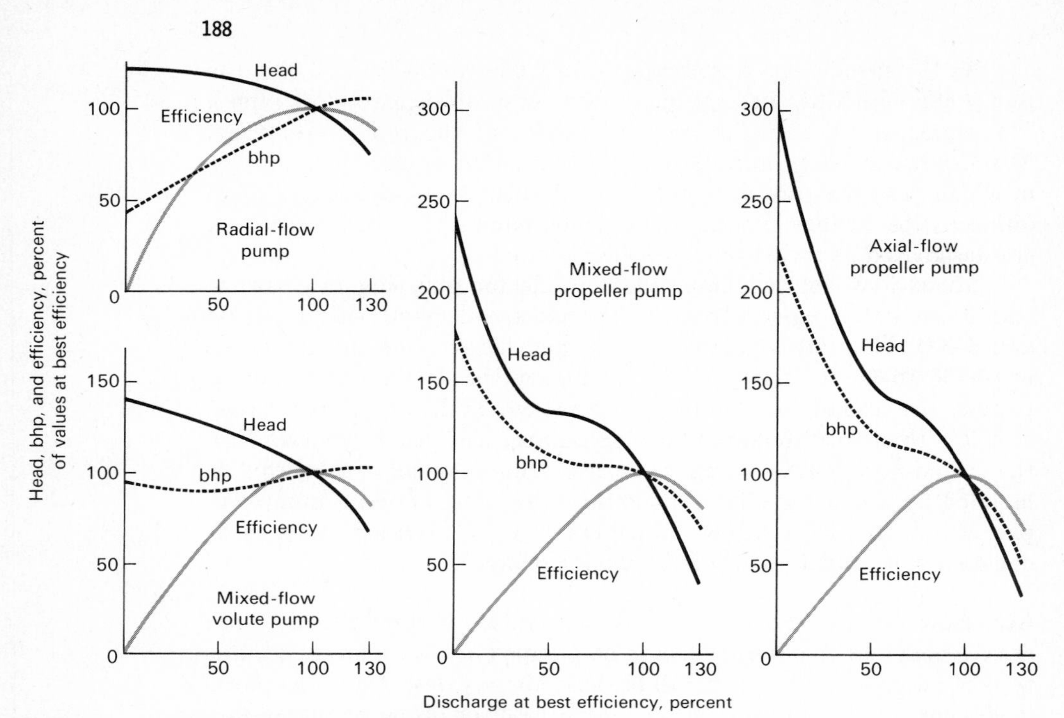

FIG. 6·3 Typical centrifugal pump characteristic curves.

total head H in ft, the efficiency E in percent, and the power input P in horsepower (hp), plotted as ordinates against the flow Q in gpm as abscissas. The general shape of these curves varies as the specific speed varies.

Characteristic curves for typical radial-flow, mixed-flow volute, mixed-flow propeller, and axial-flow centrifugal pumps are shown in Fig. 6·3. In this figure, the variables have been plotted as a percentage of their values at the best efficiency point (bep). In addition, operating characteristics for various types of pumps have been summarized and are presented in Table 6·1.

For radial-flow pumps, it can be shown theoretically that the discharge Q, head H, and power P at a particular operating point on the pump curves are related to speed N in revolutions per minute (rpm) as follows:

$$\begin{aligned} &Q \text{ varies with } N \\ &H \text{ varies with } N^2 \\ &P \text{ varies with } N^3 \end{aligned} \tag{6·1}$$

TABLE 6·1 OPERATING CHARACTERISTICS OF VARIOUS TYPES OF CENTRIFUGAL AND POSITIVE-DISPLACEMENT PUMPS

	Type of pump				
	Centrifugal			Positive-displacement	
Characteristics	Radial	Mixed-flow	Axial-flow	Rotary	Piston
Flow	Even	Even	Even	Relatively even	Pulsating
Effect of increasing head:					
on capacity	Decrease	Decrease	Decrease	Almost negligible decrease*	
on power required	Decrease	Small decrease to large increase	Large increase	Increase	Increase
Effect of decreasing head:					
on capacity	Increase	Increase	Increase	Almost negligible increase*	
on power required	Increase	Slight increase to decrease	Decrease	Decrease	Decrease
Effect of closing discharge valve:					
on pressure	Up to 30% increase	Considerable increase	Large increase	Destructive pressure developed unless relief valve is used	
on power demand	Decrease 50–60%	10% decrease to 80% increase	Increase 80–150%	Increase to destruction	
Pump valves	None†	None†	None†	None†	Inlet and discharge valves

* Slip loss increases at increased pressure differential.
† Isolating valves and check valves installed in piping separate from pump. Not required to maintain pumping action.

These relationships are known as the affinity laws and are used to determine the effect of speed changes on the discharge, head, and power. Similar laws apply to a change in impeller diameter but are less reliable. In actual practice, the new operating point, resulting from a change in speed, will be the intersection of the system curve and the new head-capacity curve.

An important relationship, which has been mentioned and used previously, is that of specific speed defined by the following equation:

$$N_s = \frac{N\sqrt{Q}}{H^{3/4}} \tag{6·2}$$

where N_s = specific speed
N = speed, rpm
Q = discharge, gpm
H = head, ft

For any pump operating at any given speed, Q and H are taken at the point of maximum efficiency. When using Eq. 6·2 for pumps having double-suction impellers, one-half of the discharge is used, unless otherwise noted. For multistage pumps, the head is the head per stage.

The computed value of specific speed has no usable physical meaning, except as a type number, but it is extremely useful because it is constant for all similar pumps and does not change with speed for the same pump. Since specific speed for a given pump is independent of both physical size and speed, it depends only on shape and is also sometimes considered a shape factor. Figure 6·4 shows the variation in maximum efficiency to be expected with variations in size (capacity) and design (specific speed). At the bottom of the figure are shown the progressive changes in impeller shape as the specific speed increases.

Pump design characteristics, cavitation parameters, and abnormal operation under transient conditions can be correlated satisfactorily with the specific speed.

Consideration of the specific-speed equation reveals the following:

1. If larger units of the same type are selected for approximately the same head, the operating speed must be reduced.
2. If higher-specific-speed units are selected for the same head and capacity, they will operate at a higher speed; and hence the complete unit, including the driver, should be less expensive.

It thus becomes obvious why large propeller-type pumps are used in irrigation practice where low-lift high-capacity service is required.

Pump Construction

Sewage pumps are constructed of cast iron with bronze or stainless-steel trim and with either cast-iron or bronze impellers. Where the sewage contains grit, longer life will be obtained from cast-iron impellers.

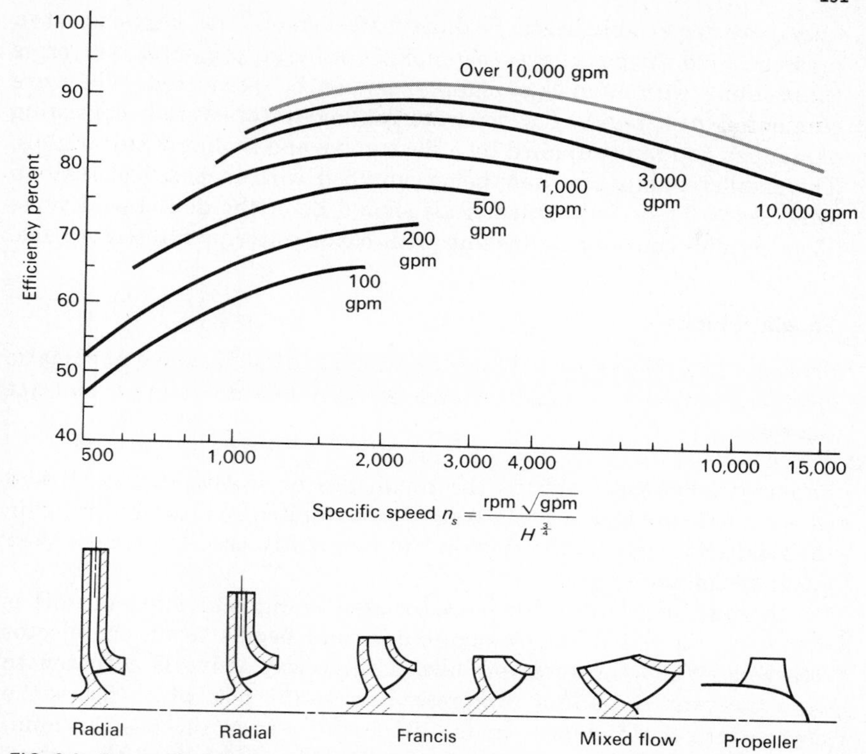

FIG. 6·4 Pump efficiency versus specific speed and pump size [from Worthington Corp.].

Larger pumps are normally provided with bronze wearing rings at the suction side of the impeller. Stainless-steel rings can be obtained if specified. Pumps smaller than 10 in. in size are usually furnished without wearing rings or with wearing rings on the casing only.

The bearings for vertical dry-pit pumps are usually of the antifriction type located in the main frame above the impeller and are commonly grease lubricated. Wet-pit pumps are usually equipped with oil-lubricated sleeve bearings.

Pump shafts should be made of high-grade forged steel and should be protected by renewable bronze or stainless-steel sleeves where the shaft passes through the stuffing box. In small pumps, the shaft may be made of stainless steel and the shaft sleeves omitted.

The stuffing boxes may be either grease sealed or water sealed. Grease seals should not be used on pumps handling wastewater containing grit if a supply of municipal water is available or can be obtained at reasonable cost. Sealing water should be supplied at a pressure 5 to 10 psi above the pump discharge pressure, and cross connections with municipal supply cannot be permitted. They are eliminated by providing a separate sealing water system consisting of an open-top tank supplied by a float valve and sealing water pumps. The smaller pumps are now being supplied with mechanical seals in some cases. These mechanical seals should be of the double-seal type. They require continuous flushing with clean water or filtered sewage.

Special Pumps

Special pumps that are used for various applications include pneumatic ejectors, torque-flow pumps, bladeless pumps, and air-lift and jet pumps.

Pneumatic Ejectors Where the quantities of sewage are small and where the future increase in sewage flow is limited to an amount within their capacity, pneumatic ejectors are frequently used because of their nonclog characteristics.

A sewage ejector with its associated equipment and controls is shown in Fig. 6·5. First, sewage enters and begins to fill the ejector reservoir. When the reservoir fills, a three-way valve is actuated to close the vent and admit compressed air to the pot, which forces the sewage into the discharge main. When the sewage reaches the minimum level, the position of the three-way valve is reversed to shut off the flow of compressed air and to pen the vent. Venting of the air within the pot permits sewage to flow again into the reservoir. The compressed air may be supplied directly by the air compressors or from an air storage tank that is maintained at the necessary pressure by the air compressors. In the latter case, the air compressors can be smaller, and the installed motor horsepower can be reduced by one-half.

Pneumatic ejectors are available in capacities varying from 20 to 600 gpm in simplex operation. Under the usual design conditions, the ejectors operate on a 1-min cycle, filling for 30 sec and discharging for 30 sec.

Sewage ejectors in municipal installations should be used in duplex. When flows are particularly small, one ejector pot may be installed initially; and space should be left for a second pot to be installed later. Large units should be alternated in operation. Smaller units, if the incoming and outgoing sewers have sufficient capacity, may be operated as two simplex units, in which case the controls are much simpler.

Pneumatic ejectors are economically feasible for flows up to 300

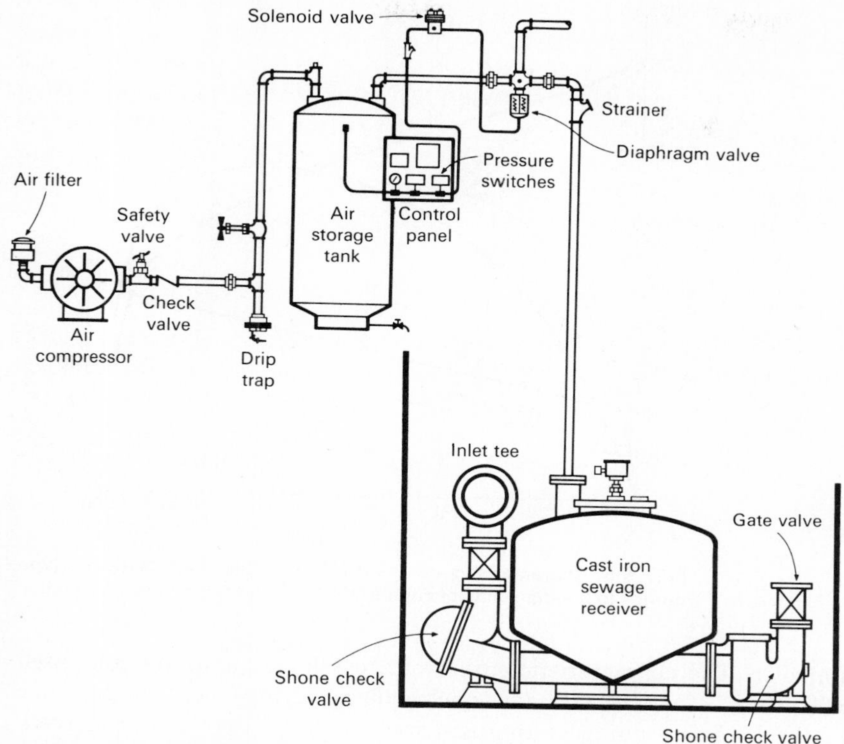

FIG. 6·5 Pneumatic ejector and associated piping [from Yeomans].

gpm, beyond which point power costs become excessive. For larger flows, the use of nonclog centrifugal sewage pumps is recommended.

Torque-Flow Pumps These pumps were developed by Wemco (trade name) for handling solid materials and have a recessed impeller in the side of the case entirely out of the flow stream. A pumping vortex is set up by viscous drag. These pumps have been installed in many sewage treatment plants for pumping raw sludge and sewage and have been practically free from clogging in locations where the ordinary nonclog pumps clogged repeatedly. On account of their high cost and lower efficiency, it is anticipated that these pumps will find their greatest usefulness in the sanitary engineering field in the pumping of sludges.

Bladeless Pumps These pumps are essentially volute-type centrifugal pumps that are fitted with a special "bladeless" or single-passage

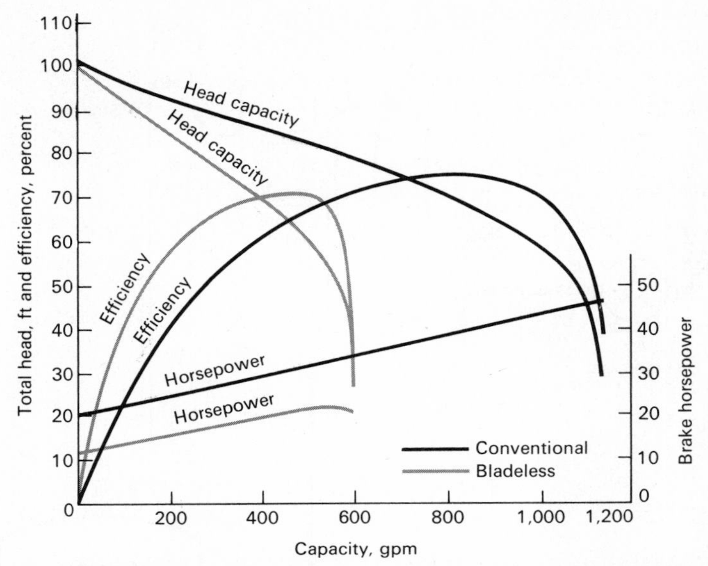

FIG. 6·6 Characteristic curves for radial-flow and bladeless-type pumps to handle solids of comparable size [from Fairbanks Morse].

impeller. The characteristic curves of a bladeless pump are compared with those of a conventional nonclog pump in Fig. 6·6. The capacity of the bladeless pump is approximately one-half that of the conventional nonclog pump. These pumps have demonstrated superior nonclogging performance and are particularly suitable for small flows. They are available in sizes up to 5 in.

Air-Lift and Jet Pumps Air-lift pumps are sometimes used in sewage treatment plants for recirculating sludge in the activated-sludge process. Jet pumps are occasionally used for priming centrifugal pumps.

PUMP DRIVE UNITS

The most commonly used drives for sewage pumps are direct-connected electric motors. Units driven by internal combustion engines are sometimes installed to ensure operation during electric power outages or where sewage gas is available for fuel.

Electric Motors

Constant-speed pumps may be driven by squirrel-cage or wound-rotor induction motors or by synchronous motors. Squirrel-cage motors will normally be selected on account of their simplicity, reliability, and economy, unless the starting current limitations of the power company make the choice of wound-rotor motors necessary. Synchronous motors may be more economical for large, slow-speed drives.

Wound-rotor motors are usually selected for variable-speed operation. Speed adjustment is obtained by control of resistance in the rotor circuit. Speed reductions up to one-third of full-load speed can normally be obtained. This is sufficient in most cases to cover the full range of capacity from zero to maximum. The motor is started with full resistance in the circuit. As magnetic contactors cut out resistances in steps, the motor speed and sewage flow are increased in steps. A common method of control is to have two accelerating steps followed by five running speeds. Stepless speed variation may be obtained by the use of liquid resistors or by magnetic drives.

The synchronous speed of an electric motor is given by the equation

$$\text{rpm} = \frac{60 \times \text{frequency}}{\text{no. of pairs of poles}} \tag{6$\cdot$3}$$

Synchronous motors operate at this speed, but the speed of squirrel-cage induction motors and the full speed of wound-rotor motors is 2 to 3 percent less due to slip. The available speeds of pumps driven by constant-speed motors operating on 60-cycle alternating current are given in Table 6·2.

Speeds above 1,750 rpm are not used for sewage pumps, and speeds below 277 rpm are uncommon. Multispeed two-winding squirrel-cage motors are available but are limited to two of the speeds shown in Table 6·2. Two additional speeds can be derived, but they must be half of the first two speeds. Since the head produced varies with the square of the speed, two-speed motors seldom can be utilized unless the static head is small in proportion to the head at the design flow or the speed of the pump is very low. In the latter case, two-speed synchronous motors are sometimes used.

Internal Combustion Engines

At sewage treatment plants, sewage pumps may be driven by sewage gas or dual-fuel engines. These pumps may be either horizontal units connected directly or through gears or vertical units driven through right-angle gears. The engine speed may be controlled manually by the operators or automatically from a float in the wet well.

Diesel engines or spark-ignition engines burning gas or gasoline may be used to drive sewage pumps; and, where pumps must operate

TABLE 6·2 OPERATING SPEEDS OF CONSTANT-SPEED MOTORS ON 60-CYCLE ALTERNATING CURRENT

Pairs of poles	Motor speed, rpm	
	Synchronous	Induction
1	3,600	3,500
2	1,800	1,750
3	1,200	1,160
4	900	870
5	720	695
6	600	580
7	514	500
8	450	435
9	400	390
10	360	350
11	327	318
12	300	290
13	277	268
14	257	249

during power outages, at least one such unit is frequently installed. In these cases, one or more dual-drive units may be installed. In this arrangement, an electric motor is mounted on top of the right-angle gear and coupled directly to the pump shaft. The engine is connected to the horizontal shaft of the right-angle gear through a clutch or quick-disconnect coupling. If the engine is to come into operation automatically, the clutch should be of the freewheeling or overrunning type.

SEWAGE PUMP SELECTION

Determination of Design Flow

The amount and variation of sewage flow (discussed in Chap. 2) must be known before the proper selection of pumps and other components of a sewage pumping station can be made. Estimates must be made of not only the initial range of flows but also the future range of flows during the life of the station. A decision will have to be made whether to (1) install equipment large enough to handle the entire range of present and future flows or (2) make suitable provisions for increasing the capacity of the station at some time in the future by installing larger impellers, larger pumps, additional units, or, in some cases, motors with larger capacities and higher speeds.

Just what the design flow should be will depend on several factors, including

1. The maximum dry weather and wet weather flows
2. The existence of overflows, their location, and the possibilities of flooding basements and streets
3. State or local regulatory agency requirements
4. If located at or pumping directly to a sewage treatment plant, the ability of the treatment process to accept large increases in flow, or if recirculation of flow has been incorporated in the plant design

If there are adequate existing overflows, which will operate without flooding the sewer system and which may be permitted to discharge to watercourses during storms, the design flow need not exceed the maximum dry weather flow by a large amount (20 percent should be sufficient). If these favorable conditions do not exist, then it may be essential to pump from 2.5 to 3.0 times the average dry weather flow to the treatment plant during storms, depending on the amount of storm water reaching the sewers. Some of this flow may be discharged through plant bypasses, either before or after partial treatment.

Pumping stations serving combined-sewer systems may contain two sets of pumps: one set to pump the dry weather sanitary flow to a treatment plant or to an interceptor leading to a treatment plant, and a second set of large storm water pumps to pump combined sewage and storm water during heavy rains to a suitable nearby discharge point on the natural drainage system of the area. The storm water pumps and their discharge conduit should have ample capacity to prevent flooding of streets and basements. Many such pumping stations have been built on combined-sewer systems, discharging storm water directly to watercourses without treatment. Recent investigations have established that such practices result in substantial pollution in many places, and research is currently under way to determine economical methods of reducing such pollution by a combination of storage and treatment.

General Considerations of Pump Selection

The small, remote station serving only a small part of the total tributary area may contain two identical units, either one capable of handling the maximum flow. The station alternates between no discharge and discharge at maximum rate. When substantial future growth of the drainage area is expected, space may be provided for one or two additional pumps. The large station, on the other hand, which pumps the entire sewage flow or major portion thereof to the sewage treatment plant, should be designed, so far as practicable, to pump continuously. The rate of outflow should change gradually in relatively small increments as the inflow to the station varies, in order that the sewage treatment plant may operate at maximum efficiency. This requires at least one pump with a capacity approximately equal to or slightly less than the minimum flow. In addition to this requirement, provision must be made for the maximum design flow, varying from 1.5 to 3.0 times the average dry weather flow. To allow for the performance of maintenance and repairs, this flow should be handled with the largest unit out of service.

To follow the variations in flow closely from minimum to maximum may require a large number of pumps of different sizes, if they are to operate at constant speed. Moreover, if the station discharges through a long force main with considerable friction loss, the smaller pumps may not be usable at maximum flow because of the increase in head. A good solution to this problem is the use of variable-speed pumps, or a combination of variable-speed and constant-speed pumps. In such cases, the number of units can be reduced by at least one and sometimes more. In addition, it is often found that the units may all be of the same size or, at most, of two sizes, which reduces the number of spare parts that must be carried in stock and may also result in more favorable pump prices initially.

Variable-speed drives are more expensive than constant-speed motors. They are also less efficient because of the slip losses. Since the horsepower required by the pump varies as the speed cubed, the actual power loss is much less than might be supposed by a consideration of the efficiency alone. The loss is not usually excessive when it is considered that variable-speed pumps result in smaller stations, smaller wet wells, and fewer starts and stops of units.

System Analysis

System analysis for a pumping plant is carried out to select the most suitable pump units and their operating points. It involves calculation of the system-head curves and the use of these curves in conjunction with the characteristic curves of available pumps.

System Head Curves The system-head curve represents the total head against which the pumps will be required to operate under various conditions of flow. It consists of the static head plus the friction losses in pipe and fittings plotted against flow. If there is negligible head loss due to friction, the system-head curve will be a straight line parallel to the x axis.

The static head is the difference in elevation between the free water surface in the wet well and the free water surface at the point of discharge. Since the water level in the wet well will vary between the float-switch settings calling for pumps to start and stop and since the discharge water surface may also vary, the static head will vary, resulting in a family of parallel system curves. Usually the maximum system curve corresponds to the lowest wet-well level, and the minimum system curve corresponds to the highest wet-well level. Friction losses in the station piping and force main may be computed by the Darcy-Weisbach or Hazen-Williams equations. Entrance and exit losses and losses in valves and fittings must be included and are most satisfactorily estimated as fractions of the velocity head by means of the equation

$$H = K \frac{v^2}{2g} \tag{6·4}$$

For appropriate values of K, the student should consult standard textbooks and reference works on hydraulics.

To compute the friction losses, the sizes of pipes and fittings must first be assumed. The design flow and the number of pumping units should be estimated (at least tentatively), and pipe sizes should usually be selected to give velocities between 3 and 8 fps. The highest velocities are used with the biggest units. Velocities ranging from 4 to 6 fps are desirable but not always economically feasible.

Operating-Point Determination for Single-Pump Operation The pump characteristic curves show the relationship between head, capacity, efficiency, and brake horsepower, over a wide range of possible operating conditions, but they do not indicate at which point on the curves the pump will operate. The operating point is found by plotting the head-capacity curve on the system curve plot. The pump will operate where the two curves cross. If too conservative a friction factor is used in computing the system curve, the pump may operate farther out on its head-capacity curve than is intended. In extreme cases, this may result in a considerable loss of efficiency, an overloaded motor, and possible cavitation. Such conditions can be anticipated and guarded against by plotting system curves using friction factors for new pipe in addition to the system curves based on design C values. The bep should lie within the family of system curves where the curves are intersected by the pump curve.

Multiple-Pump Operation In stations where two or more pumps may operate either individually or in parallel discharging into the same header and force main, an alternative computation method is recommended, as follows:

1. The friction losses in the suction and discharge piping of individual pumps are omitted from the system curve.
2. Instead, these losses are subtracted from the head-capacity curves of the individual pumps to obtain modified pump curves, which represent the head-capacity capability of the pump and its individual valves and piping combined.
3. When two or more pumps operate in parallel, the combined-pump operating curve is found by adding the capacities of the modified pump curves at the same head. The point of intersection of the combined curve with the system curve gives the total discharge of the combination of pumps and the modified head at which each operates. By entering the modified curves of each pump at this head, determinations can be made of the capacity contributed by each pump, the efficiency of each pump, and the brake horsepower required under these conditions.

To find the total head at which each individual pump will operate, one must proceed vertically at constant capacity from its modified pump curve to its actual head-capacity curve. The pump specifications or purchase order must be drawn in such a way that the pump will produce this head. Each pump can operate at several points on the head-capacity curve, the head increasing and the discharge decreasing as more pumps go into operation. An effort should be made to limit these operating points to a range of flows within ± 25 percent of the flow at the bep.

The selection of multiple constant-speed pumps and the selection of a combination of constant-speed and variable-speed pumps is illustrated in the following two examples.

EXAMPLE 6·1 *Selection of Multiple Constant-Speed Pumps*

Select two or more constant-speed pumps for installation in a hypothetical station with the following requirements. The present average flow is 20 mgd, the minimum flow is 40 percent of the average flow, the maximum flow is two times the average flow, and estimated future flows are assumed to increase 50 percent. The losses in the station header and force main are assumed to be 20 ft at future maximum design flow. The static head is assumed to be 36 ft with the wet well full and 40 ft with the wet well drawn down. Pumping is to be continuous and is to match the inflow to the station reasonably well.

Solution

A possible selection of multiple constant-speed pumps is shown in Fig. 6·7. Pump *A* cannot be used under maximum head conditions. With the selection of pumps shown, it will be necessary to install either larger impellers in all units, or larger pumps, to handle the estimated future range of flows satisfactorily. To provide a spare unit under maximum future design conditions, one additional pump of the largest size will be required. These changes can be made gradually as flows increase, or all at once, depending on the maintenance capability and financial policy of the owner.

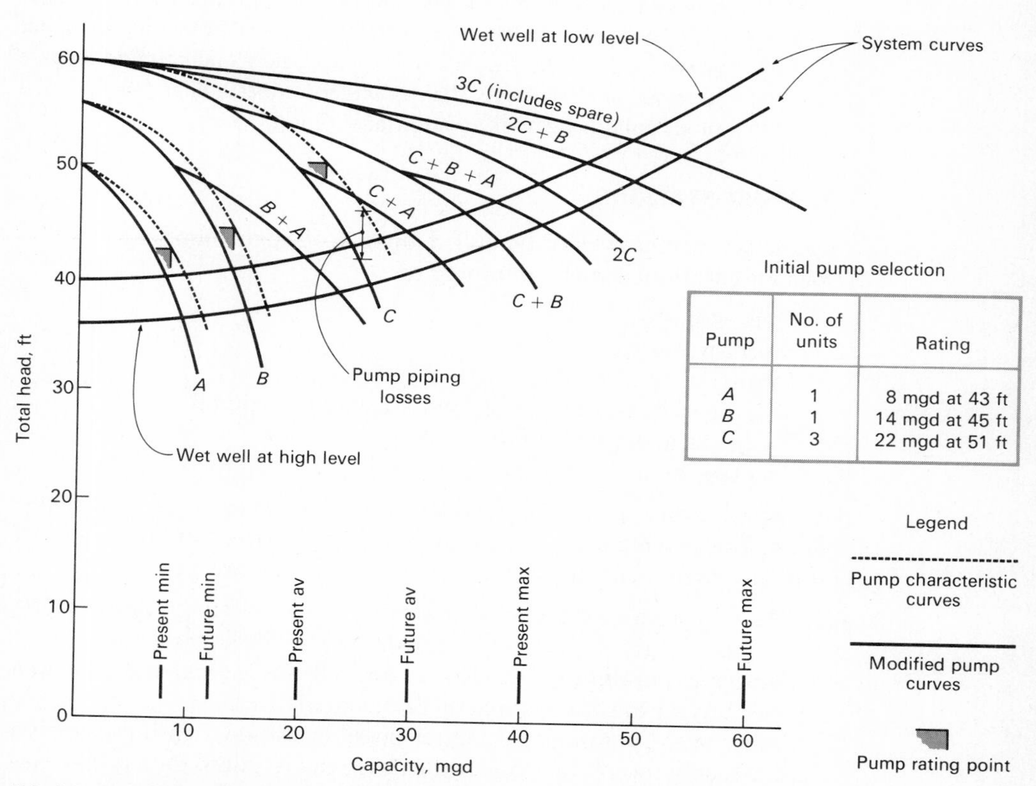

Pump	No. of units	Rating
A	1	8 mgd at 43 ft
B	1	14 mgd at 45 ft
C	3	22 mgd at 51 ft

FIG. 6·7 Selection of constant-speed pumps for a hypothetical pump station.

EXAMPLE 6·2 *Selection of a Combination of Constant- and Variable-Speed Pumps*

The A and B pumps in Example 6·1 are to be replaced with another C unit operating at variable speed, and one of the other three C units is to be made variable speed so that there will be four identical pumps, two driven by constant-speed motors and two by variable-speed motors. It is intended that one of the variable-speed pumps will be held in reserve as a spare at all times. The affinity laws can be applied to modified pump curves, since both the head developed by the pump due to a speed change and the head losses in suction and discharge piping vary as the square of the flow.

Solution

The solution is shown in Fig. 6·8. Note that the maximum capacity of the station has been increased since all pumps are usable at maximum head. One additional constant-speed pump of the same size and an increase in maximum wet-well level of 1 ft, quite possible with an increase of 50 percent in the flow in the incoming sewer, will handle the future estimated flow and provide a spare unit. As an alternative, slightly larger impellers might be installed in all units instead of an additional pump.

PUMPING STATIONS

A sanitary engineer may be called upon to design pumping installations and stations for the pumping of

1. Domestic sewage
2. Storm water runoff
3. Industrial wastes
4. Combined domestic sewage and storm water runoff
5. Sludge at a sewage disposal works
6. Treated domestic sewage

Apart from pumping facilities required at sewage treatment plants, the principal conditions and factors necessitating the use of pumping stations in the sewage collection system are as follows:

1. The elevation of the area or district to be serviced is too low to be drained by gravity to existing or proposed trunk sewers.
2. Service is required for areas that are outside the natural drainage area but within the sewage or drainage district.
3. Omission of pumping, although possible, would require excessive construction costs because of the deep cuts required for the installation of a trunk sewer to drain the area.

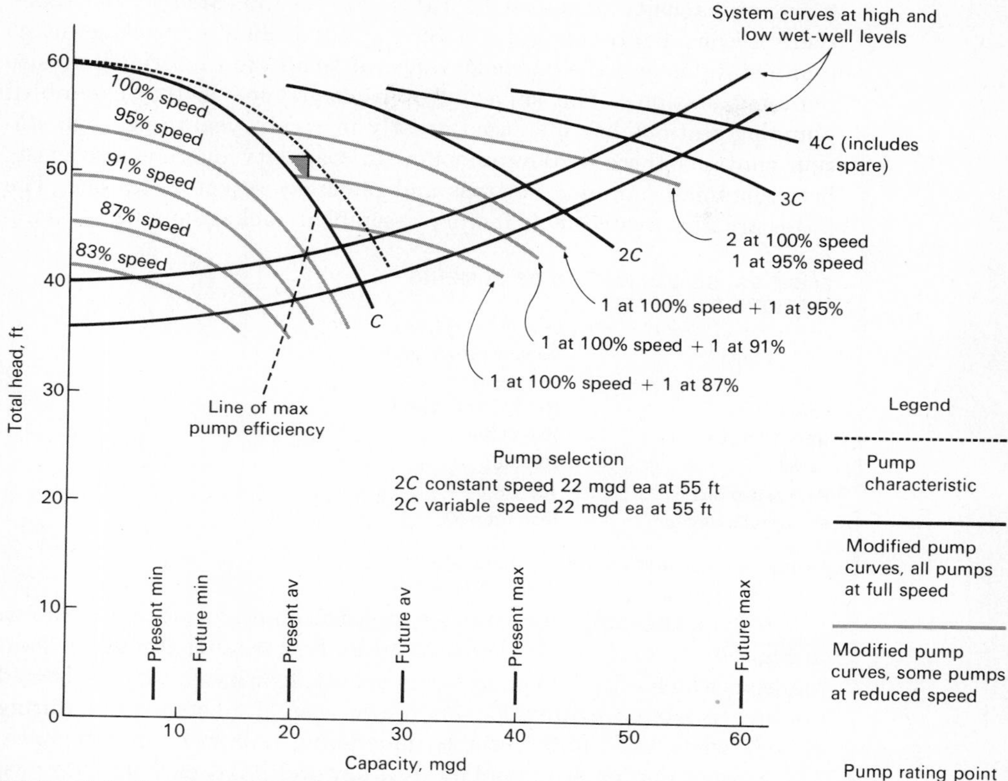

FIG. 6·8 Selection of a combination of a constant-speed and variable-speed pumps for a hypothetical pump station.

Pumping plants of modern design are almost always completely automatic in normal operation. The smaller stations are normally unattended and require little attention other than a daily check on the proper functioning and lubrication of the equipment and for removal of screenings where a hand-cleaned rack is provided instead of a comminutor. The larger stations, especially those with standby power equipment, are usually attended stations, but operating staffs are small. In most cases, one or two men per shift are sufficient for operation.

Types of Pumping Stations

Pumping stations have been classified in a variety of ways—none entirely satisfactory—such as by capacity (gpm, mgd), source of energy (diesel, steam, or electric), or method of construction. A classification based on capacity is shown in Table 6·3. As indicated in the table, there is considerable overlap in the capacity range of package-plant pump stations and the capacity range of small- and intermediate-size pumping stations. The capacity of package-type factory-assembled pumping stations has increased greatly in recent years (e.g., one station contains three 4,200-gpm pumps). Capacity overlap also exists between small pumping stations and pneumatic-ejector stations. The latter are also available as factory-assembled package plants.

TABLE 6·3 CLASSIFICATION OF PUMPING STATIONS BY CAPACITY

Class	Capacity range, gpm
Large	15 mgd and over*
Intermediate	700–10,000
Small	200–700
Package-plant	100–1,600+
Pneumatic-ejector	less than 600

* Large pump stations are rated in mgd.

A typical example of a modern sewage pumping station is shown in Fig. 6·9. Sewage enters the wet well by first passing through a comminutor, which is located on an intermediate floor level. A hand-cleaned bar screen has been provided for use in case of emergency or during periods when the comminutor is undergoing repairs or maintenance. The sewage pumps are installed in a dry well, and each has its own individual suction-pipe connection to the wet well. The motors are located at ground floor level for maximum protection against flooding and drive the pumps through vertical shafting of the universal-joint flexible type. Seepage, sealing water leakage, and washdown water are collected in a gutter and sump and pumped back to the wet well.

The design shown in Fig. 6·9 provides space for chlorination equipment, but it is not often installed. Chlorination facilities may be installed in stations that discharge through long outfalls or force mains that provide sufficient contact time for effective destruction of bacteria.

Smaller stations may have the motors mounted on the pumps or on intermediate floors. Access to both the wet well and the pump room may be through manholes or hatches and by vertical ladders. If the superstructures are omitted, they may be installed underground within the street lines, preferably with access manholes in sidewalks or grass plots, but sometimes within the traveled way.

A package plant is shown in Fig. 6·10. Access is through a 3-ft-diameter tube located in a grass plot. The tube projects above the

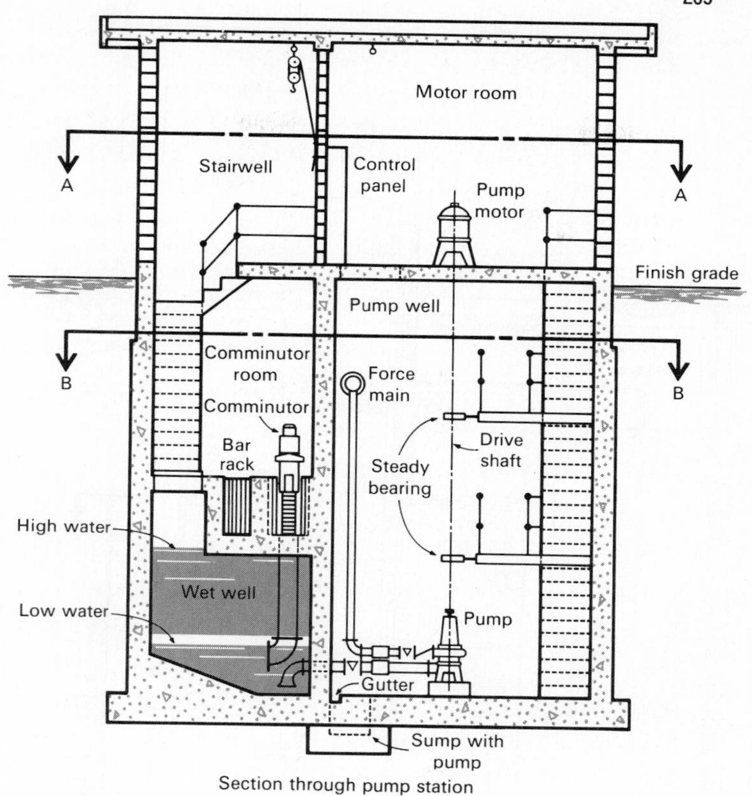

Section through pump station

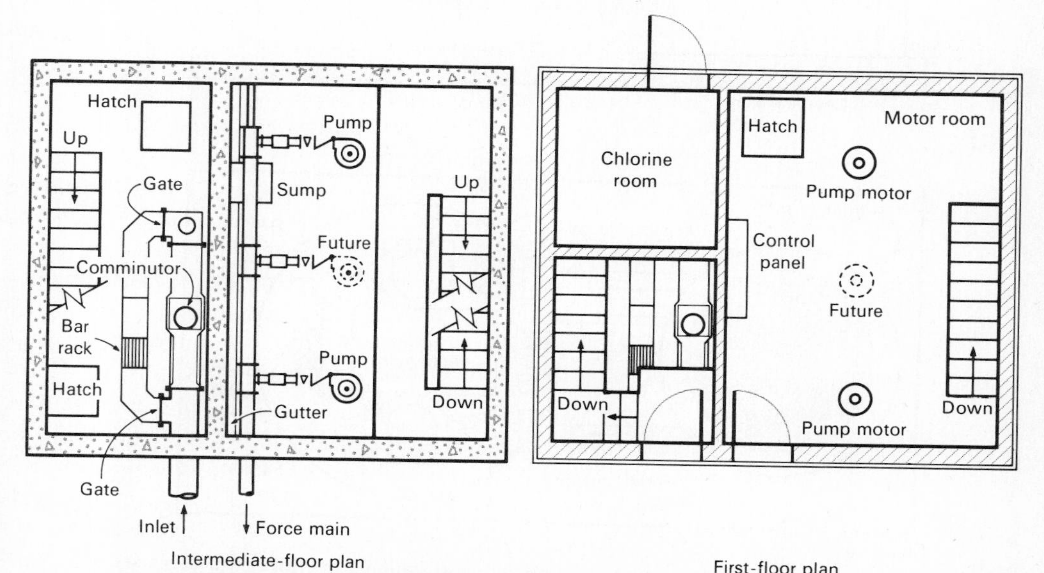

Intermediate-floor plan
SECTION B–B

First-floor plan
SECTION A–A

FIG. 6·9 Typical modern sewage pumping station.

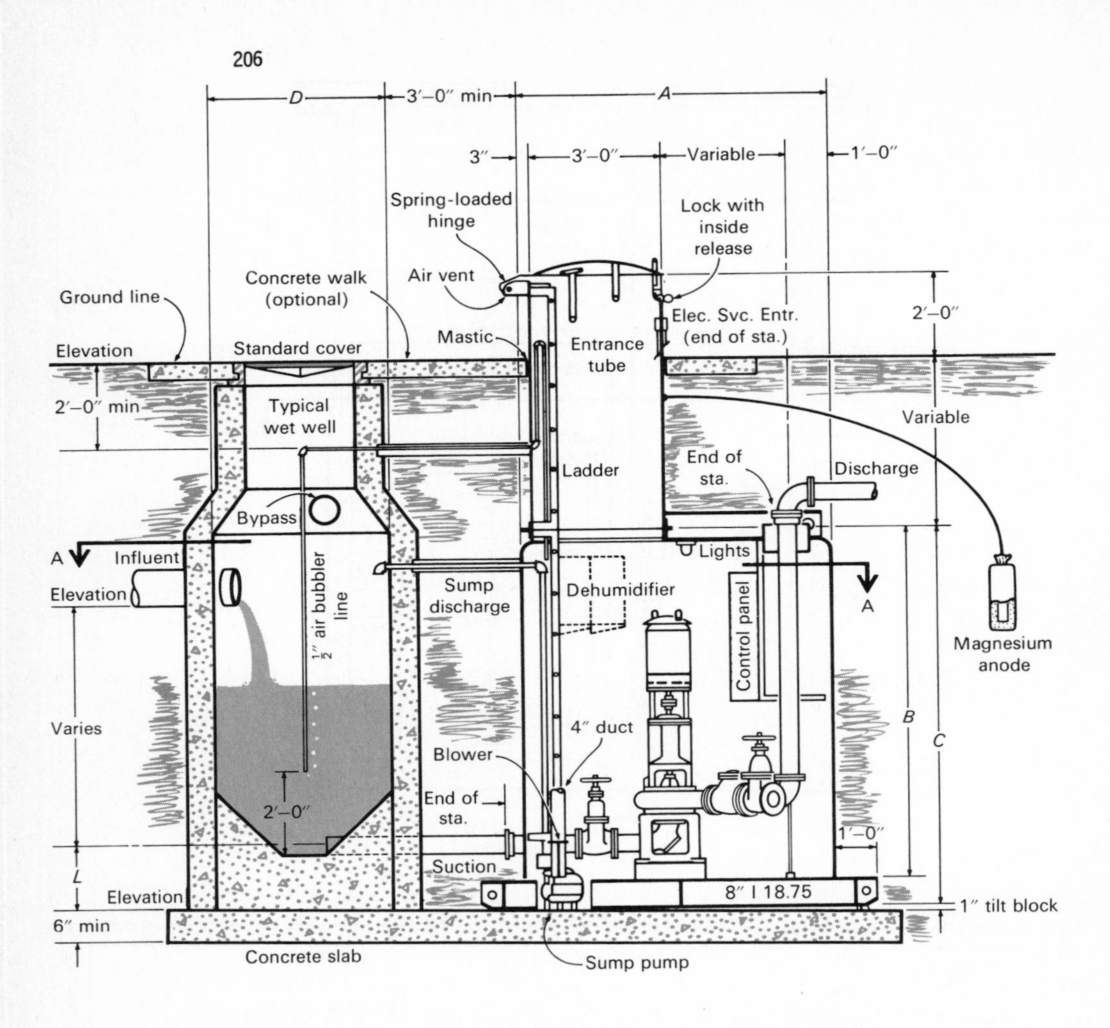

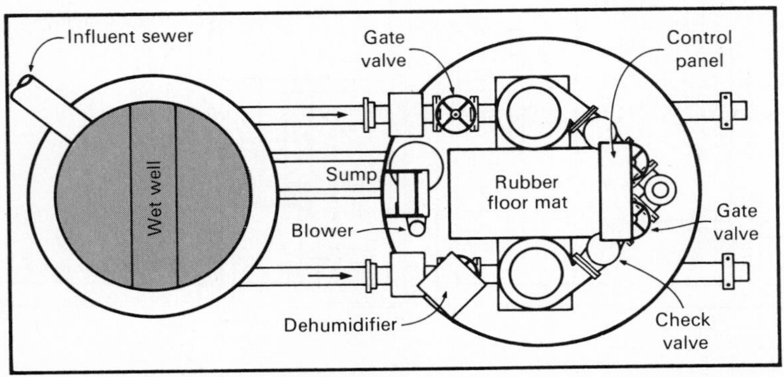

SECTION A–A

FIG. 6·10 Plan and section of a two-pump package pumping station [from Smith and Loveless].

surface to provide ventilation and to keep rainwater out. The electric control panel is located within the station, but the meter and disconnect switch are located on an electric pole nearby. An oversize manhole is constructed of masonry adjacent to the station to serve as a wet well. Magnesium anodes, as shown, are normally installed to protect the steel shell of the station from corrosion.

A typical pneumatic-ejector station that might be constructed in city streets where space is at a premium is shown in Fig. 6·11. The compressor is mounted on top of the ejector and there is no storage tank. When conditions permit, the compressors and air storage tank are installed above ground, directly over the ejector pots or in their general vicinity, with the air piped underground. In industrial installations, the ejectors may be supplied from the plant air supply.

The Chicago Pump Co. has developed a design for a two-pump station using "Flush-Kleen" sewage ejectors, in which the sewage flows into the wet well through a check valve and screen in the pump discharge and through the pump in the reverse direction. The solids in the sewage are caught on the screen and flushed out through the discharge line when the pump operates. The pumps are always installed in duplicate, with each pump sized for the maximum flow, so that when one pump is operating, the flow may find its way to the wet well through the idle pump. A screened overflow to the wet well is provided for flows in excess of the anticipated maximum. The pumps themselves are conventional nonclog pumps.

In general, wet-pit pumps are not recommended for pumping raw sewage on account of the difficulties of inspection and maintenance. In the usual case, the discharge pipe is brought up through a steel plate that supports the pump, and the entire unit, including the pump, motor, and discharge pipe, must be hoisted through a hatch in the station roof for maintenance. Wet-pit designs using submersible close-coupled pump and motor assemblies, which can be installed in underground vaults or manholes and easily hoisted above ground level for maintenance, are a recent development. These are suitable for small or temporary installations and, where applicable, reduce the cost to the absolute minimum.

Also deserving of mention are designs incorporating self-priming nonclog pumps in a building at ground level built over a wet well. This type of pump is presently available from several manufacturers.

Details of Pumping Station Design

The pumps normally will be vertical-shaft, single-suction units, installed in a dry well, with motors mounted on top of the pumps (pack-

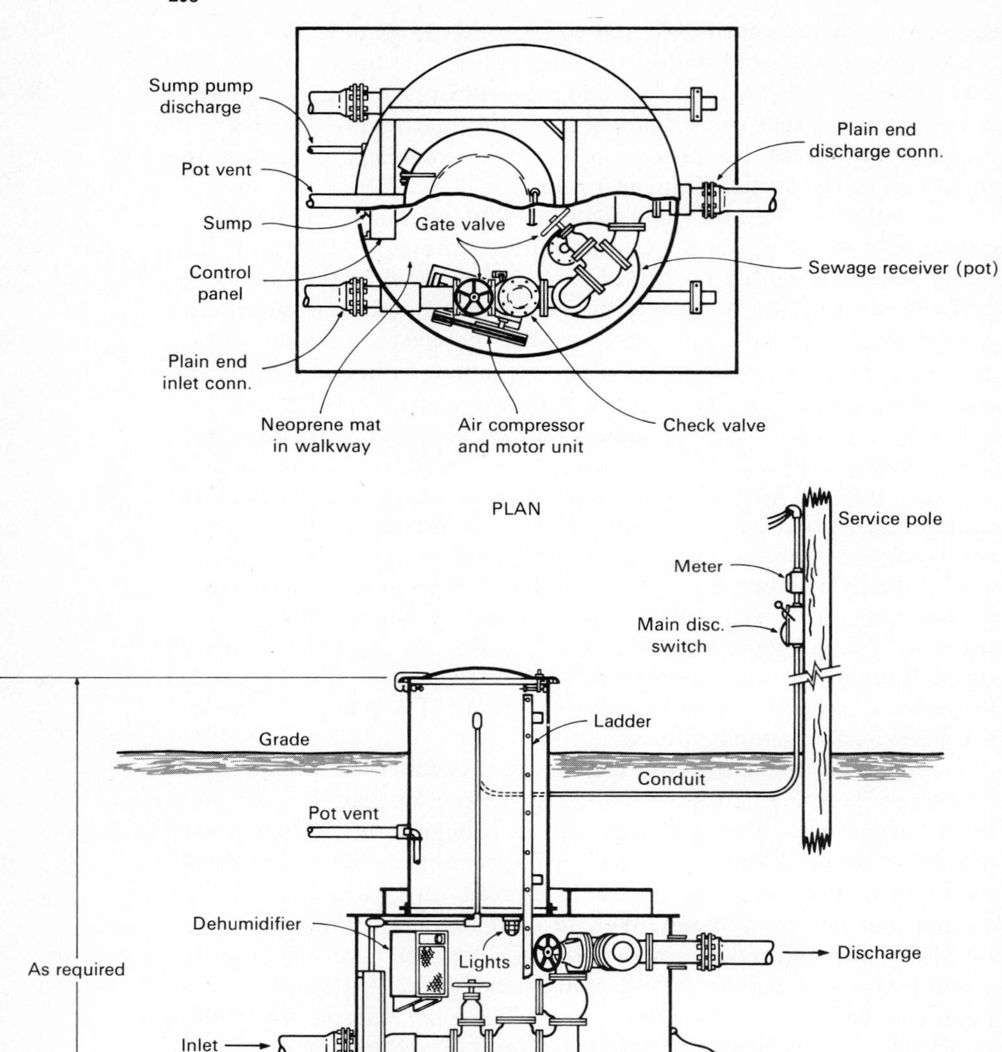

FIG. 6·11 Typical pneumatic-ejector installation in underground vault [from Zimmer and Francescon, Inc.].

age stations or very deep stations) or on an upper floor, preferably at ground level as in Fig. 6·8. Each pump will be provided with its own individual pipe connection to the wet well. Discharge normally will be to a common force main, although in the case of lift stations and stations located at treatment plants, individual discharge pipes to a gravity sewer or channel may be used.

Suction and Discharge Piping The velocity at the pump suction and discharge nozzles normally will be in the range from 10 to 14 fps. If the velocity is more or less it is highly probable that a better pump selection could be made. Pumps with much higher discharge velocities should not be accepted (but are sometimes bid). It is recommended practice for the discharge pipe to be at least one pipe size larger than the discharge nozzle and for the suction pipe to be one or two sizes larger than the suction nozzle. Most sewage pumps, however, have suction and discharge nozzles of the same size, but occasionally the suction nozzle may be one size larger.

Desirable velocities in the discharge pipe at maximum pump discharge are from 6 to 8 fps. A concentric increaser should be provided on the pump discharge, followed by a check valve and a gate valve. The gate valve should be solid wedge, outside screw and yoke (OS&Y) type preferably. The check valve will be one of the following types:

1. Swing check, preferably with outside lever and weight, with weight adjusted to assist closing. The valve should be placed in a horizontal run of pipe and in a direct line with the discharge nozzle. It should be of a type permitting the unobstructed flow of sewage in the full open position. This will be the normal arrangement for small pumps.
2. Cone, which is excellent but expensive. Cone valves may be the same size as the pump nozzle to decrease the cost, or even smaller if space is available for both a reducer and an increaser.
3. Butterfly, which is usable only with settled sewage or when screenings are removed or reliably comminuted.

Cone and butterfly valves require a reliable source of power (water or an oil accumulator system) for operation and must be correctly timed. Otherwise, with long force mains, fast closure may break the water column and result in serious water hammer when the column returns against a closed valve (see "Details of Force Main Design").

Electrically operated gate valves are sometimes used in large, attended stations. In this case, it is essential to have a reliable standby power source, as well as an auxiliary means, such as a second gate valve, for shutting off the flow in an emergency.

In large lift stations at sewage treatment plants, where the elevation of the water surface in the discharge channel does not vary greatly, the pumps may be provided with individual unvalved discharge pipes discharging over siphons equipped with vacuum breakers to prevent reverse flow. An alternative arrangement would provide for individual discharge over weirs, with the weir crest set above the maximum water level in the discharge conduit. Large low-head pumps sometimes discharge into individual chambers with provisions for stop logs, the inlet pipe being provided with a flap valve or backwater gate. These arrangements are also applicable to large storm water pumping stations.

Desirable velocities in pump suction piping are 4 to 6 fps. An eccentric reducer with the flat side uppermost is located ahead of the pump suction nozzle if the pump includes a suction elbow. If the suction elbow must be furnished as part of the piping, a reducing elbow, preferably with a long radius, may be installed under the pump. A gate valve should be installed between the pump and the wall of the wet well to permit opening the pump without flooding the pump room. The gate valve is normally solid wedge, OS&Y. If fear of flooding due to a cracked pump casing is a factor, the suction valve may be operated by a floorstand or valve box on an upper floor. In large, deep stations, the suction valves may be hydraulically operated by four-way valves located at ground floor level.

Wet-Well Design The end of the suction pipe in the wet well normally will be provided with either (1) a 90 or 45° flange and flare elbow, or (2) a 90 or 45° flanged elbow and a flange and flare fitting as shown in Fig. 6·12. If D is the diameter of the flared inlet, then the lip of the flare at mid-height should be not less than $\frac{1}{3}D$ nor more than $\frac{1}{2}D$ above the floor. An inlet flush with the wall is sometimes used where adequate submergence is available to prevent air being drawn into the suction pipe through vortices. The submergence required above a flared inlet is approximately as follows:

Velocity, at diameter D, fps	*Required submergence*, ft
2	1
5	2
7	3
11	7
15	14

The floor of the wet well should be level from the wall to a point 12 to 18 in. beyond the outermost edge of the suction bell and should

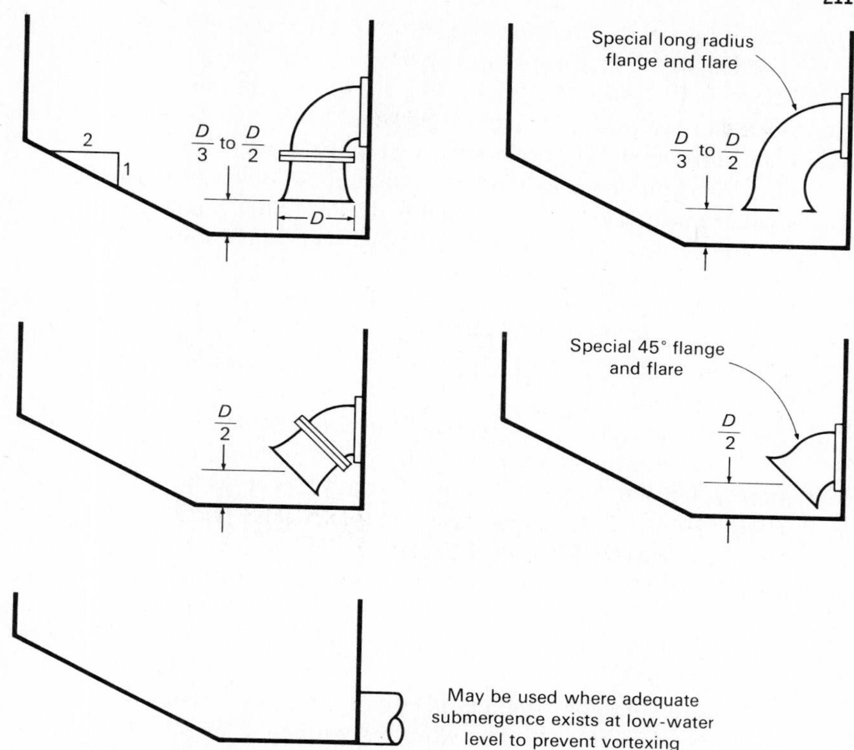

FIG. 6·12 Pump suction connections to wet well.

then slope upward to the opposite wall at a slope of approximately 2 horizontal on 1 vertical.

The volume of the wet well between start and stop elevations for a single pump or a single-speed control step for variable-speed or multi-speed operation is given by the following equation [10]:

$$V = \frac{\theta q}{4} \tag{6·5}$$

where V = required capacity, gal

θ = minimum time of one pumping cycle or time between successive starts or speed increases of a pump operating over the control range, min

q = pump capacity, gpm, or increment in pumping capacity where one pump is operating already and a second pump is started, or where pump speed is increased

The minimum cycle time for single-pump operation occurs when the inflow is exactly half the pump capacity. Under these conditions the "on" and "off" times are equal. The pump will be on a longer time and off a shorter time for larger inflows and vice versa for smaller inflows; in both cases, the cycle time will be greater.

For large pumps and motors, θ should be not less than 20 min. For smaller pumps, θ can be reduced to 10 min, but 15 min is desirable. Where this requires an undesirably large wet well in a small station containing two identical pumps, one of which is a spare, the wet-well volume can be cut in half by installing an automatic alternator in the pump control circuit. This will start and run the pumps alternately, which has the effect of making θ for the wet well half the effective θ for the pumps and motors.

Some concern has been expressed in the literature over the possibility of septic conditions developing with resultant odors due to excessive retention times in pumping station wet wells, and a maximum retention time of 30 min has been suggested [4]. This is a problem only in warm climates and then only when the time of flow in the tributary sewers is excessively long. Odors will be minimized if the floor of the wet well, including the sloping portions, is kept covered at all times by proper selection of the stop or cutout point for the first pump in sequence.

The more common problem is one of obtaining sufficient wet-well volume at reasonable cost. The effective volume V of the wet well between "on" and "off" float-switch settings includes the storage in the incoming sewers. If the "on" switch setting is below the sewer invert, no storage is available. This setting is wasteful of both the storage capacity and head available in sewers of appreciable size. It also wastes power and may require higher head pumps and larger horsepower motors. Where it is necessary to rely on the storage in the sewers to obtain adequate wet-well volume for control, backwater curves should be computed to obtain the effective volume between the various float-switch settings. This may amount to 50 percent of the volume in the wet well itself. Some stations have been built with practically no wet well; and, in the case of really large stations, the wet-well volume within the station approaches insignificance. In these cases, the only volume available for control is the storage volume in the sewers.

Screening devices normally are included in the wet well or in an adjacent screen chamber to protect the pumps against clogging. Manually raked bar racks may be used in small stations. Mechanically cleaned bar racks, screenings grinders, and comminutors are in common use in the larger stations. These should be installed in duplicate, or a manually cleaned bypass rack should be installed to permit servicing the equipment without shutting down the station. The floor of the screen chamber should be above maximum high-water level in the wet well.

Pump Settings The pumps should be set so that the high point of the casing is below the minimum level of the sewage in the wet well. This setting assures that air cannot enter the pumps during the down period and that they will be full of sewage when started automatically. It also avoids the use of a vacuum priming system, which is difficult to maintain in satisfactory operation on raw sewage.

Pumps handling settled sewage or plant effluent may be provided with an air valve or a $\frac{3}{4}$-in. vent pipe extending from the top of the volute through the wall of the wet well at a high level. This arrangement prevents the buildup of air pressure within the pump and assures equalization of the water levels in the wet well and in the pump casings. Under these conditions, the start level may be at the top of the pump casing and the stop level approximately 3 ft lower. This arrangement is sometimes used for pumping raw sewage, but the possibility of fouling of the air valve or plugging of the vent is always present, in which case the pump may lose its prime and operate dry. Unless devices are installed (such as mercury switches on the check-valve levers) that will sense failure of the pump to deliver flow, shut it down, and sound an alarm, this arrangement is not recommended.

Pumps should be lined up and equally spaced for a neat installation, and there should be ample clear space for access and maintenance. A minimum clear space between pump casings of 3 to 4 ft is recommended. If space for a future pump is provided, it should not be at the end of the station farthest from the inlet, because the end of the wet well may fill up with sewage solids. Note that in the station shown in Fig. 6·8 the space for the future pump is in the middle.

The available net positive suction head (NPSH) must be greater than the NPSH required by the pump at the operating point to avoid cavitation. With heads of 60 ft and below, this is not often a problem; but, where the head is greater than this or where the pump is operating under a suction lift or far out on its curve, the NPSH should be checked. If requested, manufacturers will plot the required NPSH on the pump characteristic curves. Preliminary estimates may be obtained from data in Refs. [2, 11, and 12]. If the pump operates at low head at a capacity considerably greater than the capacity at the bep, the following equation holds approximately:

$$\frac{\text{NPSH at operating point}}{\text{NPSH at bep}} = \frac{Q^2 \text{ at operating point}}{Q^2 \text{ at bep}} \tag{6·6}$$

The NPSH required at the bep increases with the specific speed of the pump. For high-head pumps it may be necessary to limit the speed to obtain adequate NPSH at the operating point or to deepen the station. Higher speeds also mean cheaper units and result in more wear on the equipment. Greater troubles from vibration and misalignment, particularly with rags in the pumps, may also result.

Automatic Controls The wet-well level can be sensed by (1) floats, usually installed in float tubes located in the dry well, (2) pneumatically by a bubble-tube system, or (3) electrically by probes located in the wet well. Equipment of each type is available from several manufacturers. Less difficulty in operation has been experienced with the first two systems than with the electrical probes.

Several different control schemes are available, such as constant-level, variable-level or step control, and stepless variable-speed control. Step control is the simplest and is standard for a small number of constant-speed pumps. It can also be used for step control of variable-speed or multispeed pumps.

A control range of at least 3 ft is desirable between maximum and minimum wet-well levels. The maximum wet-well level will be at the 0.8 depth of the incoming sewer, less an allowance for losses through bar racks or comminutors. High-water alarm will be 6 in. above this elevation. Minimum wet-well level will be at the top of the volute of the largest pump. Low-water cutout and alarm should be not less than 12 in. below the top of the volute nor lower than the top of the suction pipe. Manual operation of all pumps by push-button control should be possible for checking the operation and for drawing down the wet well. Manual operation should bypass the low-water cutout but not the low-water alarm.

A three-pump control scheme is shown in Fig. 6·13. The lead pump operates between elevations 0.0 and +2.0, starting at +2.0 and stopping when the wet well is pumped down to 0.0. If the flow is greater than one pump can handle, the wet well rises to +2.5 and starts the second pump. The two pumps running together will pump the wet well down to +0.5. At this point, the second pump will stop; the first pump will continue to run. A further increase in flow will start the third pump at elevation 3.0. The increment in level between start points should not be less than 3 in. and need not be more than 6 in. The drawdown (the distance between "start" and "stop" points) should not be less than 12 in. and normally should be greater to keep the area of the wet well from becoming excessively large. This area is determined by dividing the volume V from Eq. 6·5 by the drawdown. Float switches are available with a large number of speed points,

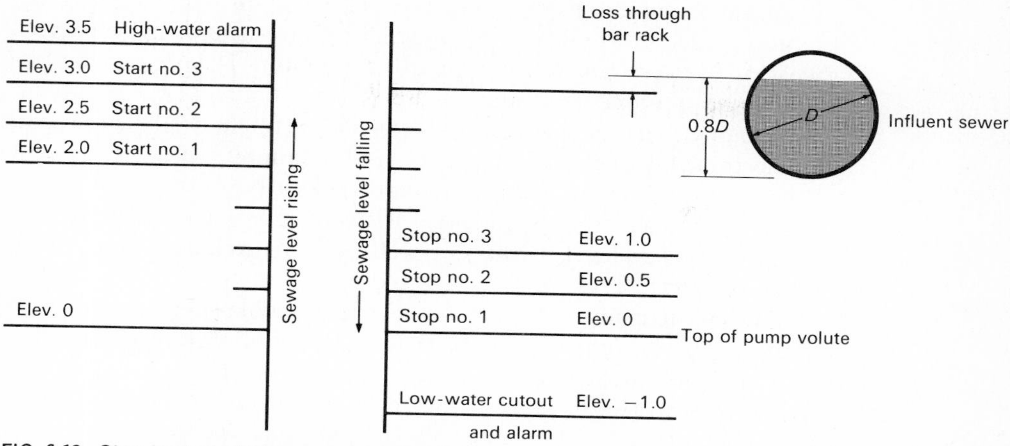

FIG. 6·13 Step control of three constant-speed pumps.

with uniform drawdown on all speed points except the first obtained by means of a lost-motion clutch, and with a separately adjustable drawdown on the first point to limit on-and-off operation of the lead pump at minimum flows.

In the constant-level system, the wet-well level is maintained between a set upper and a set lower level. These levels can be adjusted manually or automatically in proportion to flow or number of pumps in operation. While the level is between the two limits, there is no change in pump operation. When the level reaches the upper limit, a timing motor is energized to move a master controller one step, starting another pump or increasing the speed one step. At the lower level, the timing motor is reversed, decreasing the speed one step or stopping a pump. These controls are considerably more complicated and must be custom designed. They also require more maintenance. They are sometimes necessary, where a large number of control points are required, but the control range is small.

Stepless variable-speed controls operate to keep the wet well between two fixed levels by increasing the speed of the pumps gradually as the sewage level rises. At the lower level, the pump is at minimum speed and at the upper level the pump is at maximum speed. When the maximum level is exceeded by a fixed amount, usually 6 to 12 in., a second pump is started either manually or automatically. If the second pump is a constant-speed pump, control of flow and wet-well level is by speed variation of the first pump. If the second pump is a variable-speed unit, control will be by speed variation of both pumps operating over the same or different control ranges.

With magnetic drives, speed control is obtained by varying the excitation current applied to the magnet by means of a float-operated rheostat. Stepless control of wound-rotor motors is obtained by the insertion of a liquid resistance in the rotor circuit. The Flomatcher (trade name) system comprises a two-compartment tank containing the resistors and an electrolyte located in the pumping station adjacent to the electrical control board. A compressed-air bubbler system varies the level of the electrolyte in contact with the resistors in step with the change in sewage level in the wet well. A heat exchanger is required to cool the electrolyte.

Automatic control systems for several variable-speed pumps may require detailed study to ensure that pumps will start and run at a speed that will deliver flow against the existing head and that, on decreasing flow, two or more pumps do not continue to run at reduced speed when one less pump could do the job more efficiently.

Electrical Equipment Small station pumps and equipment will operate at 220 or 440 volts, three phase. The pumps in large stations may operate at 2,300 or 4,000 volts with auxiliary equipment on 440 volts. Important stations should be supplied by two separate feeders from separate substations of the power company. Motors smaller than 0.5 hp normally will operate off the lighting circuits. Power transformers will be installed in an outdoor fenced enclosure or on poles.

The motor starters and controls should be located in a factory-assembled freestanding unit-control center located at ground floor level, as shown in Fig. 6·9. This type of construction is far neater, safer, and more satisfactory than the wall-mounted assemblies of individual starters and circuit breakers used in the past. Large stations will include a separate electrical room containing the motor starters, switch gear, meters and instruments, and a control benchboard.

Electrical equipment and lights in the wet well should be explosion-proof, because of the possible danger from gasoline vapors and gases in the incoming sewage. Adequate lighting and a convenient number of equipment receptacles for power tools should be provided.

Heating and Ventilating All pumping stations except those entirely below ground or located in warm climates should be supplied with automatically controlled heaters to prevent freezing in cold weather. Thermostatically controlled strip or electric unit heaters are suitable for small stations. Gas- or oil-fired heaters may be used for larger stations. Comfortable temperatures should be maintained in attended stations, both summer and winter. Wet wells are usually unheated, especially if entirely below ground level. Under some temperature conditions, condensation on walls and fog may be severe. This can be eliminated or greatly reduced by the installation of a thermostatically controlled, explosion-proof unit heater.

Ample fresh-air ventilation is essential. Wet wells should have a positive supply of fresh air furnished by a centrifugal blower discharging across the screen room operating area and a gravity exhaust duct with low-level intake at the opposite end of the wet well discharging through the roof. The blower should have sufficient capacity to provide a 2-min air change (30 air changes per hour), based on the wet-well volume below grade and above the minimum sewage level. Such blowers normally are operated only when the attendants visit the station; hence a large capacity is required to flush out accumulated stale air and gases quickly for the operators' protection. Three-way control switches should be located just inside the door and in the screen room. As an extra precaution, the blower can be interlocked with the light switch or operated by a door switch. If the blower is equipped with a two-speed motor, it may be operated continuously on low speed and on high speed only when the station is inspected. Such operation might prevent the periodic flushing of sewage odors into the atmosphere, which could be objectionable in built-up areas.

The ventilation systems of the wet- and dry-well sides of the station should be entirely separate, and all openings for pipes or electric cables should be caulked gas-tight.

The dry well should be positively ventilated using either a supply or exhaust system with a capacity of 30 air changes per hour (based on the dry-well volume below grade) for unattended stations where operation would be intermittent. Intake or exhaust grilles of ample area, equipped with motor-operated dampers interlocked with the supply blower or exhaust fan, should be provided to ensure a positive flow of air when the fan is running. In attended stations, the system may be designed for six air changes per hour and operated continuously [4]. Additional ventilation capacity may be necessary to remove the heat generated by the pump motors or generator sets. In this case, ventilation equipment may be interlocked to operate whenever a pump is running.

Pump Station Construction Pumping station substructures should be of reinforced-concrete construction, and exterior walls below grade should be tar coated to prevent leakage. Superstructures should be designed to harmonize with the surroundings, should be of fireproof construction, and preferably should not have windows if unattended. Other items that should be installed in pumping stations include (1) adequate floor openings, doorways, or floor hatches for installation, removal and replacement of equipment; (2) eyebolts or trolley beams for hoisting equipment over motors and floor hatches (bridge cranes for servicing equipment in large stations); (3) floor drains and gutters to collect seepage and stuffing-box leakage and a small submersible sump pump to pump it back to the wet well; (4) equipment guards and railings for floor openings; and (5) rubber floor mats in front of electrical panels.

Stairs should be installed in all but the smallest stations. In these, vertical ladders or ship's ladders may be used. Ladders should be made of galvanized steel or aluminum of satisfactory rigidity.

Details of Force Main Design

Theoretically, the most economical size of force main should be selected on the basis of power costs for pumping plus annual charges for force main and pumping equipment. As a practical matter, the problem is generally resolved by selecting a force main with adequate velocity to prevent deposition of solids at minimum flow, if possible, and pumping equipment with the ability to pump the desired flows at the heads required by the force main.

The greatest difficulty is encountered in the case of long force mains. The best approach in this case is to first select the most economical size of force main with adequate carrying velocities for the entire range of present and future flows. Then various pumps are selected. At this point, it may be found that a larger force main is necessary to reduce the friction losses so that a reasonable selection of pumps can be made. Where the difference between present and future design flows is particularly great, it may be necessary to build a smaller force main initially and to install a second at some later date.

Design Criteria Force mains will generally be 8 in. in diameter or larger. In some cases, 6-in. pipe may be used for small pump stations and short force mains, and 4-in. pipe may be used for small ejector stations. The following values of C are recommended for use in the Hazen-Williams equation for computing the friction losses in force mains for design conditions:

C = 100 for unlined cast-iron pipe
C = 120 for cement-lined, cast-iron pipe, reinforced-concrete pipe, asbestos-cement pressure pipe and various types of plastic pipe
C = 110 for steel pipe, 20 in. or larger, with a bituminous or cement-mortar lining

For small- or medium-size stations serving only part of a treatment area, where flow may be pumped intermittently at any rate up to the maximum, the desirable force-main velocities are from 3.5 to 5 fps. The small station would have only two pumps, one of which would be a spare, and would discharge at maximum rate or not at all. The pump capacities required to maintain velocities of 2.0 and 3.5 fps in 6-, 8-, and 10-in. force mains are shown in Table 6·4. Solids will not settle out at a velocity of 2.0 fps, but solids in the wastewater remaining in the line when the pump stops will settle out. A velocity of 3.5 fps is desirable to ensure pickup of the deposited solids.

TABLE 6·4 PUMP CAPACITIES FOR MINIMUM FORCE-MAIN VELOCITIES

Force-main diameter, in.	Pipe area, sq ft	Pump capacity, gpm	
		$V = 2.0$ fps	$V = 3.5$ fps
6	0.196	176	308
8	0.349	313	548
10	0.545	490	860

In the small station with two pumps, it should be possible to run both pumps at once, even though only one is needed for design conditions. If flows are too small to warrant a 3.5-fps design, pumps can be selected to give 3.5-fps minimum velocity with both pumps running. In this case, both pumps should be operated together by manual control once a week for a sufficient length of time to flush out the line.

Larger stations of this type may contain three or four pumps, all of the same size, of which one is a spare. Velocities of approximately 3.0 and 5.0 fps might be selected for the three-pump stations with one and two pumps in operation and velocities of approximately 2.25, 4.0, and 5.5 fps might be selected for the four-pump station with one, two, and three pumps in operation. These velocities allow for a slight reduction in pump capacity due to greater friction losses at increased flows.

Force-main design usually becomes more complicated for stations (small, medium, or large) serving all or a major part of an area where it is desired to pump continuously, following the flow more or less closely. These pump stations may have two or three sizes of pumps, some of which may be constant-speed units and some variable-speed units. In general, the pumps should be sized to pump, in combination, the following rates of flow:

1. Initial and future minimum flows
2. 1.25 to 1.50 times the average initial and future design flows
3. Future maximum design flow

To provide for the future maximum flow, it may be necessary to plan for (1) larger impellers to be installed in the future, possibly with different speeds and different motors, (2) additional pumps, or (3) replacement of some or all of the original pumps with larger units.

The range in discharge and velocity to provide for the above rates of flow may be on the order of 7 or 8 to 1. If the maximum velocity in the force main is set at 8 fps, the initial minimum flow would then produce a velocity of only 1.0 to 1.1 fps in the force main. For continuous pumping, one pump would have to be sized for this flow. Inasmuch as 1.0 to 1.1 fps is not a self-cleansing velocity, there will be some settlement, but this can be accepted for the following reasons:

1. At minimum flow, the sewage is weakest and contains little grit. It is the grit that would be expected to settle out.
2. At daily peak flows, the pumping rate will be 1.5 to 2.0 times the daily average, resulting in a velocity of 3.8 to 5.0 fps, which will flush out any deposits daily.
3. The alternative solution of two force mains would be more expensive and operationally undesirable and is to be avoided if at all possible.
4. Pump stations and force mains designed on this basis have worked satisfactorily.

If possible, the force main should be designed without high points so that air-relief valves will not be needed. The top of the force main should be below the hydraulic grade line at minimum pumping rate.

Water Hammer in Sewage Force Mains When the power is suddenly cut off from a pump motor, the speed drops quickly, and a wave of reduced pressure travels up the pipeline to decelerate the flow according to Newton's second law. The velocity of the pressure wave is given by the following equation [7]:

$$a = \frac{4{,}720}{1 + (Kd/Ee)} \tag{6·7}$$

where a = velocity of pressure wave, fps
d = pipe diameter, in.
e = pipe-wall thickness, in.
E = modulus of elasticity of pipe material, psi
K = bulk modulus of water, taken as 300,000 psi

Parmakian [8] has plotted values of a for various types of pipe against values of d/e, and he has developed a method for determining an equivalent thickness of steel pipe for pipes of composite materials. Values of a normally lie in the range from 3,500 to 4,100 fps.

When the front of the wave reaches the end of the force main, it reverses direction and a wave of increased pressure travels back to the pumps, where reversal again takes place and a second wave of reduced pressure travels up the pipe, etc. The shape of the wave front is sloping and depends on the ratio of the decelerating torque to the WR^2 of the pump and motor. The head at the pump discharge drops rapidly to the level of the sewage in the wet well, and even below it, as sewage continues flowing through the pump by gravity. The head at the end of the force main remains constant. At intermediate points, the head is determined by the sum of the pressures of the positive and negative waves. Pipeline friction helps to decelerate the flow. Each travel of the wave along the pipe in either direction results in a decrease in the velocity V in the force main of

$$\Delta V = \frac{gH}{a} \tag{6·8}$$

where H is the difference in head at the two ends of the force main plus the friction head at the average velocity during the passage of the wave. If $\Delta t = L/a$, the time for a wave to travel the length of the pipe, $L/\Delta t$ can be substituted for a in Eq. 6·8, yielding Eq. 6·9:

$$\Delta t = \frac{L\,\Delta v}{gH} \qquad \text{or} \qquad T = \frac{LV}{gH_{av}} \tag{6·9}$$

where T = time for velocity to drop to zero

H_{av} = average decelerating head including friction

Graphical methods are recommended for computing water-hammer pressures in pump discharge lines as described by Parmakian and Stepanoff [8, 13]. A briefer description is contained in another publication by Stepanoff [12]. Computer methods are described by Streeter and Wylie [15]. These references contain sound theory but do not specifically treat the special conditions that occur in sewage force mains.

If the pump discharge piping did not contain a check valve, after the velocity dropped to zero, the unbalanced head acting on the pipe would accelerate the flow in the reverse direction, reversing the rotation of the pump and motor until equilibrium was attained with the pump operating as a turbine at runaway speed. Under these conditions, reverse flow may be from 50 to 110 percent of rated flow; transient reverse speeds, from 125 to 150 percent of rated forward speed; and transient pressures, from 150 to 175 percent of rated head.

If the pump were equipped with a power-operated check valve that failed to close, the above conditions would result. If the pump were equipped with a swing check valve that stuck open temporarily and then slammed shut under these conditions, the resultant surge pressure head H, in ft of water, could amount to

$$H = \frac{aV}{g} \tag{6·10}$$

which is equal to the surge caused by sudden valve closure. For this reason, swing check valves with outside lever and weight, with weight set to assist closure, are recommended for small- and medium-size pumps. Spring-loaded check valves may also be used. With this arrangement and long force mains, the valve disks float to their seats and surge pressures are limited to two times the static pressure at the pump discharge. With short force mains, the valves close with a clunk assisted by developing reverse flow, and pressures may be somewhat greater, but they usually are not excessive.

In large stations, check valves should be positively controlled cone valves or butterfly valves. Butterfly valves should not be used unless rags and other solids are removed or comminuted ahead of the pumps. The timing of these valves is of paramount importance, because if they close too fast, they can break the water column at the pump discharge. In this case, a vacuum or vapor cavity forms beyond the valve, which permits the water column, after reaching zero velocity, to reverse and reach almost full initial velocity before coming up against a closed valve with resultant pressures determined by Eq. 6·10, or approximately 120 ft for every 1.0 fps destroyed.

Consequently, these valves must operate slowly and should be about half closed when the velocity in the force main has dropped to zero and begins to reverse. The valve closure time can be estimated by multiplying T from Eq. 6·9 by two where V in the equation is with the maximum number of pumps in operation. The time $2T$ should be the time from the beginning of unseating to the end of rotation in the case of cone valves, not including the reseating time. The specifications should call for the operating time to be adjustable at least over the range from T to $4T$. A minimum value of 10 seconds is recommended for T.

With this scheme there will be reverse flow through the pump on power failure. Normal shutdown will be accomplished by closing the valve first, which will then trip a limit switch to stop the motor. The pumps will attain a reverse speed on power failure of 60 to 125 percent of rated speed depending on the number of pumps in the station and the number in operation at the time of power failure. Pumps and motors should be specified to withstand safely a reverse speed of 150 percent of the rated head, measured above the top of the pump volute.

The ideal force main would leave the pumping station at the same level as the pumps and run level or rise gently at a uniform slope until near the end where it would rise abruptly to the outlet. This is seldom the case, since force main profiles conform for the sake of economy to existing ground elevations and may contain both high and low points. Furthermore, at many deep stations individual pump discharges rise vertically to the main header, which then leaves the station close to ground level, or the main header itself rises just inside or outside the station and then runs approximately level near the ground surface. On power failure the water column may break and vapor cavities may form at the high points as the first negative wave moves up the pipe, provided the minimum pressure drops 32 to 34 ft below the top of the pipe at the high points. The two portions of the water column then behave independently of each other, coming to rest or slowing down and reversing with the pressure at the high point remaining at the vapor pressure. If the pump column continues its forward movement for some seconds, the cavity may not grow very large and the columns may come together with very little difference in velocity and inconsequential surge pressures. On the other hand, if the pump column comes to rest quickly, the cavity may grow large enough to cause the other column to accelerate in the reverse direction almost to its initial velocity at the moment the columns rejoin. If a swing check valve has already closed at the pump discharge, a severe water hammer, equal to the pressure caused by sudden valve closure, could result.

A possible solution would be the substitution of slow-moving, power-operated cone or butterfly valves for the swing check valves, permitting reverse flow through the pumps. Such situations require special study by a water-hammer expert. Many of the devices used to control water hammer in water pumping stations, such as surge suppressors and relief valves, are not applicable because of the solids in sewage.

PROBLEMS

6·1 A sewage pump is pumping at a rate of 7,500 gpm. It has a 14-in. discharge and a 16-in. suction. The discharge gage, located at the pump centerline, reads 18 psi. The suction gage, located 24 in. below the pump centerline, reads 1.5 psi. Determine: (*a*) the head on the pump, (*b*) the brake horsepower required, and (*c*) the kilowatt input to the motor, assuming a pump efficiency of 82 percent and a motor efficiency of 91 percent. Use Bernoulli's equation (see Eq. 3·5 in Chap. 3) to find the head on the pump.

6·2 A mixed-flow volute pump is to operate at a head of 16 ft and discharge 6 cfs. It is to be driven by a direct-coupled squirrel-cage induction motor. If the specific speed is not to exceed 6,000, what should be the operating speed? What pump efficiency could be expected, and how much brake horsepower will be required?

6·3 A variable-speed pump delivers 5,000 gpm at a speed of 1,150 rpm against a system head of 90 ft, of which 75 ft is static head and 15 ft is friction head. The pump characteristics are as follows:

Capacity, gpm	Head, ft	Efficiency, %
0	112	
2,000	111	62
3,000	109	76.5
4,000	104	86
5,000	90	88
6,000	76	83.5

Determine the speeds necessary to pump 2,500 and 3,750 gpm against the system curve and determine the heads, efficiencies, and brake horsepowers required at all three operating points. Assume that the friction losses vary as the square of the flow.

6·4 Two pumps, each with a capacity of 1,000 gpm, are installed in a lift station (no force main). The inflow varies from 200 to 1,900 gpm. If the area of the wet well is 120 sq ft, determine and show on a diagram the float-switch settings for each pump. The pumps are not to start more often than four times per hour. Assume a 6-in. difference in maximum water level with one and two pumps in operation, and a water depth of 3 ft when the first pump in sequence stops.

6·5 A pumping station containing three identical constant-speed pumps, one of which is to be a spare, is to be designed to pump sewage through a force main 2,500 ft long. The flow is expected to vary from a present minimum flow of 200 gpm to a maximum design flow of 1,800 gpm. The incoming sewer is 18 in. in diameter, and its invert is 30 ft below the water surface at the point of discharge of the force main. Allow a 6-in. loss through the bar screens, and a 6-in. difference in wet-well levels between the start levels for no. 1 and no. 2 pumps in sequence.

Select the size of force main using cement-lined cast-iron pipe, compute the system curves when no. 1 and no. 2 pumps start, and determine the head, capacity, and efficiency of each pump when it operates alone and when two pumps operate together. Assume that the pumps are radial-flow centrifugal pumps, that the shape of the pump curves will be as given in Fig. 6·3, that the efficiency at the bep will be 75 percent, and that by proper selection of pump size, speed, and impeller diameter any desired head and capacity can be obtained. Assume that the losses in the individual suction and discharge pipes of each pump amount to 3 ft when the pump capacity is equal to the capacity at the bep. What is the capacity of each pump in terms of the capacity at the bep when it operates alone and when two pumps operate together?

6·6 In Prob. 6·5, assume a drawdown of 30 in. for control of each pump. What area of wet well is required? State the basis for your answer.

REFERENCES

1. Addison, H.: *Centrifugal and Other Rotodynamic Pumps*, 3d ed., Chapman & Hall, London, 1966.

2. Church, A. H.: *Centrifugal Pumps and Blowers*, Wiley, New York, 1944.

3. Hicks, T. G., BME: *Pump Selection and Application*, McGraw-Hill, New York, 1957.

4. Joint Committee of ASCE and WPCF: *Design and Construction of Sanitary and Storm Sewers*, chap. 12, Wastewater and Stormwater Pumping Stations, WPCF Manual of Practice no. 9, New York, 1969.

5. Karassik, I. J., and R. Carter: *Centrifugal Pumps*, F. W. Dodge Corporation, New York, 1960.

6. Kristal, F. A., and F. A. Annet: *Pumps*, 2d ed., McGraw-Hill, New York, 1953.

7. Moody, L. F.: Simplified Derivation of Water Hammer Formulas, *Symposium on Water Hammer, ASME*, New York, 1933.

8. Parmakian, J.: *Water-Hammer Analysis*, Prentice-Hall, Englewood Cliffs, N.J., 1955. (Dover reprint available.)

9. Potthoff, E. O.: Motor Drives for Sewage Pumping, *Wastes Engineering*, vol. 23, no. 6, 1952.

10. Seminar Papers on Waste Water Treatment and Disposal, *Pumps, Measuring Devices, Hydraulic Controls*, Boston Society of Civil Engineers, 1961.

11. *Standards of the Hydraulic Institute*, 12th ed., New York, 1969.

12. Stepanoff, A. J.: *Centrifugal and Axial Flow Pumps*, 2d ed., Wiley, New York, 1957.

13. Stepanoff, A. J.: Elements of Graphical Solution of Water-Hammer Problems in Centrifugal Pump Systems, *Transactions ASME*, vol. 71, 1949.

14. Stratton, C. H.: Raw Sewage Pumps, *Sewage and Industrial Wastes*, vol. 26, no. 12, 1954.

15. Streeter, V. L., and E. B. Wylie: *Hydraulic Transients*, McGraw-Hill, New York, 1967.

16. Water Pollution Control Federation: *Plant Pumping Stations*, chap. 3, Sewage Treatment Plant Design, WPCF Manual of Practice no. 8, Washington, D.C., 1959.

wastewater characteristics

7

An understanding of the nature of the physical, chemical, and biological characteristics of wastewaters is essential in the design and operation of collection, treatment, and disposal facilities and in the engineering management of environmental quality. Applications of the material covered in this chapter in the design of treatment and disposal facilities are discussed in Chaps. 11 to 14. Applications in environmental-quality management are discussed in Chap. 15.

WASTEWATER ANALYSES

The analyses performed on wastewaters may be classified as physical, chemical, and biological. The principal parameters used to characterize wastewater are listed in Table 7·1. These analyses vary from precise quantitative chemical determinations to the more qualitative biological and physical determinations. Further, many of the parameters are interrelated. For example, temperature, a physical parameter, affects both the biological activity in the wastewater and the amounts of gases dissolved in the wastewater, which are classified as chemical parameters. Definitions, explanations, and applications of the parameters shown in Table 7·1

TABLE 7·1 PHYSICAL, CHEMICAL, AND BIOLOGICAL CHARACTERISTICS OF WASTEWATER

Parameter	Source
	Physical
Solids	Carriage water,* domestic and industrial wastes
Temperature	Domestic and industrial wastes
Color	Domestic and industrial wastes
Odor	Decomposing sewage, industrial wastes
	Chemical
Organic:	
Proteins	Domestic and commercial wastes
Carbohydrates	Domestic and commercial wastes
Fats, oils, and grease	Domestic, commercial, and industrial wastes
Surfactants	Domestic and industrial wastes
Phenols	Industrial wastes
Pesticides	Agricultural wastes
Inorganic:	
pH	Industrial wastes
Chlorides	Carriage water, domestic wastes, ground water infiltration
Alkalinity	Domestic wastes, carriage water, ground water infiltration
Nitrogen	Domestic and agricultural wastes
Phosphorus	Domestic and industrial wastes, natural runoff
Sulfur	Carriage water and industrial wastes
Toxic compounds	Industrial wastes, ground water infiltration
Heavy metals	Industrial wastes
Gases:	
Oxygen	Carriage water, surface water infiltration
Hydrogen sulfide	Decomposition of domestic wastes
Methane	Decomposition of domestic wastes
	Biological
Protista	Domestic wastes, treatment plants
Viruses	Domestic wastes
Plants	Open watercourses and treatment plants
Animals	Open watercourses and treatment plants

* Carriage water refers to domestic water supply.

are included in the individual sections dealing with the physical, chemical, and biological characteristics.

In discussing the various parameters reported in Table 7·1, details concerning the exact method of analysis are not presented. These details may be found in *Standard Methods* [19], the accepted reference delineating the conduct of water and wastewater analyses. *Analysis of Water and Sewage* [23], although an older reference, is still useful. As a general reference, *Chemistry for Sanitary Engineers* [15] is recommended. *Aquatic Chemistry* [21], an advanced text, should be consulted for chemical equilibrium problems, especially in natural waters. For details concerning the biology and microbiology of the various microorganisms encountered in water and wastewater, the text *The Microbial World* [20], the reference work *Fresh Water Biology* [27], and the reference work *A Treatise on Limnology* [5], are recommended. Various other specific references are given throughout this chapter.

Expression of Analytical Results

Analytical results for wastewater samples are expressed in terms of physical and chemical units of measurement. The most common of these units are reported in Table 7·2. Chemical parameters are usually expressed in the physical unit of milligrams per liter (mg/liter). For the dilute systems in which a liter weighs a kilogram, such as those encountered in natural waters and waste flows, the mg/liter unit is interchangeable with parts per million (ppm), which is the weight-to-weight ratio. For design purposes, mg/liter is converted to pounds per million gallons by multiplying by 8.34, the weight in pounds of 1 gal of water.

Dissolved gases are considered to be chemical constituents and are measured as mg/liter. Gases evolved as a by-product of sewage treatment, such as methane and nitrogen (anaerobic decomposition), are measured in terms of liters and cubic feet (cu ft). Results of tests and parameters such as temperature, odor, hydrogen ion, and biological organisms are expressed in units other than mg/liter, as explained in the sections dealing with the specific parameters.

Composition

Composition refers to the physical, chemical, and biological constituents found in sewage. Depending on the amount of these constituents, sewage is classified as strong, medium, or weak. Typical composition and concentration data for domestic sewage are reported in Table 7·3. Because both the composition and concentration will vary with the

TABLE 7·2 METHODS FOR EXPRESSING ANALYTICAL RESULTS

Basis	Application	Unit
	Physical	
Weight per unit volume	$\frac{\text{milligrams}}{\text{liter of solution}}$	mg/liter
	$\frac{\text{grams}}{\text{liter of solution}}$	g/liter
	$\frac{\text{grains}}{\text{gallon of solution}}$	grains/gal
Weight ratio	$\frac{\text{milligram}}{10^6\ \text{milligram}}$	ppm
Volume ratio	$\frac{\text{milliliters}}{\text{liter}}$	ml/liter
Density (cgs units)	$\frac{\text{mass of solution}}{\text{unit volume}}$	g/ml
Percent by weight	$\frac{\text{weight of solute} \times 100}{\text{combined weight of solute} + \text{solvent}}$	% (by wt)
Percent by volume	$\frac{\text{volume of solute} \times 100}{\text{total volume of solution}}$	% (by vol)
	Chemical	
Molarity	$\frac{\text{moles of solute}}{\text{liter of solution}}$	moles/liter
	$\frac{\text{millimoles of solute}}{\text{liter of solution}}$	millimoles/liter
Molality	$\frac{\text{moles of solute}}{\text{1,000 grams of solvent}}$	moles/kg
Normality	$\frac{\text{equivalents of solute}}{\text{liter of solution}}$	equiv/liter
	$\frac{\text{milliequivalents of solute}}{\text{liter of solution}}$	meq/liter

hour of the day, the day of the week, the month of the year, as well as with other local conditions, the data in Table 7·3 are intended only to serve as a guide and not as a basis for design.

Of equal importance to composition and concentration data are data on the mineral pickup resulting from water usage and its variation within a sewerage system. These data are especially important in evaluating the reuse potential of wastewater. Mineral pickup results from domestic usage, from the addition of highly mineralized water from private wells and ground water, and from industrial usage. Domestic and industrial water softeners also contribute to the observed mineral pickup; in some areas, they may represent the major source of mineral pickup. Occasionally, water added from private wells and ground water infiltration will, because of its high quality, serve to dilute the mineral concentration in the wastewater.

TABLE 7·3 TYPICAL COMPOSITION OF DOMESTIC SEWAGE
(All values except settleable solids are expressed in mg/liter)

Constituent	Concentration		
	Strong	Medium	Weak
Solids, total	1,200	700	350
Dissolved, total	850	500	250
Fixed	525	300	145
Volatile	325	200	105
Suspended, total	350	200	100
Fixed	75	50	30
Volatile	275	150	70
Settleable solids, (ml/liter)	20	10	5
Biochemical oxygen demand, 5-day, 20°C (BOD_5-20°)	300	200	100
Total organic carbon (TOC)	300	200	100
Chemical oxygen demand (COD)	1,000	500	250
Nitrogen, (total as N)	85	40	20
Organic	35	15	8
Free ammonia	50	25	12
Nitrites	0	0	0
Nitrates	0	0	0
Phosphorus (total as P)	20	10	6
Organic	5	3	2
Inorganic	15	7	4
Chlorides*	100	50	30
Alkalinity (as $CaCO_3$)*	200	100	50
Grease	150	100	50

* Values should be increased by amount in carriage water.

Data on the chemical composition of a typical water supply and the resultant wastewater effluent composition after treatment are shown in Table 7·4. In this case, the effect of the use of some local well water and the intensive use of water softeners on the total mineral pickup can be assessed by comparing the local incremental pickup values reported in col. 2 to the national average range given in col. 3.

TABLE 7·4 MINERAL PICKUP FROM DOMESTIC WATER USAGE [22]

Constituent	Concentration, mg/liter		
	Palo Alto water[a]	Palo Alto effluent[b]	Increment range[c]
Anions:			
Carbonate (CO_3)	2.4	0.0	
Bicarbonate (HCO_3)	45.0	251.0	
Chloride (Cl)	3.5	215.0[d]	20–50
Sulfate (SO_4)	5.8	47.5[e]	15–30
Nitrate (NO_3)	1.1	18.4	20–40[f]
Phosphate (PO_4)	0.0	21.4	20–40
Cations:			
Sodium (Na)	0.5	155.0[e]	40–70
Potassium (K)	0.8	8.8	7–15
Calcium (Ca)	10.4	49.6[d]	15–40[g]
Magnesium (Mg)	9.8	32.6[d]	15–40[g]
Other data:			
Silica (SiO_2)	5.8	14.5	
Fluoride (F)	0.8	3.8	
Manganese (Mn)	0.0	0.0	
Iron (Fe)	0.0	<0.1	
Aluminum (Al)	0.1	<3.0	
Boron (B)	0.1	0.93	0.1–0.4
Total dissolved solids (TDS)	63.8	693.0	100–300
Total alkalinity ($CaCO_3$)	39.0	206.0	100–150

[a] Provided by City of San Francisco from its Hetch Hetchy source in the Sierra.
[b] Effluent from the Palo Alto, Calif., wastewater treatment plant.
[c] Reported national average range of mineral pickup by domestic usage.
[d] Approximately 15 percent local well water used along with Hetch Hetchy water.
[e] High due to use of water softeners.
[f] Total nitrogen as N.
[g] Reported as $CaCO_3$.

Variations in Flow and Sewage Strength

The variation in sewage flowrates at a sewage plant follows a somewhat diurnal pattern, as shown in Fig. 7·1. Minimum flows occur during the early morning hours when water consumption is lowest and when the base flow consists of leakages, infiltration, and small quantities of sanitary sewage. Peak flows generally occur in the late morning when the sewage from the peak morning water demand reaches the sewage treatment plant. A second peak generally occurs in the early evening. Where infiltration rates are high or where there are drainage connections

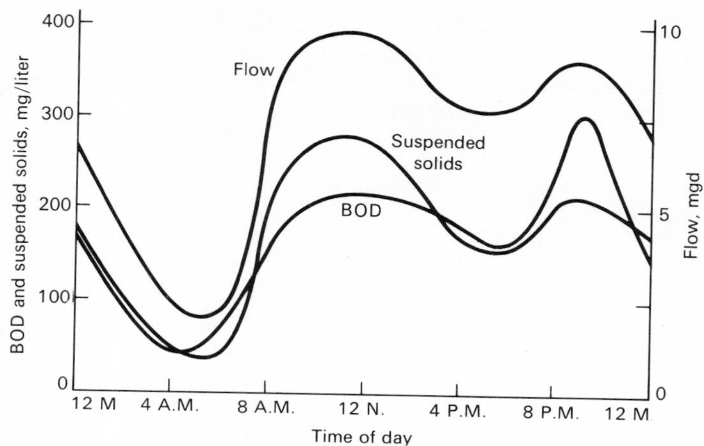

FIG. 7·1 Typical hourly variation in flow and strength of domestic sewage.

to the sewer system, rainfall patterns may have a pronounced effect on the sanitary sewage flow.

The variation in sewage strength is also shown in Fig. 7·1. The biochemical oxygen demand (BOD) variation follows the flow variation, except that the peak BOD (organic matter) concentration occurs in the evening around 9 P.M.

Sewage from combined-sewer systems contains more inorganic matter than sewage from sanitary sewer systems because of the storm drainage that enters the system. Peak flowrates and the ratio of peak flow to average flow are higher than in the sanitary sewage systems. At the beginning of a storm, the flushing action of runoff may increase the strength of combined sewage far above the strength of normal sanitary sewage, both in terms of suspended solids and BOD.

PHYSICAL CHARACTERISTICS: DEFINITION AND APPLICATION

The most important physical characteristic of wastewater is its total solids content, which is composed of floating matter, matter in suspension, colloidal matter, and matter in solution. Other physical characteristics include temperature, color, and odor.

Total Solids

The total solids in wastewater are derived from the carriage water (water supply), from domestic and industrial usage, and from local well water and ground water infiltration, as noted previously. Domestic solids include those derived from toilets, sinks, baths, laundries, garbage grinders, and water softeners. Typical data on the daily per capita quantities of dry solid material derived from these and the aforementioned sources are reported in Table 7·5.

TABLE 7·5 ESTIMATE OF THE COMPONENTS OF TOTAL SOLIDS IN WASTEWATER

Component	Dry weight, gpcd
Water supplies and ground water, assumed to have little hardness	12.7
Feces (solids, 23%)	20.5
Urine (solids, 3.7%)	43.3
Toilet (including paper)	20.0
Sinks, baths, laundries, and other sources of domestic wash waters	86.5
Ground garbage	30.0
Water softeners	*
Total for domestic sewage from separate sewerage systems, excluding contribution from water softeners	213.0
Industrial wastes	200.0‡
Total for industrial and domestic wastes from separate sewerage system	413.0
Storm water	25.0†
Total for industrial and domestic wastes from combined sewerage system	438.0

* Variable.
† Will vary with the season.
‡ Will vary with the type and size of industries,

Analytically, the total solids content of a wastewater is defined as all the matter that remains as residue upon evaporation at 103 to 105°C. Matter that has a significant vapor pressure at this temperature is lost during evaporation and is not defined as a solid. Total solids, or residue upon evaporation, can be classified as either suspended solids or filterable solids by passing a known volume of liquid through a filter. The filter is commonly chosen so that the minimum diameter of the suspended solids is about 1 micron (μ). The suspended-solids fraction includes the settleable solids that will settle to the bottom of a cone-shaped container (called an Imhoff cone) in a 60-min period. Settleable solids are an approximate measure of the quantity of sludge that will be removed by sedimentation.

The filterable-solids fraction consists of colloidal and dissolved

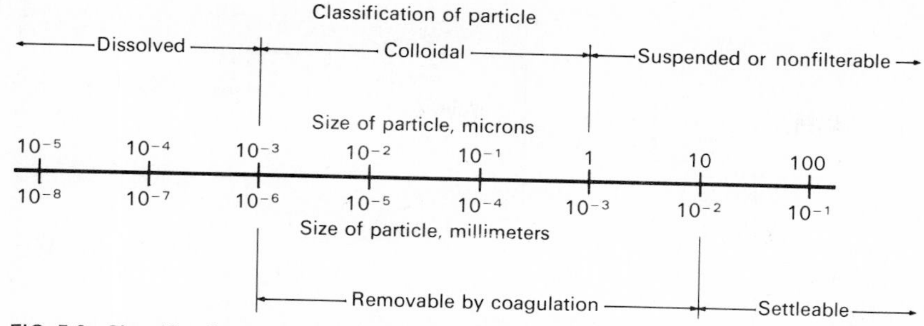

FIG. 7·2 Classification and size range of particles found in water.

solids. The colloidal fraction consists of the particulate matter with an approximate diameter range from 1 millimicron (mμ) to 1 μ (see Fig. 7·2). The dissolved solids consist of both organic and inorganic molecules and ions that are present in true solution in water. The colloidal fraction cannot be removed by settling. Generally biological oxidation or coagulation followed by sedimentation is required to remove these particles from suspension.

Each of the categories of solids may be further classified on the basis of their volatility at 600°C. The organic fraction will oxidize and be driven off as gas at this temperature, and the inorganic fraction remains behind as ash. Thus the terms "volatile suspended solids" and "fixed suspended solids" refer, respectively, to the organic and inorganic (or mineral) content of the suspended solids. At 600°C, the decomposition of inorganic salts is restricted to magnesium carbonate, which decomposes into magnesium oxide and carbon dioxide at 350°C. Calcium carbonate, the major component of the inorganic salts, is stable up to a temperature of 825°C. The volatile-solids analysis is applied most commonly to sewage sludges to measure their biological stability. The solids content of a medium-strength sewage may be classified approximately as shown in Fig. 7·3.

Turbidity, a measure of the light-transmitting properties of water, is another test used to indicate the quality of waste discharges and natural waters with respect to colloidal matter. Colloidal matter will scatter or absorb light and thus prevent its transmission.

Temperature

The temperature of wastewater is commonly higher than that of the water supply, because of the addition of warm water from households and industrial activities. As the specific heat of water is much greater

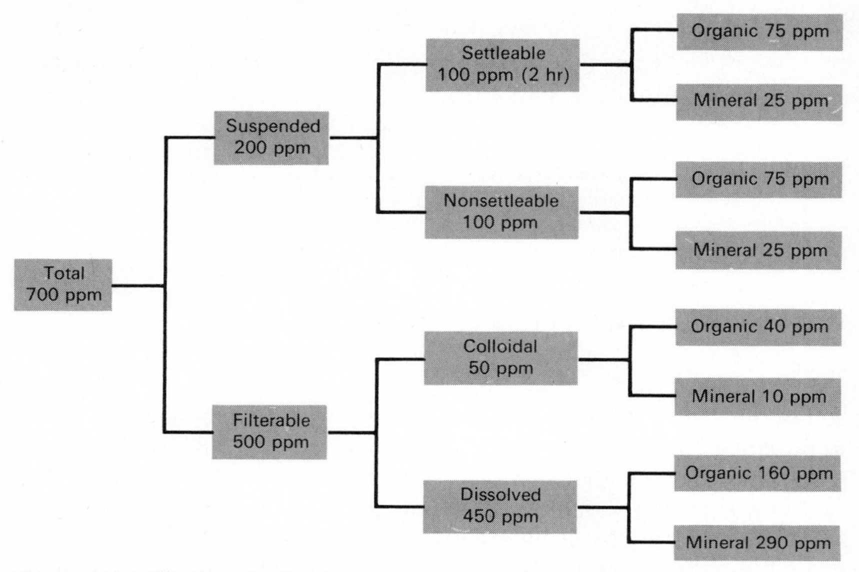

FIG. 7·3 Classification of solids found in medium-strength sewage.

than that of air, the observed wastewater temperatures are higher than the local air temperatures during most of the year and are lower only during the hottest summer months. Depending on the geographic location, the mean annual temperature of wastewater will vary from about 50 to 70°F, with 60°F being a representative value.

The temperature of water is a very important parameter because of its effect on the aquatic life, the chemical reactions and reaction rates, and the suitability of the water for beneficial uses. Increased temperature, for example, can cause a change in the species of fish that can exist in the water. Industrial establishments that use surface water for cooling water purposes are particularly concerned with the temperature of the intake water.

In addition, oxygen is less soluble in warm water than in cold water. The increase in the rate of biochemical reactions that accompanies an increase in temperature, combined with the decrease in the quantity of oxygen present in surface waters, can often cause serious depletions in dissolved oxygen concentrations in the summer months. When significantly large quantities of heated water are discharged to natural receiving waters, these effects are magnified. It should also be realized that a sudden change in temperature can result in a high rate of mortality of aquatic life. Moreover, abnormally high temperatures can foster the growth of undesirable water plants and sewage fungus.

Color

Historically, the term "condition" was used along with composition and concentration to describe sewage. Condition refers to the age of the sewage. It is determined qualitatively by its color and odor. Fresh sewage is usually grey; however, as organic compounds are broken down by bacteria, the dissolved oxygen in the sewage is reduced to zero and the color changes to black. In this condition, the sewage is said to be septic (or stale). Some industrial wastewaters may also add color to domestic wastewater.

Odors

Odors in sewage are due to gases produced by decomposition of organic matter (see "Gases"). Fresh sewage has a distinctive somewhat disagreeable odor, less objectionable than the odor of septic sewage. The most characteristic odor of stale or septic sewage is that of hydrogen sulfide produced by anaerobic microorganisms which reduce sulfates to sulfides. Industrial wastewaters may contain odorous compounds or compounds that produce odors in the process of sewage treatment. Odor control is discussed in Chap. 11.

CHEMICAL CHARACTERISTICS: DEFINITION AND APPLICATION

The material presented in this section is divided into four general categories dealing with (1) the organic matter, (2) the measurement of organic content, (3) the inorganic matter, and (4) the gases found in wastewater. The measurement of organic content is discussed separately because of its importance in both the design and operation of wastewater treatment plants and the management of water quality.

Organic Matter

In a sewage of medium strength, about 75 percent of the suspended solids and 40 percent of the filterable solids are organic in nature, as shown in Fig. 7·3. They are derived from both the animal and plant kingdoms and the activities of man as related to the synthesis of organic compounds. Organic compounds are normally composed of a combination of carbon, hydrogen, and oxygen, together with nitrogen in some cases. Other important elements, such as sulfur, phosphorus, and iron, may also be present. The principal groups of organic substances found in sewage are proteins (40 to 60 percent), carbohydrates (25 to 50 percent), and fats and oils (10 percent). Urea, the chief constituent of urine, is another important organic compound contributed to sewage. Because it decomposes so rapidly, undecomposed urea is seldom found in other than very fresh wastewater.

Along with the proteins, carbohydrates, fats and oils, and urea, wastewater contains small quantities of a large number of different synthetic organic molecules ranging from simple to extremely complex in structure. Typical examples, discussed in this section, include surfactants, phenols, and agricultural pesticides. Further, the number of such compounds is growing yearly as more and more organic molecules are being synthesized. The presence of these substances has, in recent years, complicated wastewater treatment because many of them either cannot be, or are very slowly, decomposed biologically. This factor also accounts for the renewed interest in the use of chemical precipitation followed by carbon adsorption for the complete treatment of wastewater (see Chaps. 9, 11, and 14).

Proteins Proteins are the principal constituents of the animal organism. They occur to a lesser extent in plants. All raw animal and plant foodstuffs contain proteins. The amount present varies from small percentages in watery fruits, such as tomatoes, and in the fatty tissues of meat, to quite high percentages in beans or lean meats. Proteins are complex in chemical structure and unstable, being subject to many forms of decomposition. Some are soluble in water, others are insoluble. The chemistry of the formation of proteins involves the combination or linking together of a large number of amino acids. The molecular weights of proteins are very high, ranging from about 20,000 to 20 million.

All proteins contain carbon, which is common to all organic substances, as well as hydrogen and oxygen. In addition they contain, as their distinguishing characteristic, a fairly high and constant proportion of nitrogen, about 16 percent. In many cases sulfur, phosphorus, and iron are also constituents. Urea and proteins are the chief sources of nitrogen in sewage. When the latter are present in large quantities, extremely foul odors are apt to be produced by their decomposition.

Carbohydrates Widely distributed in nature, carbohydrates include sugars, starches, cellulose, and wood fiber. All are found in sewage. They contain carbon, hydrogen, and oxygen. The common carbohydrates contain six, or a multiple of six, carbon atoms in a molecule, and hydrogen and oxygen in the proportions in which these elements are found in water. Some carbohydrates, notably the sugars, are soluble in water; others, such as the starches, are insoluble. The sugars are prone to decomposition, the enzymes of certain bacteria and yeasts setting up fermentation with the production of alcohol and carbon dioxide. The starches, on the other hand, are more stable but are converted into sugars by microbial activity as well as by dilute mineral acids. From the standpoint of bulk and resistance to decomposition, cellulose is the most important carbohydrate found in sewage. The destruction of cellulose in the soil goes on readily, largely as a result of the activity of various fungi, particularly when acid conditions prevail.

Fats, Oils, and Grease Fats and oils are the third major component of foodstuffs. The term "grease," as commonly used, includes the fats, oils, waxes, and other related constituents found in sewage. Grease content is determined by extraction of the waste sample with hexane (grease is soluble in hexane). Another group of hexane-soluble substances includes mineral oils, such as kerosene and lubricating and road oils.

Fats and oils are compounds (esters) of alcohol or glycerol (glycerin) with fatty acids. The glycerides of fatty acids that are liquid at ordinary temperatures are called oils, and those that are solids are called fats. They are quite similar, chemically, being composed of carbon, hydrogen, and oxygen in varying proportions.

Fats and oils are contributed to domestic sewage in butter, lard, margarine, and vegetable fats and oils. Fats are also commonly found in meats, in the germinal area of cereals, in seeds, in nuts, and in certain fruits.

Fats are among the more stable of organic compounds and are not easily decomposed by bacteria. Mineral acids attack them, however, resulting in the formation of glycerin and fatty acid. In the presence of alkalies, such as sodium hydroxide, glycerin is liberated, and alkali salts of the fatty acids are formed. These alkali salts are known as soaps and, like the fats, they are stable. Common soaps are made by saponification of fats with sodium hydroxide. They are soluble in water, but in the presence of hardness constituents, the sodium salts are changed to calcium and magnesium salts of the fatty acids, or so-called mineral soaps. These are insoluble and are precipitated.

Kerosene and lubricating and road oils are derived from petroleum and coal tar and contain essentially carbon and hydrogen. These oils sometimes reach the sewers in considerable volume from shops, garages, and streets. For the most part, they float on the sewage, although a portion is carried into the sludge on settling solids. To an even greater extent than fats, oils, and soaps, the mineral oils tend to coat surfaces. The particles interfere with biological action and cause maintenance problems.

As indicated in the foregoing discussion, the grease content of wastewater can cause many problems in both sewers and waste treatment plants. If grease is not removed before discharge of the waste, it can interfere with the biological life in the surface waters and create unsightly floating matter and films. Limits of 15 to 20 mg/liter of grease content and absence of iridescent oil films for wastewaters discharged to natural waters are two examples of standards that have been set by regulating agencies.

Surfactants Surfactants, or surface-active agents, are large organic molecules that are slightly soluble in water and cause foaming in wastewater treatment plants and in the surface waters into which the waste effluent is discharged. Surfactants tend to collect at the air-water interface. During aeration of wastewater, these compounds collect on the surface of the air bubbles and thus create a very stable foam.

Before 1965, the type of surfactant present in synthetic detergents, called alkyl-benzene-sulfonate (ABS) was especially troublesome because it resisted breakdown by biological means. As a result of legislation in 1965, ABS has been replaced in detergents by linear-alkyl-sulfonate (LAS), which is biodegradable [6]. Since surfactants come primarily from synthetic detergents, the foaming problem has been greatly reduced.

The determination of surfactants is accomplished by measuring the color change in a standard solution of methylene-blue dye. Another name for surfactant is methylene-blue active substance (MBAS).

Phenols Phenols and other trace organic compounds are also important constituents of water. Phenols cause taste problems in drinking water, particularly when the water is chlorinated. They are produced primarily by industrial operations and find their way to surface waters via wastewater discharges that contain industrial wastes. Phenols can be biologically oxidized at concentrations up to 500 mg/liter.

Pesticides and Agricultural Chemicals Trace organic compounds, such as pesticides, herbicides, and other agricultural chemicals, are toxic to most life forms and therefore can be significant contaminants of surface waters. These chemicals are not common constituents of sewage but result primarily from surface runoff from agricultural, vacant, and park lands. Concentrations of these chemicals can result in fish kills, in contamination of the flesh of fish that decreases their value as a source of food, and in impairment of water supplies.

The concentration of these trace contaminants is measured by the carbon-chloroform extract method, which consists of separating the contaminants from the water by passing a water sample through an activated-carbon column and then extracting the contaminant from the carbon using chloroform. The chloroform can then be evaporated and the contaminants can be weighed. Pesticides in concentrations of 1 part per billion (ppb) and less can be accurately determined by several methods, including gas chromatography and electron capture or coulometric detectors [15].

Measurement of Organic Content

Over the years, a number of different tests have been developed to determine the organic content of wastewaters. One method, discussed previously, is to measure the volatile-solids fraction of the total solids, but it is subject to many errors and is seldom used [19]. Laboratory methods commonly used today are biochemical oxygen demand (BOD), chemical oxygen demand (COD), and total organic carbon (TOC). Another recently developed test is the total oxygen demand (TOD). Complementing these laboratory tests is the theoretical oxygen demand (ThOD), which is determined from the chemical formula of the organic matter.

Other methods used in the past included (1) total, albuminoid, organic, and ammonia nitrogen, and (2) oxygen consumed. These determinations, with the exception of albuminoid nitrogen and oxygen consumed, are still included in complete sewage analyses. Their significance, however, has changed. Whereas formerly they were used almost exclusively to indicate organic matter, they are now used to determine the availability of nitrogen to sustain biological activity in industrial waste treatment processes and to foster undesirable algal growths in receiving waters.

BOD The most widely used parameter of organic pollution applied to both wastewater and surface waters is the 5-day BOD (BOD_5). This determination involves the measurement of the dissolved oxygen used by microorganisms in the biochemical oxidation of the organic matter. The BOD measurement is significant in sewage treatment and water-quality management, because it is used to determine the approximate quantity of oxygen that will be required to stabilize biologically the organic matter present. BOD data are used for sizing of waste treatment facilities and for measuring the efficiency of some treatment processes. The rate at which oxygen will be required can also be calculated from BOD data.

To ensure that meaningful results are obtained, the sample must be suitably diluted with a specially prepared dilution water so that adequate nutrients and oxygen will be available during the incubation period. Normally, several dilutions are prepared to cover the complete range of possible values. The ranges of BOD that can be measured with various dilutions based on percentage mixtures and direct pipetting are reported in Table 7·6.

TABLE 7·6 BOD MEASURABLE WITH VARIOUS DILUTIONS OF SAMPLES [15]

By using percent mixtures		By direct pipetting into 300-ml bottles	
% mixture	Range of BOD	ml	Range of BOD
0.01	20,000–70,000	0.02	30,000–105,000
0.02	10,000–35,000	0.05	12,000–42,000
0.05	4,000–14,000	0.10	6,000–21,000
0.1	2,000–7,000	0.20	3,000–10,500
0.2	1,000–3,500	0.50	1,200–4,200
0.5	400–1,400	1.0	600–2,100
1.0	200–700	2.0	300–1,050
2.0	100–350	5.0	120–420
5.0	40–140	10.0	60–210
10.0	20–70	20.0	30–105
20.0	10–35	50.0	12–42
50.0	4–14	100.0	6–21
100.0	0–7	300.0	0–7

The dilution water is "seeded" with a bacterial culture that has been acclimated, if necessary, to the organic matter present in the water. The seed culture that is used to prepare the dilution water for the BOD test is a mixed culture. Such cultures contain large numbers of saprophytic bacteria and other organisms that oxidize the organic matter. In addition, they contain certain autotrophic bacteria that oxidize noncarbonaceous matter. When the sample contains a large population of microorganisms (raw sewage, for example), seeding is not necessary.

The incubation period is usually 5 days at 20°C, but other lengths of time and temperatures can be used. The temperature, however, should be constant throughout the test. After incubation, the dissolved oxygen of the sample is measured and the BOD is calculated using Eq. 7·1*a* or 7·1*b*.

For percent mixtures:

$$\text{BOD (mg/liter)} = \left[(\text{DO}_b - \text{DO}_i)\frac{100}{\%}\right] - (\text{DO}_b - \text{DO}_s) \quad (7{\cdot}1a)$$

For direct pipetting:

$$\text{BOD} = \left[(\text{DO}_b - \text{DO}_i)\frac{\text{vol of bottle}}{\text{ml of sample}}\right] - (\text{DO}_b - \text{DO}_s) \quad (7{\cdot}1b)$$

where DO_b, DO_i = dissolved oxygen values found in blank (containing dilution water only) and dilutions of sample, respectively, at end of incubation period

DO_s = dissolved oxygen originally present in undiluted sample

As the value of DO_s approaches DO_b, or when the BOD is over 200 mg/liter, the second term in Eqs. 7·1*a* and 7·1*b* becomes negligible.

Biochemical oxidation is a slow process and theoretically takes an infinite time to go to completion. Within a 20-day period, the oxidation is about 95 to 99 percent complete, and in the 5-day period used for the BOD test, oxidation is from 60 to 70 percent complete. The 20°C temperature used is an average value for slow-moving streams in temperate climates and is easily duplicated in an incubator. Different results would be obtained at different temperatures because biochemical reaction rates are temperature-dependent.

The kinetics of the BOD reaction are, for practical purposes, formulated in accordance with first-order reaction kinetics and may be expressed as

$$\frac{dL_t}{dt} = -K'L_t \quad (7{\cdot}2)$$

where L_t is the amount of the first-stage BOD remaining in the water at time t. This equation can be integrated as

$$\ln L_t \Big|_0^t = -K't$$

$$\frac{L_t}{L} = e^{-K't} = 10^{-Kt} \quad (7{\cdot}3)$$

where L or BOD_L is the BOD remaining at time $t = 0$ (i.e., the total or ultimate first-stage BOD initially present). The relation between K' and K is as follows:

$$K = \frac{K'}{2.303}$$

FIG. 7·4 Formulation of the first stage BOD curve.

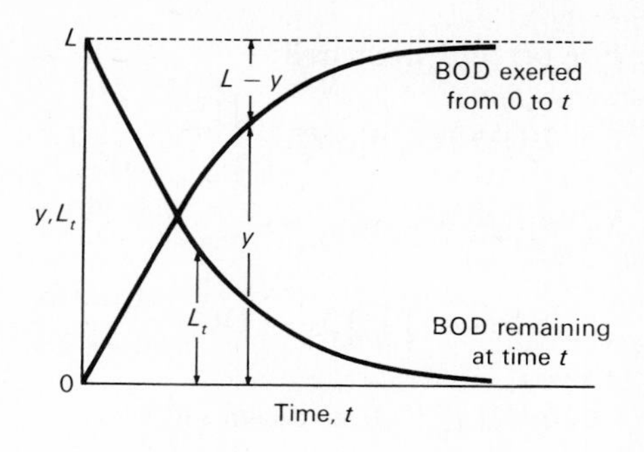

The amount of BOD remaining at any time t equals

$$L_t = L(10^{-Kt})$$

and y, the amount of BOD that has been exerted at any time t, equals

$$y = L - L_t = L(1 - 10^{-Kt}) \tag{7·4}$$

Note that the 5-day BOD equals

$$y_5 = L - L_5 = L(1 - 10^{-5K})$$

This relationship is shown in Fig. 7.4. The following example will illustrate the use of the BOD equations.

EXAMPLE 7·1 *Calculation of* BOD

Determine the 1-day BOD and the ultimate first-stage BOD for a sewage whose 5-day, 20°C BOD is 200 mg/liter. The reaction constant $K' = 0.23$.

Solution

1. Determine ultimate BOD.

$$L_t = Le^{-K't}$$
$$y_5 = L - L_5 = L(1 - e^{-5K'})$$
$$200 = L(1 - e^{-5(0.23)}) = L(1 - 0.316)$$
$$L = 293 \text{ mg/liter}$$

2. Determine 1-day BOD.

$$L_1 = Le^{-K't}$$
$$= 293(e^{-0.23(1)}) = 293(0.795) = 233 \text{ mg/liter}$$
$$y_1 = L - L_1 = 293 - 233 = 60 \text{ mg/liter}$$

For polluted water and sewage a typical value of K (base 10, 20°C) is 0.10 day^{-1}. The value of K varies significantly, however, with the type of waste. The range may be from 0.05 day^{-1} to 0.3 day^{-1} or more. For the same ultimate BOD, the oxygen uptake will vary with time and with different K values. The effect of different K values is shown in Fig. 7·5.

As mentioned, the temperature at which the BOD of a wastewater sample is determined is usually 20°C. It is possible, however, to determine the reaction constant K at a temperature other than 20°C. The following approximate equation, which is derived from the van't Hoff-Arrhenius relationship (see Eq. 9·8 in Chap. 9), may be used:

$$K_T = K_{20}\theta^{(T-20°)} \tag{7·5}$$

The value of θ has been found to vary from 1.056 in the temperature range between 20 and 30°C to 1.135 in the temperature range between 4 and 20°C [16]. A value of θ often quoted in the literature is 1.047 [12], but it has been observed that this value does not apply at cold temperatures (e.g., below 20°C) [16].

Noncarbonaceous matter, such as ammonia, is produced during the hydrolysis of proteins. Some of the autotrophic bacteria are capable of using oxygen to oxidize the ammonia to nitrites and nitrates. The nitrogenous oxygen demand caused by the autotrophic bacteria is called the second-stage BOD. The normal progression of each stage for a domestic sewage is shown in Fig. 7·6. At 20°C, however, the reproductive rate of the nitrifying bacteria is very slow. It normally takes from 6 to 10 days for them to reach significant numbers and to exert a measurable oxygen demand. The interference caused by their presence

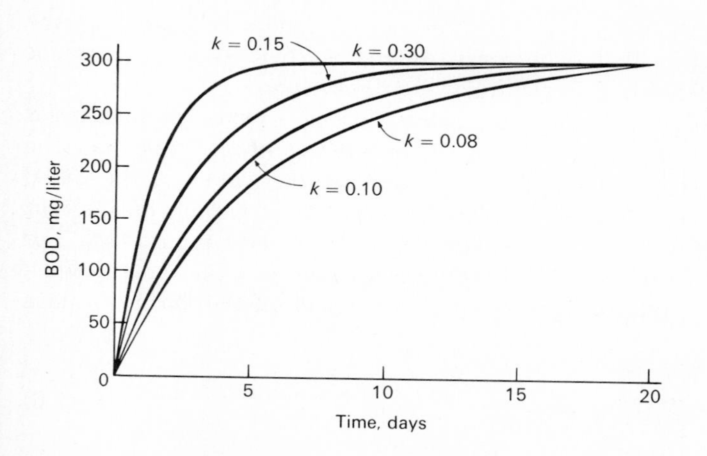

FIG. 7·5 Effect of the rate constant K on BOD (for a given L value) [15].

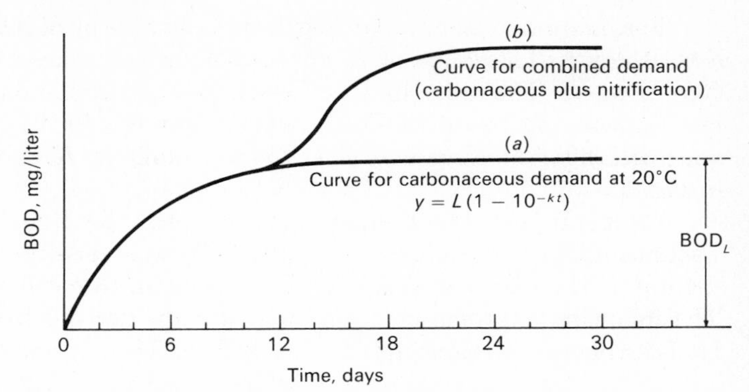

FIG. 7·6 The BOD curve [15]. (*a*) Normal curve for oxidation of organic matter. (*b*) Influence of nitrification.

can be eliminated by pretreatment of the sample or by the use of inhibitory agents.

Pretreatment procedures include pasteurization [14], chlorination, and acid treatment. Inhibitory agents are usually chemical in nature and include methylene blue, thiourea and allylthiourea, and 2-chloro-6-(trichloromethyl) pyridine [31]. For a more detailed review of these procedures and a discussion of experimental results obtained using chemical inhibitory agents, Ref. [31] should be consulted.

The value of K is needed if the BOD_5 is to be used to obtain L, the ultimate or 20-day BOD. The usual procedure followed when these values are unknown is to determine K and L from a series of BOD measurements. There are several ways of determining K and L from the results of a series of BOD measurements including (1) the least-squares method [29, 30], (2) the method of moments [9], (3) the daily-difference method [25], (4) the rapid-ratio method [17], and (5) the Thomas method [24]. The least-squares method and the Thomas method are illustrated in the following discussion.

The least-squares method involves fitting a curve through a set of data points, so that the sum of the squares of the residuals (the difference between the observed value and the value of the fitted curve) must be a minimum. Using this method, a variety of different types of curves can be fitted through a set of data points [29]. For example, for a time series of BOD measurements on the same sample, the following equation may be written for each of the various n data points:

$$\left.\frac{dy}{dt}\right|_{t=n} = K'(L - y_n) \qquad (7\cdot6)$$

In this equation both K' and L are unknown. If it is assumed that dy/dt represents the value of the slope of the curve to be fitted through all the data points for a given K' and L value, then, due to experimental error, the two sides of Eq. 7·6 will not be equal but will differ by an amount R. Rewriting Eq. 7·6 in terms of R for the general case yields

$$R = K'(L - y) - \frac{dy}{dt} \tag{7·7}$$

Simplifying and using the notation y' for dy/dt gives

$$R = K'L - K'y - y' \tag{7·8}$$

Substituting a for $K'L$ and $-b$ for K' gives

$$R = a + by - y' \tag{7·9}$$

Now, if the sum of the squares of the residuals R is to be a minimum, the following equations must hold:

$$\begin{aligned} \frac{\partial}{\partial a}\Sigma R^2 &= \Sigma 2R\frac{\partial R}{\partial a} = 0 \\ \frac{\partial}{\partial b}\Sigma R^2 &= \Sigma 2R\frac{\partial R}{\partial b} = 0 \end{aligned} \tag{7·10}$$

If the indicated operations in Eq. 7·10 are carried out using the value of the residual R defined by Eq. 7·9, the following set of equations result:

$$na + b\Sigma y - \Sigma y' = 0 \tag{7·11}$$

$$a\Sigma y + b\Sigma y^2 - \Sigma yy' = 0 \tag{7·12}$$

where n = number of data points
$K' = -b$ (base e)
$L = -a/b$

Application of the least-squares method in the analysis of BOD data is illustrated in the following example.

EXAMPLE 7·2 *Calculation of BOD Constants Using the Least-Squares Method*

Compute L and K' using the least-squares method for the following BOD data.

t, days	2	4	6	8	10
y, mg/liter	11	18	22	24	26

Solution

1. Set up a computation table and perform the indicated steps.

Time	y	y^2	y'	yy'
2	11	121	4.50	49.5
4	18	324	2.75	49.5
6	22	484	1.50	33.0
8	24	576	1.00	24.0
	Σ75	1,505	9.75	156.0

The slope y' is computed as follows:

$$\frac{dy}{dt} = y' = \frac{y_{n+1} - y_{n-1}}{2\Delta t}$$

2. Substituting the values computed in step 1 in Eqs. 7·11 and 7·12 and solving for a and b yields values of 7.5 and -0.271, respectively.

$$4a + 75b - 9.75 = 0$$
$$75a + 1505b - 156.0 = 0$$

3. Determine the values of K' and L.

$$K' = -b = 0.271 \text{ (base } e)$$
$$L = -\frac{a}{b} = \frac{7.5}{0.271} = 27.7 \text{ mg/liter}$$

4. Compare these answers to the values obtained using the Thomas method in Example 7·3.

The Thomas method, based on the similarity of two series functions, is illustrated here. It is a graphical procedure based on the function

$$\left(\frac{t}{y}\right)^{\frac{1}{3}} = (2.3KL)^{-\frac{1}{3}} + \frac{K^{\frac{2}{3}}}{3.43L^{\frac{1}{3}}}\,t \tag{7·13}$$

where y = BOD that has been exerted in time interval t
K = base 10 reaction-rate constant
L = ultimate BOD

This equation has the form of a straight line,

$$Z = a + bt$$

where $Z = (t/y)^{\frac{1}{3}}$
$a = (2.3KL)^{-\frac{1}{3}}$
$b = K^{\frac{2}{3}}/3.43L^{\frac{1}{3}}$

and Z can then be plotted as a function of t. The slope b and the intercept a of the line of best fit of the data can then be used to calculate K and L.

$$K = 2.61 \frac{b}{a} \tag{7·14}$$

$$L = \frac{1}{2.3Ka^3} \tag{7·15}$$

To use this method, several observations of y as a function of t are needed. The data observations should be limited to the first 10 days because of nitrogenous interference. The method is illustrated in the following example.

EXAMPLE 7·3 *Calculation of BOD Constants Using the Thomas Method*

Compute L and K using the Thomas method [24] for the data given in Example 7·2.

Solution

1. Determine the value $(t/y)^{\frac{1}{3}}$ for the data

t, days	2	4	6	8	10
y, mg/liter	11	18	22	24	26
$(t/y)^{\frac{1}{3}}$	0.57	0.61	0.65	0.69	0.727

2. Plot the value $(t/y)^{\frac{1}{3}}$ versus t (see Fig. 7·7).
3. From Fig. 7·7 the slope b and intercept a are

$$\text{Slope } b = \frac{0.04}{2} = 0.02$$

$$\text{Intercept } a = 0.53$$

4. Compute K and L.

$$K = 2.61 \frac{0.02}{0.53} = 0.099 \qquad K' = 0.228$$

$$L = \frac{1}{2.3(0.099)(0.53)^3} = 29.4 \text{ mg/liter}$$

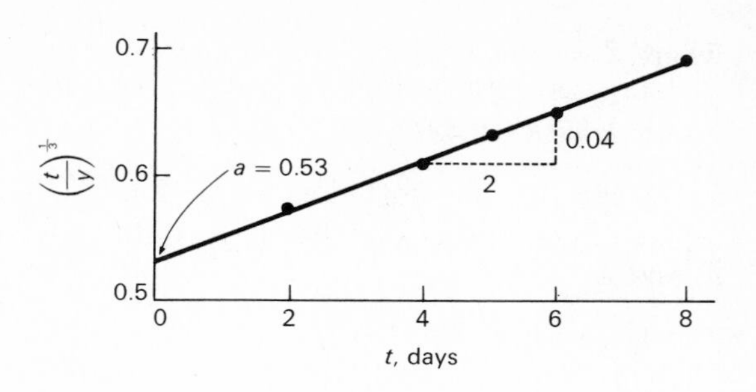

FIG. 7·7 Determination of K and L from BOD data by Thomas method.

The BOD and rate constant K determinations can be made more rapidly in the laboratory by using the Warburg respirometer, or with the aid of dissolved oxygen probes or an electrolysis cell. The Warburg apparatus consists of a constant-temperature water bath, an agitator mechanism, and a set of special flasks equipped with manometers.

Each flask has an interior well in which a small quantity of potassium hydroxide solution is placed (see Fig. 7·8). The flask is filled with

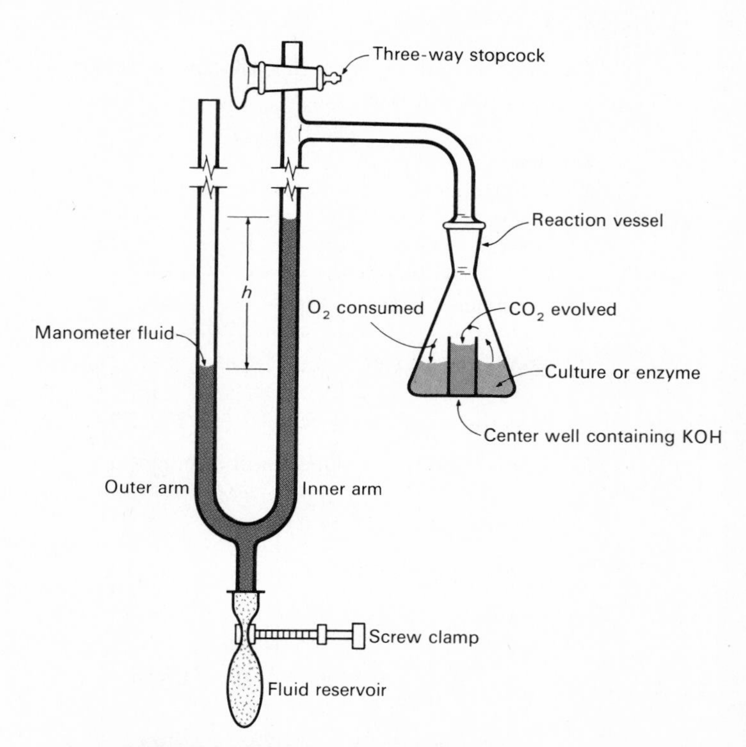

FIG. 7·8 Schematic of Warburg respirometer.

a measured quantity of waste and biological seed culture and is agitated in the water bath. After the contents are thoroughly mixed, the manometer is connected and readings are taken periodically. Biological respiration within the sample consumes oxygen and produces carbon dioxide. The carbon dioxide is absorbed by the potassium hydroxide solution. Depletion of the dissolved oxygen causes oxygen in the air space over the liquid to enter the solution, thus lowering the pressure in the flask. The quantity of oxygen consumed can then be calculated from the pressure drop as measured with the manometer.

The dissolved oxygen probe is used in conjunction with the ordinary aqueous BOD technique to reduce the amount of laboratory analyses required. Using an electric strip-chart recorder, the BOD curve can be plotted automatically as it is developed.

An electrolysis cell may also be used to obtain a continuous BOD [3, 32]. Such a cell is shown in Fig. 7·9. In it the oxygen pressure over the sample is maintained constant by continuously replacing the oxygen utilized by the microorganisms. This is accomplished by producing more oxygen by means of an electrolysis reaction in response to changes in the pressure. The BOD readings are determined by noting the length of time that the oxygen was generated and correlating it to the amount of oxygen produced by the electrolysis reaction. Advantages of the electrolysis cell over the Warburg apparatus include

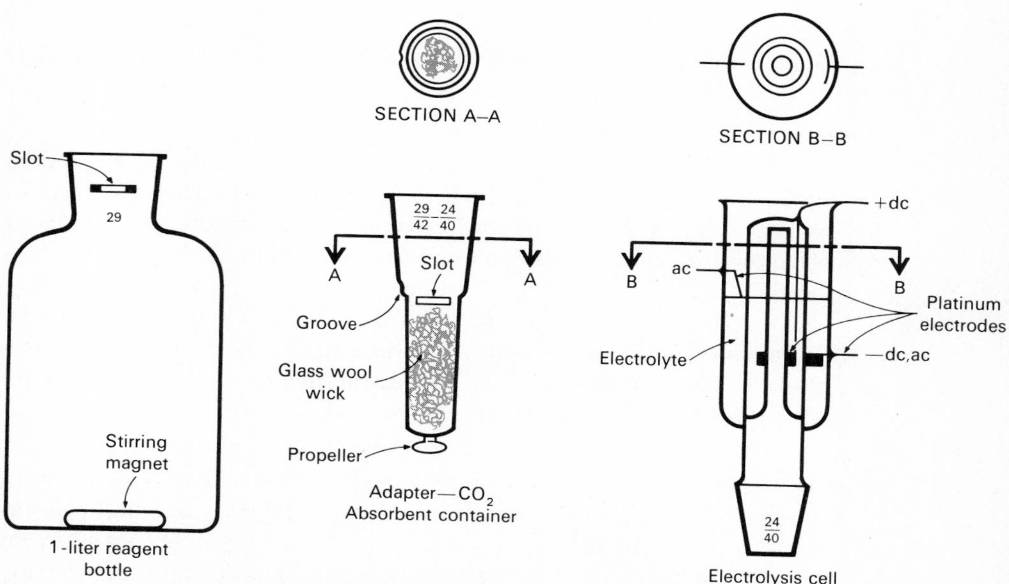

FIG. 7·9 Electrolysis cell for BOD determination [32].

the use of a large (1-liter) sample, thereby minimizing the errors of grab sampling and pipetting in dilutions, and the fact that the value of the BOD is available directly.

The limitations of the BOD determination include the necessity of having a high concentration of active, acclimated seed bacteria; the need for pretreatment when dealing with toxic wastes and the need to reduce the effects of nitrifying organisms; the arbitrary as well as long period of time required to obtain results; the fact that only the biodegradable organics are measured; and the fact that the test does not have stoichiometric validity after the soluble organic matter present in solution has been utilized.

COD The COD test is used to measure the content of organic matter of both wastewater and natural waters. The oxygen equivalent of the organic matter that can be oxidized is measured by using a strong chemical oxidizing agent in an acidic medium. Potassium dichromate has been found to be excellent for this purpose. The test must be performed at an elevated temperature. A catalyst (silver sulfate) is required to aid the oxidation of certain classes of organic compounds. Since some inorganic compounds interfere with the test, care must be taken to eliminate them. The principal reaction using dichromate as the oxidizing agent may be represented in a general way by the following unbalanced equation:

$$\text{Organic matter } (C_aH_bO_c) + Cr_2O_7^{--} + H^+ \xrightarrow[\text{heat}]{\text{catalyst}} Cr^{3+} + CO_2 + H_2O \tag{7·16}$$

The COD test is also used to measure the organic matter in industrial and municipal wastes that contain compounds that are toxic to biological life. The COD of a waste is, in general, higher than the BOD because more compounds can be chemically oxidized than can be biologically oxidized. For many types of wastes, it is possible to correlate COD with BOD. This can be very useful because the COD can be determined in 3 hr, compared with 5 days for the BOD. Once the correlation has been established, COD measurements can be used to good advantage for treatment-plant control and operation.

TOC Another means for measuring the organic matter present in water is the TOC test, which is especially applicable to small concentrations of organic matter. The test is performed by injecting a known quantity of sample into a high-temperature furnace. The organic carbon is oxidized to carbon dioxide in the presence of a catalyst. The carbon dioxide that is produced is quantitatively measured by means of an infrared analyzer. Acidification and aeration of the sample prior to analysis eliminates errors due to the presence of inorganic carbon. The test can be performed very rapidly and is becoming more popular. Certain resistant organic compounds may not be oxidized, however, and the measured TOC value will be slightly less than the actual amount present in the sample. Typical TOC values for sewage are reported in Table 7·3.

TOD Another instrumental method that can be used to measure the organic content of wastewater is the recently developed TOD test [4]. In this test, organic substances and, to a minor extent, inorganic substances are converted to stable end products in a platinum-catalyzed combustion chamber. The TOD is determined by monitoring the oxygen content present in the nitrogen carrier gas. This test can be carried out rapidly, and the results have been correlated with the COD [4].

ThOD Organic matter of animal or vegetable origin in sewage is generally a combination of carbon, hydrogen, oxygen, and nitrogen. The principal groups of these elements present in wastewater are, as previously noted, carbohydrates, proteins, fats, and products of their decomposition. The biological decomposition of the substances is discussed in Chap. 10. If the chemical formula of the organic matter is known, the ThOD may be computed as illustrated in the following example.

EXAMPLE 7·4 *Calculation of* ThOD

Determine the ThOD for glycine ($CH_2(NH_2)COOH$) using the following assumptions:

1. In the first step, the carbon atoms are oxidized to CO_2 while the nitrogen is converted to ammonia.
2. In the second and third steps, the ammonia is oxidized to nitrite and nitrate.
3. The ThOD is the sum of the oxygen required for all three steps.

Solution

1. Carbonaceous demand

$$CH_2(NH_2)COOH + \tfrac{3}{2}O_2 \rightarrow NH_3 + 2CO_2 + H_2O$$

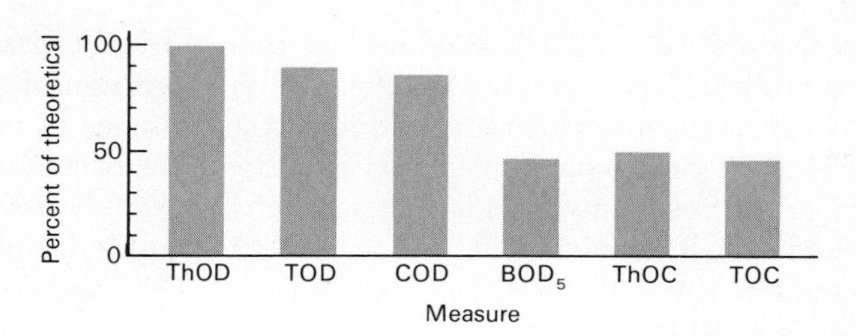

FIG. 7·10 Approximate relationship among measures of the organic content of wastewaters.

2. Nitrogenous demand

(a) $NH_3 + \frac{3}{2}O_2 \xrightarrow{\text{nitrite-forming bacteria}} HNO_2 + H_2O$

(b) $HNO_2 + \frac{1}{2}O_2 \xrightarrow{\text{nitrate-forming bacteria}} HNO_3$

$$\text{ThOD} = 3\tfrac{1}{2} \text{ moles of } O_2/\text{mole of glycine}$$
$$= 112 \text{ g } O_2/\text{mole}$$

Correlation among Measures Establishment of constant relationships among the various measures of organic content depends primarily on the nature of the wastewater and its source. In general, the relationship that exists among these parameters is reported in Table 7·3 and shown graphically in Fig. 7·10. Of all the measures, the most difficult to correlate to the others is the BOD test, because of the problems associated with biological tests (see BOD discussion). For typical domestic wastes, however, the BOD/COD and BOD/TOC ratios vary from 0.4 to 0.8 and 0.8 to 1.0, respectively. Because of the rapidity with which the COD, TOC, and TOD tests can be conducted, it is anticipated that more use will be made of these tests in the future.

Inorganic Matter

Several inorganic components of wastewaters and natural waters are important in establishing and controlling water quality. The concentrations of inorganic substances in water are increased both by the geologic formation with which the water comes in contact and by the wastewaters, treated or untreated, that are discharged to it. The natural waters dissolve some of the rocks and minerals with which they come in contact. Wastewaters, with the exception of some industrial wastes, are seldom treated for removal of the inorganic constituents that are added in the use cycle. Concentrations of inorganic constituents also are increased due to the natural evaporation process which removes some of the surface water and leaves the inorganic substance in the water. Since concentrations of various inorganic constituents can greatly affect the beneficial uses made of the waters, it is well to examine the nature of some of the constituents, particularly those added to surface water via the use cycle.

pH The hydrogen-ion concentration is an important quality parameter of both natural waters and wastewaters. The concentration range suitable for the existence of most biological life is quite narrow and critical. Wastewater with an adverse concentration of hydrogen ion is difficult to treat by biological means, and if the concentration is not altered before discharge, the wastewater effluent may alter the concentration in the natural waters.

The hydrogen-ion concentration in water is closely connected with the extent to which water molecules dissociate. Water will dissociate into hydrogen and hydroxyl ions as follows:

$$H_2O \rightleftharpoons H^+ + OH^- \tag{7·17}$$

Applying the law of mass action to this equation,

$$\frac{[H^+][OH^-]}{H_2O} = K \tag{7·18}$$

where the brackets indicate concentration of the constituents in moles/liter. Since the concentration of water in a dilute aqueous system is essentially constant, this concentration can be incorporated into the equilibrium constant K to give

$$[H^+][OH^-] = K_w \tag{7·19}$$

K_w is known as the ionization constant, or ion product, of water and is approximately equal to 1×10^{-14} at a temperature of 25°C. Equation 7·19 can be used to calculate the hydroxyl-ion concentration when the hydrogen-ion concentration is known, and vice versa.

The usual means of expressing the hydrogen-ion concentration is as pH, which is defined as the negative logarithm of the hydrogen-ion concentration.

$$\text{pH} = -\log_{10}[H^+] \tag{7·20}$$

With pOH, which is defined as the negative logarithm of the hydroxyl-ion concentration, it can be seen from Eq. 7·19 that, for water at 25°C,

$$\text{pH} + \text{pOH} = 14 \tag{7·21}$$

The pH of aqueous systems can be conveniently measured with a pH meter. Various indicator solutions that change color at definite pH values are also used. The color of the solution is compared with the color of standard tubes or disks. This method can be used only for relatively clear liquids.

Chlorides Another quality parameter of significance is the chloride concentration. Chlorides in natural water result from the leaching of chloride-containing rocks and soils with which the water comes in contact, and in coastal areas, from salt water intrusion. In addition, agricultural, industrial, and domestic wastewaters discharged to surface waters are a source of chlorides.

Human excreta, for example, contain about 6 g of chlorides per person per day. In areas where the hardness of water is high, water softeners will also add large quantities of chlorides. Since conventional methods of waste treatment do not remove chloride to any significant extent, higher-than-usual chloride concentrations can be taken as an indication that the body of water is being used for waste disposal. Infiltration of ground water into sewers adjacent to salt water is also a potential source of high chlorides as well as sulfates.

Alkalinity Alkalinity in sewage is due to the presence of the hydroxides, carbonates, and bicarbonates of elements such as calcium, magnesium, sodium, potassium, or of ammonia. Of these, calcium and magnesium bicarbonates are most common. Sewage is normally alkaline, receiving its alkalinity from the water supply, the ground water, and the materials added during domestic usage. Alkalinity is determined by titrating against a standard acid; the results are expressed in terms of calcium carbonate $CaCO_3$. The concentration of alkalinity in sewage is important where chemical treatment is to be employed (see Chap. 11) and where ammonia is to be removed by air stripping (see Chap. 14).

Nitrogen The elements nitrogen and phosphorus are essential to the growth of protista and plants and as such are known as nutrients or biostimulants. Trace quantities of other elements, such as iron, are also needed for biological growth, but nitrogen and phosphorus are, in most cases, the major nutrients of importance.

Since nitrogen is an essential building block in the synthesis of protein, nitrogen data will be required to evaluate the treatability of sewage and industrial wastes by biological processes. Insufficient nitrogen can necessitate the addition of nitrogen to make the waste treatable. Nutrient requirements for biological waste treatment are defined in Chap. 12. Where control of algal growths in the receiving water is necessary to protect beneficial uses, removal or reduction of nitrogen in wastewaters prior to discharge may be desirable (see Chap. 14).

The various forms of nitrogen that are present in nature and the pathways by which the forms are changed are depicted in Fig. 7·11.

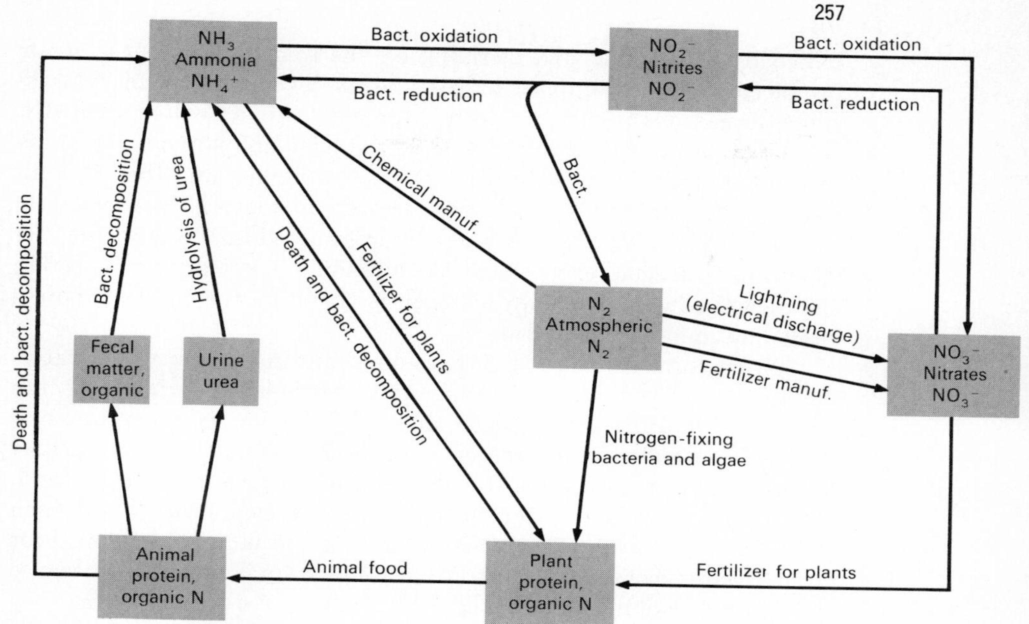

FIG. 7·11 Nitrogen cycle [15].

The nitrogen present in fresh sewage (see Table 7·3) is primarily combined in proteinaceous matter and urea. Decomposition by bacteria readily changes the form to ammonia. The age of sewage is indicated by the relative amount of ammonia that is present. In an aerobic environment, bacteria can oxidize the ammonia nitrogen to nitrites and nitrates (see Example 7·4). The predominance of nitrate nitrogen in wastewater indicates that the waste has been stabilized with respect to oxygen demand. Nitrates, however, can be used by algae and other aquatic plants to form plant protein which, in turn, can be used by animals to form animal protein. Death and decomposition of the plant and animal protein by bacteria again yields ammonia. Thus, if nitrogen in the form of nitrates can be reused to make protein by algae and other plants, it may be necessary to remove or to reduce the nitrogen that is present to prevent these growths.

Ammonia nitrogen exists in aqueous solution as either the ammonium ion or ammonia, depending on the pH of the solution, in accordance with the following equilibrium reaction:

$$NH_3 + H_2O \rightleftharpoons NH_4^+ + OH^- \tag{7·22}$$

At pH levels above 7, the equilibrium is displaced to the left; at levels below pH 7, the ammonium ion is predominant. Ammonia is determined by raising the pH, distilling off the ammonia with the steam produced when the sample is boiled, and condensing the steam that absorbs the gaseous ammonia. The measurement is made colorimetrically.

Organic nitrogen is determined by the Kjeldahl method. The aqueous sample is first boiled to drive off the ammonia, and then it is digested. During the digestion the organic nitrogen is converted to ammonia. Total Kjeldahl nitrogen is determined in the same manner as organic nitrogen, except that the ammonia is not driven off before the digestion step. Kjeldahl nitrogen is therefore the total of the organic and ammonia nitrogen.

Nitrite nitrogen is relatively unimportant in wastewater or water-pollution studies because it is unstable and is easily oxidized to the nitrate form. It is an indicator of past pollution in the process of stabilization and seldom exceeds 1 mg/liter in sewage or 0.1 mg/liter in surface or ground waters. It is determined by a colorimetric method.

Nitrate nitrogen is the most highly oxidized form of nitrogen found in wastewaters. Where secondary effluent is to be reclaimed for ground water recharge, the nitrate concentration is important because the USPHS drinking water standards [26] limit it to 45 mg/liter as NO_3^- due to its serious and occasionally fatal effects on infants. Nitrates may vary in concentration from 0 to 20 mg/liter as N in sewage effluents with a typical range from 15 to 20 mg/liter as N. The nitrate concentration is also usually determined by colorimetric methods.

Phosphorus Phosphorus is also essential to the growth of algae and other biological organisms. Because of noxious algal blooms that occur in surface waters, there is presently much interest in controlling the amount of phosphorus compounds that enter surface waters in domestic and industrial waste discharges and natural runoff. Municipal wastewaters, for example, may contain from 6 to 20 mg/liter of phosphorus as P (see Table 7·3).

The usual forms of phosphorus that are found in aqueous solutions include the orthophosphate, polyphosphate, and organic phosphate. The orthophosphates, for example, PO_4^{3+}, HPO_4^{--}, $H_2PO_4^-$, and H_3PO_4, are available for biological metabolism without further breakdown. The polyphosphates include those molecules with two or more phosphorus atoms, oxygen atoms, and in some cases, hydrogen atoms combined in a complex molecule. Polyphosphates undergo hydrolysis in aqueous solutions and revert to the orthophosphate forms; however, this hydrolysis is usually quite slow. The organically bound phosphorus is usually of minor importance in most domestic wastes, but it can be an important constituent of industrial wastes and sewage sludges.

Orthophosphate can be determined by directly adding a substance

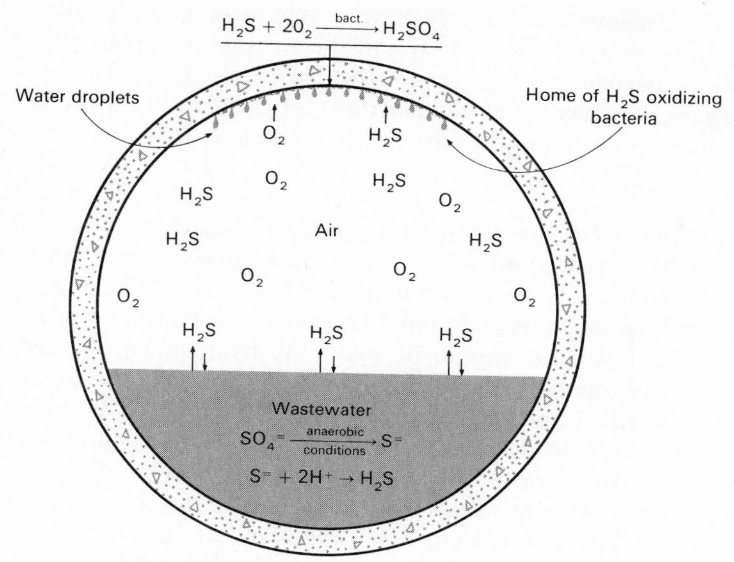

FIG. 7·12 Sewer corrosion due to H_2S oxidation [15].

such as ammonium molybdate which will form a colored complex with the phosphate. The polyphosphates and organic phosphates must be converted to orthophosphates before they can be determined in a similar manner.

Sulfur The sulfate ion occurs naturally in most water supplies and is present in sewage as well. Sulfur is required in the synthesis of proteins and is released in their degradation. Sulfates are chemically reduced to sulfides and to hydrogen sulfide (H_2S) by bacteria under anaerobic conditions, as shown in the following equations:

$$SO_4^{--} + \text{organic matter} \xrightarrow{\text{bacteria}} S^{--} + H_2O + CO_2 \qquad (7\cdot23)$$

$$S^{--} + 2H^+ \rightarrow H_2S \qquad (7\cdot24)$$

As shown in Fig. 7·12, the H_2S can then be oxidized biologically to sulfuric acid, which is corrosive to sewer pipes.

Sulfates are reduced to sulfides in sludge digesters and may upset the biological process if the sulfide concentration exceeds 200 mg/liter. Fortunately, such concentrations are rare. The H_2S gas, which is evolved and mixed with the sewage gas ($CH_4 + CO_2$), is corrosive to the gas piping and, if burned in gas engines, the products of combustion can damage the engine and severely corrode exhaust-gas heat-recovery equipment, especially if allowed to cool below the dew point. One prominent engine manufacturer limits the H_2S content to 60 grains/100 cu ft. Gas scrubbers that reduce the H_2S content to 50 grains/100 cu ft have proven satisfactory at the Miami sewage treatment plant.

Toxic Compounds Because of their toxicity, certain cations are of great importance in the treatment and disposal of wastewaters. Copper, lead, silver, chromium, arsenic, and boron are toxic in varying degrees to microorganisms and therefore must be taken into consideration in the design of a biological treatment plant. Many plants have been upset by the introduction of these ions to the extent that the microorganisms were killed and treatment ceased. For instance, in sludge digesters, copper is toxic at a concentration of 100 mg/liter; chromium and nickel are toxic at concentrations of 500 mg/liter; and sodium is also toxic at high concentrations [7]. Other toxic cations include potassium and ammonium at 4,000 mg/liter. The alkalinity present in the digesting sludge will combine with and precipitate the calcium ions before the calcium concentration approaches the toxic level.

Some toxic anions, including cyanides and chromates, are also present in industrial wastes. These are found particularly in metal-plating wastes and should be removed by pretreatment at the site of the factory rather than be mixed with the municipal sewage. Fluoride is another toxic anion. Organic compounds present in some industrial wastes are also toxic.

Heavy Metals Trace quantities of many metals, such as nickel (Ni), manganese (Mn), lead (Pb), chromium (Cr), cadmium (Cd), zinc (Zn), copper (Cu), iron (Fe), and mercury (Hg) are important constituents of most waters. Some of these metals are necessary for growth of biological life, and absence of sufficient quantities of them could limit growth of algae, for example. The presence of any of these metals in excessive quantities will interfere with many beneficial uses of the water because of their toxicity; therefore, it is frequently desirable to measure and control the concentrations of these substances.

Methods for determining the concentrations of these substances vary in complexity according to the interfering substances that may be present [19]. In addition, quantities of many of these metals can be determined at very low concentrations by such instrumental methods as polarography and atomic absorption spectroscopy. For a review of the effects of heavy metals on the environment, the reference work by McKee and Wolf [8] is recommended. Since research in this area is constantly under way, a review of the current literature is also recommended.

Gases

Gases commonly found in raw sewage include nitrogen (N_2), oxygen (O_2), carbon dioxide (CO_2), hydrogen sulfide (H_2S), ammonia (NH_3), and methane (CH_4). The first three are common gases of the atmosphere and will be found in all waters exposed to air. The latter three are derived from the decomposition of the organic matter present in sewage. Although not found in raw sewage, other gases with which the sanitary engineer must be familiar include chlorine (Cl_2) and ozone (O_3) (disinfection and odor control), and the oxides of sulfur and nitrogen (combustion processes). The following discussion is limited to those gases that are of interest in raw wastewater. Under most circumstances, the ammonia in raw wastewater will be present as the ammonium ion (see "Nitrogen"). Therefore, it will be considered further in Chap. 14 rather than here.

Dissolved Oxygen Dissolved oxygen is required for the respiration of aerobic microorganisms as well as all other aerobic life forms. However, oxygen is only slightly soluble in water. The actual quantity of oxygen (other gases too) that can be present in solution is governed by (1) the solubility of the gas, (2) the partial pressure of the gas in the atmosphere, (3) the temperature, and (4) the purity (salinity, suspended solids, etc.) of the water. The interrelationship of these variables is delineated in Chap. 8 and is illustrated in Appendix C, where the effect of temperature and salinity on dissolved oxygen concentration is presented.

Because the rate of biochemical reactions that utilize oxygen increases with increasing temperature, dissolved oxygen levels tend to be more critical in the summer months. The problem is compounded in summer months because stream flows are usually lower, and thus the total quantity of oxygen available is also lower. The presence of dissolved oxygen in wastewater is desirable because it prevents the formation of noxious odors. The role of oxygen in wastewater treatment is discussed in Chaps. 10 and 12; its importance in water-quality management is discussed in Chap. 15.

Hydrogen Sulfide Hydrogen sulfide is formed, as mentioned previously, from the decomposition of organic matter containing sulfur or from the reduction of mineral sulfites and sulfates. It is not formed in the presence of an abundant supply of oxygen. This gas is a colorless, inflammable compound having the characteristic odor of rotten eggs. The blackening of sewage and sludge is usually due to the formation of hydrogen sulfide that has combined with the iron present to form ferrous sulfide (FeS). Although hydrogen sulfide is the most important gas formed from the standpoint of odors, other volatile compounds such as indol, skatol, and mercaptans, which may also be formed during anaerobic decomposition, may cause odors far more offensive than that of hydrogen sulfide.

Methane The principal by-product of the anaerobic decomposition of the organic matter in wastewater is methane gas (see Chap. 13). Methane is a colorless, odorless, combustible hydrocarbon of high fuel value. Normally, large quantities are not encountered in wastewater because even small amounts of oxygen tend to be toxic to the organisms responsible for the production of methane (see Chap. 10). Occasionally, however, as a result of anaerobic decay in accumulated bottom deposits, methane has been produced. Because methane is highly combustible and the explosion hazard is high, manholes and sewer junctions or junction chambers where there is an opportunity for gas to collect should be ventilated with a portable blower during and before the time required for men to work in them on inspection, renewals, or repairs. In treatment plants, notices should be posted about the plant warning of explosion hazards, and plant employees should be instructed in safety measures to be maintained while working in and about the structures where gas may be present.

BIOLOGICAL CHARACTERISTICS: DEFINITION AND APPLICATION

Biological aspects with which the sanitary engineer must be familiar include knowledge of the principal groups of microorganisms found in surface and wastewaters as well as those responsible for biological treatment, knowledge of the organisms used as indicators of pollution and their significance, and knowledge of the methods used to evaluate the toxicity of treated wastewaters. These matters are discussed in this section.

Microorganisms

The principal groups of organisms found in surface and wastewater are classified as protista, plants, and animals (see Chap. 10). The category protista includes bacteria, fungi, protozoa, and algae. Seed plants, ferns, and mosses and liverworts are classified as plants. Invertebrates and vertebrates are classified as animals [20]. Viruses, which are also found in wastewater, are classified according to the host infected. Because the organisms in the various groups are discussed in detail in the subsequent chapters of this book, the following discussion is meant to serve only as a general introduction to the various groups and their importance in the field of wastewater treatment and water-quality management.

Protista As a class, protista are the most important group of organisms with which the sanitary engineer must be familiar, especially the bacteria, algae, and protozoa.

Because of the extensive and fundamental role played by bacteria in the decomposition and stabilization of organic matter, both in nature and in treatment plants, their characteristics, functions, metabolism, and synthesis must be understood. These subjects are discussed extensively in Chap. 10. Coliform bacteria are also used as an indicator of pollution by human wastes. Their significance and some of the tests used to determine their presence are discussed in a subsequent section.

Algae can be a great nuisance in surface waters because, when conditions are right, they will rapidly reproduce and cover streams, lakes, and reservoirs in large floating colonies called blooms. Algal blooms are usually characteristic of what is called a eutrophic lake, or a lake with a high content of the compounds needed for biological growth. Because effluent from wastewater treatment plants is usually high in biological nutrients, discharge of the effluent to lakes causes enrichment and increases the rate of eutrophication. The same effects can also occur in streams.

The presence of algae affects the value of water for water supply because they often cause taste and odor problems. Algae can also alter the value of surface waters for the growth of certain kinds of fish and other aquatic life, for recreation, and for other beneficial uses. Determination of the concentration of algae in surface waters involves collecting the sample by one of several possible methods and microscopically counting them. Detailed procedures for algae counts are outlined in *Standard Methods* [19]. Pictures and descriptions of common algae may be found in Refs. [5, 10, 13, 18, and 27].

One of the most important problems facing the sanitary engineering profession in terms of water-quality management is how to treat wastes of various origins so that the effluents do not encourage the growth of algae and other aquatic plants. The solution may involve the removal of carbon, the removal of various forms of nitrogen and phosphorus, and possibly the removal of some of the trace elements, such as iron and cobalt.

Protozoa of importance to sanitary engineers include amoebas, flagellates, and free-swimming and stalked ciliates. These protists feed on bacteria and other microscopic protists and are essential in the operation of biological treatment processes and in the purification of streams because they maintain a natural balance among the different groups of microorganisms. Detailed descriptions of these organisms may be found in Refs. [5, 20, and 27].

Viruses The functioning of viruses with respect to other organisms is explained in Chap. 10. Viruses that are excreted by humans may become a major hazard to public heath. For example, from experimental studies, it has been found that from 10,000 to 100,000 infectious doses of hepatitis virus are emitted from each gram of feces of a patient ill with this disease [1]. It is known that some viruses will live as long as 41 days in water or sewage at 20°C and for 6 days in a normal river. A number of outbreaks of infectious hepatitis have been attributed to transmission of the virus through water supplies. Much more study is required on the part of biologists and engineers to determine the mechanics of travel and removal of virus in soils, surface waters, and sewage treatment plants.

Plants and Animals Plants and animals of importance range in size from microscopic rotifers and worms to macroscopic crustaceans. A knowledge of these organisms is helpful in evaluating the condition of streams and lakes, in determining the toxicity of wastewaters discharged to the environment, and in observing the effectiveness of biological life in the secondary treatment processes used to destroy organic wastes. Detailed descriptions of these organisms may be found in Refs. [5, 11, and 27].

Coliform Organisms

The intestinal tract of man contains countless rod-shaped bacteria known as coliform organisms. Each person discharges from 100 to 400 billion coliform organisms per day, in addition to other kinds of bacteria. Coliforms are harmless to man and are, in fact, useful in destroying organic matter in biological waste treatment processes.

Pathogenic organisms may be discharged by humans who are infected with disease or who are carriers of a particular disease. The usual pathogenic organisms that may be excreted by man cause diseases of the gastrointestinal tract, such as typhoid fever, dysentery, diarrhea, and in certain parts of the world, cholera.

Because the numbers of pathogenic organisms present in wastes and polluted waters are few and difficult to isolate, the coliform organism, which is more numerous and more easily tested for, is used as an indicator organism. The presence of coliform organisms is taken as an indication that pathogenic organisms may also be present, and the absence of coliform organisms is taken as an indication that the water is free from disease-producing organisms.

The coliform bacteria include the genera *Escherichia* and *Aerobacter*. The use of coliforms as indicator organisms is complicated by the fact that *Aerobacter* and certain *Escherichia* can grow in soil. Thus, the presence of coliforms does not always mean contamination with human wastes. Apparently, *Escherichia coli* (*E. coli*) are entirely of fecal origin. There is difficulty in determining *E. coli* to the exclusion of the soil coliforms; as a result, the entire coliform group is used as an indicator of fecal pollution.

In recent years, tests have been developed that distinguish between total coliforms, fecal coliforms, and fecal streptococci; and all three are being reported in the literature, particularly in articles dealing with urban runoff. For a discussion and references, see USPHS *Drinking Water Standards* [26].

The usual procedure for determining the presence of coliforms consists of the presumptive and the confirmed tests. The presumptive test is based on the ability of the coliform group to ferment lactose broth, producing gas. The confirmed test consists of growing cultures of coliform bacteria on media that suppress the growth of other organisms. The completed test is based on the ability of the cultures grown in the confirmed test to again ferment the lactose broth.

There now are two accepted methods for obtaining the numbers of coliform organisms present in a given volume of water. The most probable number (MPN) technique has been used for a long time and is based on a statistical analysis of the number of positive and negative results obtained when testing multiple portions of equal volume and in portions constituting a geometric series for the presence of coliform. It is emphasized that the MPN is not the absolute concentration of organisms that are present but only a statistical estimate of that concentration. Complete MPN tables are provided in Appendix D, and use of these tables is illustrated in Examples 7·5 and 7·6.

EXAMPLE 7·5 *Calculation of* MPN

Determine the coliform density (MPN) for a surface water, the bacterial analysis of which yielded the following results for the standard confirmed test.

Size of portion, ml	No. positive	No. negative
10.0	4	1
1.0	4	1
0.1	2	3
0.01	0	5

Solution
From Appendix D, eliminating the portion with no positive tubes, as outlined in *Standard Methods* [19], the MPN per 100 ml is 47.

EXAMPLE 7·6 MPN *of Sewage Overflow*

A sample of a combined-sewer overflow is tested for coliforms by the MPN method with the following results. Determine the coliform density.

Size of portion, ml	No. positive	No. negative
0.001	5	0
0.0001	5	0
0.00001	5	0
0.000001	3	2

Solution
Eliminate the top line, as selection of the top three lines leads to no solution. Using the bottom three lines, by reference to Appendix D, if the ml of sewage in the samples were 10, 1, and 0.1, the MPN would have been 920. Since the samples are 100,000 times more dilute, the MPN per 100 ml of the original sample is 92,000,000 or 920,000 per ml.

The membrane-filter technique can also be used to determine the number of coliform organisms that are present in water. The determination is accomplished by passing a known volume of water sample through a membrane filter that has a very small pore size. The bacteria are retained on the filter because they are larger than the pores. The bacteria are then contacted with an agar that contains nutrients necessary for the growth of the bacteria. After incubation, the coliform colonies can be counted and the concentration in the original water sample determined. The membrane-filter technique has the advantage of being faster than the MPN procedure and of giving a direct count of the number of coliforms. Both methods are subject to limitations, however. Detailed procedures are given for both methods in *Standard Methods* [19].

Bioassay Tests

The results of bioassay tests are used to evaluate the toxicity of wastewaters to the biological life of the receiving waters. The specific objectives of the bioassay test are (1) to determine the concentration of a given waste that will kill 50 percent of the test organisms in a specified time period and (2) to determine the maximum concentration causing no apparent effect on the test organisms in 96 hr. These objectives are achieved by introducing fish or other appropriate test animals into test aquaria containing various concentrations of the waste in question and observing their survival with time. Observations are usually made after 24, 48, and 96 hr. General test procedures, evaluation of test results, and application of test results are described in the following discussion.

Test Procedures The procedures to be followed in the conduct of routine bioassay tests have been summarized and reported in *Standard Methods* [19]. The routine test is applied widely for detecting and evaluating acute toxicity that is not associated with excessive oxygen demand and is due to substances that are relatively stable and not extremely volatile. Death of the test organisms due to a deficiency of dissolved oxygen in polluted water should be distinguished from death due to toxicity. To detect and evaluate the direct lethality of the wastes only, adequate dissolved oxygen must be maintained during the toxicity tests.

When it is suspected that the toxicity of the test solutions declines rapidly during the course of a test, a modification of the routine procedure is recommended. Reduction of the solution's toxicity may result from reduction or removal of toxic components due to extreme volatility by oxidation, by hydrolysis, by precipitation, or by their combination with the test fish's metabolic by-products, etc.

The validity of the test depends partly on the selection of the proper species and partly on the characteristics of the receiving water where the waste is discharged. For tests relating to estuarine pollution, species such as sticklebacks, killfish, or mosquitofish (*Gambusia*) are suitable for test animals because of their tolerance to a wide variation in salinity, their abundance in many coastal waters, and their size. For freshwater tests, species such as sticklebacks, mosquitofish, minnows, trout, and sunfish have been used successfully.

For details on the size of the test aquaria, temperature, maintenance, selection of test specimens, and other related matters, *Standard Methods* [19] should be consulted.

Evaluation of Results The prescribed measure of acute toxicity is the median tolerance limit (TL_m), defined as the concentration of waste (toxicant) in which just 50 percent of the test animals are able to survive for a specified period of exposure. The method of calculating TL_m values is illustrated in Example 7·7. The TL_m values are only estimates of the acute toxicity for the waste under laboratory conditions. The values obtained do not represent the concentrations of the tested waste that may be considered harmless to the diverse aquatic biota. Waste concentrations that are not toxic to selected test organisms in a 96-hr period may be very toxic to these same animals or other economically or ecologically important species under conditions of continuous or chronic exposure. A more meaningful approach for evaluating the results of bioassay tests, based on a consideration of threshold toxicity, has been proposed by Chen and Selleck [2].

EXAMPLE 7·7 *Analysis of Bioassay Data*

Using the following hypothetical data determine the 24- and 48-hr TL_m values in percent by volume.

Concentration of waste, % by vol	No. of test animals	No. of test animals surviving	
		After 24 hr	After 48 hr
40	20	1	0
20	20	8	0
10	20	14	6
5	20	20	13
3	20	20	16

Solution

1. Plot concentration of waste in percent by volume against test animals surviving in percent.
2. Connect data points on either side of 50 percent survival for 24 and 48 hr.
3. Find waste concentration for 50 percent survival.

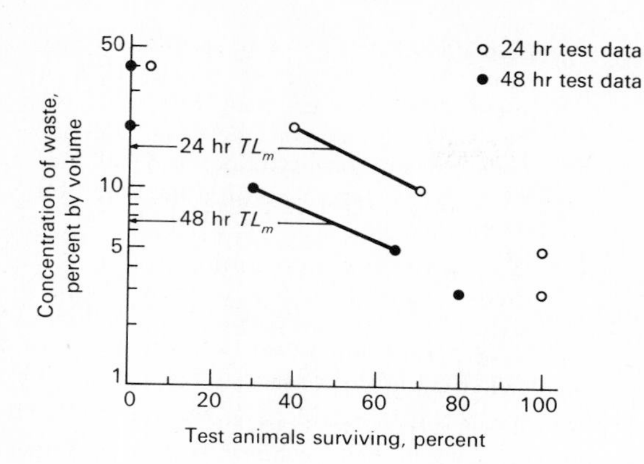

FIG. 7·13 Graphical presentation and analysis of bioassay test data.

4. The TL_m values as shown in Fig. 7·13 are 16.0 percent for 24 hr and 6.7 percent for 48 hr.

Application of Results When estimating the permissible waste dilution ratios based on acute toxicity bioassay test results, some application factor must be used. The application factor is defined as the concentration of the material or waste that is not harmful under long-term or continuous exposure divided by the 96-hr TL_m value for that waste.

In the absence of toxicity data other than the 96-hr TL_m, the National Technical Advisory Committee on Water Quality Criteria has recommended application factors varying from $\frac{1}{10}$ to $\frac{1}{100}$ of the 96-hr TL_m depending on the nature and characteristics of the waste [28]. In some cases, this recommendation is qualified to include a requirement that it be demonstrated that the safe levels prescribed on the basis of the 96-hr TL_m and the application factor do not cause decreases in productivity or diversity of the receiving water biota [28].

PROBLEMS

7·1 The variations of flow and BOD with time at a treatment plant are given in Fig. 7·14. Compute both the average and the weighted average BOD.

7·2 If a sample of sewage from a city with a population of 100,000 and an average flow of 100 gpcd contains 5 ml/liter settleable solids, how many cubic feet of sludge will be produced per day?

7·3 In a BOD determination, 6 ml of sewage are mixed with 294 ml of diluting water containing 8.6 mg/liter of dissolved oxygen. After a 5-day incubation at 20°C, the dissolved oxygen content of the mixture is 5.4 mg/liter. Calculate the BOD of the sewage. Assume the initial dissolved oxygen of sewage is zero.

7·4 The 5-day 20°C BOD of a sewage is 210 mg/liter. What will be the ultimate BOD? What will be the 10-day demand? If the bottle had been incubated at 30°C, what would the 5-day BOD have been? $K' = 0.23$.

7·5 The following BOD results were obtained on a sample of raw sewage at 20°C:

t, days	0	1	2	3	4	5
y, mg/liter	0	65	109	138	158	172

Compute the reaction constant K and ultimate first-stage BOD using both the least-squares and the Thomas methods.

7·6 Compute the carbonaceous and nitrogenous oxygen demand of a waste represented by the formula $C_9N_2H_6O_2$. (N is converted to NH_3 in the first step.)

7·7 The dissolved oxygen of a tidal estuary must be maintained at 4.5 mg/liter or more. The average temperature of the water during summer months is 24°C and the chloride concentration is 5,000 mg/liter. What percent saturation does this represent?

7·8 The results of a presumptive coliform test were 4 of 5 tubes of 10-ml portions positive, 3 of 5 tubes of 1-ml portions positive, and 0 of 5 tubes of 0.1-ml portions positive. What is the MPN per 100 ml?

7·9 Discuss the advantages and disadvantages of using the fecal coliform test to indicate bacteriological pollution.

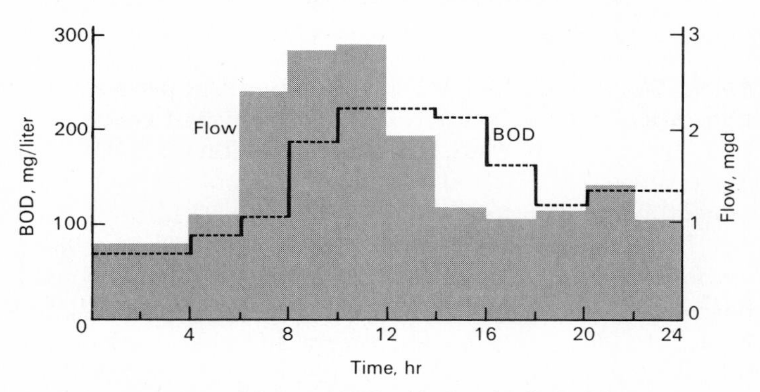

FIG. 7·14 Variations of flow and BOD with time for Prob. 7·1.

REFERENCES

1. Berg, G.: *Transmission of Viruses by the Water Route*, Wiley, New York, 1965.
2. Chen, C. W., and R. E. Selleck: A Kinetic Model of Fish Toxicity Threshold, *J. WPCF*, vol. 41, no. 8, part 2, 1969.
3. Clark, J. W.: *New Method for Biochemical Oxygen Demand*, Eng. Experimental Sta., New Mexico State University, Bulletin 11, 1959.
4. Clifford, D.: *Total Oxygen Demand—A New Instrumental Method*, American Chemical Society, Midland, Mich., 1967.
5. Hutchinson, G. E.: *A Treatise on Limnology*, vol. II, Wiley, New York, 1967.
6. Klein, S. A., and P. H. McGauhey: Degradation of Biologically Soft Detergents by Wastewater Treatment Processes, *J. WPCF*, vol. 37, no. 6, 1965.
7. McCarty, P. L.: Anaerobic Waste Treatment Fundamentals, *Public Works*, vol. 95, no. 11, 1964.
8. McKee, J. E., and H. W. Wolf: *Water Quality Criteria*, 2d ed., Report to California State Water Quality Control Board, Publication 3A, 1963.

9. Moore, E. W., H. A. Thomas, and W. B. Snow: Simplified Method for Analysis of BOD Data, *Sewage and Industrial Wastes*, vol. 22, no. 10, 1950.

10. Palmer, E. M.: *Algae in Water Supplies*, U.S. Public Health Service, Pub. 657, Washington, D.C., 1959.

11. Pennak, R. W.: *Fresh-Water Invertebrates of the United States*, Ronald Press, New York, 1953.

12. Phelps, E. B.: *Stream Sanitation*, Wiley, New York, 1944.

13. Prescott, G. W.: *The Freshwater Algae*, 2d ed., Brown Company, Dubuque, Iowa, 1970.

14. Sawyer, C. N., and L. Bradney: Modernization of the BOD Test for Determining the Efficiency of Sewage Treatment Processes, *Sewage Works Journal*, vol. 18, no. 6, 1946.

15. Sawyer, C. N., and P. L. McCarty: *Chemistry for Sanitary Engineers*, 2d ed., McGraw-Hill, New York, 1967.

16. Schroepfer, G. J., M. L. Robins, and R. H. Susag: The Research Program on the Mississippi River in the Vicinity of Minneapolis and St. Paul, *Advances in Water Pollution Research*, vol. 1, Pergamon, London, 1964.

17. Sheehy, J. P.: Rapid Methods for Solving Monomolecular Equations, *J. WPCF*, vol. 32, no. 6, 1960.

18. Smith, G. M.: *The Fresh-Water Algae of the United States*, McGraw-Hill, New York, 1950.

19. *Standard Methods for the Examination of Water and Waste Water*, 13th ed., American Public Health Association, 1971.

20. Stanier, R. Y., M. Doudoroff, and E. A. Adelberg: *The Microbial World*, 3d ed., Prentice-Hall, Englewood Cliffs, N.J., 1970.

21. Stumm, W., and J. J. Morgan: *Aquatic Chemistry*, Wiley-Interscience, New York, 1970.

22. Tchobanoglous, G., and R. Eliassen: The Indirect Cycle of Water Reuse, *Water and Wastes Engineering*, vol. 6, no. 2, 1969.

23. Theroux, F. R., E. F. Eldridge, and W. L. Mallmann: *Analysis of Water and Sewage*, McGraw-Hill, New York, 1943.

24. Thomas, H. A., Jr.: Graphical Determination of BOD Curve Constants, *Water and Sewage Works*, vol. 97, p. 123, 1950.

25. Tsivoglou, E. C.: *Oxygen Relationships in Streams*, Robert A. Taft Sanitary Engineering Center, Technical Report W-58-2, 1958.

26. U.S. Public Health Service: *Public Health Service Drinking Water Standards*, Washington, D.C., 1962.

27. Ward, H. B., and G. C. Whipple (ed. by W. T. Edmondson): *Fresh Water Biology*, 2d ed., Wiley, New York, 1959.

28. *Water Quality Criteria*, National Technical Advisory Committee, Federal Water Pollution Control Administration, Washington, D.C., 1968.

29. Waugh, A. E.: *Elements of Statistical Method*, 2d ed., McGraw-Hill, New York, 1943.

30. Young, H. D.: *Statistical Treatment of Experimental Data*, McGraw-Hill, New York, 1962.

31. Young, J. C.: Chemical Methods for Nitrification Control, *Proceedings of the 24th Industrial Waste Conference*, part 2, Purdue University, Lafayette, Indiana, 1969.

32. Young, J. C., W. Garner, and J. W. Clark: An Improved Apparatus for Biochemical Oxygen Demand, *Analytical Chemistry*, vol. 37, p. 784, 1965.

physical unit operations

8

Contaminants in wastewater are removed by physical, chemical, and biological means. Means of treatment in which the application of physical forces predominate are known as unit operations and are dealt with extensively in chemical engineering textbooks [13, 18] and reference books [1, 15]. Thus, screening, mixing, flocculation, sedimentation, flotation, elutriation, vacuum filtration, heat transfer, and drying are unit operations. Means of treatment in which the removal of contaminants is brought about by the addition of chemicals or by biological activity are known as unit processes. Precipitation, combustion, and biological oxidation are examples of unit processes.

The unit operations listed above, which are discussed in this chapter, and the chemical and biological unit processes, which are discussed in Chaps. 9 and 10, respectively, occur in various combinations in different treatment systems, but the fundamental principles of their operation do not change; hence, as in chemical engineering, it has been found advantageous to study the scientific basis of their operation as units.

The principles developed in these chapters (8 to 10) will be applied to the design of treatment facilities in Chaps. 11 and 12. The design of facilities for sludge process-

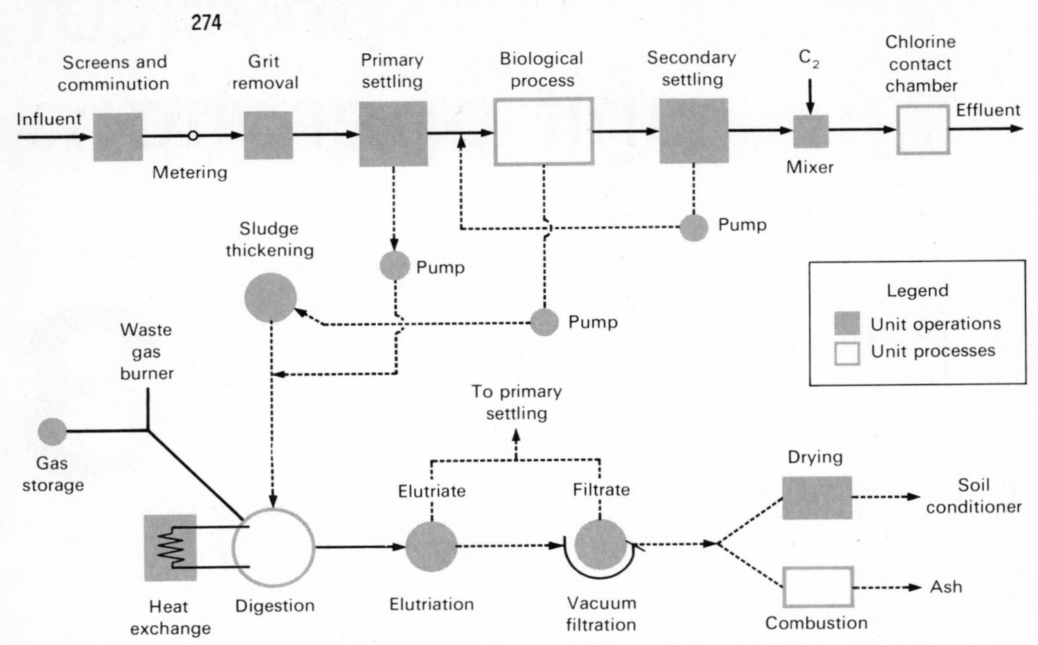

FIG. 8·1 **Wastewater treatment plant flowsheet.**

ing and disposal is discussed separately in detail in Chap. 13. Unit operations and processes associated with advanced waste treatment are discussed and applied in Chap. 14. In these latter four chapters, various unit operations and unit processes will be brought together in treatment plant flowsheets such as the one shown in Fig. 8·1.

SCREENING

The first unit operation encountered in wastewater treatment plants is the filtering operation of screening. A screen is a device with openings, generally of uniform size, used to retain coarse sewage solids. The screening element may consist of parallel bars, rods or wires, grating, wire mesh, or perforated plate, and the openings may be of any shape, generally circular or rectangular slots. A screen composed of parallel bars or rods is called a rack. Although a rack is a screening device, the use of the term "screen" should be limited to the type employing wire cloth or perforated plates. However, the function performed by a rack is called screening, and the material removed by it is known as screenings, although rakings is more convenient in some cases.

According to the method of cleaning, racks and screens are designated as hand cleaned or mechanically cleaned. According to the size of openings, screens are designated as coarse or fine. Coarse screens have openings of $\frac{1}{4}$ in. or more, and fine screens have openings of less than $\frac{1}{4}$ in.

Coarse screening devices in sewage treatment consist mainly of bar racks, which are used to protect pumps, valves, pipelines, and other appurtenances from damage or clogging by rags and large objects. Industrial waste plants may or may not need them, depending on the character of the wastes. Suspended particles greater than $\frac{1}{4}$ in. can be removed more economically by screening than by other unit operations.

Fine screens are generally of the disk or drum type. The microstrainer, discussed in Chap. 14, is a fine screening device of the drum type.

The grinding up (comminution) of screenings and their disposal is discussed in Chap. 11 (see "Racks, Fine Screens, and Comminutors").

Racks

Bar racks are made of bars of steel welded into a frame that fits across the channel. The bars run vertically or at a slope varying from 30 to 80° with the horizontal. A mechanically cleaned bar rack is shown in Fig. 8·2. Large objects are caught on the rack, carried up by traveling rakes, and scraped off and collected.

Hydraulic losses through bar racks are a function of bar shape and the velocity head of the flow between the bars. Kirschmer [10] has proposed the following equation for head loss:

$$h_L = \beta \left(\frac{w}{b}\right)^{\frac{4}{3}} h_v \sin \theta \tag{8·1}$$

where h_L = head loss, ft
β = a bar-shape factor
w = maximum cross-sectional width of bars facing direction of flow, ft
b = minimum clear spacing of bars, ft
h_v = velocity head of flow approaching rack, ft
θ = angle of rack with horizontal

Kirschmer's values of β for several shapes of bars are given in Table 8·1. The head loss calculated from Eq. 8·1 applies only when the bars are clean. Head loss increases with the degree of clogging.

Fine Screens

Fine screens of the inclined disk or drum type, whose screening media consisted of bronze or copper plates with milled slots $\frac{1}{32}$ to $\frac{3}{32}$ in. wide

FIG. 8·2 Schematic of a mechanically cleaned bar rack from REX Chain Belt].

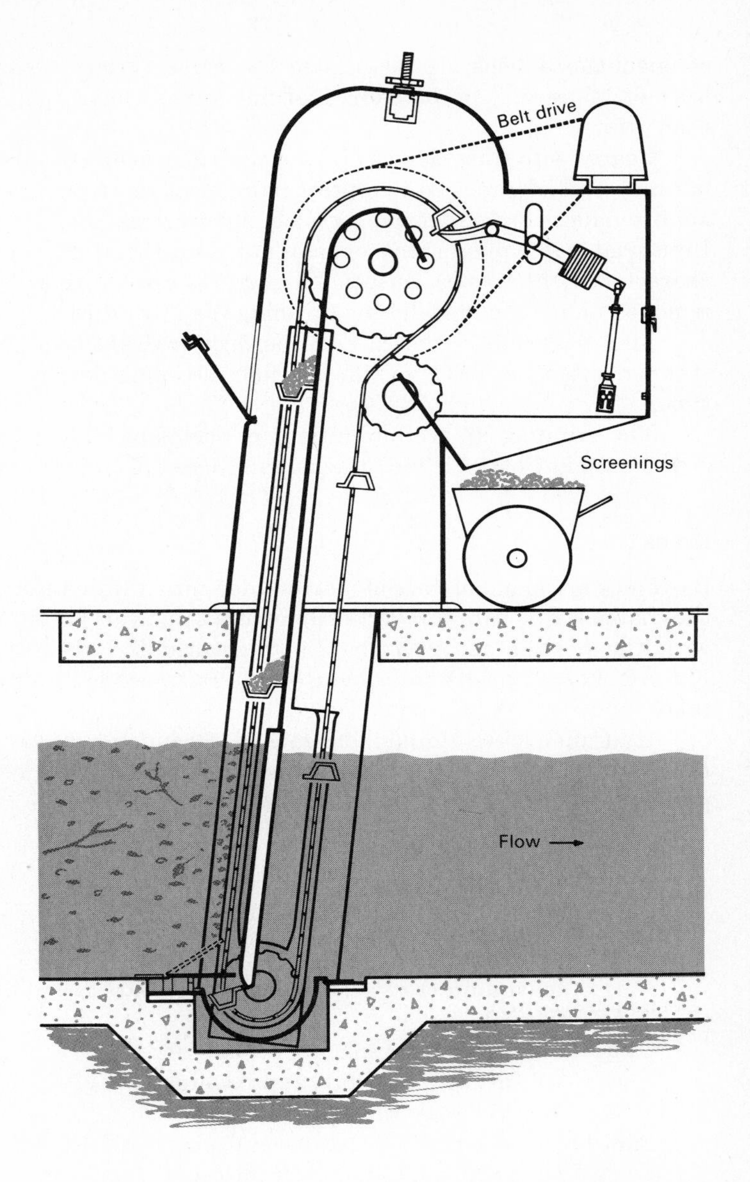

by 2 in. long, were installed in the 1920s and earlier in place of sedimentation tanks for primary treatment. Some of these screens remained in service for many years, especially where the effluent was discharged to salt water. They are no longer considered acceptable, however, as the sole means of sewage treatment before discharge.

Head loss through fine screens may be obtained from manufacturers' rating tables or may be calculated by means of the common orifice formula:

$$h_L = \frac{1}{2g}\left(\frac{Q}{CA}\right)^2 \tag{8·2}$$

where C = coefficient of discharge
Q = discharge through screen, cfs
A = effective submerged open area, sq ft
g = acceleration due to gravity, ft/sec^2
h_L = head loss, ft

Values of C and A depend on screen design factors, such as the size and milling of slots, the wire diameter and weave, and particularly the percent of open area, and must be determined experimentally. A typical value of C for a clean screen is 0.60. The head loss through a clean screen is relatively insignificant. The important determination is the head loss during operation, and this depends on the size and amount of solids in the wastewater, the size of the apertures, and the method and frequency of cleaning. Commercial screens operate partially submerged and rotate continuously or intermittently. The solids are flushed by sprays from the exposed screen surface into a collecting trough each revolution. Data based on wastewaters that have similar characteristics are invaluable for design.

TABLE 8·1 KIRSCHMER'S VALUES OF β [10]

Bar type	β
Sharp-edged rectangular	2.42
Rectangular with semicircular upstream face	1.83
Circular	1.79
Rectangular with semicircular upstream and downstream faces	1.67

MIXING

Mixing is an important unit operation in many phases of wastewater treatment where one substance must be completely intermingled with another. An example is the mixing of chemicals with sewage, as shown in the latter part of Fig. 8·1, where chlorine or hypochlorite is mixed with the effluent from the secondary settling tanks. Chemicals are also mixed with sludge to improve its dewatering characteristics before vacuum filtration. In the digestion tank, mixing is employed frequently to assure intimate contact between food and microorganisms. In the biological-process tank, air must be mixed with the activated sludge to provide the organisms with the oxygen required. In this case, diffused air is introduced in such a way as to fulfill the mixing requirements, or mechanical turbine aerator-mixers may be employed.

Types of Mixers

Liquid mixing can be carried out in a number of different ways: (1) in hydraulic jumps in open channels, (2) in Venturi flumes, (3) in pipelines, (4) in pumps, and (5) in vessels with the aid of mechanical means. In the first four of these ways, mixing is accomplished as a result of turbulence that exists in the flow regime. In the fifth, turbulence is induced through the use of rotating impellers, such as paddles, turbines, and propellers; gases (as in diffused aeration); and air and water jet lift pumps.

Paddles generally rotate slowly, as they have a large surface exposed to the liquid. They are used as flocculation devices when coagulants, such as aluminum or ferric sulfate, and coagulant aids, such as polyelectrolytes and lime, are added to wastewater or sludges. The production of a good floc usually requires a detention time of 15 to 30 min. On the other hand, a detention time of 2 to 5 min is more than adequate for the flash mixing of chemicals in tanks equipped with turbine or propeller mixers.

Turbine impellers are used to mix air with activated sludge in mechanical aeration tanks and to mix sludge in digesters. These impellers usually consist of blades mounted radially at the edges of and perpendicular to a flat plate. They rotate at moderate speeds (about 150 rpm). Propellers are used for high-speed mixing up to 2,000 rpm. They are shaped like ship propellers and generate strong axial currents that rapidly mix chemicals or gases with liquids.

Vortexing or mass swirling of the liquid must be restricted with all types of impellers. Vortexing causes a reduction in the difference between the fluid velocity and the impeller velocity and thereby decreases the effectiveness of mixing. If the mixing vessel is fairly small, vortexing can be prevented by mounting the impellers off-center or at an angle with the vertical, or by having them enter the side of the basin at an angle. The usual method in circular tanks is to install four or more vertical baffles extending approximately one-tenth the diameter out from the wall. These effectively break up the mass rotary motion and promote vertical mixing. Mixing tanks of concrete may be made square and the baffles omitted.

Theory

Mixing in sanitary engineering processes occurs generally in the regime of turbulent flow in which inertial forces predominate. As a general rule, the higher the velocity and the greater the turbulence, the more efficient the mixing. Rushton [16] has developed, on the basis of inertial and viscous forces, the following mathematical relationships for power requirements for laminar and turbulent conditions.

Laminar: $$P = \frac{k}{g_c} \mu n^2 D^3 \qquad (8\cdot3)$$

Turbulent: $$P = \frac{k}{g_c} \rho n^3 D^5 \qquad (8\cdot4)$$

where P = power requirement, ft-lb force/sec
k = constant (see Table 8·2)
g_c = Newton's law conversion factor, 32.2 ft-lb mass/sec²-lb force
μ = absolute viscosity of fluid, lb mass/sec-ft
ρ = mass density of fluid, lb mass/cu ft
D = diameter of impeller, ft
n = revolutions per second (rps)

Values of k, as developed by Rushton, are presented in Table 8·2. For the turbulent range, it is assumed that vortex conditions have been eliminated by four baffles at the tank wall, each 10 percent of the tank diameter, as shown in Fig. 8·3.

TABLE 8·2 VALUES OF k FOR MIXING POWER REQUIREMENTS [16]

Impeller	Laminar range, Eq. 8·3	Turbulent range, Eq. 8·4
Propeller, square pitch, 3 blades	41.0	0.32
Propeller, pitch of two, 3 blades	43.5	1.00
Turbine, 6 flat blades	71.0	6.30
Turbine, 6 curved blades	70.0	4.80
Fan turbine, 6 blades	70.0	1.65
Turbine, 6 arrowhead blades	71.0	4.00
Flat paddle, 6 blades	36.5	1.70
Shrouded turbine, 2 curved blades	97.5	1.08
Shrouded turbine with stator (no baffles)	172.5	1.12

Equation 8·3 applies if the Reynolds number is less than 10 and Eq. 8·4 applies if the Reynolds number is greater than 10,000. For

FIG. 8·3 Turbine mixer in baffled tank [16].

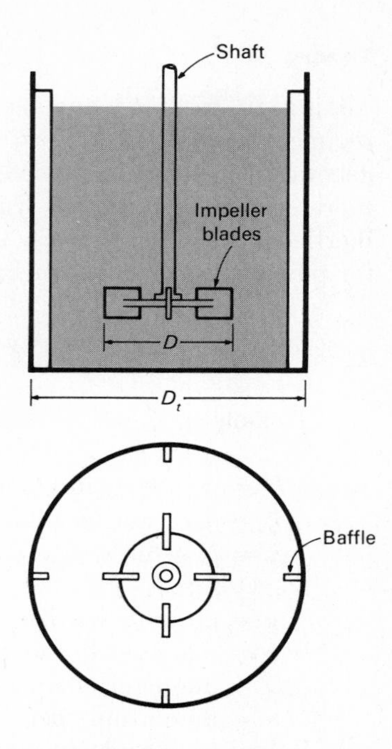

intermediate values of the Reynolds number, Ref. [13] should be consulted. The Reynolds number is given by

$$N_R = \frac{D_a{}^2 n \rho}{\mu} \tag{8·5}$$

where D_a = diameter of impeller, ft
n = rps
ρ = mass density of liquid, lb mass/cu ft
μ = viscosity, lb mass/ft-sec = centipoises $\times$ 6.72 $\times$ 10^{-4}

Analysis

The power input per unit volume is a rough measure of mixing effectiveness, based on the reasoning that more input power creates greater turbulence, and greater turbulence leads to better mixing. Mixers are selected on the basis of laboratory or pilot plant tests or similar data provided by manufacturers. No satisfactory method exists for scaling up from an agitator of one design to a unit of a different design. Geometrical similarity should be preserved and the power input per unit volume should be kept the same. A small impeller of high speed gives high turbulence but low flow and is best for dispersing gases or small amounts of chemicals in wastewater. A large, slow impeller gives high flow but low turbulence and is best for blending two fluid streams, or for flocculation.

FLOCCULATION

An essential part of any chemical or chemically aided precipitation system is stirring or agitation to increase the opportunity for particle contact (flocculation), after the chemicals have been added. Flocculation is promoted by gentle stirring with slow moving paddles. The action is sometimes aided by the installation of stationary slats or stator blades, located between the moving blades, that serve to break up the mass rotation of the liquid and promote mixing. Increased particle contact will promote floc growth; however, if the agitation is too vigorous, the shear forces that are set up will break up the floc into smaller particles. Agitation should be carefully controlled so that the floc particles will be of suitable size and will settle readily.

Numerous experiments have been performed by equipment manufacturers and plant operators to determine the optimum configuration of paddle size, spacing, and velocity. It has been found that a paddle-tip speed of approximately 2 to 3 fps achieves sufficient turbulence without breaking up the floc. Camp and Stein [3] studied the establishment and effect of velocity gradients in coagulation tanks of various types and developed the following equations for use in the design and operation of flocculation systems:

$$D = \frac{C_D A \rho v^2}{2} \tag{8·6}$$

$$P = \frac{C_D A \rho v^3}{2} \tag{8·7}$$

$$G = \sqrt{\frac{P}{\mu V}} \tag{8·8}$$

where D = drag, lb

C_D = coefficient of drag of flocculator paddles moving perpendicular to fluid

A = area of paddles, sq ft

ρ = mass fluid density, slugs/cu ft

v = relative velocity of paddles in fluid, fps, usually about 0.7 to 0.8 of paddle-tip speed

P = power requirement, ft-lb/sec

G = mean velocity gradient, ft/sec-ft = 1/sec

V = flocculator volume, cu ft

μ = absolute fluid viscosity, lb force-sec/sq ft

In Eq. 8·8, G is a measure of the mean velocity gradient in the fluid. As shown, it depends on the power input, the viscosity of the fluid, and the volume of the basin. Multiplying both sides of Eq. 8·8 by the theoretical detention time $t_d = V/Q$ yields

$$Gt_d = \frac{V}{Q}\sqrt{\frac{P}{\mu V}} = \frac{1}{Q}\sqrt{\frac{PV}{\mu}} \tag{8·9}$$

Typical values for G for a detention time of about 15 to 30 min vary from 20 to 75 sec^{-1}. Reported values of Gt_d vary from 10^4 to 10^5 [9]. The use of these equations is illustrated in the following example.

EXAMPLE 8·1 *Power Requirement and Paddle Area for Flocculator*

Determine the theoretical power requirement and the paddle area required to achieve a G value of 50 sec^{-1} in a tank with a volume of 10^5 cu ft. Assume that the water temperature is 60°F, the coefficient of drag for rectangular paddles is 1.8, the paddle-tip velocity v_p is 2 fps, and the relative velocity of the paddles v is $0.75v_p$.

Solution

1. Determine the theoretical power requirement.

$$\begin{aligned} P &= \mu G^2 V \\ &= 2.35 \times 10^{-5}(50)^2(10^5) \\ &= 5{,}890 \text{ ft-lb/sec} \end{aligned}$$

2. Convert the power to horsepower.

$$\frac{P}{\dfrac{550 \text{ ft-lb}}{\text{sec}}} = \frac{5{,}890}{550} = 10.7 \text{ hp}$$

3. Determine the paddle area.

$$\begin{aligned} A &= \frac{2P}{C_D \rho v^3} \\ &= \frac{2\left(\dfrac{5{,}890 \text{ ft-lb}}{\text{sec}}\right)}{1.8\left(\dfrac{1.938 \text{ lb-sec}^2}{\text{ft}^4}\right)\left(0.75 \times \dfrac{2 \text{ ft}}{\text{sec}}\right)^3} \\ &= 1{,}000 \text{ sq ft} \end{aligned}$$

SEDIMENTATION

Sedimentation is the separation of suspended particles that are heavier than water from water by gravitational settling. It is one of the most widely used unit operations in wastewater treatment. This operation is used for grit removal, particulate-matter removal in the primary settling basin, biological-floc removal in the activated-sludge settling basin, chemical-floc removal when the chemical coagulation process is used, and for solids concentration in sludge thickeners. In most cases, the primary purpose is to produce a clarified effluent, but it is also necessary to produce sludge with a solids concentration that can be easily handled and treated. In other processes, such as sludge thickening, the primary purpose is to produce a concentrated sludge that can be treated more economically. In the design of sedimentation basins, due consideration should be given to production of both a clarified effluent and a concentrated sludge.

On the basis of the concentration and the tendency of the particles to interact, four general classifications of the manner in which particles settle can be made. It is common to have more than one type of settling taking place at a given time during a sedimentation operation, and it is possible to have all four occurring simultaneously.

Type-1 settling refers to the sedimentation of discrete particles in a suspension of low solids concentration. Particles settle as individual entities, and there is no significant interaction with neighboring particles. A typical example is a dilute suspension of grit or sand particles. This type of settling is also called free settling.

Type-2 settling refers to a rather dilute suspension of particles that coalesce, or flocculate, during the sedimentation operation. By coalescing, the particles increase in mass and settle at a faster rate.

Type-3 settling occurs in suspensions of intermediate concentration, in which interparticle forces are sufficient to hinder the settling of neighboring particles. The particles tend to remain in fixed positions with respect to each other and the mass of particles settles as a unit. A distinct solids-liquid interface develops at the top of the settling sludge mass. This type of settling is generally called zone settling.

Type-4 settling occurs when the particles are of such concentration that a structure is formed and further settling can occur only by compression of the structure. Compression takes place due to the weight of the particles, which are constantly being added to the structure by sedimentation from the supernatant liquor. This type of settling is called compression settling and occurs usually in the lower layers of deep sludge masses.

Type-1 Settling

The settling of discrete, nonflocculating particles can be analyzed by means of the classic laws of sedimentation formulated by Newton and Stokes. Newton's law yields the terminal particle velocity by equating the effective weight of the particle to the frictional resistance, or drag. The effective weight is simply:

$$W = (\rho_s - \rho)gV \tag{8·10}$$

where ρ_s = density of particle
ρ = density of fluid
g = acceleration due to gravity
V = volume of particle

The drag per unit area depends on the particle velocity, fluid density, fluid viscosity, and particle diameter. The drag coefficient C_D (dimensionless) is defined by Eq. 8·11.

$$C_D = \frac{F_d}{\dfrac{(\rho v^2)}{2} A} \tag{8·11}$$

where F_d = drag force
v = particle velocity
A = cross-sectional or projected area of particle at right angles to v

Equating the drag force to the effective weight of the particle, for spherical particles, yields Newton's law:

$$V_c = \left[\frac{4}{3}\frac{g(\rho_s - \rho)d}{C_D\rho}\right]^{\frac{1}{2}} \tag{8·12}$$

where V_c = terminal velocity of particle
d = diameter of particle

The drag coefficient takes on different values depending on whether the flow regime surrounding the particle is laminar or turbulent. The drag coefficient is shown in Fig. 8·4 as a function of the Reynolds number. Although particle shape affects the value of the drag coefficient, for spherical particles the curve in Fig. 8·4 is approximated by the following equation (upper limit of $N_R = 10^4$) [9]:

$$C_D = \frac{24}{N_R} + \frac{3}{\sqrt{N_R}} + 0.34 \tag{8·13}$$

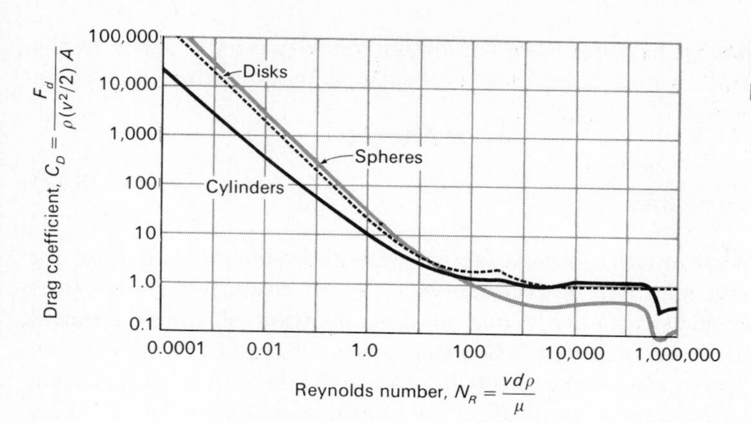

FIG. 8·4 Drag coefficients of spheres, disks, and cylinders [13].

For Reynolds numbers less than 0.3, the first term in Eq. 8·13 predominates, and substitution of this drag term into Eq. 8·12 yields Stokes' law:

$$V_c = \frac{g(\rho_s - \rho)d^2}{18\mu} \tag{8·14}$$

For laminar flow conditions, Stokes found the drag force to be

$$F_d = 3\pi\mu v d \tag{8·15}$$

Equating this force to the effective particle weight also yields Eq. 8·14.

In the design of sedimentation basins, the usual procedure is to select a particle with a terminal velocity V_c and to design the basin so that all particles that have a terminal velocity equal to or greater than V_c will be removed. The rate at which clarified water is produced is then

$$Q = AV_c \tag{8·16}$$

where A is the surface area of the sedimentation basin. Equation 8·16 yields

$$V_c = \frac{Q}{A} = \text{overflow rate, gpd/sq ft}$$

which shows that the overflow rate or surface loading rate, a common basis of design, is equivalent to the settling velocity. Equation 8·16 also indicates that for type-1 settling the flow capacity is independent of the depth.

For continuous-flow sedimentation, the length of the basin and the time a unit volume of water is in the basin (detention time) should be such that all particles with the design velocity V_c will settle to the bottom of the tank. The design velocity, detection time, and basin depth are related as follows:

$$V_c = \frac{\text{depth}}{\text{detention time}} \tag{8·17}$$

In actual practice, design factors have to be included to allow for the effects of inlet and outlet turbulence, short circuiting, sludge storage, and velocity gradients due to the operation of sludge-removal equipment. These factors will be discussed in Chap. 11. The discussion in this chapter refers to ideal settling in which the factors are omitted.

Type-1 settling in an ideal settling basin is shown in Fig. 8·5. Particles that have a velocity of fall less than V_c will not all be removed during the time provided for settling. Assuming that the particles of various sizes are uniformly distributed over the entire depth of the basin at the inlet, it can be seen from an analysis of the particle trajectory in Fig. 8·5 that particles with a settling velocity less than V_c will be removed in the ratio

$$X_r = \frac{V_p}{V_c} \tag{8·18}$$

where X_r is the fraction of the particles with settling velocity V_p that are removed.

In a typical suspension of particulate matter, a large gradation of particle sizes occurs. To determine the efficiency of removal for a given settling time, it is necessary to consider the entire range of settling velocities present in the system. This can be accomplished in two ways: (1) by use of sieve analyses and hydrometer tests combined with Eq. 8·12 or (2) by use of a settling column. With either method, a settling-velocity analysis curve can be constructed from the data. Such a curve is shown in Fig. 8·6.

FIG. 8·5 Type-1 settling in an ideal settling basin.

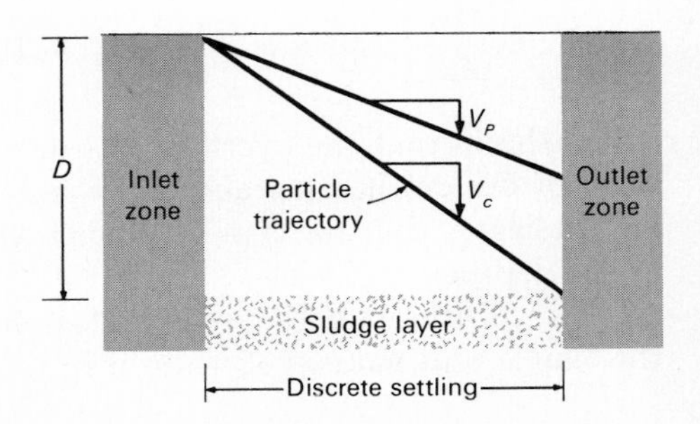

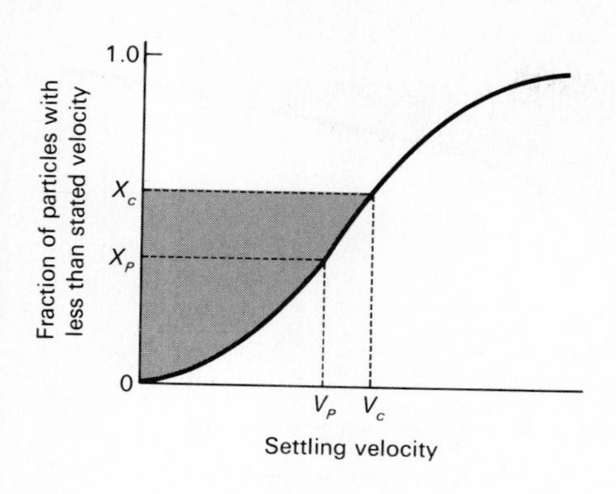

FIG. 8·6 Settling-velocity analysis curve for discrete particles.

For a given clarification rate Q where

$$Q = V_c A \tag{8·19}$$

only those particles with a velocity greater than V_c will be completely removed. The remaining particles will be removed in the ratio V_p/V_c. The total fraction of particles removed is given by Eq. 8·20.

$$\text{Fraction removed} = (1 - X_c) + \int_0^{X_c} \frac{V_p}{V_c}\,dx \tag{8·20}$$

where $1 - X_c$ = fraction of particles with velocity V_p greater than V_c

$\int_0^{X_c} \frac{V_p}{V_c}\,dx$ = fraction of particles removed with V_p less than V_c

The following example illustrates the use of Eq. 8·20.

EXAMPLE 8·2 *Removal of Discrete Particles (Type-1 Settling)*

A particle size distribution has been obtained from a sieve analysis of sand particles. For each weight fraction, an average settling velocity has been calculated. The data are

Settling velocity, feet per minute (fpm)	10.0	5.0	2.0	1.0	0.75	0.5
Weight fraction remaining	0.55	0.46	0.35	0.21	0.11	0.03

What is the overall removal for an overflow rate of 100,000 gpd/sq ft?

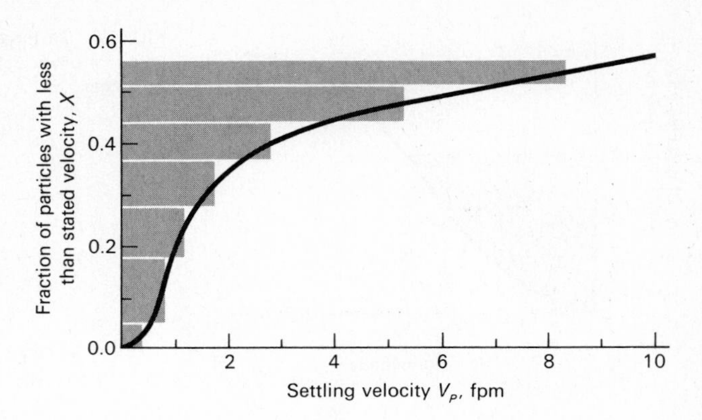

FIG. 8·7 Settling-velocity curve for Example 8·2.

Solution

Draw the settling-velocity analysis curve as shown in Fig. 8·7. Compute the settling velocity V_c of the particles that will be completely removed when the rate of clarification is 100,000 gpd/sq ft.

$$V_c = \frac{100{,}000 \text{ gpd}}{\text{sq ft}} \times \frac{1 \text{ mgd}}{10^6 \text{ gpd}} \times \frac{1.547 \text{ cfs}}{\text{mgd}} \times \frac{60 \text{ sec}}{\text{min}} = 9.3 \text{ fpm}$$

From the curve it is found that 0.55 of the particles have a settling velocity less than 9.3 fpm. The graphical integration of the second term in Eq. 8·20 is shown on the curve as a series of rectangles (shaded) and in the following tabulation. (Note that since V_c is constant it is taken outside the integral or summation sign.)

dx	V_p	$V_p\,dx$
0.05	0.4	0.020
0.13	0.8	0.104
0.10	1.1	0.110
0.09	1.7	0.153
0.07	2.8	0.196
0.07	5.2	0.364
0.04	8.3	0.332
0.55		1.279

$$\text{Fraction removed} = (1 - X_c) + \frac{1}{V_c}\Sigma V_p\,dx$$

$$= (1 - 0.55) + \frac{1.279}{9.3} = 0.588$$

Type-2 Settling

Particles in relatively dilute solutions sometimes will not act as discrete particles but will coalesce during sedimentation. As coalescence or flocculation occurs, the mass of the particle increases and it will settle faster. The extent to which flocculation occurs is dependent on the opportunity for contact, which varies with the overflow rate, the depth of the basin, the velocity gradients in the system, the concentration of particles, and the range of particle sizes. The effect of these variables can be determined only by sedimentation tests.

To determine the settling characteristics of a suspension of flocculent particles, a settling column may be used. Such a column can be of any diameter but should be equal in height to the depth of the proposed tank. Satisfactory results can be obtained with a 6-in. diameter plastic tube about 10 ft high. Sampling ports should be inserted at 2-ft intervals. The solution containing the suspended matter should be introduced into the column in such a way that a uniform distribution of particle sizes occurs from top to bottom.

Care should also be taken to ensure that a uniform temperature is maintained throughout the test to eliminate convection currents. Settling should take place under quiescent conditions. At various time intervals, samples are withdrawn from the ports and analyzed for suspended solids. The percent removal is computed for each sample analyzed, and is plotted as a number against time and depth, as elevations are plotted on a survey grid. Between the plotted points, curves of equal percent removal are drawn. A settling column and the results of a sedimentation test are shown in Fig. 8·8. The resulting curves are

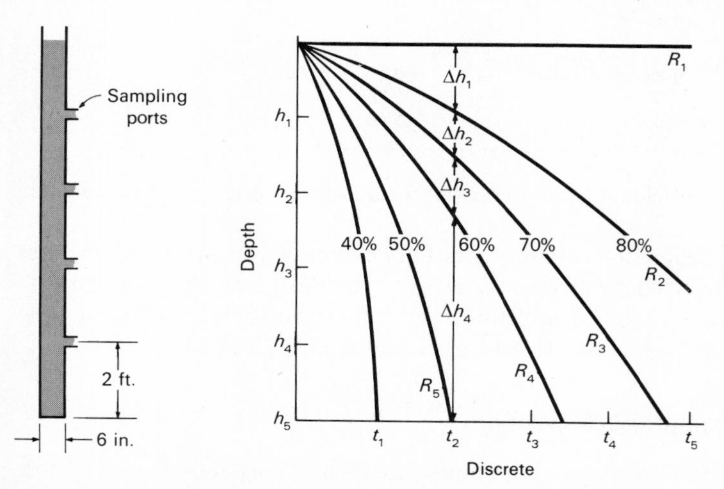

FIG. 8·8 Settling column and settling curves for flocculent particles.

shown, but the plotted numbers representing the individual samples have been omitted from the figure. Determination of removals by using Fig. 8·8 is illustrated in Example 8·3.

EXAMPLE 8·3 *Removal of Flocculent Suspended Solids (Type-2 Settling)*

Using the results of the settling test shown in Fig. 8·8, determine the overall removal of solids if the detention time is t_2 and the depth is h_5.

Solution

1. Determine the percent removal.

$$\text{Percent removal} = \frac{\Delta h_1}{h_5} \times \frac{R_1 + R_2}{2} + \frac{\Delta h_2}{h_5} \times \frac{R_2 + R_3}{2} + \frac{\Delta h_3}{h_5} \times \frac{R_3 + R_4}{2} + \frac{\Delta h_4}{h_5} \times \frac{R_4 + R_5}{2}$$

2. For the curves shown in Fig. 8·8 the computations would be

$\frac{\Delta h_n}{h_5} \times$	$\frac{R_n + R_{n+1}}{2}$	=	percent removal
0.20 ×	$\frac{100 + 80}{2}$	=	18.00
0.11 ×	$\frac{80 + 70}{2}$	=	8.25
0.15 ×	$\frac{70 + 60}{2}$	=	9.75
0.54 ×	$\frac{60 + 50}{2}$	=	29.70
1.00			65.70

yielding a total removal for quiescent settling of 65.7 percent.

In the design of a settling tank to achieve the removals comparable to those indicated by the settling tests, the design settling velocity or overflow rate should be multiplied by a factor of 0.65, and the detention times should be multiplied by a factor of 1.75 or 2.0.

Zone and Compression Settling

In systems that contain high concentrations of suspended solids, both zone settling (type-3) and compression settling (type-4) usually occur

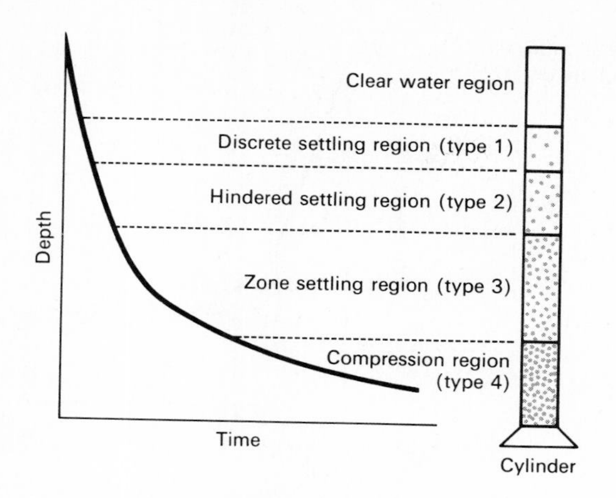

FIG. 8·9 Schematic of settling regions for activated sludge.

in addition to free settling. The settling phenomenon that occurs when a concentrated suspension, initially of uniform concentration throughout, is placed in a graduated cylinder, is shown in Fig. 8·9.

Because of the hydraulic characteristics of flow around the particles and other interparticle forces, the particles settle as a zone or "blanket," maintaining the same relative position with respect to each other. As this region settles, a volume of relatively clear water is produced above the zone-settling region. Particles remaining in this region settle as discrete or flocculated particles, as discussed previously in this chapter. A distinct interface exists between the discrete-settling region and the hindered-settling region on Fig. 8·9. The rate of settling in the hindered-settling region is a function of the concentration of solids and their condition.

As settling continues, a compressed layer of particles begins to form on the bottom of the cylinder in the compression-settling region. The particles in this region apparently form a structure in which there is physical contact between particles. As the compression layer forms, regions containing successively lower concentrations of solids than those in the compression region extend upward in the cylinder. The zone-settling region then consists of a gradation in solids concentration from that found in the hindered-settling region to that found in the compression-settling region. According to Dick and Ewing [6], the forces of physical interaction between the particles that are especially strong in the compression-settling region lessen progressively with height. They may exist to some extent in the zone-settling region.

Settling tests are usually required to determine the settling characteristics of the suspension when zone-settling and compression are important considerations. In design, the overflow rate determination

FIG. 8·10 Graphical analysis of interface settling curve [17].

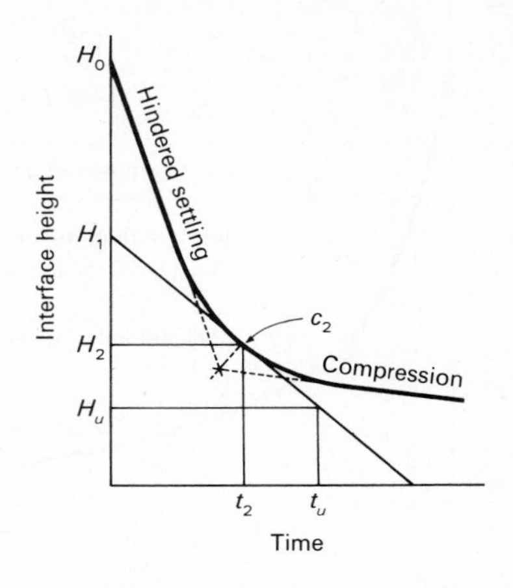

should be based on three factors: (1) the area needed for free settling in the discrete-settling region; (2) the area needed on the basis of the rate of settling of the interface between the discrete-settling region and the zone-settling region; and (3) the rate of sludge withdrawal from the compression region. The long-column settling tests as previously described can be used to determine the area needed for the free-settling region; however, the rate of zone settling is usually less than the rate of free particle settling, so that the rate of free particle settling rarely is the controlling factor. Furthermore, the sludge blanket or top mass of sludge just below the interface acts as a filter to entrap and strain out slower-settling particles that might otherwise remain in the discrete-settling region or that, in a settling tank, could be carried upward by water rising through the sludge mass to the overflow weirs.

The area requirement for zone settling is determined according to a method developed by Talmadge and Fitch [17]. A column of height H_0 is filled with a suspension of solids of uniform concentration C_0. As the suspension settles, the position of the interface as time elapses will be as given in Fig. 8·10. The rate at which the interface subsides is then equal to the slope of the curve at that point in time. According to the procedure, the critical area for thickening is given by Eq. 8·21:

$$A = \frac{Qt_u}{H_0} \tag{8·21}$$

where A = area required for sludge thickening, sq ft (see Fig. 8·10)
Q = flowrate into tank, cfs
H_0 = initial height of interface in column, ft
t_u = time to reach desired underflow concentration, sec

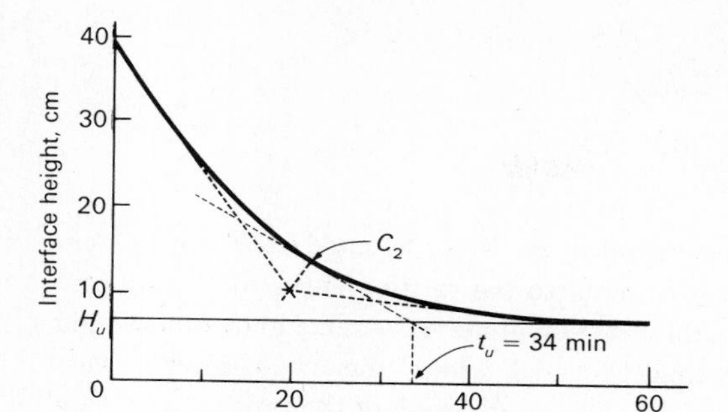

FIG. 8·11 Settling curve for Example 8·4.

The critical concentration controlling the sludge-handling capability of the tank occurs at a height H_2 where the concentration is C_2. This point is determined by extending the tangents to the free-settling and compression regions of the subsidence curve to the point of intersection and bisecting the angle thus formed, as shown in Fig. 8·10. The time t_u can be determined as follows:

1. Construct a horizontal line at the depth H_u that corresponds to the depth at which all the solids are at the desired underflow concentration C_u. Thus

$$H_u = \frac{C_0 H_0}{C_u}$$

2. Construct a tangent to the settling curve at the point indicated by C_2.
3. Construct a vertical line from the point of intersection of these two lines to the time axis to determine t_u.

This procedure is illustrated in the following example.

EXAMPLE 8·4 *Calculations for Sludge Thickeners*

In a settling cylinder 40 centimeters (cm) high, the settling curve shown in Fig. 8·11 was obtained for an activated sludge with an initial solids concentration C_0 of 4,000 mg/liter. Compute the thickener area to yield a thickened sludge concentration C_u of 24,000 mg/liter with a sludge inflow of 0.1 mgd. Determine the solids loading in lb/sq ft/day and the overflow rate.

Solution

1. Since $C_0H_0 = C_uH_u$,

$$H_u = \frac{4{,}000 \times 40}{24{,}000} = 6.67 \text{ cm}$$

In Fig. 8·11 a horizontal line is constructed at $H_u = 6.67$ cm. A tangent is constructed to the settling curve at C_2, the midpoint of the region between hindered settling and compression. Bisecting the angle formed where the two tangents meet determines point C_2. The intersection of the tangent at C_2 and the line $H_u = 6.67$ cm determines t_u. Thus $t_u = 34$ min and the required area is

$$A = \frac{Qt_u}{H_0} = \frac{0.10(1.55)(34)(60)}{40/30.48} = 241 \text{ sq ft}$$

2. This area must be adequate for clarification to occur in the discrete-settling region; therefore, determine the subsidence velocity v from the hindered-settling portion of the curve.

$$v = \frac{40 - 25}{10 \times 60} = 0.025 \text{ cm/sec} = 8.2 \times 10^{-4} \text{ fps}$$

3. The overflow rate is proportional to the volume above the sludge zone.

$$Q = 0.155 \frac{40.0 - 6.7}{40.0} = 0.129 \text{ cfs}$$

Therefore

$$A = \frac{Q}{v} = \frac{0.129}{8.2 \times 10^{-4}} = 157 \text{ sq ft}$$

4. The controlling requirement is the thickening area of 241 sq ft, since it exceeds the area required for clarification.
5. The solids loading is computed as follows:

$$\text{lb of solids/day} = 0.1(8.34)(4{,}000) = 3{,}340$$

$$\text{Solids loading} = \frac{3{,}340}{241} = 13.8 \text{ lb/sq ft/day}$$

6. $\text{The overflow rate} = \dfrac{100{,}000(33.3/40)}{241} = 345 \text{ gpd/sq ft}$

The volume required for the sludge in the compression region can also be determined by settling tests. The rate of consolidation in this region has been found to be proportional to the difference in the depth at time t and the depth to which the sludge will settle after a long period of time. This can be represented as Eq. 8·22:

$$H_t - H_\infty = (H_2 - H_\infty)e^{-i(t-t_2)} \qquad (8\cdot22)$$

where H_t = sludge height at time t

H_∞ = sludge depth after long period, say, 24 hr

H_2 = sludge height at time t_2

i = constant for a given suspension

It has been observed that stirring serves to compact sludge in the compression region by breaking up the floc and permitting water to escape. Rakes are often used on sedimentation equipment to manipulate the sludge and thus produce better compaction. Dick and Ewing [6] found that stirring would produce better settling in the zone-settling region also. With these facts in mind, it is apparent that, when appropriate, stirring should be investigated as an essential part of the settling tests if the proper areas and volumes are to be determined from the tests.

Scour Velocity

Scour velocity is important in sedimentation operations. Forces on settled particles are caused by the friction of water flowing over the particles. In sewers, velocities should be maintained high enough that solid particles will be kept from settling. In sedimentation basins, horizontal velocities should be kept low so that settled particles are not scoured from the bottom of the basin. The critical velocity is given by Eq. 8·23, which was developed by Camp [2] using the results from studies by Shields.

$$V_H = \left[\frac{8k(s-1)gd}{f}\right]^{\frac{1}{2}} \qquad (8\cdot23)$$

where V_H = horizontal velocity that will just produce scour

s = specific gravity of particles

d = diameter of particles

k = constant which depends on type of material being scoured

Typical values of k are 0.04 for unigranular sand and 0.06 or more for sticky, interlocking matter. The term f is the Darcy-Weisbach friction factor, which depends on the characteristics of the surface over which flow is taking place and the Reynolds number. Typical values of f are 0.02 to 0.03. Either English or metric units may be used in Eq. 8·23, so long as they are consistent, since k and f are dimensionless.

FLOTATION

Flotation is a unit operation used to separate solid or liquid particles from a liquid phase [7]. Separation is brought about by introducing fine gas (usually air) bubbles into the liquid phase. The bubbles attach to the particulate matter, and the buoyant force of the combined particle and gas bubbles is great enough to cause the particle to rise to the surface. Particles that have a higher density than the liquid can thus be made to rise. The rising of particles with lower density than the liquid can also be facilitated (e.g., oil suspension in water).

In wastewater treatment, flotation is used to remove suspended matter and to concentrate biological sludges (see Chap. 13). The principal advantage of flotation over sedimentation is that very small or light particles that settle slowly can be removed more completely and in a shorter time. Once the particles have been floated to the surface, they can be collected by a skimming operation.

Types of Systems

The present practice of flotation as applied to municipal wastewater treatment is confined to the use of air as the flotation agent. Air bubbles are added or caused to form in one of the following methods:

1. Aeration at atmospheric pressure (air flotation)
2. Injection of air while the liquid is under pressure, followed by release of the pressure (dissolved-air flotation)
3. Saturation with air at atmospheric pressure, followed by application of a vacuum to the liquid (vacuum flotation)

Further, in all these systems the degree of removal can be enhanced through the use of various chemical additives.

Air Flotation In this system, air bubbles are formed by introducing the gas phase directly into the liquid phase through a revolving impeller or through diffusers. Aeration alone for a short period is not particularly effective in bringing about the flotation of solids. The provision of aeration tanks for flotation of grease and other solids from normal wastewater is usually not warranted, although some success with these units has been experienced on certain scum-forming wastes.

Dissolved-Air Flotation In this system air is dissolved in the wastewater under a pressure of several atmospheres, followed by release of the pressure to the atmospheric level (see Fig. 8·12). In small pressure systems, the entire flow may be pressurized by means of a pump to 40 to 50 pounds per square inch gage (psig) with compressed air added

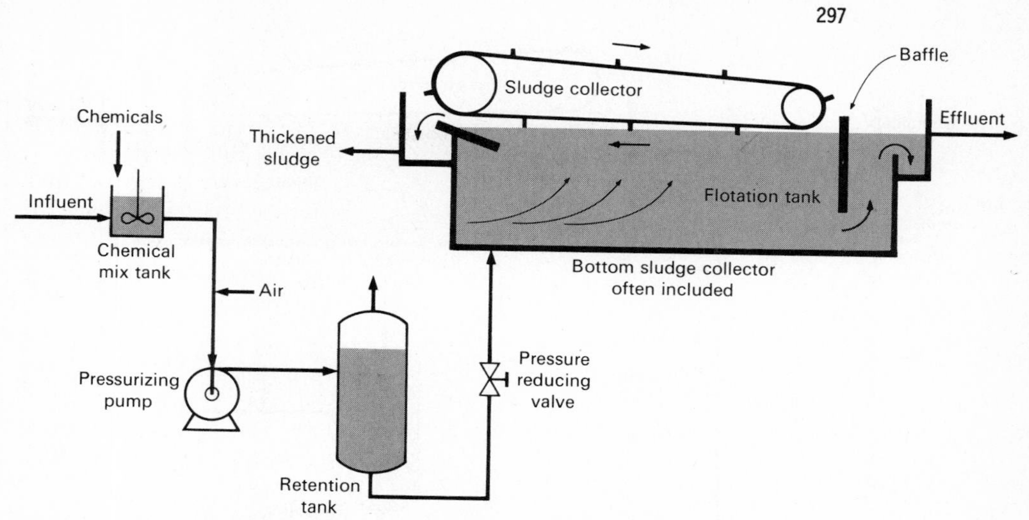

FIG. 8·12 Schematic of dissolved-air flotation tank without recycle [7].

at the pump suction. The entire flow is held in a retention tank under pressure for several minutes to allow time for the air to dissolve. It is then admitted through a pressure-reducing valve to the flotation tank where the air comes out of solution in minute bubbles throughout the entire volume of liquid.

In the larger units, a portion of the effluent (15 to 120 percent) is recycled, pressurized, and semisaturated with air (Fig. 8·13). The recycled flow is mixed with the unpressurized main stream just before admission to the flotation tank, with the result that the air comes out of solution in contact with particulate matter at the entrance to the tank. Pressure types of units have been used mainly for the treatment of industrial wastes and for the concentration of sludges.

Vacuum Flotation This process consists of saturating the wastewater with air either (1) directly in an aeration tank or (2) by permitting air to enter on the suction side of a sewage pump. A partial vacuum is applied, which causes the dissolved air to come out of solution as minute bubbles. The bubbles and the attached solid particles rise to the surface to form a scum blanket, which is removed by a skimming mechanism. Grit and other heavy solids that settle to the bottom are raked to a central sludge sump for removal. If this unit is used for grit removal and if the sludge is to be digested, the grit must be separated from the sludge in a grit classifier before the sludge is pumped to the digesters.

The unit consists of a covered cylindrical tank in which a partial

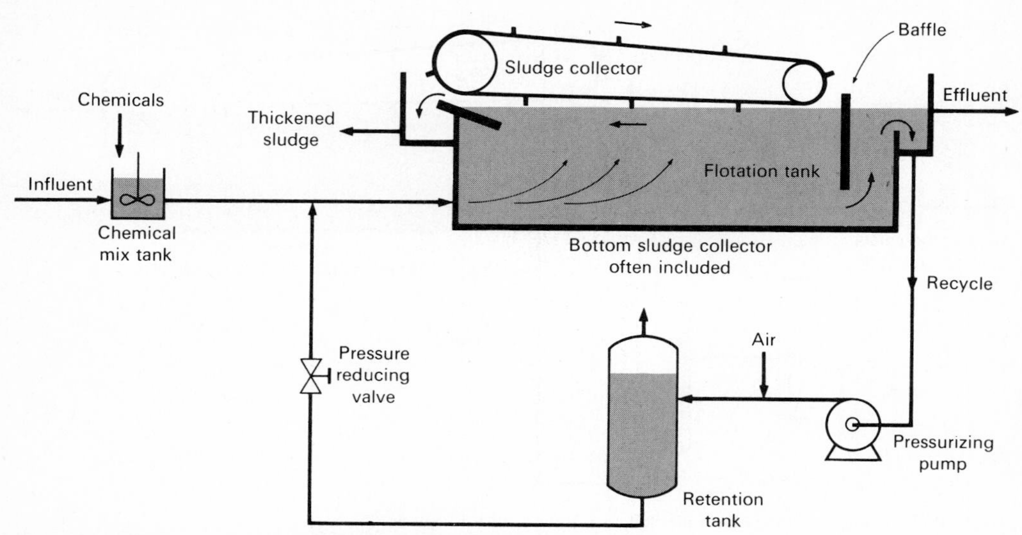

FIG. 8·13 Schematic of dissolved-air flotation tank with recycle [7].

vacuum is maintained. The tank is equipped with scum- and sludge-removal mechanisms. The floating material is continuously swept to the tank periphery, automatically discharged into a scum trough, and removed from the unit to a pump also under partial vacuum. Auxiliary equipment includes an aeration tank for saturating the wastewater with air, a short-period detention tank for removal of large air bubbles, vacuum pumps, and sludge and scum pumps.

Chemical Additives

Chemicals are commonly used to aid the flotation process. These chemicals, for the most part, function to create a surface or a structure that can easily absorb or entrap air bubbles. Inorganic chemicals, such as the aluminum and ferric salts and activated silica, can be used to bind the particulate matter together and, in so doing, create a structure that can easily entrap air bubbles. Various organic chemicals can be used to change the nature of either the air-liquid interface or the solid-liquid interface, or both. These compounds usually collect on the interface to bring about the desired changes.

Analysis

Because flotation is very dependent on the type of surface of the particulate matter, laboratory and pilot plant tests must usually be performed to yield the necessary design criteria. Factors that must be considered in the design of flotation units include the concentration of particulate matter, quantity of air used, the particle rise velocity, and the solids loading rate. With the exception of air flotation, the design of the other flotation systems is discussed in this section.

Dissolved-Air Flotation The performance of a dissolved-air flotation system depends primarily on the ratio of the pounds of air to the pounds of solids required to achieve a given degree of clarification. This will vary with each type of suspension and must be determined experimentally using a laboratory flotation cell. Typical air to solids (A/S) ratios encountered in the thickening of sludge in wastewater treatment plants vary from about 0.005 to 0.060.

The relationship between the A/S ratio and the solubility of air, the operating pressure, and the concentration of sludge solids for a system in which all the flow is pressurized is given in Eq. 8·24 [7].

$$\frac{A}{S} = \frac{1.3s_a(fP - 1)}{S_a} \tag{8·24}$$

where s_a = air solubility, cc/liter

Temp., °C	s_a, cc/liter
0	29.2
10	22.8
20	18.7
30	15.7

f = fraction of air dissolved at pressure P, usually 0.5

P = pressure, atmospheres (atm)

$= \dfrac{p + 14.7}{14.7}$

p = gage pressure, psig

S_a = sludge solids, mg/liter

The corresponding equation for a system with only pressurized recycle is

$$\frac{A}{S} = \frac{1.3s_a(fP - 1)R}{S_aQ} \tag{8·25}$$

where R = pressurized recycle, mgd

Q = mixed-liquor flow, mgd

In both of the foregoing equations, the numerator represents the weight of air and the denominator the weight of the solids. The factor 1.3 is the weight in milligrams of 1 cc of air and the term (-1) within the brackets accounts for the fact that the system is to be operated at atmospheric conditions. The use of these equations is illustrated in Example 8·5.

The required area of the thickener is determined from a consideration of the rise velocity of the solids (0.2 to 4.0 gpm/sq ft depending on the solids concentration), the degree of thickening to be achieved, and the solids-loading rate (see Table 13·3, Chap. 13).

EXAMPLE 8·5 *Flotation Thickening of Activated-Sludge Mixed Liquor*

Design a flotation thickener without and with pressurized recycle to thicken the solids in activated-sludge mixed liquor from 0.3 to about 4 percent. Assume that the following conditions apply.

1. Optimum A/S ratio = 0.008
2. Temperature = 20°C, 68°F
3. Air solubility = 18.7 cc/liter
4. Recycle system pressure = 40 psig
5. Fraction of saturation = 0.5
6. Surface-loading rate = 0.2 gpm/sq ft
7. Sludge flowrate = 0.1 mgd

Solution (without Recycle)

1. Compute the required pressure using Eq. 8·24.

$$\frac{A}{S} = \frac{1.3s_a(fP - 1)}{S_a}$$

$$0.008 = \frac{1.3(18.7)(0.5P - 1)}{3{,}000}$$

$$0.5P = 0.99 + 1$$

$$P = 3.98 \text{ atm} = 3.98 = \frac{p + 14.7}{14.7}$$

$$p = 43.8 \text{ psig}$$

2. Determine the required surface area.

$$A = \frac{\dfrac{100{,}000 \text{ gal}}{\text{day}}}{\dfrac{0.2 \text{ gal}}{\text{min-sq ft}} \times \dfrac{60 \text{ min}}{\text{hr}} \times \dfrac{24 \text{ hr}}{\text{day}}}$$

$$= 347 \text{ sq ft}$$

3. Check the solids-loading rate.

$$\text{lb/sq ft/day} = \frac{\dfrac{3{,}000 \text{ mg}}{\text{liter}} \times 8.34 \times 0.1 \text{ mgd}}{347 \text{ sq ft}}$$
$$= 7.2$$

Solution (with Recycle)

1. Determine pressure in atmospheres.

$$P = \frac{40 + 14.7}{14.7} = 3.72$$

2. Determine the required recycle rate.

$$\frac{A}{S} = \frac{1.3s_a(fP - 1)R}{S_aQ}$$
$$0.008 = \frac{1.3(18.7)[0.5(3.72) - 1]R}{3{,}000(0.1)}$$
$$R = 0.115 \text{ mgd}$$

Alternatively, the recycle flowrate could have been set and the pressure determined. In an actual design, the costs associated with recycle pumping, pressurizing system, and tank construction can be evaluated to find the most economical combination.

3. Determine the required surface area.

$$A = \frac{215{,}000}{0.2(60)(24)} = 747 \text{ sq ft}$$

Vacuum Flotation Vacuum-tank equipment is available commercially in sizes ranging from 12 to 60 ft in diameter. The water depth in the tank is fixed at about 10 ft, and the vacuum is equivalent to 9 in. Hg. For proper flotation, the air requirements are between 0.025 and 0.05 cu ft/gal of wastewater. Surface-loading rates in the unit vary, ranging from 5,000 to 10,000 gpd/sq ft, depending on the purpose of the unit. At the lower rate, the equivalent of preliminary treatment is approached. At the higher rate, the unit is used for scum and grit removal ahead of primary sedimentation.

Although limited performance data are available on vacuum-flotation units, suspended-solids reduction is reported to vary from 35 to 55 percent with surface loadings of 6,000 to 4,000 gpd/sq ft. The corresponding BOD reduction is reported to be in the vicinity of 17 to 35 percent. It is reported that at higher surface loadings, the percentage of removal of suspended solids and BOD is lowered rapidly.

ELUTRIATION

Elutriation is a unit operation in which various chemical components are leached from digested sludge. The operation consists of intimately mixing a solid, or a solid-liquid mixture, with a liquid for the purpose of transferring certain components to the liquid. A typical example is the washing of digested sewage sludge before chemical conditioning to remove certain soluble organic and inorganic components that would consume large amounts of chemicals. The cost of washing the sludge is, in general, more than compensated for by the savings that result from a lower demand for conditioning chemicals.

The usual leaching operation consists of two steps involving (1) a thorough mixing of the solid or solid-liquid mixture with the leaching liquid and (2) separation of the leaching liquid. Each combination of mixing and washing is called a stage. A stage is said to be ideal if the concentration of the component being leached is the same in the separating liquid as it is in the liquid that remains with the solids. Mixing and separating can be carried out either in the same tank or in separate tanks. In sanitary engineering, separate tanks are usually used for each stage. The mixing and thickening operations should be designed in accordance with the principles previously discussed in this chapter.

Since alkalinity is usually present in high concentrations in digested sludge, it is commonly used to measure leaching efficiency. A decrease in the quantity of chemicals required to condition sludge has been correlated, by Genter in McCabe and Eckenfelder [12], with the decrease in alkalinity that results from elutriation. Some of the flow arrangements that can be used for elutriation are outlined in Fig. 8·14.

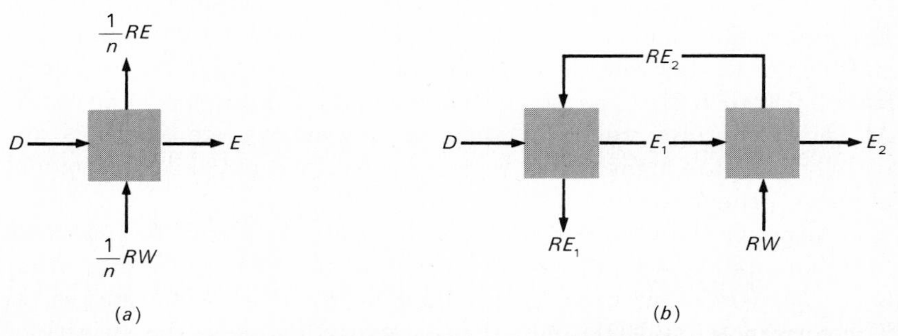

FIG. 8·14 (*a*) Single-tank, multiple-stage elutriation. (*b*) Multiple-tank countercurrent elutriation.

Single-Tank

Multiple stages may be carried out successively in a single tank, as shown in Fig. 8·14*a*, with E_n becoming D_{n+1}. Striking a mass balance in a single tank for a single stage yields

$$\frac{1}{n}RW + D = \frac{1}{n}RE + E \tag{8·26}$$

where n = number of stages
R = volume ratio of total wash water to wet sludge, all stages
W = alkalinity of wash water, mg/liter
D = alkalinity of sludge before elutriation, mg/liter
E = alkalinity of sludge after elutriation, mg/liter

The alkalinity E_n after n elutriations, assuming equal volumes of wash water R/n in each stage, is

$$E_n = \frac{D + W\left[\left(\frac{R}{n} + 1\right)^n - 1\right]}{\left(\frac{R}{n} + 1\right)^n} \tag{8·27}$$

and R may be calculated by Eq. 8·28 [9]:

$$R = n\left[\left(\frac{D - W}{E - W}\right)^{1/n} - 1\right] \tag{8·28}$$

Multiple-Tank

In multistage countercurrent operation, fresh water is introduced only into the nth stage. E_n is given by Eq. 8·29 [9]:

$$E_n = \frac{D(R - 1) + WR(R^n - 1)}{R^{n+1} - 1} \tag{8·29}$$

and R may be calculated by Eq. 8·30 [9]:

$$\frac{R^{n+1} - 1}{R - 1} = \frac{D - W}{E - W} \tag{8·30}$$

The efficiency of a multiple-tank countercurrent operation is compared to a single-tank two-stage operation in Example 8·6.

EXAMPLE 8·6 *Elutriation of Digested Sludge*

If the alkalinity of a digested sludge is 4,000 mg/liter and if it is to be elutriated with a wash water whose alkalinity is 40 mg/liter, compare the efficiency of using a single tank with two elutriations to using two tanks in a countercurrent elutriation operation. Assume in both cases that the ratio of wash water to sludge is 4:1.

Solution

$$D = 4{,}000 \text{ mg/liter} \qquad R = 4$$
$$W = 40 \text{ mg/liter} \qquad n = 2$$

1. For a single tank use Eq. 8·27:

$$E_2 = \frac{4{,}000 + 40(8)}{9} = 480 \text{ mg/liter}$$

$$\text{Efficiency, } \% = \frac{4{,}000 - 480}{4{,}000 - 40} = 88.7\%$$

2. For a multiple-tank countercurrent elutriation use Eq. 8·29:

$$E_2 = \frac{4{,}000(3) + 40(4)(4^2 - 1)}{4^3 - 1} = 228 \text{ mg/liter}$$

$$\text{Efficiency, } \% = \frac{4{,}000 - 228}{4{,}000 - 40} = 95\%$$

VACUUM FILTRATION

The function of the unit operation of vacuum filtration is to reduce the water content of sludge, whether raw, digested, or elutriated, so that the proportion of solids increases from the 5 to 10 percent range to about 30 percent. At this higher percentage, sewage sludge is a moist cake, easily handled. To visualize the amount of water to be removed, consider a ton of sludge with 5 percent solids, or 100 lb of dry solids and 1,900 lb of water. After filtration to 30 percent solids, the 100 lb of solids removed would be associated with 233 lb of water in 333 lb of sludge. Thus 1,667 lb of water would have been extracted by the vacuum filter. This represents an 83 percent reduction in the weight of sludge to be disposed of from the treatment process.

In wastewater treatment plants, vacuum filtration is a continuous operation that is generally accomplished on cylindrical drum filters. These drums have a filter medium, which may be a cloth of natural or synthetic fibers, coil springs, or a wire-mesh fabric. The drum is suspended above and dips into a vat of sludge. As the drum rotates slowly, part of its circumference is subject to an internal vacuum that draws sludge to the filter medium. Water is drawn through the porous filter cake for that sector of the circumference, as shown in Fig. 8·15. The piping arrangement within the filter permits the suction to be maintained until the release point, at which time compressed air is blown

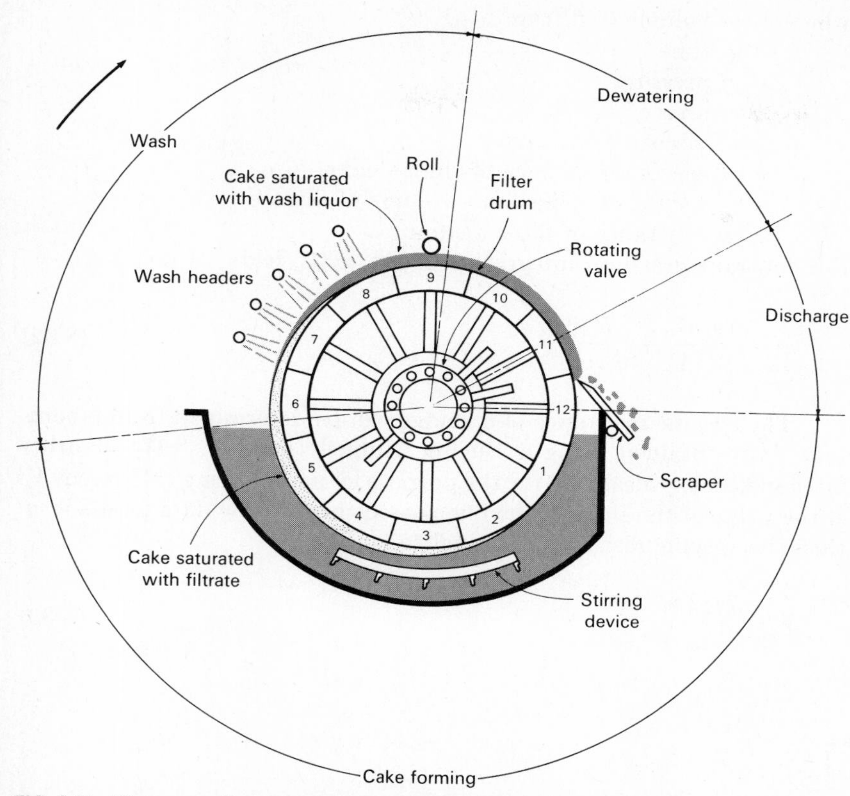

FIG. 8·15 Diagrammatic cross section of rotary drum vacuum filter [1].

through the medium to release the cake to a scraper for discharge. The filter medium may be washed in the small sector before suction begins again. The flowsheet of a vacuum filtration system is shown in Fig. 8·16.

Theory

The basic theory of the filtration process was developed by Carman [4] and extended by Coackley [5, 12] for conditions of streamline flow by application of Poiseuille's and Darcy's laws [8]. The basic filtration equation is

$$\frac{dV}{dt} = \frac{PA^2}{\mu(rcV + R_m A)} \tag{8·31}$$

where V = volume of filtrate
t = time
P = pressure
A = area
μ = viscosity of filtrate
r = specific resistance of sludge cake
c = weight of solids/unit volume of filtrate
R_m = resistance of filter medium

For constant pressure, integration of Eq. 8·31 yields

$$\frac{t}{V} = \frac{\mu r c V}{2PA^2} + \frac{\mu R_m}{PA} \tag{8·32}$$

The specific resistance of a sludge can be determined from laboratory data obtained using a Büchner funnel (see Fig. 8·17) or other filter specifically designed for the purpose [5] and plotting t/V versus V. If the value of the slope of the line passed through the data points is m, then the specific resistance is equal to

$$r = \frac{2PA^2m}{\mu c} \tag{8·33}$$

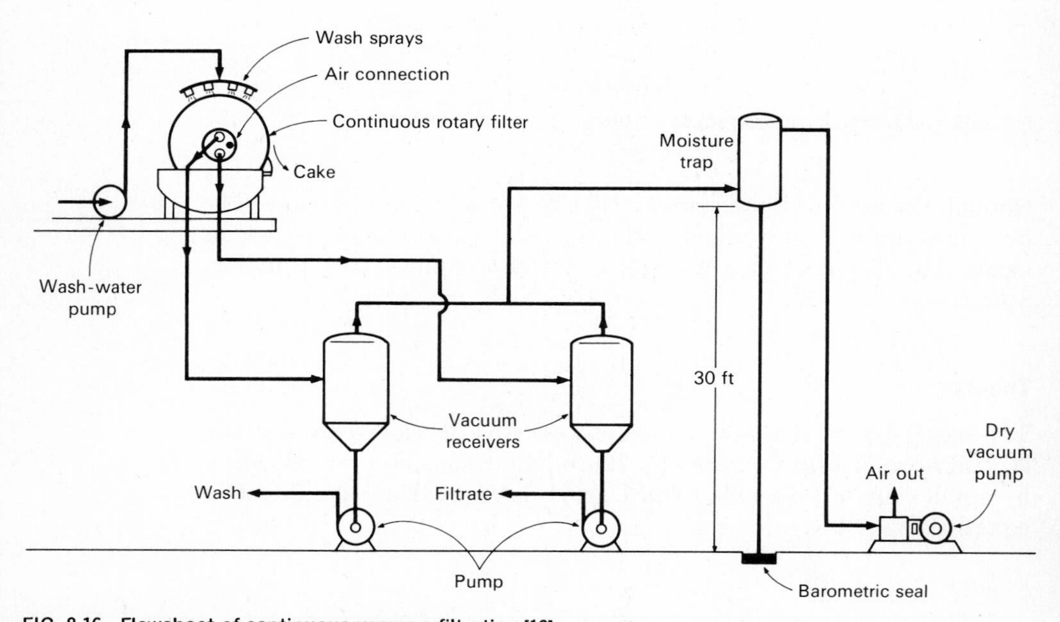

FIG. 8·16 Flowsheet of continuous vacuum filtration [13].

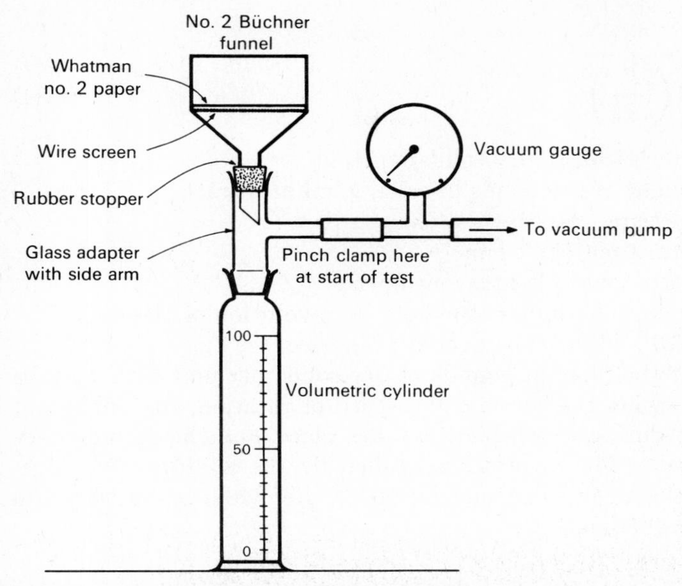

FIG. 8·17 Büchner funnel test apparatus used for the determination of the specific resistance of sludge [7].

Typical specific-resistance values for various biological sludges are reported in Table 8·3. As a point of caution, it should be noted that the

TABLE 8·3 TYPICAL SPECIFIC-RESISTANCE VALUES FOR VARIOUS SLUDGES

Sludge	Specific resistance r, sec^2/g
Primary	$1.5\text{–}5.0 \times 10^{10}$
Activated	$1\text{–}10 \times 10^{9}$
Digested	$1\text{–}6 \times 10^{10}$
Digested + coagulant	$3\text{–}40 \times 10^{7}$

literature on vacuum filtration is confusing in terms of unit inconsistencies and that the reader is advised to check units before using the various equations.

Analysis

Where the resistance of the filter medium is small compared with the resistance of the filter cake, Eq. 8·34 may be used to determine the filter yield.

$$L = 35.6 \left(\frac{xPw}{\mu R\theta}\right)^{\frac{1}{2}} \tag{8·34}$$

where L = filter yield, lb dry solids/sq ft/hr
w = weight of dry solids in cake, g/ml of filtrate
P = pressure, psi
x = form time/cycle time
μ = viscosity of filtrate, centipoises
θ = cycle time, min = time of one revolution of the drum
R = $r/10^7$, where r = specific resistance, sec^2/g

The yield of the filter in pounds of dry solids per unit time may be changed by varying the suction, the speed of rotation, the portion of the cycle time during which suction takes place, and the permeability of the filter cake. The latter is controlled by the addition of sludge-conditioning chemicals. The calculation of filter yield is illustrated in the following example.

EXAMPLE 8·7 *Vacuum-Filter Yield*

A conditioned, digested sludge with a specific resistance r of 3.1×10^7 sec^2/g is to be dewatered on a rotary drum filter under a vacuum of 25 in. Hg. The weight of filtered solids per unit volume of filtrate is to be 4 lb/cu ft. The filtration temperature of the sludge is 77°F and the cycle time, half of which is form time, is 6 min. Calculate the filter yield.

Solution

1. Compute the pressure.

$$P = 25 \text{ in. Hg} = \frac{25}{29.9} 14.7 = 12.3 \text{ psi}$$

2. Compute R.

$$R = \frac{r}{10^7} = \frac{3.1 \times 10^7}{10^7} = 3.1$$

3. Substitute the following values in Eq. 8·34 and compute the load.

$x = 0.5$ $\quad w = 0.064$ g/ml $\quad \theta = 6$ min
$\mu = 0.896$ centipoises

$$L = 35.6 \left[\frac{0.5(12.3)(0.064)}{0.896(3.1)(6)}\right]^{\frac{1}{2}} = 5.5 \text{ lb/sq ft/hr}$$

Operational results vary greatly with the characteristics of the sludge being filtered. The solids content of the sludge, among other parameters, is very important. Chemical conditioning of the sludge prior to filtration is usually practiced to control the solids content and to improve the dewatering characteristics. The optimum solids content for filtration is about 8 to 10 percent. Higher solids content makes the sludge difficult to distribute and to condition for dewatering; lower solids content requires the use of larger-than-necessary vacuum filters. Chemicals that are commonly used for conditioning sludge are lime, ferric chloride, and polyelectrolytes. Sludge from primary settling tanks, in general, requires lesser amounts of conditioning chemicals than sludge from biological-waste treatment processes. Elutriation of digested sludge, as previously discussed, reduces the chemical requirements. Laboratory tests and engineering economic studies are necessary to determine the optimum process of sludge handling to achieve minimum construction and operating costs, including labor, power, and chemicals.

HEAT TRANSFER

Heat transfer is the transmission of thermal energy from one medium to another. Unit operations that use heat transfer include sludge digestion and sludge drying. Anaerobic digesters are maintained at 90°F or above by heat transfer from heat exchangers or heating coils, and sludge drying is accomplished by intimately mixing the sludge with hot air in a dryer or by placement on open or glass-covered drying beds. Building and pipe gallery heating and summer cooling of offices, laboratories, and indoor working areas involve the installation of heat-transfer equipment. Waste heat for building and sludge heating may be recovered from dual-fuel and gas engine exhausts.

The literature on heat transfer is extensive and the amount of data available is voluminous, since the subject has been of prime concern to the American Society of Mechanical Engineers, the American Society of Heating, Ventilating and Refrigerating Engineers, and the American Institute of Chemical Engineers. What follows is no more than a brief introduction to the subject, with special attention to its application in sanitary engineering.

Theory

Heat transfer will occur by conduction, convection, and radiation. Conduction is involved in transferring heat through a metallic pipe, as in a heat exchanger where hot water is on one side and sludge is on the other.

The flow of heat by conduction is given by Fourier's law [11, 13], which is applicable to heating coils, pipes, and digester walls:

$$\frac{dQ}{d\theta} = -kA\frac{dT}{dx} \tag{8·35}$$

where $\frac{dQ}{d\theta}$ = rate of heat transfer through wall, Btu/hr

k = thermal conductivity, Btu/(hr)(sq ft)(°F/ft)

A = cross-sectional area perpendicular to x, sq ft

$\frac{dT}{dx}$ = temperature gradient across wall, °F/ft

Under steady-state conditions $dQ/d\theta = Q/\theta =$ a constant value designated as q. With a flat plate or large-radius wall, the equation is integrated to

$$q = \frac{kA}{x}\Delta T \tag{8·36}$$

where q = rate of heat transfer, Btu/hr

T = temperature drop across wall, °F

x = thickness of wall or plate, ft

For cylindrical pipes and pipe insulation, where the ratio of the radii of the outer to the inner wall is less than 2, the mean radius can be employed in calculating the area of the pipe. Where the ratio is significantly greater, as for pipe insulation, Eq. 8·35 becomes

$$q = -k2\pi L\frac{dT}{dr} \tag{8·37}$$

which integrates to

$$q = k\frac{2\pi L}{\ln(r_o/r_i)}\Delta T \tag{8·38}$$

where L = length of pipe or coil, ft

r_o, r_i = outer and inner radii, ft

Values of the thermal conductivity k vary with the ambient temperature and the type of material. Typical values in Btu/(hr)(sq ft) (°F/ft) are as follows: mild steel at 100°F, 26.2; stainless steel, 8.5; copper, 220; structural concrete, 1.0; 85 percent magnesia insulation, 0.034; and corkboard, 0.025.

Where multiple barriers are encountered, such as in a composite wall in a digester (concrete + air + brick facing), individual barrier thicknesses and thermal conductivities must be considered. Equation 8·36 can be rewritten in the form

$$\Delta T = q\frac{x}{kA} = qR \tag{8·39}$$

where $x/kA = R$ is a resistance, and the equation is analogous to the equation for the flow of electric current, $E = IR$. In the case of a composite wall, the same quantity of heat flows through several resistances in series; the temperature drops are additive. Thus

$$\Delta T = \Delta T_1 + \Delta T_2 + \Delta T_3 = \frac{x_1}{k_1A_1}q + \frac{x_2}{k_2A_2}q + \frac{x_3}{k_3A_3}q \tag{8·40}$$

which by rearrangement yields

$$q = \frac{\Delta T}{\dfrac{x_1}{k_1A_1} + \dfrac{x_2}{k_2A_2} + \dfrac{x_3}{k_3A_3}} \tag{8·41}$$

This equation cannot be used without further development because ΔT is the unknown temperature difference between the inside and outside surfaces of the wall, instead of the difference between the known ambient air and/or water temperatures on the two sides of the wall.

In general, heat is brought to one side of the wall and removed from the other side by convection, although radiation may also be a factor, as will be discussed later. Convection is heat transfer by fluid turbulence. A boundary layer film with a definite heat resistance and temperature drop exists on each side of the wall, as shown in Fig. 8·18. Since the thicknesses of the films are unknown, the resistance of each film is written $1/hA$, instead of x/kA. The general heat-flow equation then becomes

$$q = \frac{\Delta T}{\dfrac{1}{h_1A_1} + \sum\dfrac{x}{kA} + \dfrac{1}{h_2A_2}} \tag{8·42}$$

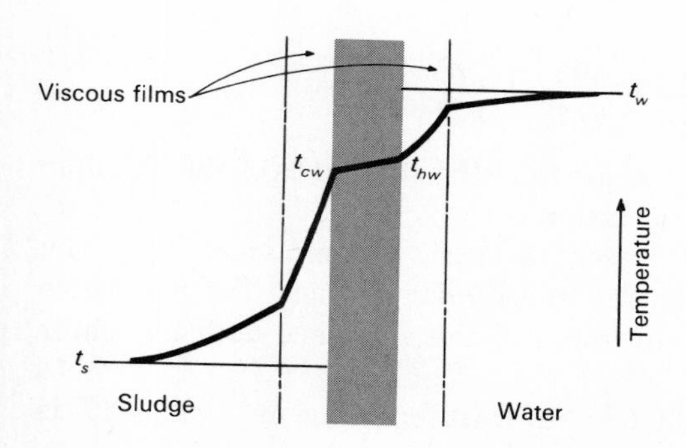

FIG. 8·18 Temperature gradients in forced convection.

If the wall is straight and all areas are the same, the numerator and denominator of the right-hand side can be multiplied by the area A, obtaining

$$q = \frac{A\,\Delta T}{\dfrac{1}{h_1} + \sum \dfrac{x}{k} + \dfrac{1}{h_2}} = UA\,\Delta T \tag{8·43}$$

$$U = \frac{1}{\dfrac{1}{h_1} + \sum \dfrac{x}{k} + \dfrac{1}{h_2}} \tag{8·44}$$

U is called the overall coefficient of heat transfer, and the problem of determining heat losses from buildings or tanks or the capacity of heat-transfer equipment consists of the estimation or computation of U and application of Eq. 8·43.

If the areas are not equal, as in the transfer of heat through pipe walls, one of the areas is selected, both numerator and denominator are multiplied by that area, and the value of U thus obtained must be used with that area. For example, if A_1 is selected in Eq. 8·42,

$$U = \frac{1}{\dfrac{1}{h_1} + \sum \dfrac{xA_1}{kA} + \dfrac{A_1}{h_2A_2}} \tag{8·45}$$

and

$$q = UA_1\,\Delta T \tag{8·46}$$

During operation, scales and films of dirt particles build up on heat-transfer surfaces of heating coils, heat exchangers, and radiators, and seriously impede the flow of heat. Accessible surfaces are cleaned periodically; where surfaces are inaccessible the films build up to a limiting thickness, and very low coefficients must be used. In the selection of equipment, scaling or fouling must be allowed for. This is done by adding additional film coefficients as h_3 and h_4, so that the equation of U becomes

$$U = \frac{1}{\dfrac{1}{h_1} + \sum \dfrac{xA_1}{kA} + \dfrac{A_1}{h_2A_2} + \dfrac{A_1}{h_3A_3} + \dfrac{A_1}{h_4A_4}} \tag{8·47}$$

Recommended fouling factors are given in the standards of the Tubular Exchanger Manufacturers Association.

In computing the heat losses through a compound wall, the resistance of an air space is handled in a similar manner. The resistance is taken equal to $1/hA$; the thickness of the air space does not enter into the computation.

In the computation of heat losses from buildings or tanks, ΔT is merely the difference between the inside and outside ambient temperatures. In the design of heat-transfer equipment, the selection of the correct value of ΔT is not so obvious. Flow may be parallel, with a large temperature difference at the inlet and a small temperature difference at the outlet. Flow may also be countercurrent, with a more or less uniform or varying temperature difference not necessarily equal at inlet and outlet. For countercurrent flow, the hot fluid may be cooled well below the outlet temperature of the cooling fluid. In commercial heat exchangers the flow may be mixed with the fluid flows crossing and recrossing each other in patterns, depending on the design of the equipment. In all three cases—parallel, countercurrent, and mixed-flow—the ΔT used is the log mean ΔT, but in the mixed-flow exchangers, correction factors must be applied [11]. The log mean temperature difference is given by the following equation:

$$\Delta T_m = \frac{\Delta T_1 - \Delta T_2}{\ln (\Delta T_1/\Delta T_2)} \tag{8·48}$$

where ΔT_1 and ΔT_2 are the temperature differences at the inlet and outlet. ΔT_1 may be taken as the greater temperature difference and ΔT_2 as the smaller.

Analysis

Digestion tanks may be heated by circulating hot water through pipe coils located inside the tank close to the wall or by pumping sludge and supernatant through external heat exchangers and back to the tank (see Fig. 8·19).

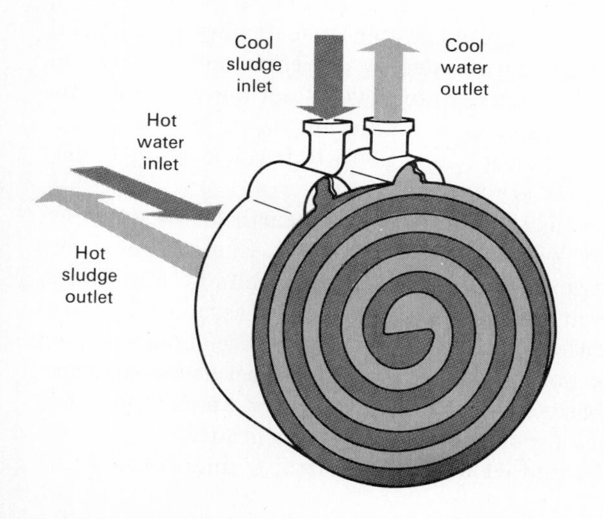

FIG. 8·19 Countercurrent spiral heat exchanger [from Dorr-Oliver].

Water temperatures are kept low, usually less than 130°F, to prevent sludge caking on the tubes. Values for the coefficient U for pipe coils installed in the tank are 8 to 15 for thick sludge, 30 for thin sludge, and 60 to 80 for thin supernatant surrounding the pipe coils. Coefficients are low because the velocity of sludge flowing across the tubes is dependent on thermally induced convection currents in viscous sludge at low temperature differences. Mixing devices, frequently installed in digesters, promote uniformity of temperature as well as composition throughout the tank.

When external heaters are installed, the sludge is pumped at high velocity through the tubes while water circulates at high velocity around the tubes. This promotes high turbulence on both sides of the heat-transfer surface and results in higher heat-transfer coefficients and better heat transfer. Another advantage of external heaters is that raw cold sludge on its way into the digesters can be both warmed and intimately blended and seeded with sludge liquor before entering the tank.

The heat requirements of digesters consist of the amount needed to raise the incoming sludge to digestion-tank temperatures plus the amount needed to compensate for the heat losses through walls, floor, and roof. Digestion-tank walls may be surrounded by earth embankments that serve as insulation or may be of compound construction consisting of approximately 12 in. of concrete, corkboard insulation, or an insulating air space, plus brick facing or corrugated aluminum facing over rigid insulation. In the latter case, the overall coefficient U varies from 0.35 for the wall with the air space as the only insulation to 0.12 for the wall with 2 in. of corkboard. The heat transfer from plain concrete walls below ground level and from floors depends on whether or not they are below ground water level. It may be assumed that the sides of the tank are surrounded by dry earth and the bottom is in saturated earth if the ground water level is not known. Since the heat losses from the tank warm up the adjacent earth, it is assumed that the earth forms an insulating blanket 5 to 10 ft thick before stable ambient earth temperatures are reached.

For plain concrete walls surrounded by dry earth below grade or in embankments, U may be taken as 0.10 to 0.12. Since in northern climates frost may penetrate to a depth of 4 ft, the ground temperature could be taken at 32°F at this depth and could be assumed to vary uniformly above this depth to the design air temperature at the surface, and below this depth to normal winter ground temperatures that are 10 to 20° higher at the base of the wall. Alternatively, an average temperature may be assumed for the entire wall below grade.

For concrete walls in contact with moist earth, U may be taken as

0.20 to 0.25. For floors in contact with moist earth, U may be taken as 0.12 to 0.15. The loss through the roof will depend on the type of construction, the absence or presence of insulation and its thickness, the presence of air space as in floating covers between the skin plate and the roofing, and whether the underside of the roof is in contact with sludge liquor or gas. The value of U for floating covers may be taken as 0.33 for covers with a 1.5-in. wood deck and built-up roofing but no insulation, and as 0.16 if 1.0-in.-thick insulating board is installed under the roofing. For fixed concrete covers that are 4 in. thick and covered with built-up roofing, U may be taken as 0.72 if uninsulated and 0.23 if insulated with 1.0-in.-thick insulating board.

Radiation from roofs and above-ground walls also contributes to heat losses. At the temperatures involved, the effect is small and is included in the coefficients normally used, such as those given in the foregoing discussion. For the theory of radiant heat transmission the reader is referred to McAdams [11] and Perry [14]. Heat requirements for a digester are determined in Example 8·8.

EXAMPLE 8·8 *Digester Heating Requirements*

A digester with a capacity of 100,000 lb of sludge per day is to be heated by circulation of sludge through an external hot-water heat exchanger. Using the data given below, find the heat required to maintain the required digester temperature. If all heat were shut off for 24 hr, what would be the average drop in temperature of the tank contents? Assume that the following conditions apply.

1. Concrete digester dimensions:

   Diameter = 60 ft

   Side depth = 20 ft

   Mid-depth = 30 ft

2. Heat-transfer coefficients:

   Dry earth embanked for entire depth, $U = 0.12$

   Floor of digester in ground water, $U = 0.15$

   Roof exposed to air, $U = 0.16$

3. Temperatures:

   Air, 15°F

   Earth next to wall, 32°F

   Incoming sludge, 50°F

Earth below floor, 42°F

Sludge contents in digester, 90°F

4. Specific heat of sludge = 1.0

Solution

1. Compute the heat requirement for the sludge.

$$Q = 100{,}000(90 - 50)(1) = 4{,}000{,}000 \text{ Btu/day}$$

2. Compute the area of the walls, roof, and floor.

$$\begin{aligned}\text{Wall area} &= \pi 60(20) = 3{,}770 \text{ sq ft}\\ \text{Floor area} &= \pi 30(30^2 + 10^2)^{\frac{1}{2}} = 2{,}980 \text{ sq ft}\\ \text{Roof area} &= \pi(30)^2 = 2{,}830 \text{ sq ft}\end{aligned}$$

3. Compute heat losses by conduction.

(*a*) Walls:

$$q = 0.12(3{,}770)(90 - 32) = 26{,}200 \text{ Btu/hr}$$

(*b*) Floor:

$$q = 0.15(2{,}980)(90 - 42) = 22{,}200 \text{ Btu/hr}$$

(*c*) Roof:

$$q = 0.16(2{,}830)(90 - 15) = 34{,}000 \text{ Btu/hr}$$

(*d*) Total losses:

$$\begin{aligned}Q &= 24(26{,}200 + 22{,}200 + 34{,}000)\\ &= 1{,}973{,}000 \text{ Btu/day}\end{aligned}$$

4. Compute the required heat-exchanger capacity.

$$\begin{aligned}\text{Capacity} &= 4{,}000{,}000 + 1{,}973{,}000\\ &= 5{,}973{,}000 \text{ Btu/day}\end{aligned}$$

5. Compute the drop in temperature.

$$\begin{aligned}\text{Digester volume} &= \pi(30)^2\,(20 + \tfrac{10}{3})7.48\\ &= 495{,}000 \text{ gal}\\ \text{Weight of sludge} &= 495{,}000(8.34) = 4{,}130{,}000 \text{ lb}\\ \text{Drop in temperature} &= \frac{5{,}973{,}000}{4{,}130{,}000} = 1.45°\text{F}\end{aligned}$$

DRYING

Sludge drying is a unit operation that involves reducing water content by vaporization of water to the air. In sludge-drying beds, vapor-pressure differences account for evaporation to the atmosphere. In mechanical drying apparatuses, auxiliary heat is provided to increase the vapor-holding capacity of the ambient air and to provide the latent heat of evaporation.

Theory

Under equilibrium conditions of constant-rate drying, mass transfer is proportional to (1) the area of wetted surface exposed, (2) the difference between water content of the drying air and saturation humidity at the wet-bulb temperature of the sludge-air interface, and (3) other factors, such as velocity and turbulence of drying air expressed as a mass-transfer coefficient. The equation reads

$$W = k_y(H_s - H_a)A \tag{8·49}$$

where W = evaporation rate, lb/hr

k_y = mass-transfer coefficient of gas phase, lb mass/hr-sq ft-unit of humidity difference (ΔH)

H_s = saturation humidity of air at sludge-air interface, lb water vapor/lb dry air

H_a = humidity of drying air, lb water vapor/lb dry air

The sludge-air interface temperature may be taken as equal to the wet-bulb temperature of the bulk volume of drying air or hot gases, provided the temperature of the air and the walls of the container are approximately the same. For extension of the theory and its application to specific types of drying equipment, refer to McAdams [11] and Walker *et al.* [18].

It is evident that drying may be accomplished most rapidly on a finely divided sludge by exposing new areas to the drying air stream. Furthermore, maximum contact between dry air and wet sludge should be obtained to assure a maximum value of ΔH. These factors must be considered in the selection of drying apparatuses for sludge disposal in a wastewater treatment plant.

Flash Drying

This operation involves pulverizing the sludge in a cage mill or by an atomized suspension technique in the presence of hot gases. The equipment should be designed so that the particles remain in contact with the turbulent hot gases long enough to accomplish mass transfer of moisture from sludge to the gases.

One operation involves a cage mill that receives a mixture of wet sludge or sludge cake and recycled dried sludge. The mixture contains

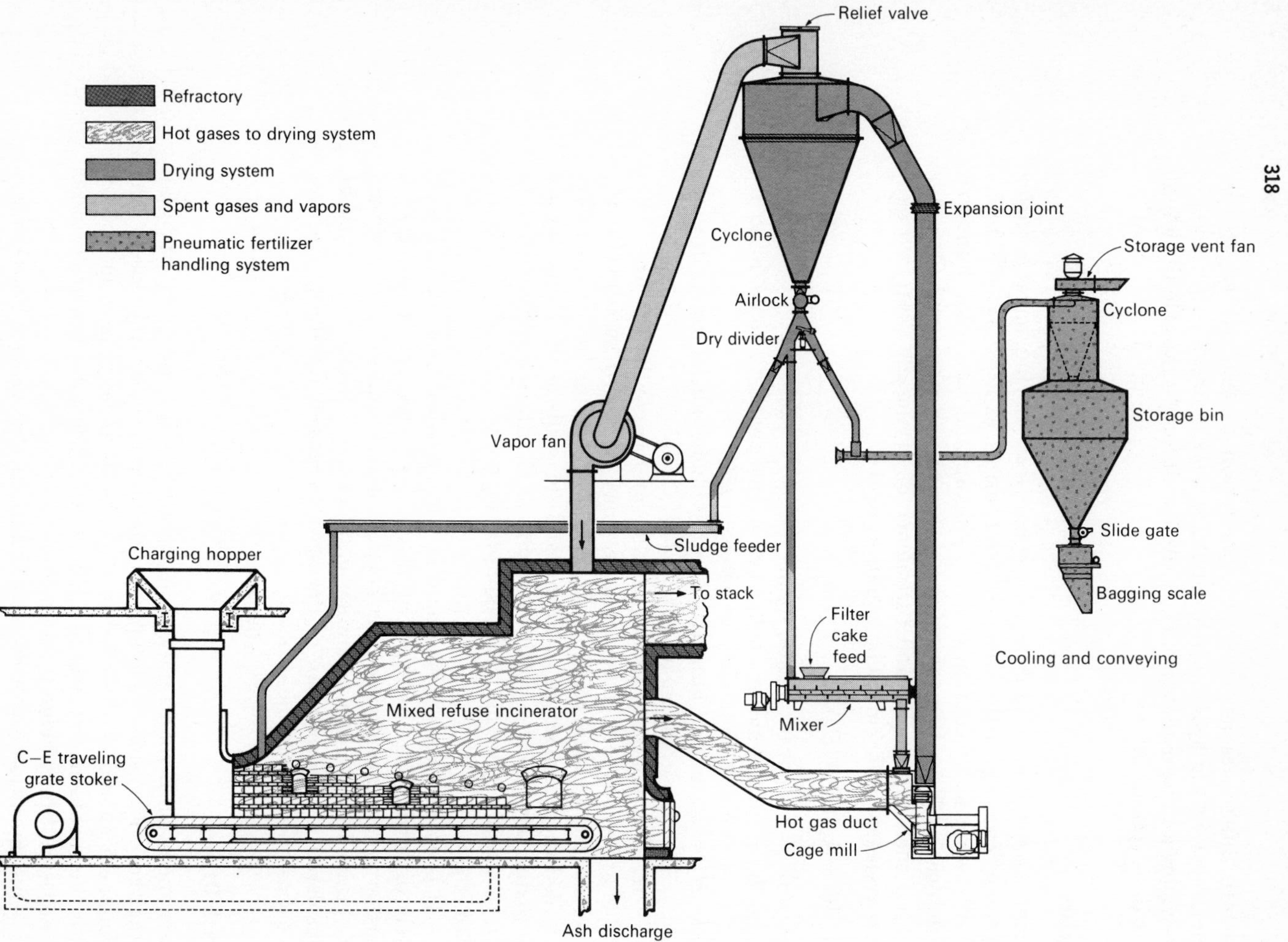

FIG. 8·20 Flash-drying system with mixed refuse incinerator [from Combustion Engineering].

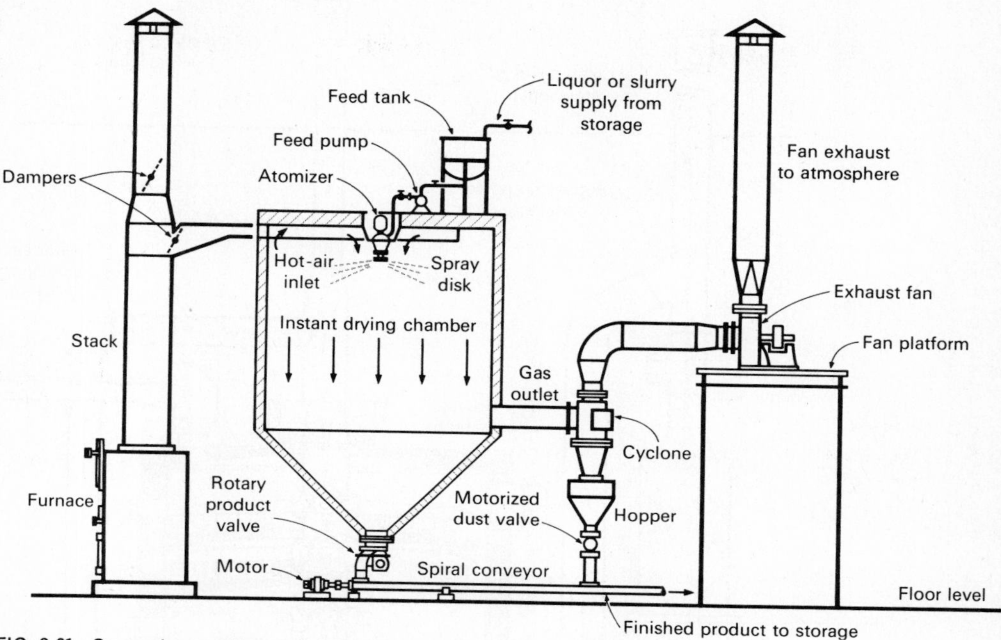

FIG. 8·21 Spray dryer with parallel flow [from Instant Drying Company].

approximately 50 percent moisture. The hot gases and sludge are forced up a duct in which most of the drying takes place and to a cyclone, which separates the vapor and solids. It is possible to achieve a moisture content of 8 percent in this operation. A schematic drawing of the entire drying process is shown on Fig. 8·20. The dried sludge may be used or sold as fertilizer or it may be incinerated in the furnace in any proportion up to 100 percent of production.

A spray dryer utilizes a high-speed centrifugal bowl into which liquid sludge is fed. Centrifugal force serves to atomize the sludge into fine particles and to spray them into the top of the drying chamber where steady transfer of moisture to the hot gases takes place. This is shown in Fig. 8·21. A nozzle may be utilized in place of the bowl if the design prevents clogging of the nozzle.

Rotary Dryers

Rotary kiln dryers have been used in several plants for the drying of sludge and for the drying and burning of municipal refuse and industrial wastes. Many different dryers have been developed for industrial processes [11, 18], including direct-heating types in which the material

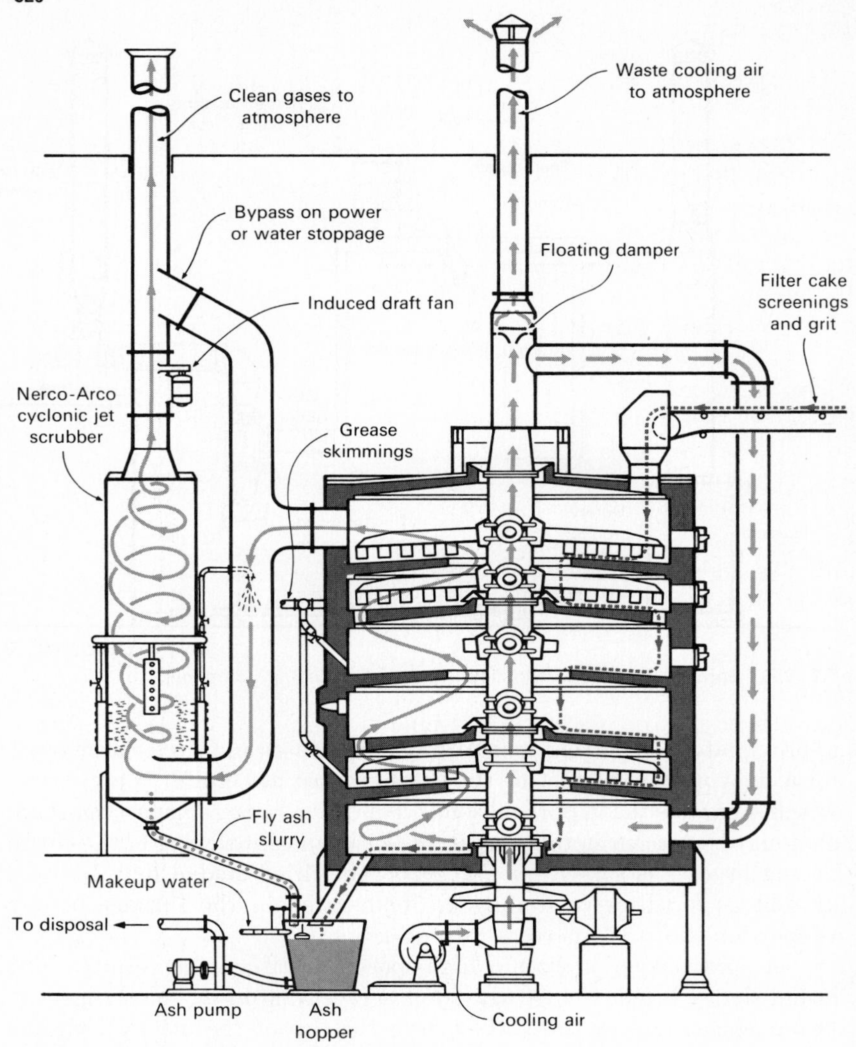

FIG. 8·22 Multiple hearth incinerator. [From Nichols Herreshoff.]

being dried is in contact with the hot gases, indirect-heating types in which the hot gases are separated from the drying material by steel shells, and indirect-direct types in which the hottest gases surround a central shell containing the material but return through it at reduced temperatures. Coal, oil, gas, municipal refuse, or the dried sludge may be used as fuel. Plows or louvres may be installed for lifting and agitating the material as the drum revolves.

Multiple Hearth Incinerator

A multiple hearth incinerator is frequently used to dry and burn sludges that have been partially dried by vacuum filtration. This is a countercurrent operation in which heated air and products of combustion pass by finely pulverized sludge that is continually raked to expose fresh surfaces.

In the unit shown on Fig. 8·22, wet sludge (70 to 75 percent moisture) is fed onto the top hearth and is slowly raked to the center. From the center, it drops to the second hearth where the rakes move it to the periphery. Here it drops to the third hearth and is again raked to the center. The hottest temperatures are on the middle hearths where the sludge burns and where auxiliary fuel is also burned as necessary to warm up the furnace and to sustain combustion. Preheated air is admitted to the lowest hearth and is further heated by the sludge as it rises past the middle hearths where combustion is occurring. The air then cools as it gives up its heat to dry the incoming sludge on the top hearths.

The highest moisture content of the air is found on the top hearths where sludge with the highest moisture is heated and some water is vaporized. This air passes twice through the furnace. Cooling air is initially blown into the central column and hollow rabble arms to keep them from burning up. A large portion of this air, after passing out the central column at the top, is recirculated to the lowest hearth, either directly or after passing through a recuperator (not shown in Fig. 8·22).

This furnace may also be designed as a dryer only. In this case, a furnace is needed to provide hot gases, and the sludge and gases both proceed downward through the furnace in parallel flow. Parallel flow of product and hot gases is frequently used in drying operations to prevent burning or scorching a heat-sensitive material.

PROBLEMS

8·1 A bar rack is inclined at a 50° angle with the horizontal. The circular bars have a diameter of 0.75 in. and a clear spacing of 1 in. Determine the head loss when the bars are clean and the velocity approaching the rack is 3.2 fps.

8·2 The contents of a mixing tank are to be mixed with a turbine impeller with six flat blades. The diameter of the turbine is 6 ft and the impeller is installed 4 ft above the bottom of the 20-ft tank. If the temperature is 30°C and the impeller is rotated at 30 rpm, what will be the power consumption? Find the Reynolds number using Eq. 8·5.

8·3 Derive Stokes' law by equating Eq. 8·15 to the effective particle weight.

8·4 Determine the settling velocity in fps of a sand particle with a specific gravity of 2.6 and a diameter of 1 mm. Assume the Reynolds number is 175.

8·5 Using the settling test curves shown in Fig. 8·23, determine the efficiency of a settling tank in removing flocculent particles if the depth is 4 ft and the detention time is 20 min.

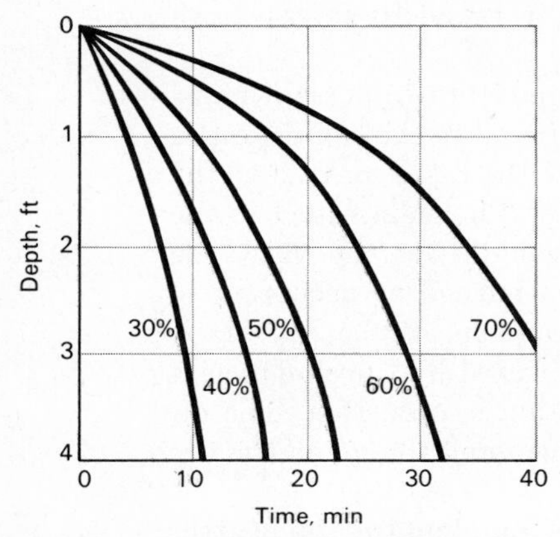

FIG. 8-23 Settling test curves for Prob. 8·5.

8·6 The curve shown in Fig. 8·24 was obtained from a settling test in a 5-ft cylinder. The initial solids concentration was 3,600 mg/liter. Determine the thickener area required for a concentration C_u of 22,500 mg/liter with a sludge flow of 0.4 mgd.

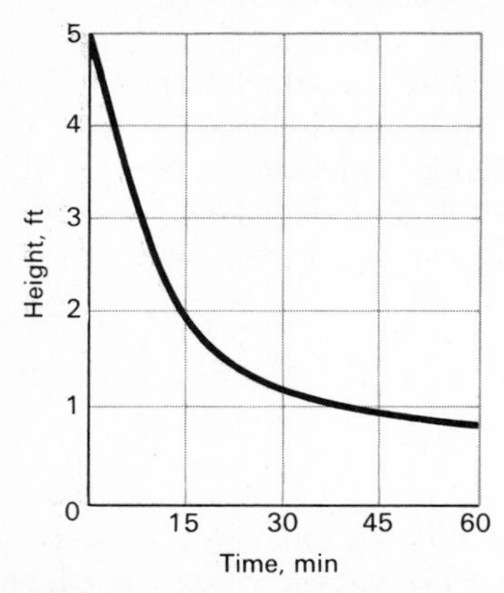

FIG. 8·24 Interface settling curve for Prob. 8·6.

8·7 A rectangular settling tank has an overflow rate of 750 gpd/sq ft and dimensions of 8 ft deep by 20 ft wide by 50 ft long. Determine whether or not particles with a diameter of 0.1 mm and a specific gravity of 2.5 will be scoured from the bottom. Use $f = 0.03$ and $k = 0.04$.

8·8 Determine the number of volumes of wash water required to elutriate a sludge containing 4,500 mg/liter of alkalinity. The effluent alkalinity is to be 150 mg/liter and the alkalinity of the wash water is 30 mg/liter. Use a two-stage countercurrent operation.

8·9 A digester is to be heated by circulation of sludge through an external hot-water heat exchanger. Using the data given below, find the heat required to maintain the required digester temperature.

1. U_x = overall heat-transfer coefficient, Btu/(hr)(sq ft)(°F)
2. $U_{air} = 0.15$, $U_{ground} = 0.12$, $U_{cover} = 0.20$.
3. Digester is a concrete tank with floating steel cover, diameter = 36 ft, side wall depth = 26 ft, 13 ft of which is above the ground surface.
4. Sludge fed to digester = 4,000 gpd at 58°F.
5. Outside temperature = 5°F.
6. Average ground temperature = 40°F.
7. Sludge in tank is to be maintained at 95°F.
8. Assume a specific heat of the sludge = 1.0.
9. Sludge contains 4 percent solids.
10. Assume cone-shaped cover with center 2 ft above digester top, and a bottom with center 4 ft below bottom edge.

REFERENCES

1. Brown, G. G., and Associates: *Unit Operations*, Wiley, New York, 1960.
2. Camp, T. R.: Sedimentation and the Design of Settling Tanks, *Transactions ASCE*, vol. 111, 1946.

3. Camp, T. R., and P. C. Stein: Velocity Gradients and Internal Work in Fluid Motion, *Journal, Boston Society of Civil Engineers*, vol. 30, p. 219, 1943.

4. Carman, P. C.: A Study of the Mechanism of Filtration, parts I–III, *Journal Soc. Chem. Ind.* (Britain), vol. 52–53, 1933–1934.

5. Coackley, P., and B. R. S. Jones: "Vacuum Sludge Filtration," *Sewage and Industrial Wastes*, vol. 28, no. 6, 1956.

6. Dick, R. I., and B. B. Ewing: Evaluation of Activated Sludge Thickening Theories, *Journal Sanitary Division*, ASCE, vol. 93, no. SA-4, 1967.

7. Eckenfelder, W. W., Jr.: *Industrial Water Pollution Control*, McGraw-Hill, New York, 1966.

8. Eckenfelder, W. W., Jr., and D. J. O'Connor: *Biological Waste Treatment*, Pergamon, New York, 1961.

9. Fair, G. M., J. C. Geyer, and D. A. Okun: *Water and Wastewater Engineering*, vol. 2, Wiley, New York, 1966.

10. Jaeger, C.: *Engineering Fluid Mechanics*, Blackie, London, 1956.

11. McAdams, W. H.: *Heat Transmission*, 2d ed., McGraw-Hill, New York, 1954.

12. McCabe, B. J., and W. W. Eckenfelder, Jr.: *Biological Treatment of Sewage and Industrial Wastes*, vol. 2, Reinhold, New York, 1958.

13. McCabe, W. L., and J. C. Smith: *Unit Operations of Chemical Engineering*, 2d ed., McGraw-Hill, New York, 1967.

14. Perry, J. H.: *Chemical Engineer's Handbook*, 4th ed., McGraw-Hill, New York, 1963.

15. Rich, L. G.: *Unit Operations of Sanitary Engineering*, Wiley, New York, 1961.

16. Rushton, J. H.: Mixing of Liquids in Chemical Processing, *Industrial and Engineering Chemistry*, vol. 44, no. 12, 1952.

17. Talmadge, W. P., and E. B. Fitch: "Determining Thickener Unit Areas," *Industrial and Engineering Chemistry*, vol. 47, no. 1, 1955.

18. Walker, W. H., W. K. Lewis, W. H. McAdams, and E. R. Gilliland: *Principles of Chemical Engineering*, 3d ed., McGraw-Hill, New York, 1937.

chemical unit processes

9

Chemical unit processes commonly used for wastewater treatment, including (1) chemical precipitation, (2) gas transfer, (3) adsorption, (4) disinfection, and (5) combustion, are delineated in this chapter. Such chemical unit processes as ion exchange and electrodialysis are covered in Chap. 14.

Reaction kinetics are discussed briefly first, because knowledge concerning the order of reactions and their determination is fundamental to the chemical unit processes covered here and to the biological unit processes discussed in Chap. 10. The fundamentals of chemistry are not presented, but a basic knowledge of chemistry is essential to the understanding of chemical processes.

REACTION KINETICS

The rate at which a reaction proceeds is an important consideration in wastewater treatment. Treatment processes may be designed, for example, on the basis of the rate at which the reaction proceeds rather than the equilibrium position of the reaction, because the reaction usually takes too long to go to completion. In this case, quantities of chemicals in excess of the stoichiometric, or exactly reacting amount, may be used to accomplish the treatment step in a reasonable period of time. This is true, for example, with disinfection and with the biological processes for BOD removal and digestion.

Reaction Order

A first-order reaction is one in which the rate of completion is observed to be directly proportional to the first power of the concentration of the reactant. For example, in the reaction

$$aA + bB + \cdots \rightarrow pP + qQ + \cdots \tag{9·1}$$

if the rate is experimentally found to be proportional to the first power of the concentration of A, then the reaction is said to be first order with respect to A. The following equations will illustrate the relationships used to identify reactions of various orders. A first-order reaction is defined as

$$\frac{dC}{dt} = k_1C \tag{9·2}$$

which integrates to

$$\ln \frac{C_t}{C_0} = k_1t \qquad \text{or} \qquad C_t = C_0e^{k_1t} \tag{9·3}$$

where C = concentration of reactant
C_0 = concentration at time zero
C_t = concentration at time t
k_1 = reaction-rate constant

A second-order reaction for the same reactant is defined as

$$\frac{dC}{dt} = k_2C^2 \tag{9·4}$$

which integrates to

$$\frac{1}{C_t} - \frac{1}{C_0} = k_2t \tag{9·5}$$

A second-order reaction for different reactants is defined as

$$\frac{dC_A}{dt} = \frac{dC_B}{dt} = k_2'C_AC_B \tag{9·6}$$

Fractional orders are defined as

$$\frac{dC_A}{dt} = \frac{dC_B}{dt} = k'C_A{}^aC_B{}^b \tag{9·7}$$

Reactions for other orders are defined in a similar manner; however, the order of the reaction is an empirical quantity and has no necessary relationship to any stoichiometric equation for the reaction. The determination of the order of a reaction and the corresponding rate constant is illustrated in Example 9·1.

EXAMPLE 9·1 *Determination of Reaction Order and Rate Constant*

Determine if the following substrate removal data can be described with a first-order reaction, and if so, determine the rate constant.

Time, hr	0	1	2	3	4	6
Substrate conc, mg/liter	50	35.6	25.8	18.5	12.8	7.3

Solution

1. Assuming $C_0 = 50$ mg/liter, compute the value of the rate constant at the various times, using Eq. 9·3.

Time	C/C_0	ln C/C_0	k, hr^{-1}
0	1.00	0.000	
1	0.71	−0.342	−0.342
2	0.52	−0.654	−0.327
3	0.37	−0.994	−0.331
4	0.27	−1.310	−0.328
6	0.14	−1.966	−0.328

2. Since the rate constant is approximately the same, it can be concluded that the reaction is first order with respect to substrate and that the average value of the rate constant is about -0.331 hr^{-1}.
3. The rate constant could also be determined by plotting log $(C_0 - x)$ versus t on arithmetic paper, where the quantity $(C_0 - x)$ represents the amount of material remaining at each time period.

Effect of Temperature

The temperature dependence of reaction-rate constants is of special importance because of the need to use constants that are determined at one temperature for systems of another temperature. For example, the reaction-rate constant determined for the BOD reaction at 20°C must often be used for systems at temperatures other than 20°C. The temperature dependence of the rate constant is given by the van't Hoff-Arrhenius equation

$$\frac{d(\ln k)}{dT} = \frac{E}{RT^2} \tag{9·8}$$

where T = temperature, °K
R = ideal gas constant, 199 calories (cal)/°K-mole
E = a constant characteristic of reaction called activation energy

Integration of Eq. 9·8 between the limits T_1 and T_2 gives

$$\ln \frac{k_2}{k_1} = \frac{E(T_2 - T_1)}{RT_1T_2} \tag{9·9}$$

With k_1 known for a given temperature and with E known, k_2 can be calculated from this equation. The activation energy E can be calculated by determining the k at two different temperatures and by using Eq. 9·9. Common values of E for waste treatment processes are in the range of 2,000 to 20,000 cal/mole.

Because most wastewater treatment operations and processes are carried out at or near the ambient temperature, the quantity E/RT_1T_2 may be assumed to be a constant for all practical purposes. If the value of the quantity is designated by C, then Eq. 9·9 can be rewritten as

$$\ln \frac{k_2}{k_1} = C(T_2 - T_1) \tag{9·10}$$

$$\frac{k_2}{k_1} = e^{C(T_2 - T_1)} \tag{9·11}$$

Replacing e^C in Eq. 9·11 with a temperature coefficient θ yields

$$\frac{k_2}{k_1} = \theta^{(T_2 - T_1)} \tag{9·12}$$

which is used commonly in the sanitary engineering field to adjust the value of the operative rate constant to reflect the effect of temperature. An alternate form of the temperature-correction equation may be obtained by expanding Eq. 9·11 as a series and dropping all but the first two terms. It should be noted, however, that although the value of θ is assumed to be constant, it will often vary considerably with temperature. Therefore, caution must be used in selecting appropriate values for θ for different temperature ranges. Typical values for various operations and processes for different temperature ranges are given, where available, in the sections in which the individual topics are discussed.

CHEMICAL PRECIPITATION

Chemical precipitation in wastewater treatment involves the addition of chemicals for the expressed purposes of improving plant performance and removing specific components contained in the wastewater. In the past (also see "Chemical Treatment" in Chap. 11), chemical precipitation was used to enhance the degree of suspended solids and BOD removal (1) where there were seasonal variations in the concentration of the sewage (wastes from canneries), (2) where an intermediate degree of treatment was required, and (3) as an aid to the sedimentation process.

More recently (1970), interest in chemical precipitation has been renewed because (1) it can be used effectively for the removal of phosphorus and (2) it can be combined with activated-carbon adsorption to provide complete wastewater treatment, bypassing the need for biological treatment and, at the same time, providing more effective removal of the organics in wastewater that are resistant to biological treatment. For example, the residual COD after chemical precipitation and carbon adsorption is about 10 to 20 mg/liter, whereas the residual COD after biological treatment is about 100 to 300 mg/liter.

Chemicals

Many different substances have been used as precipitants. The most common ones are listed in Table 9·1. The degree of clarification obtained depends on the quantity of chemicals used and the care with which the process is controlled. It is possible by chemical precipitation to obtain a clear effluent, substantially free from matter in suspension or in the colloidal state. From 80 to 90 percent of the total suspended matter, 50 to 55 percent of the total organic matter, and 80 to 90 percent of the bacteria can be removed by chemical precipitation. These figures can be compared with plain sedimentation in which 50 to 70 percent of the total suspended matter and 30 to 40 percent of the organic matter settle out.

TABLE 9·1 CHEMICALS USED IN WASTEWATER TREATMENT

Chemical	Formula	Molecular weight
Alum	$Al_2(SO_4)_3 \cdot 18H_2O$*	666.7
Ferrous sulfate (copperas)	$FeSO_4 \cdot 7H_2O$	278.0
Lime	$Ca(OH)_2$	56 as CaO
Sulfuric acid	H_2SO_4	98
Sulfur dioxide	SO_2	64
Ferric chloride	$FeCl_3$	162.1
Ferric sulfate	$Fe_2(SO_4)_3$	400

* Number of bound water molecules will vary from 13 to 18.

Chemical Reactions (Simplified)

The chemicals added to sewage in chemical precipitation react with substances that are either normally present in the sewage or are added for this purpose. The quantity of chemicals used is commonly expressed in pounds per million gallons. In this section, the reactions involved in chemical precipitation with (1) alum, (2) ferrous sulfate (copperas) and lime, (3) lime, and (4) sulfuric acid and sulfur dioxide are presented and discussed in detail. Reactions for ferric chloride, ferric chloride and lime, and ferric sulfate and lime are also presented. The reactions involved in the removal of phosphate are discussed in the following section.

Alum When alum is added to sewage containing calcium and magnesium bicarbonate alkalinity, the reaction that occurs may be illustrated as follows:

$$\underset{\text{Aluminum sulfate}}{\overset{666.7}{Al_2(SO_4)_3 \cdot 18H_2O}} + \underset{\text{Calcium bicarbonate}}{\overset{3 \times 100CaCO_3}{3Ca(HCO_3)_2}} \rightleftharpoons$$

$$\underset{\text{Calcium sulfate}}{\overset{3 \times 136}{3CaSO_4}} + \underset{\text{Aluminum hydroxide}}{\overset{2 \times 78}{2Al(OH)_3}} + \underset{\text{Carbon dioxide}}{\overset{6 \times 44}{6CO_2}} + \overset{18 \times 18}{18H_2O} \qquad (9\cdot13)$$

The insoluble aluminum hydroxide is a gelatinous floc that settles slowly through the sewage, sweeping out suspended matter as well as producing other changes. The reaction is exactly analogous when magnesium bicarbonate is substituted for the calcium salt. The numbers above the chemical formulas are the combining molecular weights of the different substances and therefore denote what quantity of each is involved. Because alkalinity is reported in terms of calcium carbonate ($CaCO_3$), the molecular weight of which is 100, the quantity of alkalinity required to react with 10 mg/liter of alum is

$$10.0 \times \frac{3 \times 100}{666.7} = 4.5 \text{ mg/liter}$$

If less than this amount of alkalinity is available, it must be added. Lime is commonly used for this purpose when necessary, but it is seldom required in sewage treatment.

Ferrous Sulfate and Lime In most cases, ferrous sulfate cannot be used alone as a precipitant since lime must be added at the same time, to form a precipitate. The reaction with ferrous sulfate alone is illustrated in Eq. 9·14.

$$\underset{\text{Ferrous sulfate}}{\overset{278}{FeSO_4 \cdot 7H_2O}} + \underset{\text{Calcium bicarbonate}}{\overset{100CaCO_3}{Ca(HCO_3)_2}} \rightleftharpoons \underset{\text{Ferrous bicarbonate}}{\overset{178}{Fe(HCO_3)_2}} + \underset{\text{Calcium sulfate}}{\overset{136}{CaSO_4}} + \overset{7 \times 18}{7H_2O} \tag{9·14}$$

If lime in the form $Ca(OH)_2$ is now added, the reaction that takes place is

$$\underset{\text{Ferrous bicarbonate}}{\overset{178}{Fe(HCO_3)_2}} + \underset{\text{Calcium hydroxide}}{\overset{2 \times 56CaO}{2Ca(OH)_2}} \rightleftharpoons \underset{\text{Ferrous hydroxide}}{\overset{89.9}{Fe(OH)_2}} + \underset{\text{Calcium carbonate}}{\overset{2 \times 100}{2CaCO_3}} + \overset{2 \times 18}{2H_2O} \tag{9·15}$$

The ferrous hydroxide is next oxidized to ferric hydroxide, the final form desired, by the oxygen dissolved in the sewage:

$$\underset{\text{Ferrous hydroxide}}{\overset{4 \times 89.9}{4Fe(OH)_2}} + \underset{\text{Oxygen}}{\overset{32}{O_2}} + \overset{2 \times 18}{2H_2O} \rightleftharpoons \underset{\text{Ferric hydroxide}}{\overset{4 \times 106.9}{4Fe(OH)_3}} \tag{9·16}$$

The insoluble ferric hydroxide is formed as a bulky, gelatinous floc similar to the alum floc. The alkalinity required for 10 mg/liter of ferrous sulfate is

$$10.0 \times \frac{100}{278} = 3.6 \text{ mg/liter}$$

The lime required is

$$10.0 \times \frac{2 \times 56}{278} = 4.0 \text{ mg/liter}$$

The oxygen required is

$$10.0 \times \frac{32}{4 \times 278} = 0.29 \text{ mg/liter}$$

Oxidation is favored by a high pH value, which is established to some extent by the lime.

In sewage treatment, lime is commonly added in excess of the amount required. Experience has shown that the best results are obtained by adding sufficient lime to produce a pink color when phenolphthalein is used as an indicator. Since the formation of ferric hydroxide is dependent on the presence of dissolved oxygen, the reaction in Eq. 9·16 cannot be completed with septic sewages or industrial wastes devoid of oxygen. Ferric sulfate may take the place of ferrous sulfate, and its use often avoids the addition of lime and the requirement of dissolved oxygen.

Lime When lime alone is added as a precipitant or is used in excess of the amount required for the precipitation of the iron in the previous reactions, the principles of clarification are explained by the following reactions:

$$\underset{\substack{\text{Calcium}\\ \text{hydroxide}}}{\overset{56CaO}{Ca(OH)_2}} + \underset{\substack{\text{Carbonic}\\ \text{acid}}}{\overset{44CO_2}{H_2CO_3}} \rightleftharpoons \underset{\substack{\text{Calcium}\\ \text{carbonate}}}{\overset{100}{CaCO_3}} + \overset{2 \times 18}{2H_2O} \tag{9·17}$$

$$\underset{\substack{\text{Calcium}\\ \text{hydroxide}}}{\overset{56CaO}{Ca(OH)_2}} + \underset{\substack{\text{Calcium}\\ \text{bicarbonate}}}{\overset{100CaCO_3}{Ca(HCO_3)_2}} \rightleftharpoons \underset{\substack{\text{Calcium}\\ \text{carbonate}}}{\overset{2 \times 100}{2CaCO_3}} + \overset{2 \times 18}{2H_2O} \tag{9·18}$$

A sufficient quantity of lime must therefore be added to combine with all the free carbonic acid and with the carbonic acid of the bicarbonates (half-bound carbonic acid) to produce calcium carbonate, which acts as the coagulant. Much more lime is generally required when it is used alone than when sulfate of iron is also used. Where industrial wastes introduce mineral acids or acid salts into the sewage, these must be neutralized before precipitation can take place.

If too much lime is used in the treatment of sewage, some of the suspended organic matter will be dissolved by the caustic calcium hydroxide, and the effluent may be worse than the untreated sewage. If too little is used, the effluent will not be well clarified.

Sulfuric Acid and Sulfur Dioxide Neither sulfuric acid nor sulfur dioxide is a floc-forming substance. When sulfuric acid is added to sewage it neutralizes the alkalinity as follows:

$$\begin{array}{lllll} 98 & 100CaCO_3 & 136 & 2 \times 44 & 2 \times 18 \\ H_2SO_4 + & Ca(HCO_3)_2 \rightleftharpoons & CaSO_4 + & 2CO_2 + & 2H_2O \\ \text{Sulfuric acid} & \text{Calcium bicarbonate} & \text{Calcium sulfate} & \text{Carbon dioxide} & \end{array} \qquad (9{\cdot}19)$$

Any excess remains in solution as sulfuric acid and only soluble substances are formed.

When sulfur dioxide is used it hydrolizes to form sulfurous acid and reacts with the alkalinity of the sewage to form bisulfites. Both of these compounds are oxidized in the presence of dissolved oxygen. The reactions may be stated as follows:

$$\begin{array}{lllll} 2 \times 82 & 100CaCO_3 & 202 & 2 \times 44 & 2 \times 18 \\ 2H_2SO_3 + & Ca(HCO_3)_2 \rightleftharpoons & Ca(HSO_3)_2 + & 2CO_2 + & 2H_2O \\ \text{Sulfurous acid} & \text{Calcium bicarbonate} & \text{Calcium bisulfite} & \text{Carbon dioxide} & \end{array} \qquad (9{\cdot}20)$$

$$\begin{array}{lll} 64 & 18 & 82 \\ SO_2 & + H_2O \rightleftharpoons & H_2SO_3 \\ \text{Sulfur dioxide} & & \text{Sulfurous acid} \end{array} \qquad (9{\cdot}21)$$

$$\begin{array}{lll} 2 \times 82 & 32 & 2 \times 98 \\ 2H_2SO_3 + & O_2 \rightleftharpoons & 2H_2SO_4 \\ \text{Sulfurous acid} & \text{Oxygen} & \text{Sulfuric acid} \end{array} \qquad (9{\cdot}22)$$

$$\begin{array}{llll} 202 & 32 & 136 & 98 \\ Ca(HSO_3)_2 + & O_2 \rightleftharpoons & CaSO_4 + & H_2SO_4 \\ \text{Calcium bisulfite} & \text{Oxygen} & \text{Calcium sulfate} & \text{Sulfuric acid} \end{array} \qquad (9{\cdot}23)$$

Here again, only soluble substances are formed. The action of these chemicals is discussed later in this chapter.

Ferric Chloride The reactions for ferric chloride are

$$\begin{array}{lll} 162.1 & 3 \times 18 & 106.9 \\ FeCl_3 + & 3H_2O \rightleftharpoons & Fe(OH)_3 + 3H^+ + 3Cl^- \\ \text{Ferric chloride} & \text{Water} & \text{Ferric hydroxide} \end{array} \qquad (9{\cdot}24)$$

$$\begin{array}{lll} 3H^+ + & 3HCO_3^- \rightleftharpoons & 3H_2CO_3 \\ & \text{Bicarbonate} & \text{Carbonic acid} \end{array} \qquad (9{\cdot}25)$$

Ferric Chloride and Lime The reactions for ferric chloride and lime are

$$\begin{array}{llll} 2 \times 162 & 3 \times 56CaO & 3 \times 111 & 2 \times 106.9 \\ 2FeCl_3 + & 3Ca(OH)_2 \rightleftharpoons & 3CaCl_2 + & 2Fe(OH)_3 \\ \text{Ferric chloride} & \text{Calcium hydroxide} & \text{Calcium chloride} & \text{Ferric hydroxide} \end{array} \qquad (9{\cdot}26)$$

Ferric Sulfate and Lime The reactions for ferric sulfate and lime are

$$\underset{\substack{\text{Ferric}\\\text{sulfate}}}{\overset{400}{Fe_2(SO_4)_3}} + \underset{\substack{\text{Calcium}\\\text{hydroxide}}}{\overset{3\times 56CaO}{3Ca(OH)_2}} \rightleftharpoons \underset{\substack{\text{Calcium}\\\text{sulfate}}}{\overset{408}{3CaSO_4}} + \underset{\substack{\text{Ferric}\\\text{hydroxide}}}{\overset{2\times 106.9}{2Fe(OH)_3}} \tag{9·27}$$

Phosphate-Removal Reactions (Simplified)

Phosphate can be removed by chemical precipitation with various multivalent metal ions. Reactions with calcium [Ca(II)], aluminum [Al(III)], and iron [Fe(III)] ions are presented in Eqs. 9·28 to 9·30. (The Roman numerals following the metal refer to the oxidation state.)

Calcium:

$$10Ca^{++} + 6PO_4^{3-} + 2OH^- \rightleftharpoons \underset{\text{Hydroxylapatite}}{Ca_{10}(PO_4)_6(OH)_2} \tag{9·28}$$

Aluminum:

$$Al^{3+} + H_nPO_4^{3-n} \rightleftharpoons AlPO_4 + nH^+ \tag{9·29}$$

Iron:

$$Fe^{3+} + H_nPO_4^{3-n} \rightleftharpoons FePO_4 + nH^+ \tag{9·30}$$

The chemistry of the removal of phosphate with lime is quite different from that of alum or iron. From the equations presented previously it will be noted that when lime is added to water it reacts with the natural bicarbonate alkalinity to precipitate $CaCO_3$. Excess calcium ions will then react with the phosphorus, as shown in Eq. 9·28, to precipitate hydroxylapatite $Ca_{10}(PO_4)_6(OH)_2$. Therefore, the quantity of lime required will, in general, be independent of the amount of phosphorus present and will depend primarily on the alkalinity of the wastewater. In the case of alum and iron, 1 mole will precipitate 1 mole of phosphate; however, these reactions are deceptively simple and must be considered in light of the many competing reactions and their associated equilibrium constants.

For example, for equimolar initial concentrations of Al(III), Fe(III), and phosphorus, the total concentration of soluble phosphorus in equilibrium with both insoluble $FePO_4$ and $AlPO_4$ is shown in Fig. 9·1. The solid lines trace the concentration of residual soluble phosphorus after precipitation. Pure metal phosphates are precipitated within the shaded area, and mixed complex precipitates are formed outside toward the higher pH values [5].

Because the theoretical details of the precipitation of phosphorus

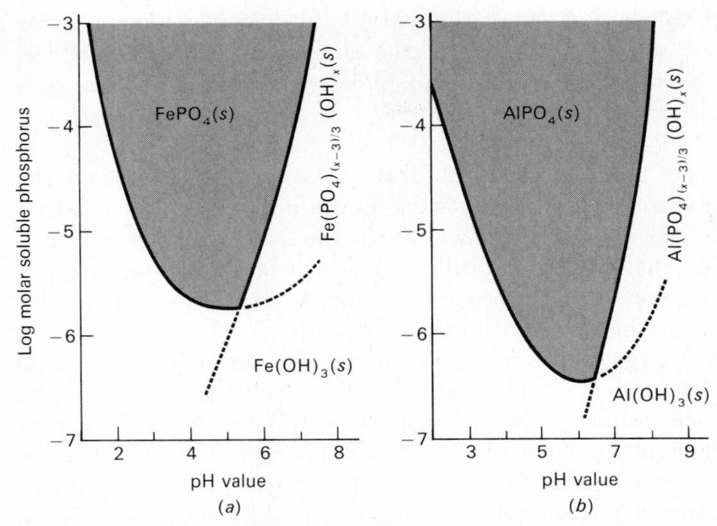

FIG. 9·1 Concentration of ferric and aluminum phosphate in equilibrium with soluble phosphorus. (*a*) Fe(III)-phosphate, (*b*) Al(III)-phosphate [5].

with metallic ions are beyond the scope of this text, the reader is advised to consult Stumm and Morgan [15]. Practical details on the removal of phosphorus, including process flowsheets and required chemical dosages, are presented in Chaps. 11 and 14.

Action of Chemical Precipitants

The theory of chemical precipitation reactions is very complex. The reactions that have been presented explain it only in part, and even they do not necessarily proceed as indicated. They are often incomplete, and numerous side reactions with other substances in sewage may take place. Therefore, the following discussion is of necessity incomplete, but it will serve as an introduction to the nature of the phenomena involved.

Nature of Particles in Wastewater There are two general types of colloidal solid particle dispersions in liquids. When water is the solvent, these are called the hydrophobic or "water-hating" and the hydrophilic or "water-loving" colloids. These two types are based on the attraction of the particle surface for water. Hydrophobic particles have relatively little attraction for water, while hydrophilic particles have a great attraction for water. It should be noted, however, that water can interact to some extent with hydrophobic particles. Some water molecules will generally adsorb on the typical hydrophobic surface, but the reaction between water and hydrophilic colloids occurs to a much greater extent.

Surface Charge An important factor in the stability of colloids is the presence of surface charge. It develops in a number of different ways, depending on the chemical composition of the medium (wastewater in this case) and the colloid. Regardless of how it is developed, this stability must be overcome if these particles are to be aggregated (flocculated) into larger particles with enough mass to settle easily.

Surface charge develops most commonly through preferential adsorption, ionization, and isomorphous replacement. For example, oil droplets, gas bubbles, or other chemically inert substances dispersed in water will acquire a negative charge through the preferential adsorption of anions (particularly hydroxyl ions). In the case of substances such as proteins or microorganisms, surface charge is acquired through the ionization of carboxyl and amino groups [14]. This can be represented as $R_{NH_2}{}^{COO^-}$ at high pH, $R_{NH_3{}^+}{}^{COOH}$ at low pH, and $R_{NH_3{}^+}{}^{COO^-}$ at the isoelectric point where R represents the bulk of the solid [5]. Charge development through isomorphous replacement occurs in clay and other soil particles, in which ions in the lattice structure are replaced with ions from solution (e.g., the replacement of Si with Al).

When the colloid or particle surface becomes charged, some ions of the opposite charge (known as counter ions) become attached to the surface. They are held there through electrostatic and van der Waals forces strongly enough to overcome thermal agitation. Surrounding this fixed layer of ions is a diffuse layer of ions, which is prevented from forming a compact double layer by thermal agitation. This is illustrated schematically in Fig. 9·2. As shown, the double layer consists of a compact layer (Stern) in which the potential drops from ψ_0 to ψ_s and a diffuse layer in which the potential drops from ψ_s to 0 in the bulk solution.

If a particle such as shown in Fig. 9·2 is placed in an electrolyte solution, and an electric current is passed through the solution, the particle, depending on its surface charge, will be attracted to one or the other of the electrodes, dragging with it a cloud of ions.

The potential at the surface of the cloud (called the surface of shear) is sometimes measured in wastewater treatment operations. The measured value is often called the zeta potential. Theoretically, however, the zeta potential should correspond to the potential measured at the surface enclosing the fixed layer of ions attached to the particle, as shown in Fig. 9·2. The use of the measured zeta potential

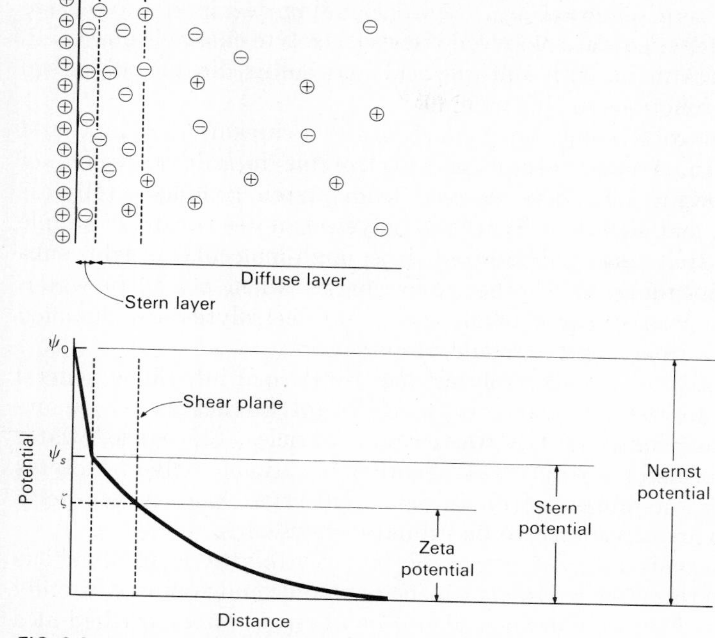

FIG. 9·2 Stern model of electrical double layer [14].

value is limited because it will vary with the nature of the solution components, and it therefore is not a repeatable measurement.

Particle Aggregation To bring about particle aggregation, steps must be taken to reduce particle charge or to overcome the effect of this charge. This can be accomplished by (1) the addition of potential-determining ions, which will be taken up by or will react with the colloid surface to lessen the surface charge, or the addition of electrolytes, which have the effect of reducing the thickness of the diffuse electric layer and thereby reduce the zeta potential; (2) the addition of long-chained organic molecules (polymers), whose subunits are ionizable and are therefore called polyelectrolytes, that bring about the removal of particles through adsorption and bridging; and (3) the addition of chemicals that form hydrolyzed metal ions.

Addition of potential-determining ions to promote coagulation can be illustrated by the addition of strong acids or bases to reduce the charge of metal oxides or hydroxides to near zero, so that coagulation can occur. Electrolytes can also be added to coagulate colloidal suspensions. Increased concentration of a given electrolyte will cause a decrease in zeta potential and a corresponding decrease in repulsive forces. Similar effects are observed if the electrolyte charge is increased. Chemical treatment with sulfuric acid and sulfur dioxide, discussed previously, functions in this manner.

Polyelectrolytes may be divided into two categories: natural and synthetic. Important natural polyelectrolytes include polymers of biological origin and those derived from starch products, cellulose derivatives, and alginates. Synthetic polyelectrolytes consist of simple monomers that are polymerized into high-molecular-weight substances. Depending on whether their charge, when placed in water, is negative, positive, or neutral, these polyelectrolytes are classified as anionic, cationic, and nonionic, respectively.

The action of polyelectrolytes may be divided into three general categories. In the first category, polyelectrolytes act as coagulants lowering the charge of the wastewater particles. Since wastewater particles normally are charged negatively, cationic polyelectrolytes are used for this purpose [18]. In this application, the cationic polyelectrolytes are considered to be primary coagulants.

The second mode of action of polyelectrolytes is interparticle bridging. In this case, polymers which are anionic and nonionic (usually anionic to a slight extent when placed in water) become attached at a number of adsorption sites to the surface of the particles found in the settled effluent. A bridge is formed when two or more particles become adsorbed along the length of the polymer. Bridged particles become intertwined with other bridged particles during the flocculation process. The size of the resulting three-dimensional particles grows until they can be removed easily by sedimentation.

The third type of polyelectrolyte action may be classified as a coagulation-bridging phenomenon, which results from using extremely high-molecular-weight cationic polyelectrolytes. Besides lowering the charge, these polyelectrolytes also will form particle bridges, as discussed previously.

In contrast with the aggregation brought about by the addition of chemicals that act as electrolytes (e.g., sulfuric acid and sulfur dioxide) and polymers, aggregation brought about by the addition of alum or ferric sulfate is a more complex process. In the past, it was thought that free Al^{3+} and Fe^{3+} were responsible for the effects observed during particle aggregation; however, it is now known that their hydrolysis products are responsible [16, 17]. Although the effect of these hydrolysis products is only now appreciated, it is interesting to note that their chemistry was first elucidated in the early 1900s by Pfeiffer (1902–1907), Bjerrum (1907–1920), and Werner (1907) [18]. For example, Pfeiffer proposed that the hydrolysis of trivalent metal salts, such as chromium, aluminum, and iron, could be represented as

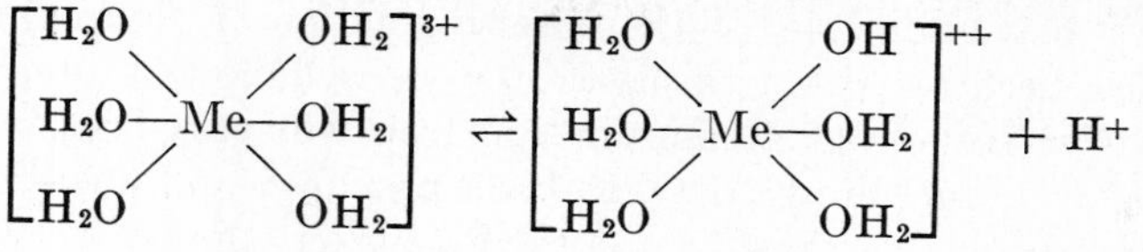

$$\left[\begin{array}{ccc} H_2O & & OH_2 \\ & \diagdown \quad \diagup & \\ H_2O & —Me— & OH_2 \\ & \diagup \quad \diagdown & \\ H_2O & & OH_2 \end{array}\right]^{3+} \rightleftharpoons \left[\begin{array}{ccc} H_2O & & OH \\ & \diagdown \quad \diagup & \\ H_2O & —Me— & OH_2 \\ & \diagup \quad \diagdown & \\ H_2O & & OH_2 \end{array}\right]^{++} + H^+ \qquad (9\cdot31)$$

with the extent of the dissociation depending on the anion associated with the metal and on the physical and chemical characteristics of the solution. Further, it was proposed that, upon the addition of sufficient base, the dissociation can proceed to produce a negative ion [18], such as

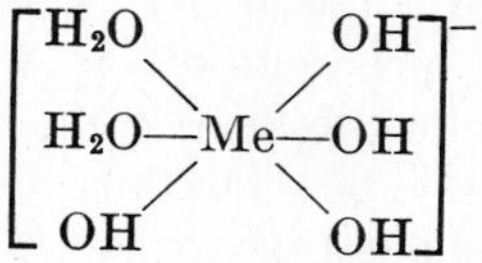

$$\left[\begin{array}{ccc} H_2O & & OH \\ & \diagdown \quad \diagup & \\ H_2O & —Me— & OH \\ & \diagup \quad \diagdown & \\ OH & & OH \end{array}\right]^{-}$$

Recently, however, it has been observed that the intermediate hydrolysis reactions of Al(III) are much more complex than would be predicted on the basis of a model in which a base is added to the solution. A hypothetical model proposed by Stumm [5] for Al(III) is shown in Eq. 9·32.

$$[Al(H_2O)_6]^{3+} \xrightarrow{OH^-} [Al(H_2O)_5OH]^{++} \xrightarrow{OH^-} [Al(H_2O)_4(OH)_2]^{+}$$

$$\xrightarrow{OH^-} [Al_6(OH)_{15}]^{3+}(aq) \quad \text{or} \quad [Al_8(OH)_{20}]^{4+}(aq) \xrightarrow{OH^-}$$

$$[Al(OH)_3(H_2O)_3]\ (s) \xrightarrow{OH^-} [Al(OH)_4(H_2O)_2]^{-} \qquad (9\cdot32)$$

Before the reaction proceeds to the point where a negative ion is produced, polymerization as depicted in the following formula will usually take place [18].

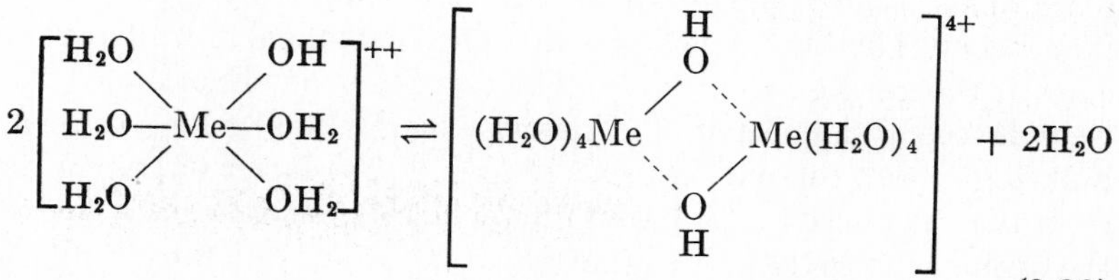

$$2\left[\begin{array}{ccc} H_2O & & OH \\ & \diagdown \quad \diagup & \\ H_2O & —Me— & OH_2 \\ & \diagup \quad \diagdown & \\ H_2O & & OH_2 \end{array}\right]^{++} \rightleftharpoons \left[\begin{array}{ccc} & \overset{H}{O} & \\ & \diagup \quad \diagdown & \\ (H_2O)_4Me & & Me(H_2O)_4 \\ & \diagdown \quad \diagup & \\ & \underset{H}{O} & \end{array}\right]^{4+} + 2H_2O \qquad (9\cdot33)$$

The possible combinations of the various hydrolysis products is endless, and their enumeration is not the purpose here. What is important, however, is the realization that one or more of the hydrolysis products may be responsible for the observed action of aluminum or iron. Further, because the hydrolysis reactions follow a stepwise process, the effectiveness of aluminum and iron will vary with time. For example, an alum slurry that has been prepared and stored will behave differently from a freshly prepared solution when it is added to a wastewater. For a more detailed review of the chemistry involved, the excellent articles on this subject by Stumm and Morgan [16] and Stumm and O'Melia [17] are recommended.

The characteristics of the three major modes of particle destabilization discussed in this section are reported in Table 9·2. In general, aggregation brought about by the hydrolysis products of alum and iron can be described best by the third mode of destabilization reported in Table 9·2, although polymer bridging may also be a factor. In addition, particles will also be removed by enmeshment as the alum and iron flocs sweep through the liquid.

GAS TRANSFER

Gas transfer is a vital part of wastewater treatment processes. For example, the functioning of aerobic processes, such as activated sludge, biological filters, and aerobic digestion, depends on the availability of sufficient quantities of oxygen. Chlorine, when used as a gas, must be transferred to solution in the water for disinfection purposes. One process for removing nitrogen compounds consists of conversion of the nitrogen to ammonia and of transferring the ammonia gas from the water to air. If the sewage entering the treatment plant is septic, air may be added before primary sedimentation, in preaeration tanks or in aerated grit chambers, to decrease odors and to improve treatability.

Theory

Of the numerous theories of mass transfer that are used to explain the mechanism of gas transfer one of the most commonly used is the two-film theory. This theory is based on a physical model in which two films exist at the gas-liquid interface. The two films, one liquid and one gas, provide the resistance to the passage of gas molecules between the bulk-liquid and the bulk-gaseous phases. For transfer of gas molecules from the gas phase to the liquid phase, slightly soluble gases encounter the primary resistance to transfer from the liquid film, while very soluble gases encounter the primary resistance from the gaseous film. Gases of intermediate solubility encounter significant resistance from both films. A sketch of these films is shown in Fig. 9·3. The rate of gas

TABLE 9·2 MODES OF DESTABILIZATION AND THEIR CHARACTERISTICS [17]

Phenomena	Physical double-layer theory (coagulation)	Chemical bridging model (flocculation)	Aggregation by adsorbable species (adsorption coagulation)
Electrostatic interaction	Predominant	Subordinate	Important
Chemical interactions, and adsorption	Absent	Predominant	Important
Zeta potential for optimum aggregation	Near zero	Usually not zero	Not necessarily zero
Addition of an excess of destabilizing species	No effect	Restabilization due to complete surface coverage	Restabilization usually accompanied by charge reversal
Relationship between optimum dosage of destabilizing species and the concentration of colloid (or concentration of colloidal surface)	CCC* virtually independent of colloid concentration	Stoichiometry, a linear relationship between flocculant dose and surface area	Stoichiometry possible but does not always occur
Physical properties of the aggregates which are produced	Dense, great shear strength but poor filtrability in cake filtration	Flocs of three-dimensional structure; low shear strength but excellent filtrability in cake filtration	Flocs of widely varying shear strength and density

* Critical coagulation concentration.

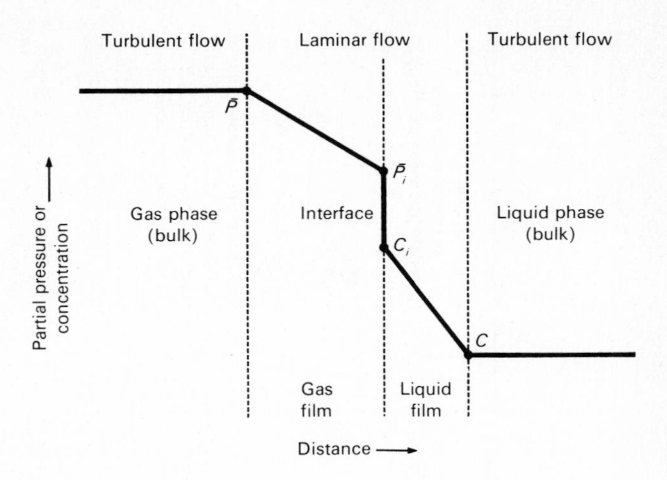

FIG. 9·3 Definition sketch for two-film gas transfer theory [8].

transfer, in general, is proportional to the difference between the existing concentration and the equilibrium concentration of the gas in solution. In equation form, this relationship can be expressed as

$$\frac{dC}{dt} = K(C_s - C) \tag{9·34}$$

where C = gas concentration
t = time
C_s = saturation concentration of gas
K = proportionality constant

K, in this case, includes the effect of the resistance of either or both films and is also a function of the area of liquid-gas interface that exists per unit volume of fluid.

The equilibrium concentration of gas dissolved in a liquid is a function of the partial pressure of the gas adjacent to the liquid. This relationship is given by Henry's law

$$p_g = Hx_g \tag{9·35}$$

where p_g = partial pressure of gas, atmospheres
H = Henry's law constant
x_g = equilibrium mole fraction of dissolved gas,

$$= \frac{\text{moles gas } (n_g)}{\text{moles gas } (n_g) + \text{moles water } (n_w)}$$

H is a function of the type, temperature, and constituents of the liquid. Values of H for various gases are tabulated in Table 9·3, and the use of Henry's law is illustrated in the following example.

TABLE 9·3 HENRY'S LAW CONSTANTS FOR SEVERAL GASES THAT ARE SLIGHTLY SOLUBLE IN WATER [Adapted from 11]

T, °C	$H \times 10^{-4}$, atm/mole fraction							
	Air	CO_2	CO	H_2	H_2S	CH_4	N_2	O_2
0	4.32	0.0728	3.52	5.79	0.0268	2.24	5.29	2.55
10	5.49	0.104	4.42	6.36	0.0367	2.97	6.68	3.27
20	6.64	0.142	5.36	6.83	0.0483	3.76	8.04	4.01
30	7.71	0.186	6.20	7.29	0.0609	4.49	9.24	4.75
40	8.70	0.233	6.96	7.51	0.0745	5.20	10.4	5.35
50	9.46	0.283	7.61	7.65	0.0884	5.77	11.3	5.88
60	10.1	0.341	8.21	7.65	0.103	6.26	12.0	6.29

EXAMPLE 9·2 *Saturation Concentration of Nitrogen in Water*

What is the saturation concentration of nitrogen in water in contact with dry air at 1 atm and 20°C?

Solution

1. Dry air contains about 79 percent nitrogen. Therefore $p_g = 0.79$.
2. From Table 9·3, $H = 8.04 \times 10^4$, and

$$x_g = \frac{p_g}{H} = \frac{0.79}{8.04 \times 10^4}$$
$$= 9.84 \times 10^{-6}$$

3. One liter of water contains 1,000/18 = 55.6 g moles, thus

$$\frac{n_g}{n_g + n_w} = 9.84 \times 10^{-6}$$
$$n_g = (n_g + 55.6)\, 9.84 \times 10^{-6}$$

Because the quantity $(n_g)\, 9.84 \times 10^{-6}$ is very much less than n_g

$$n_g \approx (55.6)\, 9.84 \times 10^{-6}$$
$$\approx 5.47 \times 10^{-4} \text{ moles/liter nitrogen}$$

4. Determine the saturation concentration of nitrogen.

$$C_s \approx \frac{5.47 \times 10^{-4} \text{ moles}}{\text{liter}} \left(\frac{28 \text{ g}}{\text{mole}}\right) \left(10^3 \frac{\text{mg}}{\text{g}}\right)$$
$$\approx 15.3 \text{ mg/liter}$$

In addition to the partial pressure, the amount of gas that can be taken up by a liquid depends on any subsequent reactions that the gas may undergo after it is dissolved. Chlorine, for example, reacts with water to form HCl and HOCl, and for this reason, large quantities of chlorine can be taken up by water. Oxygen molecules, on the other hand, apparently do not react with water molecules, and thus comparatively little oxygen can be taken up by water.

Sufficient oxygen to meet the requirements of aerobic waste treatment does not enter water through normal surface air-water interfaces. This is because of the low solubility of oxygen and the consequent low rate of oxygen transfer. To transfer the large quantities of oxygen into water that are needed, additional interfaces must be formed.

Oxygen can be supplied by means of bubbles introduced to the water to create additional air-water interfaces. In wastewater treatment plants, aeration is most frequently accomplished by dispersing air bubbles in the liquid at depths up to 15 ft by means of small orifices. Aerating devices include porous plates and tubes, perforated pipes, and various configurations of metal and plastic diffusers. Hydraulic shear devices may also be employed to create small bubbles by impinging a flow of liquid at an orifice to break up the air bubbles into smaller sizes.

A high-speed mixer, such as a turbine aerator, may be used to disperse air bubbles introduced below the center of the turbine. Mechanical aerators generally consist of high-speed turbines operating at the surface of the liquid, partially submerged. They are designed so as to intimately mix and circulate large volumes of air and water.

For a given amount of air introduced to a liquid system, the available surface through which gas transfer can take place increases with decreasing bubble size. The generalization can then be made that K in Eq. 9·34 will increase as the bubble size decreases, because of the advantageous change in the surface-area-to-volume ratio; however, this principle holds only within certain limits.

Efficient gas transfer also depends on agitation of the water. This turbulence reduces the thickness of the liquid film and lowers resistance to transfer and to dispersion of the dissolved gas once transfer has taken place. Air bubbles have a lifting effect due to viscous drag, and they promote turbulence and circulation of the liquid. Introduction of air along one side of an activated-sludge tank promotes spiral flow of the liquid mass. This flow carries the smaller bubbles across the surface, thus increasing the period of contact and also continually exposing fresh liquid to absorption of atmospheric air. The action keeps the mixed-liquor solids in suspension with a minimum use of air for agitation and provides the oxygen transfer required by the process. Ridge and furrow tanks require greater quantities of air just for agitation but they also have greater oxygenation capacities. As air bubbles rise, they tend to grow in size due to reduction of pressure and to coalescence. This latter effect increases as the air flowrate increases, resulting in larger-size bubbles.

For a given volume of water being aerated, aerators are evaluated on the basis of the quantity of oxygen transferred per unit of air introduced to the water for equivalent conditions (temperature and chemical composition of the water, depth at which the air is introduced, etc.). Usual procedures for analyzing efficiencies may involve:

1. Measuring the rate of sulfite oxidation when a sodium sulfite solution is aerated. Sulfite is rapidly oxidized by oxygen, and thus C in Eq. 9·34 is zero for this analysis.
2. Measuring the rate at which oxygen is added to the water by direct analysis for the oxygen. The oxygen content of the water is usually lowered to approximately zero prior to initiating the test. The oxygen level can be measured at known time intervals after the start of aeration. The data are plotted and then used in Eq. 9·34 to determine K.
3. Measuring the uptake of oxygen by microorganisms. In the activated-sludge system, oxygen is usually maintained at a level of 1 to 3 mg/liter. The oxygen is used by the microorganisms as rapidly as it is supplied. In equation form,

$$\frac{dC}{dt} = K(C_s - C) - U_m \tag{9·36}$$

where U_m is the quantity of oxygen used by the microorganisms. If the oxygen level is maintained at a constant level, dC/dt is zero and

$$U_m = K(C_s - C) \tag{9·37}$$

C in this case is constant also. Values of U_m can be determined in a laboratory by means of the Warburg apparatus. In this case, K can easily be determined as follows:

$$K = \frac{U_m}{C_s - C} \tag{9·38}$$

Studies of the transfer coefficient are usually made on water and then corrected for the wastewater. For instance, Sawyer and Lynch

[13] studied the effect of various synthetic detergents on K and found values as low as 40 percent of the value in pure water with 15 mg/liter of detergents. King [7] reported that, for relatively fresh sewage, values of 26 to 46 percent of the freshwater transfer coefficient were observed. This percentage will increase with time of aeration in the activated-sludge process. In addition, the saturation value of oxygen in domestic wastewater is about 95 percent of that in distilled water.

ADSORPTION

The adsorption process is, in general, one of collecting soluble substances that are in solution on a suitable interface. The interface can be between the liquid and a gas, a solid, or another liquid. Use is made of adsorption at the air-liquid interface in the flotation process. In this discussion, however, only the case of adsorption at the liquid-solid interface will be considered. The adsorption process has not been extensively used in wastewater purification to date. Demands for a better-quality effluent of treated wastewater have led to an intensive examination of the process of adsorption on activated carbon. This particular process holds much potential for meeting these demands, and thus it will be considered in some detail.

Theory

The adsorption process can be pictured as one in which molecules leave solution and are held on the solid surface by chemical and physical bonding. The molecules are called the absorbate and the solid is called the adsorbent. If the bonds that form between the adsorbate and adsorbent are very strong, the process is almost always irreversible, and chemical adsorption or chemisorption is said to have occurred. On the other hand, if the bonds that are formed are very weak, as is characteristic of bonds formed by van der Waals forces, physical adsorption is said to have occurred. The molecules adsorbed by this means are easily removed, or desorbed, by a change in the solution concentration of the adsorbate, and for this reason, the process is said to be reversible. Physical adsorption is the process that occurs most frequently in the removal of wastewater constituents by activated carbon. It should also be noted that the terms "adsorption" and "physical adsorption" are often used interchangeably. The practice will be followed in this discussion.

The quantity of adsorbate that can be taken up by an adsorbent is a function of both the concentration of adsorbate C and the temperature T. Generally, the amount of material adsorbed is determined as a function of C at a constant temperature, and the resulting function is called an adsorption isotherm. The procedures involved in obtaining liquid adsorption data for activated carbon are outlined in Chap. 16. Equations that are often used to describe the experimental isotherm data were developed by Freundlich and Langmuir.

Freundlich Isotherm Equation Derived from empirical considerations, the Freundlich isotherm is

$$\frac{X}{M} = kC^{1/n} \tag{9·39}$$

where X/M = amount adsorbed per unit weight of adsorbent (carbon)
C = equilibrium concentration of adsorbate in solution after adsorption
k, n = empirical constants

The constants in this equation can be evaluated by plotting X/M versus C on double logarithmic paper.

Langmuir Isotherm Equation Derived from rational considerations, the Langmuir adsorption isotherm is

$$\frac{X}{M} = \frac{abC}{1 + bC} \tag{9·40}$$

where X/M = amount adsorbed per unit weight of adsorbent (carbon)
a, b = empirical constants
C = equilibrium concentration of adsorbate in solution after adsorption

This equation was developed on the basis of the assumptions that (1) a fixed number of accessible sites are available on the adsorbent surface, all of which have the same energy and that (2) adsorption is reversible. Equilibrium is reached when the rate of adsorption of molecules onto the surface is the same as the rate of desorption of molecules from the surface. The rate at which adsorption proceeds, then, is proportional to the driving force, which is the difference between the amount adsorbed at a particular concentration and the amount that can be adsorbed at that concentration. At the equilibrium concentration, this difference is zero.

Correspondence of experimental data to the Langmuir equation does not mean that the stated assumptions are valid for the particular

FIG. 9·4 Sketch showing generalized form of Langmuir isotherm.

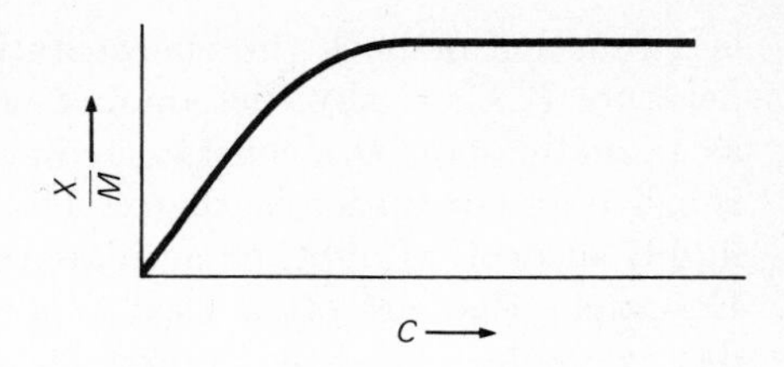

system being studied, because departures from the assumptions can have a cancelling effect. A Langmuir isotherm is shown in Fig. 9·4. The constants in the Langmuir equation can be determined by plotting $C/(X/M)$ versus C and making use of Eq. 9·40 rewritten as

$$\frac{C}{X/M} = \frac{1}{ab} + \frac{1}{a}C \tag{9·41}$$

Application of the Langmuir adsorption isotherm is illustrated in the following example.

EXAMPLE 9·3 *Analysis of Activated-Carbon Adsorption Data*

Determine if the carbon adsorption data presented below follow the Langmuir adsorption isotherm, and if they do, determine the constants a and b.

C, mg/liter	10	20	30
X/M, g/g	0.133	0.187	0.220

Solution

1. Plot $C/(X/M)$ versus C (see Fig. 9·5)

C, mg/liter	10	20	30
$\frac{C}{X/M}$	75	107	136

2. Because the data plot as a straight line, it can be concluded that they can be described with the Langmuir adsorption isotherm.

3. Determine the constant a. From Fig. 9·5,

$$\frac{1}{a} = \frac{37}{12} = 3.08$$

Therefore

$$a = \frac{1}{3.08} = 0.325$$

4. Determine the constant b. From Fig. 9·5,

$$\frac{1}{ab} = 45$$

Therefore

$$b = \frac{1}{45(0.325)} = 0.0685$$

Rate of Adsorption The adsorption process can be divided into three steps: (1) transfer of the adsorbate molecules through the film that surrounds the adsorbent, (2) diffusion through the pores if the adsorbent is porous, and (3) uptake of the adsorbate molecules by the active surface, including formation of the bonds between the adsorbate and the carbon. Step 3 is considered to be very rapid, since equilibrium on nonporous adsorbents can be accomplished in a matter of minutes. Steps 1 and 2 are generally held to be rate limiting. The thickness of the stagnant aqueous film that surrounds the adsorbent depends on the flow regime maintained in the system. The rate of adsorption then depends on the rate at which the molecules move or diffuse in solution or the rate at which the molecules can reach the available surface by diffusing through the film and the pore.

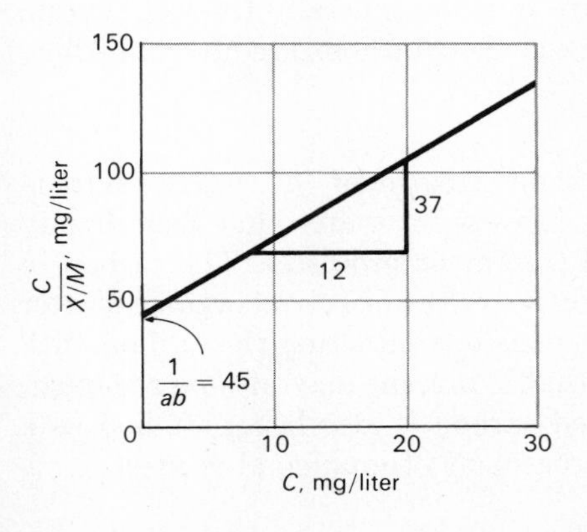

FIG. 9·5 Plot of Langmuir adsorption data for Example 9·3.

Adsorption of Mixtures In the application of adsorption to sewage treatment, mixtures of organic compounds are encountered. Weber [19] found that, although there will be a depression of the adsorptive capacity of any individual compound in a solution of many compounds, the total adsorptive capacity of the adsorbent may be larger than the adsorptive capacity with a single compound. The amount of inhibition due to competing adsorbates is related to the size of the molecules being adsorbed, their adsorptive affinities, and their relative concentrations.

Activated Carbon At the present time, activated carbon holds the most promise for efficient treatment of wastewater by adsorption. It has been used extensively in the past for removal of organic compounds that cause taste and odor in drinking water supplies. Research that has been performed on the treatment of wastewater by activated carbon has been especially encouraging.

Activated carbon is prepared by first making a char. A material such as wood or coal is brought to a red heat in a retort to drive off the hydrocarbons but with an insufficient supply of air to sustain combustion. The char particle is then activated by exposure to an oxidizing gas at high temperature. This gas develops a porous structure in the char and thus creates a large internal surface area. The surface properties that result are a function both of the initial material used and the exact preparation procedure, so that many variations are possible.

After activation, the carbon can be separated into, or prepared in, different sizes. The two general size classifications are powdered and granular. The diameter of powdered carbon is generally less than 200-mesh, while granular carbon has a diameter greater than 0.1 mm.

The size of the pore developed during activation is of importance in adsorption from the vapor phase but not generally from the liquid phase. Typical pore sizes and pore size distributions are given in Allen, Joyce, and Kasch [1].

Process Principles Activated-carbon treatment of wastewater is usually thought of as a polishing process for water that has already been treated by normal biological treatment processes. The carbon in this instance is used to remove a portion of the dissolved organic matter that remains. Depending on the means of contacting the carbon with the water, the particulate matter that is present may also be removed. Complete treatment with activated carbon is also being studied as a possible substitute for biological treatment of municipal wastes.

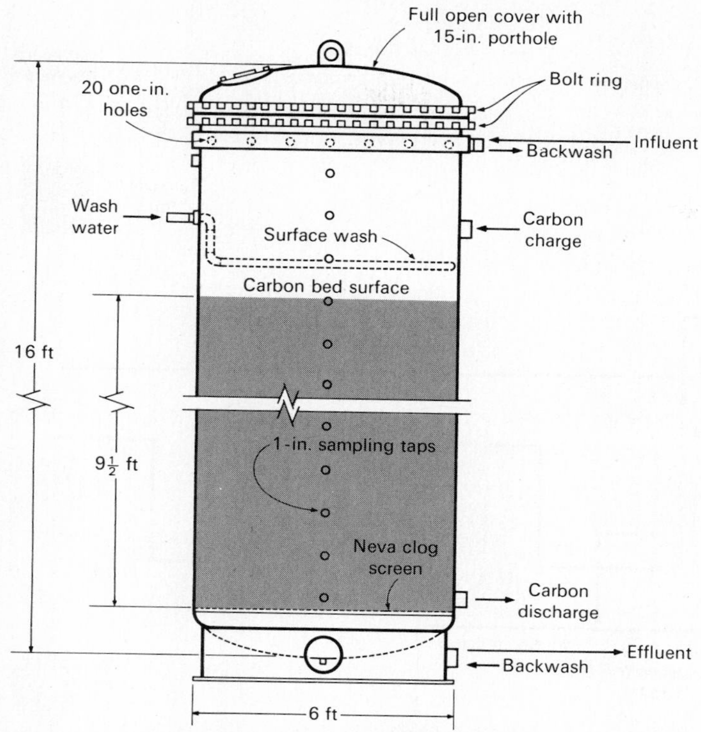

FIG. 9·6 Typical activated-carbon adsorption column [9].

A fixed-bed column is often used as a means of contacting wastewater with granular carbon. A typical column, which is being used in a pilot plant in Pomona, Calif., is shown in Fig. 9·6 [9]. The water is applied to the top of the column and withdrawn at the bottom. The carbon is held in place with a screen at the bottom of the column. Provision for backwash and surface wash is necessary, because the particulate matter in the influent averages 10 mg/liter and is almost completely removed by the filter. Backwashing is necessary to keep excessive head loss from building up. Columns such as these can be operated singly, in series, or in parallel. A typical series application is shown in Fig. 9·7.

The problem of clogging of the carbon bed can be partially overcome if an expanded bed is used. In operation, the influent is introduced at the bottom of the column and the bed is allowed to expand, much as a filter bed expands during backwash. Spent carbon is displaced continuously with fresh carbon. In such a system head loss does not build up with time after the operating point has been reached.

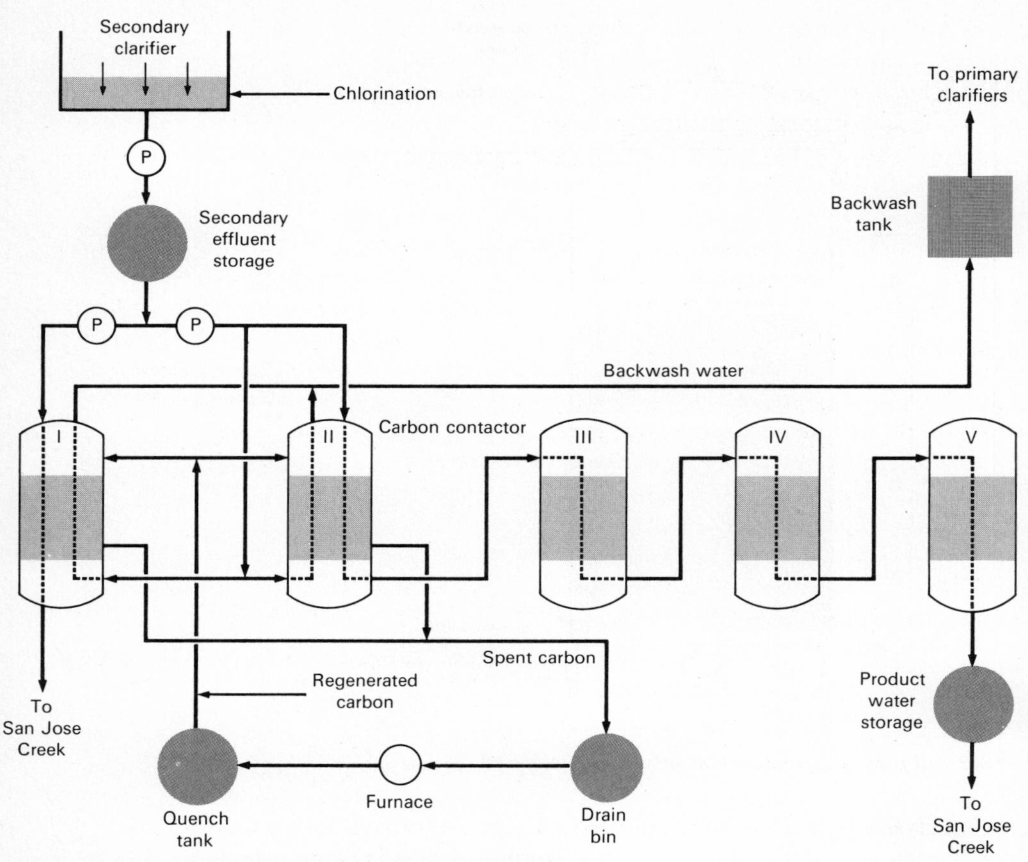

FIG. 9·7 Flowsheet of an activated-carbon pilot plant [9].

An alternate means of application that is being studied is that of adding powdered activated carbon to biological treatment effluent. The carbon, in this case, is added to the effluent in a contacting basin. After a certain amount of time for contact, the carbon is allowed to settle to the bottom of the tank and the treated water is then removed from the tank. Since the carbon is very fine, a coagulant, such as a polyelectrolyte, may be needed to aid the removal of the carbon particles, or filtration through rapid sand filters may be required.

Economical application of carbon depends on an efficient means of regenerating the carbon after its adsorptive capacity has been reached. Granular carbon can easily be regenerated by oxidizing the organic matter and thus removing it from the carbon surface in a furnace. Some of the carbon (about 10 percent) is also destroyed in this process and must be replaced with new or virgin carbon. The capacity of regenerated carbon, it should be noted, is slightly less than that of virgin carbon. The use of powdered carbon in wastewater treatment has been somewhat limited by lack of an efficient means of collecting and regenerating the spent carbon; however, research along these lines is proceeding. The use of powdered activated carbon produced from solid wastes may obviate the need to regenerate the spent carbon. Reference [8] in Chap. 13 may be consulted for a more complete discussion of the use of activated carbon for wastewater treatment.

DISINFECTION

Three categories of human enteric organisms of the greatest consequence in producing disease are bacteria, viruses, and amoebic cysts. Typical waterborne bacterial diseases are typhoid, cholera, paratyphoid, and bacillary dysentery. Diseases caused by waterborne viruses include poliomyelitis and infectious hepatitis.

Disinfection refers to the selective destruction of disease-causing organisms. All of the organisms are not destroyed during the process. This differentiates disinfection from sterilization, which is the destruction of all organisms.

Agents and Means

In the field of wastewater treatment, disinfection most commonly is accomplished through the use of (1) chemical agents, (2) physical agents, (3) mechanical means, and (4) radiation.

Chemical Agents The requirements for an ideal chemical disinfectant are reported in Table 9·4. As shown, an ideal disinfectant would have to possess a wide range of characteristics. Although such a compound may not exist, the requirements set forth in Table 9·4 should be considered in evaluating proposed or recommended disinfectants. To sanitary engineers, it is also important that the disinfectant be safe to handle and apply, and that its strength or concentration in treated waters be measurable so that the presence of a residual can be determined.

Chemical agents that have been used as disinfectants include (1) phenol and phenolic compounds, (2) alcohols, (3) iodine, (4) chlorine and its compounds, (5) bromine, (6) ozone, (7) heavy metals and related compounds, (8) dyes, (9) soaps and synthetic detergents, (10) quaternary ammonium compounds, (11) hydrogen peroxide, and (12) various alkalies and acids.

TABLE 9·4 CHARACTERISTICS OF AN IDEAL CHEMICAL DISINFECTANT [10]

Characteristic	Remarks
Toxicity to microorganism	Should have a broad spectrum of activity of high dilutions
Solubility	Must be soluble in water or cell tissue
Stability	Loss of germicidal action on standing should be low
Nontoxic to higher forms of life	Should be toxic to organisms and nontoxic to man and other animals
Homogeneity	Solution must be uniform in composition
Interaction with extraneous material	Should not be absorbed by organic matter
Toxicity at room temperature	Should be effective in environmental temperature range
Penetration	Should have the capacity to penetrate through surfaces
Noncorrosive and nonstaining	Should not disfigure metals or stain clothing
Deodorizing ability	Should deodorize while disinfecting
Detergent capacity	Should have cleansing action to improve effectiveness of disinfectant
Availability	Should be available in large quantities and reasonably priced

The most common disinfectants are the oxidizing chemicals of which chlorine is the most universally used. Bromine and iodine occasionally are used for swimming pools but have not been used for treated wastewater. Ozone is a highly effective disinfectant and even though it leaves no residual, its use is increasing. It also has been used to eliminate odors from air discharged from pumping stations, covered treatment tanks, and from enclosed sludge thickeners. Highly acid or alkaline water can also be used to destroy pathogenic bacteria, because water with a pH greater than 11 or less than 3 is relatively toxic to most bacteria.

Physical Agents Physical disinfectants that can be used are heat and light. Heating water to the boiling point, for example, will destroy the major disease-producing, non-spore-forming bacteria. Heat is commonly used in the beverage and dairy industry, but it is not a feasible means of disinfecting large quantities of wastewater because of the high cost.

Sunlight is also a good disinfectant. In particular, ultraviolet radiation can be used. Special lamps that emit ultraviolet rays have been used successfully for the sterilization of small quantities of water. The efficiency of the process depends on the penetration of the rays into water. The contact geometry between the ultraviolet-light source and the water is extremely important, because suspended matter, dissolved organic molecules, and water itself will absorb the radiation, in addition to the microorganisms. It is thus difficult to use ultraviolet radiation in aqueous systems, especially when particulate matter is present.

Mechanical Means Bacteria are also removed by mechanical means during sewage treatment. Typical removal efficiencies for various treatment processes are reported in Table 9·5. The first five processes listed may be considered to be physical. The removals accomplished are a by-product of the primary function of the process.

TABLE 9·5 REMOVAL OR DESTRUCTION OF BACTERIA BY DIFFERENT TREATMENT PROCESSES

Process	Percent removal
Coarse screens	0–5
Fine screens	10–20
Grit chambers	10–25
Plain sedimentation	25–75
Chemical precipitation	40–80
Trickling filters	90–95
Activated sludge	90–98
Chlorination of treated sewage	98–99

Radiation The major types of radiation are electromagnetic, acoustic, and particle. Gamma rays are emitted from radioisotopes, such as cobalt 60. Because of their penetration power, gamma rays have been used to disinfect (sterilize) both water and wastewater.

Factors Influencing the Action of Disinfectants

In applying the disinfection agents or means that have been described, the following factors must be considered: (1) contact time, (2) concentration and type of chemical agent, (3) intensity and nature of physical agent, (4) temperature, (5) number of organisms, (6) types of organisms, and (7) nature of suspending liquid [10].

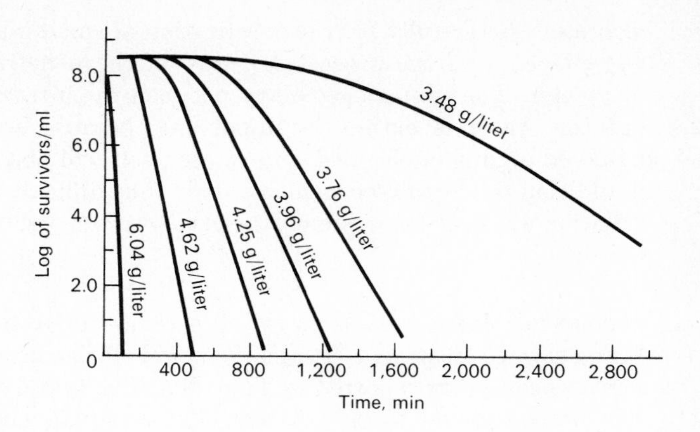

FIG. 9·8 Effect of time and concentration on survival of *E. coli* using phenol as disinfectant at 35°C [6].

Contact Time Perhaps one of the most important variables in the disinfection process is contact time. In general, as shown in Figs. 9·8 and 9·9, it has been observed that, for a given concentration of disinfectant, the longer the contact time, the greater the kill. This observation was first formalized in the literature by Chick [3]. In differential form, Chick's law is

$$\frac{dN}{dt} = -kN \tag{9·42}$$

where N = number of organisms
t = time
k = constant, time^{-1}

If N_0 is the number of organisms when t equals 0, Eq. 9·42 can be integrated to

$$\frac{N}{N_0} = e^{-kt} \tag{9·43}$$

or

$$\ln \frac{N}{N_0} = -kt$$

Departures from this rate law are common. Rates of kill have been found to increase with time in some cases and to decrease with time in other cases. To formulate a valid relationship for the kill of organisms under a variety of conditions, an assumption often made is that

$$\ln \frac{N}{N_0} = -kt^m \tag{9·44}$$

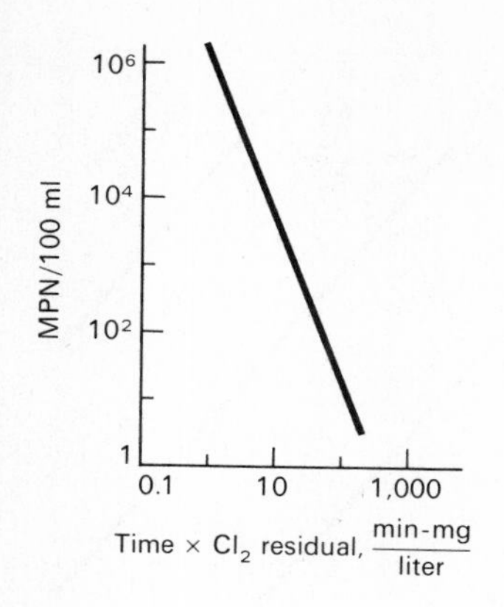

FIG. 9·9 Effect of time and concentration on survival of *E. coli* in wastewater using chlorine as disinfectant.

where m is a constant. If m is less than 1, the rate of kill decreases with time, and if m is greater than 1, the rate of kill increases with time. The constants in Eq. 9·44 can be obtained by plotting $-\ln (N/N_0)$ versus the contact time t on log-log paper. The straight line form of the equation is

$$\log \left(-\ln \frac{N}{N_0}\right) = \log k + m \log t \tag{9·45}$$

Concentration and Type of Chemical Agent Depending on the type of chemical agent, it has been observed that, within limits, disinfection effectiveness is related to concentration (see Figs. 9·8 and 9·9). The effect of concentration has been formulated empirically [4]:

$$C^n t_p = \text{constant} \tag{9·46}$$

where C = concentration of disinfectant

n = constant

t_p = time required to effect a constant percentage kill

The constants in Eq. 9·46 can be evaluated by plotting, on log-log paper, the concentration versus the time required to effect a given percentage kill. The slope of the line then corresponds to the value of $-1/n$. A plot of this relationship, as experimentally determined for a 99 percent kill of different organisms, is shown in Fig. 9·10. In general, if n is greater than 1, contact time is more important than the dosage; if n equals 1 the effect of time and dosage are about the same [4].

Intensity and Nature of Physical Agent As noted earlier, heat and light are physical agents that have been used from time to time in the disinfection of wastewater. It has been found that their effectiveness is a

FIG. 9·10 Concentration of chlorine as HOCl required for 99 percent kill of *E. coli* and three enteric viruses at 0 to 6°C [2].

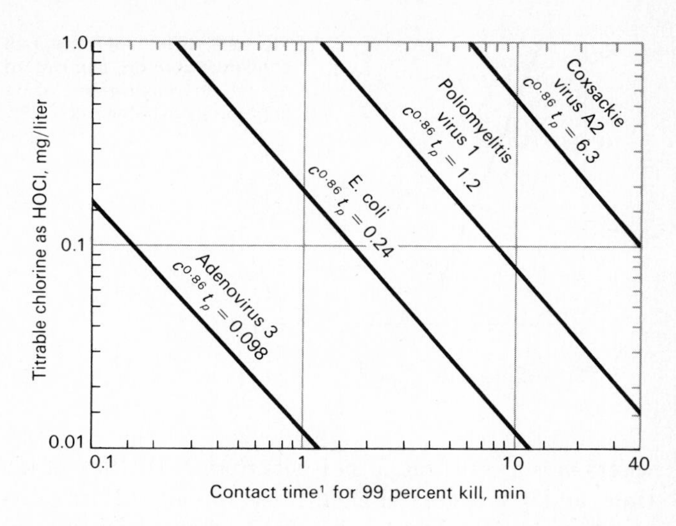

function of intensity. For example, if the decay of organisms can be described with a first-order reaction, such as

$$\frac{dN}{dt} = -kN$$

where N = number of organisms
t = time, min
k = reaction velocity of constant, 1/min

then the effect of the intensity of the physical disinfectant is reflected in the constant k through some functional relationship.

Temperature The effect of temperature on rate of kill can be represented by a form of the van't Hoff-Arrhenius relationship. Increasing the temperature results in a more rapid kill. In terms of the time t required to effect a given percentage kill, the relationship is

$$\log \frac{t_1}{t_2} = \frac{E(T_2 - T_1)}{2.303RT_1T_2} = \frac{E(T_2 - T_1)}{4.58T_1T_2} \tag{9·47}$$

where t_1, t_2 = time for given percentage kill at temperatures T_1 and T_2, °K, respectively
E = activation energy
R = gas constant (1.99 cal/°K-mole)

Typical values for the activation energy for various chlorine compounds at different pH values are reported in Table 9·6.

TABLE 9·6 ACTIVATION ENERGIES FOR AQUEOUS CHLORINE AND CHLORAMINES AT NORMAL TEMPERATURES [4]

Compound	pH	E, cal
Aqueous chlorine	7.0	8,200
	8.5	6,400
	9.8	12,000
	10.7	15,000
Chloramines	7.0	12,000
	8.5	14,000
	9.5	20,000

Number of Organisms In a dilute system such as wastewater, the concentration of organisms is seldom a major consideration. However, it can be concluded from Eq. 9·46 that the larger the organism concentration, the longer the time required for a given kill. An empirical relationship that has been proposed to describe the effect of organism concentration is [4]

$$C^q N_p = \text{constant} \tag{9·48}$$

where C = concentration of disinfectant
N_p = concentration of organisms reduced given percentage in given time
q = constant related to strength of disinfectant

Types of Organisms The effectiveness of various disinfectants will be influenced by the nature and condition of the microorganisms. For example, viable growing bacteria cells are killed easily. In contrast, bacterial spores are extremely resistant, and many of the chemical disinfectants normally used will have little or no effect. Other disinfecting agents, such as heat, may have to be used.

Nature of Suspending Liquid Besides all the above factors, the nature of the suspending liquid must be evaluated carefully. For example, extraneous organic material will react with most oxidizing disinfectants and will reduce their effectiveness. Turbidity will reduce the effectiveness of disinfectants by absorption and by protecting entrapped bacteria.

Mechanisms of Disinfectants Four mechanisms that have been proposed to explain the action of disinfectants are (1) damage to the cell wall, (2) alteration of cell permeability, (3) alteration of the colloidal nature of the protoplasm, and (4) inhibition of enzyme activity [10].

Damage or destruction of the cell wall will result in cell lysis and death. Some agents, such as penicillin, inhibit the synthesis of the bacterial cell wall.

Agents such as phenolic compounds and detergents alter the permeability of the cytoplasmic membrane. These substances destroy the selective permeability of the membrane and allow vital nutrients, such as nitrogen and phosphorus, to escape.

Heat, radiation, and highly acid or alkaline agents alter the colloidal nature of the protoplasm. Heat will coagulate the cell protein while acids or bases will denature proteins, producing a lethal effect.

Another mode of disinfection is the inhibition of enzyme activity. Oxidizing agents, such as chlorine, can alter the chemical arrangement of enzymes and inactivate the enzymes.

Disinfection with Chlorine

Of all the chemical disinfectants, chlorine is perhaps the one most commonly used throughout the world. The reason is that it satisfies most of the requirements specified in Table 9·4. Because the practice of chlorination is discussed in Chap. 11, the following presentation will be limited to a discussion of chlorine chemistry and break-point chlorination.

Chlorine Chemistry in Water When chlorine, in the form of Cl_2 gas, is added to water, two reactions take place: hydrolysis and ionization.

Hydrolysis may be defined as

$$Cl_2 + H_2O \rightleftharpoons HOCl + H^+ + Cl^- \tag{9·49}$$

The stability constant for this reaction is

$$K = \frac{(HOCl)(H^+)(Cl^-)}{(Cl_2)} = 4.5 \times 10^{-4} \text{ at } 25°C \tag{9·50}$$

Because of the large magnitude of this coefficient, large quantities of chlorine can be dissolved in water.

Ionization may be defined as

$$HOCl \rightleftharpoons H^+ + OCl^- \tag{9·51}$$

The ionization constant for this reaction is

$$K_i = \frac{(H^+)(OCl^-)}{(HOCl)} = 2.7 \times 10^{-8} \text{ at } 25°C \tag{9·52}$$

The variation in the value of K_i with temperature is reported in Table 9·7.

TABLE 9·7 VALUES OF THE IONIZATION CONSTANT OF HYPOCHLOROUS ACID AT DIFFERENT TEMPERATURES [4]

Temperature, °C	0	5	10	15	20	25
$K_i \times 10^8$, moles/liter	1.5	1.7	2.0	2.2	2.5	2.7

The quantity of HOCl and OCl^- that is present in water is called the free available chlorine. The relative distribution of these two species (see Fig. 9·11) is very important because the killing efficiency of HOCl is about 40 to 80 times that of OCl^-. The percentage distribution of HOCl at various temperatures can be computed using Eq. 9·53 and the data in Table 9·7.

$$\frac{HOCl}{HOCl + OCl} = \frac{1}{1 + OCl/HOCl} = \frac{1}{1 + K_i/H} \tag{9·53}$$

Free chlorine can also be added to water in the form of hypochlorite salts. Calcium hypochlorite, $Ca(OCl)_2$, is commonly used for this purpose. These hypochlorite ions are then subject to the reaction represented in Eq. 9·51.

Free chlorine in solution will react with ammonia in water to form

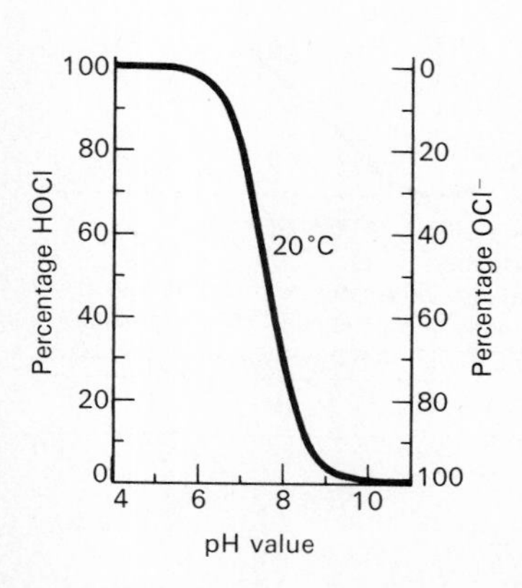

FIG. 9·11 Distribution of hypochlorous acid and hypochlorite in water at different pH values.

chloramines. The chloramines also serve as disinfectants, although they are extremely slow reacting. The reactions of importance are

$$NH_3 + HOCl \rightleftharpoons NH_2Cl + H_2O \quad (9\text{·}54)$$

$$NH_2Cl + HOCl \rightleftharpoons NHCl_2 + H_2O \quad (9\text{·}55)$$

$$NHCl_2 + HOCl \rightleftharpoons NCl_3 + H_2O \quad (9\text{·}56)$$

where monochloramine (NH_2Cl), dichloramine ($NHCl_2$), and nitrogen trichloride (NCl_3) are formed in the successive substitution reactions. The two species that predominate, in most cases, are NH_2Cl and $NHCl_2$; they are commonly called combined available chlorine.

Break-Point Chlorination The fact that free chlorine will react with ammonia, coupled with the fact that free chlorine is a strong oxidizing agent, complicates the maintenance of a residual (combined or free) for the purpose of wastewater disinfection. The phenomena that result when chlorine is added to a wastewater containing ammonia can be explained by referring to Fig. 9·12.

As chlorine is added, readily oxidizable substances such as Fe^{++}, Mn^{++}, H_2S, and organic matter react with the chlorine, reducing most of it to the chloride ion (point A in Fig. 9·12). After meeting this immediate demand, the chlorine will continue to react with the ammonia to form chloramines between points A and B. For mole ratios of chlorine to ammonia less than one, monochloramine and dichloramine will be formed. The distribution of these two forms is governed by their

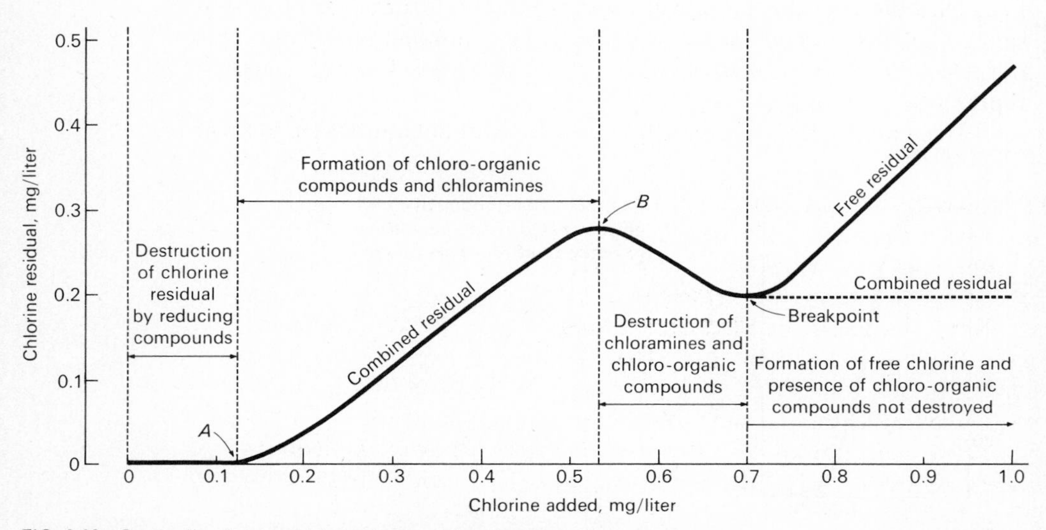

FIG. 9·12 Generalized curve obtained during break-point chlorination.

rates of formation, which are dependent on the pH and temperature [5]. Between point B and the break point some of the chloramines will be converted to nitrogen trichloride (see Eq. 9·56), while the remaining chloramines will be oxidized to nitrous oxide and nitrogen with the chlorine being reduced to the chloride ion. With continued addition of chlorine, essentially all of the chloramines will be oxidized at the break point. Possible reactions to account for the appearance of the aforementioned gases and the disappearance of chloramines are as follows (see also Eq. 9·56).

$$NH_2Cl + NHCl_2 + HOCl \rightarrow N_2O + 4HCl \quad (9{\cdot}57)$$

$$4NH_2Cl + 3Cl_2 + H_2O \rightarrow N_2 + N_2O + 10HCl \quad (9{\cdot}58)$$

$$2NH_2Cl + HOCl \rightarrow N_2 + H_2O + 3HCl \quad (9{\cdot}59)$$

$$NH_2Cl + NHCl_2 \rightarrow N_2 + 3HCl \quad (9{\cdot}60)$$

Continued addition of chlorine past the break point will result in a directly proportional increase in the free available chlorine (unreacted hypochlorite). The main reason for adding enough chlorine to obtain a free chlorine residual is that disinfection can then usually be assured. Occasionally, due to the formation of nitrogen trichloride and related compounds, serious odor problems have developed during break-point chlorination operations. In practice, the hydrochloric acid formed during chlorination will react with the alkalinity of the wastewater and, under most circumstances, the pH drop will be slight.

The presence of additional compounds that will react with chlorine may greatly alter the shape of this curve. The amount of chlorine that must be added to reach a desired level of residual is called the chlorine demand.

COMBUSTION

To reduce the weight and volume of sludge and to product an odorless and inert residue for final disposal, combustion of the organics is practiced in many of the larger municipal and industrial wastewater treatment plants. "Sludge incineration" is the term applied to dry combustion of sludge, which usually has undergone vacuum filtration for reduction of water content. This may be accomplished in a multiple hearth furnace, similar to the one shown in Chap. 8 in Fig. 8·22 in which maximum temperatures are maintained slightly above 1400°F to prevent odors. In another type of unit, dried sludge is blown into a furnace and used for fuel to heat the air which, in turn, is used for flash drying and/or incineration of the sludge. In another type of unit, the sludge is burned in a fluidized bed at high temperatures. The fuel value of sludge will range from 4,800 to 10,000 Btu/lb dry solids depending on the type of sludge and the volatile content (see Eq. 9·63). This is equivalent to some of the lower grades of coal.

Dry Combustion

Incineration of sludge implies complete combustion of all of the organic substances present. The predominant elements in the carbohydrates, fats, and proteins composing the volatile matter of sludge are carbon, oxygen, hydrogen, and nitrogen (C–O–H–N). The approximate percentages of these may be determined in the laboratory by a technique known as ultimate analysis.

Oxygen requirements for complete combustion may be determined from a knowledge of the constituents, assuming that carbon and hydrogen are oxidized to the ultimate end products CO_2 and H_2O. The formula becomes

$$C_aO_bH_cN_d + (a + 0.25c - 0.5b)O_2 \rightarrow aCO_2 + 0.5cH_2O + 0.5dN_2 \quad (9{\cdot}61)$$

The theoretical quantity of air will be 4.35 times the calculated quantity of oxygen, because air is composed of 23 percent oxygen on a weight basis. To ensure complete combustion, excess air amounting to about 50 percent of the theoretical amount will be required. A materials balance must be made to include the above compounds and the inorganic substances in the sludge, such as the inert material and moisture, and the moisture in the air. The specific heat of each of these substances and of the products of combustion must be taken into account in determining the heat required for the incineration process.

Heat requirements will include the sensible heat Q_s in the ash, plus the sensible heat required to raise the temperature of the flue gases to 1400°F, or whatever higher temperature of operation is selected for complete oxidation and elimination of odors, less the heat recovered in preheaters or recuperators. Latent heat Q_e must also be furnished to evaporate all of the moisture in the sludge. Total heat required Q may be expressed as

$$Q = \Sigma Q_s + Q_e = \Sigma C_pW_s(T_2 - T_1) + W_w\lambda \quad (9{\cdot}62)$$

where C_p = specific heat for each category of substance in ash and flue gases
W_s = weight of each substance
T_1, T_2 = initial and final temperatures
λ = latent heat of evaporation per pound
W_w = weight of moisture in sludge

It should be obvious that reduction of moisture content of the sludge is the principal way to lower heat requirements and may determine whether or not additional fuel will be needed to support combustion.

The heating value of a sludge may be determined by the conventional bomb-calorimeter test. An empirical formula, based on a statistical study of fuel values of vacuum-filtered sludges of different types, taking into account the amount of coagulant added before filtration, is as follows [5]:

$$Q = a\left[\frac{P_v(100)}{100 - P_c} - b\right]\left[\frac{100 - P_c}{100}\right] \tag{9·63}$$

where Q = fuel value, Btu/lb dry solids
a = coefficient (131 for primary sludge, raw or digested; 107 for fresh activated sludge)
b = coefficient (10 for primary sludge; 5 for activated sludge)
P_v = percent of volatile solids in sludge
P_c = percent of coagulating solids added to the sludge

A heat balance of the process must include heat losses through the walls and pertinent equipment of the incinerator, as well as losses in the stack gases and ash. Information on these amounts should be made available by the equipment manufacturer and checked by the design engineer. Often sludge will contain enough available heat to support combustion. In the case of high nonvolatile content or high moisture content, auxiliary fuel will be required. Designs should include provisions for auxiliary heat for start-up and to assure complete oxidation at the desired temperature under all conditions.

The determination of materials and heat balances in the incineration of solid wastes is illustrated in Example 9·4. Computations for a sludge incinerator would be similar.

EXAMPLE 9·4 *Materials and Heat Balances in Incineration*

Determine the available heat for steam production from a quantity of refuse with the following characteristics to be incinerated at a rate of 250,000 lb/day.

Constituent	Percent of total	lb/day
Combustible	60	150,000
Water	20	50,000
Noncombustible	20	50,000

Component	*Percent*
Carbon	28
Hydrogen	4
Oxygen	23
Nitrogen	4
Sulfur	1
Water	20
Metals and inerts	20

Assume the following conditions are applicable.

1. The as-fired heating value of the refuse is 5,000 Btu/lb.
2. Grate residue contains 5 percent carbon.
3. Temperatures:

   Entering air, 80°F

   Residue from grate, 800°F

4. Specific heat of residue, 0.25 Btu/lb-°F
5. Latent heat of water, 1,040 Btu/lb
6. Radiation loss, 0.005 Btu/Btu of total furnace input
7. All oxygen in refuse is bound as water.
8. Theoretical air requirements based on stoichiometry:

   Carbon ($C + O_2 \rightarrow CO_2$) = 11.5 lb/lb
   Hydrogen ($2H_2 + O_2 \rightarrow 2H_2O$) = 34.3 lb/lb
   Sulfur ($S + O_2 \rightarrow SO_2$) = 4.3 lb/lb

9. The heating value of carbon is 14,000 Btu/lb.
10. Moisture in the combustion air is 1 percent.

Solution

1. Compute the weights of the elements of the refuse.

Element		*lb/day*
Carbon	= 0.28(250,000)	= 70,000
Hydrogen	= 0.04(250,000)	= 10,000
Oxygen	= 0.23(250,000)	= 57,500
Nitrogen	= 0.04(250,000)	= 10,000
Sulfur	= 0.01(250,000)	= 2,500
Water	= 0.20(250,000)	= 50,000
Inerts	= 0.20(250,000)	= 50,000

2. Compute the amount of residue.

   Inerts $= 50{,}000$ lb/day

   $$\text{Carbon} = \frac{0.05(50{,}000)}{0.95} = 2{,}600 \text{ lb/day}$$

   Total residue $= 52{,}600$ lb/day

3. Compute the available hydrogen and bound water.

   $$\underset{2}{2H} + \underset{16}{O} \rightarrow \underset{18}{H_2O}$$

   Oxygen/hydrogen in bound water $= 8$

   Hydrogen in bound water $= 23\%/8 = 2.9\% = 7{,}200$ lb/day

   Available hydrogen $= 10{,}000 - 7{,}200 = 2{,}800$ lb/day

   $$\text{Bound water} = 0.23 \frac{9 \text{ lb } H_2O}{8 \text{ lb } O_2} 250{,}000 = 64{,}700 \text{ lb/day}$$

4. Compute the air required.

Element		*lb/day*
Carbon	$= (70{,}000 - 2{,}600)(11.5)$	$=$ 775,000
Hydrogen	$= 2{,}800(34.3)$	$=$ 96,000
Sulfur	$= 2{,}500(4.3)$	$=$ 11,000
Total dry theoretical air		$=$ 882,000
Total dry air including 50 percent excess		$=$ 1,323,000
Moisture 1,323,000(0.01)		$=$ 13,000
Total air		$=$ 1,336,000

5. Compute gross heat input from refuse.
   250,000 lb/day $\times$ 5,000 Btu/lb $= 1{,}250{,}000{,}000$ Btu/day
6. Compute latent heat losses.
   (*a*) Inherent moisture:

   $$50{,}000(1{,}040) = 52{,}000{,}000 \text{ Btu/day}$$

   (*b*) Moisture from oxidation of hydrogen:

   $$\frac{9 \text{ lb } H_2O}{\text{lb H}} 10{,}000(1{,}040) = 95{,}000{,}000 \text{ Btu/day}$$

7. Compute reactor losses.
   (*a*) Radiation loss:

   $$0.005 \text{ Btu/Btu}(1{,}250{,}000{,}000 \text{ Btu/day}) = 6{,}000{,}000 \text{ Btu/day}$$

(*b*) Sensible heat in residue:

$$52{,}600(0.25)(800 - 80) = 9{,}000{,}000 \text{ Btu/day}$$

(*c*) Unburned carbon:

$$2{,}600(14{,}000) = 36{,}000{,}000 \text{ Btu/day}$$
$$\text{Total losses} = 198{,}000{,}000 \text{ Btu/day}$$

8. Compute the heat available for steam production. Available heat in the hot gases:

$$1{,}250{,}000{,}000 - 198{,}000{,}000 = 1{,}052{,}000{,}000 \text{ Btu/day}$$

This is the heat above normal air temperature (assumed 80°F) in the high-temperature waste gases available at the inlet of a waste heat boiler. The amount of steam produced would depend on the efficiency of the waste heat boiler.

Wet Oxidation Process

Organic substances may be oxidized under high pressures at elevated temperatures with the sludge in a liquid state by feeding compressed

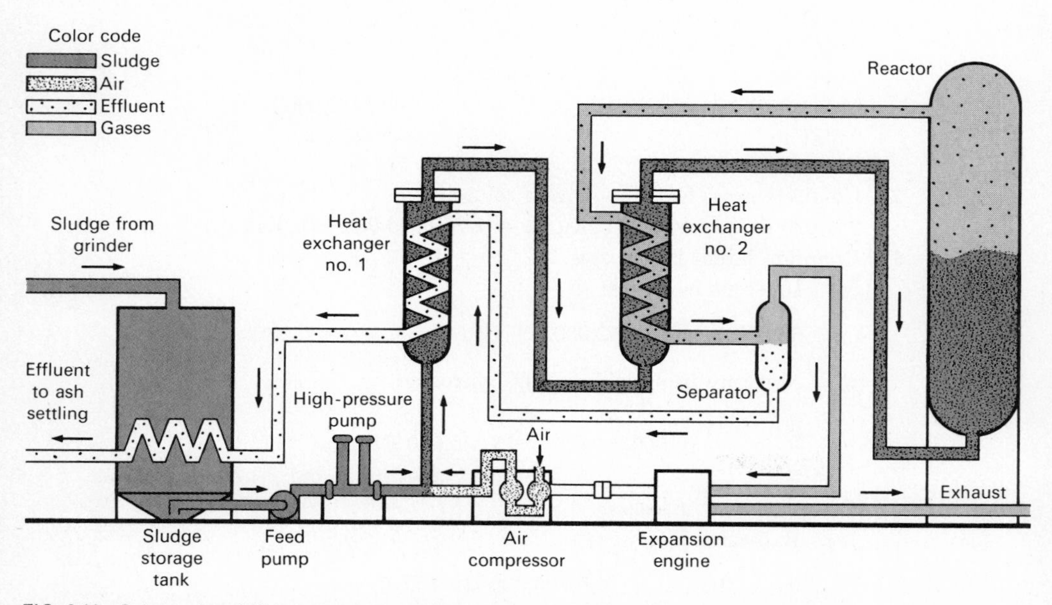

FIG. 9.13 Schematic of Zimmerman wet oxidation process [from Zimpro, Inc.].

air into the pressure vessel. The process was developed in Norway for pulp-mill wastes, but has been revised for the oxidation of raw sewage sludges pumped directly from the primary settling tank or thickener. Combustion is not complete; the average is 80 to 90 percent completion. Thus, some organic matter, plus ammonia, will be observed in the end products. For this incomplete combustion reaction, Rich [12] has suggested that

$$C_aH_bO_cN_d + 0.5(ny + 2s + r - c)\ O_2 \rightarrow nC_wH_xO_yN_z + sCO_2 + rH_2O + (d - nz)NH_3 \quad (9\cdot64)$$

where $r = 0.5[b - nx - 3(d - nz)]$
$s = a - nw$

The results obtained from this equation can also be approximated by the COD of the sludge, which is approximately equal to the oxygen required in combustion. The range of heat released per pound of air required has been found to be from 1,200 to 1,400 Btu [12]. Maximum operating temperatures for the system vary from 350 to 600°F with design operating pressures ranging from 150 to 3,000 psig. A flowsheet of the wet oxidation process known as the Zimmerman process is shown on Fig. 9·13.

PROBLEMS

9·1 Assuming that a given flocculation process can be defined by a second-order reaction ($dN/dt = -KN^2$), complete the following table:

Time, t	*No. of particles, N*
0	10
5	5
10	(?)

9·2 To aid sedimentation in the primary settling tank, 200 lb/million gal of ferrous sulfate ($FeSO_4 \cdot 7H_2O$) is added to the sewage. Determine the minimum alkalinity required to react initially with the ferrous sulfate. How many pounds of lime should be added as CaO to react with the $Fe(HCO_3)_2$ and the dissolved oxygen in the sewage to form insoluble $Fe(OH)_3$?

9·3 How much oxygen will be used up, theoretically, when 700 lb of sulfur dioxide per million gallons are added to sewage with an alkalinity of 50 mg/liter?

9·4 If 100 lb of (*a*) alum and (*b*) ferrous sulfate and lime are added per million gallons of sewage, and it is assumed that all insoluble and very slightly soluble products of the reactions, with the exception of 15 mg/liter $CaCO_3$, are precipitated as sludge, how many pounds of sludge per million gallons of sewage will result in each case?

9·5 Compute the equilibrium carbon dioxide level in a reservoir. Assume that the atmosphere contains 0.03 percent CO_2 and the temperature is 20°C.

9·6 Laboratory tests were conducted on a waste containing 50 mg/liter phenol.

Four jars containing 1 liter of the waste were dosed with powdered activated carbon. When equilibrium was reached, the contents of each jar were analyzed for phenol. The results are shown in the following table. Determine the constants *a* and *b* in the Langmuir equation and the dosage required to yield an effluent with a phenol concentration of 0.10.

Jar	Activated carbon added, g	Equilibrium conc of phenol, mg/liter
1	0.5	6.0
2	0.64	1.0
3	1.0	0.25
4	2.0	0.08

9·7 The following dosages of chlorine to a wastewater resulted in corresponding chlorine residuals. Draw a chlorine residual curve. Determine (*a*) the break-point dosage and (*b*) the design dosage to obtain a residual of 0.75 mg/liter free available chlorine.

Dosage, mg/liter	0.1	0.5	1.0	1.5	2.0	2.5	3.0
Residual, mg/liter	0.0	0.4	0.8	0.4	0.4	0.9	1.4

9·8 Discuss the advantages and disadvantages of using ozone as a disinfectant.

9·9 Ultimate analysis of a dried sludge yielded the following breakdown:

Carbon	52.1%
Oxygen	38.3%
Hydrogen	2.7%
Nitrogen	6.9%
Total	100.0%

How many pounds of air will be required for complete oxidation?

9·10 Compute the fuel value of the sludge from a primary settling tank (*a*) if no chemicals are added and (*b*) if the coagulating solids amount to 10 percent by weight of the dry sludge. The amount of volatile solids is 75 percent.

REFERENCES

1. Allen, J. B., R. S. Joyce, and R. H. Kasch: Process Design Calculations for Adsorption from Liquid in Fixed Beds of Granular Activated Carbon, *J. WPCF*, vol. 39, no. 2, 1967.
2. Berg, G.: The Virus Hazard in Water Supplies, *J. New England Water Works Association*, vol. 78, p. 79, 1964.
3. Chick, H.: Investigation of the Laws of Disinfection, *J. Hygiene*, vol. 8, p. 92, 1908.
4. Fair, G. M., *et al.*: The Behavior of Chlorine as a Water Disinfectant, *J. AWWA*, vol. 40, p. 1051, 1948.
5. Fair, G. M., J. C. Geyer, and D. A. Okun: *Water and Wastewater Engineering*, vol. 2, Wiley, New York, 1968.
6. Jordan, R. C., and S. E. Jacobs: *Journal of Hygiene*, vol. 43, p. 297, 1944.
7. King, H. R.: Oxygen Absorption in Spiral Flow Tanks, *Sewage and Industrial Wastes*, vol. 27, nos. 8–10, 1955.
8. Lewis, W. K., and W. C. Whitman: Principles of Gas Adsorption, *Industrial Engineering Chemistry*, vol. 16, p. 1215, 1924.
9. Parkhurst, J. D., F. D. Dryden, G. N. McDermott, and J. English: Pomona Activated Carbon Pilot Plant, *J. WPCF*, vol. 39, no. 10, part 2, 1967.

10. Pelczar, M. J., Jr., and R. D. Reid: *Microbiology*, 2d ed., McGraw-Hill, New York, 1965.
11. Perry, J. H.: *Chemical Engineer's Handbook*, 4th ed., McGraw-Hill, New York, 1963.
12. Rich, L. G.: *Unit Processes of Sanitary Engineering*, Wiley, New York, 1963.
13. Sawyer, C. N., and W. O. Lynch: Effects of Detergents on Oxygen Transfer in Bubble Aeration, *J. WPCF*, vol. 32, no. 1, 1960.
14. Shaw, D. J.: *Introduction to Colloid and Surface Chemistry*, Butterworth, London, 1966.
15. Stumm, W., and J. J. Morgan: *Aquatic Chemistry*, Wiley-Interscience, New York, 1970.
16. Stumm, W., and J. J. Morgan: Chemical Aspects of Coagulation, *J. AWWA*, vol. 54, no. 8, 1962.
17. Stumm, W., and C. R. O'Melia: Stoichiometry of Coagulation, *J. AWWA*, vol. 60, p. 514, 1968.
18. Thomas, A. W.: *Colloid Chemistry*, McGraw-Hill, New York, 1934.
19. Weber, W. J., Jr., and J. C. Morris: Adsorption in Heterogeneous Aqueous Systems, *J. AWWA*, vol. 56, no. 4, 1964.

biological unit processes

10

In biological treatment of wastewater the objectives are to coagulate and remove the nonsettleable colloidal solids and to stabilize the organic matter. For domestic wastewater (i.e., sewage), the major objective is to reduce the organic content. For agricultural return wastewater, the objective is to remove the nutrients, such as nitrogen and phosphorus, that are capable of stimulating the growth of aquatic plants. For industrial wastewater, the objective is to remove or reduce the concentration of organic and inorganic compounds. Because many of these compounds are toxic to microorganisms, pretreatment may be required.

In most cases, with proper environmental control, wastewater can be treated biologically. Therefore, it is the responsibility of the sanitary engineer to ensure that the proper environment is produced and effectively controlled. This can frequently be achieved by careful selection of the method of introducing the wastewater to the treatment units or by dilution of the wastewater through recirculation of a portion of the treated effluent. In view of the importance of biological treatment, it is the purpose of this chapter to

1. Present and review some of the fundamentals of wastewater microbiology.
2. Review and discuss the key factors governing biological growth and waste treatment kinetics, and the application of kinetics to waste treatment systems.
3. Review the basic biological treatment processes, emphasizing their biological characteristics and distinguishing factors.

SOME FUNDAMENTALS OF MICROBIOLOGY

Basic to the design of a biological treatment process, or to the selection of the type of process to be used, is an understanding of the form, structure, and biochemical activities of the important microorganisms. In this section, the cytology and physiology of the microorganisms commonly encountered in wastewater treatment will be presented and discussed.

Basic Concepts

In the past, microorganisms were commonly grouped into two kingdoms: plants and animals. Because of taxonomic difficulties, the recent trend is to group them into three kingdoms: protista, plants, and animals. The members of the kingdom protista are called protists. Summary data on the characteristics of the microorganisms in each kingdom are presented in Table 10·1. While the most significant differ-

TABLE 10·1 THE THREE KINGDOMS OF MICROORGANISMS

Kingdom	Representative members	Characterization
Animal	Rotifers	Multicellular, with tissue differentiation
	Crustaceans	
Plant	Mosses	
	Ferns	
	Seed plants	
Protista:		
Higher*	Algae	Unicellular or multicellular, without tissue differentiation
	Protozoa	
	Fungi	
	Slime molds	
Lower†	Blue-green algae	
	Bacteria	

* Contain true nucleus (eucaryotic cells).
† Contain no nuclear membrane (procaryotic cells).

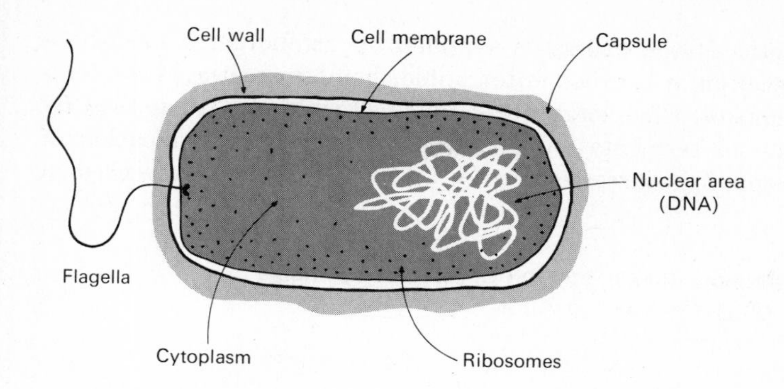

FIG. 10·1 Generalized schematic of bacterial cell [11].

ences among protists, plants, and animals are shown in the table, the three kingdoms are similar in that the cell is the basic unit of life for each, regardless of the complexity of the organism.

In general, most living cells are quite similar. As shown in Fig. 10·1, they have a cell wall, which may be either a rigid or a flexible membrane. If they are motile, they usually possess flagella or some hairlike appendages. The interior of the cell contains a colloidal suspension of proteins, carbohydrates, and other complex organic compounds, called the cytoplasm.

Each cell contains nucleic acids, the hereditary material that is vital to reproduction. The cytoplasmic area contains ribonucleic acid (RNA), whose major role is in the synthesis of proteins. Also within the cell wall is the area of the nucleus, which is rich in deoxyribonucleic acid (DNA). DNA contains all the information necessary for the reproduction of all the cell components and may be considered to be the blueprint of the cell. In some cells, the DNA is enclosed by a membrane and the nucleus is clearly defined (eucaryotic cells). In other cells, the nucleus is poorly defined (procaryotic cells). As shown in Table 10·1, bacteria and blue-green algae are examples of procaryotic cells.

To continue to produce and function properly, an organism must have a source of energy and carbon for the synthesis of new cellular material. Inorganic elements, such as nitrogen and phosphorus, and other trace elements, such as sulfur, potassium, calcium, and magnesium, are also vital to cell synthesis. Two of the most common sources of cell carbon for microorganisms are carbon dioxide and organic matter. If an organism derives its cell carbon from carbon dioxide, it is called autotrophic; if it uses organic carbon, it is called heterotrophic.

Energy is also needed in the synthesis of new cellular material. For autotrophic organisms, the energy can be supplied by the sun, as in photosynthesis, or by an inorganic oxidation-reduction reaction. If the energy is supplied by the sun, the organism is called autotrophic photosynthetic. If the energy is supplied by an inorganic oxidation-reduction reaction, it is called autotrophic chemosynthetic. For heterotrophic organisms, the energy needed for cell synthesis is supplied by the oxidation or fermentation of organic matter. A classification of microorganisms by sources of energy and cell carbon is presented in Table 10·2.

TABLE 10·2 GENERAL CLASSIFICATION OF MICROORGANISMS BY SOURCES OF ENERGY AND CARBON

Classification	Energy source	Carbon source
Autotrophic:		
Photosynthetic	Light	CO_2
Chemosynthetic	Inorganic oxidation-reduction reaction	CO_2
Heterotrophic	Organic oxidation-reduction reaction	Organic carbon

Organisms can also be classed according to their ability to utilize oxygen. Aerobic organisms can exist only when there is a supply of molecular oxygen. Anaerobic organisms can exist only in an environment that is void of oxygen. Facultative organisms have the ability to survive with or without free oxygen.

Important Microorganisms

The sanitary engineer should be familiar with the characteristics of the following microorganisms because of their importance in biological treatment processes: (1) bacteria, (2) fungi, (3) algae, (4) protozoa, (5) rotifers, (6) crustaceans, and (7) viruses.

Bacteria Bacteria are single-cell protists. They utilize soluble food and, in general, will be found wherever moisture and a food source are available. Their usual mode of reproduction is by binary fission, although some species reproduce sexually or by budding. Even though there are thousands of different species of bacteria, their general form falls into one of three categories: spherical, cylindrical, and helical. Bacteria vary widely in size. Representative sizes are 0.5 to 1.0 μ in diameter for the spherical, 0.5 to 1.0 μ in width by 1.5 to 3.0 μ in length for the cylindrical (rods), and 0.5 to 5 μ in width by 6 to 15 μ in length for the helical (spiral). The general structure of a cell was discussed earlier. For further information concerning the bacterial cell see Refs. [25 and 29].

Tests on a number of different bacteria indicate that they are about 80 percent water and 20 percent dry material, of which 90 percent is organic and 10 percent inorganic. An approximate formula for the organic fraction is $C_5H_7O_2N$ [9]. As indicated by the formula, about 53 percent by weight of the organic fraction is carbon. Compounds comprising the inorganic portion include P_2O_5 (50 percent), SO_3 (15 percent), Na_2O (11 percent), CaO (9 percent), MgO (8 percent), K_2O (6 percent), and Fe_2O_3 (1 percent). Since all these elements and compounds must be derived from the environment, a shortage of any of these substances would limit and, in some cases, alter growth.

Temperature and pH play a vital role in the life and death of bacteria, as well as in other microscopic plants and animals. It has been observed that the rate of reaction for microorganisms increases with increasing temperature, doubling with about every 10°C of rise in temperature until some limiting temperature is reached. According to the temperature range in which they function best, bacteria may be classified as cryophilic or psychrophilic, mesophilic, and thermophilic. Typical temperature ranges for bacteria in each of these categories are presented in Table 10·3. For a more detailed discussion of the organisms in the various temperature ranges, see Refs. [3, 11, and 29].

TABLE 10·3 TYPICAL TEMPERATURE RANGES FOR VARIOUS BACTERIA

Type	Temperature, °C	
	Range	Optimum
Cryophilic*	−2–30	12–18
Mesophilic	20–45	25–40
Thermophilic	45–75	55–65

* Also called psychrophilic.

The pH of a solution is also a key factor in the growth of organisms. Most organisms cannot tolerate pH levels above 9.5 or below 4.0. Generally, the optimum pH for growth lies between 6.5 and 7.5.

Metabolically, bacteria can be classified as heterotrophic or autotrophic. The most common autotrophs are chemosynthetic, but a few are able to perform photosynthesis. The purple sulfur bacteria (*Thiorhodaceae*) and the green sulfur bacteria (*Chlorobiaceae*) are representative examples of autotrophic photosynthetic bacteria. In biological wastewater treatment, the heterotrophic are usually the most important group, because of their requirement for organic compounds for cell carbon. The autotrophic and heterotrophic bacteria can be further classified as aerobic, anaerobic, or facultative, depending on their need for oxygen.

Fungi In sanitary engineering, fungi are considered to be multicellular, nonphotosynthetic, heterotrophic protists. Microbiologists are less restrictive in their definition and include bacteria as a class of fungi, but in sanitary engineering bacteria are so important that they are considered separately.

Fungi are usually classed by their mode of reproduction. They reproduce sexually or asexually, by fission, budding, or spore formation. Molds or "true fungi" produce microscopic units (hyphae), which collectively form a filamentous mass called the mycelium. In sanitary engineering, the terms "fungi" and "molds" are used synonymously. Yeasts are fungi that cannot form a mycelium and are therefore unicellular.

Most fungi are strict aerobes. They have the ability to grow under low-moisture conditions and can tolerate an environment with a relatively low pH. The optimum pH for most species is 5.6; the range is 2 to 9. Fungi also have a low nitrogen requirement and need approximately one-half as much as bacteria. The ability of the fungi to survive under low pH and nitrogen-limiting conditions makes them very important in the biological treatment of some industrial wastes and in the composting of solid organic wastes.

Algae Algae are unicellular or multicellular, autotrophic, photosynthetic protists. Algae are undesirable in water supplies because they produce bad tastes and odors. In water filtration plants, the presence of algae will shorten filter runs. The green color of most species and their ability to form mats lowers the aesthetic value of the water. In oxidation ponds, algae are valuable in that they have the ability to produce oxygen through the mechanism of photosynthesis. At night, when light is no longer available for photosynthesis, they use up the oxygen in respiration. Respiration also occurs in the presence of sunlight; however, the net reaction is the production of oxygen. Equations 10·1 and 10·2 represent simplified biochemical reactions for photosynthesis and respiration.

Photosynthesis:

$$CO_2 + 2H_2O \xrightarrow{\text{light}} \underset{\substack{\text{New}\\\text{algae}\\\text{cells}}}{(CH_2O)} + O_2 + H_2O \tag{10·1}$$

Respiration:

$$CH_2O + O_2 \rightarrow CO_2 + H_2O \tag{10·2}$$

In an aquatic environment, it can be seen that this type of metabolic system will produce a diurnal variation in dissolved oxygen. The ability of algae to produce oxygen is vital to the ecology of the water environment. For an aerobic or facultative oxidation pond to operate effectively, algae are needed to supply oxygen to aerobic, heterotrophic bacteria. This symbiotic relationship between algae and bacteria will be expanded upon in the section on aerobic oxidation ponds at the end of this chapter.

Because algae use carbon dioxide in photosynthetic activity, high pH conditions can result. In addition, as the pH increases, the alkalinity components change, and carbonate and hydroxide alkalinity tend to predominate. If the water has a high concentration of calcium, calcium carbonate will precipitate when the carbonate and calcium-ion concentrations become great enough to exceed the solubility product. This removal of the carbonate ion by precipitation will keep the pH from continuing to increase. As is the case with dissolved oxygen, there is also a diurnal variation in pH. During the day algae utilize carbon dioxide, which results in a rising pH, while at night they produce carbon dioxide, which results in a falling pH.

Like other microorganisms, algae require inorganic compounds to reproduce. The principal nutrients required, other than carbon dioxide, are nitrogen and phosphorus. Other trace elements, such as iron, copper, and molybdenum, are also very important. It is noteworthy that the problem of preventing excessive algal growth in natural waters has, to date, centered around nutrient removal. Some scientists advocate the removal of nitrogen from treatment plant effluents, while others recommend the removal of phosphorus, and still others recommend removal of both nitrogen and phosphorus. Not to be forgotten are the trace elements, which in some cases may be the limiting nutrients in the growth of algae.

Four classes of freshwater algae are of importance [28]:

1. *Green* (*Chlorophyta*). The green algae are principally a freshwater species, and can be unicellular or multicellular. A distinguishing feature of this species is that the chlorophyll and other pigments are contained in chloroplasts. Chloroplasts are the sites of photosynthesis and are membrane-surrounded structures that contain chlorophyll and other pigments. Common green algae are those of the *Chlorella* group found in stabilization ponds.
2. *Motile Green* (*Volvocales Euglenophyta*). Colonial in nature, these

algae are bright green, unicellular, and flagellated. *Euglena* is a member of this particular group of algae. *Mastigophora* that contain chlorophyll are also often included in this category.

3. *Yellow Green or Golden Brown* (*Chrysophyta*). Most forms of the *Chrysophyta* are unicellular. They are freshwater inhabitants, and their characteristic color is due to yellowish brown pigments that conceal the chlorophyll. Of this group of algae, the most important are the diatoms. They are found in both fresh and salt waters. Diatoms have shells, which are composed mainly of silica. Deposits of these shells are known as diatomaceous earth, which is used as a filter aid.
4. *Blue-Green* (*Cyanophyta*). The blue-green algae are of very simple form and are similar to bacteria in several respects. They are unicellular, usually enclosed in a sheath, and have no flagella. They differ from other algae in that their chlorophyll is not contained in chloroplasts but is diffused throughout the cell. The blue-green algae are of interest in water and wastewater engineering for numerous reasons. They have the ability to form large, dense mats on the surface of the water and as a result lower the aesthetic value of the water. At times they can impart undesirable tastes and odors to waters. An important characteristic is their ability to utilize nitrogen from the atmosphere as a nutrient in cell synthesis. Thus, the removal of nitrogenous compounds from the water will not eliminate the source of nitrogen for these species of algae.

Protozoa Protozoa are motile, microscopic protists that are usually single cells. The majority of protozoa are aerobic heterotrophs, although a few are anaerobic. Protozoa are generally an order of magnitude larger than bacteria and often consume bacteria as an energy source. In effect, the protozoa act as polishers of the effluents from biological waste treatment processes by consuming bacteria and particulate organic matter.

Protozoa are usually divided into the following five groups:

1. *Sarcodina*. The *Sarcodina* are characterized by their pseudopods or false feet, which they use for movement and the capturing of food. *Endamoeba histolytica,* the cause of an intestinal disease in man, is a member of the *Sarcodina* group.
2. *Mastigophora*. The *Mastigophora* are characterized by their flagella, which they use for motility. Some microbiologists divide the *Mastigophora* into two groups: those that do and do not contain chlorophyll. *Euglena* and its colorless counterpart *Astasia* are examples.
3. *Sporozoa*. The *Sporozoa* are spore-forming protozoa and obligate parasites. Their only real interest to sanitary engineers is that

certain *Sporozoa,* namely four species of *Plasmodium,* cause malaria.

4. *Infusoria or Ciliata.* Movement by means of cilia is characteristic of these protozoa. The cilia are hairlike extensions from the cell membrane. Besides being responsible for the organism's movement, they are also important in assisting the protozoa to capture solid food. Sanitary engineers usually consider the *Ciliata* to be divided into two types, the free-swimming and the stalked. The free-swimming type must swim after bacteria. They require a great deal of food, because they expend so much energy in swimming. The *Paramecium* is a free-swimming ciliate that is important in wastewater treatment. The stalked ciliates are attached to something solid and must catch their food as it passes by. Because their movement is limited, they require less food for energy. The *Vorticella* is a stalked ciliate that is important in biological treatment processes, especially in the activated-sludge process.
5. *Suctoria.* The *Suctoria* are protozoa that have long tentacles, which they use to capture other protozoa and then draw the protoplasm from these protozoa into their own bodies. During the early stages of their life cycle, the *Suctoria* have cilia, but later in the adult stage they obtain their tentacles.

Rotifers The rotifer is an aerobic, heterotrophic, and multicellular animal. Its name is derived from the fact that it has two sets of rotating cilia on its head which are used for motility and capturing food. Rotifers are very effective in consuming dispersed and flocculated bacteria and small particles of organic matter. Their presence in an effluent indicates a highly efficient aerobic biological purification process.

Crustaceans Like the rotifer, the crustacean is an aerobic, heterotrophic, and multicellular animal. Unlike the rotifer, the crustacean has a hard body, or shell. Crustaceans are an important source of food for fish and as such are normal occupants in most natural waters. Except for their occasional presence in underloaded oxidation ponds, they do not exist in biological treatment systems to any noticeable extent. Their presence is indicative of an effluent that is low in organic matter and high in dissolved oxygen.

Viruses A virus is the smallest biological structure containing all the information necessary for its own reproduction. Viruses are so small that they can be seen only with the aid of an electron microscope. They are obligate parasites and as such require a host in which to live. Once they have a host, they redirect its complex machinery to produce new virus particles. Eventually the host cell ruptures, releasing new virus particles, which can go on to infect new cells.

Viruses are usually classified by the host they infect. Many viruses that produce diseases in man are also known to be excreted in the feces of man. Thus in sewage treatment the sanitary engineer has the responsibility of ensuring that these viruses are effectively controlled. This is usually done by chlorination and proper disposal of the plant effluent.

Cell Physiology

Since the bacteria are the most widely occurring microorganism in biological waste treatment, the discussion of cell physiology that follows will be centered around the bacteria; however, the basic principles are applicable to all living cells.

The process by which microorganisms grow and obtain energy is complex and intricate; there are many pathways and cycles. Vital to the reactions involved in these pathways and cycles are the actions of enzymes, which are organic catalysts produced by the living cell. Enzymes are proteins or proteins combined either with an inorganic molecule or with a low-molecular-weight organic molecule. As catalysts, enzymes have the capacity to increase the speed of chemical reactions greatly without altering themselves.

There are two general types of enzymes, extracellular and intracellular. When the substrate or nutrient required by the cell is unable to enter the cell wall, the extracellular enzyme converts the substrate or nutrient to a form that can then be transported into the cell. Intracellular enzymes are involved in the synthesis and energy reactions within the cell.

Enzymes are known for their high degree of efficiency in converting substrate to end products. One enzyme molecule can change many molecules of substrate per minute to end products. Enzymes are also known for their high degree of substrate specificity. This high degree of specificity means that the cell must produce a different enzyme for every substrate it uses. An enzyme reaction can be represented by the following general equation:

$$\underset{\text{Enzyme}}{(E)} + \underset{\text{Substrate}}{(S)} \rightarrow \underset{\substack{\text{Enzyme-}\\\text{substrate}\\\text{complex}}}{(E)(S)} \rightarrow \underset{\text{Product}}{(P)} + \underset{\text{Enzyme}}{(E)} \qquad (10{\cdot}3)$$

As illustrated, the enzyme functions as a catalyst by forming a complex with the substrate, which is then converted to a product and the original enzyme. At this point, the product may be acted upon by another enzyme. In fact, a sequence of complexes and products may be formed before the final end product is produced. In a living cell, the transformation of the original substrate to the final end product is accomplished by such an enzyme system. The activity of enzymes is substantially affected by pH and temperature, as well as by substrate concentration. Each enzyme has a particular optimum pH and temperature. The optimum pH and temperature of the key enzymes in the cell are reflected in the overall temperature and pH preferences of the cell.

Along with enzymes, energy is required to carry out the biochemical reactions in the cell. Energy is released in the cell by oxidizing organic or inorganic matter or by a photosynthetic reaction. The energy released is captured and stored in the cell by certain organic compounds. The most common storing compound is adenosine triphosphate (ATP). The energy captured by this compound is used for cell synthesis, maintenance, and motility. When the ATP molecule has expended its captured energy to the reactions involved in cell synthesis and maintenance, it changes to a discharged state called adenosine diphosphate (ADP). This ADP molecule can then capture the energy released in the breakdown of organic or inorganic matter. Having done this, the compound again assumes an energized state as the ATP molecule. The ADP-ATP cellular energy system is shown schematically in Fig. 10·2.

Simplified biochemical reactions that release energy for heterotrophic and autotrophic bacteria are listed in Table 10·4. While it is true that the energy they release is used to charge ADP molecules, many steps, all catalyzed by enzyme systems, are involved. A discussion of the pathways and cycles in these energy-releasing reactions is beyond the scope of this text; however, in simple terms, the overall metabolism of bacterial cells can be thought of as consisting of two biochemical reactions: energy and synthesis. The first reaction releases energy so that the second reaction of cellular synthesis can proceed. Both reactions are the result of numerous systems within the cell, and

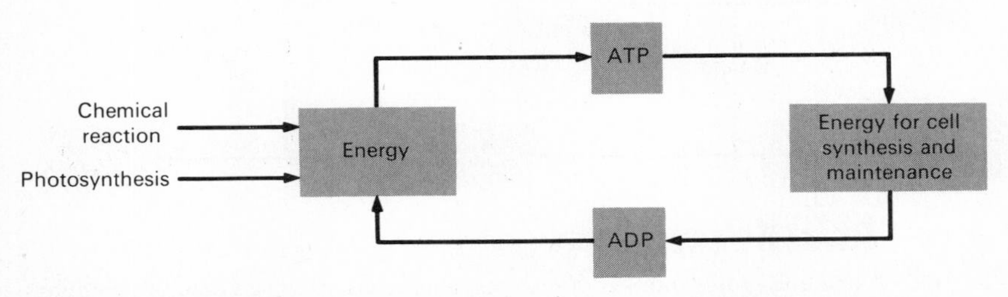

FIG. 10·2 Schematic representation of ADP-ATP cellular energy transfer system.

TABLE 10·4 TYPICAL EXOTHERMIC BIOCHEMICAL REACTIONS

Biochemical energy reaction	Nutrition of bacteria
$C_6H_{12}O_6 + 6O_2 \rightarrow 6CO_2 + 6H_2O$	Heterotrophic, aerobic
$C_6H_{12}O_6 \rightarrow 3CH_4 + 3CO_2$	Heterotrophic, anaerobic
$2NH_4^+ + 3O_2 \rightarrow 2NO_2^- + 2H_2O + 4H^+$	Autotrophic, chemosynthetic, aerobic
$5S + 2H_2O + 6NO_3^- \rightarrow 5SO_4^{--} + 3N_2 + 4H^+$	Autotrophic, chemosynthetic, anaerobic

each system consists of many enzyme-catalyzed reactions. The energy released in the "energy reaction" is captured by the enzyme-catalyzed system involving ATP and then transferred via ATP to the energy-requiring "synthesis reaction." A schematic presentation of cellular or bacterial metabolism is given in Figs. 10·3 and 10·4. These diagrams apply to aerobic, anaerobic, or facultative organisms.

It can be seen in Fig. 10·3 that, for heterotrophic bacteria, only a portion of the organic waste is converted into end products. The energy obtained from this biochemical reaction is used in the synthesis of the remaining organic matter into new cells. As the organic matter in the wastewater becomes limiting, there will be a decrease in cellular mass,

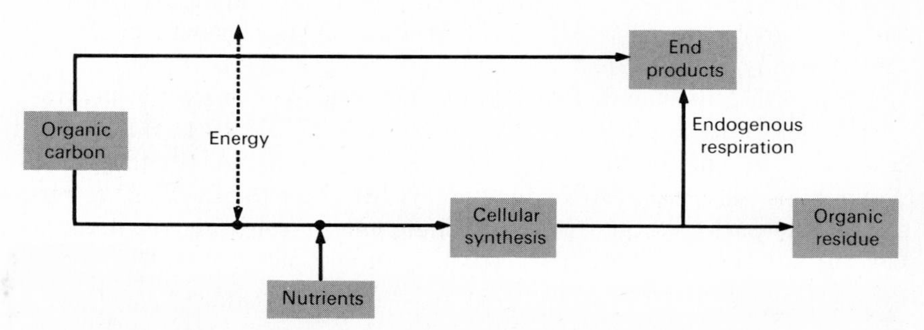

FIG. 10·3 Schematic representation of heterotrophic bacterial metabolism.

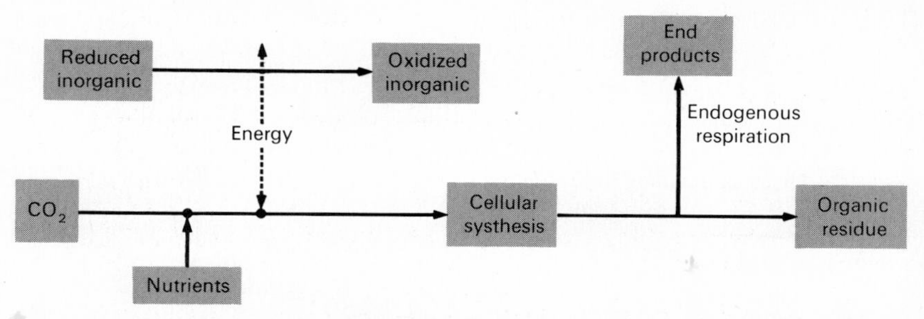

FIG. 10·4 Schematic representation of chemosynthetic autotrophic bacterial metabolism.

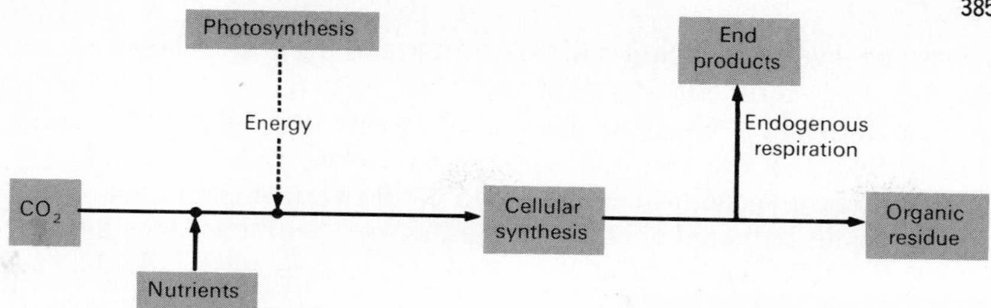

FIG. 10·5 Schematic representation of photosynthetic autotrophic bacterial metabolism.

because of the utilization of cellular material without replacement. If this situation continues, eventually all that will remain of the cell is a relatively stable organic residue. This overall process of a net decrease in cellular mass is termed endogenous respiration. Its place in the flow of energy and carbon for heterotrophic and autotrophic organisms is shown in Figs. 10·3 and 10·4.

When an autotrophic organism synthesizes new cellular material, the carbon source is carbon dioxide. The energy source for a cell synthesis is either light or the energy given off from an inorganic oxidation-reduction reaction. The flow of carbon and energy for autotrophic, chemosynthetic bacteria and for autotrophic, photosynthetic bacteria is shown in Figs. 10·4 and 10·5, respectively.

Nutrients, rather than organic or inorganic substrate in the wastewater, may at times be the limiting material for cell synthesis and growth. Bacteria, like algae, require nutrients, principally nitrogen and phosphorus, for growth. These nutrients may not always be present in sufficient quantities, as in the case of high-carbohydrate industrial wastes (e.g., sugar beets, sugar cane). Nutrient addition to the waste may be necessary for the proper growth of the bacteria and the subsequent degradation of the waste material. The position of nutrient addition in the flow of energy and carbon for heterotrophic and autotrophic organisms is shown in Figs. 10·3 to 10·5.

Aerobic and Anaerobic Cycles

To the sanitary engineer, there are two very important cycles in nature involving the growth and decay of organic matter:

1. The aerobic cycle, in which oxygen is used for the decay of the organic matter.

2. The anaerobic cycle, in which oxygen is not used for the decay of organic matter.

These two cycles are shown in Figs. 10·6 and 10·7. The elements of nitrogen and sulfur are shown as integral parts of the cycles. These are two important elements in the synthesis and decay of organic matter, but they are certainly not the only ones. Other elements are involved, and biochemical cycles could be drawn including them.

It should be noted that the title of aerobic and anaerobic applies only to the right-hand side of Figs. 10·6 and 10·7, or to the decomposition portion of the cycles. Here, dead organic matter is first broken down into initial and intermediate products, before the final stabilized products are produced. Both heterotrophic and autotrophic bacteria are involved in the many biodegradation processes required to obtain the final stabilized products. In the aerobic systems, the final products of degradation are more fully oxidized and hence at a lower energy level than the final products of the anaerobic degradation system. This accounts for the fact that much more energy is released in aerobic than in anaerobic degradation. As a result, anaerobic degradation is a much slower process.

The left-hand side of the cycle is the same in both the aerobic and anaerobic systems. This side is involved with the building or synthesis of the organic matter necessary for plant and animal life. Eventually, because of death or wastes from animal life, dead organic matter is made available to the bacterial decomposers, and the cycle is again repeated.

The decomposition portion of the cycle that occurs in nature is the concern of the sanitary engineer. By controlling the environment of the microorganisms, the decomposition of wastes is speeded up. Regardless of the type of waste, the biological treatment process consists of controlling the environment required for optimum growth of the microorganisms involved.

BACTERIAL GROWTH

Effective environmental control in biological waste treatment is based on an understanding of the basic principles governing the growth of microorganisms. Therefore, the following discussion is concerned with the growth of bacteria, the microorganisms of primary importance in biological treatment.

General Growth Patterns

As mentioned earlier, bacteria can reproduce by binary fission, by a sexual mode, or by budding. Generally, they reproduce by binary fission, that is, by dividing; the original cell becomes two new organisms.

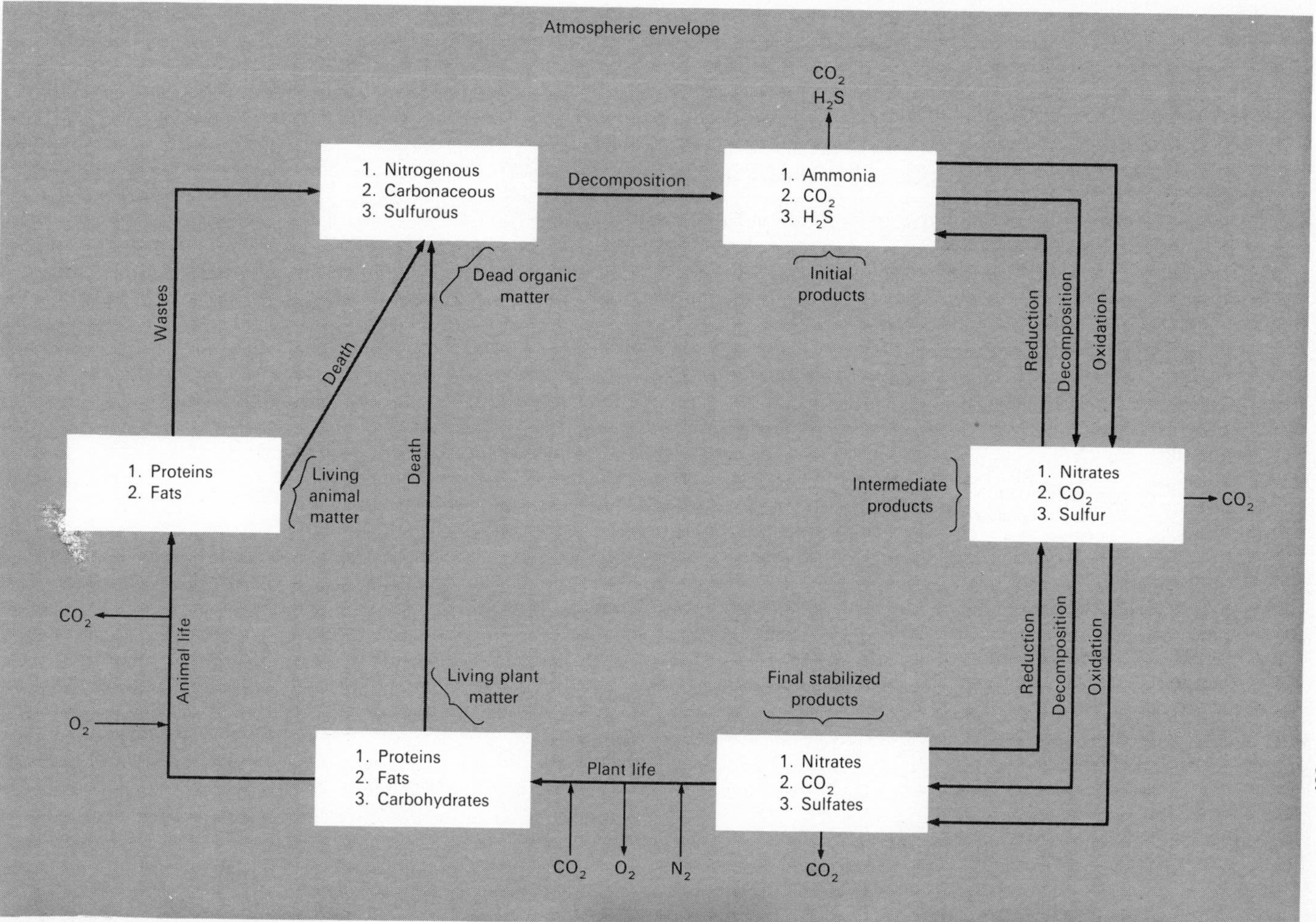

FIG. 10·6 The aerobic cycle [19].

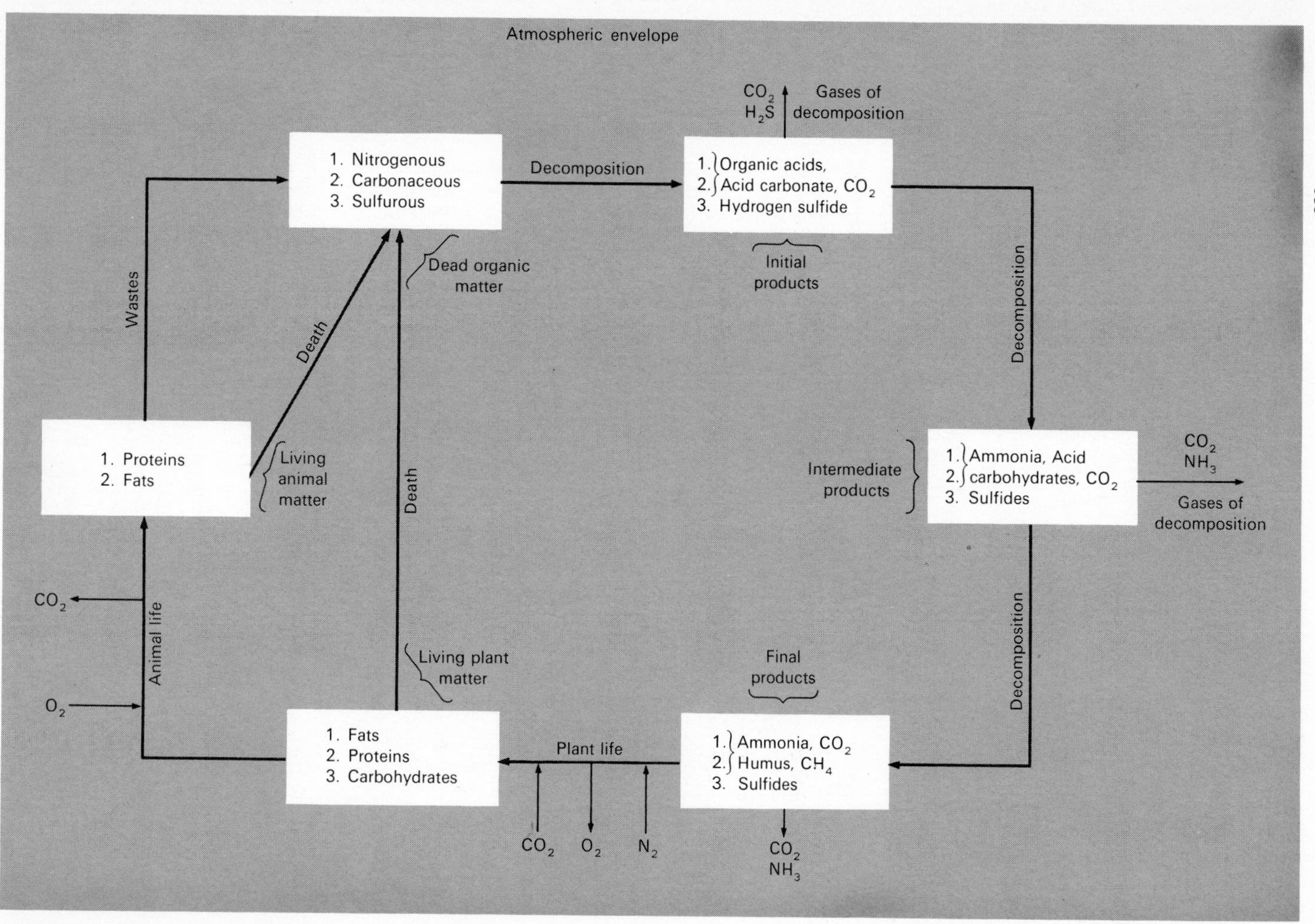

FIG. 10·7 The anaerobic cycle [19].

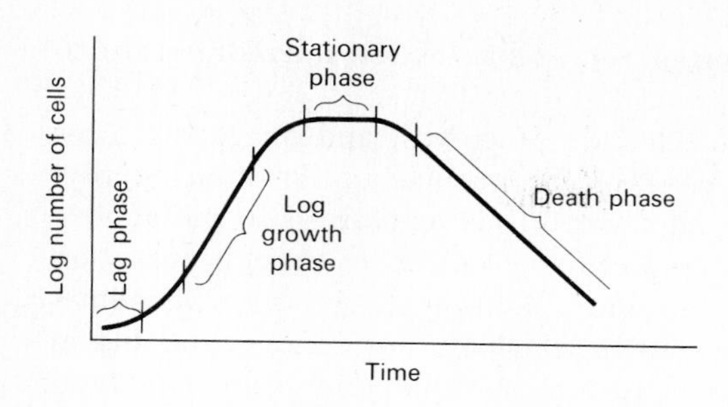

FIG. 10·8 Typical bacterial growth curve.

The time required for each fission, which is termed the generation time, can vary from days to less than 20 min. For example, if the generation time is 30 min, one bacterium would yield 16,777,216 bacteria after a period of 12 hr. This is a hypothetical figure, for bacteria would not continue to divide indefinitely because of various environmental limitations, such as substrate concentration, nutrient concentration, or even system size.

The general growth pattern of bacteria is shown in Fig. 10·8. Initially, a small number of organisms are inoculated into a culture medium, and the number of viable organisms is recorded as a function of time. The growth pattern based on the number of cells has four more or less distinct phases.

1. The lag phase. Upon addition of an inoculum to a culture medium, the lag phase represents the time required for the organisms to acclimate to their new environment.
2. The log growth phase. During this period the cells divide at a rate determined by their generation time and their ability to process food (constant percentage growth rate).
3. The stationary phase. Here the population remains stationary. Reasons advanced for this phenomenon are (*a*) that the cells have exhausted the substrate or nutrients necessary for growth, and (*b*) that the growth of new cells is offset by the death of old cells.
4. The log death phase. During this phase the bacterial death rate exceeds the production of new cells. The death rate is usually a function of the viable population and environmental characteristics. In some cases, the log death phase is the inverse of the log growth phase.

The growth pattern can also be discussed in terms of the variation of the mass of microorganisms with time. This growth pattern consists of the following three phases:

1. The log growth phase. There is always an excess amount of food surrounding the microorganisms, and the rate of metabolism and

growth is only a function of the ability of the microorganism to process the substrate.

2. Declining growth phase. The rate of growth and hence the mass of bacteria decreases because of limitations in the food supply.
3. Endogenous phase. The microorganisms are forced to metabolize their own protoplasm without replacement, since the concentration of available food is at a minimum. During this phase, a phenomenon known as lysis can occur in which the nutrients remaining in the dead cells diffuse out to furnish the remaining cells with food (known as cryptic growth).

It is important to note that the preceding discussions concerned a single population of microorganisms. Often biological treatment units are composed of complex, interrelated, mixed biological populations, with each particular microorganism in the system having its own growth curve. The position and shape of a particular growth curve in the system, on a time scale, depends on the food and nutrients available and on environmental factors, such as temperature, pH, and whether the system is aerobic or anaerobic. The variation of microorganism predominance with time in the aerobic stabilization of liquid organic waste is given in Fig. 10·9. While the bacteria are of primary importance, many other microorganisms take part in the stabilization of the organic waste. When designing or analyzing a biological treatment

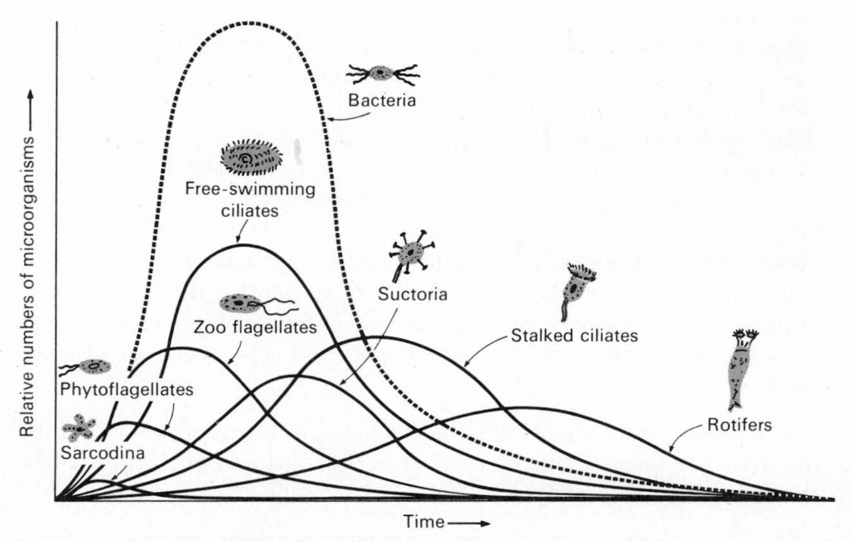

FIG. 10·9 Relative growth of microorganisms stabilizing an organic waste in a liquid environment [20].

process, the engineer should think in terms of an ecosystem, such as the one shown in Fig. 10·9, and not in terms of a "black box" that contains mysterious microorganisms. Other ecosystems that are important in biological waste treatment will be discussed more thoroughly later in this chapter (see "Aerobic-anaerobic Waste Treatment").

Kinetics of Biological Growth

The need for a controlled environment and biological community in the design of a biological waste treatment unit is stressed throughout this chapter. The microorganisms of importance in wastewater treatment have been discussed, along with their cytological and metabolic characteristics and their growth patterns. Although the characteristics of the environment needed for their growth have been described, nothing has been said about how to control the environment of the microorganisms. Environmental conditions can be controlled by pH regulation, temperature regulation, nutrient or trace element addition, oxygen addition or exclusion, and proper mixing. Control of the environmental conditions will ensure that the microorganisms have a proper medium in which to grow.

To ensure that the microorganisms will grow, they must be allowed to remain in the system long enough to reproduce. This period depends on their growth rate, which is related directly to the rate at which they metabolize or utilize the waste. Assuming that the environmental conditions are controlled properly, effective waste stabilization can be ensured by controlling the growth rate of the microorganisms.

An empirically developed relationship between biological growth and substrate utilization that is commonly used for biological systems stabilizing organic and/or inorganic wastes is [8, 29]:

$$\frac{dX}{dt} = Y\frac{dF}{dt} - k_d X \qquad (10\cdot4)$$

where $\frac{dX}{dt}$ = net growth rate of microorganisms, mass/volume-time

Y = growth-yield coefficient, mass of microorganisms/mass of substrate utilized

$\frac{dF}{dt}$ = rate of substrate utilization by microorganisms, mass/volume-time

k_d = microorganism-decay coefficient, time^{-1}

X = concentration of microorganisms, mass/volume

This equation, or one with slight modifications, has been used successfully by numerous investigators to describe both aerobic and anaerobic biological waste treatment systems [7, 8, 10, 18, 24, 29].

The rate of substrate utilization can be approximated by the following expression [18]:

$$\frac{dF}{dt} = \frac{kXS}{K_s + S} = \frac{dS}{dt} \tag{10·5}$$

where k = maximum rate of waste utilization per unit weight of microorganisms, time^{-1}

K_s = waste concentration at which rate of waste utilization per unit weight of microorganisms is one-half the maximum rate, mass/volume

S = concentration of waste surrounding the microorganisms, mass/volume

A graphical representation of Eq. 10·5 is presented in Fig. 10·10. Equation 10·5 is similar to one developed by Monod [21] to describe the relationship between the concentration of a limiting nutrient and the growth rate of microorganisms. It is also similar to the Michaelis-Menten relationship used to describe enzyme-catalyzed reactions.

Dividing both sides of Eq. 10·4 by X gives

$$\frac{dX/dt}{X} = Y\frac{dF/dt}{X} - k_d \tag{10·6}$$

In Eq. 10·6, $(dX/dt)/X$, the net growth rate, is often symbolized by mu, μ. Utilizing this and the expression for dF/dt given in Eq. 10·5, Eq. 10·6 can be rewritten to yield

$$\mu = \frac{YkS}{K_s + S} - k_d \tag{10·7}$$

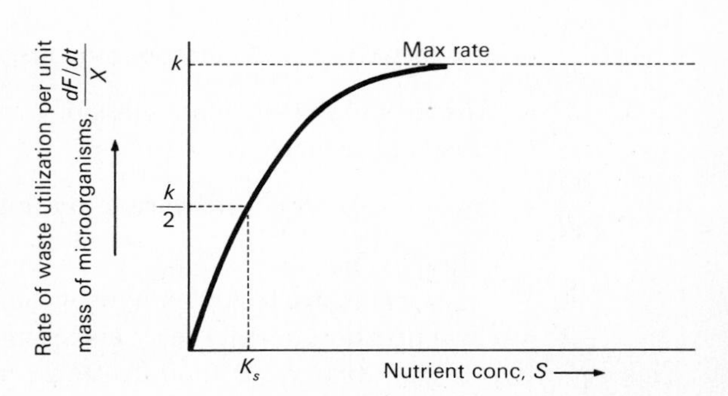

FIG. 10·10 Rate of waste utilization per unit mass of microorganisms versus concentration of a limiting nutrient.

Equation 10·7 is identical to the equation proposed by Van Uden [30] for pure culture systems. Often, $\hat{\mu}$ is used to symbolize the product of the growth yield Y and the maximum rate of waste utilization per unit weight of microorganisms k. Using this, Eq. 10·7 can be rewritten as

$$\mu = \frac{\hat{\mu} S}{K_s + S} - k_d \qquad (10\cdot8)$$

where $\hat{\mu}$ is the product of Y and k, and hence equals μ_{max}. Equation 10·8 is similar to the equation developed by Monod [21] for pure culture systems.

Rewriting and rearranging Eq. 10·6 on a finite mass and time basis, in which the subscript M refers to a definite mass of microorganisms, yields the following:

$$\frac{1}{X_M/(\Delta X/\Delta t)_M} = Y \frac{(\Delta F/\Delta t)_M}{X_M} - k_d \qquad (10\cdot9)$$

In Eq. 10·9, the term $(\Delta F/\Delta t)_M/X_M$ is commonly known as the process loading factor, the specific utilization, the substrate removal rate, or the food-to-microorganism ratio. In the following discussion, this term will be referred to as the food-to-microorganism ratio U.

$$U = \frac{(\Delta F/\Delta t)_M}{X_M} \qquad (10\cdot10)$$

In Eq. 10·10, $(\Delta F/\Delta t)_M$ represents the mass of substrate utilized by the mass of microorganisms X_M over a finite time period Δt.

The term $X_M/(\Delta X/\Delta t)_M$ on the left-hand side of Eq. 10·9 has often been referred to as the solids retention time, the sludge age, or the mean cell residence time. In this discussion, this term will be referred to as the mean cell residence time and will be symbolized by θ_c.

$$\theta_c = \frac{X_M}{(\Delta X/\Delta t)_M} \qquad (10\cdot11)$$

In Eq. 10·11, X_M equals the total active microbial mass in the treatment system, and $(\Delta X/\Delta t)_M$ equals the total quantity of microbial mass withdrawn daily from the treatment system. The quantity wasted includes those microbial solids wasted purposely, as well as those lost in the effluent. The reciprocal of θ_c, $(\Delta X/\Delta t)_M/X_M$, equals the specific or fractional growth rate [12].

Utilizing Eqs. 10·10 and 10·11, Eq. 10·9 can be rewritten as

$$\frac{1}{\theta_c} = YU - k_d \tag{10·12}$$

It can be seen from Eq. 10·12 that $1/\theta_c$, the microorganism growth rate, and U, the food-to-microorganism ratio, are related directly. In the following sections, it will be shown that a desired treatment efficiency can be obtained by the control of either θ_c or U. It will also be shown that because θ_c is easy to measure, it is the more desirable parameter for process control.

The equations and parameters developed in the foregoing discussion are basic to an understanding of biological treatment kinetics. Before these equations can be applied to basic reactor systems, assumptions inherent in the development of these equations should be noted.

1. Equation 10·4 is empirical and was developed originally from batch waste treatment studies; Eqs. 10·5 and 10·7 were developed from continuous pure culture studies. The parameters of importance in analyzing and designing biological waste treatment processes are delineated in Eqs. 10·5 and 10·7 when used in conjunction with Eq. 10·12.
2. Although known to be variable, the growth yield coefficient Y in Eq. 10·4 has been assumed to be constant. Based on laboratory studies, Ribbons [26] concluded that yield is dependent upon (1) oxidation-reduction state of the carbon source and nutrient elements, (2) degree of polymerization of the substrate, (3) pathways of metabolism, (4) growth rate, and (5) various physical parameters of cultivation. The term k_d in Eq. 10·4 accounts for factors such as death, predation, and the diversion of energy for cell maintenance reactions. An alternative approach to the analysis of net growth (Sherrard [27]) accounts for the variation of yield by incorporating k_d into an observed yield coefficient that varies with the mean cell residence time.
3. All the necessary nutrients needed for proper biological growth are present. The only limiting substance is the organic matter needed for heterotrophic growth or, in the case of autotrophic organisms, the inorganic energy source. The pH and temperature are regulated for the optimum rate of growth.
4. The equations apply to those waste treatment processes that employ bacteria as the primary organism. This assumption does not hinder the application of the equations to the mixed biological populations of biological waste treatment systems, because in the

majority of these populations, bacteria are the primary waste-stabilizing organisms.

5. The equations apply only to that portion of the waste that is soluble and biodegradable.

Application of Kinetics to Treatment Systems

The containers, vessels, or tanks in which chemical and biological reactions are carried out are commonly called reactors. Before discussing the application of the kinetic equations it is appropriate to consider the nature of reactors, especially with respect to flow.

Principal Reactor Types and Flow Characteristics The four principal types of reactors used for biological waste treatment are classified with respect to their hydraulic flow characteristics as batch, plug flow, complete mix (also known as a continuous flow stirred-tank reactor), and arbitrary flow.

In terms of flow, the batch reactor is characterized by the fact that flow is neither entering nor leaving on a continuous basis. For example, the BOD test discussed in Chap. 7 is carried out in a bottle batch reactor. Arbitrary flow is any degree of partial mixing between plug flow and complete mixing. Only plug-flow and complete-mix models will be discussed in detail in this section.

In plug flow, fluid particles pass through the tank and are discharged in the same sequence in which they enter. The particles retain their identity and remain in the tank for a time equal to the theoretical detention time. This type of flow is approximated in long tanks with a high length-to-width ratio in which longitudinal dispersion is absent. If a continuous flow of tracer were injected into such a tank to produce a concentration C_0, the appearance of the tracer in the effluent C_e would occur as shown in Fig. 10·11; t_a equals the actual time and t_d equals the theoretical detention time V/Q. Under ideal plug-flow conditions t_a equals t_d. The effect of a slug injection of tracer is also shown in Fig. 10·11.

Complete mixing occurs when the particles entering the tank are immediately dispersed throughout the tank. The particles leave the tank in proportion to their statistical population. Complete mixing can be accomplished in round or square tanks if the contents of the tank are uniformly and continuously redistributed. If a continuous flow of conservative (nonreactive) tracer were injected into the inlet at concentration C_0, the appearance of the tracer at the outlet would occur as shown in Fig. 10·11. The effluent concentration as a function of time can be determined from a materials mass balance for the tracer around

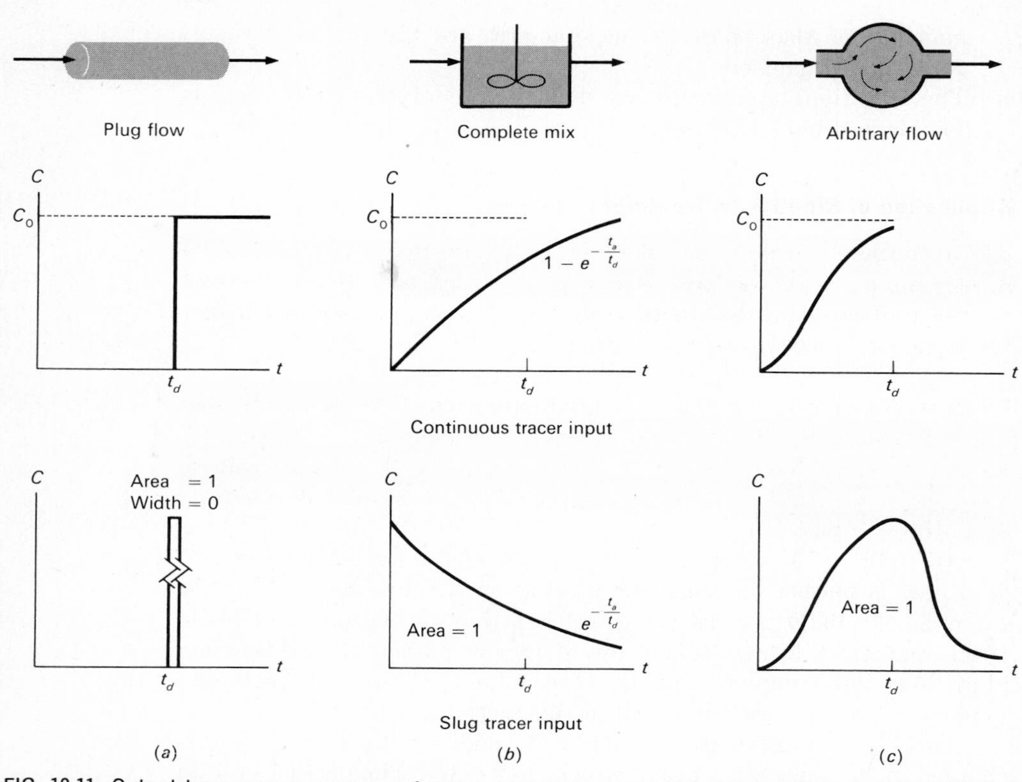

FIG. 10·11 Output tracer response curves for step and slug tracer inputs for (a) plug-flow, (b) complete-mix and (c) arbitary-flow reactors [adapted from 14].

the reactor as follows:

$$\begin{bmatrix}\text{Rate of}\\ \text{change in}\\ \text{amount of}\\ \text{tracer in}\\ \text{reactor}\end{bmatrix} = \begin{bmatrix}\text{rate of}\\ \text{tracer}\\ \text{inflow to}\\ \text{reactor}\end{bmatrix} - \begin{bmatrix}\text{rate of}\\ \text{tracer}\\ \text{outflow}\\ \text{from}\\ \text{reactor}\end{bmatrix}$$

$$\frac{dC}{dt}(V) = QC_0 - QC \tag{10·13}$$

where C = effluent concentration of tracer at any time t
V = volume of reactor
Q = flowrate
C_0 = influent concentration of tracer

Rewriting and simplifying Eq. 10·13 yields

$$\frac{dC}{dt} = \frac{Q}{V}(C_0 - C) \tag{10·14}$$

which can be integrated as

$$\int_0^C \frac{dC}{C_0 - C} = \frac{Q}{V}\int_0^t dt \tag{10·15}$$

to yield

$$C = C_0(1 - e^{-t/(V/Q)}) = C_0(1 - e^{-t/t_d}) \tag{10·16}$$

where t_d equals the detention time V/Q.

The corresponding expression for the effluent concentration from a reactor that is being purged of tracer is derived similarly, and is given by

$$C = C_0e^{-t/t_d} \tag{10·17}$$

Arbitrary flow represents any degree of partial mixing between plug and complete mixed flow. This type of flow is encountered frequently in actual aeration and settling tanks. It is also more difficult to describe mathematically (see Chap. 12). Therefore, in the mathematical treatment of the chemical and biological unit processes carried out in reactors, ideal plug-flow or completely mixed-flow models are usually assumed.

To this point, the discussion has dealt only with the hydraulic characteristics of the various reactors. In the sections that follow, kinetic equations describing the growth of microorganisms and the removal of substrate will be combined with the hydraulic analysis of these reactors. In addition, because most biological reactors are operated with some form of recycle, the effects of recycle will also be considered. The effect of reaction kinetics and other operational factors on reactor selection is considered in Chap. 12.

Complete Mix–No Recycle In this reactor scheme, shown in Fig. 10·12, the reactor unit is completely mixed and there are no organisms in the influent. For this system, the hydraulic or liquid retention time θ is

$$\theta = \frac{V}{Q} \tag{10·18}$$

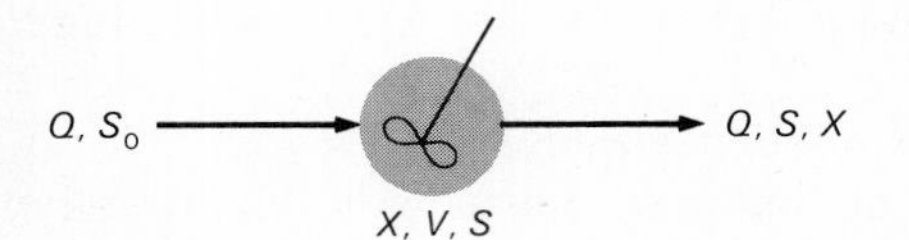

FIG. 10·12 Schematic of a complete-mix reactor without recycle.

where V is the volume of the reactor and Q is the volumetric flowrate.

The mean cell residence time θ_c is defined as

$$\theta_c = \frac{VX}{QX} \tag{10·19}$$

where X is the mass concentration of microorganisms in the reactor. By simplifying Eq. 10·19 it can be seen that

$$\theta_c = \theta \tag{10·20}$$

Thus the average retention time of the cells in the system is the same as that of the liquid. This is a very important characteristic of the complete-mix–no-recycle system.

A mass balance for the mass of microorganisms in the reactor system can be written as

$$\begin{bmatrix}\text{Rate of}\\ \text{change of}\\ \text{organism}\\ \text{concentration}\\ \text{in reactor}\end{bmatrix} = \begin{bmatrix}\text{net rate of}\\ \text{organism}\\ \text{growth in}\\ \text{reactor}\end{bmatrix} - \begin{bmatrix}\text{rate of}\\ \text{organism}\\ \text{outflow}\\ \text{from}\\ \text{reactor}\end{bmatrix}$$

$$V\left(\frac{dX}{dt}\right) = \left(Y\frac{dF}{dt} - k_d X\right)V - QX \tag{10·21}$$

At steady state, dX/dt equals 0, and Eq. 10·21 can be rewritten as

$$\frac{Q}{V} = Y\frac{dF/dt}{X} - k_d \tag{10·22}$$

Utilizing Eqs. 10·10 and 10·11, Eq. 10·22 can be rewritten to give

$$\frac{1}{\theta_c} = YU - k_d \tag{10·23}$$

Since θ_c equals θ for this reactor system, the food-to-microorganism ratio U is related directly to the hydraulic retention time of the system θ.

The relationship of θ_c and U, and therefore of θ, to the efficiency of waste stabilization in the reactor can be demonstrated as follows. The efficiency of waste stabilization can be defined as

$$E = 100\,\frac{S_0 - S}{S_0} \tag{10·24}$$

where E = efficiency of waste stabilization expressed in percentage form

S_0 = mass concentration of influent waste

S = mass concentration of influent waste not biologically degraded appearing in the effluent

To obtain an expression for the effluent substrate concentration S, Eq. 10·23 can be rewritten, utilizing Eqs. 10·5 and 10·10, to yield

$$\frac{1}{\theta_c} = Y \frac{kS}{K_s + S} - k_d \tag{10·25}$$

Solving Eq. 10·25 for S yields

$$S = \frac{K_s(1 + k_d\theta_c)}{\theta_c(Yk - k_d) - 1} \tag{10·26}$$

By comparing Eq. 10·23 with Eq. 10·25 it will be seen that

$$U = \frac{kS}{K_s + S}$$

from which the following equation is obtained:

$$S = \frac{UK_s}{k - U} \tag{10·27}$$

For a specified waste, a biological community, and a particular set of environmental conditions, the kinetic coefficients Y, k, K_s, and k_d can be defined. Consequently, the effluent waste concentration S is a direct function of either θ_c or U. Setting one of these three parameters not only fixes the other two but also specifies the efficiency of biological waste stabilization.

For a complete-mix–no-recycle system that is growth specified (specified values of S_0, Y, K_s, k, and k_d), fixing the mean cell residence time θ_c establishes the microorganism concentration in the reactor. This can be shown in the following development. The rate of food utilization dF/dt in the reactor can be evaluated on a finite time basis:

$$\frac{\Delta F}{\Delta t} = \frac{Q}{V}(S_0 - S) \tag{10·28}$$

Utilizing Eqs. 10·18, 10·20, and 10·28, Eq. 10·22 can be solved for the mass concentration of microorganisms in the reactor X to yield

$$X = \frac{Y(S_0 - S)}{1 + k_d\theta_c} \tag{10·29}$$

However, it is shown in Eq. 10·26 that the effluent waste concentration S is also a function of θ_c. Thus, for a complete-mix–no-recycle system that is growth specified, the microorganism concentration X is a function of the mean cell residence time θ_c.

Equations 10·24 and 10·26 are plotted in Fig. 10·13 for a growth-specified complete-mix–no-recycle system. As shown, the effluent con-

FIG. 10·13 Effluent waste concentration and removal efficiency versus mean cell residence time for a complete-mix reactor without recycle.

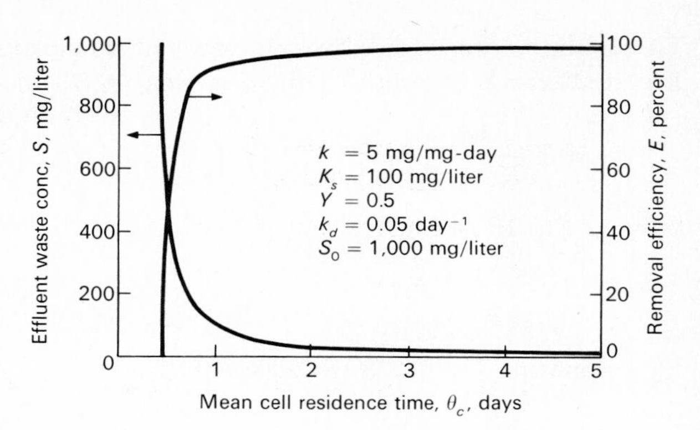

centration S and the treatment efficiency E are related directly to θ_c, which is equal to θ. In a system such as this, there is no separate control of the microorganisms, because the mean microorganism retention time θ_c and the liquid retention time θ are the same. To obtain a high treatment efficiency, θ_c must be long, which means that θ must also be long. This mode of operation is characteristic of conventional anaerobic treatment systems and some modified activated-sludge processes. The complete-mix–no-recycle model can also be used, but not without caution, to simulate oxidation ponds and lagoons. Both of these are flow-through systems with no recycle, but they are not completely mixed. Settling is an effective unit operation in the overall efficiency of these processes, and the foregoing equations did not include the effect of settling.

It can also be seen from Fig. 10·13 that there is a certain value of θ_c below which waste stabilization does not occur. This critical value of θ_c is called the minimum mean cell residence time θ_c^M. Physically, θ_c^M is the residence time at which the cells are washed out or wasted from the system faster than they can reproduce. The minimum mean cell residence time can be calculated from Eq. 10·25, since for this condition the influent waste concentration S_0 is equal to the effluent waste concentration S:

$$\frac{1}{\theta_c^M} = Y \frac{kS_0}{K_s + S_0} - k_d \tag{10·30}$$

In many situations encountered in waste treatment, S_0 is much greater than K_s so that Eq. 10·30 can be rewritten to yield

$$\frac{1}{\theta_c^M} \approx Yk - k_d \tag{10·31}$$

Use of Eqs. 10·30 and 10·31 determines the minimum mean cell residence time $\theta_c{}^M$. Obviously, biological treatment systems should not be designed with θ_c values equal to $\theta_c{}^M$. To ensure adequate waste treatment, biological treatment systems are usually designed and operated with a θ_c value from 2 to 20 times $\theta_c{}^M$.

Values for the parameters Y, k, K_s, and k_d must be available for a particular model to be used effectively. The experimental procedures involved in obtaining these values in the laboratory are delineated in Chap. 16. Presented in Tables 10·5 and 10·6 are a few of the values

TABLE 10·5 KINETIC COEFFICIENTS FOR MIXED-CULTURE AEROBIC SYSTEMS

Wastewater composition	k, days^{-1}	K_s, mg/liter	k_d, days^{-1}	Y, $\frac{\text{mg-VSS}}{\text{mg}}$	Coef basis	$\theta_c{}^M$, days	Ref
Domestic waste	...	...	0.055	0.5	BOD_5	...	8
Domestic waste	5.6	22	0.07	0.67	COD	0.27	2
Skim milk	5.1	100	0.045	0.48	BOD_5	0.42	7

TABLE 10·6 KINETIC COEFFICIENTS FOR ANAEROBIC FERMENTATION SYSTEMS

Wastewater composition	k, days^{-1}	K_s, mg/liter	k_d, days^{-1}	Y, $\frac{\text{mg-VSS}}{\text{mg}}$	Coef basis	$\theta_c{}^M$, days	Study temp., °C	Ref.
Acetic acid	3.6	2,130	0.015	0.040	COD	7.8	20	22
Municipal sludge	...	...	...	...	...	10.0	20	22
Acetic acid	4.7	869	0.011	0.054	Acetic acid	4.2	25	13
Municipal sludge	...	...	...	...	...	7.5	25	22
Synthetic milk waste	0.38	24.3	0.07	0.37	COD		20–25	5
Acetic acid	8.1	154	0.015	0.044	Acetic acid	3.1	35	13
Municipal sludge	...	...	...	...	...	2.8	35	22
Packinghouse waste	0.32	5.5	0.17	0.76	BOD		35	5

reported in the literature for various wastes treated aerobically and anaerobically. Most of the data presented in these tables pertain to the growth-yield coefficient Y and the microorganism–decay coefficient k_d.

McCarty [17] has presented a unique method of determining the growth-yield coefficient Y, based on a thermodynamic approach to substrate utilization and cell synthesis.

The growth parameters k and K_s have not been evaluated as extensively as Y and k_d. A lack of values for these parameters places some limitations on the model; however, with the data available, reasonable estimates of the growth parameters can be made.

Values of the minimum mean cell residence time are also presented in Tables 10·5 and 10·6. It would appear from Fig. 10·13 that a continual increase in θ_c would result in an increasing treatment efficiency; however, operational constraints limit the value of θ_c that can be used. These constraints result from the high microorganism concentration that, in turn, results from high values of θ_c. They include (1) difficulty in mixing the microorganisms in the reactor, (2) oxygen mass transfer limitations in an aerobic system, (3) inadequate separation of the microorganisms in the settling tank, and (4) a depletion of nutrients.

Complete Mix–Cellular Recycle In this system, shown schematically in Fig. 10·14, the reactor contents are completely mixed, and it is assumed that there are no microorganisms in the waste influent. The system contains a settling unit in which the cells from the reactor are settled and then returned to the reactor. Because of the presence of this settling unit, two additional assumptions must be made in the development of the kinetic model for this system:

1. Waste stabilization by the microorganisms occurs only in the reactor unit. This assumption leads to a conservative model, as in some systems there may be some waste stabilization in the settling unit.
2. The volume used in calculating the mean cell residence time for the system includes only the volume of the reactor unit. In effect, it is assumed that the settling tank serves as a reservoir from which solids are returned to maintain a given solids level in the aeration

FIG. 10·14 Schematic of a complete-mix reactor with cellular recycle.

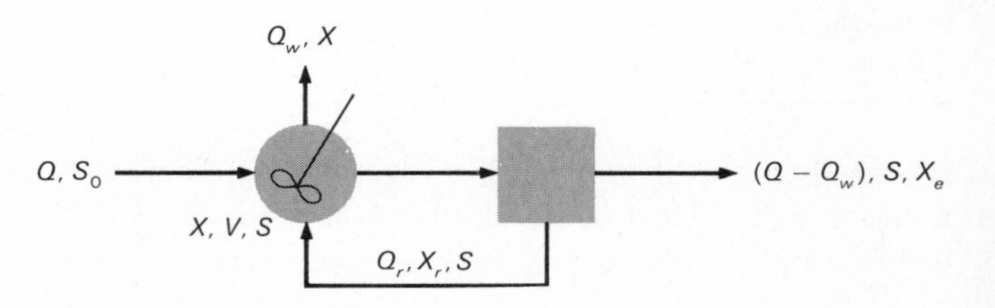

tank. If the system is such that these assumptions do not hold true, then the model should be modified.

The mean hydraulic retention time for the system θ_s is defined as

$$\theta_s = \frac{V_s}{Q} \tag{10·32}$$

where V_s = volume of reactor plus volume of settling tank
Q = influent flowrate

The mean hydraulic retention time for the reactor θ is defined as

$$\theta = \frac{V}{Q} \tag{10·33}$$

where V is the volume of the reactor.

For the system shown in Fig. 10·14, the mean cell residence time θ_c, as defined by Eq. 10·11, is

$$\theta_c = \frac{VX}{Q_wX + (Q - Q_w)X_e} \tag{10·34}$$

where Q_w = flowrate of liquid containing fraction of cells wasted from reactor
X_e = microorganism concentration in effluent from settling unit

In a system with a properly operating settling unit, the quantity of cells in the effluent is very small, and Eq. 10·34 can be simplified to yield

$$\theta_c \approx \frac{V}{Q_w} \tag{10·35}$$

Comparing Eq. 10·35 with Eqs. 10·32 and 10·33, it can be seen that, for a given reactor volume, θ_c is theoretically independent of both θ and θ_s. Practically speaking, however, θ_c cannot be completely independent of θ and θ_s. The factors relating θ_c to θ and θ_s will be discussed later.

A mass balance for the mass of microorganisms in the entire system can be written as

$$\begin{bmatrix}\text{Rate of}\\ \text{change of}\\ \text{organism}\\ \text{concentration}\\ \text{in reactor}\end{bmatrix} = \begin{bmatrix}\text{net rate of}\\ \text{organism}\\ \text{growth}\\ \text{in reactor}\end{bmatrix} - \begin{bmatrix}\text{rate of}\\ \text{organism}\\ \text{outflow}\\ \text{from}\\ \text{reactor}\end{bmatrix}$$

$$V\left(\frac{dX}{dt}\right) = \left(Y\frac{dF}{dt} - k_dX\right)V - [Q_wX + (Q - Q_w)X_e] \tag{10·36}$$

Making use of Eq. 10·34 and considering steady-state conditions, Eq. 10·36 can be simplified and rearranged to yield

$$\frac{1}{\theta_c} = Y \frac{dF/dt}{X} - k_d \tag{10·37}$$

or, considering a finite time basis and using Eq. 10·10,

$$\frac{1}{\theta_c} = YU - k_d \tag{10·38}$$

Equation 10·38 is the same as Eq. 10·23, which was developed for a complete-mix–no-recycle system. In both systems, there is a direct relationship between θ_c and U. By further developing Eq. 10·38, expressions can be obtained for the complete-mix–recycle system that duplicate Eqs. 10·25 to 10·27. Thus, in the growth-specified recycle system, the effluent waste concentration S is directly related to θ_c or U. For both recycle and nonrecycle systems, controlling θ_c or U establishes the effluent concentration.

In a complete-mix–no-recycle system, both θ_c and U are direct functions of the hydraulic retention time of the reactor θ. In a recycle system, however, θ_c and U are theoretically independent of the hydraulic retention time of the reactor θ and of the system θ_s. Thus it is possible to achieve a high θ_c, and therefore good treatment efficiency, without raising θ or θ_s.

The mass concentration of microorganisms X in the reactor can be obtained by utilizing Eqs. 10·28 and 10·33 in conjunction with Eq. 10·37, and solving for X:

$$X = \frac{\theta_c}{\theta} \frac{Y(S_0 - S)}{1 + k_d\theta_c} \tag{10·39}$$

In the analysis of the foregoing equations, the predominant theme was that in a recycle system, θ_c and U were independent of θ and θ_s, whereas in a nonrecycle system, θ_c, U, and θ were directly related. Since both θ_c and U are related directly to the effluent quality or treatment efficiency, as shown by Eqs. 10·26 and 10·27, controlling either θ_c or U in a biological treatment process will directly control the process efficiency. This can be done independently of θ or θ_s in a recycle system. The choice of which parameter to use for treatment control, θ_c or U, is a matter of ease of attainment.

To determine the food-to-microorganism ratio U (see Eq. 10·10), the food utilized and the mass of microorganisms effective in this utilization must be known. The food utilized can be evaluated by determining the difference between the influent and effluent COD or

BOD_5. The evaluation of the mass of microorganisms is usually what makes the use of U impractical as a control parameter. The most common parameter used as a measure of the biological solids is the volatile suspended solids in the treatment unit. The use of this parameter is not entirely satisfactory, because of the variability of volatile matter in the waste that is not related to active cellular material.

Using θ_c as a treatment control parameter, there is no need to determine the amount of biological solids in the system, nor is there the need to evaluate the amount of food utilized. The use of θ_c is simply based upon the fact that, to control the growth rate of microorganisms and hence their degree of waste stabilization, a specified percentage of the cell mass in the system must be wasted each day. Thus, if it is determined that a θ_c of 10 days is needed for a desired treatment efficiency, then 10 percent of the total cell mass is wasted from the system per day. In the complete-mix–recycle system, cell wastage is accomplished by wasting from the reactor or mixed-liquor tank. As shown in Eq. 10·35, by wasting cells directly from the reactor, only Q_w and V need to be known to determine θ_c. Wasting cells in this manner provides for a direct method of controlling and measuring θ_c.

In most biological treatment processes, cell wastage is accomplished by drawing off from the sludge recycle line. If this were done, Eq. 10·34 would become

$$\theta_c = \frac{VX}{Q'_w X_r + (Q - Q'_w) X_e} \tag{10·40}$$

where X_r = microorganism concentration in return sludge line
Q'_w = cell wastage rate from recycle line

Assuming that X_e is very small, Eq. 10·40 can be rewritten as

$$\theta_c \simeq \frac{VX}{Q'_w X_r} \tag{10·41}$$

Thus, wasting from the recycle line requires that both the mixed-liquor and return-sludge microorganism concentration be known.

For a complete-mix–recycle system, just as in the nonrecycle system, there is a minimum mean cell residence time $\theta_c{}^M$ below which waste stabilization cannot occur. The specific value of $\theta_c{}^M$ is a function of the waste concentration and the biological kinetic parameters Y, k, K_s, and k_d. Equations 10·30 and 10·31 are expressions that can be used to determine $\theta_c{}^M$, with or without recycle. Values of θ_c used in the design of biological process are based on the value of $\theta_c{}^M$ for the particular waste.

As was shown earlier, regardless of the location from which cells

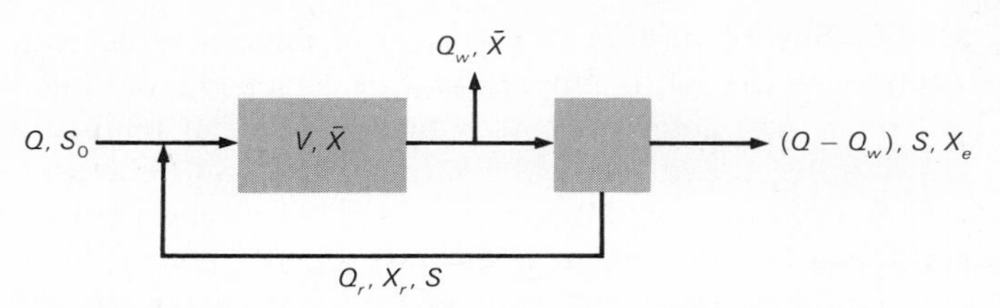

FIG. 10·15 Schematic of a plug-flow reactor with cellular recycle.

are wasted, θ_c is independent of θ. Practically speaking, however, in the successful operation of a wastewater treatment plant using biological processes, a minimum hydraulic retention time θ must be met before θ_c becomes a controlling parameter. The factors involved in selecting the proper θ for a biological treatment unit are discussed in Chap. 12. It is sufficient to note here that the most important factors tending to negate the complete independence between θ_c and θ are (1) the oxygen transfer rate in the reactor unit of an aerobic system, (2) the proper operation of the settling unit, and (3) the settling characteristics of suspended solids in the mixed liquor.

The kinetic model developed in the foregoing discussion can be used in the design and analysis of completely mixed aerobic and anaerobic processes, employing cellular recycle. It can also be used to show how parameters of the system vary under different design conditions. The final choice of design values for the system should be based on the results of the analysis with the model, in conjunction with economic and operational considerations.

Plug Flow–Cellular Recycle This system, shown schematically in Fig. 10·15, can be used to model certain forms of the activated-sludge process. The distinguishing feature of this recycle system is that the hydraulic regime of the reactor is of a plug-flow nature. In a true plug-flow model, all the particles entering the reactor will stay in the reactor an equal amount of time. Some of the particles may make more passes through the reactor because of recycle, but while the particles are in the tank, they all pass through in the same amount of time.

A kinetic model of the plug-flow system is mathematically difficult; however, Lawrence and McCarty [12] have made two simplifying assumptions that lead to a useful kinetic model of the plug-flow reactor:

1. The concentration of microorganisms in the influent to the reactor is approximately the same as in the reactor effluent. This assumption applies only if $\theta_c/\theta > 5$. The resulting average concentration of microorganisms in the reactor is symbolized as $\bar{X}$.

2. The rate of substrate utilization as the waste passes through the reactor is given by the following expression:

$$\frac{dS}{dt} = \frac{kS\bar{X}}{K_s + S} \tag{10·42}$$

Integrating Eq. 10·42 over the retention time of the waste in the tank, and after simplifying, Lawrence and McCarty [12] obtained an expression similar to the following:

$$\frac{1}{\theta_c} = \frac{Yk(S_0 - S)}{(S_0 - S) + K_s \ln (S_0/S)} - k_d \qquad \text{for } r < 1 \tag{10·43}$$

where r is the volumetric recycle ratio, and all other terms are as defined previously. Equation 10·43 is quite similar to Eq. 10·25, which applied to complete-mix systems, with or without recycle. The main difference in the two equations is that in Eq. 10·43 θ_c is also a function of the influent waste concentration S_0.

It should be noted that in Fig. 10·15 the excess microorganisms are wasted from the reactor and not from the recycle line. Thus, θ_c for the plug-flow–recycle system could also be defined as in Eq. 10·35 with the same assumptions applying. The average hydraulic retention time of the waste in the reactor θ, and the average hydraulic retention time of the waste in the plug-flow system θ_s, can be defined using Eqs. 10·32 and 10·33. The average microorganism concentration in the reactor of the plug-flow system can be obtained using Eq. 10·39 by noting that $\bar{X}$ must be substituted for X.

The true plug-flow–recycle system is theoretically more efficient in the stabilization of most soluble wastes than is the complete-mix–recycle system. This is shown graphically in Fig. 10·16. In actual

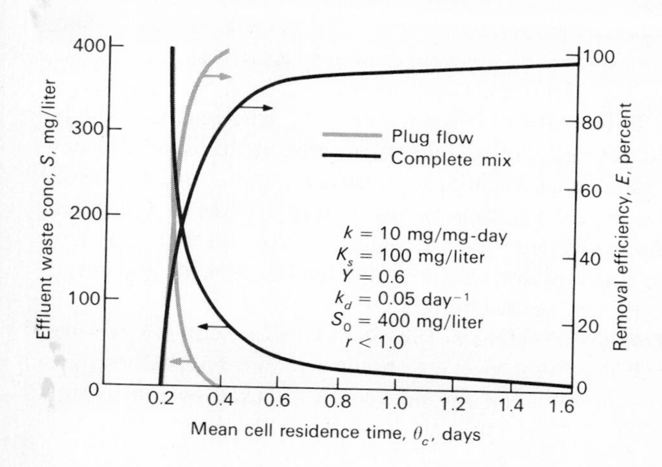

FIG. 10·16 Effluent waste concentration and removal efficiency for complete-mix and plug-flow reactors with recycle versus mean cell residence time [18].

practice, a true plug-flow regime is difficult to obtain because of longitudinal dispersion. This difficulty, plus the fact that the plug-flow system cannot handle shock loads as well as the complete-mix system, tends to reduce differences in treatment efficiency in the two models. By dividing the aeration tank into a series of complete-mix reactors, it has been shown that improvement in treatment performance can be obtained without a major loss in the ability of the system to handle shock loads [10]. Reactor selection is discussed further in Chap. 12.

BIOLOGICAL TREATMENT PROCESSES

Biological processes are classified by the oxygen dependence of the primary microorganisms responsible for waste treatment. In aerobic processes, waste stabilization is accomplished by aerobic and facultative microorganisms; in anaerobic processes, anaerobic and facultative microorganisms are used. Where aerobic, anaerobic, and facultative microorganisms are present, the processes are usually referred to as aerobic-anaerobic or facultative. In this section the aerobic, anaerobic, and aerobic-anaerobic biological treatment processes currently in use will be discussed briefly. Emphasis will be on the microbial ecology of the process; design aspects will be discussed in Chaps. 12 and 13.

Aerobic Waste Treatment

The aerobic processes considered are (1) activated sludge, (2) trickling filter, and (3) aerobic stabilization ponds. The activated-sludge process is used almost exclusively in large cities. Trickling filters are often used in smaller cities and for high-strength industrial wastes. Aerobic ponds are used for small cities where large land area is available.

Activated Sludge This process was developed in England in 1914 by Ardern and Lockett [1] and was so named because it involved the production of an activated mass of microorganisms capable of aerobically stabilizing a waste. Many versions of the original process are in use today, but fundamentally they are all similar. The system shown in Fig. 10·15 is a conventional activated-sludge system. Other activated-sludge systems are discussed in Chap. 12.

In the activated-sludge process, a waste, usually domestic sewage, is stabilized biologically in a reactor under aerobic conditions. The aerobic environment is achieved by the use of diffused or mechanical aeration. The reactor contents are referred to as the mixed liquor. After the waste is treated in the reactor, the resulting biological mass is separated from the liquid in a settling tank. A portion of the settled biological solids is recycled; the remaining mass is wasted. A portion of the microorganisms must be wasted; otherwise the mass of microorganisms would keep increasing until the system could no longer contain them. The level at which the biological mass should be kept depends on the desired treatment efficiency and other considerations related to growth kinetics. Microorganism concentrations maintained in various activated-sludge treatment systems are listed in Table 12·4 in Chap. 12.

To design and operate an activated-sludge system efficiently, it is necessary to understand the importance of the microorganisms in the system. In nature, the key role of the bacteria is to decompose organic matter produced by other living organisms. In the activated-sludge process, the bacteria are the most important microorganisms because they are responsible for the decomposition of the organic material in the influent. In the reactor, or mixed-liquor tank, a portion of the organic waste matter is used by aerobic and facultative bacteria to obtain energy for the synthesis of the remainder of the organic material into new cells, as shown in Fig. 10·3. Only a portion of the original waste is actually oxidized to low-energy compounds, such as NO_3, SO_4, and CO_2; the remainder is synthesized into cellular material. Also, many intermediate products are formed before the final end products of oxidation are obtained. Some of these intermediate products are shown in the right-hand side of Fig. 10·6.

While the bacteria are the microorganisms that actually degrade the organic waste in the influent, the metabolic activities of other microorganisms are also important in the activated-sludge system. For example, protozoa and rotifers act as effluent polishers. Protozoa consume dispersed bacteria that have not flocculated, and rotifers consume any small biological floc particles that have not settled.

Further, while it is important that bacteria decompose the organic waste as quickly as possible, it is also important that they form a satisfactory floc, which is a prerequisite for the effective separation of the biological solids in the settling unit. It has been observed that as the mean cell residence time is increased, the settling characteristics of the biological floc are enhanced. The reason for this is that, as the mean age of the cells increases, the surface charge is reduced and the microorganisms start to produce extracellular polymers, eventually becoming encapsulated in a slime layer (see Fig. 10·1). The presence of these polymers and the slime promotes the formation of floc particles that can be removed readily by gravity settling. For domestic wastes, mean cell residence times on the order of 3 to 4 days are required to achieve effective settling. Typical values of mean cell residence times used in the design and operation of various activated sludge processes are shown in Table 12·4 in Chap. 12.

Even though excellent floc formation is obtained, the effluent from the system could still be high in biological solids as a result of poor design of the secondary settling unit, poor operation of the aeration units, or because of the presence of filamentous microorganisms, such as *Sphaerotilus*, *E. coli* [23], and fungi. These subjects are discussed in detail in Chap. 12.

The foregoing discussion has been limited to the aerobic biological degradation of organic material. While this is the primary concern in the treatment of a domestic waste such as sewage, it is also often desirable to stabilize those inorganic compounds that have a BOD. The most important inorganic compound is ammonia, because its presence in the plant effluent can stimulate the lowering of the dissolved oxygen in the receiving stream through the biological process of nitrification. In nitrification, as was explained in Chap. 7, ammonia is biologically oxidized to nitrite. The nitrite is then oxidized by another group of microorganisms to nitrate. Nitrate is the final oxidation state of the nitrogen compounds, and as such represents a stabilized product.

If nitrification is to be accomplished in an activated-sludge system and thereby eliminated from the receiving stream, certain operational adjustments must be made beyond those necessary for the stabilization of the organic matter. First, additional oxygen must be provided for the nitrification process. Second, a longer mean cell residence time must be used. The bacteria that are responsible for this oxidation are strict autotrophs, and as such they are distinctly different from the heterotrophic bacteria that are responsible for the degradation of the organic matter. Nitrifying bacteria have a growth rate that is much slower than that of the heterotrophic bacteria, and they require a longer mean cell residence time to be effective. Jenkins and Garrison [10] found that, for a domestic wastewater treated by the activated-sludge process at a temperature of 21 to 22°C, a mean cell residence time of at least 10 days was needed to ensure nitrification. This subject is discussed further in Chap. 14.

Temperature dependence of the biological reaction-rate constant is very important in assessing the overall efficiency of a biological treatment process. Temperature not only influences the metabolic activities of the microbiological population, but also has a profound effect on such factors as gas transfer rates and the settling characteristics of the biological solids. The temperature effect on the reaction rate of a biological process is usually expressed in the following form:

$$\frac{K_T}{K_{20}} = \theta^{(T-20)} \tag{10·44}$$

where K_T = reaction rate at T°C
K_{20} = reaction rate at 20°C
θ = temperature-activity coefficient
T = temperature, °C

Values of θ for certain biological processes, not to be confused with values for the BOD determination, are given in Table 10·7.

TABLE 10·7 TEMPERATURE–ACTIVITY COEFFICIENTS FOR VARIOUS BIOLOGICAL PROCESSES [4]

Process	θ
Activated sludge	1.0–1.03
Trickling filters	1.02–1.04
Aerated lagoons	1.06–1.09

Trickling Filter The first trickling filter was placed in operation in England in 1893. The concept of a trickling filter grew from the use of contact filters, which were watertight basins filled with broken stones. In operation, the contact bed was filled with sewage from the top and the sewage was allowed to contact the media for a short time. The bed was then drained and allowed to rest before the cycle was repeated. A typical cycle required 12 hr with 6 hr being the time of resting. The limitations of the contact filter included a relatively high incidence of clogging, the long rest period required, and the relatively low loading that could be used.

The trickling filter is designed to overcome these limitations. The sewage is sprinkled over the stones and allowed to trickle through the bed. The trickling filter consists of a bed of highly permeable media to which microorganisms are attached and through which a liquid waste is percolated. The filter media usually consist of rocks, varying in size from 1 to 4 in. in diameter. The depth of the rock varies with each particular design, usually from 3 to 8 ft; an average depth is 6 ft. Trickling filters employing a plastic media have been build with depths of 30 to 40 ft. The filter bed is now usually circular, and the liquid waste is distributed over the top of the bed by a rotary distributor. Formerly, the bed was rectangular, and the sewage was applied through fixed spray nozzles. Each filter has an underdrain system for collecting the treated waste and any biological solids that have become detached from the media. This underdrain system is important both as a collection unit and as a porous structure through which air can circulate (see Fig. 10·17).

The organic material present in the wastewater is degraded by a population of microorganisms attached to the filter media. Organic material from the liquid is adsorbed on to the biological film or slime layer. In the outer portions of the biological slime layer (see Fig. 10·18), the organic material is degraded by aerobic microorganisms. As the microorganisms grow, the thickness of the slime layer increases and the diffused oxygen is consumed before it can penetrate the full depth of the slime layer. Thus an anaerobic environment is established near the surface of the media.

As the slime layer increases in thickness, the adsorbed organic matter is metabolized before it can reach the microorganisms near the media face. As a result of having no external organic source available for cell carbon, the microorganisms near the media face enter into an endogenous phase of growth. In this phase, the microorganisms lose their ability to cling to the media surface. The liquid then washes the slime off the media, and a new slime layer will start to grow. This phenomenon of losing the slime layer is called sloughing and is primarily a function of the organic and hydraulic loading on the filter. The hydraulic loading accounts for shear velocities, and the organic loading accounts for the rate of metabolism in the slime layer. Based on the hydraulic and organic loading rates, filters are usually divided into

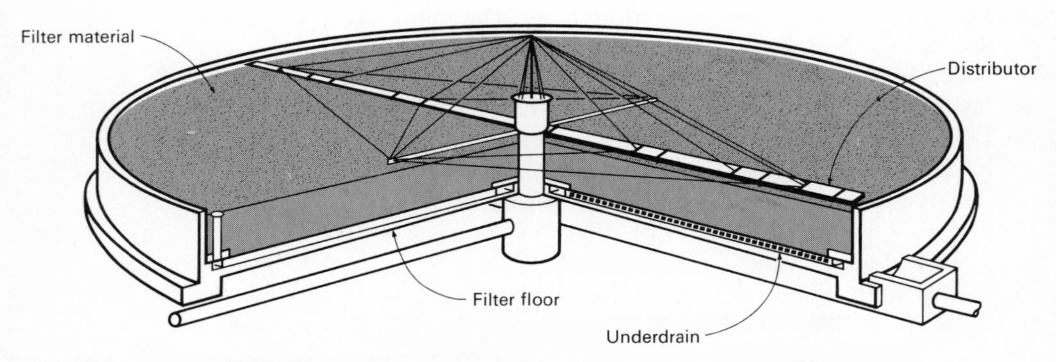

FIG. 10·17 Cutaway view of a trickling filter [from Dorr-Oliver].

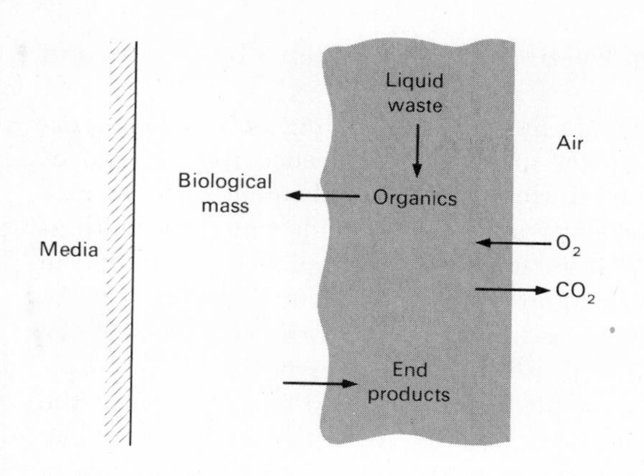

FIG. 10·18 Schematic representation of the cross section of a biological slime in a trickling filter.

two classes: low-rate and high-rate. Their relative merits are discussed in Chap. 12.

The biological community in the filter consists primarily of protists, including aerobic, anaerobic, and facultative bacteria, fungi, algae, and protozoans. Higher animals, such as worms, insect larvae, and snails, are also present.

Facultative bacteria are the predominating microorganisms in the trickling filter, and along with the aerobic and anaerobic bacteria, their role is to decompose the organic material in the wastewater. The fungi present are also responsible for waste stabilization, but their contribution is usually important only under low pH conditions or with certain industrial wastes. Algae can grow only in the upper reaches of the filter where sunlight is available. In general, algae do not take a direct part in waste degradation, but during the daylight hours they add oxygen to the percolating wastewater. From an operational standpoint, the algae are troublesome because they can cause clogging of the filter surface.

The protozoa in the filter are predominantly of the *Ciliata* group. As in the activated-sludge process, their function is not to stabilize the waste but to control the bacterial population. The higher animals, such as snails, worms, and insects, feed on the biological films in the filter and as a result help to keep the bacterial population in a state of high growth or rapid food utilization.

Variations in the individual populations of the biological community described above will occur throughout the filter depth with changes in organic loading, hydraulic loading, influent wastewater composition, pH, temperature, air availability, and other factors. The variation of the efficiency of waste removal with temperature, in the trickling-filter process, can be estimated with the use of Eq. 10·44 and Table 10·7.

As in the activated-sludge process, the settling unit is an important part of the trickling-filter process. It is needed for removal of suspended solids sloughed off during periods of unloading with low-rate filters and for removals of lesser amounts of solids sloughed continuously by high-rate filters. If recirculation is employed, some of the settled solids may be recycled and some wasted, but the recycle of the settled biological solids is not as important as in the activated-sludge process. In the trickling-filter process, the majority of the active microorganisms are attached to the filter media and do not pass out of the reactor as in the activated-sludge process. Although recirculation can help in seeding the filter, the primary purposes of recirculation are to dilute strong influent wastes and to bring the filter effluent back in contact with the biological population for further treatment. Recirculation is almost always included in high-rate trickling-filter systems.

Because of the unstable characteristics of the biological slime layer and the unpredictable hydraulic characteristics, a kinetic model of the trickling filter is very difficult to develop. Studies have been conducted on the biological kinetics of fixed-film reactors, but their application to the design of full-scale systems has not yet been demonstrated. Current trickling-filter design concepts are discussed in Chap. 12.

Other Processes Other aerobic treatment processes include aerated lagoons and photosynthetic and mechanically aerated stabilization ponds (see Chap. 12). In aerated lagoons, mechanical aeration is used to supply oxygen to the bacteria and the process is essentially the same as the activated-sludge process without recycle. In aerobic photosynthetic ponds, the oxygen is supplied by natural surface aeration and by algal photosynthesis. Except for the algal population, the biological community present in stabilization ponds is similar to that present in an activated-sludge system. The oxygen released by the algae through the process of photosynthesis is utilized by the bacteria in the aerobic degradation of organic matter. The nutrients and carbon dioxide released in this degradation are, in turn, used by the algae. This cyclic-symbiotic relationship is shown in Fig. 10·19. Higher animals, such as rotifers and protozoa, will also be present in the pond, and their main function is to polish the effluent.

Because of the presence of aeration units, algae do not have as much significance in the mechanically aerated stabilization ponds. The aeration units also serve to mix the contents of the pond and prevent

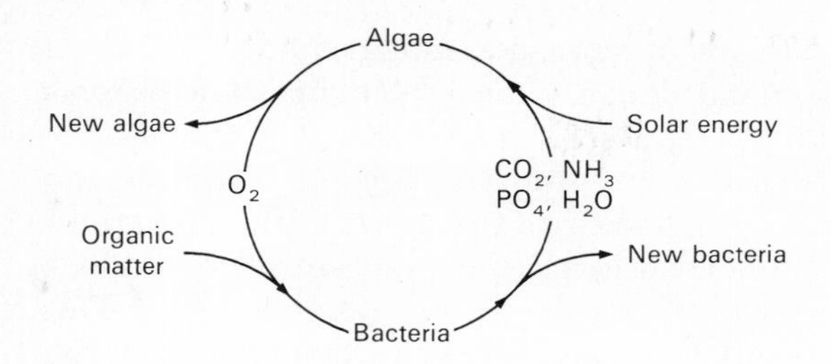

FIG. 10·19 Schematic representation of the symbiotic relationship between algae and bacteria.

the sedimentation of suspended solids. If the solids were allowed to settle out, an anaerobic sludge layer would accumulate on the bottom, and the pond would become an aerobic-anaerobic (facultative) pond.

The particular algal group, animal group, or bacterial species present in any section of an aerobic pond will depend on such factors as organic loading, degree of pond mixing, pH, nutrients, sunlight, and temperature. Temperature has a profound effect on the operation of aerobic ponds, particularly in regions having cold winters, where it is possible for ice to cover the pond surface. The water temperature under the ice cover is only slightly above freezing, and as a result the metabolic activities of the microorganisms are greatly reduced. The ice cover also greatly reduces photosynthetic activities and presents operational difficulties for the aeration units. As a result, anaerobic conditions will soon prevail throughout the pond. In the spring when the ice melts and better aeration and mixing are obtained in the pond, odorous anaerobic end products, such as hydrogen sulfide, are released to the atmosphere. Thus, for a short time during the spring, odor problems may be encountered.

Anaerobic Waste Treatment

Anaerobic waste treatment involves the decomposition of organic and/or inorganic matter in the absence of molecular oxygen. The major application is in the digestion of concentrated sewage sludges and in the treatment of some industrial wastes; however, with the introduction of the anaerobic contact process and the anaerobic filter it has been shown that dilute organic wastes can be treated anaerobically. Another application of anaerobic treatment is in anaerobic lagoons or ponds.

The usual mode of operation of an anaerobic waste treatment unit receiving a concentrated sewage sludge is by use of a complete-mix reactor system with minimum cellular recycle for the purpose of heating and mixing the tank contents. The detention time of the liquid in the reactor is from 10 to 30 days, or more, depending on how the system is operated. The operation and design of anaerobic digestion systems are discussed in Chap. 13.

The microorganisms responsible for the decomposition of the organic matter are commonly divided into two groups. The first group hydrolyzes and ferments complex organic compounds to simple organic acids, the most common of which are acetic and propionic acid. This group of microorganisms consists of facultative and anaerobic bacteria, collectively called the acid formers.

The second group converts the organic acids formed by the first group to methane gas and carbon dioxide. The bacteria responsible for this conversion are strict anaerobes and are called the methane formers. The most important bacteria of this group are the ones that degrade acetic acid and propionic acid. They have very slow growth rates; and, as a result, their metabolism is usually considered rate limiting in the anaerobic treatment of an organic waste. It is in this second step that actual waste stabilization is accomplished by the conversion of the organic acids into methane and carbon dioxide. Methane gas is highly insoluble, and its departure from solution represents actual waste stabilization.

Many other groups of anaerobic and facultative bacteria utilize the various inorganic ions present in the sludge. *Desulfovibrio* is responsible for the reduction of the sulfate ion $SO_4^=$ to the sulfide ion S^{--}. Other bacteria reduce nitrates NO_3^- to nitrogen gas N_2 (denitrification).

To maintain an anaerobic treatment system that will stabilize an organic waste efficiently, the acid formers and the methane formers must be in a state of dynamic equilibrium. To establish and maintain such a state, the reactor contents should be void of dissolved oxygen and free from inhibitory concentrations of such constituents as heavy metals and sulfides, and the aqueous environment should be in the pH range of 6.6 to 7.6. Sufficient alkalinity should be present to ensure that the pH of the system will not drop below 6.2, because the methane formers cannot function below this point. When digestion is proceeding satisfactorily, the alkalinity will normally range from 1,000 to 5,000 mg/liter and the volatile acids will be less than 250 mg/liter. A sufficient amount of nutrients, such as nitrogen and phosphorus, must also be available to ensure the proper growth of the biological community. Temperature is another important environmental parameter that must be considered. The optimum temperature ranges are the mesophilic (85 to 100°F) and the thermophilic (120 to 135°F). These environmental conditions are summarized in Table 10·8.

The advantages and disadvantages of the anaerobic treatment of

TABLE 10·8 OPTIMUM CONDITIONS FOR ANAEROBIC WASTE TREATMENT [16]

Optimum temperatures:
Mesophilic range, 85 to 100°F
Thermophilic range, 120 to 135°F
Anaerobic conditions
Sufficient biological nutrients:
Nitrogen
Phosphorus
Others
Optimum pH, 6.6 to 7.6
Absence of toxic materials

an organic waste, as compared to aerobic treatment, stem directly from the slow growth rate of the methane formers. Of particular interest are those methane formers responsible for the fermentation of acetic and propionic acids. That these methane formers have very slow growth rates can be seen by comparing their minimum mean cell residence times, $\theta_c{}^M$, of 3 to 10 days (listed in Table 10·6), with those of about 6 to 10 hr for aerobic bacteria. It can also be seen from Tables 10·5 and 10·6 that the growth-yield coefficient Y is much lower for the anaerobic bacteria and particularly for the methane formers (Y of 0.6 for aerobic bacteria and Y of 0.04 to 0.054 for anaerobic bacteria). The low growth yield signifies that only a small portion of the degradable organic waste is being synthesized into new cells. With the methane-forming bacteria, most of the organic waste is converted to methane gas, which is combustible and therefore a useful end product. If sufficient quantities are produced, as is customary with municipal sewage sludge, it can be used to run gas engines or burned to heat the digesting sludge.

Because of the low production of microorganisms in anaerobic waste treatment, there is a low requirement for biological nutrients, such as nitrogen and phosphorus, and there are fewer microorganisms to waste, when compared with aerobic processes such as activated sludge. The slow growth rate of the methane bacteria also limits the anaerobic treatment of an organic waste, in that these microorganisms react slowly to changing environmental conditions. Thus relatively long periods of time are required to establish a balanced system.

Because of the low cellular growth rate and the conversion of organic matter to methane gas and carbon dioxide, the resulting solid matter is well stabilized and frequently suitable for disposal, after drying or dewatering, in dumps or on land as a soil conditioner or humuslike material. On the other hand, the sludge solids resulting from aerobic processes must either be digested, usually anaerobically, or dewatered and incinerated, on account of the large proportion of cellular organic material. A small amount, proportionately, is heat dried and sold as fertilizer.

The high temperatures necessary to achieve adequate treatment are often listed as disadvantages of the anaerobic treatment process; however, high temperatures are necessary only when sufficiently long mean cell residence times cannot be obtained at nominal temperatures. In the anaerobic treatment system shown in Fig. 10·20, the mean cell residence time of the microorganisms in the reactor is equivalent to the hydraulic detention time of the liquid in the reactor. As the operation temperature is increased, the minimum mean cell residence time is reduced significantly. This means that at higher temperatures the system can be operated at a lower mean cell residence time. Thus, heating of the reactor contents lowers not only the mean cell residence time necessary to achieve adequate treatment but also the hydraulic detention time, and a smaller reactor volume can be used.

If the reactor volume could be made large enough, effective treatment could be obtained at nominal temperatures, but such a size would be impractical.

The anaerobic contact and anaerobic filter processes are discussed in Chap. 13. Anaerobic lagoons may be sludge lagoons as described in Chap. 13 or heavily loaded ponds. In the latter case, the organic loading applied to the pond is so high that not enough oxygen can be supplied by natural surface aeration and the entire pond becomes anaerobic. In such ponds, the facultative and anaerobic bacteria are the predominant microorganisms.

FIG. 10·20 Schematic of a complete-mix anaerobic digester.

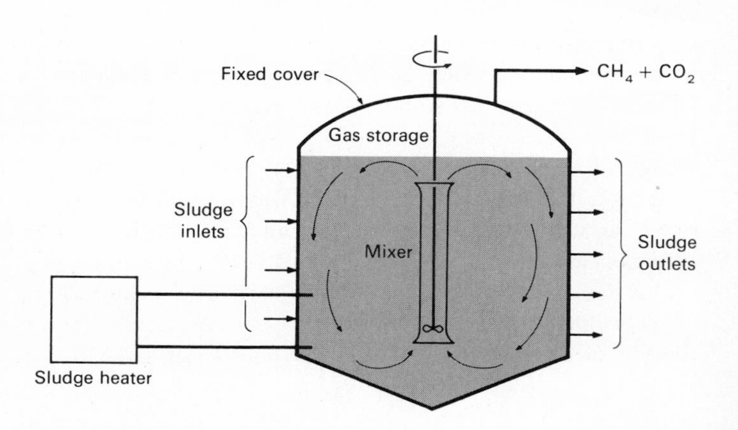

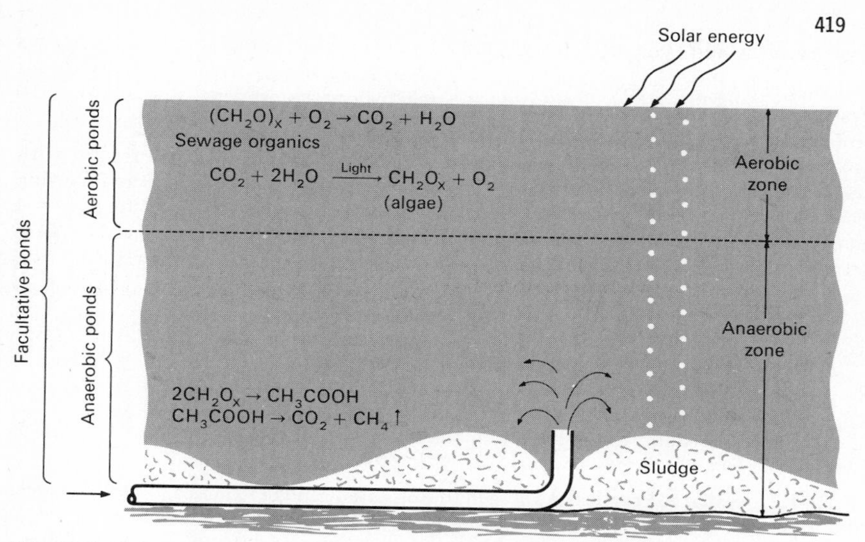

FIG. 10·21 Schematic representation of a waste stabilization pond [adapted from 6].

Aerobic-Anaerobic Waste Treatment

Ponds in which the stabilization of wastes is brought about by a combination of aerobic, anaerobic, and facultative bacteria are known as aerobic-anaerobic stabilization ponds. As shown in Fig. 10·21, such ponds have an aerobic upper layer and an anaerobic bottom layer. In practice, oxygen is maintained in the upper layer by the presence of algae or by the use of surface aerators (see Chap. 12). Where surface aerators are used, algae are not required. The biological community in the upper or aerobic layer is similar to that of an aerobic pond, while the microorganisms in the bottom layer of the pond are facultative and anaerobic bacteria. The metabolic activities of these bacteria were discussed earlier in the anaerobic treatment section.

PROBLEMS

10·1 Assuming that second-order kinetics apply ($dC/dt = -kC^2$), compare the volume of a complete-mix reactor to the volume of a plug-flow reactor in which $C_e = 0.1$, $C_0 = 1$, and $Q = 1$. (See also "Selection of Reactor Type" in Chap. 12.)

10·2 A waste is to be treated aerobically in a complete-mix reactor with no recycle. Determine $\theta_c{}^M$ using the following constants: $K_s = 50$ mg/liter; $k = 5.0$ days^{-1}; $k_d = 0.06$ days^{-1}; and $Y = 0.60$. The initial waste concentration is 200 mg/liter.

What value of θ would you set for design?

10·3 Using a design value of $\theta_c = 2$ days and the constants in Prob. 10·2, calculate the effluent substrate concentration. What is the substrate removal efficiency? Determine the food-to-microorganism ratio and the mass concentration of microorganisms in the reactor.

10·4 Discuss the advantages and disadvantages of wasting sludge from the mixed-liquor reactor versus wasting sludge from the recycle line.

10·5 Determine θ_c for a plug-flow reactor with a recycle ratio of 0.25 using the following constants: $K_s = 100$ mg/liter; $k = 4$ days^{-1}; $k_d = 0.03$ days^{-1}; $Y = 0.6$; $S_0 = 250$ mg/liter; and $S = 10$ mg/liter.

REFERENCES

1. Ardern, E., and W. T. Lockett: Experiments on the Oxidation of Sewage without the Aid of Filters, *J. Society of Chem. Ind.*, vol. 33, pp. 523, 1122, 1914.
2. Beneder, P., and I. Horvath: A Practical Approach to Activated Sludge Kinetics, *Water Research*, vol. 1, no. 10, 1967.
3. Brock, T. D.: *Biology of Microorganisms*, Prentice-Hall, Englewood Cliffs, N.J., 1970.
4. Eckenfelder, W. W., Jr.: *Industrial Water Pollution Control,* McGraw-Hill, New York, 1966.
5. Gates, W. E., *et al.:* A Rational Model for the Anaerobic Contact Process, *J. WPCF*, vol. 39, no. 12, 1967.
6. Gloyna, E. F., and W. W. Eckenfelder, Jr. (eds.): *Advances in Water Quality Improvement*, University of Texas Press, Austin, 1968.
7. Gram, A. L.: *Reaction Kinetics of Aerobic Biological Processes*, I.E.R. Series 90, Report 2, Sanitary Engineering Research Laboratory, University of California, Berkeley, 1956.
8. Heukelekian, H., H. E. Orford, and R. Manganelli: Factors Affecting the Quantity of Sludge Production in the Activated Sludge Process, *Sewage and Industrial Wastes,* vol. 23, no. 8, 1951.
9. Hoover, S. R., and N. Porges: Assimilation of Dairy Wastes by Activated Sludge, *Sewage and Industrial Wastes*, vol. 24, p. 306, 1952.
10. Jenkins, D., and W. E. Garrison: Control of Activated Sludge by Mean Cell Residence Time, *J. WPCF*, vol. 40, no. 11, part 1, 1968.
11. Kimball, J. W.: *Biology*, 2d ed., Addison-Wesley, Reading, Mass., 1966.
12. Lawrence, A. W., and P. L. McCarty: A Unified Basis for Biological Treatment Design and Operation, *Journal of the Sanitary Division,* ASCE, vol. 96, no. SA3, 1970.
13. Lawrence, A. W., and P. L. McCarty: Kinetics of Methane Fermentation in Anaerobic Treatment, *J. WPCF*, vol. 41, no. 2, part 2, 1969.
14. Levenspiel, O.: *Chemical Reaction Engineering,* Wiley, New York, 1962.
15. McCarty, P. L.: Anaerobic Treatment of Soluble Wastes, in E. F. Gloyna and W. W. Eckenfelder (eds.), *Advances in Water Quality Improvement*, University of Texas Press, Austin, 1968.
16. McCarty, P. L.: Anaerobic Waste Treatment Fundamentals, *Public Works*, vol. 95, nos. 9–12, 1964.
17. McCarty, P. L.: Energetics and Bacterial Growth, Fifth Rudolf Research Conference, Rutgers University, New Brunswick, N.J., 1969.

18. McCarty, P. L.: Kinetics of Waste Assimilation in Anaerobic Treatment, *Developments in Industrial Microbiology*, vol. 7, American Institute of Biological Sciences, Washington, D.C., 1966.

19. McGauhey, P. H.: *Engineering Management of Water Quality*, McGraw-Hill, New York, 1968.

20. McKinney, R. E.: *Microbiology for Sanitary Engineers*, McGraw-Hill, New York, 1962.

21. Monod, J.: The Growth of Bacterial Cultures, *Annual Review of Microbiology*, vol. III, 1949.

22. O'Rourke, J. T.: "Kinetics of Anaerobic Treatment at Reduced Temperatures," Ph.D. dissertation, Stanford University, Stanford, Calif., 1968.

23. Pasveer, A.: A Case of Filamentous Activated Sludge, *J. WPCF*, vol. 41, no. 7, 1969.

24. Pearson, E. A.: Kinetics of Biological Treatment, in E. F. Gloyna and W. W. Eckenfelder, Jr. (eds.), *Advances in Water Quality Improvement*, University of Texas Press, Austin, 1968.

25. Pelczar, M. J., Jr., and R. D. Reid: *Microbiology*, 2d ed., McGraw-Hill, New York, 1965.

26. Ribbons, D. W.: Quantitative Relationships Between Growth Media Constituents and Cellular Yields and Composition, in J. W. Norris and D. W. Ribbons (eds.), *Methods in Microbiology*, vol. 3A, Academic Press, London, 1970.

27. Sherrard, J. H.: Personal Communication, Davis, California, August 1971.

28. Smith, G. M.: *The Fresh-Water Algae of the United States*, 2d ed., McGraw-Hill, New York, 1950.

29. Stanier, R. Y., M. Doudoroff, and E. A. Adelberg: *The Microbial World*, 3d ed., Prentice-Hall, Englewood Cliffs, N.J., 1970.

30. Van Uden, N.: Transport-Limited Growth in the Chemostat and Its Competitive Inhibition; A Theoretical Treatment, *Archiv für Mikrobiologie*, vol. 58, 1967.

31. Weston, R. F., and W. W. Eckenfelder, Jr.: Application of Biological Treatment to Industrial Waste: I. Kinetics and Equilibria of Oxidative Treatment, *Sewage and Industrial Wastes*, vol. 27, no. 7, 1955.

design of facilities for physical and chemical treatment of wastewater

11

The purpose of this chapter is to discuss the design of the unit operations and processes used in the preliminary (also called primary) treatment of wastewater. The principal unit operations and processes and their functions as applied to the preliminary treatment of wastewater are reported in Table 11·1. As shown, physical operations are used for the removal of coarse solids, suspended and floating solids, and grease and for the pumping of sludge. Chemical processes result in the removal of suspended and colloidal solids by precipitation, disinfection of the wastewater, and control of odors. Ordinarily, chlorination is the only chemical process used in preliminary treatment. Chlorination is also used following biological treatment. Details on the design of these unit operations and processes are described in separate sections in the remainder of this chapter. Because of its importance, the processing of sludge produced from preliminary treatment operations and processes as well as that produced by biological processes is discussed in Chap. 13.

The degree of treatment afforded a wastewater depends mainly on the discharge requirements for the effluent. Requirements of settleable-solids removal and disinfection can be met by conventional preliminary treatment. A flowsheet for a plant designed to provide preliminary treatment is shown in Fig. 11·1. Require-

TABLE 11·1 UNIT OPERATIONS AND PROCESSES USED IN THE PRELIMINARY TREATMENT OF WASTEWATER

Operation or process	Function
Racks and screens	Removal by interception of coarse sewage solids
Comminutors and grinders	Grinding of sewage solids
Grit chambers	Removal of grit, sand, and gravel
Skimming and grease traps	Removal of lighter floating solids including grease, soap, cork, wood, vegetable debris, etc.
Preaeration	Improvement of hydraulic distribution, replenishment of dissolved oxygen
Flocculation	Improvement of settling of suspended solids
Sedimentation	Removal of settleable solids and floating material
Flotation	Removal of finely divided suspended solids and grease
Chemical precipitation	Removal of settleable and colloidal solids and phosphorus. First step in the complete chemical treatment of wastewater.
Sludge pumping	Removal of sludge from the bottom of sedimentation tanks. Pumping of sludge between various operations and processes.
Chlorination	Odor control, oxidation, disinfection, etc.

ments for 80 to 90 percent removal of BOD and suspended solids cannot be met by conventional preliminary treatment; thus biological treatment or other means, such as chemical precipitation followed by sand filtration and activated-carbon adsorption, must be used. Requirements for removal of nitrogen and phosphorus may be met by modifications to biological treatment processes, through chemical treatment, or through the use of advanced waste treatment operations and processes (see Chap. 14).

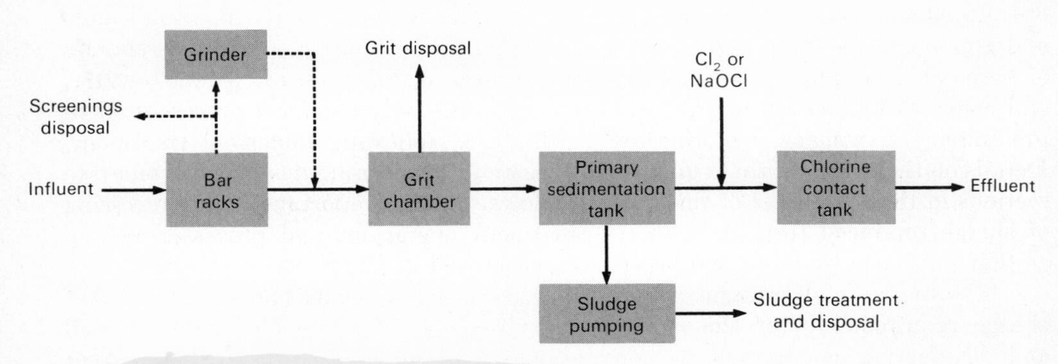

FIG. 11·1 Typical flowsheet of a preliminary treatment plant.

RACKS, FINE SCREENS, AND COMMINUTORS

The first step in the preliminary treatment of wastewater involves the removal of the coarse solids. As shown in Fig. 11·1 the usual procedure is to pass the influent wastewater through racks or fine screens. Alternatively, comminutors, which grind up the coarse solids without removing them from the flow, may be used.

Racks

Bar racks may be hand cleaned or mechanically cleaned. Characteristics of the two types are compared in Table 11·2. Details on each type

TABLE 11·2 CHARACTERISTICS OF BAR RACKS

Item	Hand-cleaned	Mechanically cleaned
Bar size:		
Width, in.	$\frac{1}{4}$–$\frac{5}{8}$	$\frac{1}{4}$–$\frac{5}{8}$
Depth, in.	1–3	1–3
Spacing, in.	1–2	$\frac{5}{8}$–3
Slope from vertical, °	30–45	0–30
Approach velocity, fps	1–2	2–3
Allowable head loss, in.	6	6

of rack are discussed below along with some of the factors that must be considered in the design of rack installations.

Hand-Cleaned Bar Racks Units of this type are frequently used in small wastewater pumping stations ahead of the pumps. In the past, they have been used at the head works to small sewage treatment plants. The tendency during recent years has been to provide mechanically cleaned racks, or comminutors, even for small installations, not only to minimize the amount of manual labor involved for cleaning the racks and for removal and disposal of the rakings, but also to reduce flooding and overflows due to clogging.

The length of the hand-cleaned rack should not exceed what can be conveniently raked by hand. The rack bars are usually not less than $\frac{3}{8}$ in. thick by 2 in. deep. They are welded to spacing bars located at the rear face, out of the way of the tines of the rake. A perforated drainage plate should be provided at the top of the rack where the rakings may be stored temporarily for drainage. A typical hand-cleaned rack is shown in Fig. 11·2.

The rack channel should be designed to prevent the accumulation of

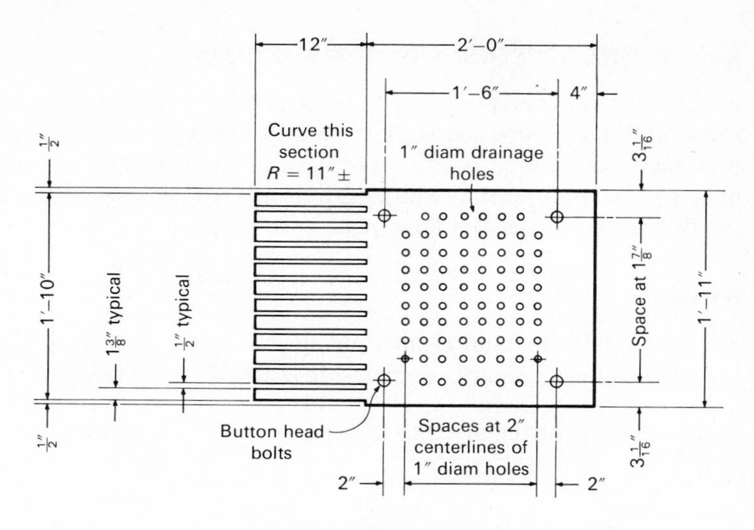

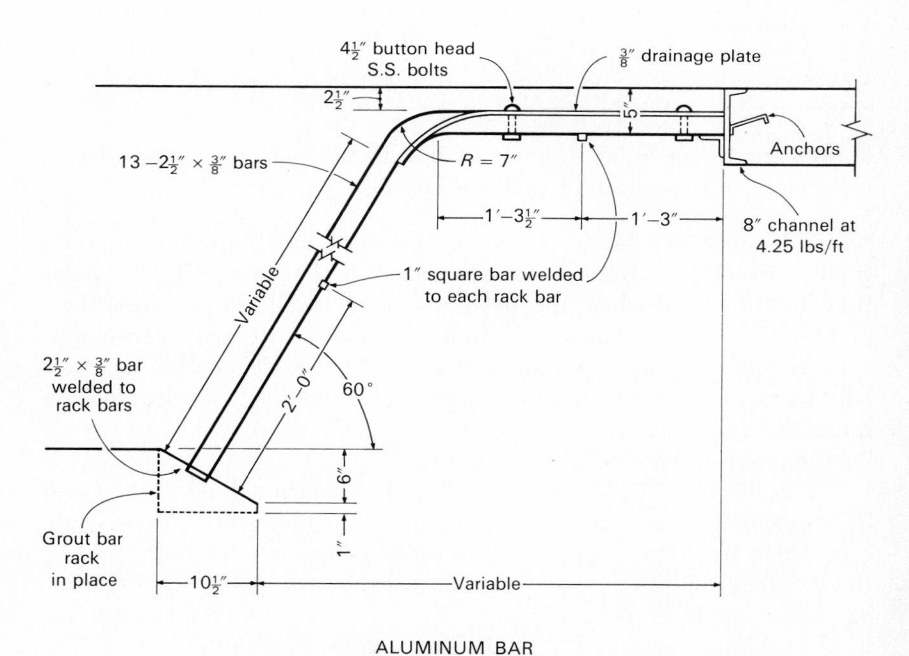

FIG. 11·2 Typical hand-cleaned bar rack.

grit and other heavy materials in the channel ahead of, and following, the rack. The channel floor should be level or should slope downward through the screen without pockets to trap solids. Fillets may be desirable at the base of the sidewalls. The channel preferably should have a straight approach, perpendicular to the bar rack, to promote uniform distribution of screenable solids throughout the flow and on the rack.

To provide adequate rack area for accumulation of screenings between raking operations, it is essential that the velocity of approach be limited to approximately 1.5 fps at average flow. Additional area to limit the velocity may be obtained by widening the channel at the rack and by placing the rack on a flatter slope. As screenings accumulate, partially plugging the rack, the head will increase, submerging new areas for the flow to pass through. The structural design of the screen should be adequate to prevent collapse if it becomes completely plugged.

Mechanically Cleaned Bar Racks Mechanically cleaned bar racks, such as the one shown in Fig. 11·3, may be purchased from a number of manufacturers. The design engineer determines in advance the type of equipment to be used, the dimensions of the rack channel, the range of depth of flow in the channel, the clear spacing between bars, and the method of control of the rack. Mechanically cleaned bar racks may be obtained either front-cleaned or back-cleaned. Each type has advantages and disadvantages. A front-cleaned model is shown in Fig. 11·3. A back-cleaned unit is shown in Chap. 8 in Fig. 8·2.

In the front-cleaned model, the mechanism is entirely in front of the rack. If solids settle out at the foot of the rack, the rakes must pull down through them at the bottom of their travel before they can start up the rack. There have been cases where obstructions have jammed the mechanism, putting it out of operation until the obstruction could be removed. Fortunately, such occurrences are infrequent.

The back-cleaned rack was developed to eliminate jamming due to obstructions at the base of the screen and has successfully overcome this problem at the expense of introducing two more. The rakes travel downward in back of the screen, free from obstructions. In one manufacturer's model they enter between the bars and project through them from behind; in another, they pass under the rack in a boot before starting up the front face of the rack (Fig. 8·2). If there are large solids at the base of the rack, the rakes enter underneath them, pulling them upward and lifting them up the rack or rolling them out of the way without jamming. In the first design mentioned above the bars are fixed at the bottom but are supported by the traveling rake teeth above. This allows the bars to move, and where round bars are used,

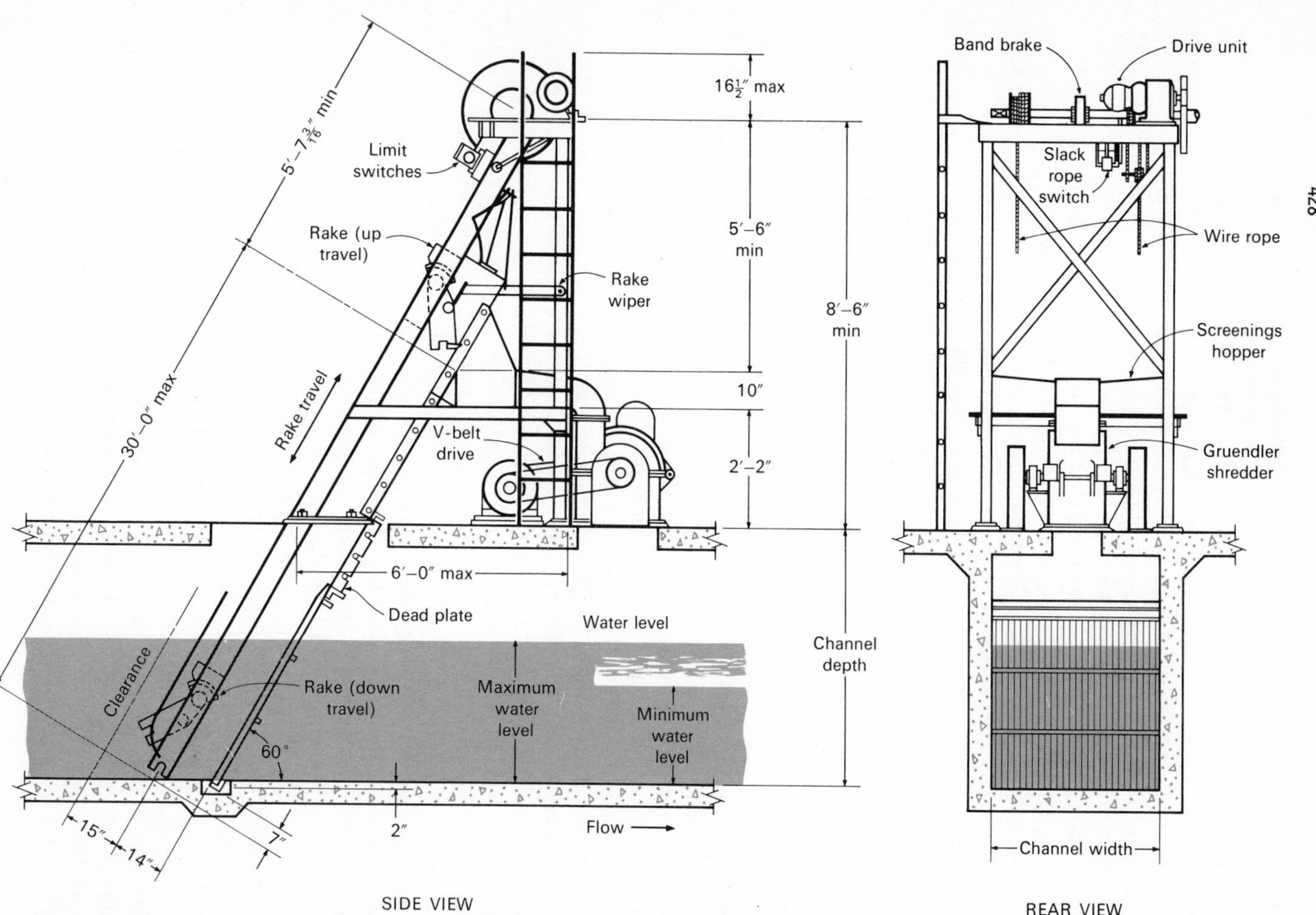

FIG. 11·3 Cross section and end views of a mechanically cleaned bar rack [from Link-Belt].

solids measuring up to two times the nominal clear spacing between the bars may pass through the rack.

Another problem with the back-cleaned racks is that, because the raking mechanism returns behind the rack, any solids remaining on the rakes return to the sewage behind the rack where they are washed off into the screened flow. This problem usually can be minimized with proper adjustment of rake wiper blades.

Design of Rack Installations Two or more units should be installed so that one unit may be taken out of service for maintenance. Stop-log grooves should be provided ahead of, and behind, each rack so that the unit can be dewatered for painting, chain or cable replacement, replacement of teeth, removal of obstructions, straightening of bent bars, etc. If only one unit is installed it is absolutely essential that a bypass channel with a manually cleaned rack be provided for emergency use. Flow through the bypass channel normally would be prevented by stop logs or a closed sluice gate.

The rack channel should be designed to prevent the settling and accumulation of grit and other heavy materials. The majority of racks utilize endless chains operating over sprockets to move the rakes. They are normally provided with "hand"-"off"-"automatic" controls. On "hand" position the rakes operate continuously. On "automatic" position they may be operated when the differential head loss increases above a certain minimum value or by a time clock. Operation by a time clock for a period, adjustable by the operator, out of a 15-min cycle is recommended with either a high-water or high-differential contact that will place the rack in continuous operation when needed.

Sheet-metal enclosures with access doors are available for the head works of the screens above the operating flow level. They neatly enclose the mechanism and the screenings hopper, but they are a nuisance to the operator and are frequently omitted in below-ground screen chambers and areas not open to the general public. Enclosures should always be provided for racks located outdoors.

Fine Screens

Fine screens, once frequently used for preliminary or primary treatment, are mechanically cleaned devices. Generally, they are equipped with a perforated bronze plate with slotted openings $\frac{1}{8}$ in. wide or less. At present, comparatively few treatment plants use such fine screens. For a description of these units, including more information on the quantity and character of screenings and data on removal efficiencies, the reader is referred to Metcalf and Eddy [10].

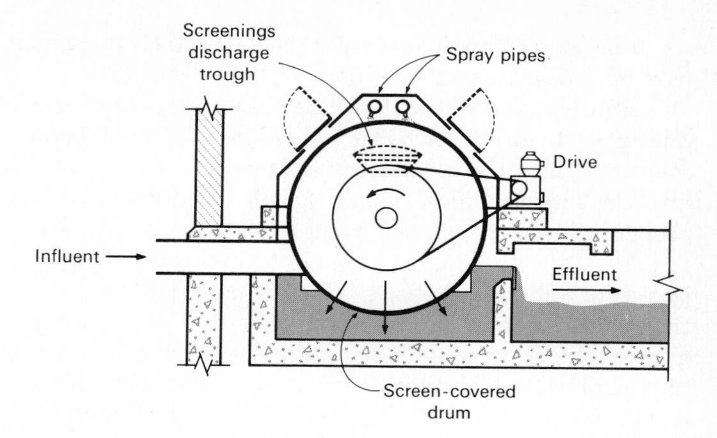

FIG. 11·4 Fine screen mounted on a drum [from Link-Belt].

Modern fine screens are of the disk or drum type, with stainless-steel or nonferrous wire-mesh screen cloth. The openings vary from 0.126 to 0.009 in. (6- to 60-mesh). The disk type has a vertical circular screening surface that rotates on a horizontal shaft set slightly above the water surface. It is available in sizes from 4 to 18 ft in diameter.

The drum type revolves at about 4 rpm around a horizontal axis and operates slightly less than half-submerged (see Fig. 11·4). The wastewater flows in one end of the drum and outward through the screen cloth. Drum screens are available in various sizes, from 39 in. to 5 ft in diameter and from 4 to 12 ft in length.

In both the disk and drum types, the solids are raised above the liquid level by rotation of the screen and are back-flushed into receiving troughs by high-pressure jets. With the finer mesh cloth, effluent may be used for spray water.

Fine wire-mesh screens, or hand-cleaned types set in stop-log grooves, have been used to a limited degree in municipal sewage treatment plants for protection of trickling-filter nozzles against clogging and for screening of rubber goods and other objects of sewage origin from plant effluents. Disk and drum types of screens have been used extensively in industrial plants for the screening of wastewater from packing plants, canneries, tanning plants, poultry-dressing plants, textile mills, paper mills, etc., before discharge.

Quantities of Screenings

The quantity of screenings collected for disposal will vary, depending on the type of rack or screen employed as well as on the sewer system and geographic location. For estimating purposes, the following values are suggested.

The quantity of screenings removed by bar racks usually varies from about 0.5 to 5 cu ft/million gal of wastewater treated; the average is about 2 cu ft/million gal [15]. In plants served by combined sewers, the quantity of screenings has been observed to increase greatly during storm flow periods. The screenings removed by fine screens have amounted to approximately 5 to 30 cu ft/million gal of wastewater treated, equivalent to 5 to 15 percent of suspended matter [10].

Disposal of Screenings

Means of disposal of screenings include (1) discharge to grinders or disintegrator pumps where they are ground and returned to the wastewater and (2) removal by hauling to disposal areas (landfill). Some engineers are of the opinion that it is undesirable to grind up the screenings and return them to the flow, because rags, which are discharged from the grinders as shredded material, frequently collect in clumps or balls to cause operating difficulties in the units that follow. Grindings also accentuate scum problems in digesters. Other engineers are of the opinion that the disposal of the ground screenings in the wastewater flow is the simplest and cleanest solution of a disagreeable disposal problem.

In small installations, screenings may be disposed of by burial on the plant site. Alternatively, they may be disposed of with the municipal refuse and garbage. In large installations, incineration may be found appropriate.

Comminutors

Devices have been developed that cut up (comminute) the material retained on the screen without removing it from the wastewater. They are available from several manufacturers [14]. The original device of this type was developed by the Chicago Pump Co. [11]. It has a vertical revolving drum screen with $\frac{1}{4}$-in. slots in small machines and $\frac{3}{8}$-in. slots in large machines. The drum, which operates nearly submerged, is furnished with stellite cutting teeth and stellite-faced shear bars (see Fig. 11·5).

Coarse material is cut by the cutting teeth and the shear bars on the revolving drum as the solids are carried past a stationary comb. The small sheared particles pass through the drum slots and out of a bottom opening through an inverted siphon and into the downstream channel. Manufacturers' data and rating tables for these units should be consulted for recommended channel dimensions, capacity ranges, upstream and downstream submergence, and power requirements.

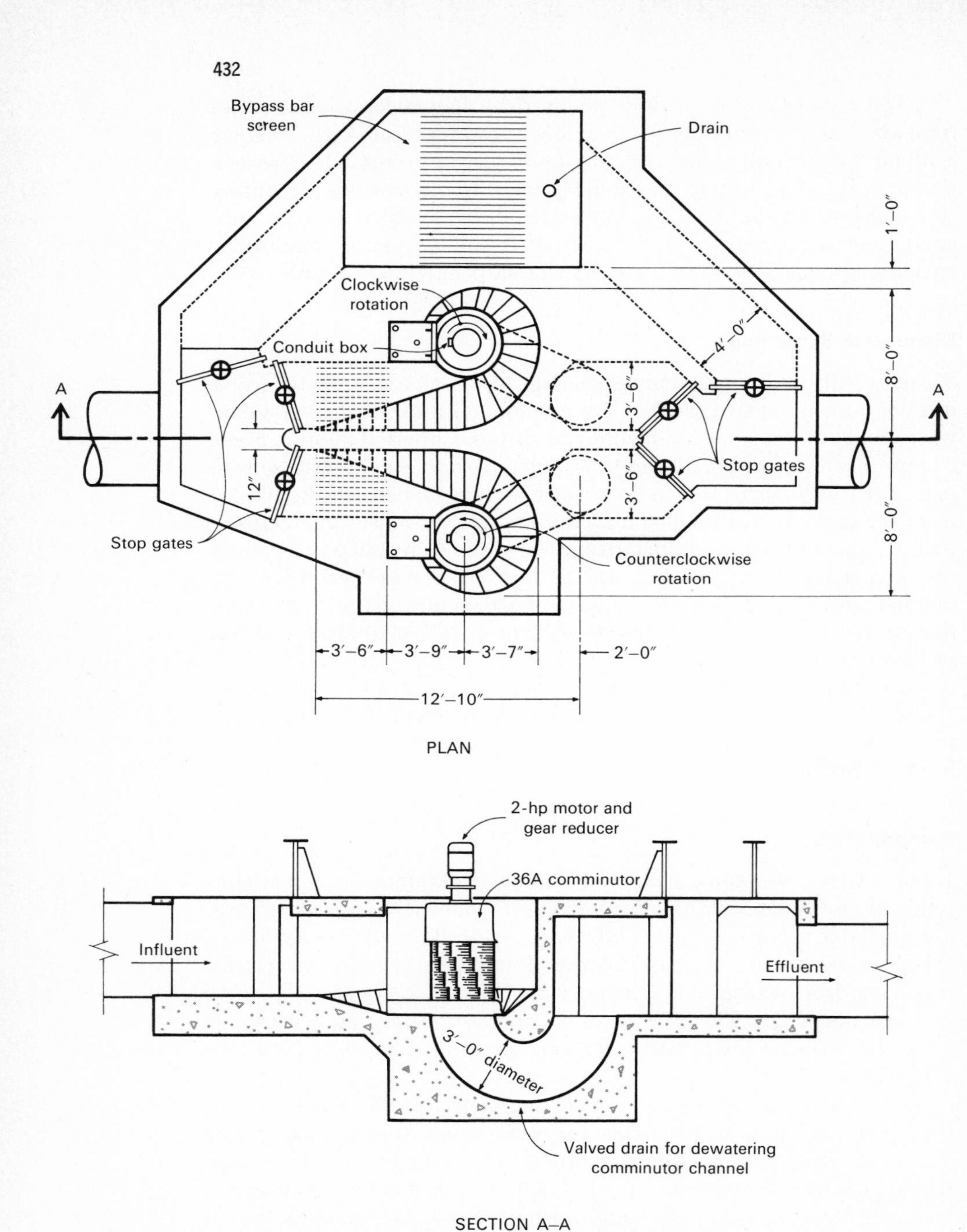

FIG. 11·5 Plan and cross-sectional views of a comminutor installation [from Chicago Pump Co.].

Another type of comminuting device consists of a stationary semicircular screen grid mounted in a rectangular channel with rotating circular cutting disks. The grid intercepts the larger solids, while smaller solids pass through the space between the grid and cutting disks. The teeth are mounted on the rotating part, and the comb is located on the stationary screen grid.

Still another type consists of a semicircular vertical stationary stainless-steel screen with horizontal slots set in a rectangular channel concave to the flow. A motor-driven vertical arm, with cutting teeth, oscillates back and forth between the slots conveying the screenings to the sides of the unit where they are shredded between the oscillating cutting teeth and stationary cutter bars.

Comminuting devices may be preceded by grit chambers to prolong the life of the equipment and to reduce the wear on the cutting surfaces and on portions of the mechanisms where there is a small clearance between moving and stationary parts. They are installed frequently in the wet well of pumping stations to protect the pumps against clogging by rags and large objects, especially in the smaller cities served by separate sanitary sewers carrying a minimum of grit. Provisions must be made to bypass comminutors in the case of flows in excess of the capacity of the comminutor or in case of power or mechanical failure.

GRIT CHAMBERS

Grit chambers are designed to remove grit, consisting of sand, gravel, cinders, or other heavy solid materials that have subsiding velocities or specific gravities substantially greater than those of the organic putrescible solids in wastewater. Grit also includes eggshells, bone chips, seeds, coffee grounds, and large organic particles, such as food wastes. Grit chambers are provided to protect moving mechanical equipment from abrasion and accompanying abnormal wear; to reduce formation of heavy deposits in pipelines, channels, and conduits; and to reduce the frequency of digester cleaning that may be required as a result of excessive accumulations of grit in such units.

Grit chambers may be located ahead of all other units in treatment plants where removal of grit would facilitate operations. However, the installation of mechanically cleaned bar racks or comminutors ahead of grit chambers facilitates operation of grit removal and cleaning facilities.

When it is desirable to locate the grit chambers ahead of wastewater pumps, this would normally involve placing them at considerable depth at added expense. It is therefore usually deemed more economical to pump the wastewater, including the grit, to grit chambers located at a convenient position ahead of the treatment plant units, recognizing that the pumps may require greater maintenance than otherwise.

The design of grit chambers will depend on the type selected, on whether or not mechanical grit-removal equipment is provided, and on the requirements of the selected grit-removal equipment. A recent study has disclosed that the majority of grit-chamber installations having a capacity in excess of 1.5 mgd have been provided with mechanical cleaning equipment [15].

Types

There are two general types of grit chambers: horizontal-flow and aerated. In the horizontal-flow type the flow passes through the chamber in a horizontal direction, the straight-line velocity of flow being controlled by the dimensions of the unit or by the use of special weir sections at the effluent end. The aerated type consists of a spiral-flow aeration tank, the spiral velocity being controlled by the dimensions and the quantity of air supplied to the unit.

Horizontal-Flow Grit Chambers Until recently, most grit chambers have been of the horizontal-flow, velocity-controlled type. These chambers have been designed so as to maintain a velocity as close to 1 fps as practical. Such a velocity will carry most organic particles through the chamber and will tend to resuspend any that settle but will permit the heavier grit to settle out.

Plants designed in the 1920s and 1930s contained up to 12 long, narrow grit channels, which were cut in and out of service to control the velocity as the flow varied. Later, the number of channels was reduced, and the velocity was maintained constant by installing proportional or Sutro-type weirs at the outlet of the channels, as shown in Fig. 11·6. Such weirs maintain a constant velocity in a rectangular channel as the depth varies (if the grit storage or space for grit-collecting equipment in the bottom of the channel is neglected), but they must have a free discharge and are wasteful of hydraulic head.

Control sections with parallel vertical sides, which dissipate less of the depth as lost head, were also used. Theoretically, these require a grit chamber of parabolic cross section for constant velocity with varying depth [3], but this can be satisfactorily approximated by a trapezoidal cross section (see Example 11·2). If desired, the control section may be made narrower at the top than at the bottom to obtain a more desirable trapezoidal cross section. If the grit chambers are followed

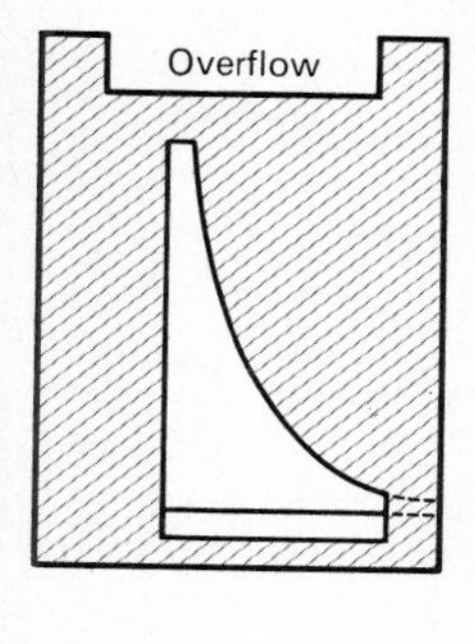

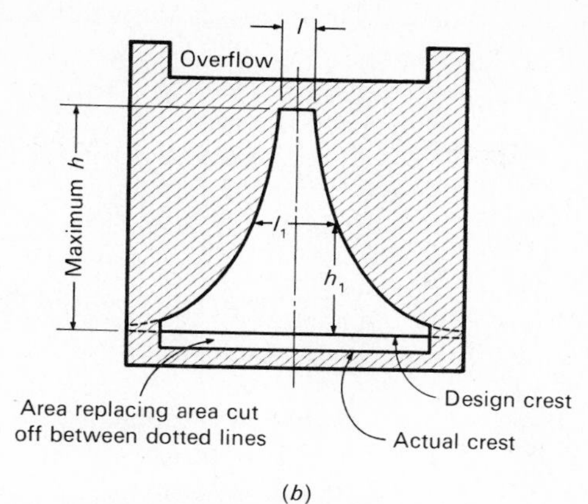

(a) (b)

FIG. 11·6 Weirs for grit-chamber control channel. (a) Sutro and (b) proportional.

immediately by a Parshall flume, the flume may be designed to control the velocity.

The head loss in the control section at any particular rate of flow amounts to about 36 percent of the depth of the flow in the grit chamber. This amount is about 1.1 times the velocity head in a control section with a well-rounded entrance. In addition, there is a considerable variation in the water level in the grit chamber and in the control section between minimum and maximum flow. Every plant should have a meter for measuring the flow. If the meter is located between the grit chamber and the sedimentation tanks, the difference in head can be utilized by the meter; otherwise it may be wasted. The effect of the control-section geometry on the geometry of the grit chamber is illustrated in Example 11·1.

EXAMPLE 11·1 *Effect of Control-Section Geometry on Grit-Chamber Geometry*

Demonstrate that the use of a rectangular control section requires an approach channel with a parabolic cross section to maintain constant velocity with varying rates of flow.

Solution

1. For a prismatic channel

$$A = \int_0^h t\, dh$$

where t = channel width at depth h

2. The discharge through the channel is given by

$$Q = AV = V \int_0^h t \, dh$$

3. For a constant velocity of 1 fps

$$Q = A = \int_0^h t \, dh$$

4. For a rectangular control section

$$Q = cwh^n$$

where c = constant
w = throat width
n = constant

5. Equating the two expressions (steps 3 and 4) for discharge yields

$$\int_0^h t \, dh = cwh^n$$

6. Differentiating the foregoing expression gives

$$t \, dh = ncwh^{n-1} \, dh$$

7. Because n is 3/2 for a rectangular section, the height of the grit chamber is given by

$$h = \left(\frac{2}{3cw}\right)^2 t^2 = kt^2$$

8. Therefore, the cross section of the grit chamber must be parabolic.

The design of horizontal-flow grit chambers should be such that, under the most adverse conditions, the lightest particle of grit will reach the bed of the channel prior to its outlet end. Normally, grit chambers are designed to remove all grit particles that will be retained on a 65-mesh screen (0.21-mm diameter), although many chambers have been designed to remove grit particles retained on a 100-mesh screen (0.15-mm diameter). It is good practice to use a settling velocity of 3.7 fpm for removal of 65-mesh material and 2.5 fpm for 100-mesh material. Where the specific gravity of grit, due to local conditions, is less than 2.65, the use of lower velocities should be considered [15].

The length of channel will be governed by the depth required by the settling velocity and the control section, and the cross-sectional area will be governed by the rate of flow and by the number of channels.

Allowance should be made for inlet and outlet turbulence. A minimum allowance of approximately twice the maximum depth of flow is recommended. A maximum allowance of 50 percent of the theoretical length is sometimes used.

There is nothing worse than accumulations of grit in digesters, because the plant is crippled while the digesters are being emptied and the grit removed. This is both a disagreeable and a hazardous undertaking. Where the plant flowsheet includes digesters, no compromise leading to less than ideal grit-removal facilities should be tolerated. On the other hand, where raw sludge is barged to sea or dewatered on vacuum filters and incinerated, grit chambers that are far less than ideal have given thoroughly satisfactory service. Grit-chamber design is illustrated in the following example.

EXAMPLE 11·2 *Horizontal-Flow–Grit-Chamber Design*

Design a grit chamber with three channels with a flow-through velocity of 1.0 fps for a plant with a maximum design flow of 30 mgd, an average flow of 15 mgd, and a minimum flow of 6 mgd. Assume that the maximum width of each channel is 6 ft and design each channel for a maximum emergency flow of 15 mgd, a normal maximum flow of 10 mgd, an average flow of 5 mgd, and a minimum flow of 2 mgd. Use a fixed-width control section with vertical sides and a well-rounded and smooth approach, so that the head loss may be assumed equal to 10 percent of the velocity head. The flow in the control section will be at critical depth and the critical-depth equations will apply [9].

Solution

1. For the control section selected, a parabolic grit-chamber cross section is required for constant velocity. For a parabola, the area $A = \frac{2}{3}HT$ where H is equal to the height and T is equal to the top width. In the final design, the parabolic section will be approximated with straight lines.

2. For the normal maximum flow of 10 mgd, $Q = 15.5$ cfs. For a 6-ft-wide channel and a velocity of 1 fps, the depth in the grit chamber is 3.88 ft.

$$Q = AV = \tfrac{2}{3}HTV$$

$$H = \frac{3}{2}\frac{Q}{TV} = \frac{3(15.5)}{2(6)(1)} = 3.88 \text{ ft}$$

Determine the velocity head and depth for the control section. Equating the upstream specific energy to that in the control section yields

$$\underset{\text{Depth}}{H} + \underset{\text{Velocity head}}{\frac{V^2}{2g}} = \underset{\text{Depth}}{d_c} + \underset{\text{Velocity head}}{\frac{V_c^2}{2g}} + \underset{\text{Head loss}}{0.1\frac{V_c^2}{2g}}$$

However, at critical depth,

$$d_c = 2\frac{V_c^2}{2g}$$

and

$$H + \frac{V^2}{2g} = 2\frac{V_c^2}{2g} + \frac{V_c^2}{2g} + 0.1\frac{V_c^2}{2g} = 3.1\frac{V_c^2}{2g}$$

The velocity head in the control section equals 1.25 ft.

$$\frac{V_c^2}{2g} = \frac{1}{3.1}\left(H + \frac{V^2}{2g}\right) = \frac{1}{3.1}\left[3.88 + \frac{(1)^2}{64.4}\right] = 1.25 \text{ ft}$$

The depth in control section $d_c = 2(1.25) = 2.5$ ft. Determine the control section width w.

$$V_c = \sqrt{2(32.2)(1.25)} = 8.95 \text{ fps}$$

$$a = \frac{Q}{V} = \frac{15.5}{8.95} = 1.73 \text{ sq ft}$$

$$w = \frac{a}{d_c} = \frac{1.73}{2.50} = 0.693 \text{ ft} = 8.3 \text{ in.}$$

3. For the average flow of 5 mgd, $Q = 7.75$ cfs.

$$a = \sqrt[3]{\frac{Q^2w}{g}} = \sqrt[3]{\frac{(7.75)^2(0.693)}{32.2}} = 1.089 \text{ sq ft}$$

$$d_c = \frac{1.089}{0.693} = 1.57 \text{ ft}$$

$$H = \frac{3.1}{2}1.57 = 2.43 \text{ ft (depth in grit chamber)}$$

$$T = \frac{3Q}{2HV} = \frac{3(7.75)}{2(2.43)(1)} = 4.8 \text{ ft}$$

4. For the minimum flow of 2 mgd, $Q = 3.1$ cfs.

$$a = \sqrt[3]{\frac{(3.1)^2(0.693)}{32.2}} = 0.607 \text{ sq ft}$$

$$d_c = \frac{0.607}{0.693} = 0.877 \text{ ft}$$

$$H = \frac{3.1}{2} 0.877 = 1.36 \text{ ft}$$

$$T = \frac{3(3.1)}{2(1.36)(1)} = 3.42 \text{ ft}$$

5. For the maximum emergency flow of 15 mgd (one chamber out of service), $Q = 23.25$ cfs.

$$a = \sqrt[3]{\frac{(23.25)^2(0.693)}{32.2}} = 2.27 \text{ sq ft}$$

$$d_c = \frac{2.27}{0.693} = 3.28 \text{ ft}$$

$$H = \frac{3.1}{2} 3.28 = 5.08 \text{ ft}$$

$$T = \frac{3(23.25)}{2(5.08)(1)} = 6.87 \text{ ft}$$

6. The parabolic cross section and the adopted cross section are shown in Fig. 11·7.
7. The length of the grit chamber required for removal of 65-mesh material with a settling velocity of 3.7 fpm under normal maximum flow conditions is

$$\frac{3.88 \text{ ft}}{3.7 \text{ fpm}} = 1.05 \text{ min flow time} \times 60 \text{ fpm} = 63 \text{ ft}$$

Under emergency maximum flow conditions the theoretical detention time required would be $5.08/3.7 = 1.37$ min equivalent to an 82-ft length.

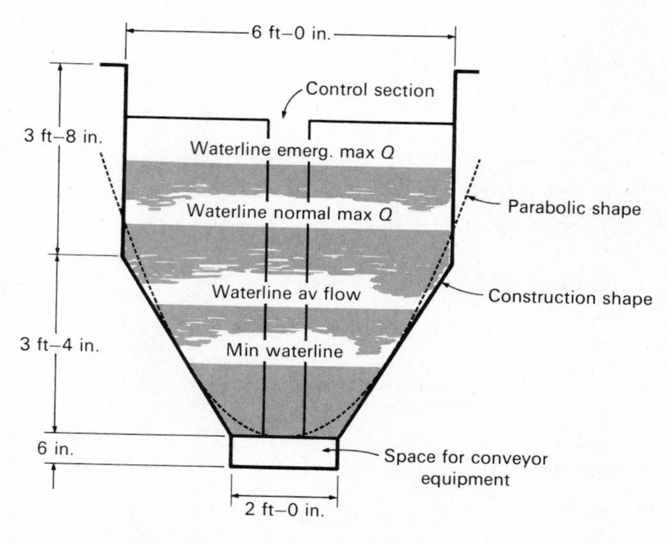

FIG. 11·7 Section through grit chamber designed in Example 11·2.

8. If the grit chambers are made 82 ft long, the extra 19 ft will provide adequately for turbulence at inlet and outlet under normal maximum flow conditions. Slightly less than maximum efficiency is acceptable under infrequent emergency operation with one chamber out of service.

Mechanical devices for grit removal from long, narrow horizontal-flow channels usually consist of conveyor-type equipment with buckets, plows, or scrapers. Elevation of the grit is usually an integral part of the collection process in smaller plants. Separate grit elevators are also used, especially in larger plants. The most commonly used grit elevators are of two types: (1) bucket and continuous chain and (2) screw conveyor.

For multiple channels equipped with collectors and elevators, belt conveyors may be provided to collect and transport the grit to one side where it is dumped into trucks for disposal, elevated to storage tanks, or discharged to grit washers before disposal. Temporary storage tanks for grit should have steep hoppers equipped with vibrators, or grit removal will be difficult.

Square grit chambers may also be used (Fig. 11·8); however, two units are advisable. These chambers are designed for approximately

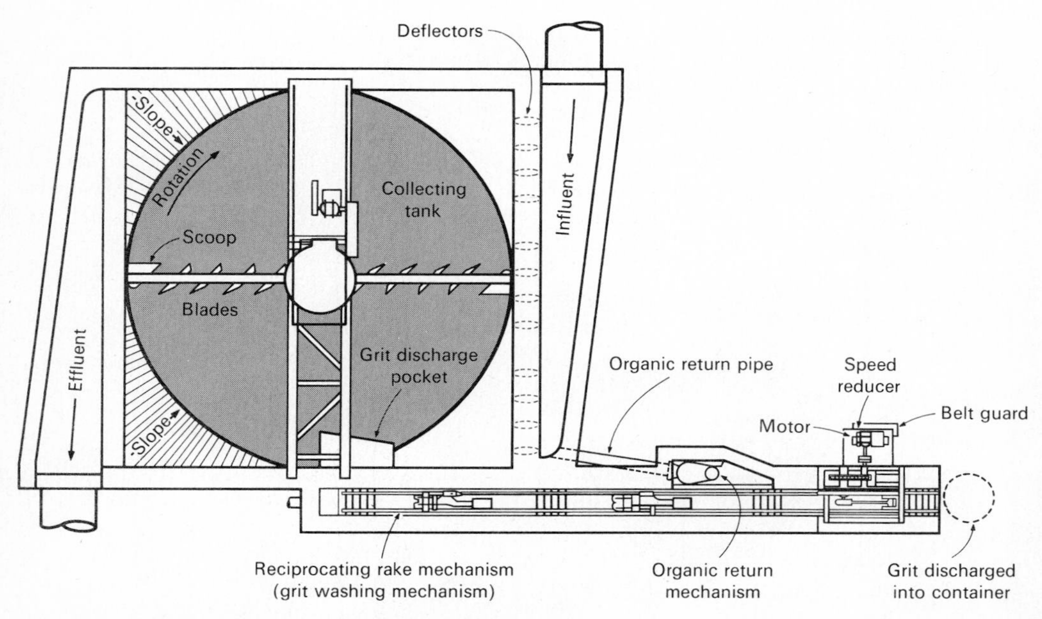

FIG. 11·8 Typical square grit chamber [from Dorr-Oliver].

1 fps velocity at maximum flow, with the water level controlled by the hydraulic gradient through the plant. This results in lower velocities and more organics in the settled grit at average and minimum flows. The solids are raked by a rotating mechanism to a sump at the side of the tank, from which they are moved up an incline by a reciprocating rake mechanism. While passing up the incline, organic solids are separated from the grit and flow back into the basin. By this method, a cleaner, dryer grit is obtained, comparable to the washed grit from separate grit washers.

The wear on removal equipment of the conveyor type, whether buckets, plows, or scrapers, has been considerable in the larger plants serving areas drained by combined sewers. For this reason, grab buckets of the clamshell type, operating on an overhead monorail system, have been installed at some plants for periodic cleaning as required. Chambers may or may not be dewatered during cleaning, but the flow should be shut off by closing the outlet gate.

Aerated Grit Chambers The discovery of grit accumulations in spiral-flow aeration tanks preceded by grit chambers led to the development of the aerated grit chamber. No doubt the excessive wear on grit-handling equipment and the necessity in most cases for separate grit-washing equipment led to the current popularity of this type of grit chamber.

These chambers are usually designed to provide detention periods of about 3 min at the maximum rate of flow. The cross section of the tank is similar to that provided for spiral circulation in activated-sludge aeration tanks, except that a grit hopper about 3 ft deep with steeply sloping sides is located along one side of the tank under the air diffusers. The diffusers are located about 18 in. or 2 ft above the normal plane of the bottom. A typical cross section is shown in Fig. 11·9.

The velocity of roll or agitation governs the size of particles of a given specific gravity that will be removed. If the velocity is too great, grit will be carried out of the chamber; and, if it is too small, organic material will be removed with the grit. Fortunately, the quantity of air is easily adjusted. With proper adjustment, almost 100 percent removal will be obtained, and the grit will be well washed. Sewage will move through the tank in a helical path and will make two to three passes across the bottom of the tank at maximum flow and more at lesser flows. Sewage should be introduced in the direction of the roll. The head loss required by this type of chamber is minimal.

Many aerated grit chambers have been provided with means for grit removal by grab buckets, traveling on monorails, centered over the grit collection and storage trough. Other installations are equipped

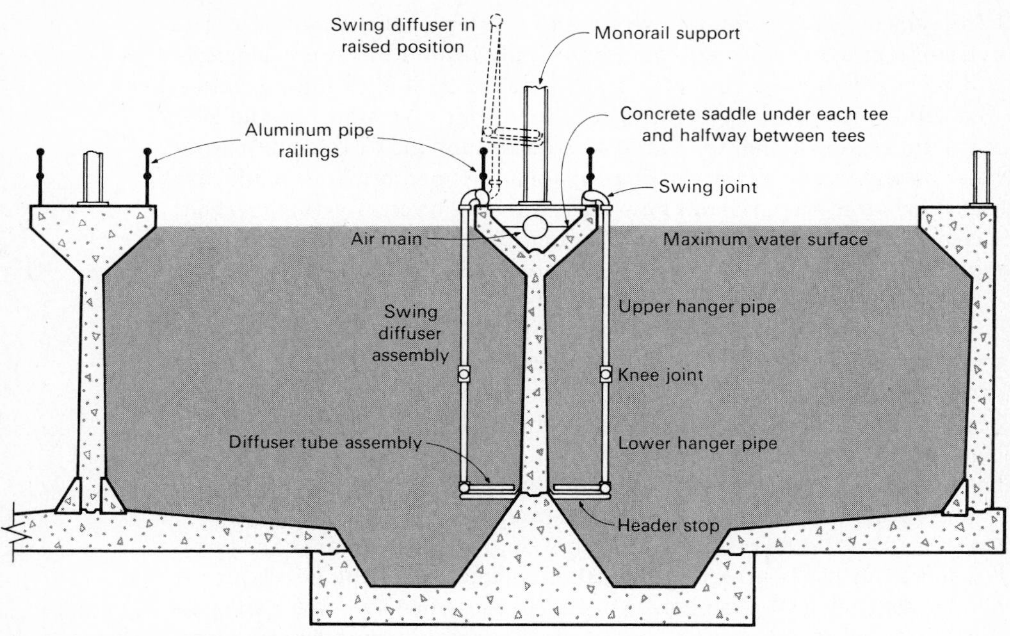

FIG. 11·9 Typical cross section of an aerated grit chamber.

with chain-and-bucket conveyors, running the full length of the storage troughs, which move the grit to one end of the trough and elevate it above the sewage level in a continuous operation. Screw conveyors, jet pumps, and air lifts have also been used [15].

Tubular conveyors are also being used for removal of grit. This type of conveyor consists of a tube, up to 12 in. in diameter, through which an endless chain moves. Between each two links in the chain, a close-fitting disk is attached, only slightly smaller than the tube. A motor-driven sprocket at one end of the installation engages the chain and pulls the disks through the tube. An open section of the tube in the storage trough admits the grit, while another open section above the liquid level permits its discharge.

The basic data for the aerated grit chambers at the Pittsburgh sewage treatment plant, designed by Metcalf & Eddy in 1954 and completed in 1959, were as follows:

Number of chambers	4
Types of diffusers	nozzles
Capacity each chamber, mgd	75
Effective depth, ft	15

Effective width, ft	20
Effective length, ft	60
Detention at maximum flow, min	2.5
Air supply, cfm/ft of length	5
Maximum amount of air available, cfm/ft of length	6.25
Grit size removed, 100 percent of grit retained on a 100-mesh sieve	
Grit removal by grab bucket	
Grit disposal by dumping	

Quantities of Grit

The quantities of grit will vary greatly from one location to another, depending on the type of sewerage system, the characteristics of the drainage area, the condition of the sewers, the frequency of street sanding to counteract icing conditions, the type of industrial wastes, the number of household garbage grinders served, and the proximity and use of sandy bathing beaches.

A review of a number of plant records indicates that the quantity of grit removed from wastewaters may range from as little as $\frac{1}{3}$ cu ft/million gal treated to as much as 24 cu ft/million gal treated [15]. In any design, particularly for systems serving combined sewers, extreme variations in grit volume and quantity should be anticipated, and a generous factor of safety should be employed in calculations concerning the actual storage, handling, or disposal of the grit.

Disposal of Grit

Possibly the most common method of grit disposal is as fill, covered if necessary to prevent objectionable conditions. In some large plants, grit is incinerated with the sludge. In New York City and in some other large coastal cities, grit and screenings are barged to sea and dumped. Generally the grit must be washed before removal.

Washing The character of grit normally collected in horizontal-flow grit chambers varies widely from what might be normally considered as clean grit to grit that includes a large proportion of putrescible organic material. Unwashed grit may contain 50 percent, or more, organic material. Unless promptly disposed of, this material may attract insects and rodents. It has a distinctly disagreeable odor.

Several types of washers are available. One type relies on an inclined submerged rake or screw that provides the necessary agitation for separation of the grit from the organic materials and, at the same time, raises the washed grit to a point of discharge above water level. Another type is a separate jig that depends on an up-and-down flow of liquid through the grit bed to wash out the organic material. Its performance is excellent but it requires an additional manually supervised plant operation.

Removal from Plant Grit is normally hauled to the dumping areas in trucks for which loading facilities are required. In larger plants, elevated grit storage facilities may be provided with bottom gates through which the trucks are loaded. Difficulties experienced in getting the grit to flow freely from the storage hoppers have been minimized by applying air beneath the grit and by the use of vibrators. Facilities for collection and disposal of drippings from the bottom gates are desirable. Grab buckets operating on a monorail system may also be used to load trucks directly from the grit chambers or from storage bins at grade.

In some plants, grit is successfully conveyed to grit disposal areas by pneumatic conveyors. This system requires no elevated storage hoppers and eliminates problems in storage and trucking, but the wear on piping, especially elbows, is considerable.

PRETREATMENT

Pretreatment is used to remove material, such as grease and scum, from sewage prior to primary sedimentation to improve treatability. Pretreatment may include skimming, grease traps, preaeration, and flocculation.

Skimming Tanks

A skimming tank is a chamber so arranged that floating matter rises and remains on the surface of the sewage until removed, while the liquid flows out continuously through deep outlets or under partitions, curtain walls, or deep scum boards. This may be accomplished in a separate tank or combined with primary sedimentation, depending on the process and nature of the wastewater.

The object of skimming tanks is the separation from the sewage of the lighter, floating substances. The material collected on the surface of skimming tanks, whence it can be removed, includes oil, grease, soap, pieces of cork and wood, and vegetable debris and fruit skins originating in households and in industry.

Most skimming tanks are rectangular or circular in shape and provide for a detention period of 1 to 15 min. The outlet, which is submerged, is situated opposite the inlet and at a lower elevation to assist in flotation and to remove any solids that may settle.

Grease Traps

Grease traps are small skimming tanks. They are situated close to the source of grease, which may be an industry, house sewer, or small treatment plant. Grease traps are employed at many manufacturing plants, garages, hospitals, and hotels. A number of proprietary tank patterns are in use. In most of them the inlet is situated below the surface and the outlet is at the bottom. To be efficient, they must be large enough to hold and, if necessary, to cool the sudden discharges of oily or greasy wastes, and they must be cleaned frequently and regularly. In many cases, neither of these requirements is met. Detention times of 10 to 30 min have been used.

Preaeration

The objectives of aerating sewage prior to primary sedimentation are: to improve its treatability; to provide grease separation, odor control, grit removal, and flocculation; to promote uniform distribution of suspended and floating solids to treatment units; and to increase BOD removals. Short-period preaeration of 3 to 5 min formerly employed does not materially improve BOD or grease removal [6].

Detention times for preaeration range from 10 to 45 min. Tank depths are generally about 15 ft, and air requirements range from 0.1 to 0.4 cu ft/gal of wastewater. Current practice when preaeration is employed frequently consists of increasing the detention period in aerated grit chambers. In this case provisions for grit removal may be provided in only the first portion of the tanks.

The use of aerated channels for distributing sewage to primary sedimentation tanks in large plants keeps the solids in suspension at all rates of flow so that the channel velocity is no longer critical. It also ensures a uniform distribution of solids to each tank, and, in conjunction with aerated grit chambers and preaeration tanks, adds dissolved oxygen to freshen the sewage and reduce odors. The amount of air required ranges between 2 and 5 cfm/linear ft of channel. Aerated channels are almost universally used for distributing mixed liquor to activated-sludge final settling tanks.

Flocculation

As mentioned in Chap. 8, the purpose of sewage flocculation is to form aggregates or flocs from the finely divided matter. Flocculation of sewage by air or mechanical agitation may be worthy of consideration when it is desired to increase the removal of suspended solids and BOD in the primary sedimentation tanks. Although not often used in sewage treatment, flocculation may be beneficial in conditioning sewage containing certain industrial wastes. When chemicals are added, the detention period at design flow should not be less than 20 min and preferably should be 30 min.

Paddles for mechanical agitation should have a peripheral speed of about 1.5 fps, with variable-speed drives permitting a 50 percent adjustment both above and below this speed. Porous tubes or diffuser nozzles are commonly used in air agitation. The amount of air required is approximately 0.08 to 0.15 cu ft/gal for a 45-min detention period. The tanks are similar to spiral-flow aeration tanks and to aerated grit chambers. If the spiral flow due to air agitation becomes highly turbulent, flocculation will not occur.

PRIMARY SEDIMENTATION TANKS

Whenever a liquid containing solids in suspension is placed in a relatively quiescent state, those solids having a higher specific gravity than the liquid will tend to settle, and those with a lower specific gravity will tend to rise. These principles are utilized in the design of sedimentation tanks for treatment of wastewaters. The objective of treatment by sedimentation is to remove readily settleable solids and floating material and thus to reduce the suspended solids content.

Primary sedimentation tanks may provide the principal degree of wastewater treatment, or they may be used as a preliminary step in the further processing of the wastewater. When used as the only means of treatment, these tanks provide for the removal of (1) settleable solids capable of forming sludge banks in the receiving waters and (2) much of the floating material. When used as a preliminary step to biological treatment, their function is to reduce the load on the biological treatment units. Efficiently designed and operated primary sedimentation tanks should remove from 50 to 65 percent of the suspended solids, and from 25 to 40 percent of the BOD_5.

Where primary sedimentation tanks precede biological treatment processes, they may be designed to provide shorter detention periods and to have a higher rate of surface loading than tanks providing the only method of treatment.

Sedimentation tanks have also been used as storm water tanks, which are designed to provide a moderate detention period (10 to 30 min) for overflows from either combined sewers or storm sewers. The purpose is to remove a substantial portion of the organic solids that otherwise would be discharged directly to the receiving water and which could form offensive sludge banks. Such tanks have also been used to provide sufficient detention periods for effective chlorination of such overflows.

Basis of Design

If all solids in wastewater were discrete particles of uniform size, uniform density, reasonably uniform specific gravity, and fairly uniform shape, the removal efficiency of these solids would be dependent on the surface area of the tank and time of detention; the depth of the tank would have little influence, providing horizontal velocities are maintained below the scouring velocity. However, the solids in most wastewaters are not of such regular character but are heterogeneous in nature, and the conditions under which they are present range from total dispersion to complete flocculation. The bulk of the finely divided solids reaching primary sedimentation tanks are incompletely flocculated but are susceptible to flocculation.

Flocculation is aided by eddying motion of the fluid within the tanks and proceeds through the coalescence of fine particles, at a rate that is a function of their concentration and of the natural ability of the particles to coalesce upon collision. As a general rule, therefore, coalescence of a suspension of solids becomes more complete as time elapses. For this reason, detention time is also a consideration in the design of sedimentation tanks. The mechanics of flocculation are such, however, that as the time of sedimentation increases, less and less coalescence of remaining particles occurs.

Normally, primary sedimentation tanks are designed to provide 90 to 150 min of detention based on the average rate of sewage flow. Tanks that provide shorter detention periods (30 to 60 min), with less removal of suspended solids, are used frequently for preliminary treatment ahead of biological treatment units.

Surface-Loading Rates Sedimentation tanks are now normally designed on the basis of a surface-loading rate at the average rate of flow, expressed as gallons per day per square foot of horizontal area. The selection of a suitable loading rate depends on the type of suspension to be separated. Typical values for various suspensions are reported in Table 11·3. Designs for municipal plants must meet the approval of state regulatory agencies, most of which have adopted standards that must be followed. For example, New York State requirements, based on the *Ten States Standards* [8] for mechanically cleaned primary settling tanks, are as follows: "Surface settling rates for primary tanks not followed by secondary treatment shall not exceed 600 gpd/sq ft for plants having a design flow of 1 mgd or less. Higher surface settling rates may be permitted for larger plants."

TABLE 11·3 RECOMMENDED SURFACE-LOADING RATES FOR VARIOUS SUSPENSIONS

Suspension	Loading rate, gpd/sq ft	
	Range	Peak flow
Untreated wastewater	600–1,200	1,200
Alum floc*	360–600	600
Iron floc*	540–800	800
Lime floc*	540–1,200	1,200

* Mixed with the settleable suspended solids in the untreated wastewater and colloidal or other suspended solids swept out by the floc.

The effect of the surface-loading rate and detention time on suspended-solids removal varies widely depending on the character of the wastewater, proportion of settleable solids, concentration of solids, and other factors. It should be emphasized that overflow rates must be set low enough to ensure satisfactory performance at peak rates of flow, which may vary from 3 times the average flow in small plants to 1.5 times the average in large plants.

When the area of the tank has been established, the detention period in the tank is governed by water depth as shown in Table 11·4.

TABLE 11·4 DETENTION TIMES FOR VARIOUS SURFACE-LOADING RATES AND TANK DEPTHS

Surface-loading rate, gpd/sq ft	Detention time, hr			
	7-ft depth	8-ft depth	10-ft depth	12-ft depth
400	3.2	3.6	4.5	5.4
600	2.1	2.4	3.0	3.6
800	1.6	1.8	2.25	2.7
1,000	1.25	1.4	1.8	2.2

Overflow rates in current use result in nominal detention periods of 2 to 2.5 hr, based on average design flow. As design flows in all cases are usually based on some future condition, the actual detention periods during the early years of operation are somewhat longer.

Weir Rates The *Ten States Standards* [8] make this statement about weirs: "Weir loadings should not exceed 10,000 gallons per linear foot per day for plants designed for average flows of 1 mgd or less. Special consideration will be given to weir loadings for plants designed for flows in excess of 1 mgd, but such loadings should preferably not exceed 15,000 gpd per linear foot."

While some authorities, in addition to the above, recommend maximum weir loadings, in actual practice there appears to have been little limitation except in the design of final tanks for activated sludge. Weir loadings on circular tanks have amounted to as much as 70,000 gpd/linear ft, and on rectangular tanks up to 215,000 gpd/linear ft. In any case, weir rates appear to have less effect on efficiencies of removal than do overflow rates [15].

Tank Type, Size, and Shape

Almost all treatment plants of any size, except for those with Imhoff tanks, now use mechanically cleaned sedimentation tanks of standardized circular or rectangular design. The selection of the type of sedimentation unit for a given application is governed by the size of the installation, by rules and regulations of local control authorities, by local site conditions, and by the experience and judgment of the engineer and his estimate of the economics involved. In some cases, alternative bids have been taken on circular and rectangular tanks.

Two or more tanks should be provided in order that the process may remain in operation while one tank is out of service for maintenance and repair work. At large plants, the number of tanks is determined largely by the limitations of size. The maximum length of rectangular tanks has been approximately 300 ft. Where widths of rectangular mechanically cleaned tanks are greater than 20 ft, multiple bays with individual cleaning equipment may be employed, thus permitting tank widths up to 80 ft or more. Circular tanks have been constructed with diameters of 12 to 200 ft, although the more common range is from 40 to 100 ft.

Rectangular Tanks A rectangular tank is shown in Fig. 11·10. Sludge-removal equipment for this type of tank is available from a number of manufacturers and usually consists of a pair of endless conveyor chains.

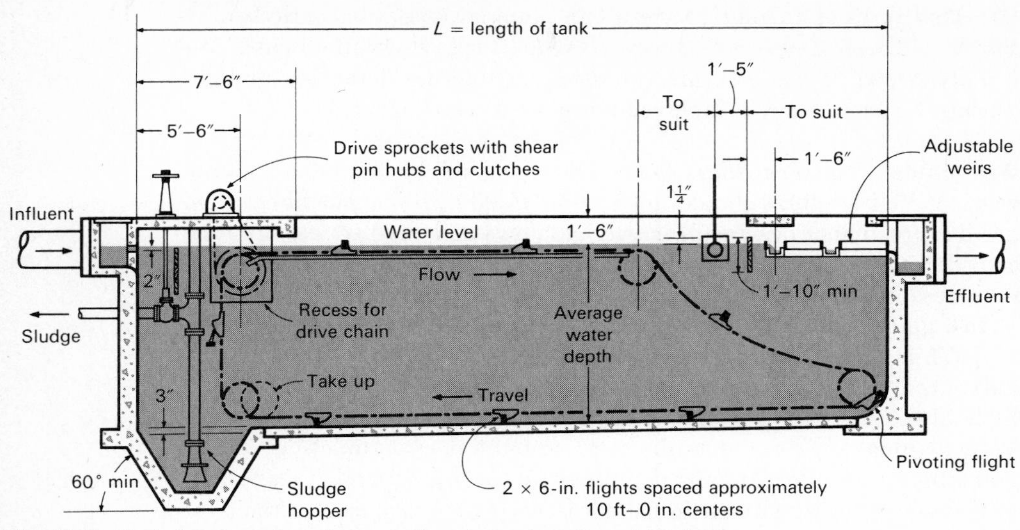

FIG. 11·10 Typical rectangular primary sedimentation tank [from Link-Belt].

Attached to the chains at intervals, usually of 10 ft, are 2-in.-thick crosspieces of wood, or flights, 6 to 8 in. deep, extending the full width of the tank or bay. Linear conveyor speeds of 2 to 4 fpm are common, although speeds of 1 fpm have been used at activated-sludge plants. The solids settling in the tank are scraped to sludge hoppers in small tanks and to transverse troughs in large tanks. These, in turn, are equipped with collecting mechanisms (cross collectors), usually of the same type as the longitudinal collectors, which convey solids to one or more sludge hoppers. In some recent designs, screw conveyors have been used for the cross collectors. Rectangular tanks may also be cleaned by a bridge-type mechanism which travels up and down the tank on rails supported on the sidewalls. One or more scraper blades are suspended from the bridge. They are lifted clear of the sludge on the return travel.

Where cross collectors are not provided, multiple hoppers must be installed. Their use for collection of sludge delivered by the longitudinal conveyors has introduced operating difficulties, notably sludge hanging on the slopes and in the corners and even arching over the sludge drawoff piping. The use of a cross collector is advisable, except possibly in the smallest tanks, since it results in the withdrawal of a more uniform and concentrated sludge.

Multiple rectangular tanks require less area than multiple circular tanks and for this reason are used where ground area is at a premium. Rectangular tanks also lend themselves to nesting with preaeration tanks and aeration tanks in activated sludge plants. They are also used generally where tank roofs or covers are required.

Influent channels should be provided across the inlet end of the tanks, and effluent channels should be provided across the effluent end of the tanks. It is also desirable to locate sludge pumping facilities close to the hoppers where sludge is collected at the ends of the tanks. One sludge pumping station can conveniently serve two tanks.

For large multiple installations of rectangular tanks, a pipe and operating gallery can be constructed integrally with the tanks along the influent end to contain the sludge pumps. This gallery can be extended as a service tunnel to sludge disposal, heating, and other plant units.

Scum is usually collected at the effluent end of rectangular tanks with the flights returning at the liquid surface. The scum is moved by the flights to a point where it is trapped by baffles before removal. The scum can also be moved by water sprays. The scum can be scraped manually up an inclined apron, or it can be removed hydraulically or mechanically, and for this process a number of means have been developed. For small installations, the most common scum drawoff facility consists of a horizontal, slotted pipe that can be rotated by a lever or a screw. Except when drawing scum, the open slot is above the normal tank water level. When drawing scum, the pipe is rotated so that the open slot is submerged just below the water level, permitting the scum accumulation to flow into the pipe. Use of this equipment results in a relatively large volume of scum liquor.

Another method for removing scum by mechanical means is a transverse rotating helical wiper attached to a shaft. By this apparatus, it is possible to draw the scum from the water surface over a short inclined apron for discharge to a cross-collecting scum trough. The scum may then be flushed to a scum ejector or hopper ahead of a scum pump. Another method of scum removal consists of a chain-and-flight type of collector that collects the scum at one side of the tank and scrapes it up a short incline for deposit in scum hoppers, whence it can be pumped to disposal units. Scum is also collected by special scum rakes in those rectangular tanks that are equipped with the carriage or bridge type of sedimentation tank equipment. Scum is usually disposed of with the sludge produced at the plant.

Circular Tanks Design of circular tanks has become standardized to the extent that comparable sludge-removal equipment is available from several manufacturers. Tanks 12 to 30 ft in diameter have the sludge-removal equipment supported on beams spanning the tank. Tanks 35 ft in diameter and larger employ a central pier that supports the mechanism and is reached by a walkway or bridge.

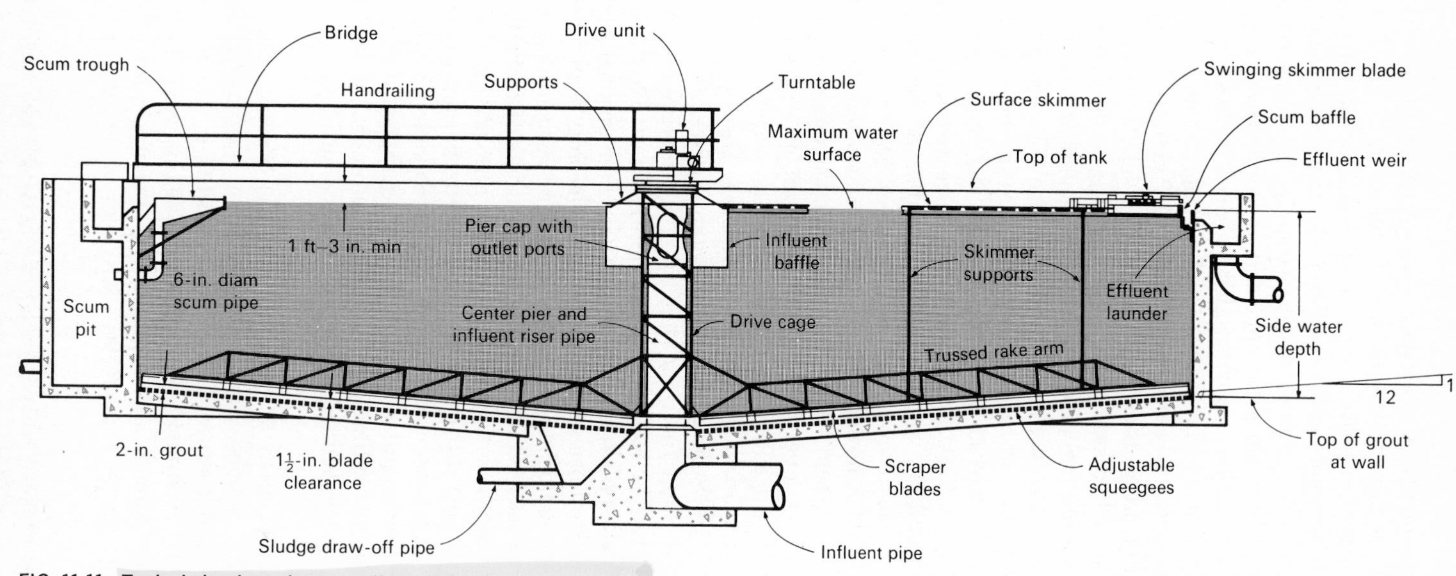

FIG. 11·11 Typical circular primary sedimentation tank [from Infilco].

In most designs, the sewage is carried to the center of the tank in a pipe suspended from the bridge, or encased in concrete beneath the tank floor. At the center of the tank, the sewage enters a circular well designed to distribute the flow equally in all directions. The removal mechanism revolves slowly and may have two or four arms equipped with scrapers. The arms also support blades for scum removal.

In another type, a suspended circular aluminum baffle a short distance from the tank wall forms an annular space into which the sewage is discharged in a tangential direction. The sewage flows spirally around the tank and underneath the baffle, the clarified liquid being skimmed off over weirs on both sides of a centrally located weir trough. Grease and scum are confined to the surface of the annular space.

A circular sedimentation tank is shown in Fig. 11·11. The bottom of the tank is sloped at about 1 in./ft to form an inverted cone, and the sludge is scraped to a relatively small hopper located near the center of the tank.

Multiple tanks are customarily arranged in groups of two or four. The flow is divided among the tanks by a control chamber located between the tanks. Sludge is usually withdrawn to the control chamber from which it is pumped to the sludge disposal units.

Quantities of Sludge

The volume of sludge produced in primary settling tanks must be known or estimated so that these tanks and subsequent sludge processing and disposal facilities can be properly designed. The volume of sludge produced will depend on (1) the characteristics of the raw sewage, including strength and freshness; (2) the period of sedimentation and the degree of purification to be effected in the tanks; (3) the condition of the deposited solids, including specific gravity, water content, and changes in volume under the influence of tank depth or mechanical sludge-removal devices; and (4) the period between sludge-removal operations. The following example and subsequent discussion will illustrate how these factors enter into the calculation of the required storage capacity.

EXAMPLE 11·3 *Sludge Volume Estimation*

Estimate the volume of primary sludge produced per million gallons from a typical medium-strength sewage. Assume that the detention time in the primary tank is 2 hr and that removal efficiency is 60 percent.

Solution

1. From Table 7·3 a medium-strength sewage is found to contain 200 mg/liter suspended solids.
2. The weight of dry solids removed per million gallons is equal to 1,000 lb.

   Dry solids = 0.60(200)(8.34)(1) = 1,000 lb
3. If the specific gravity of the sludge is 1.03 (see Table 11·5) and it contains 95 percent moisture (see Table 13·1), the volume of sludge per million gallons of sewage is 2,330 gal.

TABLE 11·5 SPECIFIC GRAVITY OF RAW SLUDGE PRODUCED FROM VARIOUS TYPES OF SEWAGE

Type of sewerage system	Strength of sewage*	Specific gravity†
Sanitary	Weak‡	1.02
Sanitary	Medium	1.03
Combined	Medium	1.05
Combined	Strong	1.07

* See Table 7·3, Chap. 7.
† Also see Table 13·1, Chap. 13.
‡ Due to infiltration or high water consumption.

$$\text{Volume} = \frac{1{,}000}{8.34(1.03)(0.05)} = 2{,}330 \text{ gal}$$

The foregoing calculation is directly applicable to the design of sludge pumping facilities for primary sedimentation tanks. Sludge should be removed by pumping at least once per shift and more frequently in hot weather to avoid deterioration of the effluent. In large plants sludge pumping may be controlled by a time clock providing continuous "on-off" operation. In primary sedimentation tanks used in activated-sludge plants, provision may be required for the excess activated sludge that may be discharged into the influent of the preliminary tanks for settlement and consolidation with the fresh sludge.

For sedimentation tanks used with trickling filters, provision may be required for the "unloading" of trickling filters and for the accumulation of sludge over longer periods than ordinarily employed in primary sedimentation tanks, if mechanical equipment for sludge removal is not provided. For sedimentation tanks in the activated-sludge process, provision must be made for light, flocculent sludge of 98 to 99.5 percent moisture and for quantities of sludge ranging from 1,500 to 10,000 mg/liter in the influent mixed liquor. Further consideration of the requirements for sedimentation tanks following trickling filters and aeration units is included in Chap. 12, which deals with these processes.

OTHER SOLIDS REMOVAL OPERATIONS AND UNITS

Flotation is a unit operation that may be used in place of primary sedimentation for removal of suspended and floating solids. Two other solids removal units are the Imhoff tank and the septic tank. These units are particularly suitable for small communities and individual households, respectively.

Flotation

Flotation is used primarily in the treatment of wastewater containing large quantities of industrial wastes that carry heavy loads of finely divided suspended solids and grease. Tannery, packinghouse, oil refinery, cannery, and laundry wastes are examples of the types of industrial wastes for which this process may be found adaptable. It is also considered to be particularly suited for treating wastes containing scum-producing materials, such as peach-processing waste, because scum can be removed and handled easily in a flotation unit. Solids having a specific gravity only slightly greater than 1.0, which would require abnormally long sedimentation times, may be removed in much less time by flotation. The theoretical aspects of the design of the various types of flotation systems are discussed in Chap. 8.

Imhoff and Septic Tanks

The removal of settleable solids and the anaerobic digestion of these solids is accomplished simultaneously in Imhoff and septic tanks.

Imhoff Tanks The Imhoff tank (see Fig. 11·12) consists of a two-story tank in which sedimentation is accomplished in an upper compartment and digestion is accomplished in a lower compartment. Settling solids pass through trapped slots into the unheated lower compartment for digestion. Scum accumulates in the sedimentation compartment and in vents adjoining the sedimentation compartments. Gas produced in the digestion process in the lower compartment escapes through the vents.

Before the use of separate heated digestion tanks, the Imhoff tank was widely used. Today, it has limited application, mostly in relatively small plants. It is simple to operate and does not require highly skilled supervision. There is no mechanical equipment to main-

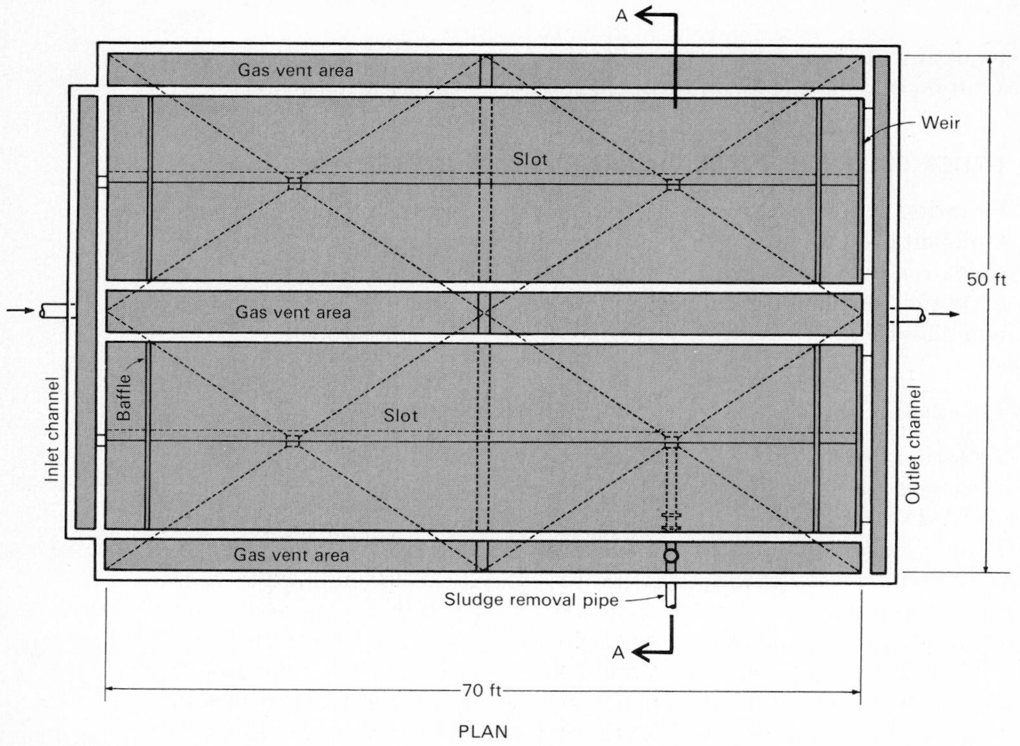

Imhoff Tank

1. Flow Q av = 1.5 mgd
2. Detention time = 2 hr
3. Surface loading = 594 gpd/sq ft
4. Wier loading = 47,000 gal/linear ft
5. Sludge digestion volume = 30,000 cu ft
6. Scum area = 20 percent of total

6″ min
6″
SLOT DETAIL

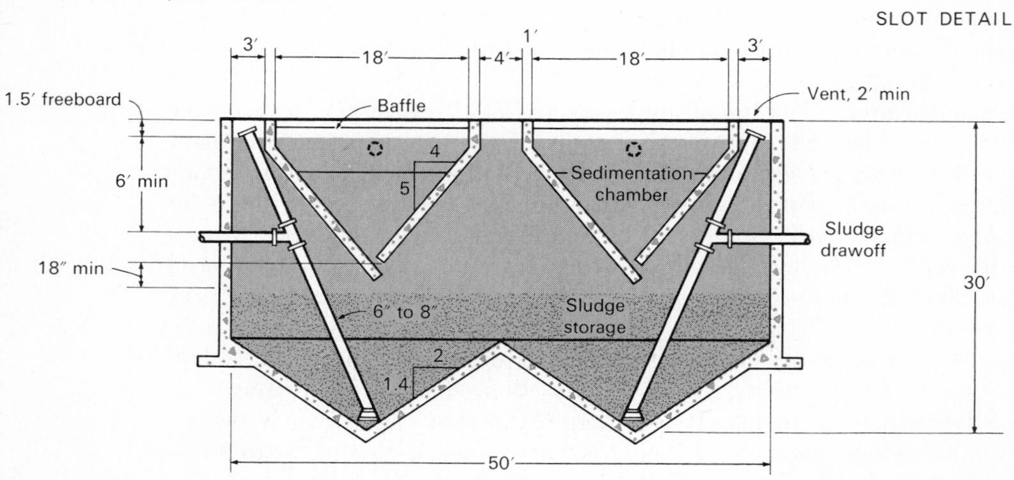

FIG. 11·12 Typical Imhoff tank.

tain, and operation consists of removing scum daily and discharging it into the nearest gas vent, reversing the flow of sewage twice a month to even up the solids in the two ends of the digestion compartment, and drawing sludge periodically to the sludge beds. Recent designs developed by manufacturers for a modified form of Imhoff tank provide means of heating the sludge compartment and mechanical removal of sludge. Conventional unheated Imhoff tanks are usually rectangular, although some small circular tanks have been used.

The settling compartments of Imhoff tanks are customarily designed to have a surface overflow rating of 600 gpd/sq ft at the average rate of flow, and a detention period of about 3 hr. The bottom of the settling compartment of the conventional unheated tank is usually sloped 1.4 vertical to 1.0 horizontal. The slot that permits solids to drop through to the digestion compartment has a minimum opening of 6 in. Several settling compartments may be installed over one digestion compartment.

The capacity of the unheated digestion compartment should provide for 6 months' storage of sludge during the cold portion of the year. For additional details regarding design and operation of Imhoff tanks, see Metcalf and Eddy [10].

The mechanized form of Imhoff tank consists of a circular sedimentation tank mounted on top of a circular sludge-digestion tank, with several gas vents rising to the surface around the periphery of the unit. The floors of both the settling and the digestion compartments are sloped slightly to the center to form flat inverted cones. Settled sludge is scraped by mechanical equipment to a trapped opening through which it is discharged to the lower or digestion compartment. Digested sludge is also scraped mechanically to a central drawoff pipe. The mechanized tank can be equipped for scum collection at the surface of the tank and for scum stirring beneath the roof of the digestion compartment. Details of the mechanized tanks may be obtained from several manufacturers. Mechanized tanks can be shallower than conventional tanks and should be cheaper to construct, especially where ground water may be a problem; however, the most desirable feature of the conventional tank, its nonmechanical nature, is lost.

Septic Tanks Septic tanks are used principally for the treatment of wastes from individual residences. In rural areas they are also used for establishments such as schools, summer camps, parks, trailer parks, and motels. Although single-chamber tanks are often used, the preferred type consists of two or more chambers in series (see Fig. 11·13). In a dual-chamber septic tank, the first compartment provides for sedimentation, sludge digestion, and sludge storage. The second compart-

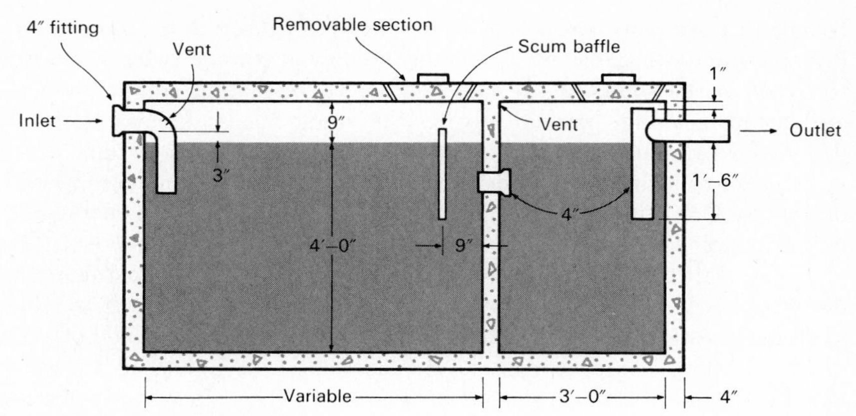

FIG. 11·13 Typical septic tank.

ment provides additional sedimentation and sludge storage capacity and thus serves to protect against the discharge of sludge and other material that might escape the first chamber. If designed for residential use, a 24-hr detention period is generally used. For larger installations serving multiple families or institutions, a shorter detention period may be permissible. In either case it is essential that adequate storage capacity be provided so that the deposited sludge may remain in the tank for a sufficient length of time to undergo decomposition or digestion before being withdrawn. In general, sludge should be removed every two to three years.

Effluent from septic tanks is normally discharged to subsurface tile or leaching fields from where it percolates into the ground. In the past, because of a lack of understanding concerning the fundamental factors governing their design and operation, a number of tile field installations have failed. However, based on the extensive studies conducted at the University of California Richmond Field Station these factors are now more clearly understood and the design of leaching fields is now a more rational undertaking. Because the details of leaching field design are beyond the scope of this text, it is recommended that the interested reader consult and review the many reports that have been issued by the University and the U.S. Public Health Service. Table 15·4 gives typical application rates.

CHEMICAL TREATMENT

Chemical precipitation, discovered in 1762, was a well-established method of sewage treatment in England as early as 1870. Lime was used as a precipitant in most cases, sometimes alone, but more often in combination with calcium chloride, magnesium chloride, alum, ferrous sulfate (copperas), charcoal, or any one of a number of other substances [10]. Chemical treatment was used extensively in the United States in the 1890s and early 1900s; but, with the development of biological treatment, the use of chemicals was abandoned and biological treatment was adopted. In the early 1930s, attempts were made to develop new methods of chemical treatment, and a number of plants were installed. Currently (1971), there is a considerable amount of renewed interest in the use of chemical treatment, and a number of processes have been developed.

The purpose of this section is, therefore, to (1) mention some of the early processes, (2) briefly review the recent applications of chemical treatment, and (3) illustrate the method used in estimating the quantity of sludge produced from chemical precipitation operations and discuss the methods and means available for its disposal.

Early Chemical Treatment Processes

In the period from 1930 to 1936 a number of different chemical treatment processes were developed to provide intermediate or complete treatment, thereby eliminating the need for biological treatment. A considerable amount of pilot plant testing was conducted and a number of full-scale plants were installed [10 and 14]. In some cases chemicals were to be added only during critical periods during the summer. In a large portion of these plants the use of chemicals was abandoned shortly after the plants were built; in others they were never used.

Processes developed during the period included the Cabrera, Diamond Alkali, Guggenheim, Laughlin, Lewis, Miller-Koller, Putnam, Scott-Darcey, Stevenson, Travers-marl, and Wright [10]. Among these, the Guggenheim and Laughlin processes, which are described below, received considerable attention. Many of the aforementioned as well as other processes are discussed in a recent article by Culp [4].

Guggenheim Process In this process, the sewage was subjected to two operations: (1) removal of suspended matter and nonbasic dissolved matter by coagulation and precipitation with an iron salt and an alkali, such as ferric sulfate and lime; and (2) removal of soluble basic compounds by passing the clarified sewage through a bed of base-exchange zeolite. The sludge was disposed of by filtration and incineration, the resulting ash being treated for the recovery of iron in the form of ferric sulfate. The zeolite was regenerated by means of a brine solution, from which ammonia was subsequently recovered [10]. A modification of the Guggenheim process involving both chemical and biological treatment was also developed and tested.

Laughlin Process In 1932, a sewage treatment plant using a chemical precipitation process developed by Laughlin was put into operation at the West Side plant in Dearborn, Mich. Substances added to the screened sewage varied from time to time, but in general they consisted of pulped waste paper, lime, and ferric chloride. The sewage, after receiving the dose of chemicals, was subjected to a short period of vigorous mixing, followed by flocculation and sedimentation for a period of 1 to 3 hr. The effluent from the two precipitation tanks underwent rapid upward filtration through a thin layer of magnetite sand, which was cleaned periodically by a traveling magnetic device. The final effluent was chlorinated, and the sludge was dewatered by vacuum filtration. Later, an incinerator was installed to dispose of the dewatered sludge [10].

Recent Applications of Chemical Treatment

The need to provide more complete removal of the organic compounds and nutrients (nitrogen and phosphorus) contained in wastewater is responsible for the renewed interest in chemical precipitation. Processes have been developed for the complete treatment of raw wastewater, including the removal of either nitrogen and phosphorus or both. Other processes have been developed to remove phosphorus by chemical precipitation and are designed to be used in conjunction with biological treatment. Still other process flowsheets have been developed to remove nitrogen and phosphorus from treated secondary effluents (see Chap. 14).

Complete Treatment A flowsheet for the complete treatment of raw wastewater, including the removal of phosphorus and nitrogen, is presented in Fig. 11·14. As shown, after first-stage lime precipitation, the wastewater is passed through a stripping tower to remove ammonia (see Chap. 14 for details). After stripping, the wastewater pH is reduced by carbonation, and calcium carbonate is removed by second-stage precipitation. The waste is then passed through a mixed-media filter to remove any residual floc and finally through carbon columns to remove the dissolved organics before discharge to the environment or reuse. As indicated in Chap. 9, the lime required will be a function of the raw wastewater alkalinity. For most wastewaters, Fig. 11·15 may be used to estimate the required dosage. Depending on the operation of the stripping tower (e.g., air-to-liquid ratio), as well as the charac-

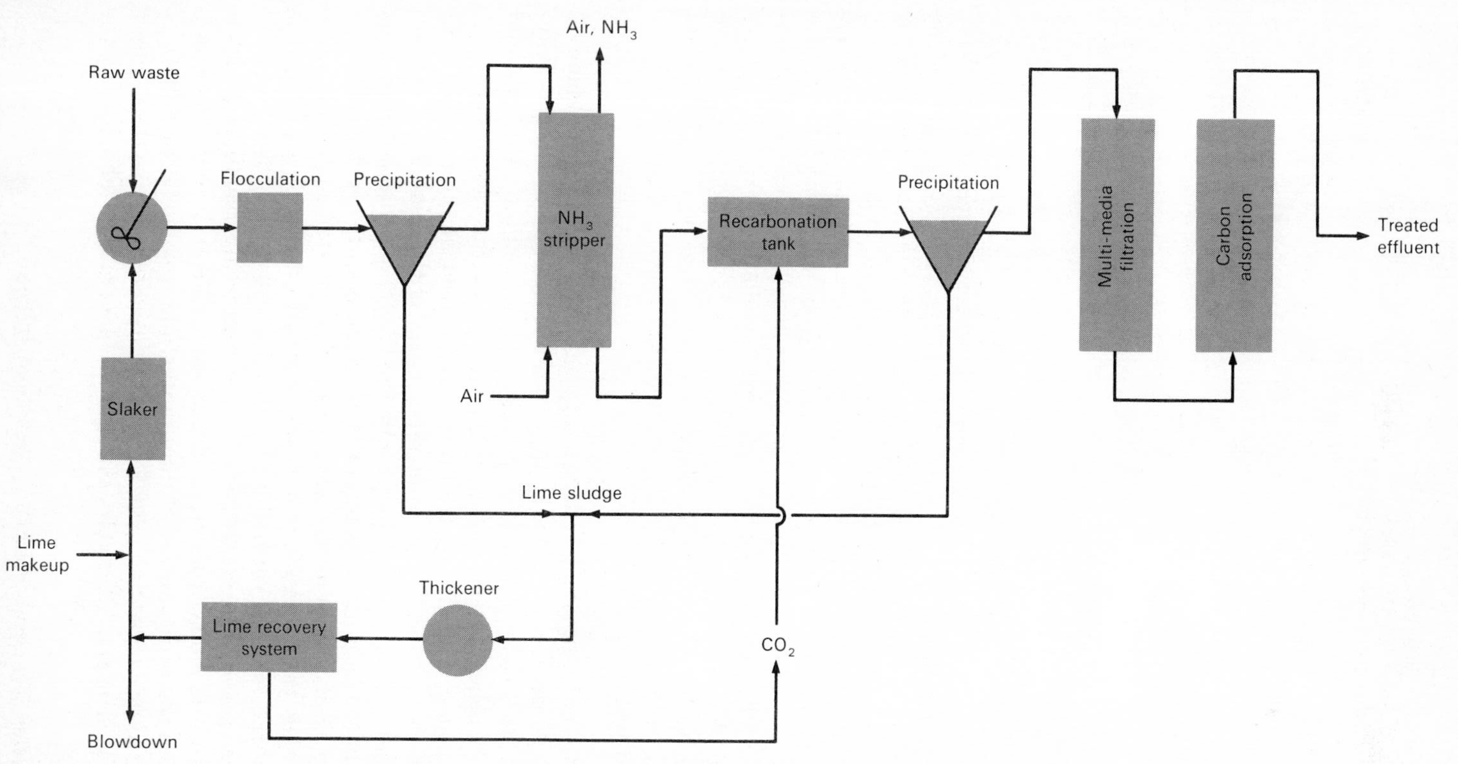

FIG. 11·14 Typical flowsheet for the complete chemical-physical treatment of raw wastewater including removal of nitrogen and phosphorus.

FIG. 11·15 Lime dosage required to raise the pH to 11 as a function of raw wastewater alkalinity.

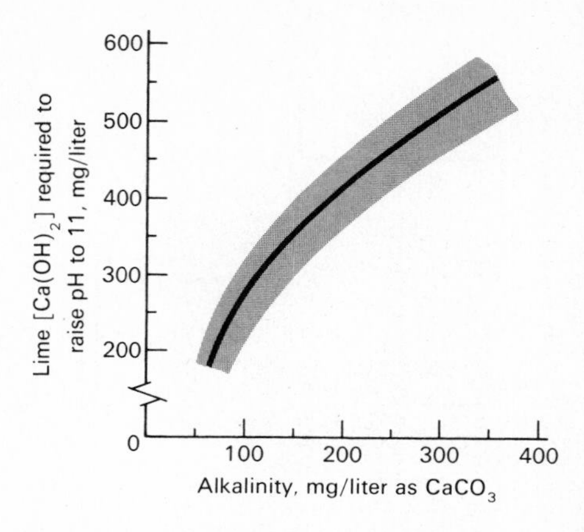

teristics of the raw sewage, it may be possible to reduce the lime dosage. Unfortunately, the required dosage can be determined only by pilot plant testing on the actual wastes. The characteristics of the sludges produced from chemical precipitation operations are reported in Table 13·1 (see Chap. 13, "Sources, Quantities, and Characteristics").

Where nitrogen removal is not a requirement, an alternative complete chemical treatment process that may be used involves the use of alum as the primary precipitant. After flocculation and sedimentation, the effluent is passed through mixed-media filters to remove any residual floc. The final step in the process involves passage of the filtered water through activated-carbon columns to remove the remaining organic material. The effluent can then be reused or discharged to the environment.

Phosphorus Removal Chemicals that have been used for the removal of phosphorus include lime, alum, and ferric chloride or sulfate. Polymers have also been used effectively in conjunction with lime and alum. Normally, these chemicals are added to the raw sewage, and phosphorus-containing precipitants are removed in the primary sedimentation tanks. Effluent from these tanks is then subjected to conventional biological treatment. Alternatively, chemicals have been added in the aeration tank or in the line leading to the activated-sludge settling tanks. Review of all the variations reported in the literature concerning the optimum point and method of chemical addition is beyond the scope of this book.

Because of the complexity of the competing reactions and the effects of alkalinity, pH, trace elements, and ligands found in wastewater on the precipitation of phosphorus, the stoichiometric equations presented previously in Chap. 9 cannot be used to estimate the required chemical dosages. Therefore, dosages are generally established on the

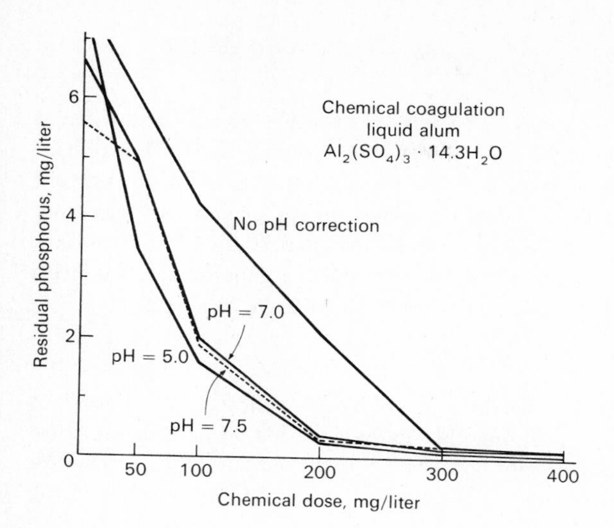

FIG. 11·16 Residual phosphorus concentration as a function of alum dosage.

basis of bench-scale tests and occasionally by full-scale tests, especially in the case of polymers. The results of a bench-scale test program to evaluate the removal of phosphorus with alum are shown in Fig. 11·16. Typical chemical dosage ranges for the removal of phosphorus with aluminum [Al(III)], iron [Fe(II)], and iron [Fe(III)] are reported in Table 11·6.

TABLE 11·6 DOSAGE RANGE FOR ALUM AND IRON FOR PHOSPHORUS REMOVAL*

Chemical	moles of chemical / moles of phosphorus
Alum [Al(III)]	1.6–2.6
Iron [Fe(II)]	1.8–2.6
Iron [Fe(III)]	1.8–2.2

* Residual phosphorus ≤ 1 mg/liter.

Quantities of Sludge

The handling and disposal of the sludge resulting from chemical precipitation was in the past and is still one of the greatest difficulties of this method of treatment. Sludge is produced in great volume from most chemical precipitation operations, often reaching 0.5 percent of the volume of sewage treated. Estimation of the quantity of sludge resulting from a typical chemical precipitation process is illustrated in the following example.

EXAMPLE 11·4 *Estimation of Sludge Volume from Chemical Precipitation of Raw Sewage*

Estimate the pounds and volume of raw sludge with and without the use of chemical precipitants from a sewage flow of 1 mgd containing 300 mg/liter of suspended solids. Assume that 60 percent of the suspended solids are removed in the primary settling tank without the addition of chemicals, and that 70 lb of ferrous sulfate and 600 lb of lime are added per million gallons (see Chap. 9), increasing the removal of suspended solids to 70 percent.

Solution

1. Compute the weight and volume of solids removed without chemicals, assuming the sludge contains 95 percent moisture and has a specific gravity of 1.02 (see Chap. 13, "Sources, Quantities, and Characteristics").

$$\text{Weight of solids removed} = 0.60(300)(8.34) = 1{,}500 \text{ lb}$$

$$\text{Volume of sludge} = \frac{1{,}500 \text{ lb}}{1.02\left(\dfrac{62.4 \text{ lb}}{\text{cu ft}}\right)(0.05)} = 470 \text{ cu ft}$$

2. Determine the weight of sewage solids removed with the addition of chemicals.

$$\text{Weight of solids removed} = (0.70)(300)(8.34) = 1{,}750 \text{ lb}$$

3. Determine the amount of ferric hydroxide formed from the addition of 70 lb of ferrous sulfate.

$$\text{Ferric hydroxide formed} = 70\,\frac{106.9}{278} = 27 \text{ lb}$$

4. Determine the amount of calcium carbonate formed from the addition of lime.

Calcium carbonate formed from reaction with ferrous sulfate $= 70\,\frac{112}{278}\,\frac{100}{56} = 50$ lb

$CaCO_3$ formed from reaction with CO_2 and bicarbonates, assuming equal molar concentrations of CO_2 and $Ca(HCO_3)_2$

$$= \tfrac{300}{112}\left(600 - 50\,\tfrac{56}{100}\right) = 1{,}530 \text{ lb}$$

Solubility of $CaCO_3$, 25 mg/liter $= 25\,(8.34) = 208$ lb

Total $CaCO_3$/million gal in sludge $= 50 + 1{,}530 - 208 = 1{,}372$ lb

5. Total solids in sludge on a dry basis = 1,750 + 27 + 1,372 = 3,149 lb/million gal
6. Assuming an overall specific gravity of 1.07 and a moisture content of 92.5 percent, compute the volume of sludge.

$$\text{Volume of sludge} = \frac{3{,}149 \text{ lb}}{1.07\left(\dfrac{62.4 \text{ lb}}{\text{cu ft}}\right)(0.075)}$$
$$= 630 \text{ cu ft}$$

Disposal of Sludge

Disposal methods that have been used for chemical sludges include (1) spreading on soil, (2) lagooning, (3) landfilling, and (4) ocean dumping. Details on these methods will be found in Chap. 13. One of the advantages of using lime or alum is the possibility of recovering calcium oxide and aluminate by burning. Development of effective pure-product recovery methods would permit recycling of the precipitant, reduction of chemical costs, and minimization of the disposal problem. It is anticipated that such methods will be developed within the next few years as the use of chemical treatment continues to expand.

SLUDGE AND SCUM PUMPING

Sludge produced in sewage treatment plants must be conveyed from point to point in the plant in conditions ranging from a watery sludge or scum to a thick sludge. For each type of sludge encountered, a different method of conveyance or pump may be needed.

Pumps

Pump types used to convey sludge include the plunger, progressing-cavity, centrifugal, and torque-flow.

Plunger Plunger pumps have been used frequently, and if rugged enough for the service, have proven to be quite satisfactory. The adantages of plunger pumps may be listed as follows:

1. Pulsating action tends to concentrate the sludge in the hoppers ahead of the pumps.
2. They are suitable for suction lifts up to 10 ft, and are self-priming.
3. Low pumping rates can be used with large port openings.

4. Positive delivery is provided unless some object prevents the ball check valves from seating.
5. They have constant but adjustable capacity, regardless of large variations in pumping head.
6. Large discharge heads may be provided for.
7. Heavy-solids concentrations may be pumped if the equipment is designed for the load conditions.

Plunger pumps come in simplex, duplex, and triplex models with capacities of 40 to 60 gpm per plunger, and larger models are available. Pump speeds should be between 40 and 50 rpm, and the pumps should be designed for a minimum head of 80 ft, since grease accumulations in sludge lines cause a progressive increase in head with use. Capacity is decreased by shortening the stroke of the plunger; however, the pumps seem to operate more satisfactorily at, or near, full stroke. For this reason, many pumps are provided with variable-pitch V-belt drives for speed control of capacity.

Progressing-Cavity The Moyno (trade name) progressing-cavity pump has been used successfully, particularly on concentrated sludge. The pump is composed of a single-threaded rotor that operates with a minimum of clearance in a double-threaded helix of rubber. It is self-priming at suction lifts up to 28 ft, is available in capacities up to 350 gpm, and will pass solids up to 1.125 in. in diameter.

Centrifugal With centrifugal pumps, the problem is to obtain a large enough pump to pass the solids without clogging and a small enough capacity to avoid pumping a sludge diluted by large quantities of the overlying sewage. Centrifugal pumps of special design have been used for pumping primary sludge in large plants. Since the capacity of a centrifugal pump varies with the head, which is usually specified great enough so that the pumps may assist in dewatering the tanks, the pumps have considerable excess capacity under normal conditions. Throttling the discharge to reduce the capacity is impractical because of frequent stoppages; hence it is absolutely essential that these pumps be equipped with variable-speed drives.

Centrifugal pumps of the bladeless impeller type have been used to some extent and in some cases have been deemed preferable to either the plunger or screw-feed types of pumps. Bladeless pumps have approximately one-half the capacity of conventional nonclog pumps of the same nominal size and consequently approach the hydraulic requirements more closely. The design of the pump makes clogging at the suction of the impeller almost impossible.

Torque-Flow This type of pump, which uses a fully recessed impeller, is very effective in conveying sludge. The size of particles that can be handled is limited only by the diameter of the suction or discharge valves. The rotating impeller develops a vortex in the sludge so that the main propulsive force is the liquid itself.

Head-Loss Determination

The head loss encountered in the pumping of sludge is primarily dependent on the flow velocity and the nature of the material being pumped. When the Reynolds number for the sludge flow is greater than about 5,000, head losses are nearly the same as those for water. For pipe sizes of 6 to 12 in. and a Reynolds number of 5,000, turbulence occurs at about 5 to 6 fps. For velocities below about 5 fps, laminar or transitional flow conditions exist and, due to the properties of the sludge, it has been found that conventional head-loss equations cannot be used. For determining the head loss in the laminar flow range, the use of Fig. 11·17 is recommended. From this figure the factor k is obtained for a given moisture content and type of sludge. The head loss when pumping sludge is determined by multiplying the head loss for water by k. For turbulent flow, head losses for digested, activated, and trickling-filter sludges are from 10 to 25 percent greater than for water, while losses for primary sludge and mixtures of primary and other sludges may be from two to four times the loss for water. It has been

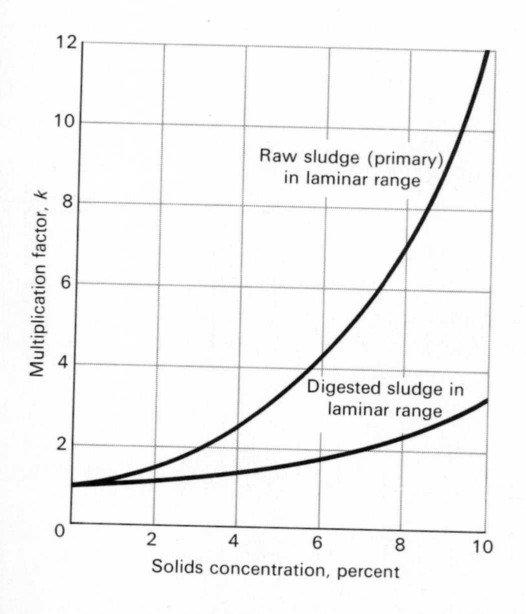

FIG. 11·17 Head-loss multiplication-factor curves for different sludge types and moisture contents.

observed that head losses increase with increased solids content, increased volatile content, and lower temperatures. When the percent volatile matter multiplied by the percent solids exceeds 600, difficulties will be encountered in pumping sludge. The calculation of head loss in pumping sludge in the laminar-flow range is illustrated in Example 11·5.

EXAMPLE 11·5 *Head Loss for Sludge Pumping*

Sludge is to be pumped from a primary settling tank to a sludge thickener. The difference in elevation between these two units is 20 ft, and 200 ft of 8-in. pipe is required to interconnect them. Determine the horsepower required to pump the sludge and the pressure that must be maintained at the outlet side of the pump if a pressure of 2 psi is maintained on the inlet side. The sludge is to be pumped at a velocity of 2.5 fps. Assume that the sludge has a moisture content of 92 percent and a specific gravity of 1.03.

Solution

1. Using the Darcy-Weisbach equation (Eq. 3·18) compute the head loss as if the sludge were water, assuming a value of $f = 0.020$.

$$h_L \text{ (water)} = 0.020 \frac{200 \ (6.25)}{0.67 \ (64.4)} = 0.58 \text{ ft}$$

2. Determine the value of k from Fig. 11·17 and compute the head loss when pumping sludge.

$$k = 7.3$$

$$h_L \text{ (sludge)} = 0.58 \times 7.3 = 4.22 \text{ ft}$$

3. Compute the total pumping head.

$$H = 20 + 4.22 = 24.22 \text{ ft of sludge}$$

To this should be added an allowance for the losses in valves, elbows, and other fittings. Assume this equals another 2 ft, making the total pumping head 26.22 ft.

4. Compute the horsepower required to pump the sludge assuming a pump efficiency of 65 percent.

$$\text{hp} = \frac{Q\gamma \text{ (specific gravity) } (h)}{(550)\text{eff}}$$

$$Q = 0.349 \ (2.5) = 0.873 \text{ cfs}$$

$$\text{hp} = \frac{0.873 \ (62.4) \ (1.03) \ (26.22)}{550 \ (0.65)} = 4.12$$

5. Compute the discharge pressure at the pump outlet.

$$\text{Pressure} = 2.0 + \frac{26.22\ (1.03)\ (62.4)}{144} = 13.7\ \text{psi}$$

An alternate method of computing the head loss for laminar flow conditions was derived by Babbitt and Caldwell [1 and 2], based on the results of experimental and theoretical studies. Presented in the form of equations, the head loss when pumping sludge is considered to depend on the plastic properties of a non-Newtonian fluid. Although this method is theoretically sound, the selection of suitable parameters for insertion in the equations presented is difficult, based on the available data. For further information on this method, the student is referred to the original work [2] and the literature.

In treatment plants, conventional sludge piping should not be smaller than 6 in. in diameter (smaller-diameter glass-lined pipe has been used successfully) and need not be larger than 8 in., unless the velocity exceeds 5 to 6 fps, in which case the pipe is sized to maintain that velocity. It is common practice to install a number of cleanouts in the form of plugged crosses instead of elbows so that the lines can be rodded if necessary. Pump connections should not be smaller than 4 in. in diameter.

Grease has a tendency to coat the inside of piping used for transporting primary sludge and scum. This results in a decrease in effective diameter and a large increase in pumping head. For this reason, low-capacity positive-displacement pumps are designed for heads greatly in excess of the theoretical. Centrifugal pumps, with their larger capacity, usually pump a more dilute sludge, often containing some sewage, and head buildup due to grease accumulations appears to occur more slowly. In some plants provisions have been made for melting out the grease by circulating hot water or steam through the main sludge lines.

In treatment plants, friction losses are low (see Example 11·5) and there is no difficulty in providing an ample factor of safety. In the design of long sludge lines transporting solids from one plant to another for further treatment, or discharging digested sludge offshore in the ocean, total friction losses may exceed 100 psi and should be estimated carefully. For solving such problems, the engineer should refer to the literature for design and operating data on long sludge lines [12].

Application of Pumps to Types of Sludge

Types of sludge that are pumped include primary, chemical, trickling-filter and activated, elutriated, thickened, and concentrated. Scum that accumulates at various points in a treatment plant must also be pumped.

Primary Sludge Ordinarily, it is desirable to obtain as concentrated a sludge as practicable from primary tanks. This is usually accomplished by collecting the sludge in hoppers and by pumping intermittently, allowing the sludge to collect and consolidate between pumping periods. The character of primary raw sludge (see Chap. 13) will vary considerably, depending on the characteristics of the solids in the wastewater, the types of units and their efficiency, and, where biological treatment follows, the quantity of solids added from

1. Overflow liquors from digestion tanks
2. Waste activated sludge
3. Humus sludge from settling tanks following trickling filters
4. Overflow liquors from sludge elutriation tanks

The character of primary sludge is such that conventional nonclog pumps cannot be used. Plunger pumps have been used frequently on primary sludge. Centrifugal pumps of the screw-feed and bladeless type, and torque-flow pumps have also been used.

Chemical Precipitation Sludge Sludge from chemical precipitation processes can usually be handled in the same manner as primary sludge.

Trickling-Filter and Activated Sludge Sludge from trickling filters is usually of such homogeneous character that it can be easily pumped with either plunger or nonclog centrifugal pumps.

Return activated sludge is dilute and contains only fine solids, so that it may readily be pumped with nonclog centrifugal pumps, which operate at slow speed because the head is low and because the flocculent character of the sludge should not be broken up.

Elutriated, Thickened, and Concentrated Sludge Plunger pumps are frequently used for concentrated sludge to accommodate the high friction-head losses in pump discharge lines. The progressing-cavity type of positive-displacement pump has been used successfully for dense sludges containing up to 20 percent solids. Because these pumps have limited clearances, it is necessary to reduce all solids to small size.

Scum Pumping Screw-feed pumps, plunger pumps, and pneumatic ejectors are frequently used for pumping scum. Bladeless or torque-flow centrifugal pumps may also be used for this service.

CHLORINATION

The chemistry of chlorine in water and wastewater has been discussed in Chap. 9 along with an analysis of how chlorine functions as a disinfectant. However, chlorine has been applied for a wide variety of objectives other than disinfection in the wastewater treatment field. Therefore, the purpose of this section is to discuss briefly (1) the various uses and required dosages, (2) the chlorine compounds most commonly used, (3) the equipment and methods used in its application, and (4) the design of chlorine contact chambers used for disinfection.

Application

To aid in the design and selection of the required facilities and equipment, the uses, including dosage ranges, to which chlorine and its compounds have been applied are discussed in this section.

TABLE 11·7 CHLORINATION APPLICATIONS IN WASTEWATER COLLECTION, TREATMENT, AND DISPOSAL

Application	Dosage range, mg/liter	Remarks
Collection:		
Slime-growth control	1–10	Control of fungi and slime-producing bacteria
Corrosion control (H_2S)	2–9*	Control brought about by destruction of H_2S in sewers
Odor control	2–9*	Especially in pump stations and long flat sewers
Treatment:		
Grease removal	2–10	Added before preaeration
BOD reduction	0.5–2†	Oxidation of organic substances
Ferrous sulfate oxidation	‡	Production of ferric sulfate and ferric chloride
Filter-ponding control	1–10	Residual at filter nozzles
Filter-fly control	0.1–0.5	Residual at filter nozzles, used during fly season
Sludge-bulking control	1–10	Temporary control measure
Digester supernatant oxidation	20–140	
Digester and Imhoff tank foaming control	2–15	
Nitrate reduction	See Chap. 14	Conversion of nitrate to ammonia
Disposal:		
Bacterial reduction	2–20	Plant overflows, storm water
Disinfection	See Table 11·8	Depends on nature of wastewater

* Per mg/liter of H_2S.
† Per mg/liter of BOD_5 destroyed.
‡ $6FeSO_4 \cdot 7H_2O + 3Cl_2 \rightarrow 2FeCl_3 + 2Fe_2(SO_4)_3 + 42H_2O$.

Uses The principal uses of chlorine and its compounds in the collection, treatment, and disposal of wastewater are reported in Table 11·7. Of the many different applications, disinfection of wastewater effluents is still the most important, although serious questions are being raised about the merits of effluent chlorination in general.

Dosages The dosages for various application, with the exception of disinfection, are reported in Table 11·7. When used for disinfection, the probable amounts of chlorine that may be required are presented in Table 11·8. A range of dosage values is given because they will vary depending on the characteristics of the wastewater. It is for this reason that whenever possible laboratory chlorination studies should be conducted to determine optimum chlorine dosages.

Chlorination capacities are generally selected to meet the specific design criteria of the state or other regulatory agencies controlling the receiving body of water (see Example 11·6). In any case where the residual in the effluent is specified or the final number of coliform bacteria is limited, the actual amount of chlorine must be determined by experiment, if at all possible. However, in the absence of more specific data the maximum values given in Table 11·8 can be used as a guide in sizing chlorination equipment.

TABLE 11·8 TYPICAL CHLORINE DOSAGES FOR DISINFECTION

Effluent from	Dosage range, mg/liter
Untreated wastewater (prechlorination)	6–25
Primary sedimentation	5–20
Chemical-precipitation plant	2–6
Trickling-fliter plant	3–15
Activated-sludge plant	2–8
Multimedia filter following activated-sludge plant	1–5

EXAMPLE 11·6 *Chlorinator Selection*

Determine the capacity of a chlorinator for a treatment plant with an average wastewater flow of 1 mgd. The peaking factor for the treatment plant is 3.0 and the maximum required chlorine dosage (set by state regulations) is to be 20 mg/liter.

Solution

1. Determine the capacity of the chlorinator at peak flow.

$$\text{lb Cl}_2\text{/day} = \frac{20 \text{ mg}}{\text{liter}} \times \frac{8.34 \text{ lb}}{\text{gal}} \times 1 \text{ mgd} \times 3.0$$
$$= 500$$

Use three 250-lb/day units. Although this capacity will not be required during most of the day, it must be available to meet the chlorine requirements at peak flow. Best practice calls for the availability of a standby chlorinator.

2. Estimate the daily consumption of chlorine. Assume an average dosage of 10 mg/liter.

$$\text{lb Cl}_2\text{/day} = 10(8.34)(1) = 83.4$$

Chlorine Compounds

The most common chlorine compounds used in wastewater treatment plants are calcium and sodium hypochlorite and chlorine gas. Calcium and sodium hypochlorite have been used in very small treatment plants, such as package plants, where simplicity and safety were far more important than cost. It is interesting to note that in 1965 New York City completed a change-over from liquid chlorine in ton containers to sodium hypochlorite, fundamentally for reasons of safety as influenced by local conditions. For a complete discussion, see Ref. [13]. In 1967, Chicago, faced with the need to install chlorination facilities to meet upgraded effluent and receiving-water quality standards, installed sodium hypochlorite storage tanks and feed equipment at its North Side and Calumet treatment plants. The decision to use hypochlorite was based solely on safety considerations.

Calcium Hypochlorite Calcium hypochlorite is available commercially in either a dry or a wet form. High-test calcium hypochlorite contains at least 70 percent available chlorine. In dry form, it is available as a powder or as granules, compressed tablets, or pellets. A wide variety of container sizes is available depending on the source. Because calcium hypochlorite granules or pellets are readily soluble in water and, under proper storage conditions, are relatively stable, they are often favored over other available forms. Because of its oxidizing potential, calcium hypochlorite should be stored in a cool, dry location away from other chemicals in corrosion-resistant containers.

Sodium Hypochlorite Sodium hypochlorite solution is available in strengths from 1.5 to 15 percent with 3 percent the usual maximum strength; thus transportation costs may limit its application. The solution decomposes more readily at high concentrations and is affected by exposure to light and heat. It must, therefore, be stored in a cool location in a corrosion-resistant tank. However, where sodium hypochlorite is available at a reasonable cost, its use should certainly be investigated by the design engineer.

Chlorine Chlorine is supplied as a liquified gas under high pressure in containers varying in size from 100- and 150-lb cylinders to ton containers, multiunit tank cars containing 15 one-ton containers, and tank cars of 16-, 30-, and 55-ton capacity. Selection of the size of chlorine pressure vessel should depend on an economic study of transportation costs, demurrage, handling charges, available space, and rate of chlorine usage. Storage and handling facilities can be designed with the aid of material developed by the Federation of Sewage and Industrial Wastes Associations [7] and the Chlorine Institute. Although all of the safety devices and precautions that must be designed into the chlorine handling facilities are too numerous to mention, the following are fundamental:

1. Chlorine gas is both very poisonous and very corrosive. Adequate exhaust ventilation at floor level should be provided because chlorine gas is heavier than air.
2. Chlorine storage and chlorinator equipment rooms should be walled off from the rest of the plant and should be accessible only from the out-of-doors. A fixed glass viewing window should be included in an inside wall. Fan controls and gas masks should be located at the room entrance.
3. Dry chlorine liquid and gas can be handled in black wrought iron piping, but chlorine solution is highly corrosive and should be handled in rubber-lined or tough plastic piping with diffusers of hard rubber.
4. Storage should be provided for a 30-day supply. Cylinders in use are set on platform scales, set flush with the floor, and the loss of weight used as a positive record of chlorine dosage.

Chlorination Apparatus and Dosage Control

In this section the apparatuses (equipment) used to inject (feed) chlorine or its related compounds into the wastewater and the methods used to control the required dosages are discussed.

Hypochlorite Feeders For very small installations serving about 10 people, it is possible to use drip-type feeders. For up to 100 people, orifice-controlled feeders using a constant-head tank fed by gravity from an overhead reservoir can be employed successfully. The only difficulty is clogging of orifices so that periodic maintenance is required. The most satisfactory means of feeding sodium or calcium hypochlorite is through the use of low-capacity proportioning pumps.

Generally the pumps are available in capacities up to 120 gpd, with adjustable stroke for any value below this. Large capacities or multiple units are available from some of the manufacturers. The pumps can be arranged to feed at a constant rate or they can be programmed by a time clock to start and stop at desired intervals. Such intervals can be determined by a totalizing meter, which will send electric impulses to the pump to feed for a selected length of time at a selected rate after a chosen number of gallons of sewage have flowed by the meter. Where the purpose of chlorination is disinfection, this method is applicable only where the flow passes into a chamber of ample size where sufficient mixing takes place to ensure continuous chlorination. The feeding of sodium and calcium hypochlorite by this method has gained acceptance over the past 15 years.

Chlorine Feeders Chlorine may be applied directly as a gas or in an aqueous solution. Some small-capacity chlorinators utilize pressure injection of the gas into the wastewater, but there are certain dangers inherent in leakages of this poisonous gas into the atmosphere of the treatment plant. Further, because of the hazards involved in handling chlorine gas, direct application is limited to large installations where adequate safety precautions can be rigidly enforced.

The most widely accepted types of chlorinators are those using vacuum-feed devices (see Fig. 11·18). An ejector or aspirator is used to create a vacuum to draw chlorine gas through an orifice or rotameter. The differential head across this measuring device may be held constant or may be varied in accordance with flow by changing the degree of vacuum. Chlorine gas is absorbed into the ejector stream and feeds into the sewage as a chlorine solution.

Dosage Control Control of dosage may be done in several ways. The simplest method is to use manual dosage with the operator changing the feed rate to suit conditions. The required dosage is usually determined by measuring the residual after 15 min of contact time and adjusting the dosage to obtain an 0.5 mg/liter residual. A second method is to use a program control that changes the feed rate to follow a preselected pattern. This is the most economical method of obtaining automatic control. A third method is to pace the chlorine flowrate to the sewage flowrate as measured by a primary meter such as a Parshall flume or flow tube. A fourth method is to control the chlorine dosage by automatic measurement of the chlorine residual. An automatic

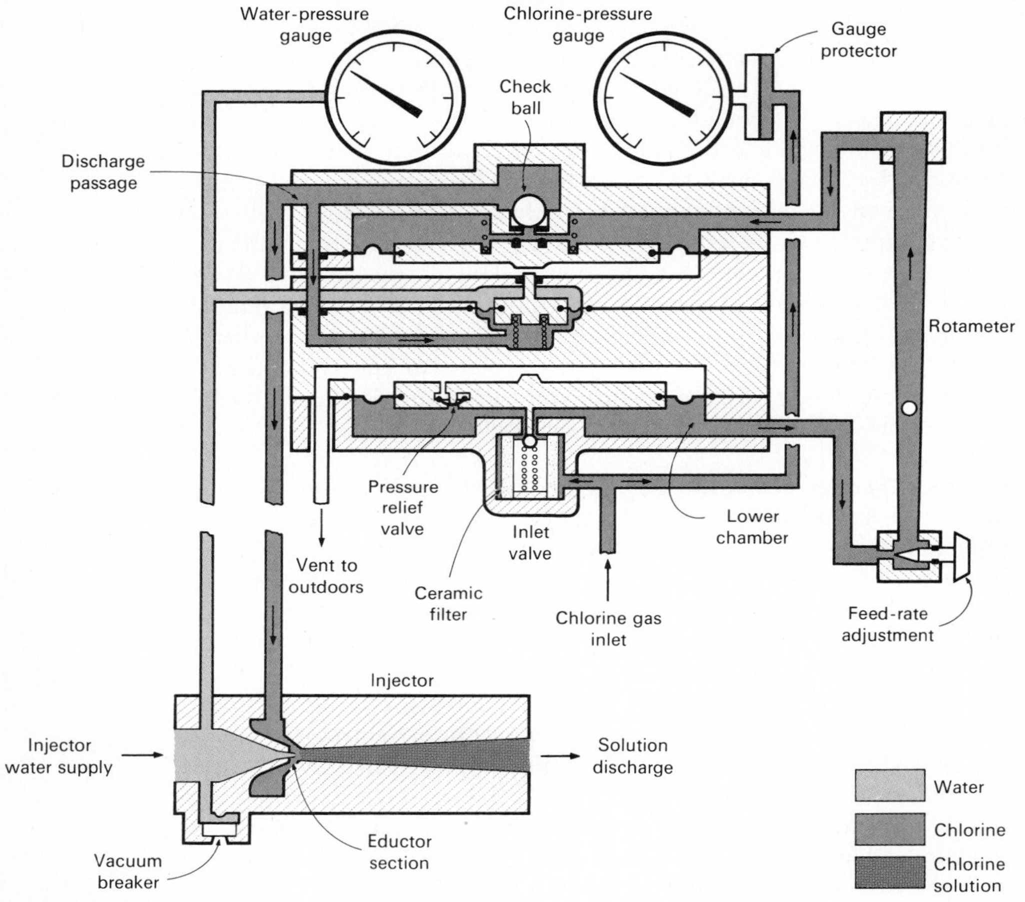

FIG. 11·18 Typical vacuum-feed chlorinator [from BIF].

analyzer with signal transmitter and recorder is required. Finally, a compound system that incorporates both methods 3 and 4 may be used. In a compound system, the control signals obtained from the sewage flow meter and from the residual recorder are superimposed to provide more precise control of chlorine dosage and residual.

Chlorine Contact Chambers

As pointed out under the theoretical discussion in Chap. 9, the time of contact is an important factor in achieving bacterial kill. The time of contact is usually specified by the regulatory agency and may range from 15 to 30 min. For instance, the *Ten States Standards* [8] state "After thorough mixing, a minimum contact period of 15 min at peak hourly flow or maximum rate of pumpage shall be provided for disinfection. In primary plants chlorine should preferably be applied ahead of sedimentation tanks."

Design Considerations Important factors that must be considered in the design of chlorine contact chambers include (1) method of chlorine addition, (2) provision for mixing, (3) avoidance of short circuiting, (4) maintenance of solids transport velocity, and (5) provision for bypassing the chamber or portions of it.

The design of the contact tank should provide for addition of chlorine solution through a diffuser, which may be a plastic or hard-rubber pipe with drilled holes through which the chlorine solution can be uniformly distributed into the path of flow of sewage, or it can flow directly to the propeller of a rapid mixer for instantaneous and complete diffusion. Mixing by hydraulic turbulence for at least 30 sec must be maintained at or near the point of addition of chlorine solution to the sewage if mechanical mixing is not used.

Because of the importance of contact time, careful attention should be given to design of the contact chamber so that at least 80 to 90 percent of the wastewater is retained in the basin for the specified contact time. This can be achieved by using a plug-flow round-the-end type contact chamber or a series of interconnected basins or compartments (a minimum of six is recommended). The advantage of using a compartmentalized chlorine contact chamber or a plug-flow reactor can be evaluated using the analysis presented in Chap. 12 in connection with the reactor selection for the activated-sludge process. If the time of travel in the outfall sewer at the maximum design flow is sufficient to equal or exceed the required time of contact, it may be possible to eliminate the chlorine contact chambers.

The horizontal velocity at minimum flow in the chamber should be sufficient to scour the bottom or at least to give a minimum deposition of sludge solids that may have passed through the settling tank. Horizontal velocities should be at least 5 to 15 fpm. Provision should be made for dewatering the chlorine contact tank and for removal of sludge by flushing or manual operation. A bypass should be provided and provision made for emergency chlorination of the effluent at such times unless multiple contact tanks are provided.

Outlet Control and MPN Measurement The flow at the end of the contact chamber may be metered by means of a V-notch or rectangular weir. Control devices for chlorination in direct proportion to the flow may be operated from these meters or from the main plant flow meter. Final determination of the success of a chlorine contact chamber must be based upon samples taken and analyzed for chlorine residual and MPN of coliform organisms. In the event that no chlorine contact chamber is provided and the outfall sewer is used for contact, the sample can be obtained at the point of chlorination, held for the theoretical detention time and the residual determined. The sample is then dechlorinated and subsequently analyzed for bacteria by normal laboratory procedures.

ODOR CONTROL

In wastewater treatment plants, the principal sources of odors are from (1) septic sewage containing hydrogen sulfide on arrival at the plant, (2) industrial wastes discharged to the collection system, (3) unwashed grit, (4) scum on primary settling tanks, (5) sludge-thickening tanks, (6) waste gas burning operations, especially if the burner goes out, (7) chemical mixing operations, (8) sludge incinerators, and (9) poorly digested sludge in drying beds.

In general, odor control is accomplished by aeration, chemical oxidation (chlorine and ozone), adsorption, or masking the odors with chemical additives and aerosols. Odor control of raw wastes is generally accomplished by preaeration or prechlorination. Preaeration has been discussed previously, and prechlorination dosages for odor control range from 10 to 20 mg/liter [1].

Ozone is a strong oxidizing agent that has been used successfully to treat the odor-causing substances in effluents from preliminary treatment. It has also been used to eliminate odors in the air emanating from aeration and sludge-thickening tanks.

Controlling odors from sludge drying beds is seldom necessary if the sludge is well digested. Open beds are generally located far from residences to minimize any possible nuisance. Methods for odor control of drying sludge include spraying aerosols from fixed fog nozzles located around the drying beds and liming the sludge.

PROBLEMS

11·1 A vertical bar rack with 1-in. openings is to screen sewage arriving at the treatment plant in a circular sewer with $d = 4$ ft, $n = 0.013$, $s = 0.00064$. The maximum carrying capacity is four times the average dry weather flow. Find the size of the steel bars comprising the rack, the number of bars in the rack, and the head loss for dry weather flow conditions. Use rectangular bars.

11·2 Design a horizontal-flow grit chamber for a plant with an average flow of 4 mgd, a minimum flow of 2 mgd, and a maximum flow of 10 mgd. The grit chamber should remove 65-mesh–size materials, have not less than two channels, and handle a flow of 7 mgd with one channel out of service. (At higher flows all channels will be placed in service.)

11·3 Design an aerated grit chamber for the same conditions as in Prob. 11·2. Determine the amount of air required and the pressure at the discharge of the blowers. Allow a 12-in. loss in the diffusers and add the submergence plus 30 percent for loss in piping and valves. Determine the horsepower required by the blower from the formula

$$\text{hp} = \frac{\text{cfm} \times \text{lb/sq ft}}{33{,}000 \times \text{efficiency}}$$

using an efficiency of 60 percent. Determine the monthly power bill, assuming a motor efficiency of 90 percent and a power cost of 2 cents/kwh.

11·4 Discuss the advantages and disadvantages of aerated grit chambers versus horizontal-flow grit chambers.

11·5 Design a skimming chamber for a treatment plant with a flow of 2.5 mgd. Use a detention time of 10 min and a rise velocity of 8 in./min. Determine the surface area, surface loading, and chamber dimensions.

11·6 Design a circular radial-flow sedimentation tank for a town with a projected population of 45,000. Assume a sewage flow of 90 gpcd. Design for 2-hr detention at 120 percent of the average flow. Determine the tank depth and diameter to produce an overflow rate of 900 gpd/sq ft for average flow. Assume standard tank dimensions to fit mechanisms that are made in diameters of whole feet and in depths of $\frac{1}{2}$ ft from 7 to 12 ft.

11·7 Contrast pressure and vacuum flotation with sedimentation discussing the following parameters:
(*a*) Detention time
(*b*) Surface loading rate
(*c*) Power input
(*d*) Efficiency
(*e*) Most favorable application for each type

11·8 Wastewater from a small plant (20 gpm) that manufactures electric storage batteries has a pH of 4.0 and contains quantities of lead, iron, and copper in excess of those permitted to be discharged into the municipal sewer. Prepare a flowsheet for a treatment system that will render the waste suitable for disposal.

11·9 Sludge is to be withdrawn by gravity from a primary settling tank for heat treatment. The available head is equal to 10 ft, and 300 ft of 6-in. pipe is to be used to interconnect the units. Determine the flowrate and velocity, assuming that the solids content of the sludge is 6 percent. Assume an f value of 0.025 for water, and that the minor losses are equal to 2 ft.

11·10 Determine the quantity of chlorine in lb/day necessary to disinfect a daily average primary effluent flow of 10 mgd. Use a dosage of 16 mg/liter and size the contact chamber for a contact time of 15 min at maximum flow, assumed two times the average.

REFERENCES

1. Babbitt, H. E., and E. R. Baumann: *Sewerage and Sewage Treatment*, 8th ed., Wiley, New York, 1958.

2. Babbitt, H. E., and D. H. Caldwell: *Laminar Flow of Sludge in Pipes*, University of Illinois Bulletin 319, 1939.

3. Camp, T. R.: Grit Chamber Design, *Sewage Works Journal,* vol. 14, p. 368, 1942.
4. Culp, G. L.: Chemical Treatment of Raw Sewage/1 and 2, *Water and Wastes Engineering,* vol. 4, nos. 7 and 10, 1967.
5. Eckenfelder, W. W., Jr.: *Industrial Water Pollution Control,* McGraw-Hill, New York, 1966.
6. Eliassen, R., and D. F. Coburn: Pretreatment—Versatility and Expandability, presented at the ASCE Environmental Engineering Conference, Chattanooga, Tenn., 1968.
7. Federation of Sewage and Industrial Wastes Associations: *Chlorination of Sewage and Industrial Wastes,* Manual of Practice 4, 1951.
8. Great Lakes–Upper Mississippi River Board of State Sanitary Engineers: *Recommended Standards for Sewage Works (Ten States Standards),* 1960.
9. King, H. W., and E. F. Brater: *Handbook of Hydraulics,* 5th ed., McGraw-Hill, New York, 1963.
10. Metcalf, L., and H. P. Eddy: *American Sewerage Practice,* vol. III, 3d ed., McGraw-Hill, New York, 1935.
11. Morgan, P. F.: The Comminution of Sewage Solids, *Sewage Works Journal,* vol. 13, no. 1, 1941.
12. Sparr, A. E.: Pumping Sludge Long Distances, *J. WPCF,* vol. 43, no. 8, 1971.
13. Steffensen, S. W., and N. Nash: Hypochlorination of Wastewater Effluents in New York City, *J. WPCF,* vol. 39, no. 8, 1967.
14. *The Sewerage Manual,* published biennially by *Public Works.*
15. Water Pollution Control Federation: *Sewage Treatment Plant Design,* Manual of Practice 8, Washington, D.C., 1967.

design of facilities for biological treatment of wastewater

12

Biological processes are used to convert the finely divided and dissolved organic matter in wastewater into flocculent settleable solids that can be removed in sedimentation tanks. Although these processes (also called secondary processes) are employed in conjunction with the physical and chemical processes used for the preliminary treatment of wastewater discussed in Chap. 11, they are not substitutes. Primary sedimentation is most efficient in removing coarse solids, whereas the biological processes are most efficient in removing organic substances that are soluble or in the colloidal-size range.

The most frequently used biological processes are the activated sludge and trickling filter. There are many modifications of these processes that can be utilized to meet specific treatment requirements. Other biological processes include aerated lagoons and stabilization ponds. The design of these processes and the physical facilities required for their implementation are discussed in detail in this chapter. Typical treatment plant flowsheets incorporating biological processes are illustrated in Figs. 12·1 to 12·3.

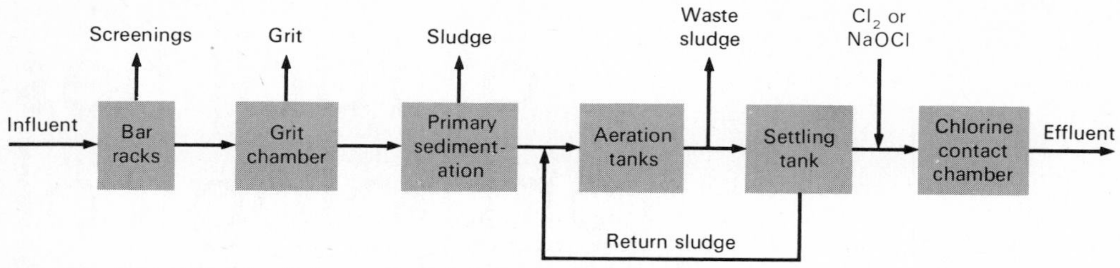

FIG. 12·1 Typical flowsheet of an activated-sludge treatment plant.

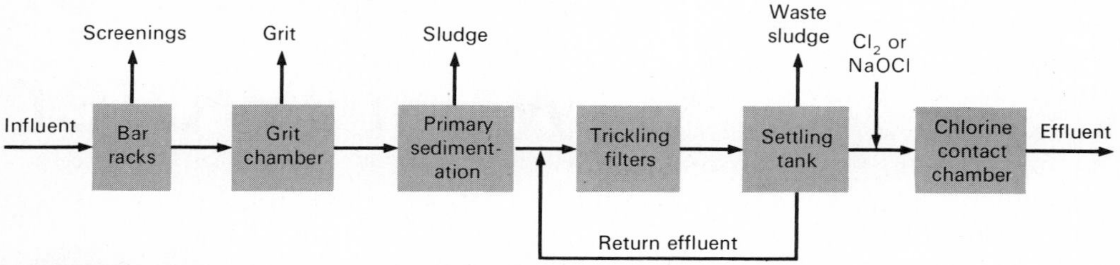

FIG. 12·2 Typical flowsheet of a trickling-filter treatment plant.

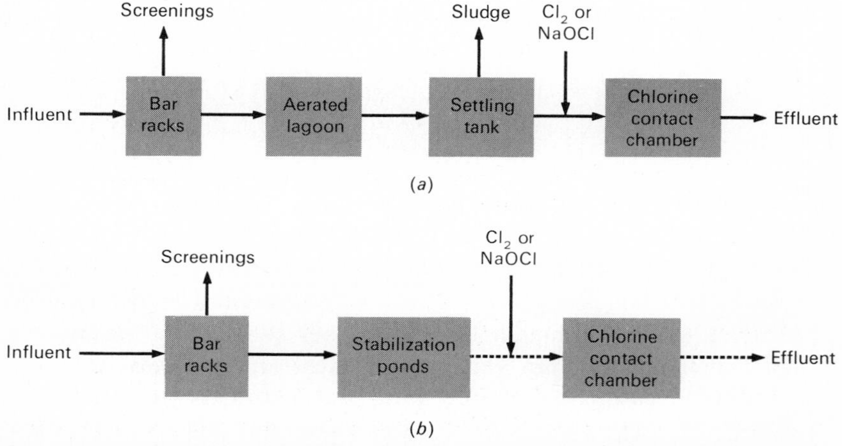

FIG. 12·3 Flowsheet of a biological treatment plant with (*a*) aerated lagoons and (*b*) stabilization ponds.

ACTIVATED-SLUDGE PROCESS

The activated-sludge process has been employed extensively in its original form as well as in several modified forms. Theoretical aspects of the process, including the microbiology, the reaction kinetics, and to some extent, the operation, have been discussed in Chap. 10. Therefore, the practical application of this process is presented and discussed in this section. The material is divided into seven major sections dealing with (1) process design considerations, (2) a discussion of the various types of activated-sludge processes and modifications, (3) the design of diffused-air aeration facilities, (4) the design of mechanical aerators, (5) the design of the aeration tanks and associated appurtenances, (6) the design of activated-sludge solids separation facilities, and (7) a discussion of the operational difficulties most often encountered with the activated-sludge process.

Process Design Considerations

Factors that must be considered in the design of the activated-sludge process include (1) loading criteria, (2) selection of reactor type, (3) sludge production, (4) oxygen requirements and transfer, (5) nutrient requirements, (6) environmental requirements, (7) solid-liquid separation, and (8) effluent characteristics. Because these factors are fundamental to process design, each one will be considered separately; and then, at the end of this section, the material presented will be incorporated in a detailed design example.

Loading Criteria Over the years, a number of both empirical and rational parameters have been proposed for the design and control of the activated-sludge process. The two rational parameters most commonly used are (1) food-to-microorganism ratio U and (2) mean cell residence time θ_c (see Chap. 10).

Typical values for the food-to-microorganism ratio reported in the literature vary from 0.2 to 0.5 with the lower value representing conservative operation. On the basis of laboratory studies and actual operating data from a number of different treatment plants throughout the United States, it has been found that mean cell residence times of about 6 to 15 days result in the production of a stable, high-quality effluent and a sludge with excellent settling characteristics.

Comparing these parameters, U can be considered to be a measure of the rate at which food (BOD) is utilized by a unit mass of organisms, while θ_c is a measure of the average residence time of the organisms in the system. The relationship between mean cell residence time θ_c and the food-to-microorganism ratio U is, as presented previously (Eq. 10·12),

$$\frac{1}{\theta_c} = YU - k_d$$

Empirical relationships based on detention time and volume have also been used. Detention time may be based on the sewage or wastes flow (sewage detention period) disregarding recirculation, or on the flow of mixed liquor that includes recirculation (mixed-liquor detention period). The sewage detention time, while greater than the hydraulic detention time, is important because it approximates the average period of contact of the sewage with the sludge due to the fact that the recirculated return sludge passes through the tanks more than once. Typically, sewage detention times in the aeration tank are in the range of 4 to 8 hr. The *Ten States Standards* require a sewage detention period of 6.0 hr for flows greater than 1.0 mgd and 7.5 hr for flows in the range of 0.2 to 0.8 mgd [10].

The volume required for treatment is thus directly proportional to the detention time and the flowrate. Organic loadings have also been expressed in terms of the pounds of BOD_5 applied daily per 1,000 cu ft of aeration tank volume. Typical values range from 20 to 40 lb/1,000 cu ft. The *Ten States Standards* limit the applied BOD_5 to not more than 35 lb/1,000 cu ft. These standards ignore the concentration of the mixed liquor, the food-to-microorganism ratio, and the mean cell residence time, which may be considered operating variables as well as design parameters; however, they do have the merit of requiring a minimum aeration-tank volume that should be adequate for satisfactory treatment, assuming proper selection of other design and operating variables.

Selection of Reactor Type One of the important considerations in the design of any biological process is the selection of the type of reactor or reactors (see Chap. 10) to be used in the treatment process. Operational factors that must be considered include (1) the reaction kinetics governing the treatment process, (2) oxygen-transfer requirements, (3) nature of the wastewater to be treated, and (4) local environmental conditions. Because the relative importance of these factors will vary with each application, they should be considered separately whenever a reactor type is to be selected.

The purpose of this section is to illustrate the type of analysis involved in considering the first of these factors and to briefly discuss the importance of these factors relative to the activated-sludge process. In practice, initial construction and operation and maintenance costs will also affect reactor selection.

The effect of reaction kinetics on reactor selection can be illustrated by considering (1) a reactor system composed of a series of identical complete-mix reactors, in which a single complete-mix and plug-flow reactor are the two extremes, and (2) a reactor with axial dispersion and arbitrary entrance and exit conditions. For the first case, the problem is approached by considering ideal flow in a series

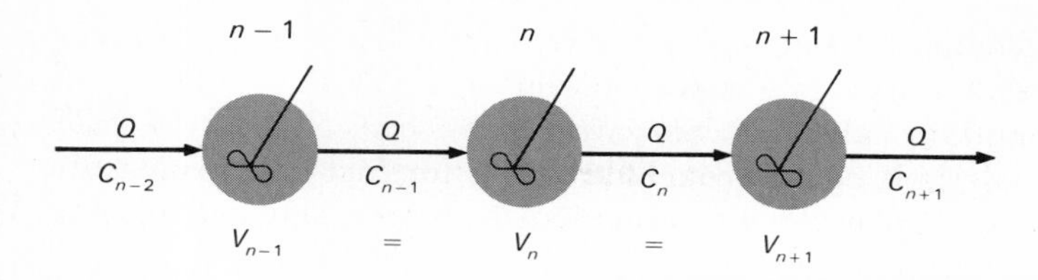

FIG. 12·4 Schematic of identical complete-mix reactors in series.

of complete-mix reactors, such as those shown in Fig. 12·4 in which the effluent from one reactor serves as the influent to the next reactor. Assuming, for the purpose of illustration, that substrate removal is governed by a first-order reaction, a materials balance around the nth reactor yields:

$$\begin{bmatrix}\text{Rate of}\\ \text{change of}\\ \text{substrate}\\ \text{in the}\\ \text{reactor}\end{bmatrix} = \begin{bmatrix}\text{rate of}\\ \text{substrate}\\ \text{inflow to}\\ \text{reactor}\end{bmatrix} - \begin{bmatrix}\text{rate of}\\ \text{substrate}\\ \text{outflow}\\ \text{from}\\ \text{reactor}\end{bmatrix} - \begin{bmatrix}\text{rate of}\\ \text{substrate}\\ \text{removal}\\ \text{in reactor}\end{bmatrix}$$

$$\frac{dC_n}{dt}(V_n) = QC_{n-1} - QC_n - kC_nV_n \tag{12·1}$$

where $\dfrac{dC_n}{dt}$ = rate of change of substrate in reactor

Q = flowrate

C_{n-1} = inflow substrate concentration

C_n = outflow substrate concentration

k = substrate removal rate

At steady state ($dC_n/dt = 0$), Eq. 12·1 becomes

$$\frac{C_n}{C_{n-1}} = \frac{1}{1 + kV_n/Q} \tag{12·2}$$

Applying Eq. 12·2 to n reactors in series results in

$$\frac{C_n}{C_0} = \frac{1}{(1 + kV/nQ)^n} \tag{12·3}$$

where V = volume of all reactors in series

n = number of reactors in series

C_0 = influent substrate concentration

Using Eq. 12·3, the total volume V, in terms of Q and k required for various removal efficiencies with 1, 2, 4, 6, 8, or 10 reactors in series, is reported in Table 12·1 and shown graphically in Fig. 12·5. In the extreme, as the number of reactors is increased, the required volume approaches that required for a plug-flow reactor, which can be computed using Eq. 12·4.

$$V_r = \frac{Q}{k}\int_{C_0}^{C_e} -\frac{dC}{C} \tag{12·4}$$

where V_r = volume of plug-flow reactor
C_0 = inflow substrate concentration
C_e = outflow substrate concentration

From this analysis it can be concluded that, for first-order substrate removal kinetics, the total volume required for a series of complete-mix

TABLE 12·1 REQUIRED REACTOR VOLUME FOR COMPLETE-MIX REACTORS FOR VARIOUS REMOVAL EFFICIENCIES*

	Required reactor volume Q/k			
No. of reactors in series	85% removal efficiency	90% removal efficiency	95% removal efficiency	98% removal efficiency
1	5.67	9.00	19.00	49.00
2	3.18	4.32	6.96	12.14
4	2.48	3.10	4.48	6.64
6	2.22	2.87	3.90	5.50
8	2.16	2.64	3.60	5.04
10	2.10	2.60	3.50	4.80
Plug flow	1.90	2.30	3.00	3.91

* Volume of individual reactors equals value in table divided by the number of reactors in series.

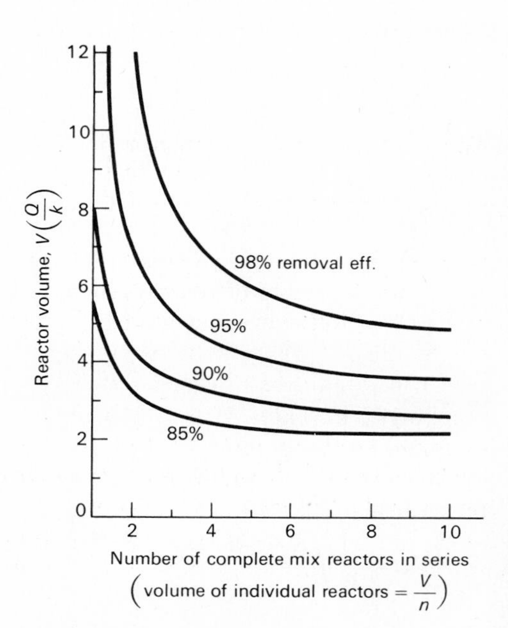

FIG. 12·5 Required reactor volume versus the number of complete-mix reactors in series for various removal efficiencies.

reactors (four or more) is considerably less than that required for a single complete-mix reactor. Further, the volume differential becomes more pronounced as the removal efficiency increases. The volume differential also increases with increasing reaction order. For example, in the previous analysis, if second-order substrate kinetics governed, the relative ratio of required volumes (single-stage complete mix to plug flow) for 90 percent removal would be equal to 10 to 1 (see Prob. 10·1 in Chap. 10). On the other hand, if zero-order substrate removal kinetics were applicable, the ratio of required volumes would be equal to 1. Thus, the form of the governing kinetic expression can greatly affect the volume requirements and must be considered in selecting reactor geometry.

In the second case, consideration of a reactor with axial dispersion (nonideal flow) provides another way of interpreting the data in Table 12·1. It can be assumed that the intermediate values in Table 12·1 represent the required volume for a plug-flow reactor with varying conditions of dispersion. For example, if the contents of a plug-flow reactor were completely dispersed, the result would be equivalent to a complete-mix reactor. Recognizing that in practice neither a plug-flow nor complete-mix reactor functions as assumed, Wehner and Wilhelm [36] derived the following equation from a reactor with axial dispersion, first-order kinetics, and arbitrary entrance and exit conditions:

$$\frac{S}{S_0} = \frac{4a \exp(1/2d)}{(1+a)^2 \exp(a/2d) - (1-a)^2 \exp(-a/2d)} \tag{12·5}$$

where S = effluent substrate concentration
S_0 = influent substrate concentration
$a = \sqrt{1 + 4ktd}$
d = dispersion factor = D/uL
D = axial-dispersion coefficient, sq ft/hr
u = fluid velocity, ft/hr
L = characteristic length, ft
k = first-order reaction constant
t = detention time, hr

To facilitate the use of Eq. 12·5, Thirumurthi, in connection with his work on stabilization ponds [32], developed Fig. 12·6, in which the term kt is plotted against S/S_0 for dispersion factors varying from zero (0) for an ideal plug-flow reactor to infinity (∞) for a complete-mix reactor. Dispersion factors for conventional plug-flow activated-sludge reactors are probably within the range from 0 to 0.2. For reactors with mechanical aerators designed to operate as completely mixed systems, values of d are probably in the range from 4.0 to ∞. Most stabilization ponds are somewhere within the range from 0.1 to 2.0.

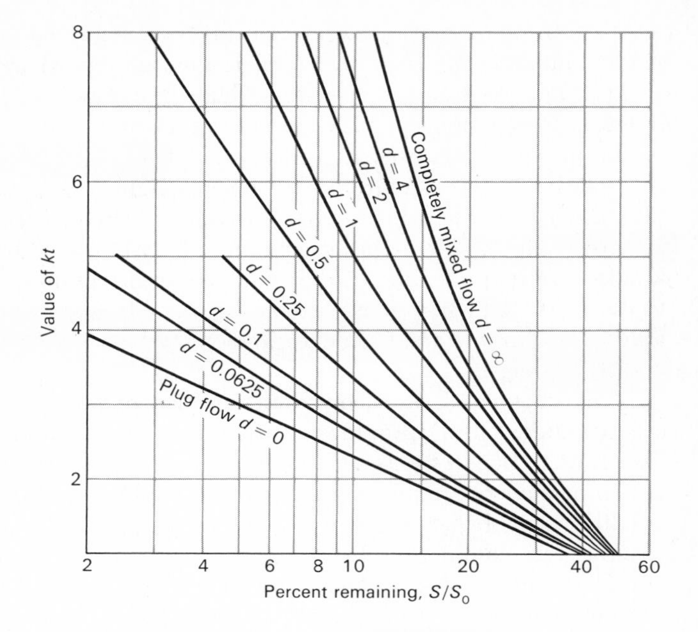

FIG. 12·6 Values of kt in the Wehner and Wilhelm equation (Eq. 12·5) versus percent remaining for various dispersion factors [32].

The relationship of the kt values derived from the dispersion model to the values given in Table 12·1 is as follows. For 90 percent removal efficiency (10 percent remaining) and a dispersion factor of 0.0625, the value of kt is 2.6 (see Fig. 12·5). If kt is rewritten as $k(V/Q)$, then the volume V is equal to $2.6(Q/k)$. Comparing this value to the data reported in Table 12·1, it is found that the performance of a plug-flow reactor with a dispersion factor of 0.0625 is equivalent to 10 complete-mix reactors in series. Thus, for a tank with a given degree of dispersion, it is possible to find its equivalent in terms of a series reactor system and to compare the volume required to that for an ideal single-stage complete-mix or plug-flow reactor. For a further analysis of these and related topics, Levenspiel [15] is recommended.

In applying the foregoing analysis to the activated-sludge process, Eq. 10·5 can be used to describe the removal of soluble substrate (see also Eqs. 10·39 and 10·43). The substrate concentration in the recycle flow can be neglected in most cases. From a practical standpoint it is interesting to note that the hydraulic detention time of many of the plug-flow and the complete-mix reactors in actual use is about the same. The reason is that the combined substrate (soluble and non-soluble) removal rate for domestic wastes is approximately zero order with respect to substrate. It is quasi-first order, however, with respect to the concentration of cells.

The second factor that must be considered in the selection of reactors for the activated-sludge process is oxygen-transfer requirements. In the past, with conventional plug-flow aeration systems, it was often found that sufficient oxygen could not be supplied to meet the oxygen requirements of the head end of the reactor (see Fig. 10·15). This condition led to the development of the following modifications of the activated-sludge process: (1) the tapered aeration process in which an attempt was made to match the air supplied to the oxygen demand, (2) the step aeration process where the incoming waste and return solids are distributed along the length of the reactor (usually at quarter points), and (3) the complete-mix activated-sludge process where the air supplied uniformly matches or exceeds the oxygen demand. Most of the past oxygen-transfer limitations have been overcome by better selection of process operational parameters and improvements in the design and application of aeration equipment. Pure oxygen instead of air can also be used to overcome this limitation.

The third factor that can influence the type of reactor selected is the nature of waste. For example, because the incoming waste is more or less uniformly dispersed in a complete-mix reactor, it can, as compared to a plug-flow reactor, more easily withstand shock loads resulting from the sludge discharge of organic and toxic materials to sewers from industrial operations. The complete-mix process has been used in a number of installations for this reason.

The fourth factor is local environmental conditions. Of these, temperature is perhaps the most important. For example, for a zero-order biological substrate removal process, if the temperature coefficient as defined by Eq. 10·44 is equal to about 1.12 and the temperature were to drop by 10°C, the required reactor volume would be three times as large, assuming that the solids level is not increased. In this situation a series of complete-mix reactors or a plug-flow reactor whose length could be reduced with stop gates could be used effectively. It should be noted, however, that in practice the conventional activated-sludge process is not significantly influenced by temperature, as reported values for the temperature coefficient only vary from 1.0 to 1.03 (see Table 10·7 in Chap. 10).

Sludge Production It is important to know the quantity of sludge to be produced per day, as it will affect the design of the sludge handling and disposal facilities. Using the mean cell residence time loading criteria, the quantity of sludge which must be disposed of per day is

$$\frac{dX}{dt} = \frac{X}{\theta_c} \qquad (12\cdot6)$$

As noted previously, the liquid quantity will depend on the volume of the reactor. Where process control is based on one of the other loading criteria, the quantity of solids to be wasted must be established on a trial-and-error basis.

Oxygen Requirements and Transfer The theoretical oxygen requirements can be computed by knowing the BOD of the waste and the amount of organisms wasted from the system per day. The reasoning is as follows. If all the BOD were converted to end products, the total oxygen demand would be computed by converting BOD_5 to BOD_L using an appropriate conversion factor. It is known that a portion of the waste is converted to new cells that are subsequently wasted from the system; therefore, if the BOD_L of the wasted cells were subtracted from the total, the remaining amount would represent the amount of oxygen that must be supplied to the system. The BOD_L of a mole of cells can be evaluated as follows:

$$\underset{\text{Cells}}{\overset{113}{C_5H_7NO_2}} + \overset{5(32)}{5O_2} \rightarrow 5CO_2 + 2H_2O + NH_3 \tag{12·7}$$

$$\frac{\text{lb } O_2}{\text{lb cells}} = \frac{160}{113} = 1.42$$

where the BOD_L of the cells is equal to

$$BOD_L = 1.42(\text{cells})$$

Therefore, the theoretical oxygen requirements for an activated-sludge system can be computed as

$$O_2\left(\frac{\text{lb}}{\text{day}}\right) = \begin{bmatrix}\text{food utilized}\\ \text{per day}\end{bmatrix} - 1.42\begin{bmatrix}\text{organisms}\\ \text{wasted per}\\ \text{day}\end{bmatrix}$$

In terms of dF/dt and dX/dt,

$$O_2\left(\frac{\text{lb}}{\text{day}}\right) = \frac{dF}{dt} - 1.42\frac{dX}{dt} \tag{12·8}$$

Then, if the oxygen-transfer efficiency of the aeration system is known or can be estimated, the actual air requirements may be determined. The air supply should be sufficient to maintain a minimum dissolved-oxygen concentration throughout the aeration tank of 1 to 2 mg/liter.

For diffused-air aeration, the amount of air used has commonly ranged from 0.5 to 2.0 cu ft/gal at different plants, with 1.0 cu ft/gal an early rule-of-thumb design factor. Because the air use depends on the strength of the sewage, the amount has become a derived quantity for recordkeeping purposes and is no longer a basic design criterion.

The air supply must be adequate to satisfy the BOD of the waste, to satisfy the endogenous respiration by the sludge organisms, and to provide adequate mixing. For food-to-microorganism ratios greater than 0.3, the air requirements for the conventional process amount to 500 to 900 cu ft/lb of BOD removed [7, 35]. At lower food-to-microorganism ratios, endogenous respiration, nitrification, and prolonged aeration periods increase air use to 1,200 to 1,800 cu ft/lb of BOD removed. The *Ten States Standards* require the air-diffusion system to be capable of delivering 150 percent of the normal requirements, which are assumed to be 1,000 cu ft of air/lb of BOD in the wastewater applied to the aeration tanks. A minimum air flow of approximately 3 cfm/ft of tank length is required for adequate mixing velocities and to avoid deposition of solids.

Nutrient Requirements If any biological system is to function properly, nutrients must be available in adequate amounts. As discussed in Chaps. 7 and 10, the principal nutrients required are nitrogen and phosphorus. Based on an average composition of cell tissue of $C_5H_7NO_2$, about 12.4 percent by weight of nitrogen will be required. The amount is based on the mass of organisms produced per day. The phosphorus requirement is usually assumed to be about one-fifth this value. These are typical values, not fixed quantities, because it has been shown that the percentage distribution of nitrogen and phosphorus in cell tissue varies with the age of the cell and environmental conditions.

Other nutrients required by most biological systems are reported in Table 12·2. In general, wastewater contains all of the nutrients required for proper cell growth. If a large portion of the wastewater is composed of industrial wastes, however, nutrient addition may be necessary.

Environmental Requirements Environmental factors of importance include temperature and pH. These factors were discussed in Chap. 10 and the reader is referred to that discussion. The effect of temperature was also discussed previously in connection with reactor selection.

Solid-Liquid Separation Perhaps the most important aspect of biological waste treatment is the design of the facilities used to separate the biological solids from the treated wastewater, for it is axiomatic that if the solids cannot be separated and returned to the aeration tank, the activated-sludge process will not function properly. Unfortunately, the importance of the separation step is little appreciated and even

TABLE 12·2 INORGANIC IONS NECESSARY FOR MOST ORGANISMS [12]

Substantial quantities	Trace quantities	
Na^{+} (except for plants)	Fe^{++}	
K^{+}	Cu^{++}	
Ca^{++}	Mn^{++}	
Mg^{++}	Zn^{++}	
PO_4^{3-}	B^{3+}	required by plants, certain protists
Cl		
SO_4^{--}	Mo^{+}	required by plants, certain protists, and animals
HCO_3^{-}	V^{++}	required by certain protists and animals
	Co^{++}	required by certain animals, protists, and plants
	I^{-}, Se^{--}	required by certain animals only

today does not receive the attention it should. Because of its importance, a separate section has been devoted to this subject.

Effluent Characteristics Organic content is a major parameter of effluent quality. The organic content of effluent from biological treatment processes is usually composed of the following three constituents:

1. Soluble biodegradable organics
   - (*a*) Organics that escaped biological treatment
   - (*b*) Organics formed as intermediate products in the biological degradation of the waste
   - (*c*) Cellular components (result of cell death or lysis)
2. Suspended organic material
   - (*a*) Biological solids produced during treatment that escaped separation with final settling tank
   - (*b*) Colloidal organic solids in the plant influent escaping treatment and separation.
3. Nonbiodegradable organics
   - (*a*) Those originally present in the influent
   - (*b*) By-products of biological degradation

The kinetic equations developed in Chap. 10 for the effluent quality apply only to the soluble organic waste that escaped biological treatment. Clearly, this is only a portion of the organic waste concen-

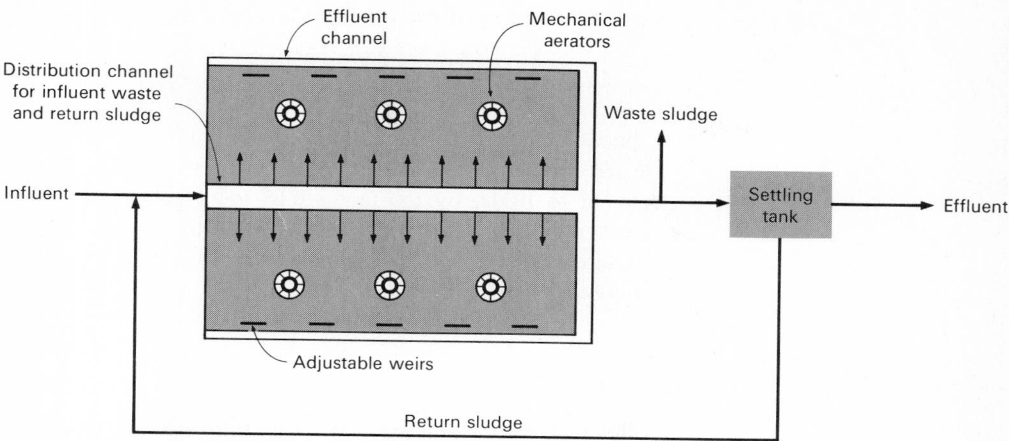

FIG. 12·7 Typical flowsheet for the complete-mix activated-sludge process.

tration in the effluent. In a well-operating activated-sludge plant treating domestic wastes the soluble carbonaceous BOD_5 in the effluent, determined on a filtered sample, will usually vary from 2 to 10 mg/liter.

Application of the aforementioned factors to the design of an activated-sludge treatment process is illustrated in Example 12·1. For purposes of the example a complete-mix system has been selected. Schematically, such a system would be depicted as shown in Fig. 12·7. Its distinguishing features are (1) uniform distribution of the inflow and return solids to the reactor (aeration tank) and (2) uniform withdrawal of mixed liquor from the reactor. Application of the principles discussed in this chapter and in Chap. 10 to other types of systems is left to the student in the problems at the end of this chapter.

EXAMPLE 12·1 *Activated-Sludge Process Design*

Design a complete-mix activated-sludge process to treat 5.0 mgd of settled sewage having a BOD_5 of 250 mg/liter. The effluent is to contain 20 or less mg/liter BOD_5. Assume that the temperature is 20°C and that the following conditions are applicable:

1. Influent volatile suspended solids to reactor are negligible.
2. Return sludge concentration = 12,500 mg/liter of suspended solids = 10,000 mg/liter volatile suspended solids.
3. Mixed-liquor volatile suspended solids (MLVSS) = 3,500 mg/liter = 0.80 total MLSS.
4. Mean cell residence time θ_c = 10 days.

5. Hydraulic regime of reactor = complete mix.
6. $Y = \frac{0.65 \text{ lb cells}}{\text{lb BOD}_5 \text{ utilized}}$ and $k_d = 0.10/\text{day}$.
   Constants are for temperatures in the range of 20–25°C.
7. Effluent contains 22 mg/liter of biological solids, of which 65 percent is biodegradable. This is converted from ultimate demand to BOD_5 by the factor 0.68 for a value of $K = 0.1$ day^{-1}.
8. Waste contains adequate nitrogen and phosphorus and other trace nutrients for biological growth.

Solution

1. Estimate the soluble BOD_5 in the effluent.

$$\text{Effluent BOD}_5 = \text{influent soluble BOD}_5 \text{ escaping treatment} + \text{BOD}_5 \text{ of effluent biological solids}$$

$$20 = S + 22(0.65)(1.42)(0.68)$$

$$S = 6 \text{ mg/liter soluble BOD}_5$$

This would mean that the biological treatment efficiency based on soluble BOD_5 would be

$$E_s = \frac{250 - 6}{250}(100) = 97.6\%$$

The overall plant efficiency would be

$$E_{\text{overall}} = \frac{250 - 20}{250}(100) = 92\%$$

2. Compute the reactor volume. The volume of the reactor can be determined using Eq. 10·39 in conjunction with Eq. 10·33 (the assumptions on which Eq. 10·39 is based are given in Chap. 10 and should be reviewed).

$$XV = \frac{YQ\theta_c(S_0 - S)}{1 + k_d\theta_c}$$

$$3{,}500V = \frac{0.65(5)(10)(250 - 6)}{1 + 0.10(10)}$$

$$V_{\text{reactor}} = 1.13 \text{ million gal}$$

3. Compute the sludge-production rate on a mass basis.

$$\frac{dX}{dt} = \frac{XV}{\theta_c} = \frac{3{,}500(8.34)(1.13)}{10}$$

$$= 3{,}300 \text{ lb VSS/day}$$

$$\text{Sludge-production rate} = \frac{3{,}300}{0.80} = 4{,}125 \text{ lb SS/day}$$

4. Compute the sludge-wasting rate if wasting is accomplished from the reactor as shown in Fig. 10·14. Neglecting the solids lost in the plant effluent, the wasting rate computed using Eq. 10·35 will be

$$Q_w \cong \frac{V}{\theta_c} \cong \frac{1.13}{10} \cong 0.113 \text{ mgd}$$

If sludge wasting is accomplished from the recycle line, then, using Eq. 10·41, the wasting rate will be

$$Q'_w \cong \frac{VX}{\theta_c X_r} \cong \frac{1.13(3{,}500)}{10(10{,}000)} \cong 0.0396 \text{ mgd}$$

Note that in either case, the weight of sludge wasted is the same (3,300 lb VSS/day), and that either wasting method will achieve a θ_c of 10 days for the system.

5. Compute the recirculation ratio (see also "Return-Sludge Requirements").

$$\begin{aligned}
\text{Aerator VSS conc} &= 3{,}500 \text{ mg/liter} \\
\text{Return VSS conc} &= 10{,}000 \text{ mg/liter} \\
3{,}500(Q + Q_r) &= 10{,}000(Q_r) \\
\frac{Q_r}{Q} &= R = 0.54
\end{aligned}$$

6. Compute the hydraulic retention time for the reactor.

$$t = \frac{V}{Q} = \frac{1.13}{5} = 0.226 \text{ days} = 5.4 \text{ hr}$$

7. Compute the oxygen requirements (neglect oxygen requirement for nitrification) based on ultimate demand, BOD_L.

$$\begin{aligned}
\text{lb } O_2\text{/day} &= \left(\frac{dF}{dt}\right)_L - 1.42\left(\frac{dX}{dt}\right) \\
&= \frac{(250 - 6)(8.34)(5.0)}{0.68} - 1.42(3{,}300) \\
&= 15{,}000 - 4{,}700 = 10{,}300 \text{ lb/day}
\end{aligned}$$

8. Check the U ratio and volumetric loading.

$$U = \frac{dF/dt}{X} \qquad \text{(Based on } BOD_5 \text{ and MLVSS)}$$

$$= \frac{(250 - 6)(8.34)(5.0)}{3{,}500(8.34)(1.13)} = 0.31$$

$$\begin{aligned}
\text{Loading} &= \frac{10{,}160(7.48)(1{,}000)}{1.13 \times 10^6} \\
&= \frac{67.3 \text{ lb } BOD_5 \text{ removed}}{1{,}000 \text{ cu ft}}
\end{aligned}$$

9. Compute the volume of air required. Oxygen requirement = 10,300 lb/day. Specific weight of air at standard temperature and pressure is 0.0750 lb/cu ft and contains 23.2 percent oxygen by weight. Then, the theoretical air requirement is

$$\frac{10{,}300}{0.0750(0.232)} = 593{,}000 \text{ cu ft/day}$$

Assume that the oxygen transfer efficiency has been computed to be 8 percent. Therefore the requirement is

$$\frac{593{,}000}{0.08} = 7{,}410{,}000 \text{ cu ft/day}$$

or

$$\frac{7{,}410{,}000}{1{,}440} = 5{,}150 \text{ cfm}$$

10. Check the air volume.

$$\frac{7.41 \times 10^6 \text{ cu ft/day}}{5 \times 10^6 \text{ gpd}} = 1.48 \text{ cu ft/gal}$$

$$\frac{7.41 \times 10^6 \text{ cu ft/day}}{10{,}160 \text{ lb BOD/day}} = 730 \text{ cu ft/lb BOD removed}$$

Types of Processes and Modifications

The activated-sludge process is very flexible and can be adapted to almost any type of biological waste treatment problem. The purpose of this section is to discuss the details of both the conventional activated processes and some of the modifications that have become standardized. The characteristics and typical removal efficiencies for these processes are listed in Table 12·3. Design parameters for these processes are shown in Table 12·4.

Conventional The conventional activated-sludge process consists of an aeration tank, a secondary clarifier, and a sludge recycle line (Fig. 12·8*a*). Sludge wasting is accomplished from the recycle or mixed-liquor line. The flow model is plug flow with cellular recycle, as described in Chap. 10. Both influent settled sewage and recycled sludge enter the tank at the head end and are aerated for a period of about 6 hr. The influent sewage and recycled sludge are mixed by the action of diffused or mechanical aeration, which is constant as the mixed liquor moves down the tank. During this period, adsorption, flocculation, and oxidation of the organic matter take place. The mixed liquor is settled in the

TABLE 12·3 OPERATIONAL CHARACTERISTICS OF ACTIVATED-SLUDGE PROCESSES

Process modification	Flow model	Aeration system	BOD removal efficiency, %	Application
Conventional	Plug-flow	Diffused-air, mechanical aerators	85–95	Low-strength domestic wastes, susceptible to shock loads
Complete-mix	Complete-mix	Diffused-air, mechanical aerators	85–95	General application, resistant to shock loads, surface aerators
Step-aeration	Plug-flow	Diffused-air	85–95	General application to wide range of wastes
Modified-aeration	Plug-flow	Diffused-air	60–75	Intermediate degree of treatment where cell tissue in the effluent is not objectionable
Contact-stabilization	Plug-flow	Diffused-air, mechanical aerators	80–90	Expansion of existing systems, package plants, flexible
Extended-aeration	Complete-mix	Diffused-air, mechanical aerators	75–95	Small communities, package plants, flexible, surface aerators
Kraus process	Plug-flow	Diffused-air	85–95	Low-nitrogen, high-strength wastes
High-rate aeration	Complete-mix	Mechanical aerators	75–90	Use with turbine aerators to transfer oxygen and control the floc size, general application
Pure-oxygen systems	Complete-mix reactors in series	Mechanical aerators	85–95	General application, use where limited volume is available, use near economical source of oxygen, turbine or surface aerators

TABLE 12·4 DESIGN PARAMETERS FOR ACTIVATED–SLUDGE PROCESSES

Process modification	Parameter					
	θ_c, days	U, lb BOD_5/lb MLVSS-day	Volumetric loading, lb BOD_5/1,000 cu ft	MLSS, mg/liter	V/Q, hr	Q_r/Q
Conventional	5–15	0.2–0.4	20–40	1,500–3,000	4–8	0.25–0.5
Complete-mix	5–15	0.2–0.6	50–120	3,000–6,000	3–5	0.25–1.0
Step-aeration	5–15	0.2–0.4	40–60	2,000–3,500	3–5	0.25–0.75
Modified-aeration	0.2–0.5	1.5–5.0	75–150	200–500	1.5–3	0.05–0.15
Contact-stabilization	5–15	0.2–0.6	60–75	(1,000–3,000)*	(0.5–1.0)*	0.25–1.0
				(4,000–10,000)†	(3–6)†	
Extended-aeration	20–30	0.05–0.15	10–25	3,000–6,000	18–36	0.75–1.50
Kraus process	5–15	0.3–0.8	40–100	2,000–3,000	4–8	0.5–1.0
High-rate aeration	5–10	0.4–1.5	100–1,000	4,000–10,000	0.5–2	1.0–5.0
Pure-oxygen systems	8–20	0.25–1.0	100–250	6,000–8,000	1–3	0.25–0.5

* Contact unit.
† Solids stabilization unit.

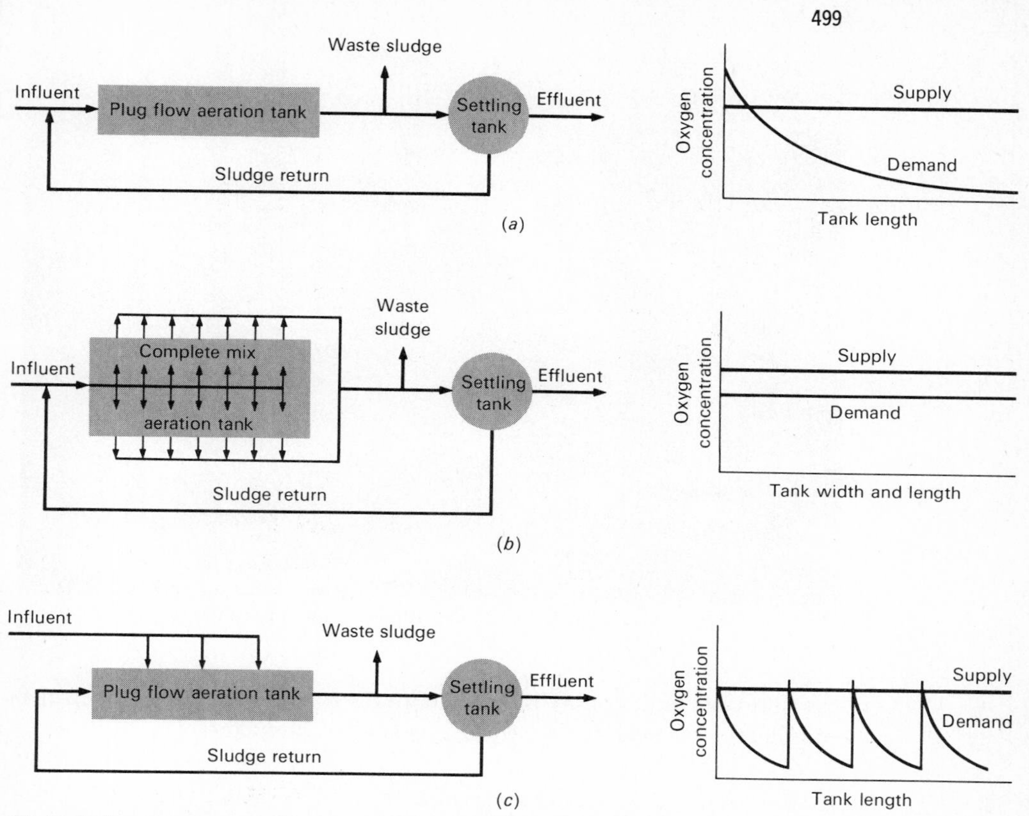

FIG. 12·8 Flowsheet and plot of oxygen demand and oxygen supply versus tank length for (a) conventional, (b) complete-mix, and (c) step-aeration activated-sludge processes.

activated-sludge settling tank, and sludge is returned at a rate of approximately 25 to 50 percent of the influent flowrate. An aerial photograph of a conventional activated-sludge wastewater treatment plant is shown in Fig. 12·9.

Complete-Mix The complete-mix process (Fig. 12·8*b*) represents an attempt to duplicate the hydraulic regime of a mechanically stirred reactor. The influent settled sewage and return sludge flow are introduced at several points in the aeration tank from a central channel. The mixed liquor is aerated as it passes from the central channel to effluent channels at both sides of the aeration tank. The aeration-tank effluent is collected and settled in the activated-sludge settling tank.

The organic load on the aeration tank and the oxygen demand are uniform from one end to the other. As the mixed liquor passes across

FIG. 12·9 Aerial view of activated-sludge treatment plant at Greenwich, Conn.

the aeration tank from the influent ports to the effluent channel, it is completely mixed by diffused or mechanical aeration.

Tapered-Aeration The objective of tapered aeration is to match the quantity of air supplied to the demand exerted by the microorganisms, as the liquor traverses the aeration tank. Tapered aeration thus affects only the arrangement of the diffusers in the aeration tank and the amount of air consumed. It is widely used and, in a strict sense, is only a modification of the conventional process.

At the inlet of the aeration tank where fresh settled sewage and return activated sludge first come in contact, the oxygen demand is very high. The diffusers are spaced close together to achieve a high oxygenation rate and thus satisfy the demand. As the mixed liquor traverses the aeration tank, synthesis of new cells occurs, increasing the number of microorganisms and decreasing the concentration of available food. This results in a lower U ratio and a lowering of the oxygen demand. The spacing of diffusers is thus increased toward the tank outlet, to reduce the oxygenation rate. Two beneficial results are obtained. Reduced oxygenation means that less air is required, reducing the size of blowers and the initial and operating costs. Avoidance of overaeration will inhibit the growth of nitrifying organisms, which can cause high oxygen demands.

Step-Aeration The step-aeration process is a modification of the activated-sludge process in which the settled sewage is introduced at several points in the aeration tank to equalize the *U* ratio, thus lowering the peak oxygen demand. The process was developed by Gould and was first applied at the Tallmans Island plant in New York City in 1939.

The aeration tank is subdivided into four or more parallel channels through the use of baffles. Each channel comprises a separate step, and the several steps are linked together in series. A typical flowsheet for the process is shown in Fig. 12·8*c*. Return activated sludge enters the first step of the aeration tank along with a portion of the settled sewage. The piping is so arranged that an increment of sewage is introduced into the aeration tank at each step. If desired, the first step can be used for reaeration of the return activated sludge alone. Flexibility of operation is one of the important features of this process.

The basic theory of the step-aeration process is the same as that of the activated-sludge process. In step aeration, however, the oxygen demand is more uniformly spread over the length of the aeration tank (see Fig. 12·8*c*), resulting in better utilization of the oxygen supplied. The multiple-point introduction of sewage maintains an activated sludge with high absorptive properties, so that the soluble organics are removed within a relatively short contact period. Higher BOD loadings are therefore possible per 1,000 cu ft of aeration-tank volume.

Modified-Aeration The flow diagram for the modified-aeration process is identical with that of the conventional or tapered-aeration process. The difference in the systems is that modified aeration uses shorter aeration times, usually 1.5 to 3 hr, and a high food-to-microorganism ratio. As shown in Table 12·4, the MLSS concentration is relatively low, whereas the organic loading is high. The resultant BOD removal is in the range of 60 to 75 percent; thus the process is not suitable where a high-quality effluent is desired. Some difficulties have been experienced with the process because of the poor settling characteristics of the sludge and the high suspended-solids concentration in the effluent [33].

Contact-Stabilization The contact-stabilization process was developed to take advantage of the absorptive properties of activated sludge. The flowsheet is shown in Fig. 12·10*a*. In some cases, primary settling is eliminated. It has been postulated that BOD removal occurs in two stages in the activated-sludge process. The first is the absorptive phase which requires 20 to 40 min. During this phase most of the colloidal, finely suspended, and dissolved organics are absorbed in the activated

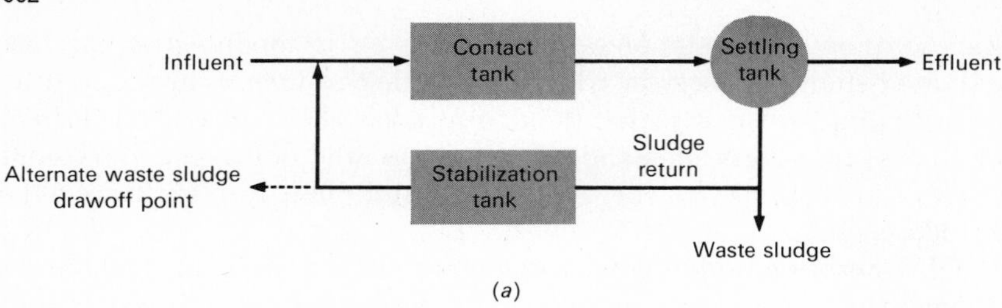

(*a*)

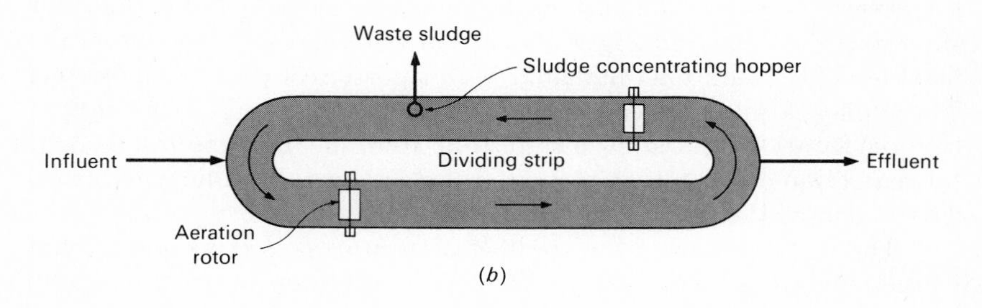

(*b*)

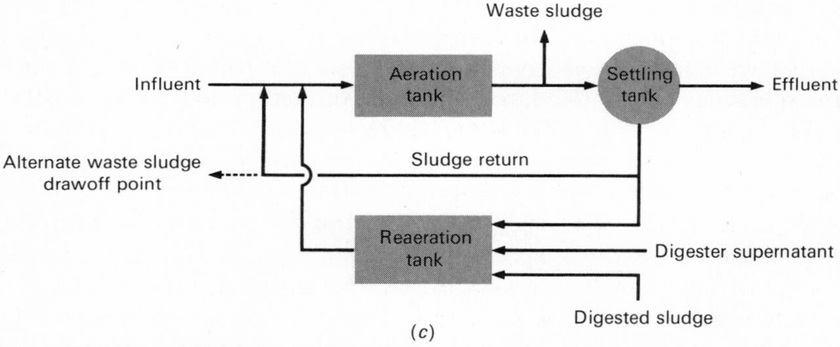

(*c*)

FIG. 12·10 Flowsheet for (*a*) contact-stabilization, (*b*) oxidation-ditch, and (*c*) Kraus activated-sludge processes.

sludge. The second phase, oxidation, then occurs, and the absorbed organics are metabolically assimilated. In the activated-sludge processes mentioned so far, these two phases occur in a single tank. In the contact-stabilization process, the two phases are separated and occur in different tanks.

The settled sewage is mixed with return activated sludge and aerated in a contact tank for 30 to 90 min. During this period, the organics are absorbed by the sludge floc. The sludge is then separated from the treated effluent by sedimentation, and the returned sludge is aerated from 3 to 6 hr in a sludge aeration tank. During this period, the absorbed organics are utilized for energy and production of new cells. A portion of the return sludge is wasted prior to recycle, to maintain a constant MLVSS concentration in the tanks.

The aeration volume requirements are approximately 50 percent of those of a conventional or tapered-aeration plant. It is thus often possible to double the plant capacity of an existing conventional plant by redesigning it to use contact stabilization. The redesign may require only changes in plant piping or relatively minor changes in the aeration system.

The contact-stabilization process has been found to work very well on domestic wastes; however, before using it on industrial wastes or mixtures of domestic and industrial wastes, laboratory tests should be performed. Its value in industrial waste treatment is limited largely to wastes in which the organic matter is not predominantly soluble.

Extended-Aeration The extended-aeration process operates in the endogenous respiration phase of the growth curve, which necessitates a relatively low organic loading and long aeration time. Thus it is generally applicable only to small treatment plants of less than 1-mgd capacity.

This process is used extensively for prefabricated package plants that provide treatment for housing subdivisions, isolated institutions, small communities, schools, etc. Although separate sludge wasting generally is not provided, it may be added where the discharge of the excess solids is objectionable. Aerobic digestion of the excess solids, followed by dewatering on open sand beds, usually follows separate sludge wasting. Primary sedimentation is omitted from the process to simplify the sludge treatment and disposal.

Oxidation Ditch Developed for small towns in the Netherlands, the oxidation ditch is an extended aeration process that has found application in many small European towns and in a variety of different applications in the United States. A schematic of a typical oxidation ditch is shown in Fig. 12·10*b*. It consists of a ring-shaped channel about 3 ft deep. An aeration rotor, consisting of a modified Kessener brush, is placed across the ditch to provide aeration and circulation. The screened sewage enters the ditch, is aerated by the rotor, and circulates at about 1 to 2 fps.

Operation of the ditch shown in Fig. 12·10*b* is intermittent. Modifications can be made for continuous operation. For intermittent operation, the cycle consists of (1) closing the inlet valve and aerating the sewage, (2) stopping the rotor and letting the contents settle, and (3) opening both inlet and outlet valves, thereby allowing the incoming sewage to displace an equal volume of clarified effluent.

Kraus Process Wastes that are deficient in nitrogen are difficult to treat by the activated-sludge process. L. S. Kraus, the operator of the treatment plant at Peoria, Ill., in about 1955, was faced with the problem of treating high-carbohydrate wastes in combination with domestic sewage. His solution was to aerate the supernatant from the sludge digesters, digested sludge, and a portion of the return sludge in a separate reaeration tank for approximately 24 hr, converting the ammonia nitrogen into nitrate, and then to mix it with the return activated sludge. This accomplished two things. The nitrate in the aerated supernatant corrected the nitrogen deficiency in the high-carbohydrate waste, and the heavy solids contained in the digested sludge improved the settleability of the mixed liquor. This process was adopted at the San Jose–Santa Clara plant in California for treating the large flow of fruit-processing wastes mixed with domestic sewage received during the canning season. The flowsheet for the process is shown in Fig. 12·10*c* [12, 13].

High-Rate Aeration High-rate aeration is a modification in which high concentrations of MLSS are combined with high volumetric loadings (see Table 12·4). This combination allows high U ratios (0.4 to 1.5) and long mean cell residence times with hydraulic detention times of 0.5 to 2 hr. Adequate mixing in the reactor to effect oxygen transfer and to control floc size is achieved through the use of turbine mixers.

Pure-Oxygen Systems Recently (1969–1970) there has been renewed interest in the use of pure oxygen as a substitute for air in the activated-sludge process. In application, the aeration tanks are covered, and the oxygen that is introduced into the aeration tank is recirculated. A schematic of a pure-oxygen system employing a series of small complete-mix reactors is shown in Fig. 12·11. Because CO_2 is released and O_2 is utilized by the microorganisms as a result of their activity, a portion of the gas must be wasted, and new oxygen must be added. Depending on the buffer capacity of the wastewater and the amount of CO_2 removed from the system, pH adjustment may be required. Using Henry's law as an approximation, if the partial pressure of the oxygen above the liquid is 0.8, then the amount of oxygen that can be put into the liquid is about four times the amount that normally could be put in with air for a given set of conditions.

A number of advantages, such as increased bacterial activity, decreased sludge volume, reduced aeration tank volume, and improved sludge settleability have been reported for this process [1]. Further testing in a variety of applications will be necessary to substantiate these findings. Quantitatively, the benefits to be derived from the use

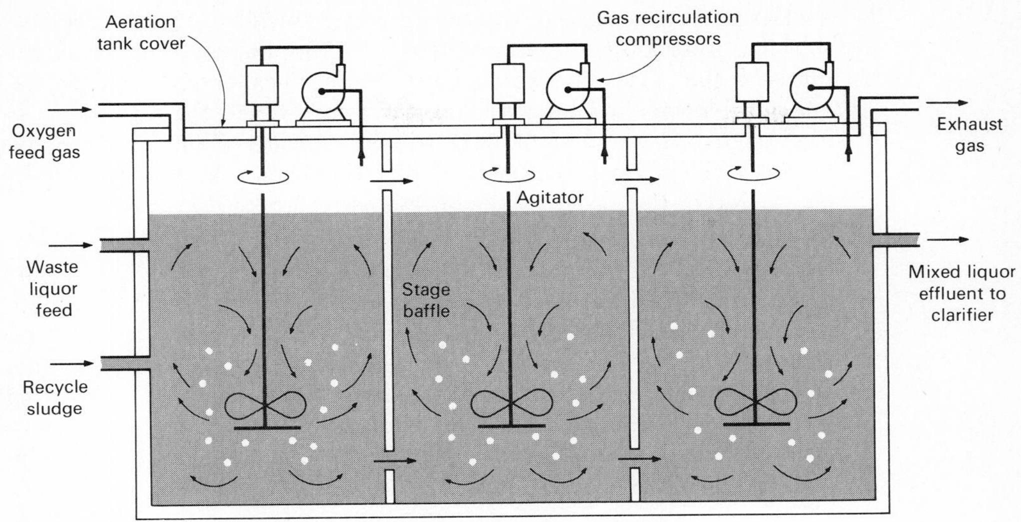

FIG. 12·11 Schematic of the pure-oxygen activated-sludge system [1].

of pure oxygen can be illustrated by considering its application in: (1) proposed and existing complete-mix systems and (2) existing conventional plug-flow aeration tanks.

In the first case, the benefits can be evaluated by considering the complete-mix activated-sludge process design example presented previously. If the mixed-liquor solids concentration in the reactor could be increased to about 5,000 mg/liter by using pure oxygen instead of air, then the reactor volume could be reduced by about 31 percent. Thus, in complete-mix systems, pure oxygen could be used to reduce the required aeration tank volume in new designs, or to extend the useful life of existing systems. Economic comparisons must include the cost of pure oxygen or an oxygen plant and its operation.

In the case of a conventional plug-flow activated-sludge system that is overloaded, conversion to pure-oxygen aeration could be used to extend the useful life of the system and might eliminate the need for additional treatment facilities. In the case of a conventional plug-flow system where sufficient oxygen is being supplied, conversion to pure oxygen will not significantly improve the performance.

Design of Diffused-Air Aeration Facilities

The two basic methods of aerating sewage are (1) to introduce air or pure oxygen into the sewage with submerged porous diffusers or air nozzles or (2) to agitate the sewage mechanically so as to promote solution of air from the atmosphere. A diffused-air system consists of diffusers that are submerged in the sewage, header pipes, air mains, and the blowers and appurtenances through which the air passes. In discussing these topics, blowers will be considered before air mains because the rise in air temperature resulting from compression must be considered in the design of the air mains.

Diffusers The diffusers used most commonly in aeration systems are designed to produce either small or relatively large bubbles. Small-orifice diffusers, designed to produce small bubbles, are tubes or plates constructed of aluminum oxide or silicon oxide grains bound together with a ceramic matrix to form a porous mass. Tubes of corrugated stainless steel wrapped with Saran cord or frames covered with porous nylon bags are also available (see Fig. 12·12). Plate diffusers are installed in concrete or aluminum plate holders, holding six or more

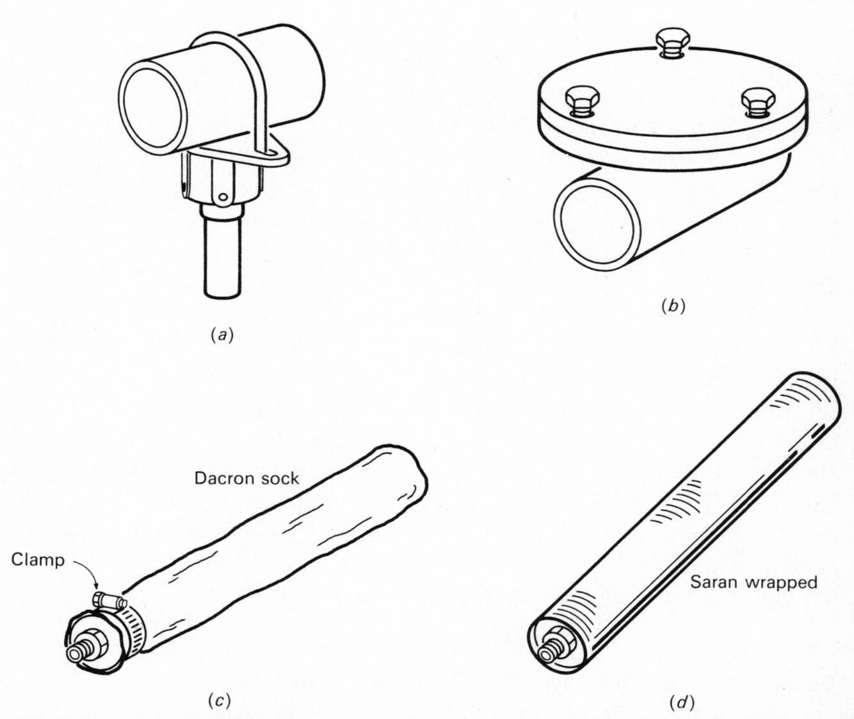

FIG. 12·12 Coarse-bubble diffusers: (*a*) Monosparj [from Walker Process Equipment] and (*b*) nonclog [from Eimco]. Fine-bubble diffusers: (*c*) Flexofuser [from Chicago Pump] and (*d*) Saran-wrapped precision diffuser tube [from Chicago Pump].

plates, which may be set in recesses or set on the bottom of the aeration tank. Groups of plate holders are connected to the air supply piping at intervals along the tank length, each group controlled by a valve. Tube diffusers are screwed into air manifolds, which may run the length of the tank close to the bottom along one side of the tank, or short manifold headers may be mounted on movable drop pipes. With the movable drop pipes, it is possible to raise a header out of the water without interrupting the process and without dewatering the tank. The diffusers can then be removed for cleaning or replacement.

With porous diffusers, it is essential that the air supplied be clean and free of dust particles that might clog the diffusers. Air filters are commonly employed and often consist of viscous impingement and dry types in series ahead of the blowers. Precoat bag-type filters and electrostatic filters have also been used. The dust problem can be alleviated by using diffusers that have large orifice openings.

Several types of coarse-bubble diffusers are available, such as the sparjer, the Deflectofuser, the Discfuser, and others [30, 35] (see Fig. 12·12). All these diffusers produce larger bubbles than the porous type of diffuser and consequently have a slightly lower aeration efficiency; however, the advantage of lower cost, less maintenance, and the absence of stringent air-purity requirements offsets the slightly lower efficiency. Water containing iron salts has caused serious clogging of porous diffusers at several plants. All diffusers will clog, however, if sufficient pressure is not maintained on the air headers at all times to prevent sewage from accumulating in the bottom of the headers.

Diffuser Performance Standard porous diffuser tubes are designed to deliver from 4 to 15 cfm per unit. Typical data for Saran-wrapped tubes (small-bubble) and sparjers are given in Figs. 12·13 and 12·14. The efficiency of oxygen transfer depends on the type and porosity of the diffuser, the size of the bubbles produced, the depth of submersion, and other factors that were explained in Chap. 9. In general, it varies from 5 to 15 percent, with 8 percent probable from porous tube diffusers and 6 percent probable from coarse-bubble diffusers [2].

The amount of air used per pound of BOD removed varies greatly from one plant to another, and there is risk in comparing the air use at different plants, not only because of the factors mentioned above, but also because of different loading rates, control criteria, and operating procedures. Extra high air flowrates applied along one side of a tank reduce the efficiency of oxygen transfer and may even reduce the net oxygen transfer by increasing circulating velocities. The result would be shorter residence time of the air bubbles as well as larger bubbles with less transfer surface [20].

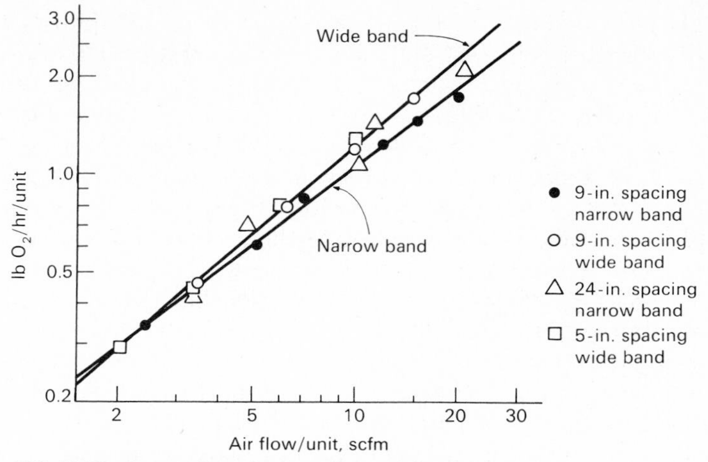

FIG. 12·13 Oxygen transfer from Saran tubes in water [5].

Blowers There are two types of blowers in common use: centrifugal and rotary positive-displacement (see Fig. 12·15). Centrifugal blowers are almost universally used where the unit capacity is equal to, or greater than, 15,000 cfm of free air. Rated discharge pressures vary from 7 to 9 psi. The preferred type is a two-stage unit direct-connected to a 3,580-rpm squirrel-cage induction motor. Units of this type are made by Elliott, Ingersoll-Rand, Allis Chalmers, Roots-Connersville,

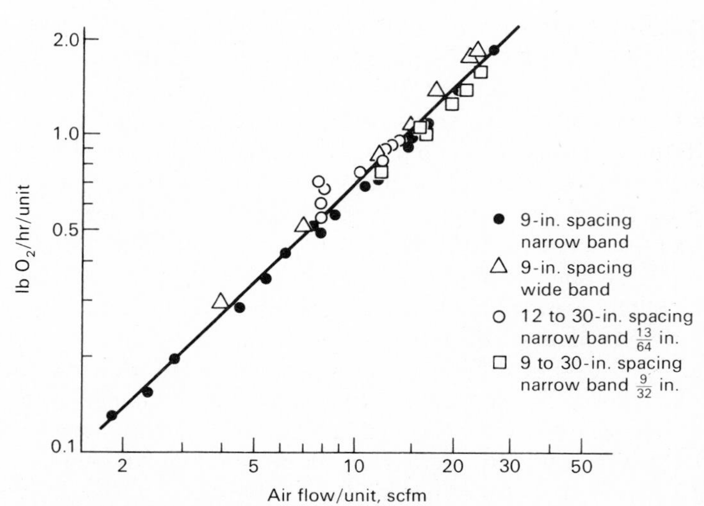

FIG. 12·14 Oxygen transfer from sparjers in water [5].

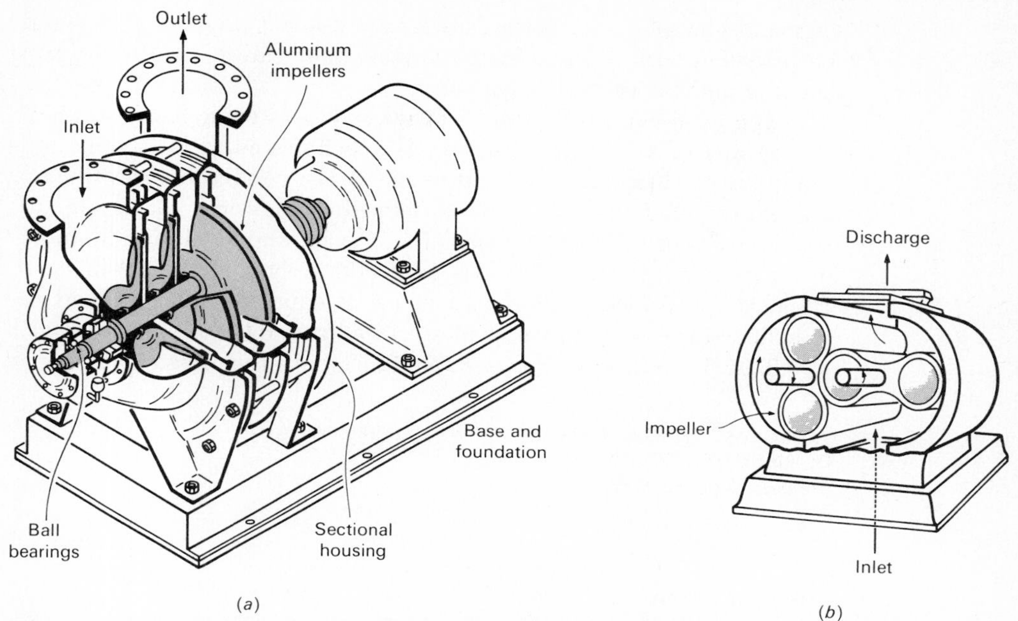

FIG. 12·15 Typical blowers used in diffused-air systems. (*a*) Centrifugal [from Hoffman] and (*b*) positive-displacement [from Roots-Connersville].

DeLaval, and Brown-Boveri of Switzerland. However, not all of these manufacturers will be competitive at any given rating. The blowers emit a high-pitched whine which is transmitted to the piping and can be very disagreeable, especially with lightweight piping, unless both inlet and discharge silencers are installed.

Centrifugal blowers have a head-capacity curve similar to that of a low-specific-speed centrifugal pump. The discharge pressure rises from shutoff to a maximum at about 50 percent of capacity and then drops off. The operating point is determined, as for a centrifugal pump, by the intersection of the head-capacity curve and the system curve.

It should be noted that surging (a phenomenon in which the blower alternately operates at zero capacity and 100 percent capacity accompanied by vibration and overheating) will result if the blower is throttled on the discharge side or by adjustment of the air valves at the aeration tanks to a capacity equal to or less than that at the peak of the head-capacity curve. Surging is avoided by installing a bypass to atmosphere arranged to be open at starting and at all times when the air flow is less than that at the surge point.

Blowers may be obtained with adjustable inlet guide vanes, at extra cost, which permit economical control of capacity by a lowering of the discharge curve. If the blowers are not so equipped, they should be provided with a valve in the suction that can be throttled during starting and for control of capacity.

The head-capacity curve (feet of air versus cubic feet per minute of free air) of a wide-open blower is a definite curve: the same as a pump curve. Free air is defined as air at the conditions prevailing at the blower inlet. Standard air is defined in blower work as air at a temperature of 68°F, a pressure of 14.7 psia, and a relative humidity of 36 percent. Standard air has a specific weight of 0.0750 lb/cu ft. Because the pressure developed in psi is equal to $\gamma_a H/144$ and the specific weight of air γ_a varies at sea level from 0.0776 lb/cu ft at 50°F to 0.0724 lb/cu ft at 86°F (see Table 12·5), both the pressure developed

TABLE 12·5 TYPICAL VALUES FOR THE SPECIFIC WEIGHT OF AMBIENT AIR*
(Pounds per cubic foot)

		Temperature		
Elevation, ft	Pressure, psi	10°C (50°F)	20°C (68°F)	30°C (86°F)
0	14.7	0.0776	0.0750	0.0724
1,000	14.2	0.0750	0.0724	0.0701
2,000	13.7	0.0722	0.0697	0.674
4,000	12.7	0.0670	0.0648	0.0625

* Relative humidity 36 percent. At other conditions of pressure and temperature, γ_a is given by the perfect gas law.

and the weight flow of air (or cubic feet per minute at standard conditions) will vary with the inlet conditions. Consequently, blowers must be selected to have adequate capacity on a hot summer day and must be provided with a motor of adequate horsepower for the coldest winter weather. The maximum horsepower may be reduced by throttling in extreme weather.

The brake horsepower (bhp) for adiabatic compression is given by Eq. 12·9:

$$\text{bhp}_{\text{ad}} = \frac{wRT_1}{550ne}\left[\left(\frac{p_2}{p_1}\right)^n - 1\right] \tag{12·9}$$

where w = weight flow of air, lb/sec
R = gas constant (53.5)
T_1 = absolute inlet temperature, °R

p_1 = absolute inlet pressure, pounds per square inch absolute (psia)
p_2 = absolute outlet pressure, psia
$n = (\kappa - 1)/\kappa = 0.283$ for air
$\kappa = 1.395$ for air
e = efficiency (usual range for compressors is 70 to 80)

The application of Eq. 12·9 is illustrated in the following example.

EXAMPLE 12·2 *Compressor Horsepower Requirements*

Determine the blower horsepower and free air volume required to supply 3,400 standard cubic feet per minute (scfm) of air at a discharge pressure of 8 psi. The ambient air temperature is 30°C (86°F), and the plant is located at sea level (760 mm Hg). Assume the compressor efficiency is 70 percent.

Solution

1. Determine the weight flow of air at 30°C.

$$\text{Volume of air at 30°C} = \frac{460 + 86}{460 + 68} 3{,}400 = 3{,}520 \text{ cfm}$$

From Table 12·5, the density of air at 30°C and 14.7 psi is 0.0724 lb/cu ft. Therefore, the weight flow of air is

$$w = \frac{3{,}520 \text{ cu ft}}{\text{min}} \times \frac{1}{\dfrac{60 \text{ sec}}{\text{min}}} \times \frac{0.0724 \text{ lb}}{\text{cu ft}}$$

$$= 4.25 \text{ lb/sec} \qquad \left(\text{note: } \frac{3{,}400}{60} 0.0750 = 4.25\right)$$

2. Determine the required horsepower using Eq. 12·9.

$$\text{hp} = \frac{4.25(53.5)(460 + 86)}{550(0.283)(0.70)} \left[\left(\frac{22.7}{14.7}\right)^{0.283} - 1\right] = 150$$

For capacities smaller than 15,000 cfm per unit, rotary positive-displacement blowers of the Roots type (Fig. 12·15*b*) are generally used. These machines operate at speeds from 360 to 900 rpm and may be direct-driven by sewage-gas engines. They produce a pulsating flow, since compression of the air is caused by backflow from the discharge main as the lobes clear the discharge port. Rugged suction and discharge silencers are essential. Discharge volume decreases only slightly with the buildup of pressure; therefore, if adequately powered, pressure buildup due to clogged diffusers is less critical than with centrifugal compressors. The units cannot be throttled, but capacity control can be obtained by multiple units and by speed change. Wound-rotor motors are uneconomical because these are constant-torque machines, and there is no saving in power. Multispeed motors may be used. Engine drive is ideal because the speed can be varied, and the units can be direct-connected without gears. The adiabatic efficiency of large blowers of this type varies from about 67 to 74 percent.

Small quantities of air for aerated grit chambers or for small plants may be supplied by small rotary positive-displacement blowers driven by variable-pitch V-belt drives or supplied with V-belt drives with interchangeable sheaves. Small multistage centrifugal blowers operating at 1,750 or 3,500 rpm are also available for this service in ratings up to 4,600 cfm at 8 psi requiring 250 hp. Units with a capacity on the order of 1,500 cfm have been used in a number of plants. These units are less efficient than the rotary positive-displacement blowers, but are less expensive in initial cost, simple to install, and smooth in operation.

Air Piping The air piping consists of the mains, valves, meters, and other fittings that transport compressed air from the blowers to the air diffusers located in the aeration tanks. Because the pressures are low (less than 10 psi), lightweight piping can be used.

Piping is usually sized on a velocity basis. The piping should be sized so that losses in tank headers and diffuser manifolds are small in comparison to the losses in the diffusers. Valves should be provided for flow regulation. Typical air velocities are given in Table 12·6.

TABLE 12·6 TYPICAL AIR VELOCITIES IN HEADER PIPES

Pipe diameter, in.	Velocity, fpm*
1–3	1,200–1,800
4–10	1,800–3,000
12–24	2,700–4,000
30–60	3,800–6,500

* Standard air.

Because of the high temperature of the air discharged by blowers (120–160°F), condensation in the air piping is not a problem, except where piping is submerged in the sewage. It is essential, however, that provisions be made for pipe expansion and contraction. Where fine-bubble diffusers are used, pipes must be of nonscaling materials or must be lined with a material that will not corrode. The practice of the Metropolitan Sanitary District of Greater Chicago, based on 30 years of experience, is to use steel pipe, bare inside, for mains 24 in. and larger; galvanized spiral-weld pipe for mains 8 to 24 in.; and standard-weight pipe, galvanized, for mains 6 in. and smaller. Piping submerged in the sewage or immediately above it, which is 3 in. and larger, should be cast iron, galvanized; piping which is $2\frac{1}{2}$ in. and smaller should be standard-weight steel, galvanized [7, 30]. Clad steel, Fiberglas, and plastic pipe are also receiving consideration by consulting engineers for current projects. Some organic coatings, such as coaltar dips, are not suitable because they will deteriorate with age, and fine particles may dislodge and clog the diffuser pores.

It is recommended that the piping losses be computed for maximum summer temperatures to be equaled or exceeded 1 percent of the time. The theoretical adiabatic temperature rise during compression is given by the following formula

$$\Delta T_{\text{ad}} = T_1 \left[\left(\frac{p_2}{p_1} \right)^n - 1 \right] \tag{12·10}$$

where the terms are the same as those defined for Eq. 12·9. The actual temperature rise is approximated by dividing ΔT_{ad} by the blower efficiency. Between the blowers and the aeration tanks, air temperatures will probably drop not more than 10 to 20°, but will quickly approach the temperature of the sewage in submerged piping.

Friction losses in the piping can be calculated using the Darcy-Weisbach equation written in the following form:

$$h_L = f \frac{L}{D} h_i \tag{12·11}$$

where h_L = friction loss, in. of water head
f = dimensionless friction factor obtained from Moody diagram (Fig. 3·10), based on relative roughness from Fig. 3·11. (It is recommended that f so obtained be increased by at least 10 percent to allow for an increase in friction factor with age.)
L/D = length of pipe in diameters
h_i = velocity head of air, in. of water

In dealing with air flow, it is customary to express velocities in feet per minute and volume rates of flow in cubic feet per minute. Lowercase letters v and q will be used for these quantities, and will refer to actual velocities and volumes under the pressure and temperature conditions prevailing in the pipe. With the subscript s attached, v_s and q_s will refer to velocities and flowrates at standard conditions.

In determining the value of the friction factor f using Fig. 3·10, the Reynolds number N_R is most conveniently found from the following formula:

$$N_R = \frac{28.4q_s}{d\mu} \tag{12·12}$$

where d = inside diameter, in.
μ = viscosity of air, centipoises

The viscosity can be approximated by using the following equation in the range of temperatures from 0 to 200°F:

$$\mu = (161 + 0.28t) \times 10^{-4} \tag{12·13}$$

where t = temperature, °F. The moderate pressures that exist in aeration piping have little effect on viscosity.

The velocity head h_i in inches of water can be computed using the following equation:

$$h_i = \left(\frac{v}{1{,}096}\right)^2 \gamma_a \tag{12·14}$$

where h_i = velocity head, in. of water
v = air velocity, fpm
γ_a = specific weight of air, lb/cu ft (see Table 12·5)

Application of the above equations is illustrated in the following example.

EXAMPLE 12·3 *Head Loss in Air Piping*

Determine the head loss in 1,000 ft of 15-in.-diameter galvanized spiral-weld pipe designed to carry 3,400 scfm of air. Assume the conditions of Example 12·2 are applicable.

Solution

1. Determine the temperature rise during compression using Eq. 12·10.

$$\Delta T = \frac{460 + 86}{0.70}\left[\left(\frac{22.7}{14.7}\right)^{0.283} - 1\right]$$
$$= 102°\text{F}$$

Therefore the air temperature at the blower discharge is 188°F (86 + 102°F).

2. Compute the Reynolds number using Eqs. 12·12 and 12·13. Assume the average temperature in the pipe is 160°F.

$$\mu = [161 + 0.28(160)] \times 10^{-4}$$
$$= 205.8 \times 10^{-4}$$
$$N_R = \frac{28.4(3{,}400)}{15(205.8 \times 10^{-4})} = 3.13 \times 10^5$$

3. Determine the friction factor f from Fig. 3·10, using the curve for commercial steel ($\epsilon = 0.00015$). The value of ϵ/D is 0.00012. Entering Fig. 3·10 with an ϵ/D value of 0.00012 and N_R of 3.13×10^5, the value of f is found to be 0.0155. Add 10 percent and use an f value of 0.017 for design.

4. Determine the air flowrate in the pipe.

$$\text{Flowrate} = 3{,}400 \frac{14.7}{22.7} \frac{460 + 160}{460 + 68}$$
$$= 2{,}580 \text{ cfm}$$

5. Determine the velocity in the pipe.

$$v = \frac{2{,}500 \text{ cu ft/min}}{\frac{3.14}{4}(1.25)^2} = 2{,}040 \text{ fpm}$$

6. Determine the specific weight of air at a pressure of 22.7 psia and a temperature 160°F.

$$\gamma_a = \frac{22.7(144)}{53.5(460 + 160)} = 0.0988 \text{ lb/cu ft}$$

7. Determine the velocity head using Eq. 12·14.

$$h_i = \left(\frac{2{,}100}{1{,}096}\right)^2 0.0988 = 0.362 \text{ in. of water}$$

8. Determine the head loss using Eq. 12·11.

$$h_L = 0.017 \frac{1{,}000}{1.25} 0.362$$
$$= 4.92 \text{ in. of water}$$

Losses in elbows, tees, valves, etc., can be computed as a fraction of the velocity head using K values in standard hydraulic texts. Meter losses can be estimated as a fraction of the differential head, depending on the type of meter. Losses in air filters, blower silencers, and check valves should be obtained from the manufacturers for estimating purposes. Air-filter losses will range from 0.5 to 2.0 in. but may run higher with certain types of filters. Check-valve losses will vary from 0.8 in. to 6.0 to 8.0 in., depending on the type of construction and installation. Losses in silencers will vary from 0.5 to 1.0 in. per silencer for centrifugal blowers and from 6.0 to 8.0 in. per silencer for those used with rotary positive-displacement blowers.

The discharge pressure at the blowers will be the sum of the above losses, the depth of water over the diffusers, and the loss through the diffusers. Diffuser losses are determined from rating curves furnished by the manufacturer and vary from 4 to 18 in. To ensure uniform air distribution, the diffuser loss at minimum air flow should be not less than 4 in. With tubes, a control orifice is installed in the tube inlet to increase the loss. An allowance of 6 to 8 in. should be added for clogging of the tubes and coarse-bubble diffusers. For plate diffusers this should be increased to 0.5 to 1.0 psi [7].

Design of Mechanical Aerators

Two types of mechanical aerators commonly used are surface and turbine (see Fig. 12·16). In the former, oxygen is entrained from the atmosphere, whereas in the latter, oxygen is entrained from air or is introduced in the tank bottom. In either case, the action of the aerator and that of the turbine help to keep the contents of the aeration tank or basin mixed.

Surface Aerators Mechanical surface aerators offer the utmost in simplicity in an aeration system. They may be obtained in sizes from 1 to 100 hp. They consist of submerged or partially submerged impellers, which are centrally mounted in the aeration tank and which agitate the sewage vigorously, entraining air in the sewage and causing a rapid change of the air-water interface to facilitate solution of the air (see Fig. 12·16*a* and *b*). For detailed descriptions, pictures, mention of major installations, and operating performance at a few locations, see Ref. [30].

Impellers are fabricated from steel, cast iron, noncorrosive alloys, and Fiberglas-reinforced plastic. In most instances, the impeller is driven through a reduction gear by an electric motor. The motor and gearbox are mounted on a platform that is supported by piers extending to the bottom of the tank or by beams that span the tank. Floating mechanical aerators have been devised for use in ponds or lagoons where the water surface elevation fluctuates, or where a rigid support would be impractical. Mechanical aerators can also be driven by internal combustion engines on a continuous or standby basis.

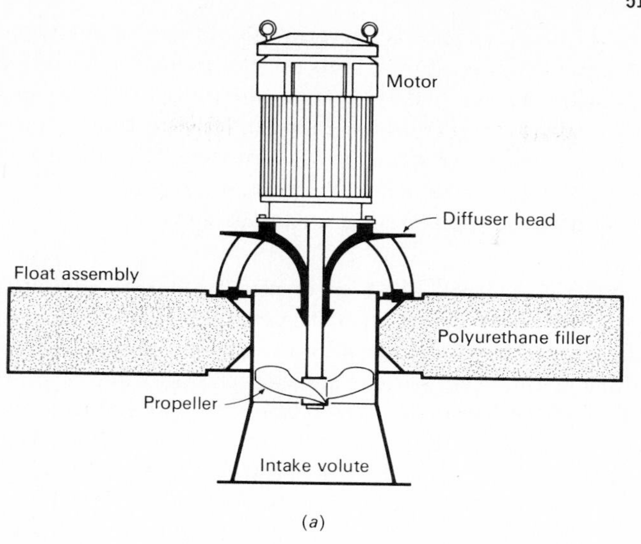

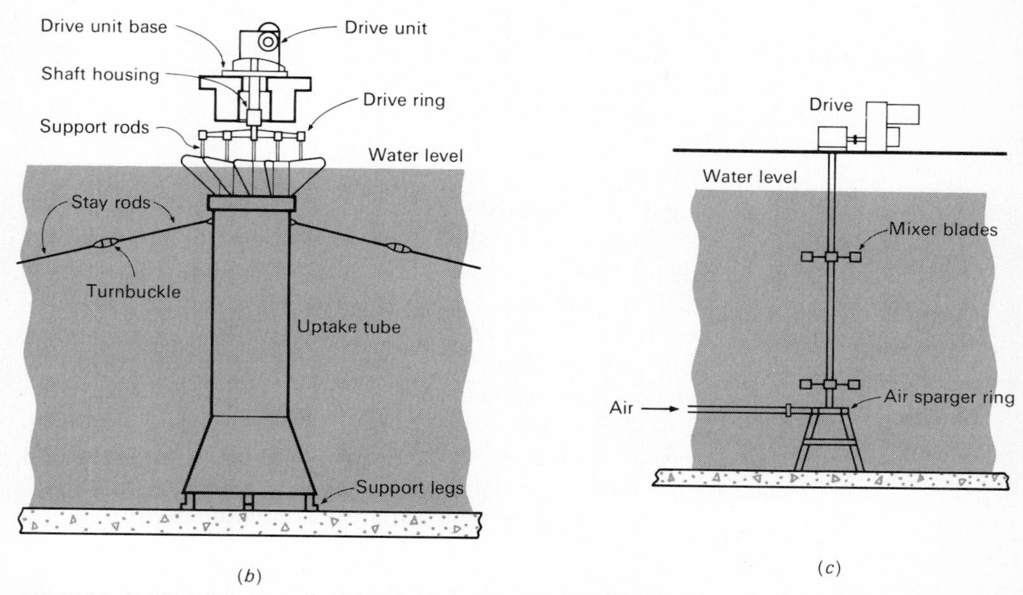

FIG. 12·16 Mechanical aerators. (*a*) Surface aerator [from Aqua-Jet], (*b*) Simplex cone [from Yeomans], and (*c*) turbine aerator [from Permutit Co.].

The Kessener brush aerator, an old device very popular in Europe, recently has been adapted to provide both aeration and circulation in oxidation ditches [30, 31]. It is horizontally mounted just above the water surface and consists of a cylinder with bristles of steel protruding from its perimeter into the wastewater. The drum is rotated rapidly by an electric motor drive, spraying sewage across the tank, promoting circulation, and entraining air in the sewage.

Turbine Aerators Most mechanical aerators are upflow types that rely on violent agitation of the surface and air entrainment for their efficiency. However, air or pure oxygen may also be introduced by diffusion into the sewage beneath the impeller of downflow aerators, the impeller being used to disperse the air bubbles and mix the tank contents. A draft tube may be utilized with either upflow or downflow models to control the flow pattern of the circulating liquid within the aeration tank. The draft tube is a cylinder with flared ends mounted concentrically with the impeller, and extending from just above the floor of the aeration tank to just beneath the impeller.

Aerator Performance Surface aerators are rated in terms of their oxygen-transfer rate as pounds of O_2 per horsepower-hour (hp-hr) at standard conditions. Standard conditions exist when the temperature is 20°C, the dissolved oxygen is 0.0 mg/liter, and the test liquid is tap water. Testing and rating is normally done under non-steady-state conditions using fresh water, deaerated with sodium sulfite. Efficiencies of up to 7 lb O_2/hp-hr have been claimed for various surface aerators, and high efficiencies can be obtained from small pilot plant aerators. Commercial-size surface aerators available today (1971) range in efficiency from 2 to 4 lb O_2/hp-hr. Efficiencies for turbine aerators range from 2 to 3 lb O_2/hp-hr. Efficiency claims should be accepted by the design engineer only when supported by actual test data for the actual model and size of aerator under consideration [30]. For design purposes, the standard performance data must be adjusted to reflect anticipated field conditions. This is accomplished using the following equation. The term within the brackets represents the correction factor.

$$N = N_0 \left[\frac{\beta C_{\text{walt}} - C_L}{9.17} (1.024)^{T-20} \alpha \right] \tag{12·15}$$

where N = lb O_2/hp-hr transferred under field conditions

N_0 = lb O_2/hp-hr transferred in water at 20°C, and zero dissolved oxygen

β = salinity-surface tension correction factor, usually 1

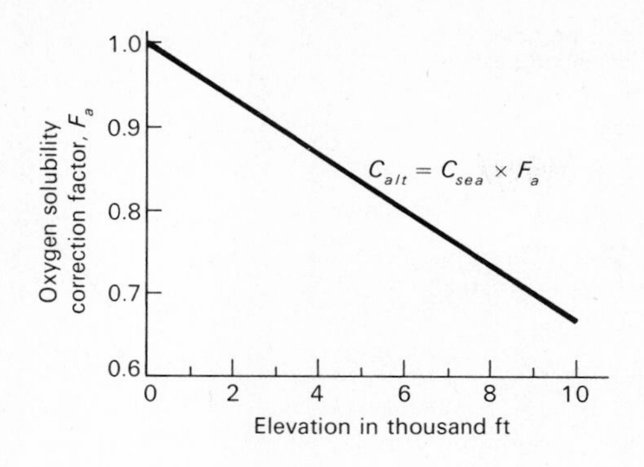

FIG. 12·17 Oxygen solubility correction factor versus elevation.

C_{walt} = oxygen-saturation concentration for waste at given temperature and altitude (see Fig. 12·17)
C_L = operating oxygen concentration
T = temperature, °C
α = oxygen-transfer correction factor for waste, usually 0.8 to 0.85 for wastewater

The application of this equation is illustrated in a subsequent section dealing with the design of aerated lagoons.

Energy Requirement for Mixing As with diffused-air systems, the size and shape of the aeration tank is very important if good mixing is to be achieved. Aeration tanks may be square or rectangular and may contain one or more units. Water depth may vary from 4 to 12 ft when using surface aerators. Depths up to 17 ft have been used with draft-tube mixers. Variable-speed drives and water-level control by adjustable weirs have been used to vary the degree of mixing and aeration. In diffused-air systems the air requirement to ensure good mixing will vary from 20 to 30 scfm/1,000 cu ft of tank volume.

Typical power requirements for maintaining a completely mixed flow regime with mechanical aerators will vary from 0.50 to 1.0 hp/1,000 cu ft. These requirements will vary with the design of the aerator and the tank, lagoon, or basin geometry. For the aerated lagoon, it is extremely important that the mixing power requirement be checked because in most instances it will be the controlling factor.

Design of Aeration Tanks and Appurtenances

Once the activated-sludge process and the aeration system have been selected and a preliminary design has been made, the next step involves the design of the aeration tanks and support facilities.

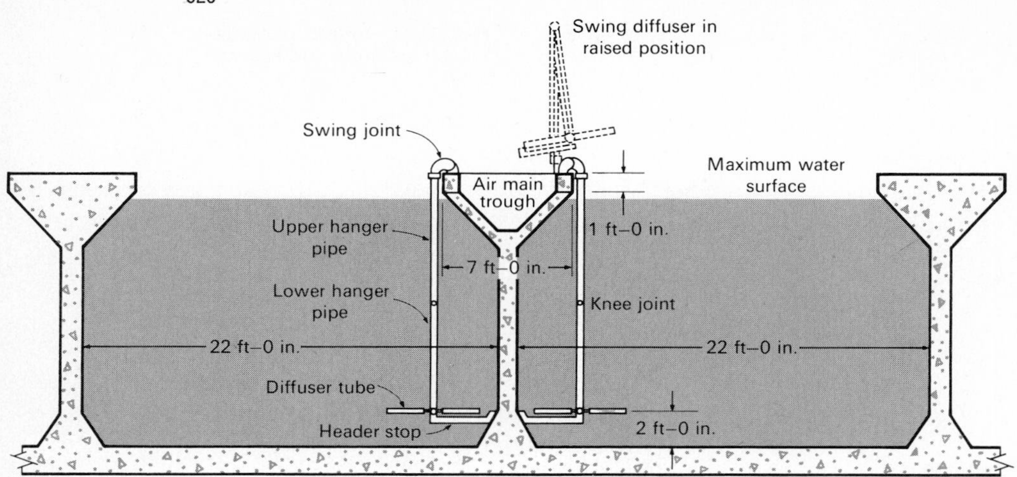

FIG. 12·18 Cross section of a typical activated-sludge aeration tank with fine-bubble diffuser aeration system.

Aeration Tanks Aeration tanks are constructed of reinforced concrete and are left open to the atmosphere. Figure 12·18 shows a cross section of a typical aeration tank. They are rectangular in shape, which permits common-wall construction for multiple tanks. The total aeration-tank volume required should be divided among two or more units capable of independent operation, if the total capacity exceeds 5,000 cu ft. The total capacity required to the water surface elevation should be determined from the biological process design. Although the air bubbles dispersed in the sewage will occupy perhaps 1 percent of the total volume, no allowance is made for this in tank sizing. The volume occupied by submerged piping is usually negligible.

If the tank is to be aerated with diffused air, the proportioning of the tank may significantly affect the aeration efficiency and the amount of mixing obtained unless the type, number, and location of diffusers are properly chosen [2]. The depth of sewage in the tank should be between 10 and 15 ft so that the diffusers can work efficiently. Freeboard from 1 to 2 ft above the waterline should be provided. The width of the tank in relation to its depth is important if spiral-flow mixing is to be used. The width-to-depth ratio for such tanks may vary from 1.0 to 1 to 2.2 to 1. This limits the width of a tank channel to between 15 and 33 ft.

In large plants, the channels become quite long, sometimes exceeding 500 ft in total length per tank. Tanks may consist of one to four channels with round-the-end flow in multiple-channel tanks. Large plants should contain not less than four tanks, preferably from six to eight or more. Some of the largest plants contain from 30 to 40 tanks arranged in several groups or batteries.

For tanks using diffusers on both sides or with diffusers in the center of the tank, greater widths are permissible. The important point is to restrict the width of the tank so that "dead spots" or zones of inadequate mixing are avoided. The dimensions and proportions of each independent unit should be such as to maintain adequate velocities so that deposition of solids will not occur. Triangular baffles or fillets may be placed longitudinally in the corners of the channels to eliminate dead spots and to deflect the spiral flow.

Individual tanks should have inlet and outlet gates or valves so that they may be removed from service for inspection and repair. The common walls of multiple tanks must therefore be able to withstand the full hydrostatic pressure from either side. Aeration tanks must have adequate foundations to prevent settlement; and, in saturated soil, they must be designed to prevent flotation when the tanks are dewatered. This is done by thickening the floor slab or installing relief valves connected by graded gravel trenches beneath the floor. Drains or sumps for aeration tanks are desirable for dewatering. In large plants where tank dewatering might be more common, it may be desirable to install mud valves in the bottoms of all tanks. These should be connected to a central dewatering pump or pumping station or to a plant drain discharging to the wet well of the plant pumping station. For small plants, portable contractors' pumps are suitable for dewatering service. It should be possible to empty a tank in 16 hr.

Froth-Control Systems Sewage normally contains soap, detergents, and other surfactants that will produce foam when the sewage is aerated. If the concentration of MLSS is high, the foaming tendency is minimized. Large quantities of foam may be produced during start-up of the process, when the MLSS are low, or whenever high concentrations of surfactants are present in the sewage. The foaming action produces a froth that contains sludge solids, grease, and large numbers of sewage bacteria. The wind may lift the froth off the tank surface and blow it about, contaminating whatever it touches. The froth, besides being unsightly, is a hazard to workmen, because it is very slippery, even after it collapses. In addition, once the froth has dried, it is difficult to clean off.

It is essential, therefore, to have some method for controlling froth formation. A commonly employed system consists of a series of spray nozzles mounted along the top edge of the aeration tank opposite the air diffusers. Screened effluent or clear water is sprayed through these nozzles either continuously or on a time clock–controlled program, and this physically breaks down the froth as it forms. Another approach is to meter a small quantity of antifoaming chemical additive into the inlet of the aeration tank or preferably into the spray water [31].

Return-Sludge Requirements The purpose of sludge return is to maintain a sufficient concentration of activated sludge in the aeration tank, so that the required degree of treatment can be obtained in the time interval desired. The return of activated sludge from the clarifier to the inlet of the aeration tank is the essential feature of the process. Sludge should be removed from settling tanks as fast as it forms. Excessive detention to form a dense sludge to minimize pumping rates is inadvisable, as it may result in sludge deterioration. Ample return-sludge pump capacity should be provided and is essential if sludge solids are not to be lost in the effluent. The reason for this is the solids have a tendency to form a sludge blanket in the bottom of the tank that varies in thickness from time to time and may fill the entire depth of the tank at peak flows if the return-sludge pump capacity is inadequate. Return-sludge pump capacities of 20 to 30 percent formerly provided have proved inadequate, and current designs for large plants provide capacity of 50 to 75 percent of the sewage flow and 100 percent for small plants.

In general, the sludge-return pumps should be set so that the return flow is approximately equal to the percentage ratio of the volume occupied by the settleable solids from the aeration tank effluent to the volume of the clarified liquid (supernatant) after settling for 30 min in a 1,000-ml graduated cylinder. This ratio should not be less than 15 percent at any time. For example, if, after 30 min of settling, the settleable solids occupied a volume of 150 ml, the percentage volume would be equal to 17.7 percent [(150 ml/850 ml) $\times$ 100]. If the plant flow were 2 mgd, the sludge return rate should be 0.35 mgd (0.177 $\times$ 2 mgd).

Another method often used to control the rate of return sludge pumping as well as plant operation is based on an empirical measurement known as the sludge volume index (SVI). This index is defined as the volume in milliliters occupied by one gram of activated-sludge mixed-liquor solids, dry weight, after settling for 30 min in a 1000-ml graduated cylinder (see Ref. [19] in Chap. 7). In practice it is taken to be the percentage volume occupied by the sludge in a mixed-liquor sample (taken at the outlet of the aeration tank) after 30 min of settling (P_v), divided by the suspended solids concentration of the mixed liquor expressed as a percentage (P_w). If the SVI is known, then the required percentage return sludge in terms of the recirculation ratio Q_r/Q to maintain a given percentage mixed-liquor solids concentration in the aeration tank is: $100Q_r/Q = 100/[(100/P_w\text{SVI}) - 1]$. For example, to maintain a mixed-liquor concentration of 0.3 percent (3,000 mg/liter), the percentage of sludge that must be returned when the SVI is 100 is equal to $100/[(100/0.30 \times 100) - 1]$, or 43 percent.

The index can also be used as an indication of the settling characteristics of the sludge. However, because the value of the index that is characteristic of a good settling sludge will vary with the mixed-liquor solids concentration, observed values at a given plant should not be compared with other values reported in the literature unless operating conditions are known to be similar. For example, if the solids did not settle at all but occupied the entire 1,000 ml at the end of 30 min, the maximum value of the index would be obtained and would vary from 1,000 for a mixed-liquor solids concentration of 1,000 mg/liter down to a value of 100 for a mixed-liquor concentration of 10,000 mg/liter. For such conditions, the computation has no meaning other than the determination of limiting values.

Typical values of the SVI for good settling sludges for diffused-air aeration plants operating with mixed-liquor suspended-solids concentrations of 800 to 3,500 mg/liter range from 150 to 35. With a fixed and limited return-sludge pump capacity, as the SVI increases, the solids concentration that can be maintained in the mixed liquor without loss of solids in the effluent is reduced [14]. For example, with a return-sludge pump capacity of 30 percent, an SVI of 50 would permit a mixed-liquor solids concentration of 3,000 to 4,000 mg/liter, an SVI of 100 would not permit a concentration greater than 2,200 mg/liter, and an SVI of 200 would require limiting the solids concentration to not over 1,100 mg/liter. A rising value of the SVI is indicative of trouble ahead, and prompt action should be taken to bring it under control [14]. Control measures are discussed in the section titled Operation Difficulties.

In some installations operation of the sludge-return pumps is controlled by a series of photoelectric cells or a sonic system located in the settling tank. The operation of these devices is based on the fact that the top of the sludge blanket usually forms a distinct interface with clarified liquid above it.

Sludge Wasting Excess activated sludge should be wasted as required to maintain a constant level of MLSS and a constant mean cell residence time. This can best and most accurately be accomplished by withdrawing mixed liquor directly from the aeration tank or the aeration-tank effluent pipe where the concentration of solids is uniform. The waste mixed liquor can then be discharged to a sludge-thickening

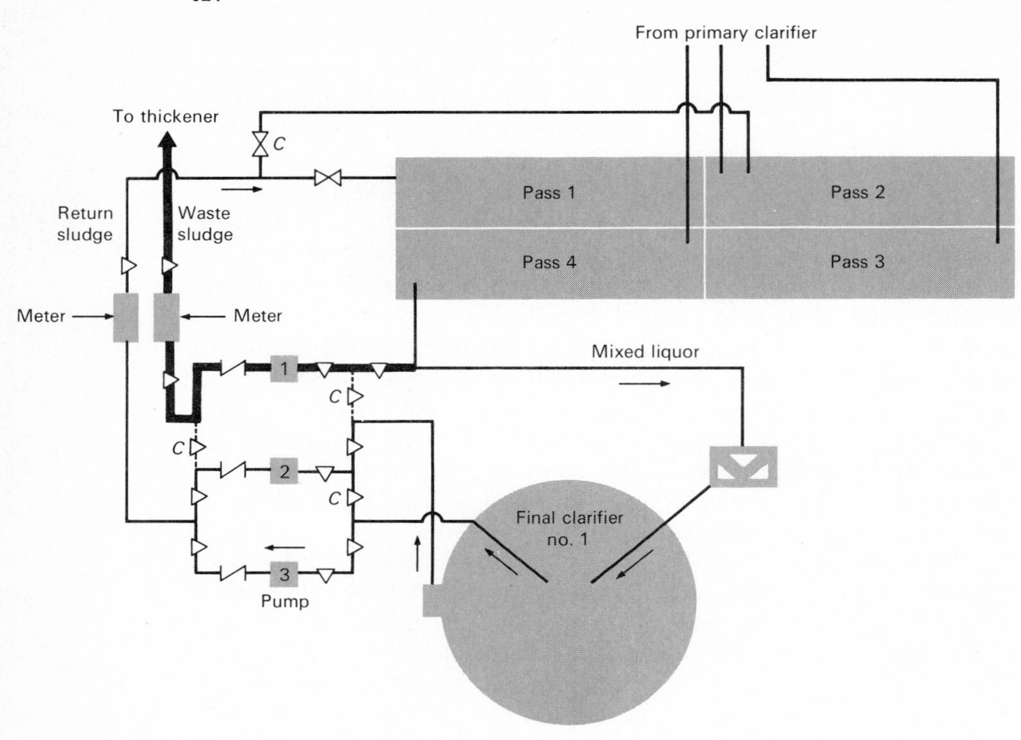

FIG. 12·19 Sludge wasting from mixed-liquor line at Wahiawa Sewage Treatment Plant, Oahu, Hawaii.

tank or to the primary tanks where the sludge settles and mixes with the raw primary sludge (see Fig. 12·19). In the past it has been more common practice to thicken the sludge in the activated-sludge settling tanks and to waste from the sludge-return line (see Fig. 12·20). The waste sludge is discharged to the primary tanks, to thickening tanks, or to other sludge-disposal facilities. Although commonly used this method of sludge wasting is not recommended. If this method must be used, provision for wasting mixed liquor should also be included.

Design of Solids Separation Facilities

The function of the activated-sludge settling tank is to separate the activated-sludge solids from the mixed liquor. This constitutes the final step in the production of a well-clarified, stable effluent low in BOD and suspended solids and, as such, represents a critical link in the operation of an activated-sludge treatment process. Although much of the material discussed in Chap. 11 in connection with the design of

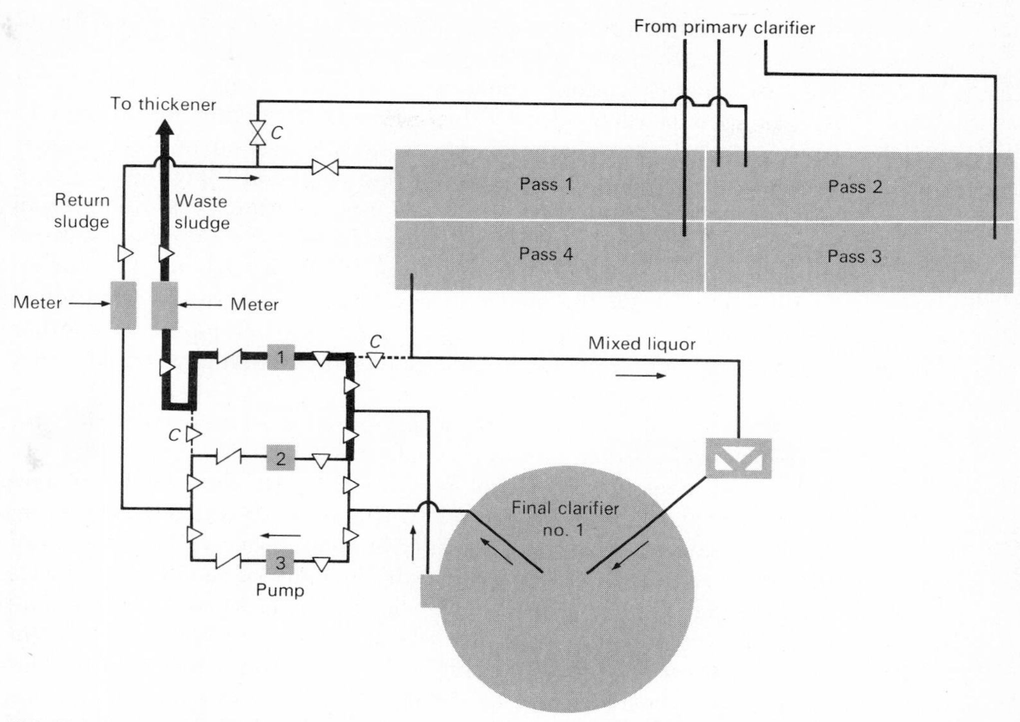

FIG. 12·20 Sludge wasting from return-sludge line, Wahiawa Sewage Treatment Plant, Oahu, Hawaii.

primary sedimentation tanks is applicable, the presence of the large volume of flocculent solids in the mixed liquor requires that special consideration be given to the design of activated-sludge settling tanks. These solids tend to form a sludge blanket in the bottom of the tank that will vary in thickness. This blanket may fill the entire depth of the tank and overflow the weirs at peak flowrates if the return-sludge pump capacity or the size of the settling tank is inadequate. Further, the mixed liquor, on entering the tank, has a tendency to flow as a density current interfering with the separation of the solids and the thickening of the sludge. To cope successfully with these characteristics, factors that must be considered in the design of these tanks include: (1) type of tank to be used, (2) surface loading rate, (3) solids loading rate, (4) flow-through velocities, (5) weir placement and loading rates, and (6) scum removal.

Tank Types Activated-sludge settling tanks may be either circular or rectangular. Square tanks are now seldom used. Circular tanks have been constructed with diameters ranging from 12 to 200 ft, although the more common range is from 30 to 100 ft. The tank radius should preferably not exceed five times the sidewater depth. Basically, there are two types of circular tanks to choose from: the center-feed and the rim-feed clarifier. Both types utilize a revolving mechanism to transport and remove the sludge from the bottom of the clarifier. Mechanisms are of two types: those that scrape or plow the sludge to a center hopper similar to the types used in primary sedimentation tanks, and those that remove the sludge directly from the tank bottom through suction orifices that serve the entire bottom of the tank in each revolution. Of the latter, in one type the suction is maintained by reduced static head on the individual drawoff pipes. In another patented suction system, sludge is removed through a manifold either hydrostatically or by pumping.

Rectangular tanks must be proportioned so that good distribution of incoming flow is achieved and horizontal velocities are not excessive. It is recommended that where possible the maximum length of rectangular tanks should not exceed 10 times the depth, but lengths up to 300 ft have been used successfully in large plants. Where widths of rectangular tanks exceed 20 ft, multiple pairs of chains and flights may be employed to permit tank widths up to 80 ft. For very long tanks, it is desirable to use two sets of chains and flights in tandem with a central hopper to receive the sludge. Tanks in which mechanisms move the sludge to the effluent end in the same direction as the density current have shown superior performance in some instances.

Regardless of tank shape, the sludge collector selected should be able to meet the following two operational conditions: (1) the collector should have a high capacity so that when a high sludge-recirculation rate is desired channeling of the overlying liquid through the sludge will not result and (2) the mechanism should be sufficiently rugged to be able to transport and remove the very dense sludges that could accumulate in the settling tank during periods of mechanical breakdown or power failure.

Overflow Rate When industrial wastes are to be treated by the activated-sludge process it is recommended that pilot plant studies be conducted to evaluate the settling characteristics of the mixed liquor. These are also desirable in the case of the more familiar municipal wastes where the process variables, such as the concentration of the mixed-liquor suspended solids, mean cell residence time, etc., are outside the range of common experience. If such studies are performed, it is important that they be conducted over a temperature range that is representative of both the average and coldest temperatures to be encountered.

Activated-sludge solids have a specific gravity so near that of water that the increased density and viscosity of the sewage under winter conditions affect the settling properties of the sludge adversely. In addition, the settling properties of the sludge may vary from time to time because of changes in the amount and specific gravity of the suspended solids passing through the primary settling tanks, the character and amount of industrial wastes contained in the sewage, and the composition of the microbial life of the floc. For these reasons, conservative design criteria are necessary if overflow of sludge solids is not to occur intermittently.

Because of the large amount of solids that may be lost in the effluent if design criteria are exceeded, effluent overflow rates should be based on peak flow conditions. The overflow rates given in Table 12·7 are recommended for use in activated-sludge plants with the exception of high-solids pure-oxygen systems.

TABLE 12·7 RECOMMENDED PEAK FLOW OVERFLOW RATES FOR VARIOUS MIXED-LIQUOR SUSPENDED-SOLIDS CONCENTRATIONS

	Recycle, percent	
MLSS	25	50
500	1,400	1,400
1,000	1,400	1,400
1,500	1,200	1,200
2,000	1,200	1,200
2,500	1,150	960
3,000	960	800
3,500	823	685

The overflow rates in Table 12·7 are based on the gallons per day of sewage flow per square foot of tank surface instead of on the mixed-liquor flow. The reason for this is that the overflow rate is equivalent to an upward flow velocity, whereas the return-sludge flow is drawn off the bottom of the tank and does not contribute to the upward flow velocity. Further, above a mixed-liquor suspended-solids concentration of 2,000 mg/liter, the overflow rates in Table 12·7 are based on a solids loading of 1.25 lb/sq ft/hr (solids loading is discussed in the following section). At mixed-liquor solids concentrations below 2,000 mg/liter, the overflow rate has been limited to fixed values consistent with satisfactory operating experience.

On the basis of limited experience with pure-oxygen systems (1971), it has been observed that the settling properties of the resulting sludge will vary with the characteristics of the wastewater being treated, its temperature, and the degree of pretreatment. From the reported results of the study conducted at Batavia, N.Y. [1] it would appear that higher overflow rates than would be indicated by an extension of Table 12·7 can be used, whereas based on pilot plant studies conducted at Washington, D.C. it was found that even lower values may be required to achieve effective solids separation. Therefore, until more information becomes available it is recommended that pilot plant tests be conducted on the waste to be treated before designing the settling tanks to be used with pure-oxygen aeration systems.

Solids-Loading Rate At some point when the concentration of mixed-liquor solids increases above 2,000 mg/liter, the area of the tank may depend on the properties of the sludge in hindered settling and the solids loading (see Example 8·4) instead of on the overflow rate. With some industrial wastes the solids loading may be limiting at much lower suspended-solids concentrations. The solids-loading rate on an activated-sludge settling tank may be computed by dividing the total solids applied by the surface area of the tank. The preferred units are pounds per square foot per hour, although units of pounds per square foot per day are common in the literature. The former is favored because the solids-loading factor should be evaluated both at peak and average flow conditions. If peaks are of short duration, average 24-hr values may govern; if peaks are of long duration, peak values should be assumed to govern to prevent the solids from overflowing the tank. In terms of the mixed-liquor flow, the solids-loading factor may be computed using Eq. 12·16.

$$\text{lb/sq ft/hr} = \frac{\dfrac{\text{MLSS mg}}{\text{liter}} \times \dfrac{\text{surface-loading rate gal}}{\text{day-sq ft}} \times \dfrac{3.785 \text{ liter}}{\text{gal}}}{\dfrac{10^3 \text{ mg}}{\text{g}} \times \dfrac{454 \text{ g}}{\text{lb}} \times \dfrac{24 \text{ hr}}{\text{day}}}$$

$$= \frac{\text{MLSS} \times \text{surface-loading rate}}{120{,}000 \times 24} \qquad (12\cdot16)$$

In effect the solids-loading rate represents a characteristic value for the suspension under consideration. It has been observed that, in a settling tank of fixed surface area, the effluent quality will deteriorate if the solids loading is increased beyond the characteristic value for the suspension [18]. Typical solids-loading values reported in the litera-

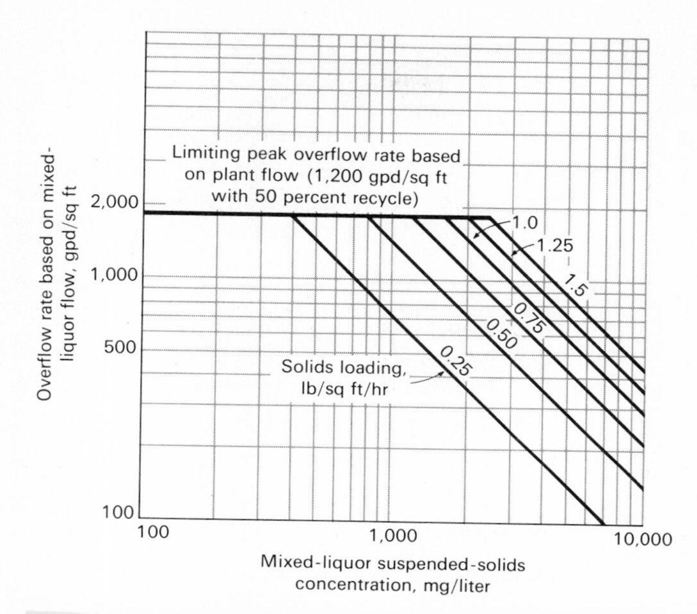

FIG. 12·21 Relationships among surface-loading rates based on mixed-liquor flow in gpd/sq ft, mixed-liquor suspended-solids concentration, and solids loadings for activated-sludge settling tanks.

ture for activated-sludge mixed-liquor vary from about 0.5 to 1.25 lb/sq ft/hr [26, 35]. Although higher rates may be observed under optimum conditions, they should not be used for design without extensive experimental work covering all seasons and operating variables.

The relationship among solids loading, mixed-liquor suspended-solids concentration, and overflow rates based on mixed-liquor flow is shown in Fig. 12·21. As shown, for a constant solids loading, the overflow rate must be reduced as the MLSS concentration increases. Experimental data obtained by Pflanz [26] in Germany are presented in Fig. 12·22. As shown, to maintain a constant effluent quality of 20 to 30 mg/liter, the solids loading should be kept more or less the same as the MLSS concentration increases. Suggested peak and average solids loadings of 1.25 and 0.6 lb/sq ft/hr, respectively, are also plotted. The data on this figure should not be taken to indicate that circular tanks are more efficient than rectangular, because in any given case the efficiency will depend on the details of design and either type could be more efficient depending on the design.

Flow-through Velocity To avoid troubles due to density currents, and scouring of sludge already deposited, horizontal velocities should be limited. For rectangular tanks, horizontal flow-through velocities based on the maximum mixed-liquor flow should not exceed a value of about 100 fph. In a circular center-feed tank, the inlet baffle should have a diameter of 15 to 20 percent of the tank diameter and should

FIG. 12·22 Relationships among surface-loading rates based on mixed-liquor flow in ft/hr, mixed-liquor suspended-solids concentration, and solids loadings for activated-sludge settling tanks.

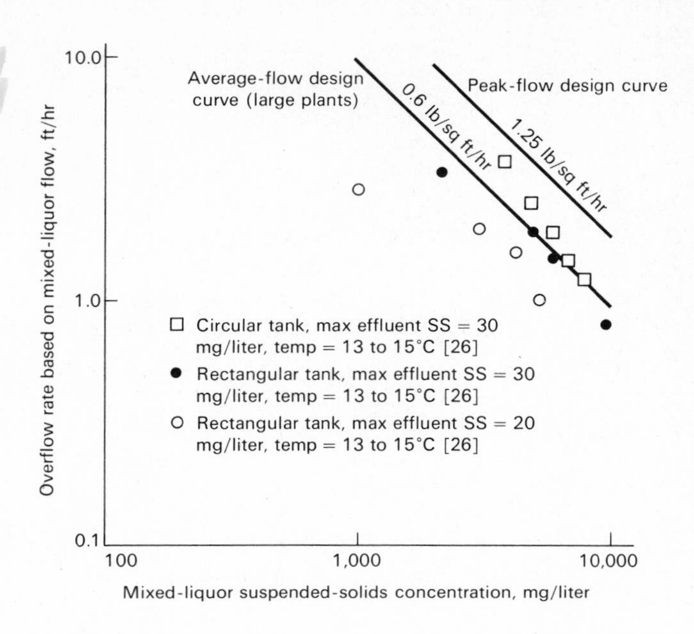

not extend more than 3 ft below the surface to avoid scouring of deposited sludge from the sludge drawoff sump.

Weir Placement and Loading Activated-sludge mixed liquor entering the tank will flow along the tank bottom as a density current until it encounters a countercurrent pattern or an end wall. Unless this is designed against, solids may be discharged over the effluent weir [28]. Anderson's experimental work at Chicago on tanks 126 ft in diameter indicated that a circular weir trough placed at $\frac{2}{3}$ to $\frac{3}{4}$ of the radial distance from the center was in the optimum position to intercept well-clarified effluent [28]. With low surface loadings and weir rates, the placement of the weirs in small tanks does not significantly affect the performance of the clarifier. Circular clarifiers are manufactured with overflow weirs located near both the center and the perimeter of the tank. The minimum water depth below effluent weirs so located should be 10 ft to prevent overflow of density currents. If weirs are located at the tank perimeter or at end walls in rectangular tanks, the minimum depth should be 12 ft.

Weir loading rates in large tanks should preferably not exceed 30,000 gpd/linear ft of weir at maximum flow when located away from the upturn zone of the density current, or 20,000 gpd/linear ft when located within the upturn zone. In small tanks the weir-loading rate should not exceed 10,000 gpd/linear ft at average flow or 20,000 gpd/

linear ft at maximum flow. The upflow velocity in the immediate vicinity of the weir should be limited to about 12 to 24 fph. This can be used to determine the spacing of multiple weirs in rectangular tanks.

Scum Removal Whether or not to make provision for skimming the final tanks will depend on the characteristics of the incoming wastewater, the extent of preliminary treatment, and the type of treatment process employed. With the modified aeration process in which the primary settling tanks are usually omitted, skimming of the final tanks is essential. Skimming facilities are now required on all federally funded projects.

Operational Difficulties

Two of the most common problems encountered in the operation of an activated-sludge plant are rising sludge and bulking sludge. Because few plants have escaped this affliction, it is appropriate to discuss the nature of these problems and methods and techniques that can be used for their control.

Rising Sludge Occasionally sludge that has good settling characteristics will be observed to rise or float to the surface after a relatively short settling period. The cause of this phenomenon is denitrification, in which the nitrites and nitrates in the wastewater are converted to nitrogen gas (see Chap. 14). As nitrogen gas is formed in the sludge layer, much of it is trapped in the sludge mass. If enough gas is formed, the sludge mass becomes buoyant and rises or floats to the surface. Rising sludge can be differentiated from bulking sludge by noting the presence of small gas bubbles attached to the floating solids.

Rising-sludge problems can be overcome by (1) increasing the rate of return activated-sludge pumping from the activated-sludge settling tank, (2) decreasing the rate of flow of aeration liquor into the offending tank if the sludge depth cannot be reduced by increasing the return activated-sludge withdrawal rate, (3) where possible, increasing the speed of the sludge-collecting mechanism in the settling tanks, and (4) decreasing the mean cell residence time by increasing the sludge-wasting rate.

Bulking Sludge A bulked sludge is one that has poor settling characteristics and poor compactability. The two principal types of sludge bulking have been identified. One is caused by the growth of filamentous organisms or organisms that can grow in a filamentous form under adverse conditions (see Ref. [23] in Chap. 10), and the other is caused by bound water in which the bacterial cells composing the floc swell through the addition of water to the extent that their density is reduced and they will not settle. The causes of sludge bulking that are most commonly cited in the literature are related to (1) the physical and chemical characteristics of the wastewater, (2) treatment plant design limitations, and (3) plant operation.

Wastewater characteristics that can affect sludge bulking include fluctuations in flow and strength, pH, temperature, staleness, nutrient content, and the nature of the waste components. Design limitations include air-supply capacity, clarifier design, return-sludge pumping-capacity limitations, and short circuiting or poor mixing. Operational causes include low dissolved oxygen in the aeration tank, organic waste overloading of aeration tanks, and final clarifier operation. In almost all cases, all the aforementioned conditions represent some sort of adverse operating condition.

In the control of bulking, where a number of variables are possible causes, it is well to have a check list of things to investigate. The following are recommended: (1) dissolved-oxygen content, (2) process loading, (3) return-sludge pumping rate, (4) internal plant overloading, (5) wastewater characteristics, and (6) clarifier operation.

Limited dissolved oxygen has been noted by more people than any other cause of bulking. If the problem is due to limited oxygen, it can be established by operating the air blowers at full capacity. Under these conditions, the blowers should have adequate capacity to maintain at least 2 mg/liter of dissolved oxygen in the aeration tank, a generally accepted value. If this level of oxygen cannot be maintained, solution to the problem may require the installation of additional blowers.

The food-to-microorganism ratio should be checked to see that it is within the range of generally accepted values (0.2 per day to 0.4 per day). When plant operation is controlled by the mean cell residence time parameter, the food-to-microorganism ratio need not be checked. However, the mean cell residence time should be checked to see that it is within the range normally found to provide efficient treatment (5 to 15 days). If not within this range, the sludge-wasting rate should be adjusted as discussed previously.

The recommendations presented previously should be followed with regard to the return-sludge pumping rate.

To avoid internal plant overloading check to see that the decanted liquid from the elutriation tanks or the liquid removed from the sludge during vacuum dewatering or other similar operations is not being returned to the plant flow during times of peak hydraulic and organic loading.

The nature of the components found in wastewater or the absence of certain components can lead to the development of a bulked sludge. If it is known that industrial wastes are being introduced into the system either intermittently or continuously, the quantity of nitrogen and phosphorus in the wastewater should be checked, because limitations of both or either are known to favor bulking. Wide fluctuations in pH are also known to be detrimental in plants of conventional design. Wide fluctuations in organic waste loads due to batch-type operations can also lead to bulking and should be checked.

If bulking conditions continue to persist after all the aforementioned factors have been checked, a critical study of clarifier behavior should be made. This is particularly true of center-feed circular tanks where sludge is removed from the tank directly under the point where the mixed liquor enters the tank. Explorations in the sludge blanket may show that a large part of the sludge is actually retained in the tank for many hours rather than the desired 30 min. If this is the case, then the design is at fault, and changes must be made in facilities.

In an emergency situation or while the aforementioned factors are being investigated, chlorination may be used to provide temporary help. Chlorination of wastewater or of return sludge has been practiced quite extensively as a means of controlling bulking. Although chlorination is effective in controlling bulking caused by filamentous growths, it is ineffective when bulking is due to light floc containing bound water. Chlorination of return sludge should be based upon its dry solids content. A reasonable range is between 0.2 and 1.0 percent by weight. Chlorination normally results in the production of a turbid effluent until such time as the sludge is freed of the filamentous forms. Chlorination of a nitrifying sludge will also produce a turbid effluent, due to death of the nitrifying organisms.

TRICKLING FILTERS

Starting with the first municipal installation in the United States in Reading, Pa., in 1908, trickling filters have been widely used to provide biological sewage treatment. The microbiology of trickling filters is described in Chap. 10. The material in this section includes filter classification, process design considerations, and design of physical facilities.

Filter Classification

Trickling filters are classified by hydraulic or organic loading as high-rate or low-rate. The range of loadings normally encountered and other operational characteristics for the high-rate and low-rate filters are shown in Table 12·8.

TABLE 12·8 COMPARISON OF LOW-RATE AND HIGH-RATE TRICKLING FILTERS

Factor	Low-rate filter	High-rate filter
Hydraulic loading, mgad	1 to 4	10 to 40
Organic loading, lb BOD_5/acre-ft-day	300 to 1,000	1,000 to 5,000
Depth, ft	6 to 10	3 to 8
Recirculation	None	1:1 to 4:1
Rock volume	5 to 10 times	1
Power requirements	None	10 to 50 hp/mg
Filter flies	Many	Few, larvae are washed away
Sloughing	Intermittent	Continuous
Operation	Simple	Some skill
Dosing interval	Not more than 5 min (generally intermittent)	Not more than 15 sec (continuous)
Effluent	Fully nitrified	Nitrification at low loadings

A low-rate filter is a relatively simple device and is highly dependable, producing a consistent effluent quality with varying influent strength. A large population of nitrifying bacteria is prevalent; thus the effluent is low in ammonia and high in nitrite and nitrate. Head loss through the filter may be 5 to 10 ft, which may be objectionable if the site is too flat to permit gravity flow. With a favorable gradient, the ability to use gravity flow is a distinct advantage. Low-rate filters are not without drawbacks. Odors are a common problem, especially if the sewage is stale or septic or if the weather is warm. Filters should not be located where the odors would create a nuisance. Filter flies (Psychoda) may breed in the filters unless control measures are employed.

In a high-rate filter, recirculation of filter effluent or final effluent permits the application of higher organic loadings. Flow diagrams for various high-rate trickling-filter configurations are shown in Fig. 12·23. Recirculation of effluent from the trickling-filter clarifier permits the high-rate filter to achieve the same removal efficiency as the low- or standard-rate filter. Recirculation of filter effluent around the filter (first flow-sheet in Fig. 12·23*a* and *b*) results in the return of viable organisms. It has been observed that this method of operation often improves treatment efficiency. Recirculation also aids in preventing ponding in the filter and in reducing the nuisance due to odors and flies.

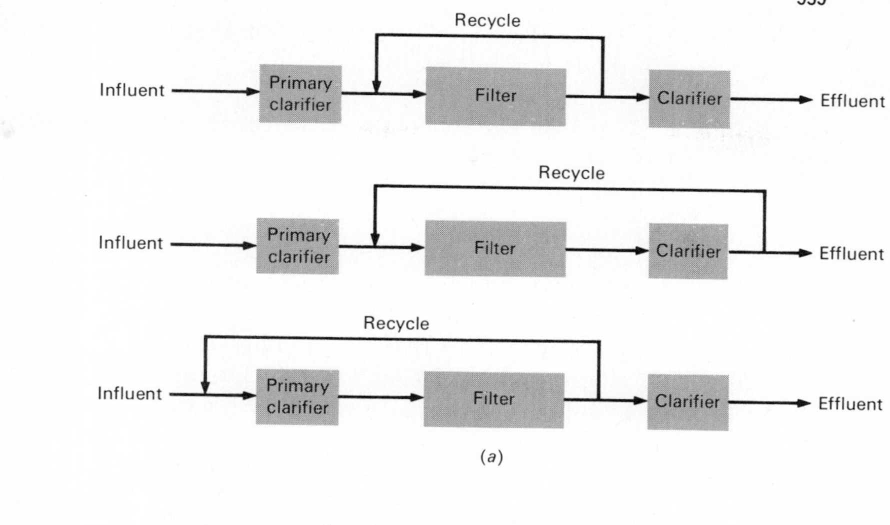

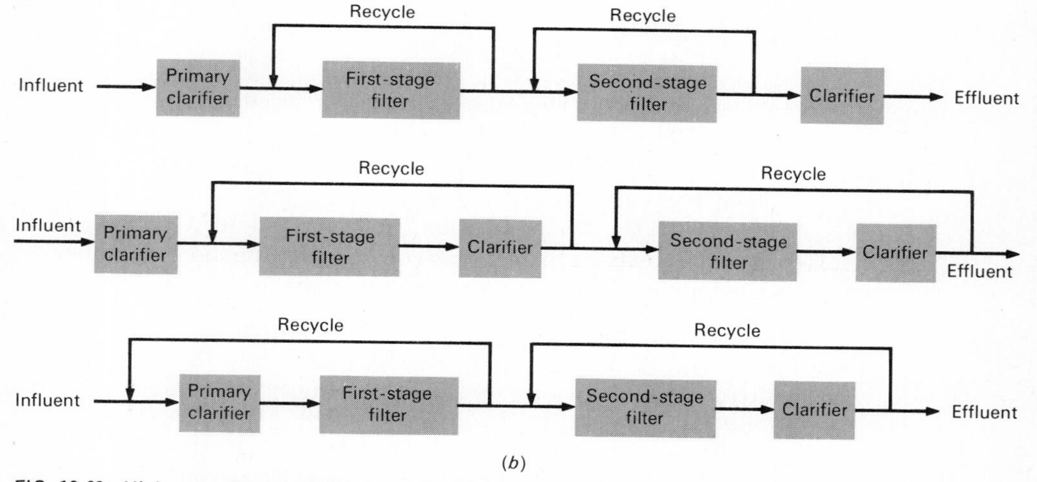

FIG. 12·23 High-rate trickling-filter flowsheets with various recirculation patterns. (*a*) Single-stage filters, (*b*) two-stage filters.

Process Design Considerations

In the design of trickling filters, the engineer must consider the organic and hydraulic loadings and the degree of purification required. Over the years, a number of investigators have proposed equations to describe the removals observed using trickling filters. Representative equations have been developed by Velz [34], The National Research Council (NRC) [22], Fairall [6], Rankin [27], Eckenfelder [4], and

Galler and Gotaas [8]. The Velz and NRC equations are presented and discussed in the following sections.

Velz Equation The Velz equation relates treatment efficiency to the depth of the filter as given by

$$\frac{L_D}{L} = 10^{-KD} \tag{12·17}$$

where L = applied BOD_L which is removable, not over $0.90L_0$ where L_0 is equal to the applied BOD_L
L_D = removable portion of BOD_L remaining at depth D
K = rate of removal (0.175 for low-rate filters; 0.15 for high-rate filters)
D = depth, ft

When recirculation is used, the applied BOD_L may be calculated from Eq. 12·18.

$$L_a = \frac{L_0 + RL_e}{1 + R} \tag{12·18}$$

where L_a = applied BOD_L after dilution by recirculation
L_0 = BOD_L of untreated wastewater
L_e = effluent BOD_L
R = recirculation ratio, Q_r/Q

The use of the Velz equation is illustrated in Example 12·4. Eckenfelder has expanded the Velz coefficient to include a number of factors, including the specific surface area and the time of contact [3].

EXAMPLE 12·4 *Determination of Depth of Trickling Filter*

Determine the filter depth for a low-rate filter using the Velz equation if the settled sewage ultimate demand L_0 is 250 mg/liter and the effluent BOD_5 is to be 30 mg/liter or better. Assume that the effluent $BOD_5 = 0.7\ BOD_L$ and that $L = 0.9L_0$.

Solution

1. Compute the BOD_L of the effluent.

$$BOD_L = \frac{30}{0.7} = 43 \text{ mg/liter}$$

2. Compute the BOD that is not removable.

$$\text{Amount of BOD not removable} = 250(0.1) = 25 \text{ mg/liter}$$

3. Compute L_D and L.

$$L_O = 43 - 25 = 18 \text{ mg/liter}$$
$$L = 250(0.9) = 225 \text{ mg/liter}$$

4. Compute the required filter depth D.

$$\frac{L_D}{L} = \frac{18}{225} = 0.08$$
$$0.08 = 10^{-0.175D}$$
$$12.5 = 10^{0.175D}$$
$$D = \frac{1.097}{0.175} = 6.3 \text{ ft}$$

NRC Equations The NRC equations for trickling-filter performance are empirical expressions developed from an extensive study of the operating records of trickling-filter plants serving military installations [22]. The formulas are applicable to single-stage and multistage systems, with varying recirculation rates (see Fig. 12·23). For a single-stage or first-stage filter, the equation is

$$E_1 = \frac{1}{1 + 0.0085\sqrt{W/VF}} \tag{12·19}$$

where E_1 = fractional efficiency of BOD removal for process, including recirculation and sedimentation
W = BOD loading to filter, lb/day
V = volume of filter media, acre-ft
F = recirculation factor

The recirculation factor is calculated using Eq. 12·20:

$$F = \frac{1 + R}{(1 + R/10)^2} \tag{12·20}$$

where R = recirculation ratio Q_r/Q. The recirculation factor represents the average number of passes of the influent organic matter through the filter. The term $R/10$ takes into account the experimental observation that the removability of organics appears to decrease as the number of passes increases [22]. For the second-stage filter (see Fig. 12·23*b*), the equation is

$$E_2 = \frac{1}{1 + \dfrac{0.0085}{1 - E_1}\sqrt{\dfrac{W'}{VF}}} \tag{12·21}$$

where E_2 = fractional efficiency of BOD removal for second-stage filtration process, including recirculation and settling
W' = BOD loading to second-stage filter, lb/day

The use of the NRC equations is illustrated in Example 12·5.

EXAMPLE 12·5 *Trickling-Filter Design Using NRC Equations*

An industrial waste having a BOD_5 of 600 mg/liter is to be treated by a two-stage trickling filter. The desired effluent quality is 50 mg/liter of BOD_5. If the filter depths are 6 ft and the recirculation ratio is 4:1, find the required filter diameters. $Q = 2$ mgd. Assume $E_1 = E_2$.

Solution

1. Compute E_1 and E_2.

$$\text{Overall efficiency} = \frac{600 - 50}{600} \times 100 = 91.7\%$$

$$E_1 + E_2(1 - E_1) = 0.917$$

$$E_1 = E_2 = 0.713$$

2. Compute the recirculation factor.

$$F = \frac{1 + R}{(1 + R/10)^2} = \frac{1 + 4}{(1.4)^2} = 2.55$$

3. Compute the BOD loading for the first filter.

$$W = 600(8.34)(2) = 10{,}000 \text{ lb BOD}_5\text{/day}$$

4. Compute the volume for the first stage.

$$E_1 = \frac{1}{1 + 0.0085\sqrt{W/VF}}$$

$$0.713 = \frac{1}{1 + 0.0085\sqrt{10{,}000/2.55V}}$$

$$V = 1.75 \text{ acre-ft}$$

5. Compute the diameter of the first filter.

$$A = \frac{V}{d} = \frac{1.75}{6} = 0.292 \text{ acre} = 12{,}700 \text{ sq ft}$$

$$d = 127 \text{ ft}$$

6. Compute the BOD loading for the second filter.

$$W' = (1 - E_1)W = 0.287(10{,}000) = 2{,}870 \text{ lb BOD}_5\text{/day}$$

7. Compute the volume of the second stage.

$$E_2 = \frac{1}{1 + \dfrac{0.0085}{1 - E_1}\sqrt{\dfrac{W'}{VF}}}$$

$$0.713 = \frac{1}{1 + \dfrac{0.0085}{1 - 0.713}\sqrt{\dfrac{2{,}870}{2.55V}}}$$

$$V = 6.08 \text{ acre-ft}$$

8. Compute the diameter of the second filter.

$$A = \frac{V}{d} = \frac{6.08}{6} = 1.013 \text{ acres} = 44{,}200 \text{ sq ft}$$

$$d = 237 \text{ ft}$$

Because a filter 237 ft in diameter is rather large, it is recommended that it be replaced with two filters, each 168 ft in diameter.

Design of Physical Facilities

Factors that must be considered in the design of trickling filters include (1) the type and dosing characteristics of the distribution system, (2) the type of filter media to be used, (3) the configuration of the underdrain system, (4) provision for adequate ventilation be it either natural or forced air, and (5) the design of the required settling tanks.

Distribution Systems The rotary distributor for trickling filtration has become a standard for the process because of its reliability and ease of maintenance. It consists of two or more arms that are mounted on a pivot in the center of the filter and revolve in a horizontal plane. The arms are hollow and contain nozzles through which the sewage is discharged over the filter bed. The distributor assembly may be driven by the dynamic reaction of the sewage discharging from the nozzles, or it may be driven by an electric motor. The speed of revolution will vary with the flowrate for the reaction-driven unit, but it should be in the range of one revolution in 10 min or less for a two-arm distributor. Clearance of 6 to 9 in. should be allowed between the bottom of the distributor arm and the top of the bed. This will permit the sewage streams from the nozzles to spread out and cover the bed uniformly, and it will prevent ice accumulations from interfering with the distributor motion during freezing weather.

Distributor arms may be of constant cross section for small units, or they may be tapered to maintain minimum transport velocity. Nozzles are spaced unevenly so that greater flow per unit of length is achieved near the periphery than at the center. For uniform distribution over the area of the filter, the flowrate per unit of length should be proportional to the radius from the center. The head loss through the distributor will be in the range of 2 to 5 ft. Important features that should be considered in selecting a distributor are the ruggedness of construction, ease of cleaning, ability to handle large variations in flow while maintaining adequate rotational speed, and corrosion resistance of the material and its coating system. Distributors are manufactured for beds with diameters up to 200 ft.

Dosing tanks providing intermittent operation or recirculation by pumping may be employed to ensure that the minimum flow will be sufficient to rotate the distributor and discharge wastes from all nozzles. Four-arm distributors may be provided with weir boxes confining the flow to two arms at minimum flowrates.

Fixed-nozzle distribution systems consist of a series of spray nozzles located at the points of equilateral triangles covering the filter bed. A system of pipes placed in the filter distributes the sewage uniformly to the nozzles. Special nozzles having a flat spray pattern are used, and the head is varied systematically so that the spray falls first at a maximum distance from the nozzle and then at a decreasing distance as the head slowly drops. In this way, a uniform dose is applied over the whole area of the bed. Half-spray nozzles are used along the sides of the filter. Twin dosing tanks, with sloping bottoms that provide more volume at the higher head (required by the greater spray area), supply the nozzles by discharging through automatic siphons and are arranged to fill and dose alternately. At maximum flow, there should be a minimum rest period of 30 sec between doses; the rest period increases as the rate of flow decreases. The head required, measured from the surface of the filter to the maximum water level in the dosing tank, is normally 8 to 10 ft. For a complete description of trickling filters with fixed nozzles and the hydraulic computations involved, the reader is referred to Metcalf and Eddy [21].

Filter Media The ideal filter medium is a material that has a high surface area per unit of volume, is low in cost, has a high durability, and does not clog easily. The most suitable material is generally a locally available crushed stone or gravel graded to a uniform size within the range of 1 to 3 in. Trap rock is particularly satisfactory. Other materials, such as slag, cinders, or hard coal, have also been used. Stones less than 1 in. in diameter do not provide sufficient pore space between the stones to permit free flow of sewage and sloughed solids. Plugging of the media and "ponding" inside the filter or at the surface will result. Large-diameter stones avoid the ponding problem but have a relatively small surface area per unit volume; thus they cannot support as large a biological population. Size uniformity is a way of ensuring adequate pore space and specifications within the 1- to 3-in. size range are usually more restrictive such as 1 to 2 in. or $1\frac{1}{4}$ to $2\frac{3}{4}$ in.

An important characteristic of the filter medium is its strength and durability. Durability may be determined by the sodium sulfate test, which is used to test the soundness of concrete aggregates. Detailed specifications for filter media can be found in the ASCE Manual 13, "Filtering Materials for Sewage Treatment Plants."

Recently developed synthetic media have been used in the treatment of strong industrial wastes with success. The media consist of interlocking sheets of plastic, which are arranged like a honeycomb to produce highly porous and clog-resistant media. The sheets are shipped unassembled, requiring very little space in comparison with the final volume of the assembled media. The sheets are corrugated, so that when the media are assembled, a strong lightweight grid is formed. Subassemblies of the media form modules that can be arranged to fit any filter configuration. Filters as deep as 20 ft have been constructed [31]. The high hydraulic capacity and resistance to plugging can best be utilized in a high-rate filter.

Underdrains A trickling filter has three primary systems: distribution, treatment (the media), and collection. The collection system catches the filtered sewage and solids discharged from the filter media and conveys it to a conduit leading to the final sedimentation tank. It consists of the filter floor, collection channel, and the underdrains. The underdrains are specially designed vitrified-clay blocks with slotted tops that admit the sewage and support the media. The body of the block consists of two or three channels with curved inverts that form the underdrain channels when laid end to end and cover the entire floor of the filter. They are available from members of the Trickling Filter Floor Institute [11].

The underdrains are laid directly on the filter floor, which is sloped toward the collection channel at a 1 to 2 percent gradient. Underdrains may be open at both ends, so that they may be inspected easily and flushed out if they become plugged. The underdrains also ventilate the filter, providing the air for the microorganisms that live in the filter slime, and they should at least be open to a circumferential channel for ventilation at the wall as well as to the central collection channel.

Ventilation Normal ventilation occurs by gravity within the filter. Because there is usually a temperature differential between the sewage temperature and the ambient air temperature, a heat-exchange process will occur within the filter bed. The change in temperature of the air within the filter will cause a density change, and a convection current will be established. The direction of the flow will depend on the relative temperatures of the air and sewage. The flow of air through a filter is downward if the air temperature is greater than the sewage temperature. During cold weather when the air temperature is low, the flow of air will be upward.

Natural ventilation has proved adequate for trickling filters, provided that the following precautions are taken [35]:

1. Underdrains and collecting channels are designed to flow no more than half-full to provide a passageway for the air.
2. Ventilating manholes with open-grating types of covers should be installed at both ends of the central collection channel.
3. Large-diameter filters should have branch collecting channels with ventilating manholes or vent stacks installed at the filter periphery.
4. The open area of the slots in the top of the underdrain blocks should not be less than 15 percent of the area of the filter.
5. One square foot gross area of open grating in ventilating manholes and vent stacks should be provided for each 250 sq ft of filter area.

In extremely deep or heavily loaded filters there may be some advantage in forced ventilation if it is properly designed, installed, and operated. Such a design should provide for an air flow of 1 cfm/sq ft of filter area in either direction.

It may be necessary during periods of extremely low air temperature to restrict the flow of air through the filter to keep it from freezing. The amount of air required by a filter is in the vicinity of 0.1 cfm/sq ft of filter area.

Filters should be designed so that the entire media can be flooded with sewage and then drained without causing any overflows. Flooding is an effective method for flushing a filter to correct ponding and to control filter fly larvae.

Trickling-Filter Settling Tanks The function of settling tanks that follow trickling filters is to produce a clarified effluent. They differ from activated-sludge settling tanks in that sludge recirculation that is essential to the activated-sludge process is lacking. All of the sludge from trickling-filter settling tanks is removed to sludge-processing facilities. The design of these tanks is similar to the design of primary settling tanks, except that the surface-loading rate is based on the plant flow plus the recycle flow (see Fig. 12·23) minus the underflow (often neglected). The overflow rate at peak flow should not exceed 1,200 gpd/sq ft.

AERATED LAGOONS

An aerated lagoon is a basin in which wastewater is treated on a flow-through basis. Oxygen is supplied usually by means of surface aerators or diffused aeration units. The action of the aerators and that of the rising air bubbles from the diffuser is used to keep the contents of the

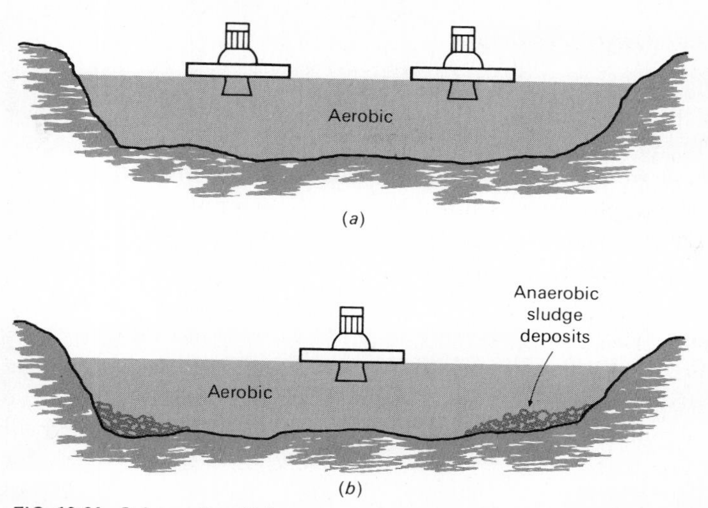

FIG. 12·24 Schematic of (a) an aerated lagoon and (b) an aerobic-anaerobic lagoon.

basin in suspension. Depending on the amount of mixing, lagoons are often classified as either aerobic or aerobic-anaerobic (see Fig. 12·24).

Types of Aerated Lagoons

As shown in Fig. 12·24, the contents of an aerobic lagoon are completely mixed, and both the incoming solids and the biological solids produced from waste conversion do not settle out. In effect the essential function of this type of lagoon is waste conversion. Depending on the detention time, the effluent will contain about one-third to one-half the value of the incoming BOD in the form of cell tissue. Before the effluent can be discharged, however, the solids must be removed by settling (a settling tank is a normal component of most lagoon systems). It will be noted that if the solids are returned to the lagoon, there is no difference between this process and a modified activated-sludge process. Solids return is often used to improve performance during winter months.

In the case of the aerobic-anaerobic lagoon the contents of the basin are not completely mixed, and a large portion of the incoming solids and the biological solids produced from waste conversion settles to the bottom of the lagoon. As the solids begin to build up, a portion will undergo anaerobic decomposition. Thus the effluent from this type of lagoon will be more highly stabilized. Because of the similarity between aerobic-anaerobic lagoons and aerobic-anaerobic (facultative) stabilization ponds, both will be discussed in a subsequent section. The following discussion is therefore limited to aerobic lagoons.

Process Design Considerations

Factors that must be considered in the process design of aerated lagoons include (1) BOD removal, (2) effluent characteristics, (3) oxygen requirements, (4) temperature effects, and (5) energy requirement for mixing. The first four factors are considered in the following discussion, and their application is illustrated in an example problem. The energy required for mixing was discussed previously in the section dealing with mechanical aerators.

BOD Removal Because an aerated lagoon can be considered to be a complete-mix reactor without recycle, the mean cell residence time basis of design outlined in Chap. 10 can be applied to its design. The basic approach involves selection of a mean cell residence time that will ensure (1) that the suspended microorganisms will bioflocculate for easy removal by sedimentation and (2) that an adequate factor of safety is provided when compared to the mean cell residence time of washout. Typical values of θ_c for aerated lagoons used for treating domestic wastes vary from about 3 to 6 days. Once the value of θ_c has been selected, the effluent soluble substrate concentration can be estimated using Eq. 10·26. The removal efficiency can then be computed using Eq. 10·24. This method of design is somewhat limited presently by the lack of specific data for the kinetic constants and the variation of these constants with temperature.

An alternative approach that has been used is to assume that the observed BOD_5 removal (either overall including soluble and suspended solids contribution or soluble only) can be described in terms of a first-order removal function. The required analysis for a complete-mix reactor has been outlined previously in this chapter in connection with the selection of reactor types for the activated-sludge process. Based on that analysis, the pertinent equation for a single aerated lagoon is

$$\frac{S}{S_0} = \frac{1}{1 + k(V/Q)} \qquad (12\cdot22)$$

where S = effluent BOD_5 concentration, mg/liter
S_0 = influent BOD_5 concentration, mg/liter
k = overall first-order BOD_5 removal rate constant, days^{-1}
V = volume, million gal
Q = flowrate, mgd

Reported overall k values vary from 0.25 to 1.0. Removal rates for soluble BOD_5 would be higher. Application of this equation is illustrated in Example 12·7 presented at the end of this section. The corresponding equation derived from a consideration of soluble substrate removal kinetics, as given by Eq. 10·5, is

$$\frac{S}{S_0} = \frac{1}{1 + \frac{kX}{K_s + S}\left(\frac{V}{Q}\right)} \tag{12·23}$$

The terms in Eq. 12·23 are as defined previously in Chap. 10.

Effluent Characteristics The characteristics of the effluent from an aerated lagoon that are of importance include the BOD_5 and the suspended-solids concentration. The effluent BOD_5 will be made up of those components previously discussed in connection with the activated-sludge process and may occasionally contain the contribution of small amounts of algae. The solids in the effluent are composed of a portion of the incoming suspended solids, the biological solids produced from waste conversion, and occasionally small amounts of algae. The solids produced from the conversion of soluble organic wastes can be estimated using Eq. 10·29.

Oxygen Requirement The oxygen requirement is computed as previously outlined in the activated-sludge process design section. Based on operating results obtained from a number of industrial and domestic installations, the amount of oxygen required has been found to vary from 0.7 to 1.4 times the amount of BOD_5 removed.

Temperature Because aerated lagoons are installed and operated in locations with widely varying climatic conditions, the effects of temperature change must be considered in their design. The two most important effects of temperature are (1) reduced biological activity and treatment efficiency and (2) the formation of ice.

The effect of temperature on biological activity is evaluated in Chap. 10. From a consideration of the influent wastewater temperature, air temperature, surface area of the pond, and wastewater flowrate, the resulting temperature in the aerated lagoon can be estimated using the following equation developed by Mancini and Barnhart [16]:

$$(T_i - T_w) = \frac{(T_w - T_a)fA}{Q} \tag{12·24}$$

where T_i = influent waste temperature, °F
T_w = lagoon water temperature, °F
T_a = ambient air temperature, °F
f = proportionality factor
A = surface area, sq ft
Q = wastewater flowrate, mgd

The proportionality factor incorporates the appropriate heat-transfer coefficients and includes the effect of surface area increase due to aeration, wind, and humidity. A typical value for the eastern United States is 12×10^{-6} [16]. To compute the lagoon temperature, Eq. 12·24 is rewritten as

$$T_w = \frac{AfT_a + QT_i}{Af + Q} \tag{12·25}$$

Alternatively, if climatological data are available, the average temperature of the lagoon may be determined from a heat-budget analysis by assuming the lagoon is completely mixed.

Where icing may be a problem, its effects on the operation of lagoons may be minimized by increasing the depth of the lagoon or by altering the method of operation. The effect of reducing the surface area is illustrated in Example 12·6. As computed, reducing the area by one-half increases the temperature about 7°F, which corresponds to about a 50 percent increase in the rate of biological activity. As the depth of the lagoon is increased, however, maintenance of a completely mixed flow regime becomes more difficult. If the depth is increased beyond about 12 ft, draft-tube aerators must be used.

EXAMPLE 12·6 *Effect of Pond Surface Area on Liquid Temperature*

Determine the effect of reducing the surface area of an aerated lagoon from 100,000 to 50,000 sq ft by doubling the depth for the following conditions:

1. Wastewater flowrate = 1 mgd
2. Wastewater temperature = 60°F
3. Air temperature = 20°F
4. Proportionality constant = 12×10^{-6}

Solution

1. Determine the lagoon water temperature for a surface area of 100,000 sq ft using Eq. 12·25.

$$T_w = \frac{100{,}000(12 \times 10^{-6})(20) + 1(60)}{100{,}000(12 \times 10^{-6}) + 1}$$
$$= 38.2°\text{F}$$

2. Determine the lagoon water temperature for a surface area of 50,000 sq ft.

$$T_w = \frac{50{,}000(12 \times 10^{-6})(20) + 1(60)}{50{,}000(12 \times 10^{-6}) + 1}$$
$$= 45°\text{F}$$

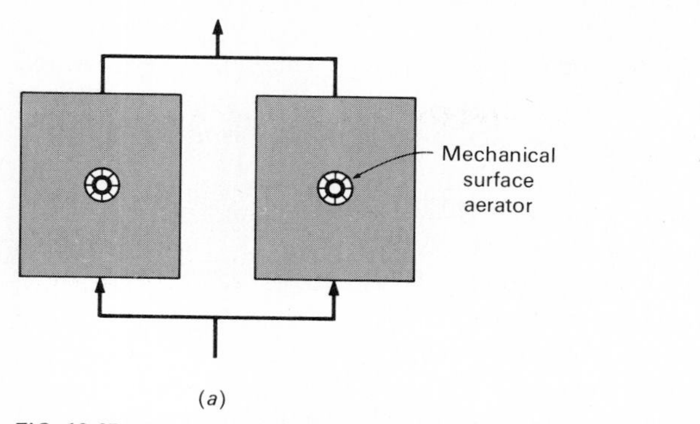

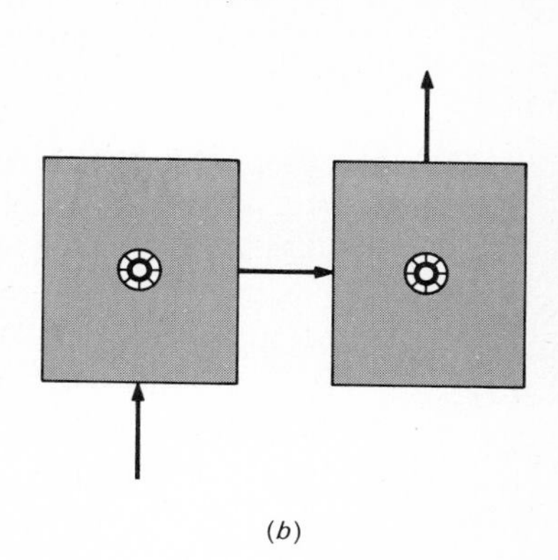

FIG. 12·25 Parallel and series operation of aerated lagoons [29].

Another approach that can be used to mitigate the effects of cold weather is to modify the operation pattern during the winter months. To do this, two aerated lagoons are required. During the summer months, the lagoons are operated in parallel, as shown in Fig. 12·25. As the cold weather begins, parallel operation is discontinued and the lagoons are operated in series [29]. This method of operation conserves heat. Both aerators are operated until ice formation forces the shut-down of the second aerator. Thus the second lagoon functions as an anaerobic pond during the winter months. Because it is covered with ice, odor problems are minimized. In spring when the ice melts, the parallel method of operation is again adopted. With this method of operation it is possible to achieve a 60 to 70 percent removal of BOD_5 even during the coldest winter months [29]. Still another method that can be used to improve performance during the winter months is to recycle a portion of the solids removed by settling.

EXAMPLE 12·7 *Aerated Lagoon Design*

Design an aerated lagoon to treat a wastewater flow of 1 mgd, including the number of surface aerators and their horsepower rating. Assume that the following conditions and requirements apply.

1. Influent suspended solids = 200 mg/liter
2. Influent suspended solids are not biologically degraded
3. Influent soluble BOD_5 = 200 mg/liter
4. Effluent soluble $BOD_5 \leq 20$ mg/liter
5. Effluent suspended solids after settling ≤ 20 mg/liter

6. Growth constants $Y = 0.65$, $K_s = 100$ mg/liter, $k = 6.0$ days^{-1}, $k_d = 0.07$ days^{-1}
7. Total solids produced are equal to computed VSS divided by 0.80
8. First-order soluble BOD_5-removal-rate constant $k = 2.5$ days^{-1} at 20°C
9. Summer temperature = 30°C, 86°F
10. Winter temperature = 10°C, 50°F
11. Wastewater temperature = 60°F
12. Temperature coefficient $\theta = 1.06$
13. Aeration constants $\alpha = 0.85$, $\beta = 1.0$
14. Elevation = 2,000 ft
15. Oxygen concentration to be maintained in liquid = 1.5 mg/liter
16. Lagoon depth = 10 ft
17. Mean cell residence time = 4 days

Solution

1. On the basis of a mean cell residence time of 4 days, determine the surface area of the lagoon.

$$\text{Volume } V = Qt = 1 \text{ mgd} \times 4 \text{ days}$$
$$= 4 \text{ million gal} = 535{,}000 \text{ cu ft}$$
$$\text{Surface area} = \frac{535{,}000 \text{ cu ft}}{10 \text{ ft}} = 53{,}500 \text{ sq ft} = 1.23 \text{ acres}$$

2. Estimate the summer and winter liquid temperatures using Eq. 12·25.

Summer:

$$T_w = \frac{53{,}500(12 \times 10^{-6})(86) + 1(60)}{53{,}500(12 \times 10^{-6}) + 1}$$
$$= 70°\text{F} = 21.1°\text{C}$$

Winter:

$$T_w = \frac{53{,}500(12 \times 10^{-6})(50) + 1(60)}{53{,}500(12 \times 10^{-6}) + 1}$$
$$= 56°\text{F} = 13.3°\text{C}$$

3. Estimate the soluble effluent BOD_5 during the summer using Eq. 10·26.

$$S = \frac{K_s(1 + k_d\theta_c)}{\theta_c(Yk - k_d) - 1}$$
$$= \frac{100[1 + 0.07(4)]}{4[0.65(6) - 0.07] - 1}$$
$$= 9.0 \text{ mg/liter}$$

This value was computed using kinetic-growth constants derived for the temperatures in the range from 20 to 25°C. Thus, during the summer months, the effluent requirement of 20 mg/liter or less will be met easily. Because there is no reliable information on how to correct these constants for the winter temperature of 13.3°C, an estimate of the effect of temperature can be obtained using the first-order soluble BOD_5 removal-rate constant.

4. Estimate the effluent BOD_5 using Eq. 12-22.

Summer (21.1°C):

$$k_{21.1} = 2.5(1.06)^{21.1-20} = 2.66$$

$$\frac{S}{200} = \frac{1}{1 + 2.66(4)}$$

$$S = \frac{200}{1 + 10.6} = 17.2 \text{ mg/liter}$$

Winter (13.3°C):

$$k = 2.5(1.06)^{13.3-20} = 1.7$$

$$\frac{S}{200} = \frac{1}{1 + 1.7(4)} = \frac{1}{7.8}$$

$$S = \frac{200}{7.8} = 25.6 \text{ mg/liter}$$

$$\text{Ratio of } \frac{S_{\text{winter}}}{S_{\text{summer}}} = \frac{25.6}{17.2} = 1.5$$

Applying the ratio to the soluble effluent BOD_5, computed using the kinetic-growth constants, yields a value of about 13.5 mg/liter. Using the ratio of the removal-rate constants yields approximately the same value. The foregoing calculations were presented only to illustrate the method. The value of the removal-rate constant must be evaluated for the waste in question, as discussed in Chap. 16.

5. Estimate the concentration of biological solids produced using Eq. 10·29.

$$X = \frac{Y(S_0 - S)}{1 + k_d\theta_c} = \frac{0.65(200 - 9)}{1 + 0.07(4)}$$

$$= 97 \text{ mg/liter VSS}$$

An approximate estimate of the biological solids produced can be obtained by multiplying the assumed growth-yield constant (BOD_5 basis) by the BOD_5 removed.

6. Estimate the suspended solids in the lagoon effluent before settling.

$$SS = 200 + \frac{97}{0.80} = 321 \text{ mg/liter}$$

Using the peak flow overflow rate recommended in connection with the design of the activated-sludge settling tanks should result in an effluent containing less than 20 mg/liter of suspended solids.

7. Estimate the oxygen requirements using Eq. 12·8.

$$\frac{\text{lb } O_2}{\text{day}} = \frac{(200 - 9)8.34(1)}{0.68} - 1.42(97)8.34(1)$$
$$= 2{,}340 - 1{,}150 = 1{,}190 \text{ lb/day}$$

8. Compute the ratio of oxygen required to BOD_5 removed.

$$\frac{O_2 \text{ required}}{BOD_5 \text{ removed}} = \frac{1{,}190}{(200 - 9)8.34(1)} = 0.75$$

9. Determine the correction factor for surface aerators under summer conditions using Eq. 12·15.

Oxygen-saturation concentration at 21.1°C = 8.97 mg/liter

Oxygen-saturation concentration at 21.1°C corrected for altitude (see Fig. 12·17) = 8.97 × 0.94 = 8.43 mg/liter

$$\text{Correction factor} = \frac{C_{walt} - C_L}{9.17}(1.024)^{T-20}\alpha$$
$$= \frac{8.43 - 1.5}{9.17}(1.024)^{21.1-20}0.85$$
$$= 0.66$$

In most situations the summer conditions will control the required aerator horsepower to meet oxygen requirements.

10. Determine surface-aerator horsepower requirements, based on the assumption that the aerators to be used are conservatively rated at 3.0 lb O_2/hp-hr. Using Eq. 12·15, the field transfer rate N is equal to 1.98 lb O_2/hp-hr.

$$N = N_0(0.66) = 3.0(0.66) = 1.98 \text{ lb } O_2\text{/hp-hr}$$

The amount of O_2 transferred per day per unit is equal to 47.5 lb O_2/hp. The total horsepower required is

$$\text{hp} = \frac{1{,}190 \text{ lb } O_2}{47.5 \text{ lb } O_2} = 25$$

11. Check energy requirements for mixing. Assume that, for a complete-mix flow regime, the power requirement is 0.5 hp/1,000 cu ft.

    Lagoon volume = 535,000 cu ft
    hp required = 535(0.5) = 267.5

    Use eight 35-hp surface aerators. For installations designed for the treatment of domestic wastewater, the energy requirement for mixing will usually be the controlling factor in sizing the aerator.

STABILIZATION PONDS

A stabilization pond is a relatively shallow body of water contained in an earthen basin of controlled shape, which is designed for the purpose of treating wastewater. The term "oxidation pond," often used, is synonymous. Ponds have become very popular with small communities because their low construction and operating costs offer a significant financial advantage over other recognized treatment methods. Ponds are also used extensively for the treatment of industrial wastes and mixtures of industrial wastes and domestic sewage that are amenable to biological treatment. Installations are now serving such industries as oil refineries, slaughterhouses, dairies, poultry-processing plants, and rendering plants.

Because of their popularity, they will be discussed in detail in this section which is divided into four parts. The purpose is to (1) describe the various types of ponds that have been used, (2) discuss briefly their application, (3) discuss the design of these ponds from the standpoint of the fundamental processes taking place, and (4) mention briefly some of the important factors that must be considered in the design of the required physical facilities.

Pond Classification

Stabilization ponds are usually classified according to the nature of the biological activity that is taking place. Other classification schemes that have been used are based on the type of influent (untreated, screened, or settled-wastewater or activated-sludge effluent); on the pond overflow condition (nonexistent, intermittent, or continuous); and on the method of oxygenation (photosynthesis, atmospheric surface transfer, or mechanical aerators) [19]. In terms of biological activity, stabilization ponds are classified as aerobic, aerobic-anaerobic, and anaerobic. In addition, a variety of different pond systems have been developed and applied to meet specific treatment objectives (see "Application").

Aerobic Ponds An aerobic stabilization pond contains bacteria and algae in suspension, and aerobic conditions prevail throughout its depth. The symbiotic relationship that exists between the aerobic bacteria and the algae has been discussed in Chap. 10. The reactions that take place are shown schematically (in Fig. 10·20).

There are basically two types of aerobic ponds. In the first, the objective is to maximize the production of algae. These ponds are usually limited to a depth of about 6 to 18 in. The application of such ponds to the treatment of irrigation return water is discussed in Chap. 14. In the second, the objective is to maximize the amount of oxygen produced, and here, pond depths up to 5 ft have been used. In both types, oxygen in addition to that produced by algae enters the liquid through atmospheric diffusion.

To achieve best results with aerobic ponds, their contents must be mixed periodically with pumps or surface aerators. In operation, the pond loading is adjusted to reflect the amount of oxygen available from photosynthesis and atmospheric reaeration. The efficiency of BOD_5 conversion in aerobic ponds is high, ranging up to 95 percent; however, it must be remembered that, although the soluble BOD_5 has been removed from the influent wastewater, the pond effluent will contain an equivalent or larger concentration of algae, which may ultimately exert a higher BOD_5 than the original waste.

The estimated characteristics of the effluent from flow-through aerobic ponds, expressed in terms of the influent BOD_5 and effluent suspended solids, are reported in Table 12·9. As shown, a range of values is given because the effluent composition will vary with pond location and operation. It is evident from Table 12·9 that, to achieve a true BOD_5 removal, the suspended algae and the microorganisms must be separated from the pond effluent.

Aerobic-Anaerobic Ponds Three zones exist in an aerobic-anaerobic pond. They are (1) a surface zone where aerobic bacteria and algae exist in a symbiotic relationship, as previously discussed; (2) an anaerobic bottom zone in which accumulated solids are actively decomposed by anaerobic bacteria; and (3) an intermediate zone that is partly aerobic and partly anaerobic, in which the decomposition of organic wastes is carried out by facultative bacteria. Because of this, these ponds are often referred to as facultative ponds.

In these ponds, the suspended solids in the wastewater are allowed to settle to the bottom. As a result, the presence of algae is not

TABLE 12·9 APPLICATION AND EFFLUENT CHARACTERISTICS OF VARIOUS TYPES OF STABILIZATION PONDS AND POND SYSTEMS

Type of pond or pond system	Application	Effluent characteristics: Suspended solids, mg/liter*: Algae $(BOD_5)_i$	Micro-organisms $(BOD_5)_i$	Other $(SS)_i$	BOD_5, mg/liter†: Soluble $(BOD_5)_i$	Suspended $(SS)_e$
Aerobic (6–18 in. deep)	Nutrient removal, treatment of soluble organic wastes, production of algal cell tissue	0.5–1.2	0.2–0.5	Low	0.02–0.1	0.3–1.2
Aerobic (up to 60 in. deep)	Treatment of soluble organic wastes and secondary effluents	0.4–1.0	0.2–0.5	Low	0.02–0.1	0.3–1.0
Aerobic-anaerobic (oxygen source: algae)	Treatment of untreated screened or primary settled wastewater and industrial wastes	0.2–0.8	0.2–0.5	0.1–0.4	0.02–0.1	0.3–1.0
Aerobic-anaerobic with and without effluent recirculation (oxygen source: surface aerators)	Treatment of untreated screened or primary settled wastewater and industrial wastes	0.02–0.1	0.2–0.5	0.1–0.4	0.02–0.1	0.3–0.8
Anaerobic	Treatment of domestic and industrial wastes	...	0.1–0.3	0.3–0.5	0.05–0.2	0.3–0.8
Anaerobic + aerobic-anaerobic with recirculation from aerobic-anaerobic to anaerobic	Complete treatment of wastewater and industrial wastes	...	0.2–0.5	0.05–0.15	0.05–0.1	0.3–0.8
Anaerobic + aerobic-anaerobic + aerobic pond system with recirculation from aerobic to anaerobic	Complete treatment of wastewater and industrial wastes with high bacterial removals	0.05–0.1	0.02–0.05	0.03–0.1	0.02–0.1	0.3–1.0

* Effluent suspended solids are composed of algae and other microorganisms, which are estimated in terms of influent $(BOD_5)_i$ and a fraction of the influent suspended solids $(SS)_i$.

† Effluent BOD_5 is composed of a fraction of the soluble influent BOD_5 $(BOD_5)_i$ plus a contribution from the effluent suspended solids $(SS)_e$.

required. The maintenance of the aerobic zone serves to minimize odor problems, because many of the liquid and gaseous anaerobic decomposition products, carried to the surface by mixing currents, are utilized by the aerobic organisms.

The advantage of using surface aerators is that higher organic loads can be applied. However, the organic load must not exceed the amount of oxygen that can be supplied by the aerators without completely mixing the pond contents, or the benefits to be derived from anaerobic decomposition will be lost. For example, if significant mixing begins to occur at 0.1 hp/1,000 cu ft, and a typical aerator will produce about 48 lb of oxygen per day per horsepower [(2.0 lb O_2/hp-hr) × 24 hr/day], then the maximum equivalent oxygen loading that can be applied is about 4.8 lb BOD_L/1,000 cu ft.

Anaerobic Ponds Anaerobic ponds are anaerobic throughout their depth except for an extremely shallow surface zone. To conserve heat energy and to maintain anaerobic conditions, these ponds have been constructed with depths up to 20 ft. Stabilization is brought about by a combination of precipitation and the anaerobic conversion of organic wastes to CO_2, CH_4, other gaseous end products, organic acids, and cell tissues. BOD_5 conversion efficiencies up to 70 percent are obtainable routinely. Under optimum operating conditions, removal efficiencies up to 85 percent are possible.

Application

Stabilization ponds have been applied singly or in various combinations to the treatment of both domestic and industrial wastes. Typical applications are reported in Table 12·9. As shown, aerobic ponds are used primarily for the treatment of soluble organic wastes and effluents from wastewater treatment plants. When more effective means are developed for separating, drying, and disposing of the algae, it is anticipated that they will become more popular. The aerobic-anaerobic ponds are the most commonly used type and have been applied to the treatment of domestic wastewater and a wide variety of industrial wastes. Anaerobic ponds are especially effective in bringing about the rapid stabilization of strong organic wastes. Usually, anaerobic ponds are used in series with aerobic-anaerobic ponds to provide complete treatment. Series-parallel operation, the application of recirculation, and state regulations are discussed in this section.

Series-Parallel Operation Stabilization ponds may be employed in parallel or series arrangement to achieve special objectives. Series operation, as noted previously (see "Selection of Reactor Type"), is beneficial where a high level of BOD or coliform removal is important. The effluent from aerobic-anaerobic ponds in series operation has a much lower algal concentration than that obtained in parallel operation, with a resultant decrease in color and turbidity. Many serially operated multiple-unit installations have been designed to provide complete treatment or to provide complete retention of the sewage, the liquid being evaporated into the atmosphere or percolated into the ground.

Parallel units provide better distribution of settled solids. Smaller units are conducive to better circulation and have less wave action. The additional cost of equipping units for both series and parallel operation is usually nominal. In some instances, actual savings can be demonstrated because of the lesser volume of earthmoving needed to adapt two or more smaller units to the topography.

Recirculation Recirculation of pond effluent has been used effectively to improve the performance of pond systems operated in series. Occasionally, internal recirculation is used. If three aerobic-anaerobic ponds are used in series, the normal mode of operation involves recirculating effluent from either the second or third pond to the first pond. The same situation applies if an anaerobic pond is substituted for the first aerobic-anaerobic pond. Recirculation rates varying from 0.5 to 2.0 Q (plant flow) have been used. If recirculation is to be considered, it is recommended that the pumps have a capacity of at least Q.

State Regulations Where ponds are commonly used, most states have regulations governing their design, installation, and management (operation). For example, the original North Dakota recommendations for complete treatment were based on an equivalent loading of 100 persons per acre of water surface area, or about 20 lb BOD/acre. During recent years, loadings of 200 persons per acre have been approved, and up to 68 lb BOD/acre/day have been employed successfully. The current Texas design criteria are based on permissible loading on stabilization ponds of 50 lb BOD/acre/day. This is approximately equal to 450 persons per acre, assuming 35 percent removal in the primary treatment.

A minimum of 60 days' detention is often required for flow-through lagoons receiving untreated wastewater. Even higher detention times (90 to 120 days) have been specified frequently. A high degree of coliform removal is assured even with a 30-day detention. When a pond (6 ft deep) is designed on the basis of 100 persons per acre per day, the detention period is approximately 200 days. When using the equations presented in the next section, it will be found that the computed detention times are considerably less than those required by most state control agencies. However, it is anticipated that, as understanding of processes improves, the detention times required by the states will be decreased.

Process Design Considerations

Of all the designs of biological treatment processes, the design of stabilization ponds is perhaps the least well defined. Numerous methods have been proposed in the literature, yet when the results obtained are correlated, a wide variance is usually found. Typical data on the parameters for the different types of ponds previously discussed are reported in Table 12·10. Data for aerated lagoons are also included for comparison. Most of the data in Table 12·10 have been derived from operating experience with a wide variety of individual ponds and pond systems. Some of the methods that have been proposed for their design, including a consideration of sludge buildup, will be described and illustrated in the following discussion.

Aerobic Ponds Although a number of different approaches [9, 17, 19] have been used for the design of aerobic stabilization ponds, only two will be considered here. In the first, the biological principles developed in Chap. 10 will be applied in connection with the reactor analysis discussed previously in this chapter. In the second, the oxygen resources of the pond will be equated to the applied organic load.

Because the bacteria in an aerobic pond are responsible for the removal (conversion) of BOD_5, the principles developed in Chap. 10 can be applied and lead to the substrate removal shown in Eq. 12·23. Since kinetic values are not now available, a first-order removal rate constant as given in Eq. 12·22 is frequently assumed. As mentioned, the rate constant must be related functionally to the temperature, solar radiation intensity, organic loading, and the nature of the waste. Further, on the basis of the first-order removal equation developed by Wehner and Wilhelm for a reactor with an arbitrary flow-through pattern (between complete-mix and plug-flow), the term kt must be related to the hydraulic characteristics of the pond. Because the contents of aerobic ponds must be mixed to achieve best performance, it is estimated that a typical value for the pond dispersion factor would be about 1.0. Typical values for the overall first-order BOD_5 removal rate constant k vary from about 0.05 to 1.0 per day depending on the operational and hydraulic characteristics of the pond [32]. One of the prob-

TABLE 12·10 DESIGN PARAMETERS FOR STABILIZATION PONDS

Parameter	Type of pond				Aerated lagoons
	Aerobic*	Aerobic-anaerobic	Aerobic-anaerobic	Anaerobic	
Flow regime	Intermittently mixed	....	Mixed surface layer	....	Completely mixed
Pond size, acres	<10 multiples	2–10 multiples	2–10 multiples	0.5–2.0 multiples	2–10 multiples
Operation†	Series or parallel	Series or parallel	Series or parallel	Series	Series or parallel
Detention time, days†	10–40	7–30	7–20	20–50	3–10
Depth, ft	3–4	3–6	3–8	8–15	6–20
pH	6.5–10.5	6.5–9.0	6.5–8.5	6.8–7.2	6.5–8.0
Temperature range, °C	0–40	0–50	0–50	6–50	0–40
Optimum temperature, °C	20	20	20	30	20
BOD_5 loading, lb/acre/day‡	60–120	15–50	30–100	200–500	
BOD_5 conversion	80–95	80–95	80–95	50–85	80–95
Principal conversion products	Algae, CO_2, bacterial cell tissue	Algae, CO_2, CH_4, bacterial cell tissue	CO_2, CH_4, bacterial cell tissue	CO_2, CH_4, bacterial cell tissue	CO_2, bacterial cell tissue
Algal concentration, mg/liter	80–200	40–160	10–40		
Effluent suspended solids, mg/liter§	140–340	160–400	110–340	80–160	260–300

* Conventional aerobic ponds designed to maximize the amount of oxygen produced rather than the amount of algae produced.
† Depends on climatic conditions.
‡ Typical values (much higher values have been applied at various locations). Loading values are often specified by state control agencies.
§ Includes algae, microorganisms, and residual influent suspended solids. Values are based on an influent soluble BOD_5 of 200 mg/liter and, with the exception of the aerobic ponds, an influent suspended-solids concentration of 200 mg/liter.

lems in comparing the values reported in the literature is that the basis (for example, soluble BOD_5 conversion, total BOD_5 including the contribution of suspended solids) on which they were derived often is not clearly stated. It is hoped that this condition will be remedied in the future.

The second method of design to be considered was developed by Oswald and his coworkers at the University of California [19, 24, 25]. In this method, the oxygen resources of the pond are equated to the applied organic loading. The principal source of oxygen in an aerobic stabilization pond is photosynthesis which is governed by the solar energy. On the basis of results from numerous studies, it has been found that the yield of oxygen can be estimated with the following equation:

$$Y_{O_2} = 0.25\,FS \tag{12·26}$$

where Y_{O_2} = oxygen yield, lb O_2/acre/day
F = oxygenation factor
S = solar radiation, cal/cm²-day

The oxygenation factor represents the ratio of the weight of oxygen produced to the BOD_L to be satisfied in the pond. The relationship between the oxygenation factor and BOD removal in a pond is shown in Fig. 12·26. For oxygenation factors lower than about 1.5 to 1.6, the removal efficiency decreases, probably because of the reduced oxygen concentration. As shown in Fig. 12·26, for a BOD_L removal of about 90 percent, the oxygenation factor should be about 1.6. If a pond is designed for this condition during the winter months, it will be somewhat overdesigned because of the increased amount of oxygen produced during the summer months. The reason for the decrease in performance is that when the oxygenation factor is increased beyond about 1.8, an accumulation of hydroxyl ion (high pH) interferes with the bacterial conversion of the incoming wastewater. The hydroxyl-ion concentration develops when the bicarbonate in the wastewater is utilized as a source of CO_2 for algal cell tissue synthesis.

The solar radiation varies principally with the time of year and the latitude. Typical values for each month for various latitudes are reported in Table 12·11. The average solar radiation value is obtained using Eq. 12·27.

$$S_{av} = S_{min} + p(S_{max} - S_{min}) \tag{12·27}$$

where p = actual hours of sunlight divided by total possible hours of sunlight

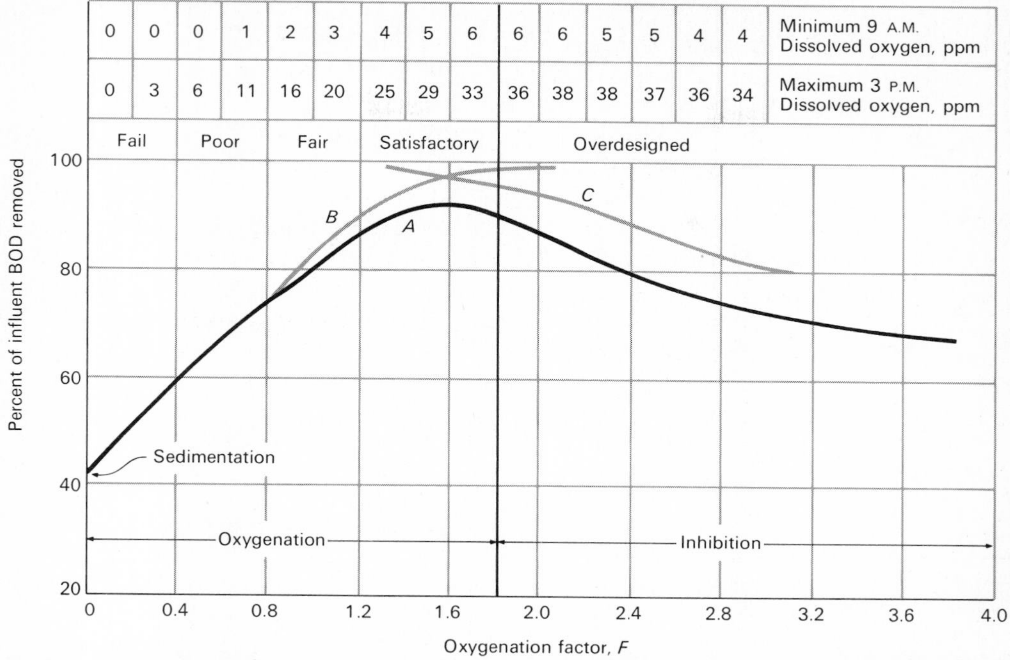

FIG. 12·26 Relationship between oxygenation factor and BOD removal in an aerobic stabilization pond [19].

The organic surface loading expressed in pounds of BOD per acre per day can be obtained using the following equation:

$$L_0 = C\left(\frac{d}{t}\right)\text{BOD}_L \tag{12·28}$$

where L_0 = organic loading, lb BOD/acre/day
C = conversion factor, 0.226
d = pond depth, in.
t = detention time, days
BOD_L = ultimate BOD, mg/liter

If the oxygen yield in Eq. 12·26 is set equal to the organic load in Eq. 12·28, the result is Eq. 12·29, which can be used for the design of aerobic ponds.

$$0.226\left(\frac{d}{t}\right)\text{BOD}_L = 0.25FS$$

$$\frac{d}{t} = 1.1\frac{FS}{\text{BOD}_L} \tag{12·29}$$

TABLE 12·11 PROBABLE VALUES OF VISIBLE SOLAR ENERGY AS A FUNCTION OF LATITUDE AND MONTH [19]

Latitude, deg N or S	Month											
	Jan.	Feb.	Mar.	Apr.	May	Jun.	Jul.	Aug.	Sep.	Oct.	Nov.	Dec.
0 max	225*	266	271	266	249	236	238	252	269	265	256	253
min	210	219	206	188	182	103	137	167	207	203	202	195
10 max	223	244	264	271	270	262	265	266	266	248	228	225
min	179	184	193	183	192	129	158	176	196	181	176	162
20 max	183	213	246	271	284	284	282	272	252	224	190	182
min	134	140	168	170	194	148	172	177	176	150	138	120
30 max	136	176	218	261	290	296	289	271	231	192	148	126
min	76	96	134	151	184	163	178	166	147	113	90	70
40 max	80	130	181	181	286	298	288	258	203	152	95	66
min	30	53	95	125	162	173	172	147	112	72	42	24
50 max	28	70	141	210	271	297	280	236	166	100	40	26
min	10	19	58	97	144	176	155	125	73	40	15	7
60 max	7	32	107	176	249	294	268	205	126	43	10	5
min	2	4	33	79	132	174	144	100	38	26	3	1

* Values of solar radiation S in cal/cm^2/day.
To determine average value of S: $S_{av} = S_{min} + p(S_{max} - S_{min})$ in which p is total hours sunshine divided by total possible hours sunshine.
To determine yield of algal cell material, in lb/acre/day: $Y_c = 0.15FS$.
To determine yield of oxygen in lb/acre/day: $O_2 = 0.25FS$.

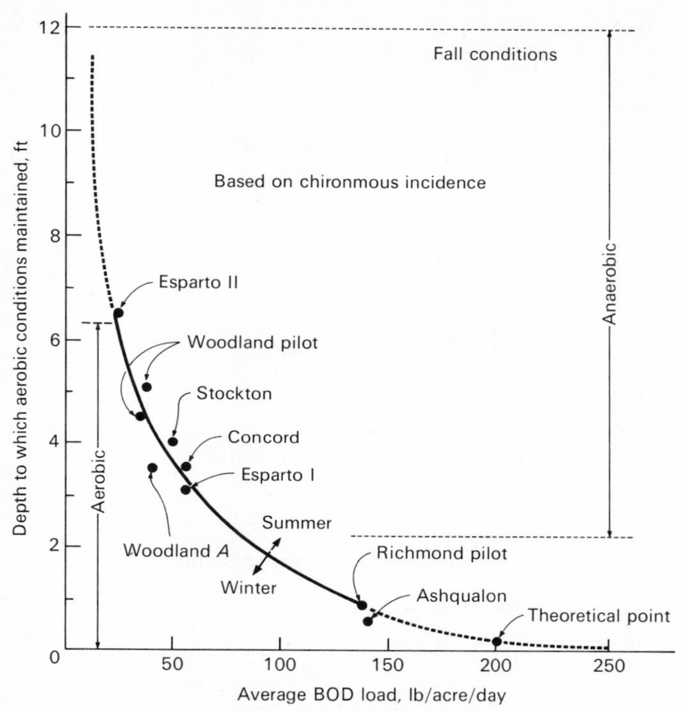

FIG. 12·27 Relationship of BOD loading to depth of aerobic zone in stabilization ponds [23].

The use of this equation is illustrated in Example 12·8. To maximize the production of algae, the depth, as noted previously, is maintained between 6 and 18 in. Where oxygen production is the main objective, depths up to 60 in. have been used. From a study of a number of different ponds, Oswald has developed a curve (see Fig. 12·27) relating the BOD loading in pounds per acre per day to the depth to which aerobic conditions were maintained. Developed for fall conditions, this curve can be used to check computed depths for both aerobic and, with a slight modification, for aerobic-anaerobic ponds to determine if aerobic conditions will exist.

EXAMPLE 12·8 *Aerobic Stabilization Pond Design*

Design an aerobic stabilization pond to treat a waste flow of 1 mgd with a soluble BOD_L of 100 mg/liter. Assume the following conditions apply.

1. Influent suspended solids are negligible
2. BOD_L removal (conversion) = 90 percent
3. First-order soluble BOD_L–removal-rate constant = 0.25 at 20°C and 40°N.
4. Temperature coefficient θ = 1.06 at 20°C
5. Latitude = 40°N
6. Pond temperature in summer = 32°C
7. Pond temperature in winter = 10°C
8. Maximum individual pond area = 10 acres
9. Maximum pond depth = 3 ft
10. Pond dispersion factor = 1.0

Solution: Method 1

1. From Fig. 12·6 determine the value of kt for the pond for a dispersion factor of 1.0 and a removal efficiency of 90 percent.

   $kt = 5$

2. Determine the detention time for winter conditions.

$$\begin{aligned} k_{10} &= 0.25(1.06)^{10-20} = 0.14 \\ 0.14t &= 5 \\ t &= 35.7 \text{ days} \end{aligned}$$

3. Determine time required for summer conditions.

$$\begin{aligned} k_{32} &= 0.25(1.06)^{32-20} = 0.5 \\ 0.5t &= 5 \\ t &= 10 \text{ days} \end{aligned}$$

4. Determine the pond surface area requirements and configuration.

$$\text{Surface area} = \frac{\dfrac{10^6 \text{ gal}}{\text{day}} \times \dfrac{1}{\dfrac{7.48 \text{ gal}}{\text{cu ft}}} \times 35.7 \text{ days}}{3 \text{ ft} \times \dfrac{43{,}560 \text{ sq ft}}{\text{acre}}}$$
$$= 36.5 \text{ acres}$$

   Use four 9.2-acre ponds. If maintenance were required, one of the ponds could be taken out of service during the summer.

Solution: Method 2

1. Determine average minimum and maximum solar radiation for a latitude of 40°N using data from Table 12·11.

Minimum, December:

$$S_{av} = S_{min} + p(S_{max} - S_{min})$$
$$= 24 + \tfrac{9}{12}(66 - 24)$$
$$= 55.5 \text{ cal/cm}^2\text{/day}$$

Maximum, June:

$$S_{av} = 173 + \tfrac{9}{12}(298 - 173)$$
$$= 267 \text{ cal/cm}^2\text{/day}$$

2. Determine the detention times during winter and summer conditions. Assume $F = 1.6$ (see Fig. 12·26).

Winter:

$$\frac{d}{t} = 1.1 \frac{FS}{\text{BOD}_L}$$

$$t = \frac{36(100)}{1.1(1.6)(55.5)} = 36.9 \text{ days}$$

Summer:

$$t = \frac{36(100)}{1.1(1.6)(267)} = 7.7 \text{ days}$$

3. Determine the pond surface area requirements and configuration.

$$\text{Surface area} = \frac{\dfrac{10^6 \text{ gal}}{\text{day}} \times \dfrac{1}{\dfrac{7.48 \text{ gal}}{\text{cu ft}}} \times 36.9 \text{ days}}{3 \text{ ft} \times \dfrac{43{,}560 \text{ sq ft}}{\text{acre}}}$$
$$= 37.8 \text{ acres}$$

Use four 9.5-acre ponds.

4. Determine the organic surface loading.

Winter:

$$L_0 = \frac{100(8.34)(1)}{38} = 21.9 \text{ lb BOD/acre/day}$$

Summer:

$$L = \frac{100(8.34)(1)}{19} = 43.9 \text{ lb BOD/acre/day}$$

5. Check to see that aerobic conditions will be maintained throughout the pond depth. From Fig. 12·27 for fall conditions,

the loading should be below about 65 lb BOD/acre/day. Because the computed winter value is only 20.8 lb BOD/acre/day, it can be assumed that aerobic conditions will prevail throughout the pond depth.

Aerobic-Anaerobic Ponds The design of aerobic-anaerobic ponds closely follows that of the first method previously outlined in connection with the design of aerobic ponds. Because of the method of operation (for example, maintenance of quiescent conditions to promote the removal of suspended solids by sedimentation), it is anticipated that dispersion factors for such ponds would vary from 0.3 to 1.0.

Another factor that must be considered is sludge accumulation, which is important in terms of the oxygen resources and the overall performance of the pond. For example, in cold climates, a portion of the incoming BOD_5 is stored in the accumulated sludge during the winter months. As the temperature increases in the spring and summer, the accumulated BOD_5 is anaerobically converted, and the oxygen demand of gases and acids produced may exceed the oxygen resources of the aerobic surface layer of the pond. In situations in which it is anticipated that BOD_5 storage will be a problem, surface aerators are recommended. If the design is based on BOD_5, the aerators should have a capacity adequate to satisfy from 175 to 225 percent of the incoming BOD_5. Another problem caused by the accumulation of sludge is a reduction in performance of the pond as measured by the suspended-solids content of the effluent. The rate of buildup of sludge in the bottom of aerobic-anaerobic ponds can be estimated using the sludge volume relationships outlined in Chap. 13. The design of an aerobic-anaerobic pond using surface aerators is illustrated in Example 12·9.

EXAMPLE 12·9 *Aerobic-Anaerobic Stabilization Pond Design*

Design an aerobic-anaerobic stabilization pond to treat a wastewater flow of 1 mgd. Because the ponds are to be installed near a residential area, surface aerators will be used to maintain oxygen in the upper layers of the pond. Assume the following conditions apply.

1. Influent suspended solids = 200 mg/liter
2. Influent BOD_5 = 200 mg/liter
3. Summer liquid temperature = 25°C = 78°F
4. Winter liquid temperature = 15°C = 59°F
5. Overall first-order BOD_5 removal-rate constant = 0.25 $days^{-1}$ at 20°C

6. Temperature coefficient $\theta = 1.06$
7. Pond depth = 6 ft
8. Pond dispersion factor = 0.5
9. Overall BOD_5 removal efficiency = 80%

Solution

1. From Fig. 12·6 determine the value of kt for a dispersion factor of 1.0 and a BOD_5 removal efficiency of 80 percent.

   $kt = 2.9$

2. Determine the detention time for winter conditions.

$$k_{15} = 0.25(1.06)^{15-20} = 0.186$$
$$0.186t = 2.9$$
$$t = 15.6 \text{ days}$$

3. Determine the detention time for summer conditions.

$$k_{25} = 0.25(1.06)^{25-20} = 0.336$$
$$0.336t = 2.9$$
$$t = 8.6 \text{ days}$$

4. Determine the pond volume and surface area requirements.

Winter:

$$\text{Volume} = \frac{10^6 \text{ gal}}{\text{day}} \times \frac{1}{\dfrac{7.48 \text{ gal}}{\text{cu ft}}} \times 15.6 \text{ days}$$
$$= 2.09 \times 10^6 \text{ cu ft}$$
$$\text{Surface area} = \frac{2.09 \times 10^6 \text{ cu ft}}{6 \text{ ft} \times \dfrac{43{,}560 \text{ sq ft}}{\text{acre}}}$$
$$= 8.0 \text{ acres}$$

Summer:

Volume = 1.15×10^6 cu ft   Area = 4.4 acres

5. Determine the volumetric loadings.

Winter:

$$\text{lb BOD}_5/1{,}000 \text{ cu ft} = \frac{200 \times 8.34}{2.88 \times 10^3} = 0.58$$

Summer:

$$\text{lb BOD}_5/1{,}000 \text{ cu ft} = \frac{200 \times 8.34}{1.59 \times 10^3} = 1.05$$

6. Determine the horsepower requirements for the surface aerators. Assume that the capacity of the aerators in terms of oxygen transferred is to be twice the value of the BOD_5 applied per day and that a typical aerator will transfer about 48 lb O_2/day/hp.

$$\text{lb } O_2\text{/day required} = 2(200)(8.34)(1) = 3{,}330 \text{ lb}$$

$$\text{hp} = \frac{3{,}330 \text{ lb/day}}{48 \text{ lb/day/hp}} = 69.4 \text{ hp}$$

Use five 15-hp units.

7. Check the horsepower input to determine if mixing will occur.

$$\text{hp}/10^3 \text{ cu ft} = \frac{75}{2.88 \times 10^3}$$

$$= 0.026 \quad \text{(winter)}$$

$$= 0.047 \quad \text{(summer)}$$

Regardless of how the ponds are operated (for example, series, parallel), the horsepower required to keep the surface aerated will not be sufficient to mix the pond contents (about 0.1 to 0.2 hp/1,000 ft is minimum required).

8. Estimate the effluent quality using the data from Table 12·9.

$(SS)_e$	$=$	$0.02(BOD_5)_i$	$+$	$0.2(BOD_5)_i$	$+$	$0.1(SS)_i$
Effluent suspended solids	=	algae	+	bacteria and other micro-organisms	+	fraction of influent suspended solids

$$= 0.02(200) + 0.2(200) + 0.1(200)$$

$$= 64 \text{ mg/liter}$$

$(BOD_5)_e$	$=$	$0.04(BOD_5)_i$	$+$	$0.5(SS)_e$
Effluent BOD_5	=	fraction of soluble influent BOD_5	+	contribution of effluent suspended solids

$$= 0.04(200) + 0.5(64)$$

$$= 40 \text{ mg/liter}$$

Anaerobic Ponds The design of anaerobic stabilization ponds follows the principles presented in Chap. 10 and previously in this chapter and therefore will not be illustrated in terms of a design example. Further, because anaerobic ponds are similar to anaerobic digesters, with the exception of mixing, the process design methods outlined in Chap. 13 should be reviewed.

Pond Systems Pond systems such as those previously discussed are designed by applying the aforementioned equations sequentially, taking into account recirculation where it is used. The general method to be followed is illustrated in the following example.

EXAMPLE 12·10 *Analysis of Pond Systems*

Determine the bacterial reduction brought about by a series of three completely mixed aerobic ponds, each having a detention time of 20 days. Assume the following conditions apply.

1. Initial bacterial concentration = 10^6 organisms/ml.
2. Organism removal rate constant (first order) = 2 days^{-1} [17].

Solution

1. For three completely mixed ponds of equal volume, Eq. 12·22 can be written as

$$\frac{N_3}{N_0} = \frac{1}{(1 + kt)^3} \qquad (12\cdot30)$$

where N_3 = effluent concentration of organisms, organisms/ml

N_0 = initial concentration of organisms, organisms/ml

2. Estimate the effluent organism concentration.

$$N_3 = \frac{10^6}{[1 + 2(20)]^3} = 14.5 \text{ organisms/ml}$$

3. Determine percent reduction.

$$\% \text{ reduction} = \frac{(10^6 - 14.5)100}{10^6} = 99.998$$

Design of Physical Facilities

Although the process design for ponds is imprecise, careful attention must be given to the design of the physical facilities to ensure optimum performance. Factors that should be considered include (1) location of influent lines, (2) outlet structure design, (3) dike construction, (4) liquid depth, (5) treatment of lagoon bottom, and (6) control of surface runoff.

Influent Lines For small ponds, a center inlet is preferred, while for ponds of 10 acres or more, the inlet can be installed 400 ft from the dike. For large aerobic-anaerobic ponds, multiple inlets are desirable to distribute the settleable solids over a larger area. For increased flexibility, movable inlets are being used more frequently.

Outlet Structure(s) The outlet structure(s) should permit lowering the water level at a rate of less than 1 ft/week while the facility is receiving its normal load. It should be large enough to permit easy access for normal maintenance. Provision for complete draining of the pond is desirable. During ice-free periods, discharge should be taken just below the water surface. This releases effluent of the highest quality and ensures retention of floating solids. For flow-through ponds, the maximum rate of effluent discharge is considerably less than the rate of peak sewage flow, because of pond losses and the leveling out of peak flows. Overflow structures generally comparable to a sewer manhole are most frequently employed, with selected level discharge facilitated through valved piping or other adjustable overflow devices. Overflow lines should be vented if the design would otherwise permit siphoning.

Dikes Dikes should be constructed in a way that minimizes seepage. Compaction afforded by the use of conventional construction equipment is usually adequate. Vegetation should be removed, and the area upon which the embankment is to be placed should be scarified. It is generally unnecessary to key the dikes into impervious subsoil, but this precaution may be advisable for sandy topsoils.

The dike should be wide enough to accommodate mowing machines and other maintenance equipment. A width of 8 ft is generally considered adequate, and narrower dikes may be satisfactory for small installations. Slopes are influenced by the nature of the soil and the size of the installation. For outer slopes, a 3 horizontal to 1 vertical is satisfactory. Inner slopes are generally from 1 vertical to 3 to 4 horizontal, although slopes exceeding 1 to 5 for larger installations and 1 to 3 for smaller installations are sometimes specified.

The freeboard is to some extent influenced by the size and shape of the installation, as wave heights are greater on larger bodies of water. Three feet above maximum liquid level is usually specified as minimum freeboard, but 2 ft is considered adequate by some states, particularly for installations of 6 acres or less not exposed to severe winds.

Liquid Depth Optimum liquid depth for adequate circulation is influenced to some extent by the pond area, greater depth being allowed for larger units. Shallow ponds encourage the growth of emergent vegetation and consequently may foster mosquito breeding.

There is a distinct advantage for facilities that permit operation at selected depths up to 5 ft, and provision for additional depth may be desirable for large installations. Facilities for adjusting pond levels can be provided at small cost. For ponds 30 acres or larger, provision for periodic operation at depths greater than 5 ft may be advantageous.

Pond Bottom The bottom of aerobic and most aerobic-anaerobic ponds should be made as level as possible except around the inlet. The finished elevation should not vary more than 6 in. from the average elevation of the bottom. An exception to this is where the bottom of an aerobic-anaerobic pond is designed specifically to retain the settleable solids in hoppered compartments or cells. The bottom should be well compacted to avoid excessive seepage. Where excessive percolation is a problem, increased hydraulic loading or partial sealing may merit consideration to maintain a satisfactory water level in the lagoon. Porous areas, such as gravel pockets, should receive particular attention.

Surface Runoff Control Ponds should not receive significant amounts of surface runoff. If necessary, provision should be made for diverting surface water around the ponds. For new installations, where maintenance of a satisfactory water depth is a problem, the diversion structure may be designed to admit surface runoff to the lagoon when necessary.

PROBLEMS

12·1 In Example 12·1 compute the required quantities of nitrogen and phosphorus if the nitrogen requirement is 0.12 dX/dt and the phosphorus requirement is one-fifth of the nitrogen requirement. In what forms should these nutrients be added?

12·2 Prepare a one-page abstract of the following article: R. E. McKinney and W. J. O'Brien: Activated Sludge—Basic Design Concepts, *Journal Water Pollution Control Federation,* vol. 40, no. 11, part 1, 1968. Prepare a one-page summary of the more important design criteria given in the article.

12·3 Prepare a one-page abstract of the following article: M. T. Garrett, Jr.: Hydraulic Control of Activated Sludge Growth Rate, *Sewage and Industrial Wastes,* vol. 39, no. 3, 1958.

12·4 A conventional activated-sludge plant is to treat 1 mgd of waste having a BOD_5 of 200 mg/liter after settling. The process loading is 0.30 lb BOD/day/lb of MLVSS. The detention time is 6 hr and the recirculation ratio is 0.33. Determine the value of MLVSS.

12·5 A conventional activated-sludge plant is operated at a mean cell residence time of 10 days. The reactor volume is 2 million gal, and the MLSS concentration is 3,000 mg/liter. Determine (*a*) the sludge production rate, (*b*) the sludge-wasting flowrate when wasting from the reactor, and (*c*) the sludge-wasting flowrate when wasting from the recycle line. Assume 10,000 mg/liter of suspended solids in the recycle.

12·6 Design an activated-sludge process to treat 10 mgd of domestic sewage having a BOD_5 of 210 mg/liter, and produce an effluent containing less than 15 mg/liter BOD_5. Assume a complete-mix reactor. Use the following data in design.

X_r = 10,000 mg/liter
MLVSS = 3,000 mg/liter
θ_c = 8 days Y = 0.60
k_d = 0.05
Effluent SS = 10 mg/liter

Determine:
(*a*) Biological treatment efficiency
(*b*) Reactor volume
(*c*) Sludge-production rate
(*d*) Recirculation ratio
(*e*) Hydraulic retention time
(*f*) Oxygen requirements
(*g*) Food-to-microorganism ratio

12·7 The step-aeration activated-sludge system shown in Fig. 12·28 is to be analyzed as a series of complete-mix reactors. Using the design basis given, determine the MLVSS concentration in each tank.

Design:
V = 0.06 mg
S_1 = 4 mg/liter S_2 = 6 mg/liter
S_3 = 8 mg/liter S_4 = 10 mg/liter
S_0 = 250 mg/liter
Q_0 = 1 mgd Q_r = 0.2 mgd
X_r = 10,000 mg/liter
Y = 0.65 k_d = 0.05

12·8 Find the theoretical diameter of the trickling filters shown in Fig. 12·29 for an installation with the following characteristics and requirements.
(*a*) A flow of 6 mgd having a BOD_5 of 300 mg/liter.
(*b*) The effluent BOD_5 must be equal to, or less than, 21 mg/liter in order to maintain stream standards.
(*c*) The filters are to have equal diameters, and the depth of both filters is to be 5 ft.
(*d*) The recirculation ratio chosen should result in a hydraulic loading of 30 mgad for each filter.
(*e*) The primary settling tank provides a BOD removal of 30 percent.
(*f*) Use NRC two-stage trickling-filter loading criteria.

12·9 Determine the temperature of the contents of an 8-acre aerated lagoon. Sewage is discharged to the lagoon at a rate of 0.5 mgd. Use a typical f value of 75×10^{-6} for the midwestern United States. The temperature of the air is 50°F and the temperature of the sewage is 65°F.

12·10 Design an aerated lagoon to treat 2.5 mgd of wastewater under the following conditions.
(*a*) Influent soluble BOD and suspended solids = 150 mg/liter
(*b*) Overall first-order BOD_5 removal rate constant = 2.0 $days^{-1}$ at 20°C
(*c*) Summer temperature = 80°F
(*d*) Winter temperature = 45°F
(*e*) Wastewater temperature = 60°F
(*f*) Temperature coefficient = 1.07
(*g*) $\alpha = 0.85$, $\beta = 1.0$
(*h*) Elevation = 4,000 ft
(*i*) Oxygen concentration to be maintained = 2.0 mg/liter
(*j*) Lagoon depth = 8 ft
(*k*) Hydraulic residence time = 10 days
(*l*) Temperature proportionality constant $f = 12 \times 10^6$

Determine the surface area, summer and winter temperatures in the lagoon, and the effluent BOD_5 in summer and winter. If the growth yield is approximately 0.5 (BOD_5 basis), determine the biological solids concentration in the lagoon, the oxygen requirements, and the horsepower requirements for summer and winter conditions. Use surface aerators rated at 2.5 lb O_2/hp-hr.

12·11 Design an aerobic stabilization pond to treat 2.5 mgd of wastewater with a BOD_5 removal efficiency of 90 percent under the following conditions.
(*a*) Influent BOD_5 = 250 mg/liter
(*b*) Overall first-order BOD_5 removal rate constant = 0.2 $days^{-1}$ at 20°C

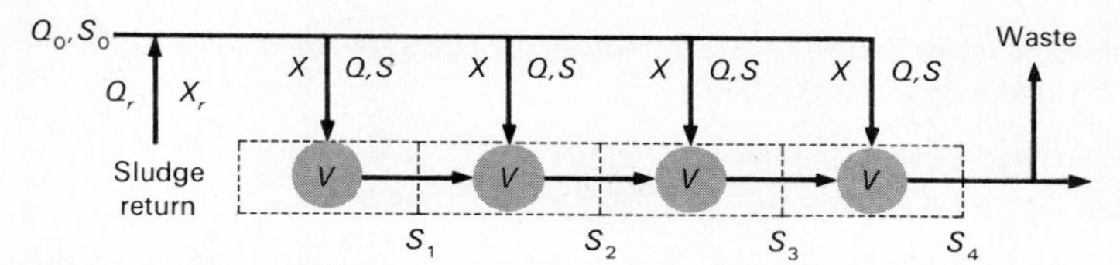

FIG. 12·28 Flowsheet for step-aeration system used in Prob. 12·7.

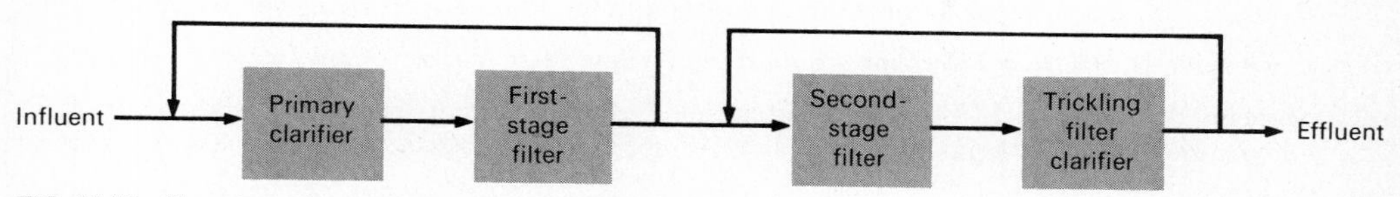

FIG. 12·29 Flowsheet for trickling-filter system used in Prob. 12·8.

(*c*) Pond temperature in summer = 30°C
(*d*) Pond temperature in winter = 12°C
(*e*) Temperature coefficient = 1.06
(*f*) Maximum pond area = 10 acres
(*g*) Maximum pond depth = 5 ft.
(*h*) Pond dispersion factor = 0.5

Determine the detention times and area requirements for summer and winter conditions.

12·12 Design the aerobic pond in Prob. 12·11 by Oswald's method for a latitude of 30°N. Use $p = 8/12$ and $F = 1.7$.

REFERENCES

1. Albertsson, J. G., *et al.*: *Investigation of the Use of High Purity Oxygen Aeration in the Conventional Activated Sludge Process*, Water Pollution Control Research Series 17050DNW05/70, 1970.
2. Bewtra, J. K., and W. R. Nicholas: Oxygenation from Diffused Air in Aeration Tanks, *J. WPCF*, vol. 36, no. 10, 1964.
3. Eckenfelder, W. W., Jr.: *Industrial Water Pollution Control*, McGraw-Hill, New York, 1966.
4. Eckenfelder, W. W., Jr.: Trickling Filtration Design and Performance, *Transactions ASCE*, vol. 128, 1963.
5. Eckenfelder, W. W., Jr., and D. L. Ford: New Concepts in Oxygen Transfer and Aeration, in E. F. Gloyna and W. W. Eckenfelder, Jr. (eds.), *Advances in Water Quality Improvement*, University of Texas Press, Austin, 1968.
6. Fairall, J. M.: Correlation of Trickling Filter Data, *Sewage and Industrial Wastes*, vol. 28, no. 9, 1956.
7. Federation of Sewage and Industrial Wastes Associations: *Air Diffusion in Sewage Works*, Manual of Practice 5, 1952.
8. Galler, W. S., and H. B. Gotaas: Optimization Analysis for Biological Filter Design, *Journal of the Sanitary Division*, ASCE, vol. 92, no. SA1, 1966.
9. Gloyna, E. F.: Basis for Waste Stabilization Pond Designs, in E. F. Gloyna and W. W. Eckenfelder, Jr. (eds.), *Advances in Water Quality Improvement*, University of Texas Press, Austin, 1968.

10. Great Lakes–Upper Mississippi River Board of State Sanitary Engineers: *Recommended Standards for Sewage Works* (*Ten State Standards*), 1960.

11. *Handbook of Trickling Filter Design*, Public Works Journal Corporation, Ridgewood, N.J., 1968.

12. Kimball, J. W.: *Biology*, 2d ed., Addison-Wesley, Reading, Mass., 1968.

13. Kraus, L. S.: A Rugged Activated Sludge Process, presented at the Annual Meeting of the Ohio Sewage and Industrial Waste Treatment Conference, 1955.

14. Kraus, L. S.: Operating Practices for Activated Sludge Plants, *J. WPCF*, vol. 37, no. 5, 1965.

15. Levenspiel, O.: *Chemical Reaction Engineering*, Wiley, New York, 1962.

16. Mancini, J. L. and E. L. Barnhart: Industrial Waste Treatment in Aerated Lagoons, in E. F. Gloyna and W. W. Eckenfelder, Jr. (eds.), *Advances in Water Quality Improvement*, University of Texas Press, Austin, 1968.

17. Marais, G. V. R., and V. A. Shaw: A Rational Theory for the Design of Sewage Stabilization Ponds in Central and South Africa, *Transactions, The Civil Engineer in South Africa*, Johannesburg, vol. 3, p. 205, 1961.

18. McCabe, B. J., and W. W. Eckenfelder, Jr.: *Biological Treatment of Sewage and Industrial Wastes*, vol. 2, Reinhold, New York, 1958.

19. McGauhey, P. H.: *Engineering Management of Water Quality*, McGraw-Hill, New York, 1968.

20. McKinney, R. E., and W. J. O'Brien: Activated Sludge—Basic Design Concepts, *J. WPCF*, vol. 40, no. 11, part 1, 1968.

21. Metcalf, L., and H. P. Eddy: *American Sewerage Practice*, vol. III, 3d ed., McGraw-Hill, New York, 1935.

22. National Research Council: Trickling Filters (in Sewage Treatment at Military Installations), *Sewage Works Journal*, vol. 18, no. 5, 1946.

23. Oswald, W. J.: Advances in Anaerobic Pond System Design, in E. F. Gloyna and W. W. Eckenfelder, Jr. (eds.), *Advances in Water Quality Improvement*, University of Texas Press, Austin, 1968.

24. Oswald, W. J.: Fundamental Factors in Stabilization Pond Design, *Advances in Waste Treatment*, Pergamon, New York, 1963.

25. Oswald, W. J., and H. B. Gotaas: Photosynthesis in Sewage Treatment, *Transactions ASCE*, vol. 122, 1957.

26. Pflanz, P.: Performance of (Activated Sludge) Secondary Sedimentation Basins, *Proceedings of the Fourth International Conference*, International Association on Water Pollution Research, Prague, 1969, p. 569.

27. Rankin, R. S.: Evaluation of the Performance of Biofiltration Plants, *Transactions ASCE*, vol. 120, 1955.

28. Sawyer, C. N.: Final Clarifiers and Clarifier Mechanism, in B. J. McCabe and W. W. Eckenfelder, Jr. (eds.), *Biological Treatment of Sewage and Industrial Wastes*, vol. 1, Reinhold, New York, 1956.

29. Sawyer, C. N.: New Concepts in Aerated Lagoon Design and Operation, in E. F. Gloyna and W. W. Eckenfelder, Jr. (eds.), *Advances in Water Quality Improvement*, University of Texas Press, Austin, 1968.

30. Technical Practice Committee, Subcommittee on Aeration in Wastewater Treatment: Aeration in Wastewater Treatment: Manual of Practice 5, *WPCF*, Washington, D.C., 1971.

31. *The Sewerage Manual*, published biennially by *Public Works*.

32. Thirumurthi, D.: Design Principles of Waste Stabilization Ponds, *Journal of the Sanitary Division*, ASCE, vol. 95, no. SA2, 1969.

33. Torpey, W. N., and A. H. Chasick: Principles of Activated Sludge Operation, in B. J. McCabe and W. W. Eckenfelder, Jr. (eds.), *Biological Treatment of Sewage and Industrial Wastes*, vol. 1, Reinhold, New York, 1956.

34. Velz, C. J.: A Basic Law for the Performance of Biological Beds, *Sewage Works Journal*, vol. 20, no. 4, 1948.

35. Water Pollution Control Federation: *Sewage Treatment Plant Design*, Manual of Practice 8, Washington, D.C., 1967.

36. Wehner, J. F., and R. H. Wilhelm: Boundary Conditions of Flow Reactor, *Chemical Engineering Science*, vol. 6, p. 89, 1958.

design of facilities for treatment and disposal of sludge

13

The solids removed by a variety of methods in wastewater treatment plants, which include grit, screenings, and sludge, constitute the most important by-product of the treatment processes. Sludge is by far the largest in volume, and its processing and disposal is perhaps the most complex problem with which the engineer is faced in the field of wastewater treatment. It is for this reason that a separate chapter has been devoted to this subject. The disposal of grit and screenings was discussed in Chap. 11, where it was noted that screenings are often ground up and disposed of together with the sludge; therefore, this chapter will be focused primarily on the processing and disposal of sludge.

The problems of dealing with sludge are complicated, especially when it is remembered (1) that it is composed largely of the substances responsible for the offensive character of untreated sewage; (2) that the portion of sludge produced from biological treatment requiring disposal is composed of the organic matter contained in the raw sewage, but in another form, and that it, too, will decompose and become offensive; and (3) that only a small part of the sludge is solid matter. Therefore, the main purpose of this chapter is to delineate the operations and processes that are used to reduce the water and organic content of sludges. These include (1) concentration (thickening), (2) digestion, (3) conditioning, (4) dewatering and drying, and (5) incineration and wet oxidation. Of these, digestion, incineration, and wet oxidation are used primarily for the treatment of the organic material in the sludge; concentration, conditioning, dewatering, and drying are used primarily for the removal of moisture from sludge. To make the study of these operations and processes more meaningful, the first two sections of this chapter are devoted to a presentation of sludge treatment flowsheets and a discussion of the sources, quantities, and characteristics of sludge. The last section deals with the ultimate disposal of the sludge or ash after processing and treatment.

SLUDGE TREATMENT PROCESS FLOWSHEETS

A generalized flowsheet incorporating the various unit operations and processes to be discussed in this chapter is presented in Fig. 13·1. As shown, numerous combinations are possible. In practice, the most commonly used process flowsheets for the treatment of sludge may be divided into two general categories, depending on whether or not biological treatment is involved; therefore, it is important that the sanitary engineer be familiar with representative flowsheets in each category.

Flowsheets incorporating biological processing are presented in Fig. 13·2. Depending on the source of the sludge, either gravity or air flotation thickeners are used; in some cases, both may be used in the same plant. Following digestion, any of the three methods shown may be used to dewater the sludge, the choice depending on local conditions.

Because the presence of industrial and other toxic wastes has presented problems in the operation of biological digesters, a number of plants have been designed with other means for sludge treatment. Three representative process flowsheets without biological treatment are shown in Fig. 13·3.

SLUDGE SOURCES, QUANTITIES, AND CHARACTERISTICS

To design sludge processing, treatment, and disposal facilities properly, the sources, quantities, and characteristics of the sludge to be handled must be known. Therefore, the purpose of this section is to present background data and information on these topics, which will serve as a basis for the material to be presented in the subsequent sections of this chapter.

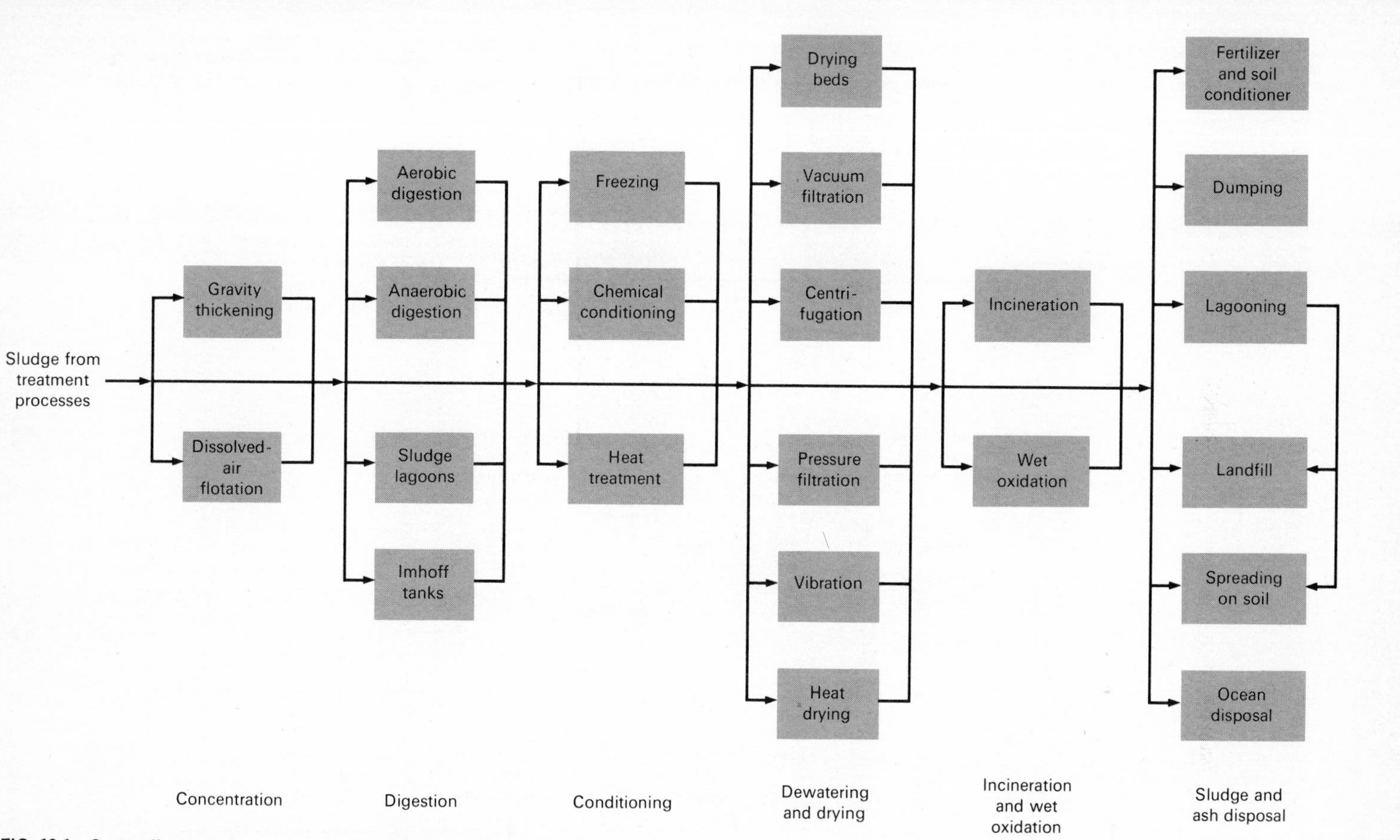

FIG. 13·1 Generalized sludge processing and disposal flowsheet.

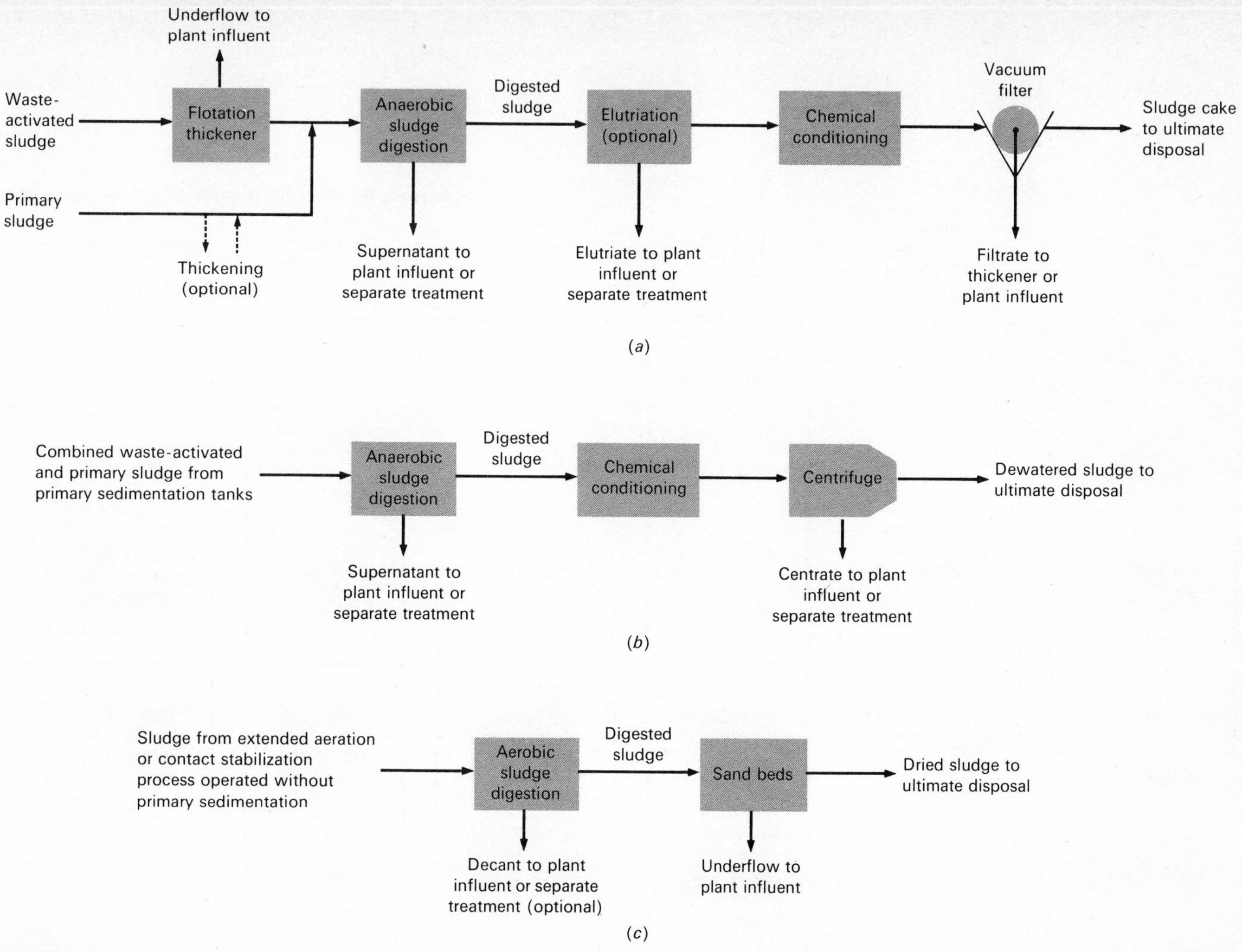

FIG. 13·2 Typical sludge treatment flowsheets with biological digestion and three different sludge dewatering processes: (*a*) vacuum filtration, (*b*) centrifugation, and (*c*) drying beds.

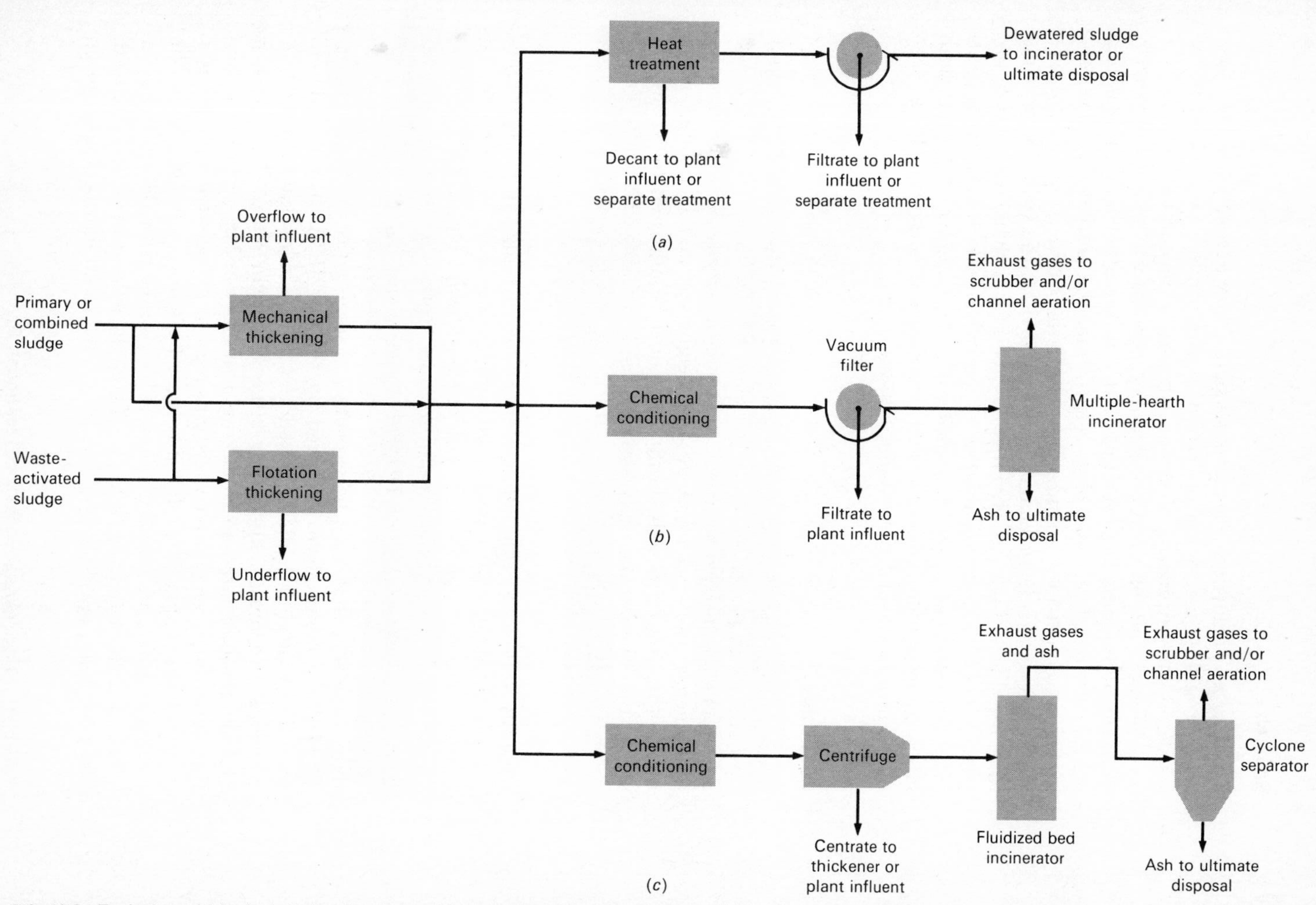

FIG. 13·3 Typical nonbiological sludge treatment flowsheets: (*a*) heat treatment with vacuum-filter dewatering, (*b*) multiple hearth incineration, and (*c*) fluidized-bed incineration.

Sources

The sources of sludge in a treatment plant vary according to the type of plant and its method of operation. For example, in a complete-mix activated-sludge process, if sludge wasting is accomplished from the mixed-liquor line or aeration chamber, sludge will not be produced from the activated-sludge settling tank. On the other hand, if wasting is accomplished from the solids return line, the activated-sludge settling tank constitutes a sludge source. If the sludge from the mixed-liquor line or aeration chamber is returned to the primary settling tank for thickening, this obviates the need for a thickener and therefore reduces by one the number of sludge sources in the treatment plant. Processes employed for thickening, digesting, conditioning, and filtering the sludge produced from primary and activated-sludge settling tanks also constitute sludge sources.

Quantities

Data on the quantities of sludge produced from various processes and operations are presented in Table 13·1. The volume and weight per 1,000 persons on either an actual or population-equivalent basis are the fundamental units; the volume and weight per million gallons, more meaningful from a design standpoint, must be adjusted to reflect the strength of the sewage. Although the data in Table 13·1 are useful as presented, it should be noted that the quantity of sludge produced from primary sedimentation tanks will vary widely.

Quantity Variations The quantity of solids entering the sewage treatment plant daily may be expected to fluctuate over a wide range. To ensure capacity capable of handling these variations, the designer of sludge-disposal facilities should consider both (1) the average and maximum rates of sludge production and (2) the potential storage capacity of the treatment units within the plant. The variation in daily quantity that may be expected in large cities is shown in Fig. 13·4. The general curve, prepared by the late R. A. Allton, is based on a study of peak loads obtained at Chicago, Cleveland, Columbus, Syracuse, Rochester, and other large American cities. The curve is characteristic of large cities having a number of large sewers laid on flat slopes [5]; even greater variations may be expected at small plants.

A limited quantity of solids may be stored temporarily in the sedimentation and aeration tanks. This storage provides capacity for equalizing short-term peak loads. Where digestion tanks are employed, their large storage capacity provides a substantial dampening effect on

TABLE 13·1 NORMAL QUANTITIES OF SLUDGE PRODUCED BY DIFFERENT TREATMENT PROCESSES*

Treatment process	Normal quantity of sludge			Moisture, %	Specific gravity of sludge solids	Specific gravity of sludge	Dry solids	
	Gal/ million gal of sewage	Tons/ million gal of sewage	Cu ft/ 1,000 persons daily				Lb/ million gal of sewage	Lb/ 1,000 persons daily
Primary sedimentation:								
Undigested	2,950	12.5	39.0	95	1.40	1.02	1,250	125
Digested in separate tanks	1,450	6.25	19.0	94	...	1.03	750	75
Digested and dewatered on sand beds	...	0.94	5.7	60	...	...	750	75
Digested and dewatered on vacuum filters	...	1.36	4.3	72.5	...	1.00	750	75
Trickling filter	745	3.17	9.9	92.5	1.33	1.025	476	48
Chemical precipitation	5,120	22.0	68.5	92.5	1.93	1.03	3,300	330
Dewatered on vacuum filters	...	6.0	19.3	72.5	...	...	3,300	330
Primary sedimentation and activated sludge:								
Undigested	6,900	29.25	92.0	96	...	1.02	2,340	234
Undigested and dewatered on vacuum filters	1,480	5.85	20.0	80	...	0.95	2,340	234
Digested in separate tanks	2,700	11.67	36.0	94	...	1.03	1,400	140
Digested and dewatered on sand beds	...	1.75	18.0	60	...	...	1,400	140
Digested and dewatered on vacuum filters	...	3.5	11.7	80	...	0.95	1,400	140
Activated sludge:								
Wet sludge	19,400	75.0	258.0	98.5	1.25	1.005	2,250	225
Dewatered on vacuum filters	...	5.62	19.0	80	...	0.95	2,250	225
Dried by heat dryers	...	1.17	3.0	4	...	1.25	2,250	225
Septic tanks, digested	900	...	12.0	90	1.40	1.04	810	81
Imhoff tanks, digested	500	...	6.7	85	1.27	1.04	690	69

* Based on a sewage flow of 100 gpcd and 300 ppm, or 0.25 lb per capita daily, of suspended solids in sewage.

FIG. 13·4 Peak sludge load as a function of the average daily load [5].

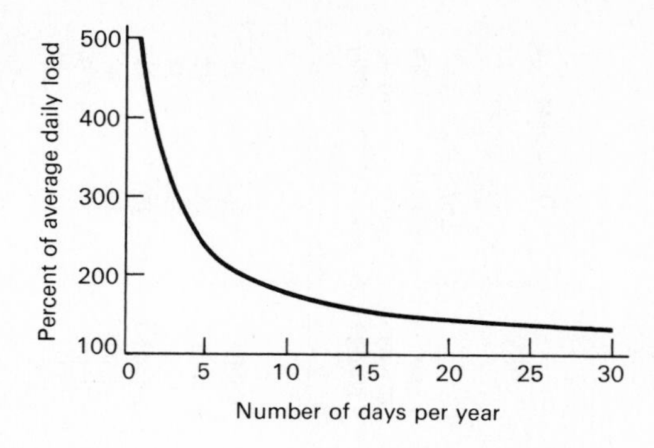

peak digested-sludge loads. Sludge-disposal systems that utilize digestion tanks may be designed on the basis of maximum monthly loadings; those that do not should be capable of handling the solids production of the maximum week.

Volume-Weight Relationships The volume of sludge depends mainly on its water content and only slightly on the character of the solid matter. The water content is commonly expressed by weight. A 90 percent sludge, for example, contains 90 percent water by weight. If the solid matter is composed of fixed (mineral) solids and volatile (organic) solids, the specific gravity of all of the solid matter can be computed using Eq. 13·1.

$$\frac{W_s}{S_s\gamma} = \frac{W_f}{S_f\gamma} + \frac{W_v}{S_v\gamma} \tag{13·1}$$

where W_s = weight of solids
S_s = specific gravity of solids
γ = weight of water, lb/cu ft
W_f = weight of fixed solids (mineral matter)
S_f = specific gravity of fixed solids
W_v = weight of volatile solids
S_v = specific gravity of volatile solids

Therefore, if one-third of the weight of the solid matter in a sludge containing 90 percent water is composed of fixed mineral solids with a specific gravity of 2.5, and two-thirds of the weight is composed of volatile solids with a specific gravity of 1.0, then the specific gravity of all solids S_s would be equal to 1.25, as follows:

$$\frac{1}{S_s} = \frac{0.33}{2.5} + \frac{0.67}{1} = 0.802$$

$$S_s = \frac{1}{0.802} = 1.25$$

If the specific gravity of the water is taken to be 1.0, as it can be without appreciable error, the specific gravity of the sludge is 1.02, as follows:

$$\frac{1}{S} = \frac{0.1}{1.25} + \frac{0.9}{1.0} = 0.98$$

$$S = \frac{1}{0.98} = 1.02$$

The volume of a sludge with a specific gravity of 1.02, with 90 percent moisture, and containing 1 lb of solid matter is 1.17 gal $\left[\frac{1}{8.33}\left(\frac{1}{1.25} + \frac{9}{1}\right)\right]$. If the water content is 95 percent, however, the volume of the sludge would be 2.38 gal $\left[\frac{1}{8.33}\left(\frac{1}{1.25} + \frac{19}{1}\right)\right]$, or slightly more than twice as much as that of a 90 percent sludge. Thus, for approximate calculations for a given solids content, it is simple to remember that the volume varies inversely with the percent of solid matter contained in the sludge as given by

$$\frac{V_1}{V_2} = \frac{P_2}{P_1} \quad \text{(approximate)} \tag{13·2}$$

where V_1, V_2 = sludge volumes
P_1, P_2 = percent of solid matter

The relationship of weight and volume of sludge is illustrated in Example 13·1.

EXAMPLE 13·1 *Volume of Raw and Digested Sludge*

Determine the liquid volume before and after digestion for 1,000 lb (dry basis) of primary sludge with the following characteristics:

	Primary	Digested
Solids, %	5	10
Volatile matter, %	60	60 (destroyed)
Specific gravity of fixed solids	2.5	2.5
Specific gravity of volatile solids	≈1.0	≈1.0

Solution

1. Compute the average specific gravity of all the solids in the primary sludge.

$$\frac{1}{S_s} = \frac{0.4}{2.5} + \frac{0.6}{1.0} = 0.76$$

$$S_s = \frac{1}{0.76} = 1.31 \qquad \text{(primary solids)}$$

2. Compute the volume of the primary sludge.

$$V = \frac{1{,}000}{62.4}\left(\frac{1}{1.31} + \frac{19}{1.0}\right)$$

$$= \frac{1{,}000}{62.4}(19.76) = 317 \text{ cu ft}$$

3. Compute the percentage of volatile matter after digestion.

$$\% \text{ volatile matter} = \frac{0.4(600)100}{400 + 0.4(600)} = 37.5$$

4. Compute the average specific gravity of all the solids in the digested sludge.

$$\frac{1}{S_s} = \frac{0.625}{2.5} + \frac{0.375}{1} = 0.625$$

$$S_s = \frac{1}{0.625} = 1.6 \qquad \text{(digested solids)}$$

5. Compute the volume of digested sludge.

$$V = \frac{400 + 0.4(600)}{62.4}\left(\frac{1}{1.6} + \frac{9}{1}\right)$$

$$= \frac{640}{62.4}(9.625) = 98.8 \text{ cu ft}$$

6. Determine the specific gravity of the sludge after digestion.

$$\frac{1}{S} = \frac{0.10}{1.6} + \frac{0.90}{1} = 0.9625$$

$$S = \frac{1}{0.9625} = 1.04$$

7. Using the value for specific gravity determined in step 6, compute the volume of sludge by an alternate method.

$$V = \frac{640 \text{ lb}}{1.04(62.4)(0.10)} = 98.8 \text{ cu ft}$$

Characteristics

The characteristics of sludge vary depending on its origin, the amount of aging that has taken place, and the type of processing to which it has been subjected. Sludge from primary sedimentation tanks is usually grey and slimy and, in most cases, has an extremely offensive odor. It can be readily digested under suitable conditions of operation.

Sludge from chemical precipitation tanks is usually black, though its surface may be red if it contains much iron. Its odor may be objectionable, but not as bad as odor from primary sedimentation sludge. While it is somewhat slimy, the hydrate of iron or aluminum in it makes it gelatinous. If it is left in the tank, it undergoes decomposition like the sludge from primary sedimentation but at a slower rate. It gives off gas in substantial quantities and its density is increased by standing.

Activated sludge generally has a brown flocculent appearance. If the color is quite dark, the sludge may be approaching a septic condition. If the color is lighter than usual there may have been underaeration with a tendency for the solids to settle slowly. The sludge, when in good condition, has an inoffensive characteristic odor. It has a tendency to become septic rather rapidly and then has a disagreeable odor of putrefaction. It will digest readily alone or mixed with fresh sewage solids.

Trickling-filter humus is brownish, flocculent, and relatively inoffensive when fresh. It generally undergoes decomposition more slowly than other undigested sludges, but when it contains many worms it may become offensive quickly. It is readily digested.

Digested sludge is dark brown to black and contains an exceptionally large quantity of gas. When thoroughly digested, it is not offensive, its odor being relatively faint and like that of hot tar, burnt rubber, or sealing wax. When drawn off on porous beds in layers 6 to 10 in. deep, the solids first are carried to the surface by the entrained gases, leaving a sheet of comparatively clear water below them which drains off rapidly and allows the solids to sink down slowly on to the bed. As the sludge dries, the gases escape, leaving a well-cracked surface with an odor resembling that of garden loam.

Sludge from septic tanks is black and, unless well digested by long storage, is offensive on account of the hydrogen sulfide and other gases it gives off. The sludge can be dried on porous beds if spread out in thin layers, but objectionable odors are to be expected while it is draining unless it has been well digested.

Typical data on the chemical composition of raw and digested sludges are reported in Table 13·2. Many of the chemical constituents,

TABLE 13·2 TYPICAL CHEMICAL COMPOSITION OF RAW AND DIGESTED SLUDGE

Item	Raw primary sludge		Digested sludge	
	Range	Typical	Range	Typical
Total dry solids (TS), %	2.0–7.0	4.0	6.0–12.0	10.0
Volatile solids (% of TS)	60–80	65	30–60	40.0
Grease and fats (ether soluble, % of TS)	6.0–30.0	...	5.0–20.0	
Protein (% of TS)	20–30	25	15–20	18
Nitrogen (N, % of TS)	1.5–4.0	2.5	1.6–6.0	3.0
Phosphorus (P_2O_5, % of TS)	0.8–2.8	1.6	1.5–4.0	2.5
Potash (K_2O, % of TS)	0–1.0	0.4	0.0–3.0	1.0
Cellulose (% of TS)	8.0–15.0	10.0	8.0–15.0	10.0
Iron (not as sulfide)	2.0–4.0	2.5	3.0–8.0	4.0
Silica (SiO_2, % of TS)	15.0–20.0	...	10.0–20.0	
pH	5.0–8.0	6.0	6.5–7.5	7.0
Alkalinity (mg/liter as $CaCO_3$)	500–1,500	600	2,500–3,500	3,000
Organic acids (mg/liter as HAc)	200–2,000	500	100–600	200
Thermal content (Btu/lb)	6,800–10,000	7,600*	2,700–6,800	4,000†

* Based on 65 percent volatile matter.
† Based on 40 percent volatile matter.

including nutrients, are of importance in considering the ultimate disposal of the processed sludge and the liquid removed from the sludge during processing. The fertilizer value of sludge, evaluated where the sludge is to be used as a soil conditioner, is based primarily on the content of nitrogen, phosphorus, and potash. The measurement of pH, alkalinity, and organic acid content is important in process control of anaerobic digestion.

The thermal content of sludge is important where incineration or some other combustion process is considered, and accurate bomb-calorimeter tests should be conducted so that a heat balance can be made for the combustion system. The thermal content of raw primary sludge is the highest, especially if it contains appreciable amounts of grease and skimmings. Where garbage grinders are used, the volatile and thermal content of the sludge will also be high. The heat content reported in Table 13·2 was calculated using Eq. 9·63.

SLUDGE CONCENTRATION

Concentration is used mainly to thicken waste activated sludge or mixtures of primary and waste activated sludges. Occasionally, separate concentration of primary or digested sludges has been practiced, although these sludges tend to compact quite well in settling tanks. Primary sludge is sometimes concentrated separately as an aid to the clarification process or when it is desirable to withdraw a relatively thin sludge from the primary clarifier for pumping with subsequent sludge concentration for volume reduction.

Separate Concentration

Although separate sludge concentration is frequently not considered as an aid to the sewage treatment process, certain benefits can be realized from its use. The major aim of the settling or clarification process in waste treatment is to obtain maximum efficiency of suspended solids removal. This is best achieved when the settled sludge is rapidly removed from the tank. On the other hand, the aim of sludge concentration is to produce as thick a sludge as possible. When an attempt is made to achieve both aims in the same tank, the long residence time required for good concentration tends to hinder solids-removal efficiency. With separate sludge concentration, the concentrating process is physically removed from the settling process so that both can be operated under optimum conditions.

Separate sludge concentration can be quite beneficial to the operation of activated-sludge treatment plants, as it makes feasible the direct removal of aeration-tank mixed liquor for excess sludge wasting rather than following the more usual practice of wasting the more concentrated return sludge. By removing a given volume of the mixed liquor each day for concentration and disposal, the sludge age or solids retention time, upon which the efficiency and operational characteristics of the activated-sludge process depend, can be closely maintained. This method of control is consistent with most current theories on activated-sludge operation. The wasting of mixed liquor is also beneficial to the separate sludge concentration process, since it has been shown that higher solids concentrations can be achieved when dilute mixed liquor, rather than the more concentrated return sludge, is applied to the concentrating unit.

Volume Reduction

The reduction in sludge volume that can be obtained by sludge concentration can be approximated from Eq. 13·2. For example, if a sludge is thickened from 1 to 4 percent solids, the volume will be reduced to 25 percent of the original volume. It is seldom desirable to achieve a solids concentration in excess of 10 percent, as such sludges are viscous and very difficult to pump. Raw activated sludges, at one extreme, characteristically have solids concentrations of only 0.5 to 1.0 percent, and the reduction in volume that can be effected by concentration (possibly to 3 or 4 percent) is very significant. Such sludges are still quite fluid and can be pumped easily. At the other extreme, raw primary sludges characteristically have solids concentrations of 2.5 to 5.0 percent, and here the potential benefit of separate sludge concentration is reduced.

Reduction in sludge volume may result in reduction in pipe size and pumping costs on large projects when sludges must be transported a significant distance, such as to a separate plant for processing or through ocean outfalls. On small projects, the requirements of a minimum practicable pipe size and minimum velocity may necessitate the pumping of significant volumes of sewage in addition to sludge which diminishes the value of volume reduction. Volume reduction is very desirable when liquid sludge is transported by tank trucks for direct application to land as a soil conditioner. The volume reduction obtained by sludge concentration can also be beneficial to subsequent treatment processes, such as digestion, dewatering, drying, and combustion, from the following standpoints: (1) capacity of tanks and equipment required, (2) quantity of chemicals required for sludge conditioning, and (3) amount of heat required by digesters and amount of auxiliary fuel required for heat drying and/or incineration.

The solids content of primary, activated, trickling filter, or mixed sludge (e.g., primary plus activated) varies considerably depending on

TABLE 13·3 CONCENTRATIONS OF UNTHICKENED AND THICKENED SLUDGES AND SOLIDS LOADINGS FOR MECHANICAL THICKENERS

Type of sludge	Sludge, percent solids		Solids loading for mechanical thickeners, lb/sq ft/day
	Unthickened	Thickened	
Separate sludges:			
Primary	2.5–5.5	8–10	20–30
Trickling filter	4–7	7–9	8–10
Modified aeration	2–4	4.3–7.9	7–18
Activated	0.5–1.2	2.5–3.3	4–8
Combined sludges:			
Primary and trickling filter	3–6	7–9	12–20
Primary and modified aeration	3–4	8.3–11.6	12–20
Primary and activated	2.6–4.8	4.6–9.0	8–16

the characteristics of the sludge, the sludge removal and pumping facilities, and the method of operation. Representative values of percent total solids are shown in Table 13·3, together with representative values for the same sludges after thickening. The range of solids loading in pounds per square foot per day for mechanical thickeners is also shown (see following section). The data in Table 13·3 have been assembled from various sources, including plant-scale tests in the New York City plants.

Design of Thickeners

Mechanical and dissolved-air flotation thickeners are commonly used to thicken sludges from various sources in wastewater treatment plants. Solids concentration of 5 to 6 percent or more can generally be obtained from mixtures of primary and waste activated sludge. At activated-sludge plants where sludge digestion is used, thickening of the waste activated sludge or combined raw sludge is most desirable for construction and operating economy. Important considerations in designing a thickener are to provide adequate capacity to meet peak demands, and to prevent septicity with its attendant odor problems during the thickening process.

Mechanical Thickeners In a mechanical thickener, dilute raw primary or waste activated sludge is fed into the thickening tank continuously. Conventional sludge-collecting mechanisms with deep trusses (see Fig. 13·5) or vertical pickets are employed to stir the sludge gently, thereby opening up channels for water to escape and promoting densification. The continuous supernatant flow that results is returned to the primary settling tank. The thickened sludge that collects on the bottom of the tank is pumped to the digesters or dewatering equipment as required; thus storage space must be provided for the sludge. Time-clock control of pumping may be desirable.

Mechanical thickeners are designed on the basis of hydraulic surface loading and solids loading. The principles that apply are the same as those used in designing sedimentation tanks, discussed in Chaps. 8, 11, and 12. Typical surface-loading rates are 400 to 900 gpd/sq ft. Solids loadings are shown in Table 13·3. To maintain aerobic conditions in gravity thickeners, provisions should be made for adding aerated mixed liquor or final effluent to the thickening tank.

In operation, a sludge blanket is maintained on the bottom of the thickener to aid in concentrating the sludge. An operating variable is the sludge volume ratio (SVR), which is the volume of the sludge blanket held in the thickener divided by the volume of thickened

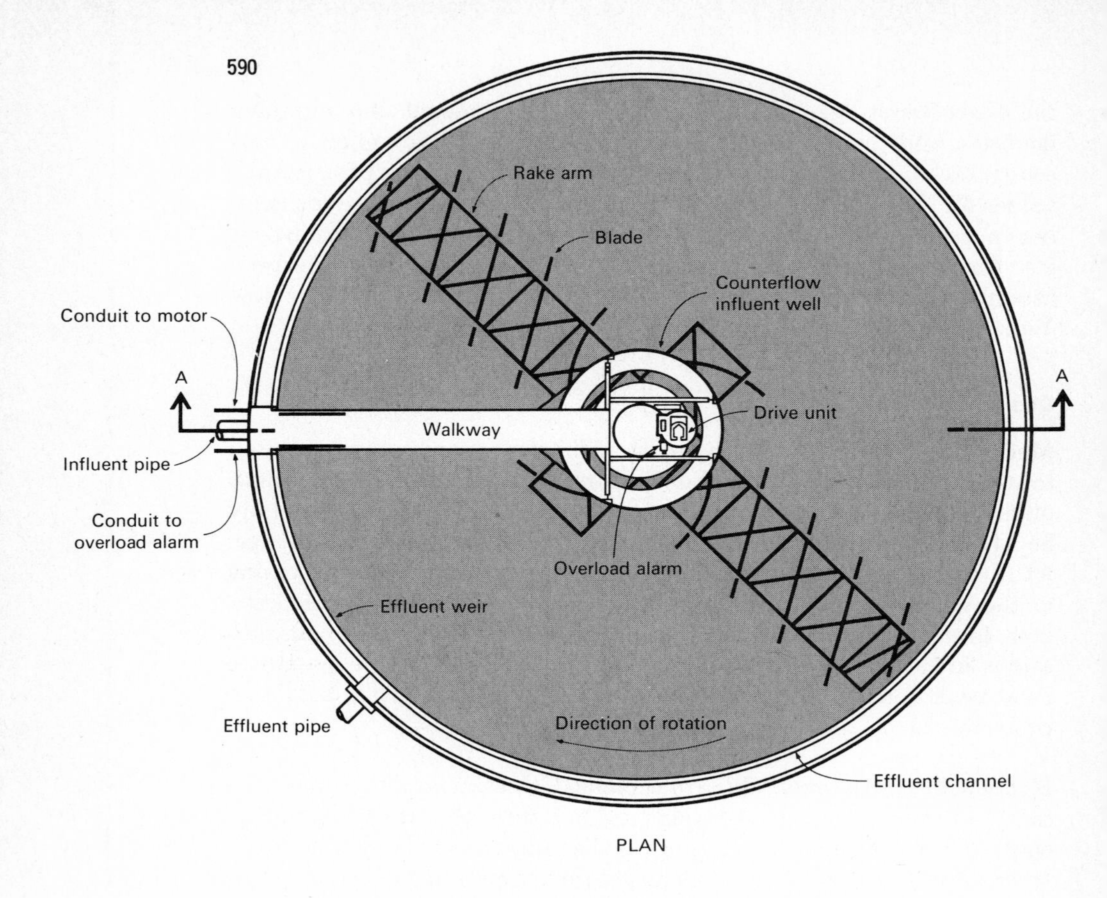

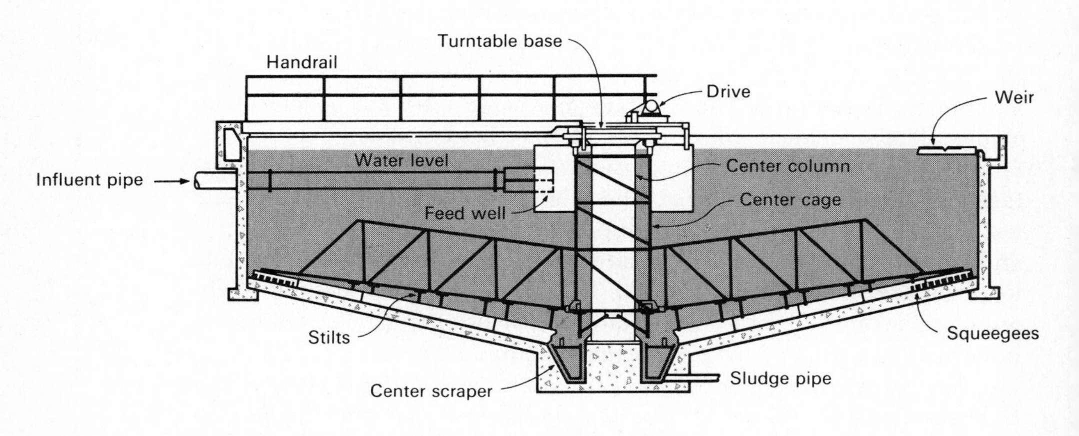

FIG. 13·5 Schematic of a mechanical thickener (from Dorr-Oliver).

sludge removed daily. Values of SVR normally range between 0.5 and 2.0 days; the lower values are required during warm weather.

Flotation Thickeners Flotation thickeners are used primarily with waste activated sludge and normally will produce a sludge with approximately 4 percent solids and 85 percent solids recovery without the use of chemicals. Concentrations averaging 6 percent and ranging as high as 8 percent have been obtained with mixtures of waste activated and primary sludges.

The use of polyelectrolytes as flotation aids may or may not be effective in increasing solids loadings and the concentration of thickened sludge, but it does appear to be effective in increasing the solids recovery in the floated sludge from 85 percent to 98 or 99 percent. At $1/lb, the addition of polyelectrolytes can increase the cost by $2 to $9/ton of dry solids [7].

Because of the rapid separation of solids from the sewage, higher loadings can be used than are permissible with gravity thickeners. Flotation thickeners may be operated at the solids loadings given in Table 13·4 [7]. For design, the minimum loadings should be used; the maximum should be 20. In general, the use of the higher solids loadings results in lower concentrations of thickened sludge. The pressure requirements and air-solids ratio requirements for dissolved-air flotation thickeners are discussed in Chap. 8.

TABLE 13·4 LOADING OF DISSOLVED-AIR FLOTATION UNITS [7]

Type of sludge	Loading, lb/sq ft/day
Activated (mixed liquor)*	5–15
Activated (settled)	10–20
50% primary + 50% activated (settled)	20–40
Primary only	to 55

* Not in [7].

Primary-tank effluent or plant effluent is recommended as the source of air-charged water rather than flotation-tank effluent, except when chemical aids are used, because of the possibility of fouling the air-pressure system with solids.

ANAEROBIC SLUDGE DIGESTION

The history of sludge digestion and its precursors can be traced from the 1850s with the development of the first tank designed to separate and retain solids. The first unit used to treat settled sewage solids was known as the Mouras automatic scavenger, which was developed by Louis H. Mouras of Vesoul, France, in about 1860 after it had been observed that if the solids were kept in a closed vault (cesspool) they were converted to a liquid state [8]. The first to recognize that a combustible gas containing methane was produced when sewage solids were liquefied was Donald Cameron, who built the first septic tank for the city of Exeter, England, in 1895. He collected and used the gas for lighting in the vicinity of the plant [6]. In 1904, the first dual-purpose tank incorporating sedimentation and sludge treatment was installed at Hampton, England. It was known as the Travis hydrolytic tank and continued in operation until 1936 [6]. Experiments on a similar unit, called a Biolytic tank, were carried out in the United States between 1909 and 1912. In 1904, a patent was issued to Dr. Karl Imhoff in Germany for a dual-purpose tank now commonly known as the Imhoff tank. One of the first installations in the United States employing separate digestion tanks was the sewage treatment plant in Baltimore, Md. Three rectangular digestion tanks were built as part of the original plant in 1911, 16 circular digestion tanks were added in 1914, and an additional rectangular tank was added in 1921 [12].

In the period from 1920 to 1935, the anaerobic digestion process was studied extensively. Heat was applied to separate digestion tanks, and major improvements were made in the design of the tanks and associated appurtenances. It is interesting to note that the same practice is being followed today, some 40 years later, but great progress has been made in the fundamental understanding and control of the process, the sizing of tanks, and the design and application of equipment. At the same time, engineers have become aware of its limitations and are learning when not to use it. Although most sludge digestion is anaerobic, the aerobic process has been growing in popularity for use in smaller installations.

Process Description

The microbiology of anaerobic digestion and the optimum environmental conditions for the microorganisms involved are discussed in Chap. 10 (see "Anaerobic Waste Treatment"). The operation and physical facilities for anaerobic digestion in conventional and high-rate digesters are described in this section.

Conventional Digestion Conventional (standard-rate) sludge digestion is carried out as either a single-stage (see Fig. 13·6) or two-stage (see Fig. 13·7) process. The sludge is normally heated (see "Heat Transfer" in Chap. 8) by means of coils located within the tanks or an external

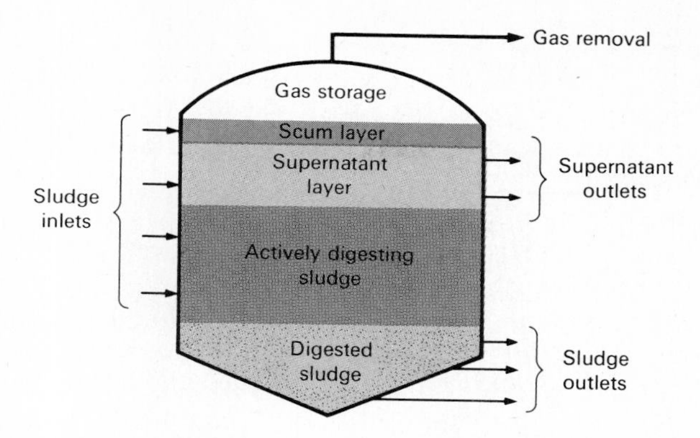

FIG. 13·6 Schematic of conventional digester used in the single-stage process.

heat exchanger. In the single-stage process, the functions of digestion, sludge thickening, and supernatant formation are carried out simultaneously. A cross section of a typical standard-rate digester is shown in Fig. 13·8. Operationally, in a single-stage process, raw sludge is added in the zone where the sludge is actively digesting and the gas is being released. As gas rises to the surface it lifts sludge particles and other materials, such as grease, oils, and fats, ultimately giving rise to the formation of a scum layer (see Fig. 13·6). As a result of digestion, the sludge becomes more mineralized (for example, percentage of fixed solids increases), and it thickens due to gravity. In turn, this leads to the formation of a supernatant layer above the digesting sludge. The biochemistry of the reactions taking place in the digesting zone is described in Chap. 10. Due to the stratification and the lack of intimate mixing, the volume of a standard-rate single-stage digester is not more than 50 percent utilized. Recognizing these

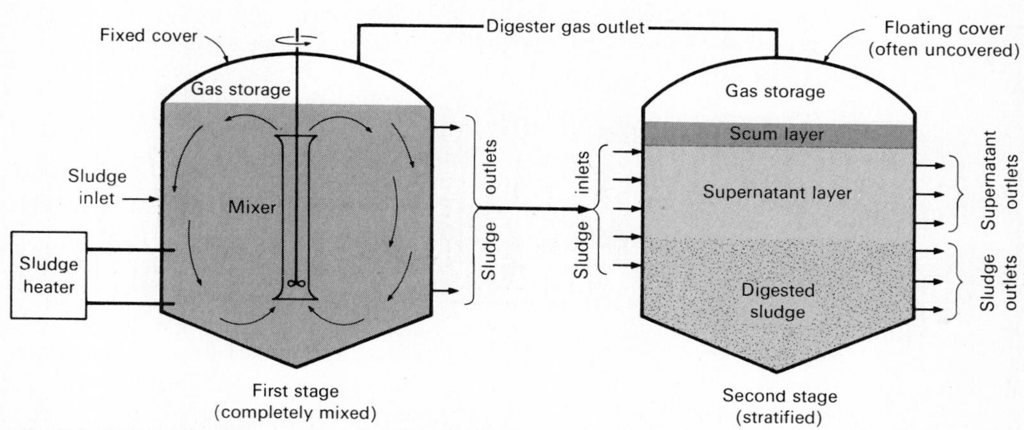

FIG. 13·7 Schematic of two-stage digestion process.

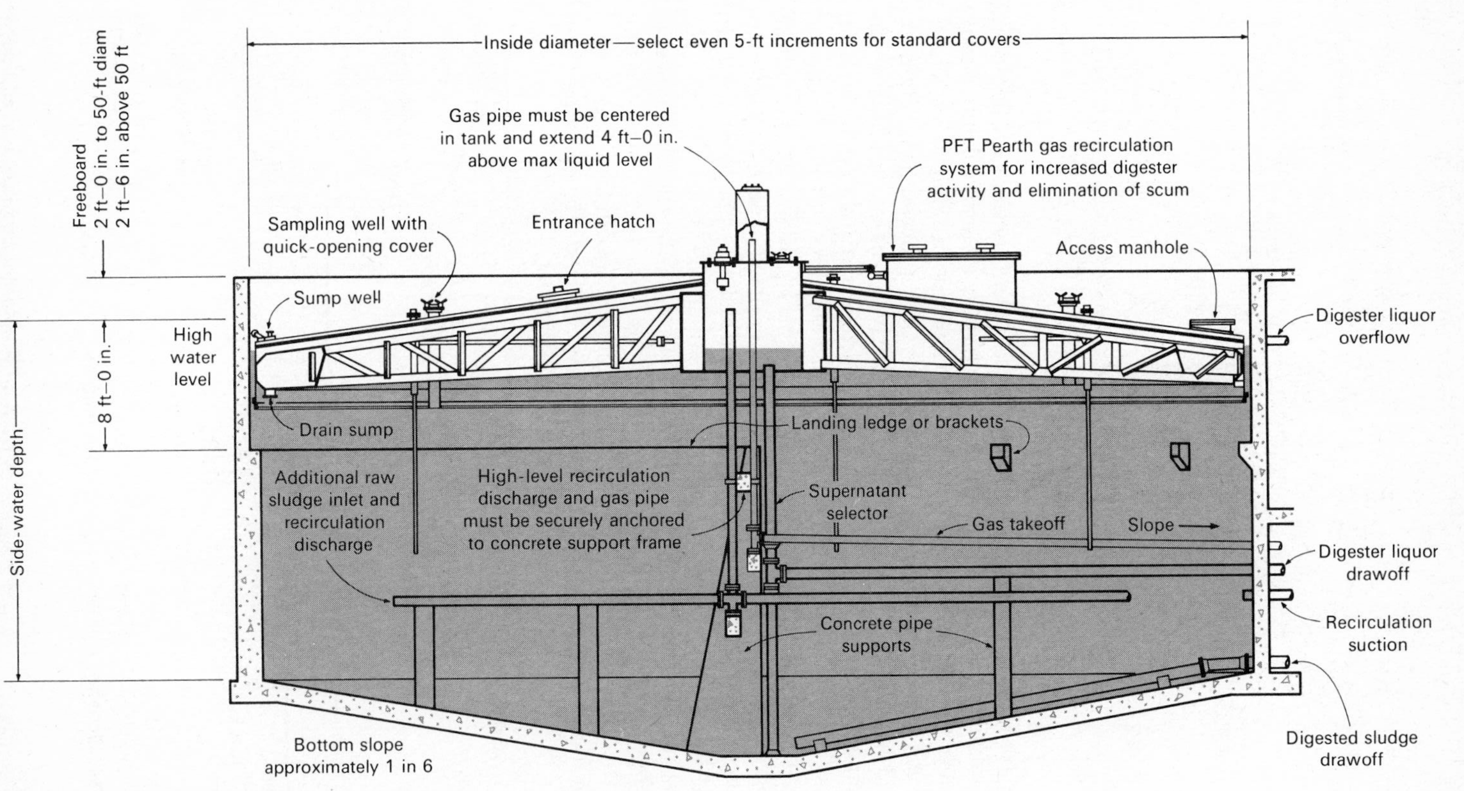

FIG. 13·8 Cross section through a typical standard-rate digester [from PFT].

limitations, most conventional digestion operations (plants having a capacity of 1 mgd or more) are carried out as a two-stage process (see Fig. 13·7).

In the two-stage process the first tank is used for digestion. It is heated and is equipped with mixing facilities consisting of one or more of the following: (1) sludge recirculation pumps; (2) gas recirculation using short mixing tubes, one or more deep-draft tubes or bottom-mounted diffusers; (3) mechanical draft-tube mixers, and (4) turbine and propeller mixers. The second tank is used for storage and concentration of digested sludge and for formation of a relatively clear supernatant. Frequently, the tanks are made identical, in which case either one may be the primary. In other cases, the second tank may be an open tank, an unheated tank, or a sludge lagoon. Tanks may have fixed roofs or floating covers. Any or all of the floating covers may be of the gasometer type. Alternatively, gas may be stored in a separate gas holder or compressed and stored under pressure. Tanks are usually circular and are seldom less than 20 ft or more than 115 ft in diameter. They should have a water depth of not less than 25 ft at the center and may be as deep as 45 ft or more. The bottom should slope to the sludge drawoff in the center, with a minimum slope of 1 vertical to 4 horizontal.

High-Rate Digestion The high-rate digestion process differs from the conventional single-stage process in that the solids-loading rate is much greater (see "Process Design"). The sludge is intimately mixed by gas recirculation, pumping, or draft-tube mixers (separation of scum and supernatant does not take place), and the sludge is heated to achieve optimum digestion rates. With the exception of higher loading rates and improved mixing, there are only a few differences between the primary digester in a conventional two-stage process and a high-rate digester. The mixing equipment should have greater capacity and should reach to the bottom of the tank, the gas piping will be somewhat larger, fewer multiple sludge drawoffs replace the supernatant drawoffs, and the tank should be deeper, if practicable, to aid the mixing process for the high-rate digester.

Sludge should be pumped to the digester continuously or by time clock on a 30-min to 2-hr cycle. The incoming sludge displaces digested sludge to a holding tank with capacity determined by subsequent disposal methods or to a second digester for supernatant separation and residual gas extraction. Because there is no supernatant separation in the high-rate digester, and the total solids are reduced by 45 to 50 percent and given off as gas, the digested sludge is about half as concentrated as the raw sludge feed.

Process Design

Ideally, the design of anaerobic sludge-digestion processes should be based on an understanding of the fundamental principles of biochemistry and microbiology discussed in Chap. 10. Because these principles have not been appreciated fully in the past, a number of empirical methods have also been used in the design of digesters. Therefore, the purpose of this section is to illustrate the various methods that have been used to design digesters in terms of size. These methods are based on (1) the concept of mean cell residence time, (2) the use of volumetric loading factors, (3) observed volume reduction, and (4) loading factors based on population.

Mean Cell Residence Time Digester design based on mean cell residence time involves application of the principles discussed in Chap. 10. To review briefly, the respiration and oxidation end products of anaerobic digestion are methane gas and carbon dioxide. The quantity of methane gas can be calculated using Eq. 13·3 [10].

$$C = 5.62\left(eF - 1.42\frac{dX}{dt}\right) \tag{13·3}$$

where C = CH_4 produced per day at standard conditions, cu ft
e = efficiency of waste utilization
F = BOD_L added, lb/day
$\frac{dX}{dt}$ = net growth rate of microorganisms, lb/day

The value 5.62 is the theoretical methane production from stabilization of 1 lb of BOD_L and is derived in Example 13·2. The efficiency of waste utilization e normally ranges from 0.60 to 0.90 under satisfactory operating conditions. Subtraction of the term $1.42(dX/dt)$ in Eq. 13·3 accounts for that portion of the substrate that is converted to new cells rather than end products.

EXAMPLE 13·2 *Conversion of BOD to Methane Gas*

Determine the cubic feet of methane produced from 1 lb of glucose.

Solution

1. Under anaerobic conditions glucose is converted to carbon dioxide and methane as follows:

$$\underset{180}{C_6H_{12}O_6} \rightarrow \underset{132}{3CO_2} + 3CH_4$$

2. The amount of oxygen required to oxidize methane to carbon dioxide and water is:

$$\overset{48}{3CH_4} + \overset{192}{6O_2} \rightarrow 3CO_2 + 6H_2O$$

3. The BOD_L of 1 lb of glucose is $\frac{192}{180}$ lb, and 1 lb of glucose yields $\frac{48}{180}$ lb of methane, so that

$$\frac{\text{lb } CH_4}{\text{lb } BOD_L} = \frac{\frac{48}{180}}{\frac{192}{180}} = 0.25$$

$$1 \text{ lb } BOD_L = 0.25 \text{ lb} \times \frac{454 \text{ g}}{\text{lb}} \times \frac{1 \text{ mole}}{16 \text{ g}} \times \frac{22.4 \text{ liters}}{\text{mole}} \times \frac{1 \text{ cu ft}}{28.32 \text{ liter}}$$

$= 5.62$ cu ft of CH_4 at standard conditions (32°F and 1 atm)

For a complete-mix high-rate digester without recycle, the mass of biological solids synthesized daily dX/dt can be estimated using Eq. 13·4, which is a modified form of Eq. 10·4.

$$\frac{dX}{dt} = \frac{Y \frac{dF}{dt}}{1 + k_d \theta_c} \tag{13·4}$$

where $\frac{dX}{dt}$ = net growth rate, lb/day

Y = growth-yield coefficient, lb/lb

$\frac{dF}{dt}$ = rate of substrate utilization, lb/day

k_d = decay coefficient, $days^{-1}$

θ_c = mean cell residence time, days

Values for Y and k_d as found for various types of waste are given in Table 13·5. Typical values for θ_c for various temperatures are reported in Table 13·6.

TABLE 13·5 GROWTH CONSTANTS AND ENDOGENOUS RESPIRATION RATES FOR VARIOUS SUBSTRATES [10, 11]

Substrate	Growth constant Y	Decay coefficient k_d
Fatty acid	0.054	0.038
Carbohydrate	0.240	0.033
Protein	0.076	0.014

TABLE 13·6 SUGGESTED MEAN CELL RESIDENCE TIMES FOR USE IN THE DESIGN OF COMPLETE-MIX DIGESTERS [10, 11]

Operating temperature, °F	θ_c^M, days	θ_c, days suggested for design
65	11	28
75	8	20
85	6	14
95	4	10
105	4	10

The application of Eqs. 13·3 and 13·4 in the process design of a high-rate digester is illustrated in the following example.

EXAMPLE 13·3 *Estimation of Digester Volume and Performance*

Estimate the size of digester required to treat the sludge from a preliminary treatment plant designed to treat 10 mgd of sewage. Check the volumetric loading and estimate the percent stabilization and the amount of gas produced per capita. For the sewage to be treated, it has been found that the quantity of dry solids and BOD_L removed per million gallons is equal to 1,200 and 1,150 lb, respectively. Assume that the sludge contains about 95 percent moisture and has a specific gravity of 1.02. Other pertinent design assumptions include

1. Hydraulic regime of reactor = complete mix.
2. $\theta_c = 10$ days at 35°C (see Table 13·6).
3. Efficiency of waste utilization, $e = 0.80$.
4. Waste contains adequate nitrogen and phosphorus for biological growth.
5. $Y = \dfrac{0.05 \text{ lb cells}}{\text{lb BOD}_L \text{ utilized}}$ and $k_d = 0.03/\text{day}$.
6. Constants are for a temperature of 35°C.

Solution

1. Compute the daily sludge volume and BOD_L loading.

$$\text{Sludge volume} = \frac{\dfrac{1{,}200 \text{ lb}}{10^6 \text{ gal}} \times 10 \text{ mgd}}{\dfrac{62.4 \text{ lb}}{\text{cu ft}} 1.02 \times 0.05}$$

$$= 3{,}770 \text{ cu ft/day}$$

$$\text{BOD}_L \text{ loading} = \frac{1{,}150 \text{ lb}}{10^6 \text{ gal}} \times 10 \text{ mgd}$$
$$= 11{,}500 \text{ lb/day}$$

2. Compute the digester volume.

$$\frac{V}{Q} = \theta = \theta_c \qquad \text{where } Q = \text{sludge flowrate}$$

$V = Q\theta_c = 3.770(10) = 37{,}700$ cu ft (also volume of first-stage digester in two-stage process)

3. Compute the volumetric loading.

$$\text{lb BOD}_L/\text{cu ft/day} = \frac{11{,}500}{37{,}700} = 0.305$$

4. Compute the pounds of volatile solids produced per day.

$$\frac{dX}{dt} = \frac{Y \dfrac{dF}{dt}}{1 + k_d\theta_c}$$

$$\frac{dF}{dt} = eF = 0.8(11{,}500) = 9{,}200 \text{ lb/day}$$

$$\frac{dX}{dt} = \frac{0.05(9{,}200)}{1 + 0.03(10)} = 354 \text{ lb/day}$$

5. Compute the percent stabilization.

$$\text{Percent stabilization} = \frac{eF - 1.42 \dfrac{dX}{dt}}{F} 100$$
$$= \frac{0.8(11{,}500) - 1.42(354)}{11{,}500} 100$$
$$= 76\%$$

6. Compute the volume of methane produced per day.

$$C = 5.62\left(eF - 1.42 \frac{dX}{dt}\right)$$
$$= 5.62[0.8(11{,}500) - 1.42(354)]$$
$$= 48{,}800 \text{ cu ft/day}$$

7. Estimate the per capita gas production. Because digester gas is about $\frac{2}{3}$ methane, the total volume of gas produced is:

$$\text{Total gas volume} = \frac{48{,}800}{0.67} = 73{,}200 \text{ cu ft/day}$$

If it is assumed that the efficiency of the treatment process is 50 percent and the per capita contribution of solids is 0.25 lb/day, then the volume of gas produced per capita is:

$$\text{Volume/capita/day} = \frac{\dfrac{73{,}200 \text{ cu ft}}{\text{day}}}{\dfrac{12{,}000 \text{ lb/day}}{0.5\ (0.25 \text{ lb/capita-day})}} = 0.76 \text{ cu ft}$$

Loading Factors One of the most common methods used to size digesters is to determine the required volume on the basis of a loading factor. Although a number of different factors have been proposed, the two that seem most favored are based on (1) the pounds of volatile solids added per day per cubic foot of digester capacity and (2) the pounds of volatile solids added per day per pound of volatile solids in the digester. From the material presented in Chap. 10, the similarity between these loading factors and the food-to-microorganism ratio is apparent. In applying these loading factors, another factor that should also be checked is the hydraulic detention time, because of its relationship to organism growth and washout (see Table 13·6) and to the type of digester used (for example, only 50 percent or less of the capacity of a conventional single-stage digester is effective).

Ideally, the conventional single-stage digestion tank is stratified into three layers with the supernatant at the top, the active digestion zone in the middle, and the thickened sludge at the bottom. Because of the storage requirements for the digested sludge, the supernatant, and the excess capacity for daily fluctuations in sludge loading, the volumetric loading for standard-rate digesters is low. Detention times based on gallons of raw sludge pumped vary from 30 to more than 90 days for this type of tank. The recommended solids loadings for standard-rate digesters are from 0.03 to 0.10 lb of volatile solids per cubic foot per day.

For high-rate digesters, loading rates of 0.10 to 0.40 lb of volatile solids per cubic foot per day and hydraulic detention periods of 10 to 20 days are practicable. The six high-rate digestion tanks at the Newtown Creek Plant of New York City are designed for a volatile solids loading of 0.214 lb/cu ft/day and a detention period of 17.6 days, with a raw sludge concentration of 8 percent solids. The tanks are also designed so that four tanks can handle the entire load and the other two can be used for storage and residual gas extraction. Under these conditions, the volatile solids loading becomes 0.32 lb/cu ft/day and the detention period 11.7 days [2]. Four draft-tube mixers in each tank are designed to turn over the entire tank contents in 30 min. The effect of sludge concentration and hydraulic detention time on the volatile-solids-loading factor is reported in Table 13·7. The determination of digester volumes using the volatile-solids-loading factor is illustrated in Example 13·4.

TABLE 13·7 EFFECT OF SLUDGE CONCENTRATION AND HYDRAULIC DETENTION TIME ON VOLATILE-SOLIDS-LOADING FACTOR*
(Pounds per cubic foot per day)

Sludge concentration, %	Hydraulic detention time, days			
	10	12	15	20
4	0.191	0.159	0.127	0.096
5	0.238	0.198	0.159	0.119
6	0.277	0.238	0.191	0.143
7	0.333	0.278	0.222	0.167
8†	0.381	0.318	0.254	0.191
9	0.428	0.357	0.286	0.214
10	0.476	0.396	0.318	0.238

* Based on 75 percent volatile content of sludge, and a sludge specific gravity of 1.02 (concentration effects neglected).
† The volatile-solids-loading factor at 11.7 and 17.6 days is 0.326 and 0.217 lb/cu ft/day.

EXAMPLE 13·4 *Estimation of Required Digester Volume Using Volatile-Solids-Loading Factor*

Determine the required volume of a high-rate digester to treat the primary sludge in Example 13·3. Assume a volatile-solids-loading factor of 0.238 based on a hydraulic detention time of 10 days and a solids concentration of 5 percent (see Table 13·7).

Solution

1. Compute the daily quantity of volatile solids based on the assumption that 75 percent of the total solids are volatile.

$$\text{Volatile solids} = \frac{1{,}200 \text{ lb}}{10^6 \text{ gal}} \times 10 \text{ mgd} \times 0.75$$
$$= 9{,}000 \text{ lb/day}$$

2. Compute the digester volume.

$$\text{Volume} = \frac{\dfrac{9{,}000 \text{ lb}}{\text{day}}}{\dfrac{0.238 \text{ lb}}{\text{day-cu ft}}}$$

$= 37{,}800$ cu ft (also volume of first-stage digester in two-stage process)

The degree of stabilization obtained is also often measured by the percent reduction in volatile solids. This can be related either to the mean cell residence time or to the detention time based on the raw sludge feed. Since the raw sludge feed can be measured easily, this method is used more commonly. In plant operation, this calculation should be made routinely as a matter of record whenever sludge is drawn to processing equipment or drying beds. The results of this calculation can also be used as a guide for scheduling the withdrawal of sludge.

In calculating the volatile-solids reduction, it is assumed that the ash content of the sludge is conservative, that is, the number of pounds of ash going into the digester is equal to that being removed. The following example illustrates a typical calculation of volatile-solids reduction.

EXAMPLE 13·5 *Determination of Volatile-Solids Reduction*

Determine the total volatile-solids reduction achieved during digestion from the following analysis of raw and digested sludge. It is assumed that (1) the weight of fixed solids in the digested sludge equals the weight of fixed solids in the raw sludge and (2) that the volatile solids are the only constituent of the raw sludge lost during digestion.

Sludge	Volatile solids, %	Fixed solids, %
Raw	70	30
Digested	50	50

Solution

1. Because the quantity of fixed solids remains the same, the weight of the digested solids based on 1.0 lb of dry raw sludge

is, as computed below, equal to 0.6 lb.

$$\text{Fixed solids raw sludge, } 30\% = \frac{0.3(100)}{0.3 + 0.7}$$

Let x equal weight of volatile solids after digestion. Then

$$\text{Fixed solids digested sludge, } 50\% = \frac{0.3(100)}{0.3 + x}$$

$$x = \frac{0.3(100)}{50} - 0.3 = 0.3 \text{ lb volatile solids}$$

$$\text{Weight of digested solids} = 0.3 + x = 0.6 \text{ lb}$$

2. Therefore, the reduction in total and volatile solids is 40 and 57.2 percent, respectively.

$$\text{Percent reduction of total solids} = \frac{(1.0 - 0.6)100}{1.0} = 40\%$$

$$\text{Percent reduction of volatile solids} = \frac{(0.7 - 0.3)100}{0.7} = 57.2\%$$

Volume Reduction It has been observed that as digestion proceeds, if the supernatant is withdrawn and returned to the head end of the treatment plant, the volume of the remaining sludge decreases approximately exponentially. If a plot is prepared of the remaining volume versus time, the required volume of the digester is represented by the area under the curve. It can be computed using Eq. 13·5.

$$V = [V_f - \tfrac{2}{3}(V_f - V_d)]t \tag{13·5}$$

where V = volume of digester, cu ft
V_f = volume of fresh sludge per day, cu ft
V_d = volume of digested sludge per day, cu ft
t = digestion time, days

The use of Eq. 13·5 in determining digester volume is illustrated in Example 13·6.

EXAMPLE 13·6 *Estimation of Required Digester Volume Using Volume-Reduction Method*

Determine the required digester volume to treat the sludge in Example 13·3 using the volume-reduction method. Assume that 60 percent of the volatile matter is destroyed and that the final sludge has a concentration of 8 percent with a specific gravity of 1.04.

Solution

1. Compute the volume of fresh sludge.

$$V_f = 3{,}770 \text{ cu ft/day} \qquad \text{(see Example 13·3)}$$

2. Compute the volume of digested sludge. Assume the volatile solids content of the sludge is equal to 75 percent.

$$V_d = \frac{3{,}000 + 0.40(9{,}000)}{1.04(62.4)(0.08)} \approx 1{,}275 \text{ cu ft/day}$$

3. Compute the required digester volume assuming a detention time of 25 days at a temperature of 90°F.

$$V = [3{,}770 - \tfrac{2}{3}(3{,}770 - 1{,}275)]25$$
$$= [3{,}770 - 1{,}660]25 = 52{,}800 \text{ cu ft (also volume of first-stage digester in two-stage process)}$$

Population Basis Digestion tanks are also designed on a volumetric basis by allowing a certain number of cubic feet per capita. Detention times of 35 to 45 days are recommended for design based on total tank volume, plus additional storage volume if sludge is dried on beds and weekly sludge drawings are curtailed due to inclement weather.

Based on 0.20 lb of suspended solids per capita in the raw sewage, these requirements translate into the number of cubic feet per capita shown in Table 13·8, which includes also the requirements of the *Ten States Standards* for comparison. These requirements are for heated tanks and are applied where analyses and volumes of sludge to be digested are not available. For unheated tanks, capacities must be in-

TABLE 13·8 DIGESTION-TANK CAPACITY REQUIREMENTS*

	Wet sludge			Volume required	
Type of plant	Dry solids, lb/capita/day	Percent solids	cu ft/ capita/day	35–45 days detention, cu ft/capita	*Ten States Standards,* cu ft/capita
Primary	0.12	5	0.038	1.3–1.7	2–3
Primary + trickling filter	0.18	4	0.072	2.5–3.2	4–5
Primary + activated sludge	0.19	3	0.100	3.5–4.5	4–6

* Based on 0.20 lb of suspended solids per capita per day in raw sewage.

creased, depending on local climatic conditions and storage volume required. The capacities shown in Table 13·8 should be increased 60 percent in a municipality where the use of garbage grinders is universal and should be increased on a population-equivalent basis to allow for the effect of industrial wastes. The sizing of digesters on a population basis is illustrated in Example 13·7.

TABLE 13·9 TIME REQUIRED FOR DIGESTION AT VARIOUS TEMPERATURES* [Adapted from 4]

Item	Mesophilic digestion					Thermophilic digestion				
Temperature, °F	50	60	70	80	90	100	110	120	130	140
Digestion period, days†	75	56	42	30	25	24	26	16	14	18

* Laboratory data for conventional digestion of primary sludge.
† Time to obtain 90 percent of ultimate gas production.

The effect of temperature on the time required for digestion, based on work of Fair and Moore [4], is shown in Table 13·9. The data in this table may be of assistance in the sizing of unheated tanks.

EXAMPLE 13·7 *Estimation of Required Digester Volume Using Volumetric Per Capita Allowance*

Estimate the volume of the digester required to treat the primary sludge in Example 13·3. Assume that the digester is to be heated.

Solution

1. Estimate the contributing population.

$$\text{Population} = \frac{\dfrac{1{,}200 \text{ lb}}{10^6 \text{ gal}} \times 10 \text{ mgd}}{\dfrac{0.12 \text{ lb}}{\text{capita-day}}}$$
$$= 100{,}000 \text{ people}$$

2. Compute the required digester volume.

$$\text{Volume} = 100{,}000 \frac{1.3 \text{ cu ft}}{\text{capita}} \qquad \text{(see Table 13·8)}$$
$$= 130{,}000 \text{ cu ft}$$ (total volume of first- and second-stage digesters in two-stage process)

Comparison of Design Methods The various design methods described were derived under different circumstances. The volume-reduction and population-allowance methods for computing the required tank capacity were developed for conventional digesters and are therefore not directly comparable to the methods used for the design of the newer complete-mix (one complete volume turnover in 30 to 45 min) high-rate digesters. However, where specific codes or standards exist, the prescribed digester volumes will usually be considerably greater than the volumes computed with the above methods.

Gas Production, Collection, and Utilization

Sewage gas contains about 65 to 70 percent CH_4 by volume, 25 to 30 percent CO_2, and small amounts of N_2, H_2, and other gases. It has a specific gravity of approximately 0.86 referred to air. Because production of gas is one of the best measures of the progress of digestion and because it can be used as fuel, the designer should be familiar with its production, collection, and utilization.

Gas Production The volume of methane gas produced during the digestion process can be estimated using Eq. 13·3, which has been discussed previously. For example, the volume of methane gas produced from 1 lb of cells is equal to about 8.0 cu ft. This value is obtained by multiplying the pounds of cells by 1.42 to convert to ultimate BOD and then by multiplying by 5.62 (see Example 13·3) to obtain the cubic feet of methane.

Total gas production is usually estimated from the volatile-solids loading of the digester or from the percentage of volatile-solids reduction. Typical values are from 8 to 12 cu ft/lb of volatile solids added, and from 12 to 18 cu ft/lb of volatile solids destroyed. Gas production can fluctuate over a wide range, depending on the volatile-solids content of the sludge feed and the biological activity in the digester. Excessive gas-production rates sometimes occur during start-up and may cause foaming and escape of foam and gas from around the edges of floating digester covers. If stable operating conditions have been achieved and the foregoing gas-production rates are being maintained, the operator can be assured that the result will be a well-digested sludge.

Gas production can also be estimated on a per capita basis. The normal yield is 0.6 to 0.8 cu ft/capita in primary plants treating normal domestic sewage. In secondary treatment plants this is increased to about 1.0 cu ft/capita.

Gas Collection Floating covers fit on the surface of the digester contents and allow the volume of the digester to change without allowing air to enter the digester. Gas and air must not be allowed to mix, or an explosive mixture may result. Explosions have occurred in sewage treatment plants. Gas piping and pressure-relief valves must include adequate flame traps. Gasometer covers may also be installed to act as gas holders that store a small quantity of gas under pressure and act as reservoirs. This type of cover can be used for single-stage digesters or in the second stage of two-stage digesters.

Fixed covers provide a free space between the roof of the digester and the liquid surface. Gas storage must be provided so that when the liquid volume is changed, gas, and not air, will be drawn into the digester and that gas will not be lost by displacement. Gas can be stored at low pressure in gas holders that use floating covers or at high pressure if gas compressors are used. Gas not used should be burned in a flare. Gas meters should be installed to measure gas produced and gas used or wasted.

Gas Utilization One cubic foot of methane at standard temperature and pressure has a net heating value of 960 Btu. Since digester gas is only 65 percent methane, the low heating value of digester gas is approximately 600 Btu/cu ft. By comparison, natural gas, which is a mixture of methane, propane, and butane, has a low heating value of approximately 1,000 Btu/cu ft.

In large plants, digester gas may be used as fuel for boiler and internal combustion engines which are, in turn, used for pumping sewage, operating blowers, and generating electricity. Hot water from heating boilers or from engine jackets and exhaust-heat boilers may be used for sludge heating and for building heating, or gas-fired sludge-heating boilers may be used.

Process Variations

Along with conventional and high-rate digesters, lagoons and Imhoff tanks (precursors of modern digesters) are also used for the digestion of sludge. Although not used specifically for the treatment of sludge, the anaerobic contact process and anaerobic filter are also discussed briefly in this section.

Sludge Lagoons The sludge lagoon is essentially a large unheated shallow digester. For small plants and where large areas are available, sludge lagooning offers an economical solution to sludge disposal. Lagoons do not permit recovery of methane gas or the continuous removal of digested sludge. When the lagoon becomes filled with digested sludge, it must be either abandoned or drained and the digested sludge excavated.

Imhoff Tanks Imhoff tanks receive primary or primary plus waste activated sludge but are generally not heated. They should be provided with 3 to 4 cu ft of volume per capita for primary treatment, and with 4 to 6 cu ft of volume per capita where biological treatment is included. The capacity is measured below a horizontal plane at the point of slot overlap (approximately 15 in. below the slots) and above planes sloping upward 30° from the end of the sludge withdrawal pipe. Due to the shape of the tank, considerable volume above the sludge is available for supernatant liquor and scum collection that is not included in the computation of the sludge volume. Some early Imhoff tanks were designed for gas collection, but this is no longer recommended (see discussion in Chap. 11).

Anaerobic Contact Process Some industrial wastes that are high in BOD can be stabilized very efficiently by anaerobic treatment. In the anaerobic contact process, raw wastes are mixed with recycled sludge solids and then digested in a mixed digestion chamber. After digestion, the mixture is separated in a clarifier or vacuum flotation unit, and the supernatant is discharged as effluent. Settled sludge is then recycled to seed the incoming waste [10]. Because of the low synthesis rate of anaerobic microorganisms, the excess sludge that must be disposed of is minimal. This process has been used successfully for the stabilization of meat-packing and other high-strength soluble wastes.

Anaerobic Filter A recent development in the field of waste treatment, the anaerobic filter is a column filled with various types of solid media, $1\frac{1}{2}$- to 2-in. rocks being the most common [9]. The waste flows upward through the column contacting the media on which anaerobic bacteria grow and are retained. Because the bacteria are retained on the media and not washed off in the effluent, mean cell residence times on the order of 100 days can be obtained. Because large values of θ_c can be obtained with short hydraulic retention times, the anaerobic filter appears to be applicable to the treatment of low-strength wastes at room temperature. A schematic of an anaerobic filter is shown in Fig. 13·9.

AEROBIC DIGESTION

Aerobic digestion is an alternative method of treating the organic sludges produced from various treatment operations. Aerobic digesters

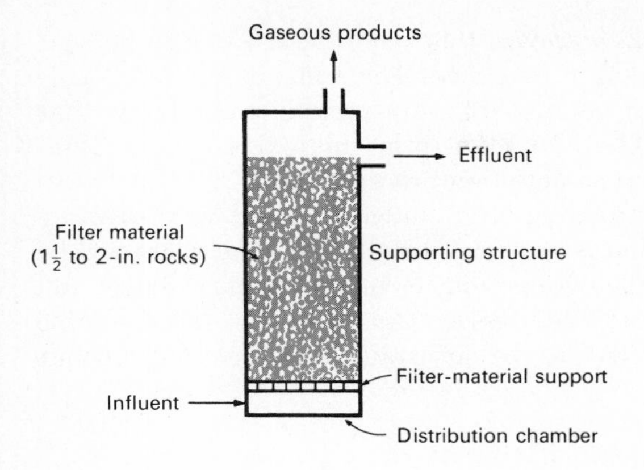

FIG. 13·9 Schematic of an anaerobic filter.

may be used to treat only waste activated sludge, mixtures of waste activated sludge or trickling-filter sludge and primary sludge, or waste sludge from activated-sludge treatment plants designed without primary settling. To date aerobic digestion has been used primarily in small plants, particularly extended aeration and contact stabilization.

Advantages claimed for aerobic digestion as compared to anaerobic digestion include (1) volatile-solids reduction approximately equal to that obtained anaerobically, (2) lower BOD concentrations in supernatant liquor, (3) production of an odorless, humus-like, biologically stable end product that can be disposed of easily, (4) production of a sludge with excellent dewatering characteristics, (5) recovery of more of the basic fertilizer values in the sludge, (6) fewer operational problems, and (7) lower capital cost [1]. The major disadvantage of the aerobic digestion process appears to be the higher power cost associated with supplying the required oxygen. That a useful by-product such as methane is not recovered may also be a disadvantage.

Comparing the advantages and disadvantages, it appears that aerobic digestion is an alternative that should be considered more often. As more reliable information on process kinetics and economics is developed, this method will probably be used more.

Process Description

Aerobic digestion is similar to the activated-sludge process. As the supply of available substrate (food) is depleted, the microorganisms will begin to consume their own protoplasm to obtain energy for cell-maintenance reactions. When this occurs the microorganisms are said to be in the endogenous phase (see Chap. 10). As shown in Eq. 12·7, cell tissue is aerobically oxidized to carbon dioxide, water, and ammonia. It should be noted, however, that only about 75 to 80 percent of the cell tissue can actually be oxidized. The remaining 20 to 25 percent is composed of inert components and organic compounds that are not biodegradable. The ammonia from this oxidation is subsequently oxidized to nitrate as digestion proceeds.

Where activated or trickling-filter sludge is mixed with primary sludge and the combination is to be aerobically digested, there will be both direct oxidation of the organic matter in the primary sludge and endogenous oxidation of the cell tissue. Operationally, most aerobic digesters can be considered to be arbitrary-flow reactors without recycle.

Process Design

Factors that must be considered in designing aerobic digesters include: hydraulic residence time, process loading criteria, oxygen requirements, energy requirements for mixing, environmental conditions, and process operation. Although there is currently little information on these factors, the following discussion is meant to serve as an introduction to the subject. The design of an aerobic digester is illustrated at the end of this section.

Hydraulic Residence Time Limited experimental data show that the amount of volatile solids in sludge is reduced more or less linearly up to a value of about 40 percent at a hydraulic detention time of about 10 to 12 days [3]. Although volatile-solids removal continues with increasing detention time, the rate of removal is reduced considerably. Depending on the temperature, the maximum reduction will range between 45 and 70 percent. The required time and degree of volatile-solids removal will also vary with the characteristics of the sludge. Typically, volatile-solids reductions will vary from about 35 to 45 percent in 10 to 12 days at temperatures equal to or above 20°C. Based on these findings, recommended hydraulic detention times for aerobic digesters are given in Table 13·10. Temperature effects are discussed under environmental conditions.

Loading Criteria Limited information is currently available on appropriate loading criteria to use for this process. Typical values in terms of pounds of volatile solids per cubic foot per day are given in Table 13·10. Because the hydraulic and mean cell residence times are nominally equivalent for this process (see "Process Operation"), loading criteria based on mean cell residence time would appear to be

TABLE 13·10 DESIGN CRITERIA FOR AEROBIC DIGESTERS

Parameter	Value
Hydraulic detention time, days at 20°C*	
Activated sludge only	12–16
Activated sludge from plant operated without primary settling	16–18
Primary plus activated or trickling-filter sludge	18–22
Solids loading, lb volatile solids/cu ft/day	0.1–0.20
Oxygen requirements, lb/lb destroyed	
Cell tissue†	≃2
BOD_5 in primary sludge	1.6–1·9
Energy requirements for mixing	
Mechanical aerators, hp/1,000 cu ft	0.5–1.0
Air mixing, scfm/1,000 cu ft	20–30
Dissolved oxygen level in liquid, mg/liter	1–2

* Detention times should be increased for temperatures below 20°C. If sludge cannot be withdrawn during certain periods (e.g., weekends, rainy weather) additional storage capacity should be provided.
† Ammonia produced during carbonaceous oxidation oxidized to nitrate.

most satisfactory. The maximum solids concentration would be governed by oxygen transfer and mixing requirements.

If only waste activated sludge is to be aerobically digested, the mean cell residence time required to achieve a given volatile-solids reduction can be estimated using Eq. 12·5 and Fig. 12·6 or Eq. 12·2. Representative values for the decay coefficient k may be about 0.05 to 0.07/day.

Oxygen Requirements The oxygen requirements that must be satisfied during aerobic digestion are those of the cell tissue and, with mixed sludges, the BOD_5 in the primary sludge. The oxygen requirement for cell tissue, computed using Eq. 12·7 and assuming that the ammonia produced is oxidized to nitrate (see Example 7·4), is equal to 7 moles per mole of cells or about 2 lb per lb of cells. The oxygen requirement for the complete oxidation of the BOD_5 contained in primary sludge will vary from about 1.7 to 1.9 lb per lb destroyed.

On the basis of operating experience it has been found that if the dissolved oxygen concentration in the digester is maintained at 1 to 2 mg/liter and the detention time is greater than 10 days, the sludge dewaters well [3].

Energy Requirements for Mixing To ensure proper operation, the contents of the aerobic digester should be well mixed. In general, because of the amount of air that must be supplied to meet the oxygen requirement, mixing should be achieved; nevertheless, horsepower mixing requirements should be checked (see Table 13·10).

Environmental Conditions Of the many environmental factors considered in Chaps. 10 and 12, temperature and pH will play an important role in the operation of aerobic digesters. It has been observed that the operation of aerobic digesters is temperature dependent, especially at temperatures below 20°C. On the basis of extremely limited data it appears that a temperature coefficient in the range of 1.08 to 1.10 might be appropriate for adjusting the hydraulic detention time for temperatures below 20°C for hydraulic residence times on the order of 15 days. As the hydraulic detention time is increased to about 60 days, the effect of temperature is negligible. In extremely cold climates consideration should be given to heating the sludge or the air supply and/or covering the tanks.

Depending on the buffering capacity of the system, the pH may drop to a rather low value (5.5 ±) at long hydraulic detention times. Reasons advanced for this include the increased presence of nitrate ions in solution and the lowering of the buffering capacity due to air stripping [3]. Although this does not seem to inhibit the process, the pH should be checked periodically and adjusted if found to be excessively low.

Process Operation Aerobic digestion is normally conducted in unheated tanks similar to those used in the activated-sludge process. In some cases, existing anaerobic digesters have been converted and are being used as aerobic digesters. Aerobic digesters should be equipped with decanting facilities so that they may also be used to thicken the digested solids before discharging them to subsequent thickening facilities or sludge-drying beds. If the digester is operated so that the incoming sludge is used to displace supernatant and the solids are allowed to build up, the mean cell residence time will not be equal to the hydraulic residence time.

EXAMPLE 13·8 *Aerobic Digester Design*

Design an aerobic digester to treat the waste sludge produced by the activated-sludge treatment plant in Example 12·1. Assume that the waste sludge is concentrated using a mechanical thickener to 3 percent (see Table 13·3). Assume that the temperature is 20°C, and use a hydraulic detention time of 15 days.

Solution

1. Compute the volume of sludge to be disposed of per day assuming a specific gravity of 1.03.

$$V = \frac{4{,}125 \text{ lb}}{1.03(62.4)(0.03)} = 2{,}140 \text{ cu ft}$$

2. Determine the volume of the aerobic digester.

$$V = 2{,}140 \times 15 \text{ days} = 32{,}100 \text{ cu ft}$$

3. Check the solids loading.

$$\text{lb volatile solids/cu ft/day} = \frac{3{,}300}{32{,}100} = 0.103$$

4. Determine the oxygen requirement, assuming that 40 percent of the cell tissue is oxidized completely.

$$\text{lb } O_2\text{/day} = 3{,}300\ (0.4)(2.0) = 2{,}640 \text{ lb/day}$$

5. Compute the volume of air required at standard conditions (see Example 12·1).

$$\frac{2{,}640}{0.0750(0.232)} = 151{,}500 \text{ cu ft/day}$$

Assuming that the oxygen transfer efficiency is 10 percent, the air requirement in cfm is

$$\frac{151{,}500}{0.10(1{,}440)} \simeq 1{,}050 \text{ cfm}$$

6. Compute the air requirement per 1,000 cu ft of digester volume.

$$\frac{1{,}050}{32.1} = 32.7 \text{ cfm/1,000 cu ft}$$

7. Check the mixing requirement. Because the air requirement computed in step 6 is greater than the maximum value given in Table 13·10, adequate mixing should prevail.

SLUDGE CONDITIONING

Conditioning of sludge is performed for the express purpose of improving its dewatering characteristics. The two methods most commonly used involve the addition of chemicals and heat treatment. Freezing and irradiation have also been investigated. Elutriation, a physical washing operation, is employed to reduce conditioning-chemical requirements.

Elutriation

Elutriation, which involves mixing of digested sludge with water and resettling, does not improve dewatering characteristics, but it does reduce the requirement for coagulating chemicals. The initial and operating cost for providing elutriation must therefore be justified by the resulting savings in chemical costs. The theoretical aspects of elutriation are discussed in Chap. 8.

Batch or Continuous Operations Elutriation can be performed as a batch or a continuous operation. The batch process permits close control and requires a minimum of equipment. The continuous process provides more flexibility and can be easily operated in stages, but requires more equipment. Plant effluent is commonly used for the diluting water, but well water or river water may also be used.

Tank Sizing The sizing of the elutriation tank is dependent on several factors [13]:

1. Nature of the sludge solids before digestion (primary, activated, or mixed)
2. Percentage of volatiles (low-volatile sludges are denser)
3. Percentage of solids in sludge (governs volume to be treated)
4. Method of elutriation (batch versus countercurrent)
5. Elutriation ratio (governs volume of water to remove and size of tank for adequate settling)
6. Operating schedule (sludge-storage volume required for filter operation)

Sludge and water are mixed in a tank with slow agitation. The time required for mixing is short (less than 1 min), but 3 to 4 hr are required for sludge resettling and densification. In circular tanks, thickening paddles and collectors are used to promote densification and to transport the sludge to the discharge sump. Rectangular tanks with fully submerged sludge collectors are also used. Surface loadings must be selected so that excessive fine particles will not be lost in the wash water. Approximately 600 gpd/sq ft is the maximum that can be tolerated with digested mixed primary plus activated sludge.

Chemical Conditioning

The use of chemicals to condition sludge for dewatering is economical because of the increased yields and greater flexibility obtained. Chemical conditioning results in coagulation of the solids and release of the absorbed water. Conditioning is used in advance of vacuum filtration and centrifugation. Chemicals used include ferric chloride, lime, alum, and organic polymers.

Chemicals are most easily applied and metered in the liquid form. Dissolving tanks are needed if the chemicals are received as dry powder. These tanks should be large enough for at least one day's supply of chemicals and should be furnished in duplicate. They must be fabricated or lined with corrosion-resistant material. PVC, polyethylene, and rubber are suitable materials for tank and pipe linings for acid solutions. Metering pumps must be corrosion resistant. These pumps are generally of the positive-displacement type with variable-speed or variable-stroke drives to control the flowrate. Another metering system consists of a constant-head tank supplied by a centrifugal pump. A rotameter and throttling valve are used to meter the flow.

Dosage The chemical dosage required for any sludge is determined in the laboratory. Filter-leaf test kits are available for determining chemical doses, filter yields, and the suitability of various filtering media. These kits have several advantages over the Büchner funnel procedure [13]. Typical dosages are presented in Table 13·11. Actual dosages in any given case may vary considerably from these figures.

TABLE 13·11 DOSAGE OF CHEMICALS FOR VARIOUS TYPES OF SLUDGES [13]
(Conditioners in Percentage of Dry Sludge)

	Fresh solids		Digested		Elutriated digested	
Description	$FeCl_2$	CaO	$FeCl_2$	CaO	$FeCl_2$	CaO
Primary	1–2	6–8	1.5–3.5	6–10	2–4	..
Primary and trickling filter	2–3	6–8	1.5–3.5	6–10	2–4	..
Primary and activated	1.5–2.5	7–9	1.5–4	6–12	2–4	..
Activated (alone)	4–6	..	..	..	..	..

Sludge Mixing Intimate admixing of sludge and coagulant is essential for proper conditioning. The mixing must not break the floc after it has formed, and the detention is kept to a minimum so that sludge reaches the filter as soon after conditioning as possible. Mixing tanks are generally of the vertical type for small plants and of the horizontal type for larger plants. They are ordinarily built of welded steel and lined with rubber or other acid-proof coating. A typical layout for a mixing or conditioning tank provides a horizontal agitator driven by a variable-speed motor to provide a shaft speed of 4 to 10 rpm. Overflow from the tank is adjustable to vary the detention period. Vertical cylindrical tanks with propeller mixers are also used.

Heat Treatment

Heat treatment is a conditioning process that involves heating the sludge for short periods of time under pressure. Heat treatment results in coagulation of the solids, breakdown of the gel structure, and a reduction of the water affinity of sludge solids. As a result the sludge is sterilized, practically deodorized, and is dewatered readily on vacuum filters or filter presses without the addition of chemicals [1]. Processes used for heat treatment include the Porteus and the low-pressure Zimpro system (see Fig. 9·13).

In the Porteus process sludge is preheated by passing it through a heat exchanger before it enters the reactor vessel. Steam is injected into the vessel to bring the temperature to within the range of 290 to 390°F under pressures of 150 to 200 psi. After a 30-min detention in the reactor the sludge is discharged through the heat exchanger and into a decant tank. The thickened sludge can be filtered to a solids content of 40 to 50 percent. Filter yields up to 20 lb/sq ft/hr have been obtained.

In the low-pressure Zimpro system the sludge is treated as in the Porteus process except that air is injected into the reactor vessel with the sludge. The reactor vessel is heated by steam to temperatures in the range of 300 to 400°F under pressures varying from 150 to 300 psi. Heat released during oxidation increases the operating temperature to a range of 350 to 600°F. The partially oxidized sludge may be dewatered by filtration, centrifugation, or drainage on beds. Solids content of the dewatered sludge can range from 30 to 50 percent depending on the degree of oxidation desired. Essentially complete oxidation of volatile solids (approximately 90 percent reduction) can be accomplished with higher pressures and temperatures (see incineration and wet oxidation).

The supernatant or filtrate liquid obtained from both of these processes contains high concentrations of short-chain water-soluble organic compounds that are amenable to biological treatment. This liquid can be either returned to the main sewage flow for biological treatment (if the treatment plant has been designed to handle this added load) or treated separately.

Other Processes

Freezing and irradiation have also been investigated as sludge conditioning methods. Laboratory investigations indicate that freezing of sludge is more effective than chemical conditioning in improving sludge filterability. Much remains to be done, however, before this method can be applied effectively. Although irradiation has been shown to be effective in improving sludge filterability, it is not considered economically competitive at present (1971).

SLUDGE DEWATERING AND DRYING

Dewatering and drying are physical (mechanical) unit operations used to reduce the moisture content of sludge so that it can be handled and processed as a semisolid instead of as a liquid.

Dewatering

Methods commonly used for dewatering sludge include spreading on drying beds, vacuum filtration, centrifugation, and pressure filtration. Mechanical or sonic vibration has also been used. The choice among these methods depends on the characteristics of the sludge, the method of final disposal, the availability of land, and the economics involved.

Drying Beds Sludge drying beds are used to dewater digested sludge. Sludge is placed on the beds in an 8- to 12-in. layer and allowed to dry. After drying, the sludge is removed and disposed of in a landfill, or ground for use as a fertilizer. A typical sludge drying bed is shown in Fig. 13·10. The economical use of sludge drying beds is generally limited to small and medium-sized communities. For populations over 20,000, consideration should be given to alternative means of sludge dewatering. The initial cost, the cost of removing the sludge and replacing sand, and the large area requirements preclude the use of drying beds in large municipalities.

The drying area is partitioned into individual beds, approximately 20 ft wide by 20 to 100 ft long, of a convenient size so that one or two beds will be filled by a normal withdrawal of sludge from the digesters. The interior partitions commonly consist of two or three creosoted planks, one on top of the other, to a height of 15 to 18 in. stretching between slots in precast concrete posts. The outer boundaries may be of similar construction or earthen embankments for open beds, but concrete foundation walls are required if the beds are to be covered.

Open beds are used where adequate area is available sufficiently

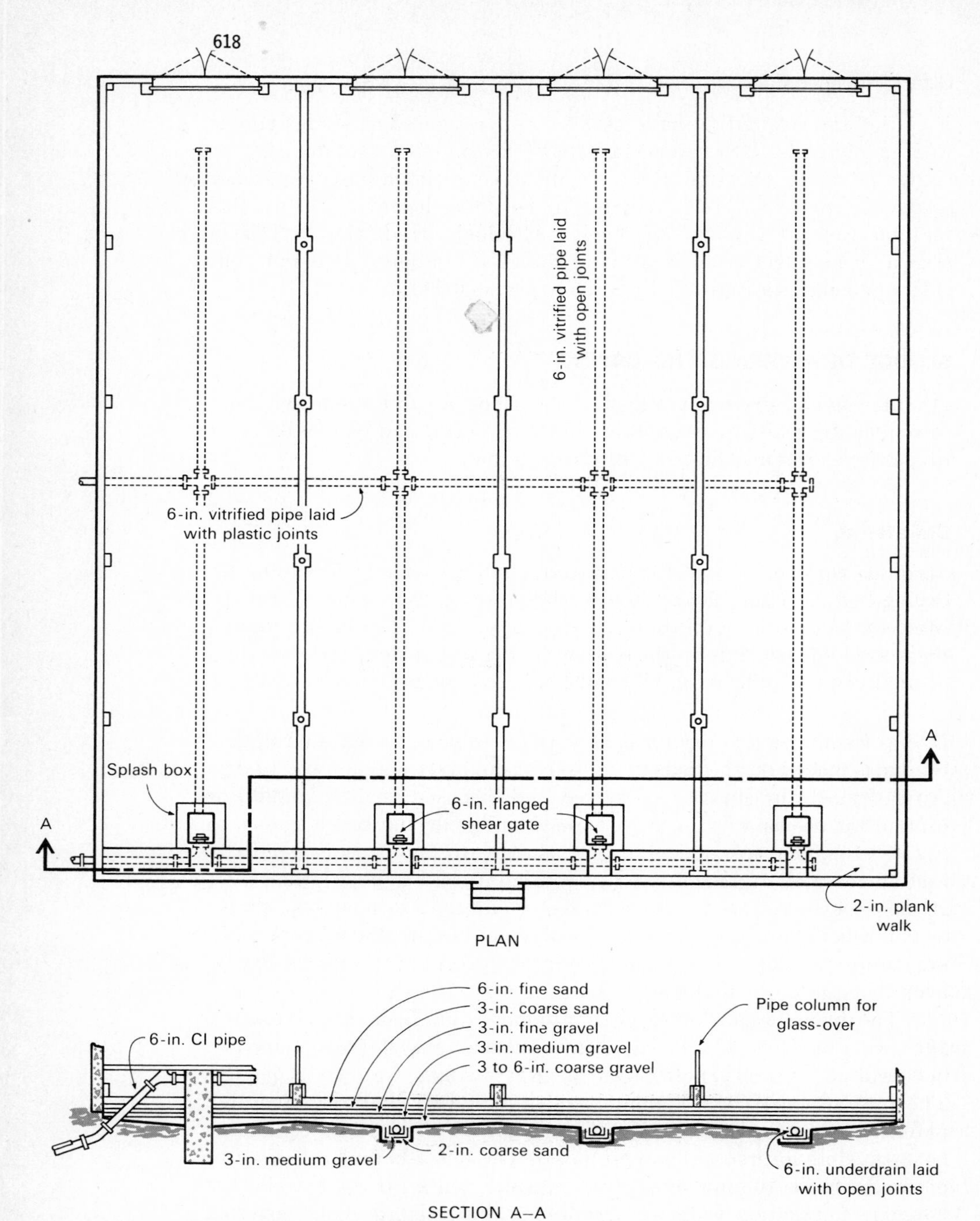

FIG. 13·10 Plan and section of a typical sludge drying bed.

isolated to avoid complaints caused by occasional odors. Covered beds with green-house types of enclosures are used where it is necessary to dewater sludge continuously throughout the year regardless of the weather, and where sufficient isolation does not exist for the installation of open beds. Well-digested sludge discharged to drying beds should present no odor problem, but to avoid nuisance from poorly digested sludge, sludge beds should be located at least 200 ft from dwellings.

Sludge-bed loadings are computed on a per capita basis or on a unit loading of pounds of dry solids per square foot per year. Typical data for various types of sludge are shown in Table 13·12. Typical

TABLE 13·12 AREA REQUIRED FOR SLUDGE DRYING BEDS IN THE NORTHERN UNITED STATES [13]
(Square feet per capita)

Type of sludge	Open beds	Covered beds
Primary digested	1.0–1.5	0.75–1.0
Primary and humus digested	1.25–1.75	1.0–1.25
Primary and activated digested	1.75–2.5	1.25–1.5
Primary and chemically precipitated digested	2.0–2.5	1.25–1.5

solids-loading rates vary from 10 to 25 lb/sq ft/year for open to 12 to 40 lb/sq ft/year for covered drying beds. With covered drying beds, more sludge drawings per year can be accommodated because of the protection from rain and snow.

Sludge dewaters by drainage through the sludge mass and supporting sand and by evaporation from the surface exposed to the air. Most of the water leaves the sludge by drainage; thus the provision of an adequate underdrainage system is essential. Drying beds are equipped with lateral drainage tiles (vitrified-clay pipe laid with open joints) spaced 8 to 20 ft apart. The tiles should be adequately supported and covered with coarse gravel or crushed stone (see Fig. 13·10). The sand layer should be from 9 to 12 in. deep with an allowance for some loss from cleaning operations. Deep sand layers will retard the draining process. Sand should have a uniformity coefficient of not over 4.0 and an effective size of 0.3 to 0.75 mm.

Piping to the sludge beds should drain to the beds and be designed for a velocity of at least 2.5 fps. Cast-iron pipe is frequently used. Arrangements should be made to flush the lines if necessary and to prevent their freezing in cold climates. Distribution boxes are required to divert the sludge flow into the bed selected. Splash plates are placed in front of the sludge outlets to spread the sludge over the bed and to prevent erosion of the sand.

Sludge can be removed from the drying bed after it has drained and dried sufficiently to be spadable. Dried sludge has a coarse, cracked surface and is black or dark brown. The moisture content is approximately 60 percent after 10 to 15 days under favorable conditions. Sludge removal is accomplished by manual shoveling into wheelbarrows or trucks or by a scraper or front-end loader. Provisions should be made for driving a truck onto or along the bed to facilitate loading.

Vacuum Filtration Vacuum filtration is probably the most widely used method of dewatering sludge by mechanical means. A uniformly dewatered sludge cake can be produced from either raw or digested sludge at any time, regardless of weather conditions. Vacuum filters and their operation are described in Chap. 8. The dewatered sludge may be hauled away for ultimate disposal, sold, or given away for use as a soil conditioner or low-grade fertilizer, or subjected to heat drying and/or incineration. The filtrate contains a high concentration of fine suspended solids and is discharged to the sludge elutriation tank or mixed with the raw sewage entering the plant.

Vacuum filters are made by several manufacturers with surface areas ranging from 50 to more than 300 sq ft and can be equipped with various types of filtering cloth. Filter cloths made of cotton, wool, nylon, dacron, and other synthetic materials are available in a variety of weaves having different porosities. Cloths made of woven stainless-steel wire are sometimes used. A unique filter media manufactured by the Komline-Sanderson Co. consists of stainless-steel springs that are wrapped around the filter drum in a double layer.

Conditioning of wet sludges is necessary to achieve satisfactory yields from vacuum filters. Conditioning coagulates the sludge particles and allows the water to drain freely. As a result, a thicker filter cake is produced and the drum can be rotated at a higher speed.

The number and size of filters is based on the type of sludge to be filtered and the number of hours of operation. At small plants, 30 hr/week may be assumed, whereas at large plants, 20 hr/day may be necessary. The additional hours in the day are used for conditioning, clean-up, and possible delays. A plant may be designed for one-shift operation initially, and for two- to three-shift operation of the same filters when the plant is expanded to provide for future or ultimate conditions.

The performance of a vacuum filter is measured in terms of the yield of solids on a dry weight basis expressed as pounds per square foot per hour. The quality of the filter cake is measured by its moisture content on a wet-weight basis expressed as a percent. Filters are operated to obtain the maximum production consistent with the desired cake quality. Where the cake is to be heat dried or incinerated, the moisture content is a critical item, since all the water remaining in the cake must be evaporated to steam. If the cake is conveyed into a truck and hauled to a disposal site, moisture content is not as important, although it does affect the tonnage that must be hauled. In such cases, the drum can be operated at the highest speed that will produce a cake that will separate easily from the filter. Moisture content normally varies from 70 to 80 percent, but filters may be operated to produce a cake of 60 to 70 percent moisture when the cake is to be heat dried or incinerated. Typical yields are shown in Table 13·13. A design rate of

TABLE 13·13 EXPECTED PERFORMANCE OF VACUUM FILTERS HANDLING PROPERLY CONDITIONED SLUDGE [13]

Type of sludge	Yield, lb/sq ft/hr
Fresh solids:	
Primary	4–12
Primary + trickling filter	4–8
Primary + activated	4–5
Activated (alone)	2.5–3.5
Digested solids (with or without elutriation):	
Primary	4–8
Primary + trickling filter	4–5
Primary + activated	4–5

3.5 lb/sq ft/hr is frequently used when the quality of the sludge must be estimated and the type of filter to be furnished is based on an open specification.

Centrifugation The centrifugation process is widely used in industry for separating liquids of different density and for thickening slurries or for removing solids. The process is applicable to the dewatering of sewage sludges and has been used with varying degrees of success in both the United States and Europe. The machines used in sewage treatment plants are of the solid-bowl type with electric-motor drive. Sludge is fed into the rotating bowl at a constant flowrate where it separates into a dense cake containing the solids and a dilute stream called centrate. The centrate contains fine, low-density solids, and is returned to the raw-sludge thickener or primary clarifier. The sludge cake containing approximately 75 to 80 percent moisture is discharged from the bowl by a screw feeder into a hopper or onto a conveyor belt. Depending on the type of sludge, solids concentration in the cake will vary from 15 to 40 percent, but reductions below 25 percent are not usually feasible economically. The cake can then be disposed of by incineration or by hauling to a sanitary landfill.

The operation of centrifuges is simple, clean, and relatively inexpensive; and it normally does not require chemical conditioning. Special consideration must be given to providing sturdy foundations and soundproofing because of the vibration and noise that result from centrifuge operation. Adequate electric power must also be provided since large motors are required.

The major difficulty encountered in the operation of centrifuges has been the disposal of the centrate, which is relatively high in suspended, nonsettling solids. The return of these solids to the sewage treatment units could result in a large recirculating load of these fine solids through the sludge and primary settling system and in reduced effluent quality. Two methods can be used to control the fine-solids discharge and to increase the capture. Longer residence of the liquid stream is accomplished by reducing the feed rate or by using a centrifuge with a larger bowl volume. Particle size can be increased by coagulating the sludge prior to centrifugation with ferric chloride and lime or organic polymers. Solids capture may be increased from a range of 50 to 80 percent to a range of 80 to 95 percent of influent solids.

The addition of lime will also aid in the control of odors that may develop when centrifuging raw sludge. Raw primary sludge can usually be dewatered to a lower moisture content than digested sludge, because it has not been subjected to the liquefying action of the digestion process, which reduces particle size. Some type of chemical conditioning is usually desirable when dewatering combined primary and waste activated sludge, regardless of whether or not it has been digested.

The area required for a centrifuge installation is less than that required for a vacuum filter of equal capacity, and the initial cost is lower. Higher power costs will partially offset the lower initial cost.

Selection of units for plant design is dependent on manufacturer's rating and performance data. Several manufacturers have portable pilot plant units, which can be used for field testing if sludge is available. Wastewater sludges from supposedly similar treatment processes but in different localities may differ markedly from each other. For this reason, pilot plant tests should be run whenever possible before final design decisions are made.

Pressure Filtration A variety of different types of filter presses have been used to dewater sludge. One type used for sludge dewatering consists of a series of rectangular plates, recessed on both sides, that are supported face to face in a vertical position on a frame with a fixed and movable head. A filter cloth is hung or fitted over each plate. The plates are held together with sufficient force to seal them to withstand the pressure applied during the filtration process. Hydraulic rams or powered screws are used to hold the plates together.

In operation, chemically conditioned sludge is pumped into the space between the plates and pressure (60 to 180 psi) is applied and maintained for 1 to 3 hr, forcing the liquid through the filter cloth and plate outlet ports. The plates are then separated and the sludge is removed. The filtrate normally is returned to the headworks of the treatment plant. The sludge cake thickness will vary from about 1 to $1\frac{1}{2}$ in. and the moisture content will vary from 55 to 70 percent. The filtration cycle time will vary from 3 to 8 hr and includes the time required to fill the press, the time the press is under pressure, the time to open the press, the time required to wash and discharge the cake, and the time required to close the press. To reduce the amount of labor to a minimum most modern presses are mechanized.

The most significant costs associated with this method of dewatering are those for chemical conditioning and maintenance and replacement of filter cloths. Although not commonly used in the United States, pressure filtration has been used extensively in England and other parts of the world.

Vibration Dewatering of sludges and screenings by mechanical or sonic vibration has been investigated using systems ranging from a simple vibrating screen to more complex systems consisting of a coarse screen, a sonic screen, a sonic filter, and a roller press. Sludges with a solids content of 35 to 40 percent have been produced using such systems; however, dewatering efficiency is largely dependent on the characteristics of the material.

Heat Drying

The purpose of heat drying is to remove the moisture from the wet sludge so that it can be incinerated efficiently or processed into fertilizer. Drying is necessary in fertilizer manufacture to permit grinding of the sludge, to reduce its weight, and to prevent continued biological action. The moisture content of the dried sludge is less than 10 percent. For heat drying of sludge, the C. E. Raymond Flash Drying System is most frequently used (see Fig. 8·20 and discussion in following section). Multiple hearth incinerators and rotary kiln driers are also used for heat drying of sludge (see Chap. 8). Spray drying equipment has been in industrial use for many years but its use for drying sewage sludge has been extremely limited.

In the drum or kiln dryer, wet sludge enters at one end of a slowly rotating drum, through which hot gases are passed. The sludge solids give up their moisture to the gases as they are agitated and transported through the drum. The dried solids emerge at the opposite end of the drum where the gases enter, thus employing the countercurrent flow principle. This type of dryer may be fuel fired, or it may use the exhaust gases from an incinerator to provide the heat. The gases that emerge from the drum contain fine sludge particles, which must be removed with a cyclone or scrubber.

The two most important control measures associated with heat drying of sludge are fly-ash collection and odor control. Cyclone separators having efficiencies of 75 to 80 percent are suitable for vent gas temperatures up to 650 or 700°F. Wet scrubbers have higher efficiencies and will condense some of the organic matter in the vent gas, but may carry over water droplets.

Sludge drying occurs at temperatures of approximately 700°F, whereas 1200 to 1400°F is required for complete incineration. To achieve destruction of odors, the exhaust gases must reach approximately 1350°F. Thus, if the gases evolved in the drying process are reheated in an incinerator to a minimum of 1350°F, odors will be eliminated. At lower temperatures, partial oxidation of odor-producing compounds may occur, resulting in an increase in the intensity or disagreeable character of odor produced.

INCINERATION AND WET OXIDATION

Incineration and wet oxidation are used to reduce the organic content and the volume of sludge (see Table 13·13). They are often incorrectly referred to as disposal methods. However, depending on the method of operation, both produce an ash or sludge requiring ultimate disposal.

Incineration

The incineration process is a natural extension of the drying process, converting the sludge into an inert ash, which can be disposed of easily. With adequate dewatering (to approximately 30 percent solids), the process is usually self-sustaining without the need for supplemental

TABLE 13·14 SLUDGE VOLUME REDUCTION AND DISPOSAL METHODS

Description	Remarks
	Volume reduction methods
Incineration	Concentration of sludge is needed. Ash must be disposed of.
Wet combustion	Heat value may be recovered for use. Ash must be disposed of.
Product recovery	Depends on waste characteristics and recovery technology and costs.
	Disposal methods
Spreading on soil	Sludge may be pretreated to aid dewatering or to remove objectionable components.
Lagooning	Provisions must be made to prevent ground water contamination.
Dumping	Can be used only with completely stabilized sludge or inert solids such as grit.
Landfill	Mixed with refuse to increase compacted density of landfill.
Ocean discharge	May not be allowed in the future.

fuel, except for initial warm-up and heat control. Where incineration is used for volume reduction of refuse and other solid wastes having a low moisture content, facilities for volume reduction of sewage sludge can be conveniently incorporated. If incineration is to be used only for sludge volume reduction, it may be preferable to incinerate raw rather than digested sludge because of its higher Btu value.

Multiple Hearth The multiple hearth furnace is one of the most successful devices available and can be used for sludge drying as well as incineration. The furnace is a circular steel cylinder containing several hearths arranged in a vertical stack (see Fig. 8·22). The operation is described in Chap. 8 (see "Drying").

In designing an incinerator for sludge volume reduction, a detailed heat balance must be prepared (see Example 9·4). Approximately 1,800 to 2,500 Btu will be required to evaporate each pound of water in the sludge. Heat is obtained from the combustion of the volatile matter in the sludge and from the burning of auxiliary fuels. For raw primary sludge incineration, the auxiliary fuel is needed only for warming up the incinerator and for maintaining the desired temperature when the volatile content of the sludge is low. Fuels such as oil, natural gas, or excess digester gas are suitable. Raw sludge has a heat content ranging from 6,500 to 9,500 Btu/lb of dry solids, while digested sludge has a heat content ranging from 2,500 to 5,500 Btu/lb.

Flash-Drying System The C. E. Raymond flash-drying system (see Fig. 8·20) can be used for either sludge drying or incineration and can do both simultaneously. A portion of the dried material is mixed with the incoming sludge cake, and the mixture is dried in a stream of hot gases. After separation from the gas stream in a cyclone, the dried sludge is divided, part going to the mixer, part to the furnace for incineration, and/or if desired, part discharged as dried sludge. The vapor from the cyclone is returned to the furnace for deodorization.

Fluidized Bed The Dorr-Oliver FS (Fluo-Solids) system utilizes a fluidized bed of sand as a heat reservoir to promote uniform combustion of sludge solids. The sludge is dewatered with a centrifuge or with a vacuum filter before it enters the fluidized-bed reactor. The fluidized bed must be preheated, using fuel oil or gas, to approximately 1200°F before the sludge is introduced. Inside the reactor, the sludge is dried and oxidized at a temperature of 1500°F. The resultant combustion gases, ash, and water vapor exit through a wet scrubber, where the ash is removed, and are then exhausted through a stack. The ash is separated from the scrubber water in a cyclone separator. The scrubber water is discharged with the plant effluent. The combustion process is controlled by varying the sludge feed rate and the air flow to the reactor to oxidize completely all of the organic material and to eliminate the need for sludge digestion. If the process is operated continuously or with shutdowns of short duration on raw sludge, there is no need for auxiliary fuel after start-up.

Dorr-Oliver has also developed a fuel-fired, horizontal, spiral-flow incinerator especially adapted to the needs of small sewage plants. In this process, thickened sludge is atomized as it enters the combustion chamber by a strong blast of compressed air. The sludge particles are oxidized rapidly and exit as ash with the combustion gases. The ash is then separated and disposed of in the same manner as with the Fluo-Solids process.

Wet Oxidation

The Zimmerman (see Fig. 9·13) process involves wet oxidation of raw sludge at an elevated temperature and pressure. The process is the same as that discussed under heat treatment except that higher pressures and temperatures are required to oxidize the volatile solids more completely. Raw sludge is ground and mixed with a specified quantity of compressed air. The mixture is pumped through a series of heat exchangers and then enters a reactor, which is pressurized to keep the water in the liquid phase at the reactor operating temperature of 350 to 600°F. High-pressure units can be designed to operate at pressures up to 3000 psig. A mixture of gases, liquid, and ash leave the reactor.

The liquid and ash are returned through heat exchangers to heat the incoming sludge and then pass out of the system through a pressure-reducing valve. Gases released by the pressure drop are separated in a cyclone and released to the atmosphere. In large installations, it may be economical to expand the gases through a turbine to recover power. The liquid and stabilized solids are cooled by passing through a heat exchanger and are then separated in a lagoon or settling tank or on sand beds. The liquid is returned to the primary settling tank and the solids are disposed of by landfill. The process can be designed to be thermally self-sufficient when utilizing raw sludge. When additional heat is needed steam is injected into the reactor vessel.

FINAL SLUDGE AND ASH DISPOSAL

The solids removed as sludge from preliminary and biological treatment are concentrated and stabilized by biological and thermal means and are reduced in volume in preparation for final disposal. The method of final disposal determines the type of stabilization and the amount of volume reduction that is needed. At present, as shown in Table 13·14, two possibilities are available for the ultimate disposal of sludge and ash: (1) disposal on land and (2) disposal in the sea.

Land Disposal

The most common methods of land disposal include spreading on soil, lagooning, dumping, and landfill.

Spreading on Soil Wet digested sewage sludge may be disposed of by spreading over farm lands and plowing under after it has dried. The humus in the sludge conditions the soil, improving its moisture retentiveness. Several cities have disposed of dried sludge by bagging it and selling it as a soil conditioner. The digested sludge may be heat dried, ground in a mill, and fortified with nitrogen to give it some fertilizer value. Air-dried sludge may also be sold or given away, for use as a soil conditioner, but the demand is usually seasonal. The major problem in disposing of sludge in this manner lies in the economical marketing of the product. Heat-dried activated sludge is sold as fertilizer by several large cities, notably Chicago, Milwaukee, and Houston. The Milwaukee sludge is distributed nationally under the trade name Milorganite.

Lagooning Lagooning of sludge is another popular disposal method because it is simple and economical if the treatment plant is in a remote location. A lagoon is an earth basin into which raw or digested sludge is deposited. Raw-sludge lagoons stabilize the organic solids by anaerobic and aerobic decomposition, which may give rise to objectionable odors. The stabilized solids settle to the bottom of the lagoon and accumulate. Excess liquid from the lagoon, if there is any, is returned to the plant for treatment. Lagoons should be located away from highways and dwellings to minimize possible nuisance conditions and should be fenced to keep out unauthorized persons. They should be relatively shallow (4 to 5 ft) if they are to be cleaned by scraping. If the lagoon is used only for digested sludge, the nuisances mentioned should not be a problem. Sludge may be stored indefinitely in a lagoon, or it may be removed periodically after draining and drying.

Dumping Dumping, such as in an abandoned mine quarry, is a suitable disposal method only for sludges that have been stabilized so that no decomposition or nuisance conditions will result. Digested sludge, clean grit, and incinerator residue can be disposed of safely by this method.

Landfill A sanitary landfill can be used for disposal of sludge, grease, and grit whether it is stabilized or not, if a suitable site is convenient. The economies of hauling sludge will usually indicate that dewatering for volume reduction will result in justifiable savings. The sanitary landfill method is most suitable if it is also used for disposal of the refuse and other solid wastes of the community. In a true sanitary landfill, the wastes are deposited in a designated area, compacted in place with a tractor or roller, and covered with a 12-in. layer of clean soil. With daily coverage of the newly deposited wastes, nuisance conditions, such as odors and flies, are minimized.

After several years' time, during which the wastes are decomposed and compacted, the land can be used for recreational or other purposes where gradual subsidence would not be objectionable. In selecting a site for a dump or landfill, consideration must be given to the nuisance and health hazards that they may cause. Trucks carrying wet sludge and grit should be able to reach the site without passing through heavily populated areas or business districts. The site should have good drainage so that runoff would not create boggy conditions that would interfere with vehicular movement. Drainage from the site that would cause pollution of ground water supplies or surface streams must also be guarded against.

Ocean Disposal

Disposal of digested sludge at sea is used by many large coastal cities. The sludge may be carried offshore in barges or sludge vessels (New York City) and dumped, or it may be pumped to deep water through a submarine outfall (Boston, Los Angeles). It may also be mixed and discharged with the plant effluent. If an outfall is used, it should terminate where adequate current velocities will provide sufficient mixing for dilution and will prevent the formation of sludge banks. Because mixtures of sewage and sludge have a lower specific gravity than seawater, they may rise to the surface under certain conditions (see Chap. 15). Digested sludge discharged through a separate outfall should remain submerged. The disposal site must also be chosen carefully to avoid currents that would transport the sludge back to the seacoast. This method of disposal may not be allowed in the future.

PROBLEMS

13·1 The water content of a sludge is reduced from 98 to 95 percent. What is the percent reduction in volume by the approximate method and by the more exact method, assuming that the solids contain 70 percent organic matter of specific gravity 1.00 and 30 percent mineral matter of specific gravity 2.00? What is the specific gravity of the 98 and the 95 percent sludge?

13·2 A 10-mgd activated-sludge plant has 60 percent removal in the primary settling tank of the 200 mg/liter suspended solids in the raw sewage, and if primary sludge alone were pumped, it would be 5 percent solids. Assume that 0.1 mgd of waste activated sludge containing 0.5 percent solids is to be wasted to the digester. If the waste activated sludge is combined with the primary sludge, the resulting mixture is 4 percent solids. Calculate the reduction in volume pumped to the digester daily by mixing the sludge.

13·3 A preliminary sewage treatment plant providing for separate sludge digestion receives an influent sewage with the following characteristics:

Average flow	2 mgd
Suspended solids removed by primary sedimentation	200 mg/liter

Volatile matter in settled solids	75%
Water in raw sludge	96%
Specific gravity of mineral solids	2.60
Specific gravity of organic solids	1.30

(*a*) Determine the required digester volume using a mean cell residence time of 12 days.
(*b*) Determine the minimum digester capacity using the recommended loading parameters of pounds of volatile matter per cubic foot per day and cubic feet per capita.
(*c*) Assuming 90 percent moisture in the digested sludge and a 60 percent reduction in volatile matter during digestion at 90°F, determine the minimum theoretical digester capacity for this plant based on parabolic reduction in sludge volume during digestion and a digestion period of 25 days.

13·4 Consider an industrial waste consisting mainly of carbohydrates in solution. Pilot plant experiments using a complete-mix anaerobic digester without recycle yielded the following data:

Run	BOD_L influent, lb/day	X reactor, lb	dX/dt effluent, lb/day
1	1,000	428	85.7
2	500	115	46

Assuming a waste-utilization efficiency of 80 percent, estimate the percentage of added BOD_L that can be stabilized when treating a waste load of 10,000 lb/day. Assume the design sludge retention time (θ_c) to be 10 days.

13·5 A digester is loaded at a rate of 600 lb BOD_L/day. Using a waste-utilization efficiency of 75 percent, what is the volume of gas produced when $\theta_c = 40$ days? $Y = 0.10$ and $k_d = 0.02$.

13·6 Volatile acid concentration, pH, or alkalinity should not be used alone to control a digester. How should they be correlated to predict most effectively how close to failure a digester is at any time? (See Chap. 10, also references [8, 12].)

13·7 Prepare a one-page abstract of each of the four articles: P. L. McCarty: Anaerobic Waste Treatment Fundamentals, *Public Works,* vol. 95, nos. 9, 10, 11, and 12, 1964.

REFERENCES

1. Burd, R. S.: *A Study of Sludge Handling and Disposal,* U.S. Department of the Interior, Publication WP-20-4, 1968.
2. Cunetta, J., and R. Feuer: Design of the Newtown Creek Water Pollution Control Project, *J. WPCF,* vol. 40, no. 4, 1968.
3. Dreier, D. E., and C. A. Obma: *Aerobic Digestion of Solids,* Walker Process Equipment Co., Bulletin no. 26-S-18194, Aurora, Ill., 1963.
4. Fair, G. M., and E. W. Moore: Heat and Energy Relations in the Digestion of Sewage Solids, *Sewage Works Journal,* vol. 4, pp. 242, 428, 589, and 728, 1932.
5. Fisher, A. J.: The Economics of Various Methods of Sludge Disposal, *Sewage Works Journal,* vol. 8, no. 2, 1936.
6. Isaac, P. C. G. (ed.): *Waste Treatment,* Pergamon, New York, 1960.
7. Katz, W. J., and A. Geinopolos: Sludge Thickening by Dissolved-Air Flotation, *J. WPCF,* vol. 39, no. 6, 1967.

8. McCabe, B. J., and W. W. Eckenfelder, Jr.: *Biological Treatment of Sewage and Industrial Wastes,* vol. 2, Reinhold, New York, 1958.

9. McCarty, P. L.: Anaerobic Treatment of Soluble Wastes, in E. F. Gloyna and W. W. Eckenfelder, Jr. (eds.), *Advances in Water Quality Improvement,* University of Texas Press, Austin, 1968.

10. McCarty, P. L.: Anaerobic Waste Treatment Fundamentals, *Public Works,* vol. 95, nos. 9–12, 1964.

11. Speece, R. E., and P. L. McCarty: Nutrient Requirements and Biological Solids Accumulation in Anaerobic Digestion, *Proceedings of First International Conference on Water Pollution Research,* Pergamon, London, 1964.

12. Wagenhals, H. H., E. J. Theriault, and H. B. Hommon: *Sewage Treatment in the United States,* Public Health Bulletin 132, Treasury Department, Government Printing Office, Washington, D.C., 1925.

13. Water Pollution Control Federation: *Sewage Treatment Plant Design,* Manual of Practice 8, 1959.

14. Water Pollution Control Federation: *Sludge Dewatering,* Manual of Practice 20, 1969.

advanced wastewater treatment

14

Many of the substances found in wastewater are not affected or are little affected by conventional treatment operations and processes. These substances range from relatively simple inorganic ions, such as calcium, potassium, sulfate, nitrate, and phosphate, to an ever-increasing number of highly complex synthetic organic compounds. As the effects of these substances on the environment become more clearly understood, it is anticipated that treatment requirements will become more stringent in terms of the allowable concentration of many of these substances in the effluent from wastewater treatment plants. In turn, this will require advanced wastewater treatment facilities, which are not now used extensively. An alternative to the addition of advanced wastewater treatment is the development of totally different process flowsheets.

The purpose of this chapter is, therefore, to delineate the unit operations and processes that have been proposed for this purpose. The chapter contains a brief summary of some of the effects caused by substances that may be found in wastewater; an overview of the available types of unit operations and processes and their selection; a more detailed analysis of the physical unit operations and the biological and chemical unit processes; and a discussion dealing with the ultimate disposal of contaminants.

EFFECTS OF CHEMICAL CONSTITUENTS IN WASTEWATER

The typical composition of domestic wastewater has been reported previously in Table 7·3 in Chap. 7. Although not routinely measured, most domestic wastewaters will also contain a wide variety of trace compounds and elements. If industrial wastewater is discharged to domestic sewers, the distribution of the various constituents will vary considerably from that reported in Table 7·3. Some of the substances found in wastewater and the concentrations that may cause problems when discharged to the environment are reported in Table 14·1. This

TABLE 14·1 TYPICAL CHEMICAL CONSTITUENTS THAT MAY BE FOUND IN WASTEWATER AND THEIR EFFECTS

Constituent	Effect	Critical concentration, mg/liter
	Inorganic	
Ammonia	Increases chlorine demand	Any amount
	Toxic to fish	2.5
	Can be converted to nitrates	Any amount
Calcium and magnesium	Increase hardness	Over 100
Chloride	Imparts salty taste	250
	Interferes with industrial processes	75–200
Mercury	Toxic to humans and aquatic life	0.005
Nitrate	Stimulates algal and aquatic growth	0.3*
	Can cause methemoglobinemia in infants (blue babies)	10†
Phosphate	Stimulates algal and aquatic growth	0.015*
	Interferes with coagulation	0.2–0.4
	Interferes with lime-soda softening	0.3
Sulfate	Cathartic action	600–1,000
	Organic	
DDT	Toxic to fish and other aquatic life	0.001
Hexachloride	May cause taste and odor problems in water	0.02
Petrochemicals	May cause taste and odor problems in water	0.005–0.1
Phenolic compounds	May cause taste and odor problems in water	0.0005–0.001
Surfactants	Cause foaming and may interfere with coagulation	1.0–3.0

* For quiescent lakes [17].
† USPHS Recommended Drinking Water Standards, 1962.

listing is not meant to be exhaustive, but rather it is meant to highlight the fact that a wide variety of substances must be considered and that they will vary with each treatment application.

Because of their actual and suspected importance in promoting aquatic growths, compounds containing available nitrogen and phosphorus have received considerable attention since the mid-1960s. In some cases, the culprit may not be either one of these substances, but some trace element, such as cobalt, molybdenum, or vanadium.

UNIT OPERATIONS AND PROCESSES

The purpose of this section is to present an overview of the treatment methods that have been applied and studied for the removal of the substances reported in Table 14·1 as well as other compounds and substances.

Classification

Unit operations and processes that have been applied to the further treatment of wastewater are classified as (1) physical (2) chemical, and (3) biological (see Chaps. 8, 9, and 10). To facilitate a general comparison of the various operations and processes, data on (1) the types of wastewater to be treated, (2) the types of constituents affected and the degree to which they are removed, and (3) the form of the ultimate wastes to be disposed of are reported in Table 14·2.

Although listed individually in Table 14·2, in application almost all these processes and operations are used in conjunction with other unit processes and operations. For example, using ion exchange for the removal of nitrogen or phosphorus from treated wastewater, the exchange unit would normally be preceded by some form of filtration.

Process Selection

Selection of a given operation, process, or combination thereof depends on (1) the use to be made of the treated effluent, (2) the nature of the wastewater, (3) the compatibility of the various operations and processes, (4) the available means for disposing of the ultimate contaminants, and (5) the economic feasibility of the various combinations. In some cases, because of extreme conditions, economic feasibility may not be a controlling factor in the design of an advanced wastewater treatment system.

TABLE 14·2 TYPICAL APPLICATION DATA FOR ADVANCED WASTEWATER TREATMENT OPERATIONS AND PROCESSES [12, 34]

Description	Type of wastewater treated*	Removal efficiency, %								Waste for ultimate disposal
		SS	BOD	COD	NH_3	Org N	NO_3	PO_4	TDS	
Physical unit operations										
Air stripping of ammonia	EBT	...	...	...	85–98	...	...	...	...	None
Filtration:										
Multimedium	EBT	80–90	50–70	40–60	...	20–40	...	...	...	Liquid and sludge
Diatomite bed	EBT	95–99	...	...	...	...	...	...	...	Sludge
Microstrainers	EBT	50–80	40–70	30–60	...	20–40	...	...	...	Sludge
Distillation	EBT nitrified + filtration	~99	98–99	95–98	...	90–98	~99	~99	95–99	Liquid
Flotation	EPT, EBT	60–80	...	...	...	20–30	...	...	...	Sludge
Foam fractionation	EBT	75–90	~70	60–70	...	...	...	...	...	Liquid
Freezing	EBT + filtration	95–98	95–99	90–99	...	90–99	~99	~99	95–99	Liquid
Gas-phase separation	EBT	...	...	...	50–70	...	...	...	...	None
Land application	EPT, EBT	95–98	90–95	80–90	60–80	80–95	5–15	60–90	...	None
Reverse osmosis	EBT + filtration	95–98	95–99	90–95	95–99	95–99	95–99	95–99	95–99	Liquid
Sorption	EBT	...	~50	~40	...	...	...	~99	~10	Liquid and sludge

Chemical unit processes										
Carbon adsorption	EPT, EBT	80–90	70–90	60–75	...	50–90	...	...	...	Liquid
Chemical precipitation	EBT	60–80	75–90	60–70	5–15	30–50	...	90–95	~20	Sludge
Chemical precipitation in activated sludge	EPT	80–95	90–95	85–90	30–40	30–40	30–40	30–40	~10	Sludge
Ion exchange	EBT + filtration	...	40–60	30–50	85–98	80–95	80–90	85–98	†	Liquid
Electrochemical treatment	Raw	80–90	50–60	40–50	80–85	80–85	...	80–85	...	Liquid and sludge
Electrodialysis	EBT + filtration + carbon adsorption	...	...	...	30–50	...	30–50	30–50	~40	Liquid
Oxidation (chlorine)	EBT	...	80–90	65–70	50–80	...	...	...	...	None
Reduction	EBT	...	...	...	...	...	$NO_3 \rightarrow NH_3$	...	...	None
Biological unit processes										
Bacterial assimilation	EPT	80–95	75–95	60–80	30–40	30–40	30–40	10–20	...	Sludge
Denitrification	Agricultural return water	...	...	...	...	...	60–95	...	...	None
Harvesting of algae	EBT	...	50–75	40–60	50–90	50–90	50–90	~50	...	Algae
Nitrification-denitrification	EPT, EBT	...	...	...	...	...	60–95	...	...	None

* EPT is effluent from preliminary treatment and EBT is effluent from biological treatment.
† Varies with type of resin.

PHYSICAL UNIT OPERATIONS

In this section and in the subsequent sections dealing with chemical and biological processes, emphasis will be given to the ones that have been used successfully or that hold the most promise. Other operations and processes that fall within the same category will be discussed briefly. Of the many physical operations that have been used in advanced wastewater treatment (see Table 14·2) only the removal of ammonia by air stripping and filtration are discussed in detail.

Air Stripping of Ammonia

The removal of ammonia from wastewaters by air stripping recently has received considerable attention as part of the work that has been carried out in the field of advanced waste treatment under the sponsorship of the federal government. Perhaps the best-known example is the work conducted at Lake Tahoe, Calif. [8, 9, 36]. This process also has received considerable attention in the industrial waste treatment field [31].

Theory The following discussion has been adapted from Tchobanoglous [38]. The air stripping of ammonia from wastewater is a modification of the aeration process used for the removal of gases dissolved in water. Ammonium ions in wastewater exist in equilibrium with ammonia, as shown in Eq. 14·1:

$$NH_3 + H_2O \rightleftharpoons NH_4^+ + OH^- \tag{14·1}$$

As the pH of the wastewater is increased above 7, the equilibrium is shifted to the left; and the ammonium ion is converted to ammonia, which may be removed as a gas by agitating the wastewater in the presence of air. This is usually accomplished in a packed tray tower equipped with an air blower. An analysis of this process, including the data required for design, is presented in the following discussion.

Because the reaction in Eq. 14·1 is pH dependent, the percentage distribution of ammonia and ammonium ion can be computed using the following equation:

$$NH_3,\ \% = \frac{NH_3 \times 100}{NH_3 + NH_4} = \frac{100}{1 + \dfrac{NH_4}{NH_3}} = \frac{100}{1 + \dfrac{K_b(\mathrm{H})}{K_w}} \tag{14·2}$$

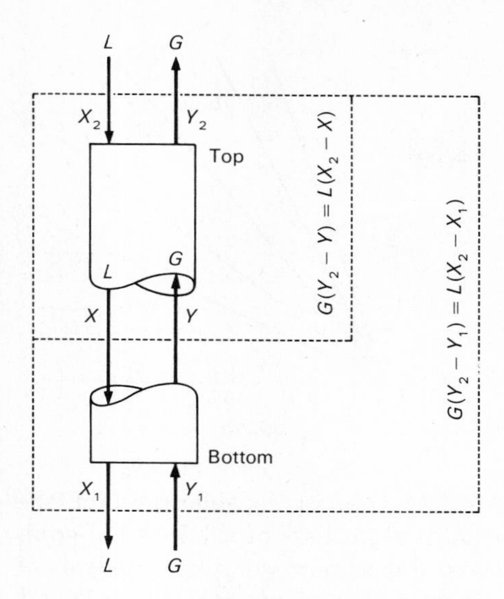

FIG. 14·1 Definition sketch for an ammonia-stripping column.

For example, at 25°C and pH 10 the percentage distribution of NH_3 is

$$\frac{100}{1 + \dfrac{1.8 \times 10^{-5} \times 10^{-10}}{10^{-14}}} = 85\%$$

Knowing that at 25°C and pH 11 the percentage distribution of ammonia is about 98 percent, the amount of base that must be added to a wastewater can be established. Typical data on the amount of lime [$Ca(OH)_2$] required to raise the pH to 11 as a function of the wastewater alkalinity are presented in Fig. 11·15 in Chap. 11.

The next step is to determine the amount of air required to remove the ammonia from the liquid. Removal is usually accomplished in a stripping tower. A process definition sketch is presented in Fig. 14·1. An overall steady-state materials balance equation for a stripping tower is given by

$$G(Y_2 - Y_1) = L(X_2 - X_1) \tag{14·3}$$

where G = moles of incoming gas per unit time
L = moles of incoming liquid per unit time
Y_1 = concentration of solute in gas at bottom of tower, moles of solute per mole of solute-free gas
Y_2 = concentration at top
X_1 = concentration of solute in liquid at bottom of tower, moles of solute per mole of liquid
X_2 = concentration at top

FIG. 14·2 Equilibrium curves for ammonia in water.

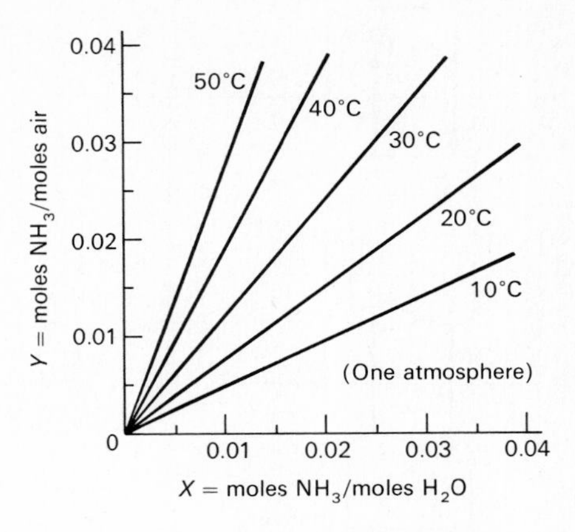

Because Eq. 14·3 was derived solely from a consideration of the equality of input and output, it holds regardless of the internal equilibria that may control the process. For a more complete analysis of this operation and other mass transfer operations, see Ref. [42] and Ref. [23] in Chap. 8.

Equilibrium curves for ammonia in water for various temperatures and 1 atm of pressure are given in Fig. 14·2. Equilibrium curves are also often presented in terms of mole fractions [moles NH_3/(moles H_2O + moles NH_3)]. The theoretical amount of air required per gallon of wastewater can be computed using Eq. 14·3 and Fig. 14·2. If it is assumed that the liquid leaving and the air entering the bottom of the tower contain no ammonia, Eq. 14·3 can be written as

$$\frac{G}{L} = \frac{X_2}{Y_2} \tag{14·4}$$

where G/L is the ratio of the gas (air) to liquid (wastewater) required for stripping the ammonia from the wastewater.

The theoretical minimum amount of air would be required if the ammonia in the air leaving the tower were in equilibrium with the ammonia in the incoming liquid. The minimum air-to-liquid ratio in moles of air per mole of water is given by the reciprocal of the slope of the equilibrium curves in Fig. 14·2. Calculation of this ratio is illustrated in the following example.

EXAMPLE 14·1 *Air Requirements for Ammonia Stripping*

Determine the minimum air-to-liquid ratio required at 20°C for complete stripping.

Solution

1. Determine the ratio X/Y at 20°C using Fig. 14·2.

$$\frac{X}{Y} = \frac{\text{moles NH}_3/\text{moles H}_2\text{O}}{\text{moles NH}_3/\text{moles air}} = \frac{0.02 \text{ moles air}}{0.015 \text{ moles H}_2\text{O}} = \frac{1.33 \text{ moles air}}{\text{mole H}_2\text{O}}$$

2. Convert the moles of air to cubic feet of air.

$$1.33 \text{ moles} \times \frac{29 \text{ g}}{\text{mole}} \times \frac{1}{\frac{454 \text{ g}}{\text{lb}}} \times \frac{1}{\frac{0.0808 \text{ lb}}{\text{cu ft}}} = 1.055 \text{ cu ft}$$

3. Convert the moles of water to gallons of water.

$$1.0 \text{ mole} \times \frac{18 \text{ g}}{\text{mole}} \times \frac{1}{\frac{454 \text{ g}}{\text{lb}}} \times \frac{1}{\frac{8.34 \text{ lb}}{\text{gal}}} = 0.00476 \text{ gal}$$

4. Determine the required air-to-liquid ratio.

$$\frac{G}{L} = \frac{1.055 \text{ cu ft}}{0.00476 \text{ gal}} = 223 \text{ cu ft/gal}$$

The required air-to-liquid ratio for various temperatures for stripping ammonia is given in Fig. 14·3. The theoretical ratio is derived by assuming the process to be 100 percent efficient with a stripping tower of infinite height—obviously unachievable in practice. Therefore, in computing the actual amount of air required, the theoretical value is

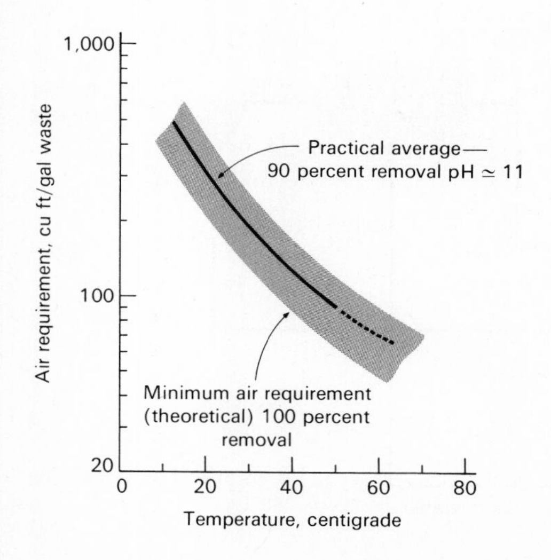

FIG. 14·3 Effect of temperature on air requirements for ammonia stripping [38].

often multiplied by a factor varying from 1.5 to 2.0. Practical air-to-liquid ratios are reported in Fig. 14·3.

The foregoing analysis is reasonable even if the concentration of ammonia in the liquid leaving the tower were on the order of 1 to 2 mg/liter. In this case, the value of X_1 would be very close to zero and, for all practical purposes, the minimum air requirement would be defined by the slope of the equilibrium line. In all practical cases, the ammonia concentration in the air leaving the tower will be less than the equilibrium value and the air-to-liquid ratio will plot below the equilibrium curves in Fig. 14·2. If practically complete removal of ammonia is obtained under these conditions, air in excess of the theoretical amount is being applied as shown in Fig. 14·3. Data on air requirements from the Lake Tahoe study for 90 percent ammonia removal at 20°C using a 24-foot high stripping tower lie approximately on the practical average curve given in Fig. 14·3 [8]. In general, removal efficiency depends on the temperature, size, and proportions of the facility, and the efficiency of the air-water contact. If the removal of ammonia is unsatisfactory, the tower has not been designed correctly or is overloaded. In this case additional air volume may improve operation.

Application The application of air stripping in the treatment of raw wastewater and digester supernatant is shown in Figs. 14·4 and 14·5, respectively. In the flowsheet shown in Fig. 14·4, lime is added to the

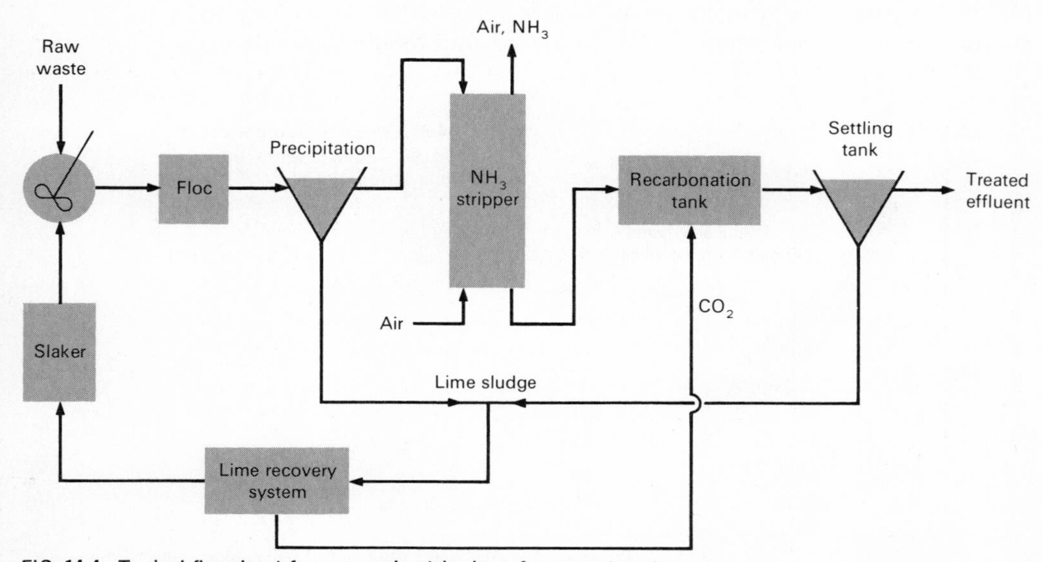

FIG. 14·4 Typical flowsheet for ammonia stripping of raw wastewater.

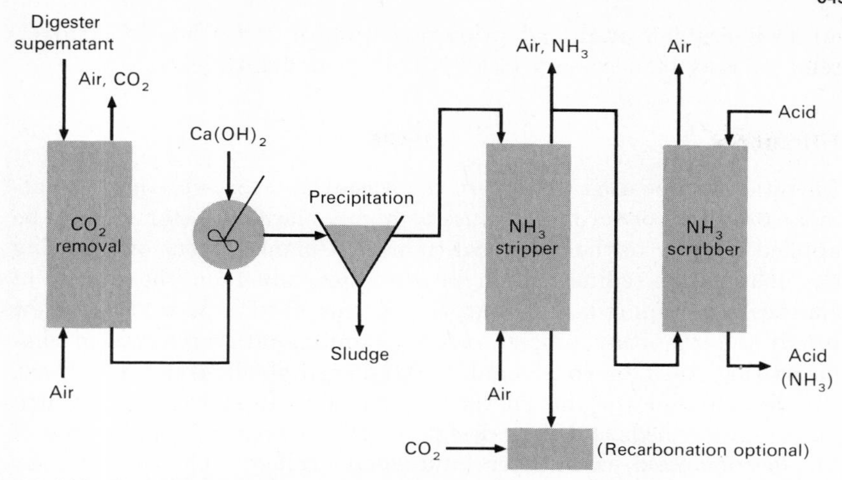

FIG. 14·5 Typical flowsheet for the stripping of ammonia from digester supernatant [10].

raw wastewater to precipitate phosphorus. The settled wastewater is then passed through a countercurrent ammonia-stripping tower. The stripped liquid is then recarbonated and resettled. A lime recovery system is also included. This flowsheet is similar to the one used at Lake Tahoe for the treatment of secondary effluent.

The flowsheet for the treatment of digester supernatant was developed by the FMC Corporation under the sponsorship of the Federal Water Pollution Control Administration [10]. The process is similar to those discussed previously, with the exception that CO_2 is stripped before adding lime, and an acid ammonia scrubber has been added where the stripped ammonia cannot be discharged to the atmosphere.

In most cases, where ammonia stripping has been applied, a number of problems have developed, such as calcium carbonate scaling within the tower and feed lines and poor performance during cold weather operation. The amount and nature (soft to extremely hard) of the calcium carbonate scale formed will vary with the characteristics of the wastewater and local environmental conditions. As the temperature decreases the amount of air required increases significantly for the same degree of removal (see Fig. 14·3). Under conditions of icing the liquid-air contact geometry in the tower is altered, which further reduces the overall efficiency. If high removal efficiencies are not required, some of the operational difficulties can be avoided by using conventional aeration basins.

Apart from operational difficulties, the discharge of ammonia to the atmosphere near large bodies of water or snow-covered areas can create serious nitrogen pollution problems. In such situations use of an acid scrubber or the adoption of an alternate method of nitrogen removal may be necessary (see "Nitrification-denitrification").

Filtration

Filtration can be used to prepare the wastewater for subsequent treatment processes or for direct reuse as highly clarified water. It may be applied directly to the biological treatment plant effluent or following the coagulation-sedimentation process, depending on the degree of clarification required. The purpose of this section is to (1) review briefly the important process variables and operative removal mechanisms that must be considered in the design of filtration operations, (2) discuss some of the characteristics of dual-medium and multimedium filter beds and their design, and (3) discuss the application of the diatomaceous-earth filters and microstrainers. The material for the following discussion has been taken from Refs. [39 and 40].

Process Variables and Removal Mechanisms Filtration process variables are summarized in Table 14·3. The variables shown are used (1) to characterize the filter material (items 1, 2, and 3); (2) as design parameters in estimating clear-water head loss (items 1, 2, 4, 5, and 7); and (3) to characterize the influent material to be filtered (item 8).

TABLE 14·3 FILTRATION PROCESS VARIABLES AND PARTICLE REMOVAL MECHANISMS [39]

Process variables	Removal mechanisms
1. Filter-media grain size, shape, density, and composition	1. Straining (*a*) Mechanical (*b*) Chance contact
2. Filter-media porosity	2. Sedimentation
3. Media head loss, characteristics	3. Inertial impaction
4. Filter-bed depth	4. Interception
5. Filtration rate	5. Chemical adsorption (*a*) Bonding (*b*) Chemical interaction
6. Allowable head loss	6. Physical adsorption (*a*) Electrostatic forces (*b*) Electrokinetic forces (*c*) Van der Waals forces
7. Fluid characteristics	7. Adhesion and adhesion forces
8. Influent characteristics (*a*) Suspended-solids concentration (*b*) Floc or particle size and distribution (*c*) Floc strength (*d*) Floc or particle charge	8. Coagulation-flocculation
	9. Biological growth

Over the past 40 years, a number of theories have been advanced to describe the manner or mechanism by which suspended matter is removed within a filter. The most prominent theories are presented in Table 14·3. The first four mechanisms may be classified as mechanical or physical and are related to physical parameters, such as grain size, porosity, and bed depth. With the exception of coagulation-flocculation and biological growth, which strictly speaking are not removal mechanisms, the remaining mechanisms are related to the chemical and surface characteristics of the suspended matter and filter bed. Flocculation within the filter will bring about the growth of large particles which may be removed by one or more of the mechanisms discussed previously. Bacterial growth will reduce the pore volume and may thus enhance removal.

Dual-Medium and Multimedium Filter Beds In recent years, increasing reference has been made in the literature to the use of dual-medium and multimedium filter beds. Some dual-medium filter beds that have been used are composed of (1) anthracite and sand, (2) activated carbon and sand, (3) resin beds and sand, and (4) resin beds and anthracite. Multimedium beds that appear to have promise are composed of (1) anthracite, sand, and garnet; (2) activated carbon, anthracite, and sand; (3) weighted spherical resin beads (charged and uncharged), anthracite, and sand; and (4) activated carbon, sand, and garnet.

Typical data on the depth and characteristics of the filtering materials used most commonly in dual-medium and tri-medium filters are presented in Table 14·4. Because filter performance is related directly to the characteristics of the liquid to be filtered, it is recommended that pilot plant studies be conducted to determine the optimum combination of filter materials. However, where this is not possible, the data in Table 14·4 can be used as a guide. Additional information may be found in Culp and Culp [8].

Filter Operation The objective of filtration is to produce an effluent that consistently meets the established treatment criteria at minimum cost. To do this, trade-offs must be established among the degree of pretreatment, the allowable head loss, the desired length of filter run, and filter cleaning requirements. Although a complete discussion of all these factors is beyond the scope of this section, some of them will be touched upon in the following discussion. For a more complete discussion of these and other topics, including the development of equations to describe the time-space removal of particulate matter in a filter, Refs. [8, 39 and 40] are recommended.

In the application of filtration to advanced wastewater treat-

TABLE 14·4 TYPICAL DATA FOR DUAL–MEDIUM AND MULTIMEDIUM FILTERS

Characteristic	Value	
	Range	Typical
Dual-medium		
Anthracite:		
Depth, in.	8–24	18
Effective size, mm	0.8–2.0	1.2
Uniformity coefficient	1.4–1.8	1.5
Sand:		
Depth, in.	10–24	12
Effective size, mm	0.3–0.8	0.5
Uniformity coefficient	1.2–1.6	1.4
Filtration rate, gpm/sq ft	2–10	6
Multimedium		
Anthracite:		
Depth, in.	8–20	15
Effective size, mm	1.0–2.0	1.4
Uniformity coefficient	1.4–1.8	1.5
Sand:		
Depth, in.	8–16	12
Effective size, mm	0.4–0.8	0.6
Uniformity coefficient	1.2–1.6	1.4
Garnet:*		
Depth, in.	2–4	3
Effective size, mm	0.2–0.6	0.3
Uniformity coefficient		~1.0
Filtration rate, gpm/sq ft	2–12	6

* Garnet becomes intermixed with sand and anthracite.

ment, it has been found that the nature of the particulate matter in the influent to be filtered and the size of the filter material or materials are perhaps the most important of the process variables (Table 14·3, items 1 and 8). Influent characteristics of greatest importance include the suspended-solids concentration, particle size and distribution, and floc strength.

Typically, the suspended solids in the effluent from an activated-sludge plant will vary between 6 and 20 mg/liter. Because the concentration of suspended solids is usually the principal parameter of

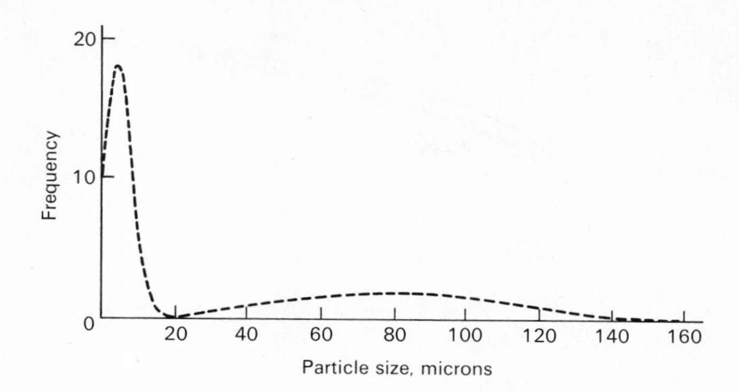

FIG. 14·6 Typical particle size distribution found in settled sewage effluent [39].

concern, turbidity is often used as a practical means of monitoring the filtration process. Within limits, it has been shown that the suspended-solids concentrations found in treated wastewater can be correlated to turbidity measurements [40].

Typical data on the size and distribution of the particles in the effluent from a pilot-scale activated-sludge plant operated at a mean cell residence time of 10 days are shown in Fig. 14·6. As illustrated, the particles fell into two distinct size ranges, small particles varying in areal size (equivalent circular diameter) from 1 to 15 μ and large particles varying in size from 50 to 180 μ.

The most significant observation relating to particle size is the fact that the distribution of sizes was found to be bimodal. This is important, as it will influence the removal mechanisms that may be operative during the filtration process. For example, it seems reasonable to assume that the mechanism of removal for particles 1.0 μ in size would be different from that for particles 80 μ in size or larger. The bimodal particle-size distribution has also been observed in water treatment plants [6].

In terms of filter performance and operation, the most important influent characteristic is floc strength, which will vary not only with the type of process, but also with the mode of operation. For example, the residual floc from the chemical precipitation of biologically processed wastewater may be considerably weaker than the residual biological floc before precipitation. Further, the strength of the biological floc will vary with the mean cell residence time, increase with longer mean cell residence times.

Different removal mechanisms will be operative, depending on the strength of the floc. In the case of a strong biological or chemical floc, straining is often the principal removal mechanism [40]. For a weak floc, a combination of removal mechanisms (see Table 14·3) will be operative.

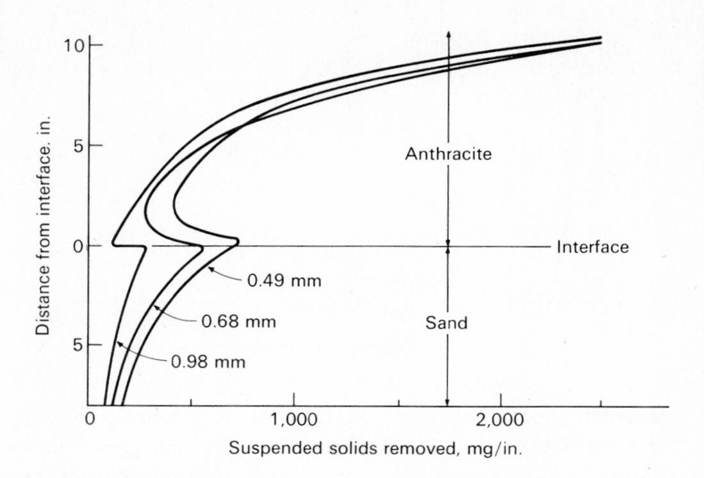

FIG. 14·7 Suspended solids removed per inch of filter depth when filtering settled sewage effluent [39].

Typical suspended-solids removal data obtained when filtering activated-sludge effluent where straining is the operative removal method are shown in Fig. 14·7. The corresponding experimental-run data are reported in Table 14·5. As shown, all three of the filters had a 12-in. layer of anthracite and an 8-in. layer of uniform sand of varying size.

The suspended solids removed in each 1-in. increment of filter bed after 48 hr are shown in Fig. 14·7. Most of the suspended material is removed in the surface layers of the filter bed. The amount removed decreases with further penetration into the bed until the zone of intermixing between the sand and anthracite is reached. At this point, the amount removed starts to increase due to the reduced pore size and continues to increase into the first layer of sand below the interface. Below this point the removal again tapers off.

The effect of sand size on both intermixing and suspended-solids removal is clearly illustrated in Fig. 14·7. Starting with the largest sand size, it is evident from the shape of the suspended-solids removal curve that there was no intermixing. In the case of the next two sand sizes, it can be concluded from the suspended-solids removal curves that there was intermixing in both cases. For the 0.68-mm sand, intermixing took place up to about 3.5 in. above the interface. This distance was approximately 5 in. for the 0.49-mm sand.

Analysis of the data shown in Fig. 14·7 is given under the heading of performance data in Table 14·5. The first item listed under performance data is the total amount of material removed by the filter and represents the area under the curves given in Fig. 14·7. The numbers in the next row represent the amount of material that would have been removed if the sand had been replaced with anthracite. The dif-

TABLE 14·5 SUMMARY DATA ON ACTIVATED–SLUDGE EFFLUENT FILTRATION WITHOUT CHEMICALS [35]

	Filter number		
	1	2	3
Filter-bed characteristics			
Anthracite:			
Depth, in.	12	12	12
Effective size, mm	1.08	1.08	1.08
Uniformity coefficient	1.42	1.42	1.42
Sand:			
Depth, in.	8	8	8
Effective size, mm	0.98*	0.68*	0.49*
Uniformity coefficient	~1.0	~1.0	~1.0
Filtration-process variables			
Filtration rate, gpm/sq ft	5.15	5.15	5.15
Influent suspended solids, mg/liter (average during run)	14	14	14
Length of run, hr	48	48	48
Performance data			
Total suspended solids removed, mg	14,460	15,460	17,430
Equivalent removal, sand replaced with anthracite, mg	13,875	13,875	13,875
Additional removal due to sand, %	0.4	13.2	20.4

* Equivalent diameters.

ference between these values represents the additional removal accomplished by the sand layer. The presence of the sand layer is more important when the floc is weak than when it is strong. With a weak floc, the sand layer does a considerably greater amount of work (as measured by the amount of material removed) in retaining the material that breaks away or that is not removed in the upper layers of the filter.

Comparing the results obtained using dual- and tri-medium filters to the results obtained using a single medium such as sand, the advantages of the former are readily apparent. In the case of a single-medium (sand) filter, because of the bimodal distribution of particle sizes often found in the wastewater to be filtered, the larger particles will be strained out at or near the surface of the filter resulting in short filter runs and poor utilization of the filter capacity. In the case of the dual- or tri-medium filter beds, the material removed will be distributed more effectively throughout the filter depth resulting in longer filter runs. Additional information and details on the application of the filtration process may be found in Culp and Culp [8].

Diatomaceous-Earth Filters Diatomite filters, designed originally for water filtration, were not designed to take shock loads. Thus, although capable of producing an extremely high-quality effluent, the occurrence of varying amounts of suspended solids in activated-sludge effluents has led to erratic operation of these filters in advanced wastewater treatment applications. New designs must be developed, with automated body feed and backwash equipment to take account of load variations in the activated-sludge effluent. Another factor that has reduced the application of diatomite filters is the high cost compared to other types of filters or solids removal operations.

Microstrainers Microstraining involves the use of variable low-speed (up to 4 to 7 rpm), continuously backwashed, rotating drum filters operating under gravity conditions (see Fig. 14·8). The principal filtering fabrics have openings of 23 or 35 μ and are fitted on the drum

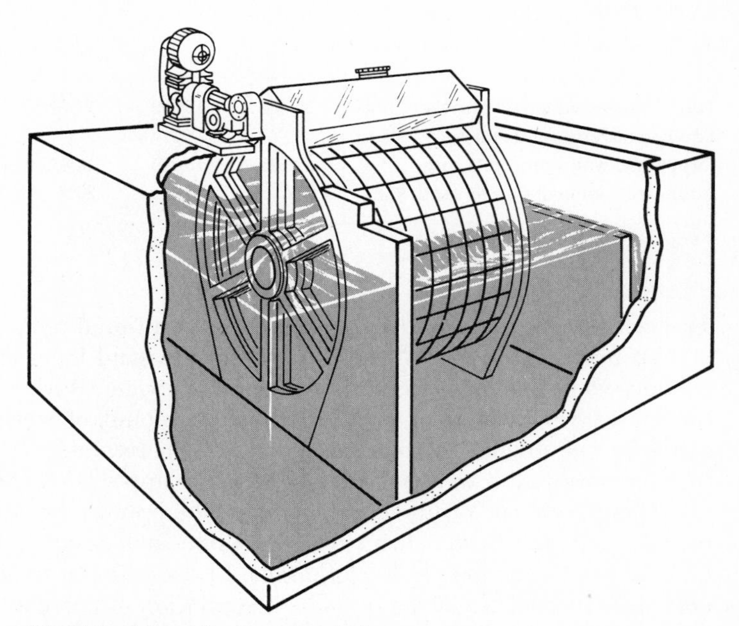

FIG. 14·8 Schematic of a microstrainer.

periphery. The wastewater enters the open end of the drum and flows outward through the rotating screening cloth. The collected solids are backwashed by high-pressure jets into a trough located within the drum at the highest point of the drum.

Microstrainers have been used to remove algae from water in reservoirs before water treatment and to polish activated-sludge effluents. Problems encountered with microstrainers include incomplete solids removals and their inability to handle solids fluctuations. Reducing the speed of rotation of the drum and less frequent flushing of the screen has resulted in increased removal efficiencies but reduced capacity.

Other Operations

As reported in Table 14·2, a number of different physical operations have been investigated in various advanced wastewater treatment applications [1]. Although many of them have proven to be technically feasible, other factors such as cost, operational requirements, and aesthetic considerations have not been favorable. Nevertheless, it is important that sanitary engineers be familiar with these operations so that in any given situation they can consider all treatment possibilities.

Distillation Distillation is a unit operation in which the components of a liquid solution are separated by vaporization and condensation. If entrainment is kept low, almost all of the nonvolatile contaminants can be removed in a single evaporation step. Volatile contaminants, such as ammonia gas and low-molecular-weight organic acids, may be removed in a preliminary evaporation step, but if their concentration is so small that their presence in the final product is not objectionable, this step with its added cost can be eliminated. Of the many distillation processes studied by the USPHS, multistage flash evaporation, multiple-effect evaporation, and vapor-compression distillation appear most feasible for the purification of municipal wastewaters [1].

Flotation The removal of the finely divided colloidal and suspended matter found in treated wastewater can often be accomplished by flotation. Its use in this and similar applications is increasing, especially in conjunction with the use of polymers. The theoretical aspects of flotation are discussed in Chap. 8, and design aspects are presented in Chaps. 11 and 13.

Foam Fractionation Foam fractionation involves the separation of colloidal and suspended material by flotation and dissolved organic matter by adsorption. When air is bubbled through the wastewater, foam is produced or induced by the addition of chemicals. Because most water-soluble organic compounds are surface-active to some extent, they tend to concentrate at the gas-liquid interface and are removed with the foam.

Freezing Freezing is a physical separation operation similar to distillation. The wastewater is sprayed into a chamber operating under a vacuum. A portion of the wastewater evaporates and the cooling effect produces contaminant-free ice crystals in the remaining liquid. The ice is then removed and melted by using the heat of condensation of the vapors from the evaporation stage. Butane and other refrigerants have been used in freezing experiments [1].

Gas-Phase Separation A promising method for the removal of ammonia as a gas is the development of selective permeable gas-phase membranes [5, 32]. The wastewater is passed through tubes lined with, or made from, selective gas-phase membranes. The gas passes through the membrane and remains attached to the tube exterior. The attached gas is removed by blowing air or nitrogen tangentially along the tubes. At the present time, more developmental work is required before this process can be applied.

Land Application Because of the abundance of nutrients in municipal wastewater treatment plant effluent, land disposal of the effluent has been considered by a number of investigators [30]. In this process, the wastewater is distributed evenly over the ground surface, which serves as a low-rate filter. Suspended particles are strained out, colloids and organic matter are absorbed by the soil particles, nutrients are utilized by vegetation, and more complex organic materials are decomposed to simple inorganic compounds by soil bacteria.

Ground water can be replenished by this process when the rate of application is not in excess of the rate at which nutrients can be utilized. Applied in excess, nitrates in the wastewater will travel unimpeded through most soil systems, which may ultimately lead to ground water contamination.

Reverse Osmosis (Hyperfiltration) Reverse osmosis is a process in which water is separated from dissolved salts in solution by filtering through a semipermeable membrane at a pressure greater than the osmotic pressure caused by the dissolved salts in the wastewater [25]. With existing membranes and equipment, operating pressures vary from atmospheric to 1,500 psi.

As an advanced wastewater treatment process, reverse osmosis has its limitations. Under existing manufacturing processes, a semipermeable membrane, such as cellulose acetate, is subjected to a temperature treatment that sets the rate at which water can be produced. Colloidal and organic matter, by fouling of the membrane surface, can cause high losses in the quantity of water that can be produced. Salts with low solubility levels will precipitate on the membrane and will reduce the quantity of product water. Pretreatment of wastewater, such as activated-carbon adsorption or chemical precipitation followed by some form of filtration, may be necessary to prevent these losses.

Sorption Conventional alum treatment for phosphate removal will increase the concentration of sulfate ion in solution. Sorption is a process developed to remove various forms of phosphate without increasing the sulfate concentration [45]. Activated alumina is used to sorb phosphates by passing a stream of water through the sorption column. Regeneration of the activated alumina for reuse is accomplished by using small amounts of caustic and nitric acid.

CHEMICAL UNIT PROCESSES

As noted in Table 14·2, a variety of advanced chemical unit processes have been applied to the treatment of wastewater. Some of the processes, such as chemical precipitation, have been applied to both treated and untreated wastewater; other processes, such as electrodialysis and ion exchange, have been applied more commonly to treated effluents.

Carbon Adsorption

The basic principles and applications of activated-carbon adsorption are discussed in Chap. 9. The conduct of activated-carbon adsorption studies is delineated in Chap. 16. Following biological treatment, adsorption has been accomplished in fixed- and expanded-bed columns of granular carbon and in tanks using powdered carbon.

A difficulty with the fixed-bed columns has been the leakage of certain organic fractions through the beds. The exact chemical explanation of this phenomenon is a matter of controversy. It has been postulated that the eluted organic fraction is composed of nonadsorbable cell fragments and small organic molecules that are hydrolyzed by biological treatment and are rendered more soluble and less subject to adsorption [43]. A detailed discussion of the application of activated carbon may be found in Culp and Culp [8].

Chemical Precipitation

As discussed in Chaps. 9 and 11, precipitation of phosphorus in wastewater is usually accomplished by the addition of coagulants, such as alum, lime, or iron salts, and polyelectrolytes. Commercial processes involving the use of metal ions and polyelectrolytes have also been used. Chemical precipitation may be carried out in primary or activated-sludge settling tanks or as a separate operation [26].

In the preliminary stage, most of the phosphorus is removed in the primary sedimentation tank. The remaining phosphorus is normally removed by assimilation in the biological treatment phase. When precipitation is to be accomplished in the biological stage of treatment, chemicals are usually added to the aeration tank. The precipitates formed in the biological reactor are removed in the activated-sludge settling tank.

Chemical Precipitation in Biological Treatment

It has been observed that the degree of phosphorus removal at some activated-sludge treatment plants is considerably higher than would be predicted on the basis of the requirements for organism growth (see "Bacterial Assimilation"). Two different theories have been proposed to account for this observation.

The first theory is that the removal of phosphate is brought about by chemical precipitation, as described by Menar and Jenkins [23]. Required conditions include (1) hydrolysis of complex phosphate to orthophosphates, (2) decreasing CO_2 production as the waste passes through a plug-flow reactor, (3) an increase in pH due to the fact that less CO_2 is being produced and more is being removed by aeration, and (4) the development of conditions favoring the precipitation of calcium phosphate (see Chap. 9). As noted from the above conditions a long plug-flow reactor would be required.

The second theory is that the removal is accomplished by biological means. Under certain ideal conditions it is thought that the microorganisms in the activated-sludge mixed liquor are able to remove an excess amount of phosphorus over that required for growth. This phenomenon has been termed "luxury uptake." It is not clear if the phosphorus is incorporated (stored) within the cell or adsorbed on the bacterial cells or if it is a combination of these two. Conditions favoring the luxury uptake of phosphorus are given in Levin and Shapiro [18].

Ion Exchange

Ion exchange is a unit process by which ions of a given species are displaced from an insoluble exchange material by ions of a different species in solution. Ion-exchange operations are either batch or continuous. In a batch process, the resin is simply stirred with the water to be treated in a reactor until the reaction is complete. The spent resin is removed by settling and subsequently is regenerated and reused. In a continuous process, the exchange material is placed in a bed or a packed column, and the water to be treated is passed through it. Theoretical and operational aspects may be found in Chapman [4].

The chemistry of the ion-exchange process may be represented by the following equilibrium equations.

Reaction:

$$RH + Na^{+} \rightleftharpoons RNa + H^{+} \tag{14·5}$$

$$RNa_2 + Ca^{++} \rightleftharpoons RCa + 2Na^{+} \tag{14·6}$$

Regeneration:

$$RNa + HCl \rightleftharpoons RH + NaCl \tag{14·7}$$

$$RCa + 2NaCl \rightleftharpoons RNa_2 + CaCl_2 \tag{14·8}$$

where R represents the resin.

Equations 14·5 and 14·6 represent the reactions involved in the removal of sodium and calcium ions from water using a synthetic cationic-exchange resin. Regeneration of the exhausted resin is represented by Eqs. 14·7 and 14·8. The extent of completion of the removal reactions shown depends on the equilibrium that is established between the ions in the aqueous phase and those in the solid phase. For the removal of sodium, this equilibrium is defined by the following expression:

$$\frac{[H]\, X_{RNa}}{[Na]\, X_{RH}} = K_{H \rightarrow Na} \tag{14·9}$$

where $K_{H \rightarrow Na}$ = selectivity coefficient
[ ] = concentration in solution phase
X_{RH} = mole fraction of hydrogen on exchange resin
X_{RNa} = mole fraction of sodium on exchange resin

The selectivity coefficient depends primarily on the nature and valence of the ion, the type of resin and its saturation, and the ion concentration in wastewater. In fact, for a given series of similar ions, exchange resins have been found to exhibit an order of selectivity or affinity for the ions. For synthetic cationic- and anionic-exchange resins, typical series are

$$Li^+ < H^+ < Na^+ < K^+ < Rb^+ < Ag^+$$
$$Mg^{++} < Zn^{++} < Cu^{++} < Co^{++} < Ca^{++} < Sr^{++} < Ba^{++}$$
$$OH^- < F^- < HCO_3^- < Cl^- < Br^- < I^- < NO_3^- < ClO_4^-$$

In practice, the selectivity coefficients are determined by measurement in the laboratory and are valid only under the conditions under which they were measured. At low concentrations, the value of the selectivity coefficient for the exchange of monovalent ions by divalent ions is, in general, larger than the exchange of monovalent ions by monovalent ions. This fact has, in many cases, limited the use of synthetic resins for the removal of certain substances in wastewater, such as ammonia in the form of the ammonium ion. There are, however, certain natural zeolites that favor NH_4^+ or Cu^{++}.

Representative listings of cationic- and anionic-exchange resins produced by the major manufacturers in the United States may be found in Applebaum [2]. Reported exchange capacities vary with the type and concentration of regenerant used to restore the resin. Exchange capacities for resins often are expressed in terms of kilograins as $CaCO_3$ per cubic foot of resin (kgr/cu ft) or gram equivalents per cubic foot (g equiv/cu ft). Conversion between these two units is accomplished using the following expression:

$$\frac{1 \text{ g equiv}}{\text{cu ft}} = \frac{0.77 \text{ kgr as } CaCO_3}{\text{cu ft}}$$

$$= \frac{1 \text{ g equiv} \times \dfrac{50 \text{ g } CaCO_3}{\text{g equiv}} \times \dfrac{7{,}000 \text{ grains}}{\text{lb}}}{\dfrac{454 \text{ g}}{\text{lb}} \times 10^3}$$

$$= 0.77 \text{ kgr as } CaCO_3$$

Calculation of the required resin volume for the ion-exchange process is illustrated in Example 14·2.

EXAMPLE 14·2 *Ion-Exchange Resin Volume Requirements*

Determine the volume of a cationic resin with an exchange capacity of 20 kgr as $CaCO_3$/cu ft required to treat 1 million gal of water containing 18 mg/liter of ammonium ion.

Solution

1. In terms of $CaCO_3$, 18 mg/liter of ammonium ion is equal to 50 mg/liter. Therefore, the required exchange capacity is equal to

$$\frac{(50 \text{ mg/liter})\ 10^6 \text{ gal}}{\left(\dfrac{17.1 \text{ mg/liter}}{\text{grains/gal}}\right)\left(\dfrac{10^3 \text{ grains}}{\text{kgr}}\right)} = 2{,}920 \text{ kgr as } CaCO_3$$

2. Compute the volume of resin.

$$\frac{2{,}920 \text{ kgr}}{20 \text{ kgr/cu ft}} = 146 \text{ cu ft}$$

3. In practice, because of leakage and other operational and design limitations, the required volume of resin will usually be about 1.1 to 1.4 times that computed on the basis of exchange capacity.

Although both natural and synthetic ion-exchange resins are available, synthetic resins are used more widely because of their durability. Nevertheless, some natural resins (zeolites) have found application in the removal of ammonia from wastewater. Of the natural zeolites that have been investigated, Hector Clinoptilolite has proven to be the most effective [24]. One of the novel features of this zeolite is the regeneration system employed. Upon exhaustion, the zeolite is regenerated with lime $Ca(OH)_2$. The ammonium ion removed from the zeolite is converted to ammonia because of the high pH. At this point, the regenerating solution is passed through a stripping tower for removal of the ammonia. The stripped liquid is collected in a storage tank for subsequent reuse [24]. The advantage of this system is that there is no process waste containing ammonia for which ultimate disposal must be provided. A flowsheet for this process is shown in Fig. 14·9.

A pressing problem that must be solved is the formation of calcium carbonate precipitates within the zeolite exchange bed and in the stripping tower and piping appurtenances. As indicated in Fig. 14·9, the zeolite bed is equipped with backwash facilities to remove the carbonate deposits that form within the filter.

A serious problem associated with application of ion exchange to the treatment of wastewater effluents is resin binding caused by the residual organic matter found in effluent from biological treatment. This problem has been solved partially by prefiltering the wastewater (see Fig. 14·9) or by using scavenger exchange resins before application to the exchange column [2].

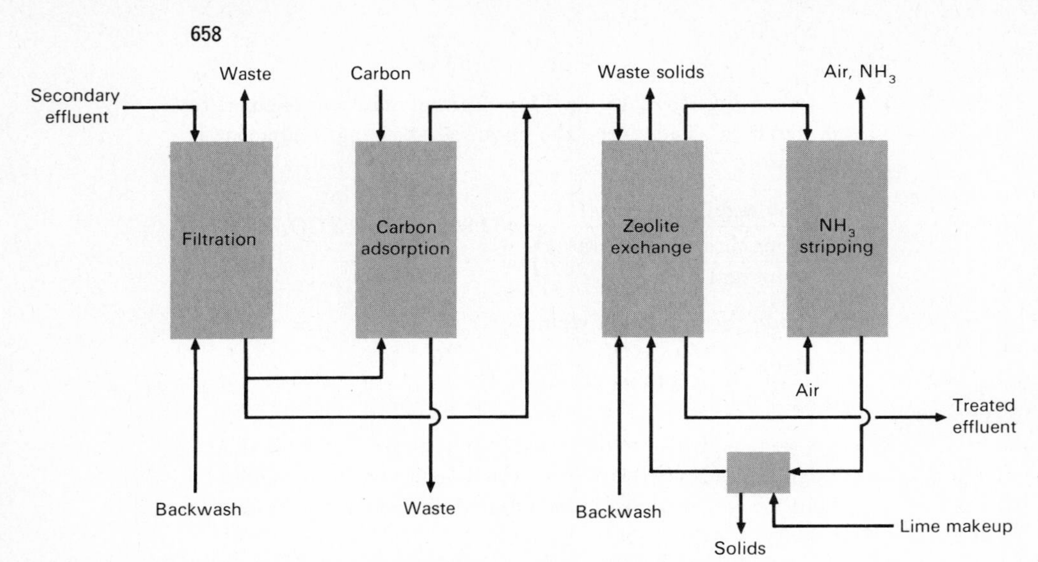

FIG. 14·9 Flowsheet for the removal of ammonia by zeolite exchange [24].

To make ion exchange economical for advanced waste treatment, it would be desirable to use regenerants and restorants that would remove both the inorganic anions and the organic material from the spent resin. Chemical and physical restorants found to be successful in the removal of organic material from resins include sodium hydroxide, hydrochloric acid, methanol, and bentonite [11, 13].

Other Processes

Other chemical processes that have been investigated include electrochemical treatment, electrodialysis, oxidation, and reduction. Here again, as in the case of the less commonly used physical operations, factors other than technical feasibility have limited the application of these processes.

Electrochemical Treatment In this process, wastewater is mixed with seawater and passed into a single cell containing carbon electrodes. Because of the relative densities of the seawater and the seawater-wastewater mixture, the former accumulates in the anode area at the bottom of the cell and the latter accumulates at the cathode area near the top of the cell. The current raises the pH at the cathode, thus precipitating the phosphorus and ammonia as $Ca_3(PO_4)_2$ and $MgNH_4PO_4$ along with $Mg(OH)_2$. Hydrogen bubbles, generated at the cathode, lift the sludge to the surface, where it is skimmed and disposed of by normal means of sludge handling. Chlorine developed at the anode of the cell provides for disinfection of the effluent. The remaining sea-water-wastewater mixture is then discharged to the ocean [14].

Electrodialysis Ionic components of a solution are separated through the use of semipermeable ion-selective membranes in the electrodialysis process. Application of an electrical potential between the two electrodes causes an electric current to pass through the solution, which, in turn, causes a migration of cations toward the negative electrode and a migration of anions toward the positive electrode. Because of the alternate spacing of cation- and anion-permeable membranes, cells of concentrated and dilute salts are formed.

Problems associated with the electrodialysis process for wastewater renovation include chemical precipitation of salts with low solubility on the membrane surface and clogging of the membrane by the residual colloidal organic matter in wastewater treatment plant effluents. To reduce membrane fouling, activated-carbon pretreatment, possibly preceded by chemical precipitation and some form of multimedium filtration, may be necessary.

Oxidation In advanced wastewater treatment applications, chemical oxidation can be used to remove ammonia, to reduce the concentration of residual organics, and to reduce the bacterial and viral content of wastewaters. At present, one of the few processes for the removal of ammonia nitrogen that has been found operationally dependable is chlorination. Ammonia can be removed chemically by adding chlorine or hypochlorite to form monochloramine and dichloramine as intermediate products and nitrogen gas and hydrochloric acid as end products. The equations proposed by Griffin to describe the process with chlorine are as follows [15]:

$$2NH_3 + 2Cl_2 \rightarrow 2NH_2Cl + 2HCl \qquad (14{\cdot}10a)$$

$$NH_2Cl + Cl_2 \rightarrow NHCl_2 + HCl \qquad (14{\cdot}10b)$$

$$NH_2Cl + NHCl_2 \rightarrow N_2 + 3HCl \qquad (14{\cdot}10c)$$

$$2NH_3 + 3Cl_2 \rightarrow N_2 + 6HCl \qquad (14{\cdot}11)$$

Theoretically, on the basis of Eq. 14·11, about 6.3 mg/liter of chlorine are required per mg/liter of ammonia. On the basis of actual operating data, Griffin found that the requirement was about 10 mg/liter of chlorine per mg/liter of ammonia, although higher ratios were not uncommon. Best results were obtained with a reaction time of 2 hr when the temperature was about 45 to 48°F and the pH was between 7 and 9 [15]. One problem in applying this method of treatment to wastewater is the presence of various organic and inorganic compounds that will exert a chlorine demand. Chemical oxidation of the organic material in wastewater with chlorine has been found to be enhanced by the use of ultraviolet radiation [22].

Ozone has been employed as a deodorant, dechlorant, and disinfectant in water treatment for a number of years. The use of ozone to reduce the organic, bacterial, and pathogenic content of treated wastewater has been studied recently by Huibers, et al. [16]. Although technically feasible in certain applications, the costs associated with this form of treatment are prohibitive compared to costs of other treatment methods for achieving the same objectives. For further details references [8, 16] should be consulted.

Reduction Nitrate may be reduced electrolytically and by the use of strong reducing agents [19]. When reducing agents are used the reaction must usually be catalyzed. The reduction of nitrate by ferrous hydroxide can be described by the following two reactions:

$$KNO_3 + 2Fe(OH)_2 + H_2O \rightarrow KNO_2 + 2Fe(OH)_3 \qquad (14\cdot12)$$

$$KNO_2 + 6Fe(OH)_2 + 5H_2O \rightarrow 6Fe(OH)_3 + NH_3 + KOH \qquad (14\cdot13)$$

Other two-step processes using different reducing agents and catalysts have been tried. However, these methods usually are limited by the availability of chemicals at low cost and the fact that the treated effluent and the waste sludge may contain toxic compounds derived from the chemicals used for catalyzing the various reactions.

BIOLOGICAL UNIT PROCESSES

The specific biological processes that will be discussed in detail in this section include (1) bacterial assimilation, (2) harvesting of algae, and (3) nitrification-denitrification. For the most part, these processes have been used principally for the removal of nitrogen in various forms and indirectly for the removal of phosphorus.

Bacterial Assimilation

The removal of nitrogen and phosphorus by biological treatment depends on the fact that the continued and normal growth of microorganisms requires the availability of certain elements and nutrients as well as a source of energy. If the gross composition of the biological cell tissue is represented as $C_5H_7NO_2$, then about 0.13 lb of nitrogen will be required for each pound of cells produced. The amount of phosphorus required is about one-fifth that of nitrogen. If the food source is properly selected and adjusted, it should be possible to convert all soluble forms of nitrogen and phosphorus into organic forms contained in bacterial cells.

Although the complete removal of these elements has been demonstrated in the laboratory, the conversion of the compounds of nitrogen and phosphorus in wastewater to bacterial protoplasm usually requires the addition of carbohydrates or similar materials as a source of food and energy [33]. Without fortification, only about 30 to 40 percent of the nitrogen found in domestic wastewater will be transformed to organic nitrogen in cell tissue by conventional biological treatment processes.

Using methanol as the supplemental energy and food source, the assimilation of ammonia into cell tissue in a conventional biological treatment process can be represented with the following two equations [20]:

Energy reaction:

$$CH_3OH + \tfrac{3}{2}O_2 \rightarrow CO_2 + 2H_2O \qquad (14{\cdot}14)$$

Synthesis:

$$5CH_3OH + NH_3 + \tfrac{5}{2}O_2 \rightarrow C_5H_7NO_2 + 8H_2O \qquad (14{\cdot}15)$$

The total quantity of methanol required will depend on the ratio of the amount required for energy to that required for synthesis. For example, if the ratio is 1 to 3, then the overall methanol requirement can be estimated using the following equation [20]:

$$20CH_3OH + 15O_2 + 3NH_3 \rightarrow 3C_5H_7NO_2 + 5CO_2 + 34H_2O \qquad (14{\cdot}16)$$

The aforementioned ratio used in deriving Eq. 14·16 will vary depending on the mean cell residence time θ_c (see Chap. 10). The removal of ammonia nitrogen by bacterial assimilation is illustrated in Example 14·3.

EXAMPLE 14·3 *Ammonia Nitrogen Assimilation*

The excess ammonia nitrogen present in wastewater that is not now assimilated in an activated-sludge plant treating a flow of 1 mgd is 18 mg/liter as nitrogen. Estimate the concentration of methanol required and the amount of cell tissue produced to bring about the assimilation of the excess ammonia. Assume that the ratio of the methanol required for energy to that for synthesis is equal to 2 to 3.

Solution

1. Derive an overall equation for the specified methanol distribution ratio based on Eqs. 14·14 and 14·15.

$$
\begin{array}{r}
10CH_3OH + \frac{30}{2}O_2 \rightarrow 10CO_2 + 20H_2O \\
15CH_3OH + 3NH_3 + \frac{15}{2}O_2 \rightarrow 3C_5H_7NO_2 + 24H_2O \\
\hline
25CH_3OH + \frac{45}{2}O_2 + 3NH_3 \rightarrow 3C_5H_7NO_2 + 10CO_2 + 44H_2O
\end{array}
$$

2. Determine the total amount of methanol required using the equation derived in step 1.

$$\frac{\text{Methanol, mg/liter}}{\text{Ammonia nitrogen, mg/liter}} = \frac{25(32)}{3(14)} = \frac{800}{42}$$

$$\text{Concentration of methanol} = \frac{800}{42}18 = 343 \text{ mg/liter}$$

3. Determine the amount of cell tissue produced using the equation derived in step 1.

$$\frac{\text{Cells, lb}}{\text{Methanol, lb}} = \frac{3(113)}{25(32)} = \frac{339}{800}$$

$$\text{Methanol added, lb/day} = 1(8.34)(343) = 2{,}860$$

$$\text{Cells, lb/day} = \frac{339}{800}2{,}860 = 1{,}210 \text{ lb/day}$$

Harvesting of Algae

Nitrogen in wastewater may be removed by harvesting algae. The concept involved is the same as that previously described for the biological removal of nitrogen by assimilation through the transformation of soluble and colloidal nitrogen to algal cell tissue. The growth of cell tissue in such a system may be represented by an equation of the following type [37]:

$$106CO_2 + 81H_2O + 16NO_3^- + HPO_4^{--} + 18H^+ + \text{light} \rightarrow C_{106}H_{181}O_{45}N_{16}P + 150O_2 \quad (14\cdot17)$$

Although this process is theoretically feasible, where large quantities of nitrate are to be removed it may be necessary to supplement the untreated waste with carbon dioxide to achieve complete nitrogen removal. At the present time, the major disadvantages of this process are the large land-area requirements and the problems and costs associated with the harvesting and disposal of the algae.

Nitrification-Denitrification

Of the biological treatment methods proposed for the removal of nitrogen, the nitrification-denitrification process appears to be the most promising. The reasons for this are (1) high potential removal efficiency, (2) process stability and reliability, (3) easy process control, (4) land-area requirements, and (5) moderate cost [20]. The removal of nitrogen with this process is carried out in one or two steps, depending on the nature of the wastewater. If the wastewater to be treated contains nitrogen in the form of ammonia, two steps are required. In the first step, the ammonia is aerobically converted to the nitrate NO_3^- form (nitrification). In the second step, the nitrates are anaerobically converted to nitrogen gas (denitrification). If the nitrogen in the wastewater is already in the form of nitrate, as in the case of irrigation return water, only the denitrification step is required.

Nitrification Nitrogen in the form of the ammonium ion is converted to nitrate in two steps by nitrifying autotrophic bacteria, as summarized by the following reactions.

Step 1:

$$NH_4^+ + \tfrac{3}{2}O_2 \xrightarrow{\text{Nitrosomonas}} NO_2^- + 2H^+ + H_2O \tag{14·18}$$

Step 2:

$$NO_2^- + \tfrac{1}{2}O_2 \xrightarrow{\text{Nitrobacter}} NO_3^- \tag{14·19}$$

Overall energy reaction:

$$NH_4^+ + 2O_2 \rightarrow NO_3^- + 2H^+ + H_2O \tag{14·20}$$

Along with obtaining energy, however, some of the ammonium ion is assimilated into cell tissue. A representative synthesis reaction for this autotrophic assimilation is as follows [20].

Synthesis:

$$4CO_2 + HCO_3^- + NH_4^+ + H_2O \rightarrow C_5H_7NO_2 + 5O_2 \tag{14·21}$$

On the basis of the results of both laboratory studies and theoretical calculations, the following overall reaction has been proposed to describe the autotrophic conversion of ammonium ion to nitrate [20].

$$22NH_4^+ + 37O_2 + 4CO_2 + HCO_3^- \rightarrow C_5H_7NO_2 + 21NO_3^- + 20H_2O + 42H^+ \tag{14·22}$$

As noted in Chaps. 10 and 12, nitrification will occur in most aerobic biological treatment processes when the operating and environmental conditions are suitable. In the activated-sludge process, one of the important controlling variables is the mean cell residence time θ_c (see Ref. [10] in Chap. 10).

Denitrification As early as 1860, it was observed that in fermentations taking place in the presence of nitrates, it was common for nitrite, nitrous oxide, and nitrogen gas to be produced. By 1909 it was recognized that the reduction of nitrate (denitrification) involved the use of the nitrate radical as a hydrogen acceptor and that the requirements for this reaction included a source of combined hydrogen and the lack of free oxygen (anaerobic conditions) [41]. Using methanol as the carbon source, the energy reaction may be represented as shown by the following equations.

Energy reaction, step 1:

$$6NO_3^- + 2CH_3OH \rightarrow 6NO_2^- + 2CO_2 + 4H_2O \qquad (14{\cdot}23)$$

Energy reaction, step 2:

$$6NO_2^- + 3CH_3OH \rightarrow 3N_2 + 3CO_2 + 3H_2O + 6OH^- \qquad (14{\cdot}24)$$

Overall energy reaction:

$$6NO_3^- + 5CH_3OH \rightarrow 5CO_2 + 3N_2 + 7H_2O + 6OH^- \qquad (14{\cdot}25)$$

A typical synthesis reaction as given by McCarty [21] is:

Synthesis:

$$3NO_3^- + 14CH_3OH + CO_2 + 3H^+ \rightarrow 3C_5H_7O_2N + H_2O \qquad (14{\cdot}26)$$

In practice 25 to 30 percent of the amount of methanol required for energy is required for synthesis. On the basis of experimental laboratory studies, McCarty [21] developed the following empirical equation to describe the overall nitrate removal reaction.

Overall nitrate removal:

$$NO_3^- + 1.08CH_3OH + H^+ \rightarrow 0.065C_5H_7O_2N + 0.47N_2 + 0.76CO_2 + 2.44H_2O \qquad (14{\cdot}27)$$

If all of the nitrogen is in the form of nitrate, the overall methanol requirement can be determined using Eq. 14·27. However, biologically processed wastewater to be denitrified may contain some nitrite and dissolved oxygen. The methanol requirement where nitrate, nitrite, and dissolved oxygen are present can be computed using the following empirically derived equation [21].

$$C_m = 2.47N_0 + 1.53N_1 + 0.87D_0 \qquad (14{\cdot}28)$$

where C_m = required methanol concentration, mg/liter
N_0 = initial nitrate nitrogen concentration, mg/liter
N_1 = initial nitrite nitrogen concentration, mg/liter
D_0 = initial dissolved oxygen concentration, mg/liter

Because it would take less oxygen to convert ammonia to nitrite and approximately half as much methanol to convert nitrite to nitrogen gas, experimental work is currently under way to define the operational parameters for such a system [34].

Application Assuming sufficient air can be supplied, nitrification generally can be assured in a conventional activated-sludge reactor by increasing the mean cell residence time beyond about 10 days. Although nitrification can be accomplished in the activated-sludge reactor, problems can develop with the operation of the settling tank due to rising sludge (see Chap. 12). To avoid this problem, the nitrification step may be carried out in a separate reactor. This method of operation allows greater process flexibility, and each process (carbonaceous oxidation and nitrification) can be operated independently to achieve optimum performance. Plug-flow reactors, complete-mix reactors, and rock filters (see Fig. 13·9) have been used. Pilot scale nitrification studies are currently (1971) under way at Stanford University using pure oxygen in conjunction with a rock filter.

Denitrification studies have been conducted using plug-flow reactors [28], complete-mix reactors [27], and rock filters [21, 35]. In addition to rock, a variety of different matrix materials on which bacteria can grow have been investigated and have proven to be successful [35].

The nitrification-denitrification flowsheet that has received considerable recent attention for the removal of nitrogen from domestic wastewater is shown schematically in Fig. 14·10. This flowsheet was developed originally on the basis of extensive bench scale laboratory studies [3]. Typical design parameters (preliminary based on limited experience) for each of the processes shown in Fig. 14·10 are given in Table 14·6. The flowsheet shown in Fig. 14·10 can also be used for the removal of phosphorus by adding alum to the raw waste and precipitating it in a preliminary settling tank. Alternatively, alum could be added and precipitated in the activated-sludge settling tank. In addition to removing phosphorus, this technique would also overcome the difficulties of separating organisms growing in the dispersed growth phase. Effluent from the denitrification step could be filtered or precipitated with alum for the removal of residual phosphorus and suspended solids and then filtered. The treated effluent would be aerated before discharge.

As shown in Fig. 14·10 and reported in Table 14·6, a complete-mix reactor would be used for the activated-sludge process and plug-flow–mixed reactors would be used for nitrification and denitrification. Mixing in the denitrification reactor would be accomplished using

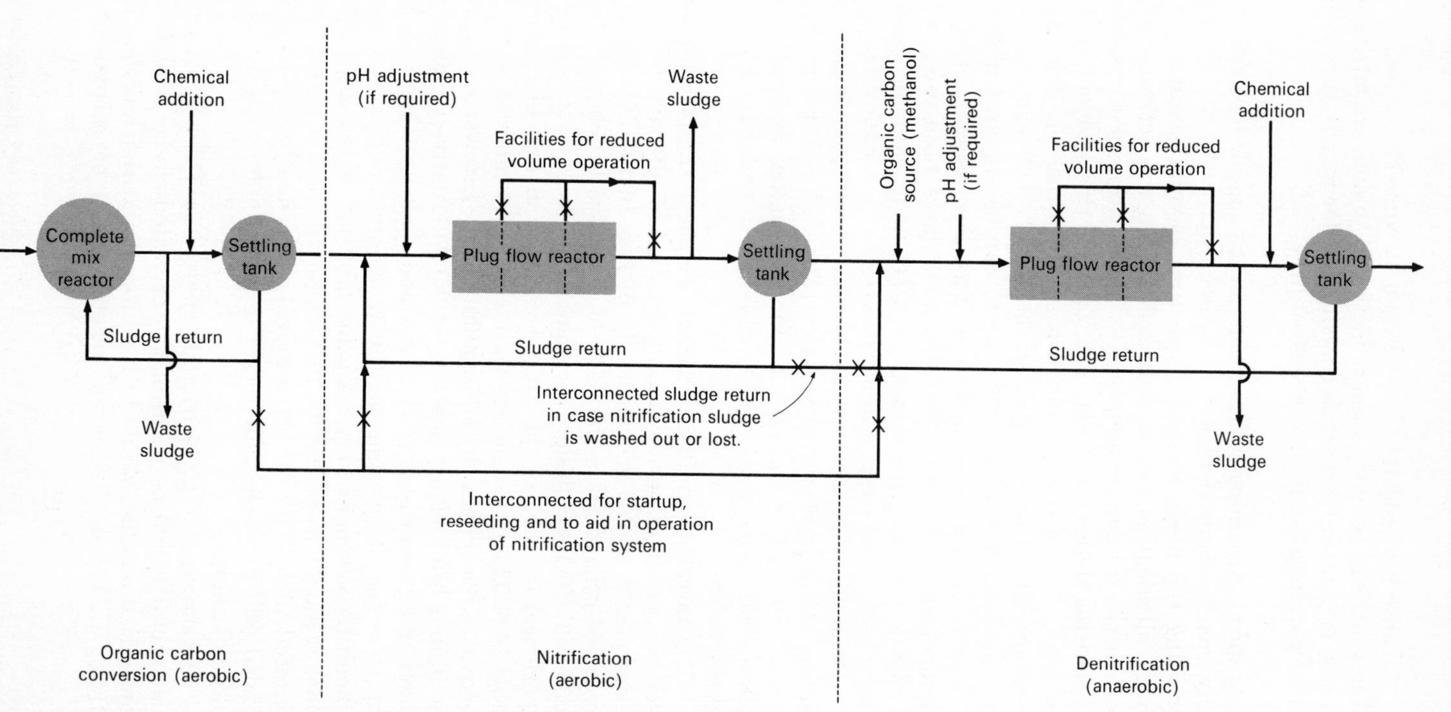

FIG. 14·10 Flowsheet for a three-stage biological treatment process for nitrogen removal.

TABLE 14·6 DESIGN PARAMETERS FOR A THREE-STAGE BIOLOGICAL TREATMENT PROCESS TO REMOVE NITROGEN FROM DOMESTIC WASTEWATER

Treatment				Design parameter				
Process	Description	Type of reactor	Aeration system	θ_c, days[a]	θ, hr[a]	MLVSS, mg/L	pH	Temperature coefficient[b]
Organic carbon conversion (modified biological treatment)	Aerobic conversion of organic carbon to end products and cell tissue	Complete mix	Air or pure oxygen	2–5	1–3	See Table 12·4, Chap. 12	6.5–8.0	1.00–1.03
Nitrification	Aerobic conversion of ammonia and nitrite to nitrate	Plug flow	Air or pure oxygen[c]	10–20	0.5–3	1,000–2,000[d]	7.4–8.6[e]	1.08–1.10[f]
Denitrification[g]	Anaerobic conversion of nitrate to nitrogen gas	Plug flow[h]	....	1–5	0.2–2	1,000–2,000[d]	6.5–7	1.14–1.16[i]

[a] Indicated values for θ_c and θ are for 20°C.
[b] Temperature coefficient to be used in the equation $K_T = K_{20}\theta^{(T-20)}$.
[c] The theoretical oxygen requirement for nitrification can be estimated using Eq. 14·22 and adding to it the value obtained by multiplying the applied BOD by a factor of 1.5.
[d] Higher values may be observed depending on the degree of solids carryover.
[e] Lower pH values have been reported.
[f] Estimated from data developed by Wild, et al. [44] and Mulbarger [28].
[g] Methanol requirement can be computed using Eq. 14·28.
[h] Covered reactors may be used to minimize air-liquid contact.
[i] Estimated from data developed by Mulbarger [28].

submerged paddles [26]. Because of the affect of temperature on the rate of nitrification and denitrification, special attention should be devoted to the design of these tanks and appurtenant facilities. Additional details on the design of the required facilities may be found in ref. [26].

Mean cell residence time, as shown in Table 14·6, is the principal process control factor, and the given values of θ_c are design values based on a temperature of 20°C. Although little full-scale data are available, recommended θ_c values for complete-mix or plug-flow nitrification and denitrification systems vary from 12 to 16 and 2 to 4 days [29], respectively.

The MLVSS in the nitrification reactor will be composed of those organisms responsible for the conversion of organic carbon (BOD) and those responsible for nitrification. The distribution between the two if at present unknown. The total MLSS in the nitrification reactor will normally be 50 to 100 percent higher than the MLVSS and may include residual chemical precipitates, if alum precipitation is used for phosphorus removal. In denitrification reactors the MLVSS have been observed to be about 40 to 70 percent of the MLSS [29 and 34].

The effect of pH on the nitrification reaction in laboratory culture systems has been observed for more than 50 years and has been studied recently for systems treating wastewater. With nonacclimated cultures the optimum pH range, as reported by Wild, et al. [44], was found to vary from about 8.2 to 8.6. On either side of this range performance dropped off rapidly. Although lower operational pH values (6.5–7.0) have been reported, pH adjustment may be necessary. For the denitrification reaction the optimum pH range with nonacclimated cultures has been found to be between 6.5 and 7.5 [28]. Adjustment of the pH of the effluent from the nitrification reactor may also be required if the amount of CO_2 produced in the oxidation of the carbonaceous matter in the denitrification reactor is insufficient to lower it to the desired range.

Another factor that will significantly affect performance is temperature. It cannot be overstressed that, unless properly considered in the design of both nitrification and denitrification systems, effluent quality (measured in terms of the amount of ammonia or nitrate in the effluent) will deteriorate at low temperatures. For example, using a temperature coefficient of 1.12 (see Table 14·6), the reactor volume at 10°C would be approximately three times the size of that required at 20°C. From a design standpoint this could be accomplished using a plug-flow–mixed reactor whose length can be reduced. Alternatively, the solids in the system could be increased to accommodate cold weather operation.

ULTIMATE DISPOSAL OF CONTAMINANTS

Various methods of ultimate disposal have been studied and a summary of the more important ones is given in Table 14·7. From a review of Table 14·7, it can be concluded that there are three general possibilities: (1) land and/or ocean disposal, (2) conversion and land and/or ocean disposal, and (3) conversion, product recovery, and land and/or ocean disposal. Of these methods the last, although the most desirable, is at present the least obtainable.

TABLE 14·7 METHODS FOR THE ULTIMATE DISPOSAL OF CONCENTRATED CONTAMINANTS RESULTING FROM ADVANCED WASTEWATER TREATMENT

Disposal method	Remarks
	Liquid
Evaporation ponds	Provisions must be made to prevent ground water contamination.
Spreading on soil	Provisions must be made to prevent ground water contamination.
Shallow-well injection	Provisions must be made to prevent ground water contamination.
Deep-well injection	Porous strata, natural or artificial cavities should be available.
Landfill	Liquid used as a wetting agent to increase compaction.
Controlled evaporation	Depends on liquid volume, power costs, and local conditions.
Ocean discharge	Truck, rail hauling, or pipeline needed for transportation.
	Sludge
Spreading on soil	Sludge may be pretreated to aid dewatering or to remove objectionable components.
Lagooning	Provisions must be made to prevent ground water contamination.
Landfill	Sludge used as a wetting agent to increase compaction.
Recovery of products	Depends on sludge characteristics, recovery technology, and costs.
Wet combustion	Heat value may be recovered for use. Disposal of ash required.
Incineration	Concentration of sludge is needed. Disposal of ash required.
Ocean discharge	May not be allowed in the future.
	Ash
Landfill	Mixed with refuse to increase compacted density of landfill.
Soil conditioner	Depends on waste characteristics.
Ocean discharge	May not be allowed in the future.

The cost of ultimate disposal for the methods reported in Table 14·7 varies over an extremely wide range depending on geographic location and local conditions. Consequently, ultimate disposal methods must be considered carefully because this cost component may be the controlling factor governing the overall feasibility of an advanced wastewater treatment plan.

PROBLEMS

14·1 A treated wastewater contains 25 mg/liter of ammonia nitrogen which is completely stripped from the wastewater with 800 cu ft of air per gal of wastewater in a stripping tower. The temperature of operation is 15°C. Evaluate the efficiency of this operation in terms of air-to-liquid ratio. What changes could be made to reduce the air-to-liquid ratio?

14·2 Prepare a one-page abstract of each of the following articles: (*a*) G. Tchobanoglous: Filtration Techniques in Tertiary Treatment, *J. WPCF*, vol. 42, no. 4, 1970; (*b*) G. Tchobanoglous and R. Eliassen: Filtration of Treated Sewage Effluent, *Journal of the Sanitary Division*, ASCE, vol. 96, no. SA2, 1970.

14·3 A quantity of sodium-form ion-exchange resin (5 g) is added to a water containing 2 meq of potassium chloride and 0.5 meq of sodium chloride. Calculate the residual concentration of potassium if the exchange capacity of the resin is 4.0 meq/g of dry weight and the selectivity coefficient is equal to 1.46.

14·4 An anionic resin is to be used to remove nitrate ions by ion exchange. If the resin has an exchange capacity of 25 kgr as $CaCO_3$/cu ft, determine the theoretical and design volumes of resin to treat 0.25 mgd of wastewater containing 20 mg/liter of nitrate nitrogen. Assume the leakage factor is 1.3.

14·5 An activated-sludge plant produces an effluent containing 10 mg/liter of ammonia nitrogen. If the plant flowrate is 2.5 mgd, determine the methanol requirement and cell production in pounds per day using Eq. 14·16 for bacterial assimilation of ammonia.

14·6 Design a three-stage biological treatment process to remove nitrogen by nitrification followed by dentrification. After preliminary treatment, 5 mgd of wastewater containing 150 mg/liter of BOD_5 and 30 mg/liter of ammonia nitrogen is to be treated. Give the detention times and volumes for the three reactors. Using Eqs. 14·22 and 14·7, determine the oxygen and methanol requirements for nitrification and denitrification. Determine the sludge produced in each reactor.

14·7 A conventional activated-sludge plant treating 1 mgd of wastewater is to be operated to produce a nitrified effluent. How would this be done? Assuming a nitrified effluent is produced containing 15 mg/liter of nitrate nitrogen, 1.5 mg/liter of nitrite nitrogen, and 2.0 mg/liter of dissolved oxygen, compute the methanol requirement for denitrification. How will the activated-sludge effluent BOD affect the methanol requirement?

REFERENCES

1. *Advanced Waste Treatment Research*—1, *Summary Report June* 1960–*December* 1961, U.S. Department of Health, Education, and Welfare, 1962.
2. Applebaum, S. B.: *Demineralization by Ion Exchange*, Academic, New York, 1968.

3. Barth, E. F., R. C. Brenner, and R. F. Lewis: Chemical-Biological Control of Nitrogen and Phosphorus in Wastewater Effluents, *J. WPCF*, vol. 40, no. 12, 1968.

4. Chapman, R. F. (ed.): *Separation Processes in Practice*, Reinhold, New York, 1961.

5. Cole, C. A., and E. J. Genetelli: Pervaporation of Volatile Pollutants from Water Using Selective Hollow Fibers, presented at the 42nd Annual Conference of the Water Pollution Control Federation, Dallas, 1969.

6. Craft, T. F.: Comparison of Sand and Anthracite for Rapid Filtration, *J. AWWA*, vol. 63, no. 1, 1971.

7. Culp, R. L.: Wastewater Reclamation by Tertiary Treatment, *J. WPCF*, vol. 35, no. 6, 1963.

8. Culp, R. L., and G. L. Culp: *Advanced Wastewater Treatment*, Van Nostrand Reinhold, New York, 1971.

9. Culp, G. L., and A. Slechta: *Nitrogen Removal from Sewage*, Final Progress Report, USPHS Demonstration Grant-26-01, 1966.

10. *Development of a Process to Remove Carbonaceous, Nitrogenous, and Phosphorus Materials from Anaerobic Digester Supernatant and Related Process Streams*, Progress Report: Phase 1, prepared for Federal Water Pollution Control Administration, Environmental Engineering Laboratories, FMC Corporation, Santa Clara, 1969.

11. Eliassen, R., and G. E. Bennett: Anion Exchange and Filtration Techniques for Wastewater Renovation, *J. WPCF*, vol. 39, no. 10, part 2, 1967.

12. Eliassen, R., and G. Tchobanoglous: Removal of Nitrogen and Phosphorus from Waste Water, *Environmental Science and Technology*, vol. 3, no. 6, 1969.

13. Eliassen, R., B. M. Wyckoff, and C. D. Tonkin: Ion Exchange for Reclamation of Reusable Supplies, *J. AWWA*, vol. 57, no. 9, 1965.

14. Föyn, E.: Removal of Sewage Nutrients by Electrolytic Treatment, *Intern. Ver. Theoret. Angew. Limnol., Verhandl.*, 15, 1962. Published in English, 1964.

15. Griffin, A. E.: Removal of Ammonia by Chlorination, *Proceedings Fifth Annual Water Conference*, The Engineers' Society of Western Pennsylvania, 1944.

16. Huibers, D. T. A., R. McNabney, and A. Halfon: *Ozone Treatment of Secondary Effluents from Wastewater Treatment Plants*, Federal Water Pollution Control Administration, Robert A. Taft Water Research Center, Cincinnati, Ohio, 1969.

17. Lackey, J. B., and C. N. Sawyer: Plankton Productivity of Certain Southeastern Wisconsin Lakes as Related to Fertilization, *Sewage Works Journal*, vol. 17, p. 573, 1945.

18. Levin G. V., and J. Shapiro: Metabolic Uptake of Phosphorus by Wastewater Organisms, *J. WPCF*, vol. 37, no. 6, 1965.

19. Maddock, A. G., and A. G. Sharpe (eds.): *Recent Aspects of the Inorganic Chemistry of Nitrogen*, Special Publication 10, The Chemical Society, Burlington House, W. I. London, 1967.

20. McCarty, P. L.: Biological Processes for Nitrogen Removal: Theory and Application, *Proceedings Twelfth Sanitary Engineering Conference*, University of Illinois, Urbana, 1970.

21. McCarty, P. L., L. Beck, and P. St. Amant: Biological Denitrification of Wastewaters by Addition of Organic Materials, *Proceedings of the 24th Purdue Industrial Waste Conference*, Lafayette, Ind., 1969.

22. Meiners, A. F., *et al.*: *An Investigation of Light-Catalyzed Chlorine Oxidation for Treatment of Wastewater*, Federal Water Pollution Control Administration, Robert A. Taft Water Research Center, Cincinnati, Ohio, 1968.

23. Menar, A. B., and D. Jenkins: The Fate of Phosphorus in Waste Treatment Processes: The Enhanced Removal of Phosphate by Activated Sludge, *Proceedings of the 24th Purdue Industrial Waste Conference,* Lafayette, Ind., 1969.

24. Mercer, B. W., *et al.*: Ammonia Removal from Secondary Effluents by Selective Ion Exchange, *J. WPCF,* vol. 42, no. 10, 1969.

25. Merten, U.: *Desalination by Reverse Osmosis,* M.I.T., 1966.

26. Metcalf & Eddy, Inc.: *Design of Nitrification and Denitrification Facilities,* Prepared for Environmental Protection Agency Symposium on Design of Wastewater Treatment Facilities, Cleveland, Ohio, April 1971.

27. Moore, S. F., and E. D. Schroeder: An Investigation of the Effects of Residence Time on Anaerobic Bacterial Denitrification, *Water Research,* vol. 4, no. 10, 1970.

28. Mulbarger, M. C.: Modification of the Activated Sludge Process for Nitrification and Denitrification, *J. WPCF,* vol. 43, no. (in press), 1971.

29. Mulbarger, M. C.: Personal Communication, May 1971.

30. Nesbitt, J. B.: *Removal of Phosphorus from Municipal Sewage Plant Effluents,* Engineering Research Bulletin B-93 College of Engineering, Pennsylvania State University, University Park, 1966.

31. Prather, B. F.: Waste-Water Aeration, *The Oil and Gas Journal,* vol. 57, no. 49, 1959.

32. Rickles, R. N.: *Membranes: Technology and Economics,* Noyes Development Corporation' Park Ridge, N.J., 1967.

33. Sawyer, C. N.: Biological Engineering in Sewage Treatment, *Sewage Works Journal,* vol. 16, no. 9, 1944.

34. Schroeder, E.: Personal Communication, May 1971.

35. Seidel, D. F., and R. W. Crites: Evaluation of Anaerobic Denitrification Processes, *Journal of the Sanitary Division,* ASCE, vol. 96, no. SA2, 1970.

36. Slechta, A. F., and G. L. Culp: Water Reclamation Studies at the South Tahoe Public Utility District, *J. WPCF,* vol. 39, no. 5, 1967.

37. Stumm, W., and M. W. Tenney: Waste Treatment for the Control of Heterotrophic and Autotrophic Activity in Receiving Waters, *Proceedings of the 12th Municipal and Industrial Waste Conference,* Raleigh, N.C., 1964.

38. Tchobanoglous, G.: Physical and Chemical Processes for Nitrogen Removal—Theory and Application, *Proceedings Twelfth Sanitary Engineering Conference,* University of Illinois, Urbana, 1970.

39. Tchobanoglous, G.: Filtration Techniques in Tertiary Treatment, *J. WPCF,* vol. 42, no. 4, 1970.

40. Tchobanoglous, G., and R. Eliassen: Filtration of Treated Sewage Effluent, *Journal of the Sanitary Division,* ASCE, vol. 96, no. SA2, 1970.

41. Thimann, K. V.: *The Life of Bacteria,* 2d ed., Macmillan, New York, 1963.

42. Treybal, R. E.: *Mass-Transfer Operations,* McGraw-Hill, 2d ed., New York, 1967.

43. Weber, W., Jr., C. B. Hopkins, and R. Bloom, Jr.: Physicochemical Treatment of Wastewater, *J. WPCF,* vol. 42, no. 1, 1970.

44. Wild, H. E., C. N. Sawyer, and T. C. McMahon: Factors Affecting Nitrification Kinetics, *J. WPCF,* vol. 43, no. (in press), 1971.

45. Yee, W. C.: Selective Removal of Mixed Phosphates by Activated Alumina, *J. AWWA,* vol. 58, no. 2, 1966.

water-pollution control and effluent disposal

15

The sanitary engineer can design a treatment plant to accomplish as much removal of pollutants as may be required. Ultimate disposal of wastewater effluents will be by dilution in receiving waters; by discharge on land; or, in some cases in desert areas, by evaporation into the atmosphere as well as seepage into the ground. Disposal by dilution (after preliminary or biological treatment) in larger bodies of water, such as lakes, rivers, estuaries, or oceans, is by far the most common method.

The fundamental thesis governing the disposal of effluents and the regulation of pollution has been to make the treatment plants do part of the work and to let nature complete it. The amount of natural or self-purification that occurs in the receiving water depends on its flow or volume, its oxygen content, and its ability to reoxygenate itself. The proportion of the self-purification capacity, sometimes called the assimilative capacity, that can be safely utilized in rivers, lakes, or estuaries depends on the uses to which the water is subjected elsewhere, the desires of the people, and the total economy of the receiving-water system. Sanitary and public health engineers must be employed by government regulatory agencies at all levels, and it is also desirable that other public officials understand the self-purification characteristics of receiving waters and the effects of the constituents in wastewater on their quality so that they can set appropriate standards regulating waste discharges. The sanitary engineer should also understand these principles so that he can select the degree of treatment and type of plant required, based on the applicable receiving-water or effluent standards.

Treated wastewater has been reused, either directly or indirectly, at a number of places. The practice has definite advantages and is growing in areas of the country where rainfall is light and water is scarce and must be conserved.

WATER-POLLUTION CONTROL

Water-pollution control includes the development of methods and means of achieving and maintaining the desired water quality, which depends on the uses to be made of the water. Thus there are water-quality criteria for domestic water supply, industrial water supply, agricultural water supply, recreation, aesthetics, fish, other aquatic life, and wildlife. The first National Technical Advisory Committee on Water Quality Criteria has collected data and recommended criteria for various beneficial uses [12]. Water-quality criteria for industrial use were suggested by a committee of the New England Water Works Association in 1940 and are still generally applicable [4].

Receiving-Water Standards

Once the criteria necessary for the protection of the various beneficial uses have been established, it is possible to set standards for surface waters with the stipulation that no discharge shall create conditions that violate them. These standards are known as receiving-water or stream standards. As an example, the standards established by the Ohio River Sanitation Commission (ORSANCO) for surface waters to be used for various purposes in the Ohio River Valley are given in Appendix E. Some state and interstate commissions have taken the approach of classifying streams in several categories in accordance with the highest beneficial use to be made of the stream. This use is based, to a certain extent, on existing conditions. Recently, all states were ordered by the federal government to adopt standards, subject to federal approval, that would maintain or enhance the existing quality of receiving waters; and all have now complied.

Effluent Standards

A difficulty in enforcing receiving-water standards arises when the combined load of several dischargers exceeds the self-purification capacity of the receiving waters. In this case, it may be difficult for the regulatory agency to allocate the blame. It is also possible for the discharger farthest upstream in a river basin to preempt most of the assimilative capacity for his own use, leaving little or none for those located downstream, which is unfair. To avoid these problems, regulatory agencies sometimes set effluent standards, which are relatively easy to enforce compared with receiving-water standards. Effluent standards should be strict enough to protect the quality of the receiving waters, should treat all dischargers fairly, but in so far as possible should be tailored to the character and volume of wastes at each point of discharge. A greater degree of treatment may logically be required of large waste flows than of small or insignificant waste flows.

Setting of Standards

Although the setting of standards comes under the jurisdiction of state and local regulatory agencies, there are fundamental considerations applicable to the setting of standards. It should be obvious that oil, grease, and floating solids, especially those of sewage origin, should be removed from wastes before discharge to receiving waters. Solids that may settle and form sludge banks should also be removed. These requirements should apply to all wastes and all receiving waters.

Degradable organic matter, through exertion of its BOD, uses the dissolved oxygen of the receiving waters to stabilize the wastes. The dissolved oxygen is replenished mainly by atmospheric reaeration, photosynthesis of algal life, and additional downstream dilution (see "Oxygen Resources in Rivers"). The minimum dissolved oxygen necessary for fish life must be assured by the setting of proper standards. A common minimum requirement is 5.0 mg/liter of dissolved oxygen. In canals utilized exclusively for drainage, where fish are not a factor, the dissolved oxygen must not be allowed to drop to zero or a nuisance due to foul odors may develop.

Where water supplies, bathing beaches, livestock, or water-contact sports may be affected, the bacterial content of the receiving waters, as measured by the number of coliform organisms, is of prime importance. Bacterial numbers are reduced by initial dilution and also by a relatively rapid death rate with the passage of time and distance from the point of discharge.

Conservative pollutants (those that do not decay), such as chemical constituents, are reduced in concentrations mainly by dilution. The dilution necessary for the discharge of toxic compounds may be assessed by the bioassay test (see Chap. 7).

The discharge of nutrients into small lakes and reservoirs and also into sluggish streams consisting of a series of pools or slack-water reaches, may promote excessive growths of algae and other forms of microscopic life. The result of this growth is undesirable turbidity and floating scum while the organisms are growing. As the organisms die, odors are produced, oxygen is consumed in the decay of cells, and the nutrients are returned to the watercourse. Where such conditions can develop, reduction of the nutrient content of wastewaters before discharge may be necessary. Fortunately, however, such conditions are not general.

Heated-water discharges, after initial mixing, should not increase the temperature of the main body of the receiving waters above 95°F if fish life is to be preserved. An excessive increase in the temperature of the wastewater–diluting water mixture above that of the diluting water should not be permitted. Increases should be limited to 3°F to 5°F, according to the report of the National Technical Advisory Committee [12]. At the present time, the permissible temperature increase is a matter of controversy. Detailed discussion of thermal pollution may be found in Krenkel and Parker [8, 13].

EFFLUENT DISPOSAL BY DILUTION

To describe the effects of a waste discharge on a body of water, the sanitary engineer must understand the physical phenomena that are occurring in that body of water. Any mathematical relationship, no matter how complicated, is simply an attempt to describe the physical phenomena. Therefore, a physical rather than purely mathematical understanding of a wastewater system is a prerequisite to intelligent analysis. The subjects discussed in this section include the special conditions applying to disposal by dilution in lakes, rivers, estuaries, and oceans, in that order, and the development of mathematical relationships that can be applied to the majority of cases arising in practice.

Disposal into Lakes

Lakes and reservoirs are often subject to significant mixing due to wind-induced currents; therefore, it is often reasonable to assume that small lakes and reservoirs are completely mixed. The theoretical model based upon this assumption is shown in Fig. 15·1. To simplify the calculations, it is assumed that flowrates are constant and that the decay of pollutants follows the first-order reaction law. Writing a mass balance through the lake:

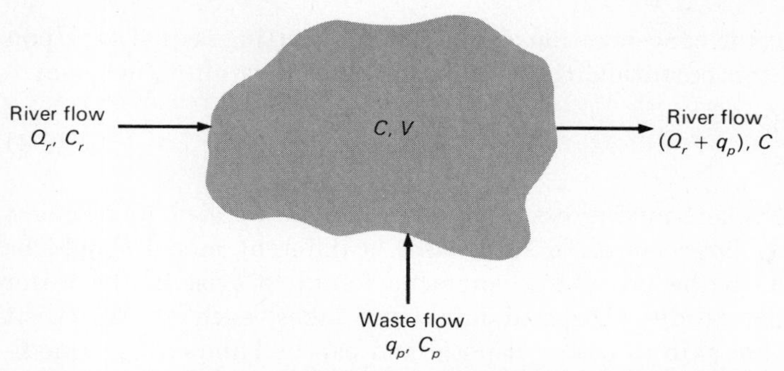

FIG. 15·1 Schematic of complete-mix model for small lakes and reservoirs.

$$\begin{array}{ccccccccc} \text{Change in storage} & = & \text{inflow} & - & \text{outflow} & + & \text{sources} & - & \text{sinks} \\ V\,dC & = & (Q_rC_r\,dt + q_PC_P\,dt) & - & (Q_r + q_P)C\,dt & + & 0 & - & VCK\,dt \end{array} \tag{15·1}$$

where V = lake volume
dC = change in pollutant concentration in the lake
Q_r = river flowrate into lake
C_r = concentration of pollutant in river
q_P = wastewater flowrate into lake
C_P = concentration of pollutant in wastewater
K = first-order decay constant

Now let

$$W = Q_rC_r + q_PC_P$$

and the detention time

$$t_0 = \frac{V}{Q}$$

where Q is the total flow into the lake $Q_r + Q_P$. Substituting the foregoing values into Eq. 15·1 yields

$$\frac{dC}{dt} + C\left(\frac{1}{t_0} + K\right) = \frac{W}{V} \tag{15·2}$$

This is a first-order linear differential equation, which, upon integration, yields

$$C = \frac{W}{\beta V}(1 - e^{-\beta t}) + C_0e^{-\beta t} \tag{15·3}$$

where $\beta = 1/t_0 + K$
C_0 = concentration in the lake at time $t = 0$

The equilibrium concentration can be found by letting t equal ∞. Upon performing this operation, the equilibrium concentration C_e becomes

$$C_e = \frac{W}{\beta V} \tag{15·4}$$

The complete-mixing assumption cannot be applied to stagnant or extremely large lakes. In such cases, a different model should be used, based on the physical phenomena found to exist in the water system under study. (Disposal into large lakes, such as the Great Lakes, is more akin to ocean disposal and can be handled by a modification of the method of Example 15·4.)

Particularly significant is the vertical stratification common during certain seasons of the year. A complete-mix model would not be a good representation of a stratified lake because waste would not normally distribute itself over the entire lake volume. However, a soluble waste might be considered completely mixed in the upper layers (epilimnion) of a small, deep lake.

Stratification in lakes is the result of an increase in water density with depth caused by a decrease in temperature. The maximum density occurs at 4°C. During the spring, most lakes are of nearly uniform temperature and therefore are easily mixed by wind currents. As summer approaches, the upper waters are warmed, their density is thereby decreased, and a stable stratification is produced.

Three zones are normally present in a stratified lake: epilimnion, thermocline, and hypolimnion. These zones are identified in Fig. 15·2. The epilimnion may be 30 to 50 ft deep and is fairly uniform in temperature because of mixing by wind action. The thermocline is a zone of significant temperature change and is extremely resistant to mixing. During the fall, temperatures drop, decreasing the amount of stratification until wind action may again completely mix the lake waters.

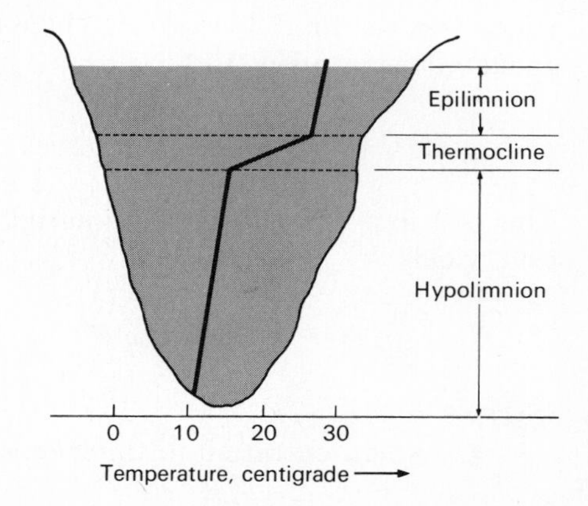

FIG. 15·2 Schematic representation of summer stratification in a lake.

This phenomenon is known as the fall turnover. In colder climates where the temperature drops below 4°C, a winter stratification may also occur because water density again decreases as the temperature falls below 4°C. It is also important to evaluate such parameters as wind, waves, and currents, because most mixing in a lake or reservoir body is the result of these forces. The determination of equilibrium waste concentration in small lakes is illustrated in Example 15·1.

EXAMPLE 15·1 *Radioactive Waste Concentration in a Lake*

A radioactive waste with a half-life of 1 day, a concentration of 10 mg/liter, and a flow of 5 cfs is continually discharged to a small lake with an average length of 300 ft, an average width of 150 ft, and an average depth of 10 ft. Wind currents are such that the lake contents are completely mixed. Determine the equilibrium concentration of radioactive waste in the lake.

Solution

1. Determine the decay constant for the waste.

$T_{\frac{1}{2}}$ = half-life = 1 day

$L_t = L_0 e^{-Kt}$ (first-order decay rate)

When $t = 1$ day,

$$L_t = \tfrac{1}{2} L_0$$
$$\tfrac{1}{2} L_0 = L_0 e^{-K(1.0)}$$
$$\ln(\tfrac{1}{2}) = -K(1.0)$$
$$K = 0.693 \text{ per day}$$

2. Determine the lake volume.

$$V = 300(150)(10) = 450{,}000 \text{ cu ft}$$

3. Determine the equilibrium concentration of the waste. Assume the waste flow to represent the total inflow to the lake.

$$t_0 = \frac{V}{Q} = \frac{450{,}000}{5(86{,}400)} = 1.04 \text{ days}$$

$$W = q_p C_p = (5 \text{ cfs})(10 \text{ mg/liter}) = 432{,}000 \text{ cu ft/day } (10 \text{ mg/liter})$$

$$\beta = \frac{1}{t_0} + K = \frac{1}{1.04} + 0.693 = 1.653 \text{ per day}$$

$$C_e = \frac{W}{\beta V} = \frac{\left(\dfrac{432{,}000 \text{ cu ft}}{\text{day}}\right)(10 \text{ mg/liter})}{\left(\dfrac{1.653 \times 1}{\text{day}}\right)(450{,}000 \text{ cu ft})}$$

$$C_e = 5.8 \text{ mg/liter}$$

Disposal into Rivers

The late Earle B. Phelps, one of the early leaders in the field of stream sanitation, wrote: "A stream is something more than a geographic feature, a line on a map, a part of the fixed permanent terrain. It cannot be adequately portrayed in terms of topography and geology. A stream is a living thing, a thing of energy, of movement, of change" [15]. Streams or rivers are subject to much natural pollution because they serve as drainage channels for large areas of the countryside. In addition, rivers are capable of absorbing some man-made pollution because they possess the ability to purify themselves through the action of living organisms that consume organic matter and the sedimentation process that contributes to the river bottom.

Oxygen Resources in Rivers Aside from the oxygen contained in tributaries, surface drainage, and ground water inflow, the sources of oxygen replenishment in a river are reaeration from the atmosphere and photosynthesis of aquatic plants and algae. The amount of reaeration is proportional to the dissolved-oxygen deficiency, while the amount of oxygen supplied by photosynthesis is a function of the size of the algal population and the amount of sunlight reaching the algae. There is more incident radiation when the sun is high in the sky than when it is near the horizon; therefore, the rate of photosynthesis is assumed to be sinusoidal. Respiration, on the other hand, is assumed to be constant because it does not depend on light radiation. Where large populations of algae are present, a diurnal variation in dissolved-oxygen concentration occurs, as shown in Fig. 15·3.

The reaeration constant K_2 is defined in Eq. 15·5

$$-\frac{dD}{dt} = K_2 D \qquad (15\cdot5)$$

where D = oxygen deficit
K_2 = reaeration constant

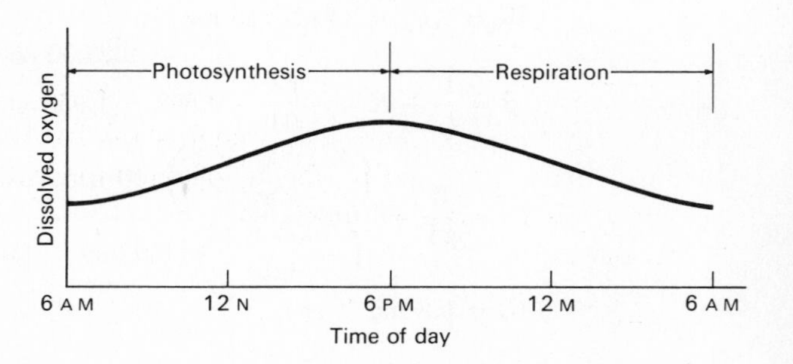

FIG. 15·3 Diurnal variation of dissolved oxygen in water containing large algal populations.

Values of K_2 have been estimated for various water bodies by the Engineering Board of Review for the Sanitary District of Chicago [17]. These values are given in Table 15·1.

TABLE 15·1 REAERATION CONSTANTS [17]

Water body	Ranges of K_2 at 20°C (base 10)*
Small ponds and backwaters	0.05–0.10
Sluggish streams and large lakes	0.10–0.15
Large streams of low velocity	0.15–0.20
Large streams of normal velocity	0.20–0.30
Swift streams	0.30–0.50
Rapids and waterfalls	>0.50

* For values to the base e, multiply by 2.3.

The concept of biological oxidation of organic matter was introduced in Chap. 10. The amount of oxygen required to stabilize a waste is normally measured by the BOD test; BOD is therefore the primary source of oxygen depletion or utilization in a waterway. Matter that is too heavy to remain in suspension will settle out, forming a sludge deposit or benthal layer on the bottom of the stream. Sludge deposits in the bottoms of slow-moving rivers can exert a significant oxygen demand on the overlying water. Although most of the sludge will be undergoing anaerobic decomposition, which is a relatively slow process, aerobic decomposition can take place at the interface between the sludge and the flowing water.

Rates of deposition and scour vary with the velocity and turbulence of the river. At times, sedimentation may reduce the BOD load in the river, if settleable solids are being discharged or if coagulation of colloidal matter takes place. At other times, scour will increase the BOD load by returning these particles to the river. Each reach of river must be analyzed for benthal load to determine its importance in the total oxygen balance. In many rivers, it is a factor that can be eliminated from consideration.

Where organic mud and sludge deposits are appreciable, their effect can be evaluated by means of an empirical equation developed by Fair, Moore, and Thomas [5]:

$$Y_m = 3.14(10^{-2}y_0)C_T w \left(\frac{5 + 160w}{1 + 160w}\right) \sqrt{t_a} \qquad (15\cdot6)$$

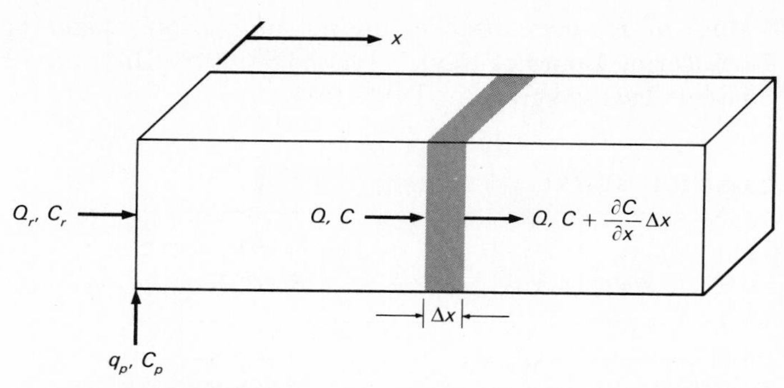

FIG. 15·4 Definition sketch for plug-flow model used in river analysis.

where Y_m = maximum daily benthal oxygen demand, gm/sq m
y_0 = BOD_5 of benthal deposit, gm/kg of volatile matter at 20°C
w = daily rate of volatile-solids deposition, kg/sq m
t_a = time during which settling takes place, days
C_T = temperature correction factor
= BOD_5 @ T/BOD_5 @ 20°C $= (1 - e^{-5K_T'})/(1 - e^{-5K_{20}'})$

Mathematical Model In most river analyses, it is assumed that the wastes are evenly distributed over the cross section of the river. This may be far from the truth in the immediate vicinity of the outlet, but the validity of the assumption improves in most cases as the wastes proceed downstream. It can also be assumed that no mixing occurs along the axis of the river, which is reasonable provided the river is not extremely turbulent. Thus the river model reduces to the plug-flow model shown in Fig. 15·4. Over any incremental volume, the following mass balance can be written, assuming a first-order decay rate for the waste under consideration and no additional sources of pollutant:

$$\text{Change in storage} = \text{inflow} - \text{outflow} + \text{sources} - \text{sinks}$$

$$V\,\Delta C = QC\,\Delta t - Q\left(C + \frac{\partial C}{\partial x}\Delta x\right)\Delta t + 0 - VK\bar{C}\,\Delta t$$

$$V\frac{\Delta C}{\Delta t} = -Q\frac{\partial C}{\partial x}\Delta x - KV\bar{C} \tag{15·7}$$

Let V equal $A\,\Delta x$, where A is the cross-sectional area of the stream. Then

$$\frac{\Delta C}{\Delta t} = -\frac{Q}{A}\frac{\partial C}{\partial x} - K\bar{C}$$

Let Δx and Δt approach zero, and let Q/A equal v, the stream velocity. Then $\bar{C}$ approaches C, and

$$\frac{\partial C}{\partial t} = -v\frac{\partial C}{\partial x} - KC$$

Under steady-state conditions $\partial C/\partial t$ equals zero at any particular point. Hence

$$v\frac{dC}{dx} = -KC$$

Integration, by separation of the variables, yields

$$C = C_0 e^{-K(x/v)} \tag{15·8}$$

This equation applies to both instantaneous and continuous discharges of pollutant at x equals zero. C_0 is the waste concentration in the river-waste mixture at x equals zero. It is given by the following equation:

$$C_0 = \frac{Q_r C_r + q_p C_p}{Q_r + q_p} = \frac{W}{Q} \tag{15·9}$$

The same approach can be used to determine the dissolved-oxygen distribution over the reach of a river downstream from a source of BOD loading. Consider that the dissolved-oxygen content is affected by the following:

$K_2(C_s - C)$ = reaeration
P = photosynthesis
$K(\text{BOD}_L)$ = BOD
R = algal respiration
S = sludge deposits

Performing the mass balance with C as the dissolved-oxygen concentration, the following equation is derived after dividing by $V\,\Delta t$ and allowing Δx and Δt to approach zero as a limit:

$$\frac{\partial C}{\partial t} = -v\frac{\partial C}{\partial x} + K_2(C_s - C) + P - K(\text{BOD}_L) - R - S \tag{15·10}$$

where C_s = saturation concentration of dissolved oxygen
v = flow velocity
x = distance

At steady state, $\partial C/\partial t$ equals zero, and because $C_s - C$ equals the dissolved-oxygen deficit D, Eq. 15·10 can be integrated to find D, using

$$\text{BOD}_{Lx} = \text{BOD}_L e^{-K'(x/v)} \tag{15·11}$$

where BOD_{Lx} = BOD_L at a distance x
BOD_L = ultimate BOD at $x = 0$
K' = BOD-rate constant (base e)

The integrated form of Eq. 15·10 is given by Eq. 15·12.

$$D = \frac{K'BOD_L}{K_2' - K'}\left(e^{-K'(x/v)} - e^{-K_2'(x/v)}\right) + D_0 e^{-K_2'(x/v)} + \frac{S + R - P}{K_2'}\left(1 - e^{-K_2'(x/v)}\right) \quad (15\cdot12)$$

where K_2' = reaeration constant (base e)
D_0 = initial oxygen deficit

The use of this equation requires the evaluation of many parameters, in particular S, R, and P. The magnitude of the effects of algae and sludge deposits on the oxygen economy of a river can be determined only from detailed testing and analysis of the river in question. If these effects are not significant, the last term in Eq. 15·12 drops out and Eq. 15·12 becomes

$$D = \frac{K'BOD_L}{K_2' - K'}\left(e^{-K'(x/v)} - e^{-K_2'(x/v)}\right) + D_0 e^{-K_2'(x/v)} \quad (15\cdot13)$$

This is the classic Streeter-Phelps equation, which is most commonly used in river analysis. It must be used with caution, however, in light of the foregoing discussion. It applies for channels of uniform cross section where effects of algae and sludge deposits are negligible.

The form of the Streeter-Phelps equation is shown in Fig. 15·5, which shows the dissolved-oxygen-sag curve for a river. Active biological decomposition begins immediately after discharge. This decomposition utilizes oxygen. Because atmospheric reaeration is pro-

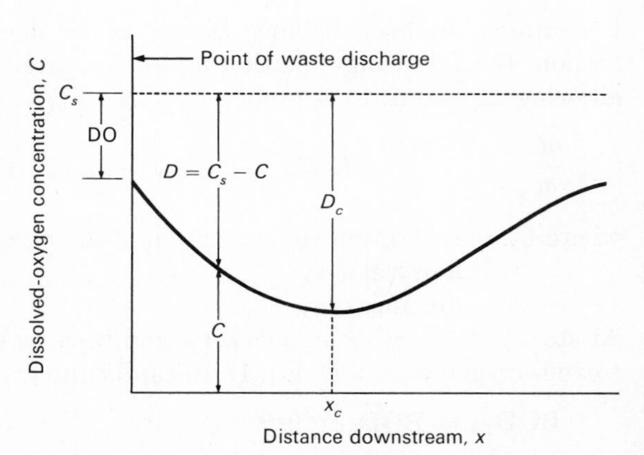

FIG. 15·5 Characteristic oxygen-sag curve obtained using the Streeter-Phelps equation.

portional to the dissolved-oxygen deficit, its rate will increase with increasing deficit. Finally a point is reached at which the rate of oxygen utilized for waste decomposition equals the rate of atmospheric reaeration. This is the point x_c in Fig. 15·5. Downstream from this point, the rate of reaeration is greater than the rate of utilization and the dissolved oxygen begins to increase. Eventually, the stream will show no effects due to the waste discharge. This is the phenomenon of natural stream purification.

The critical dissolved-oxygen deficit D_c at the point x_c is of engineering significance. This critical deficit can be determined because the rate of oxygen utilization equals the rate of reaeration at that point.

$$D_c = \frac{K'}{K'_2} \mathrm{BOD}_L \, e^{-K'(x_c/v)} \tag{15·14}$$

The value of x_c can be determined by differentiating Eq. 15·13 with respect to x and setting dD/dx equal to zero.

$$x_c = \frac{v}{K'_2 - K'} \ln \frac{K'_2}{K'} \left[1 - \frac{D_0(K'_2 - K')}{K'\mathrm{BOD}_L} \right] \tag{15·15}$$

and

$$t_c = \frac{x_c}{v}$$

where t_c equals the time required to reach the critical point. The use of the Streeter-Phelps equation in determining the dissolved-oxygen sag in a river is illustrated in Example 15·2. Additional details on the oxygen-sag method of analysis may be found in Ref. [19].

EXAMPLE 15·2 *Dissolved-Oxygen Sag in a River*

A city discharges 30 mgd of sewage into a stream whose minimum rate of flow is 300 cfs. The velocity of the stream is about 2 mph. The temperature of the sewage is 20°C, while that of the stream is 15°C. The 20°C BOD_5 of the sewage is 200 mg/liter, while that of the stream is 1.0 mg/liter. The sewage contains no dissolved oxygen, but the stream is 90 percent saturated upstream of the discharge. At 20°C, K' is estimated to be 0.30 per day while K'_2 is 0.7 per day. Determine the critical-oxygen deficit and its location. Also estimate the 20°C BOD_5 of a sample taken at the critical point. Use temperature coefficients of 1.135 for K' and 1.024 for K'_2. The dissolved-oxygen-sag curve is plotted on Fig. 15·6.

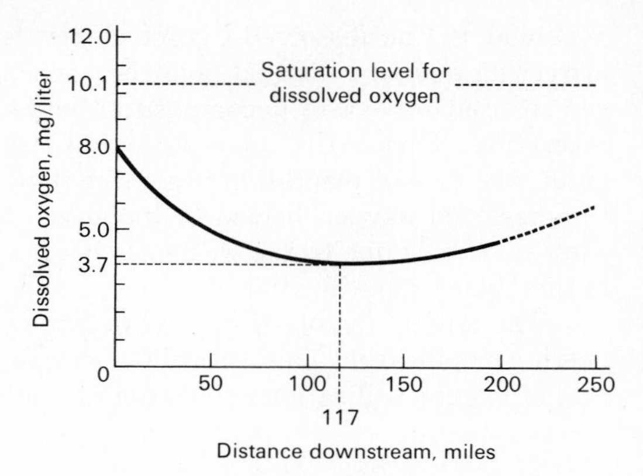

FIG. 15·6 Oxygen-sag curve for Example 15·2.

Solution

1. Determine the dissolved oxygen in the stream before discharge.

Saturation concentration,

$$15°\text{C (see Appendix C)} = 10.2 \text{ mg/liter}$$

$$\text{Dissolved oxygen in stream} = 0.9(10.2) = 9.2 \text{ mg/liter}$$

2. Determine the temperature, dissolved oxygen, and BOD of the mixture.

$$\text{Temperature of mixture} = \frac{30(1.55)(20) + 300(15)}{30(1.55) + 300} = 15.7°\text{C}$$

$$\text{Dissolved oxygen of mixture} = \frac{30(1.55)(0) + 300(9.2)}{30(1.55) + 300} = 8.0 \text{ mg/liter}$$

$$\text{BOD}_5 \text{ of mixture} = \frac{30(1.55)(200) + 300(1)}{30(1.55) + 300} = 27.7 \text{ mg/liter}$$

$$\text{BOD}_L \text{ of mixture} = \frac{27.7}{1 - e^{-0.3(5)}} = 35.6 \text{ mg/liter}$$

3. Correct the rate constants to 15.7°C.

$$K' = 0.3(1.135)^{15.7-20} = 0.174 \text{ per day}$$

$$K_2' = 0.7(1.024)^{15.7-20} = 0.63 \text{ per day}$$

4. Determine t_c and x_c. Saturation concentration, 15.7°C = 10.1

$$D_0 = 10.1 - 8.0 = 2.1 \text{ mg/liter}$$

$$t_c = \frac{1}{K_2' - K'} \ln \frac{K_2'}{K'} \left[1 - \frac{D_0(K_2' - K')}{K'\text{BOD}_L}\right]$$

$$= \frac{1}{0.63 - 0.174} \ln \frac{0.63}{0.174} \left[1 - \frac{2.1(0.63 - 0.174)}{0.174(35.6)}\right]$$

$$= 2.45 \text{ days}$$

$$x_c = vt_c = 2(24)(2.45) = 117 \text{ miles}$$

5. Determine D_c.

$$D_c = \frac{0.174}{0.63}(35.6)[e^{-0.174(2.45)}] = 6.4 \text{ mg/liter}$$

$$DO_c = 10.1 - 6.4 = 3.7 \text{ mg/liter}$$

6. Determine the BOD_5 of a sample taken at x_c.

$$L_t = 35.6\, e^{-0.174(2.45)} = 23.3 \text{ mg/liter}$$

$$20°\text{C BOD}_5 = 23.3[1 - e^{-0.3(5)}] = 18.1 \text{ mg/liter}$$

Disposal into Estuaries

An estuary can roughly be defined as the zone in which a river meets the sea. There are between 850 and 900 estuary systems along the coastal periphery of the United States. They are becoming increasingly important to the sanitary engineer because nearly 50 million people in this country live near estuaries.

The analysis of estuaries is, in general, more complicated than the analysis of rivers or lakes. The ebb and flow of tides may cause significant lateral mixing in the reaches of rivers near the estuary. Indeed, the incoming tide often reverses the direction of flow in the section of river near the ocean. Usually, estuarine waters are vertically stratified. Sea water is heavier than fresh water; therefore, a layered system is often encountered, with fresh water riding above seawater.

Some estuaries, such as San Francisco Bay, are so large that complete-mix models are clearly inadequate. Where the physical processes are extremely complicated, it may be necessary to resort to physical rather than purely mathematical models of the basin. In any case, a physical understanding of the flow processes in an estuary is necessary before any rational analysis.

In many estuarine channels, tidal action merely increases the amount of mixing and dispersion of the waste along the length of the channel. Figure 15·7 shows the effect of this phenomenon. If a dye

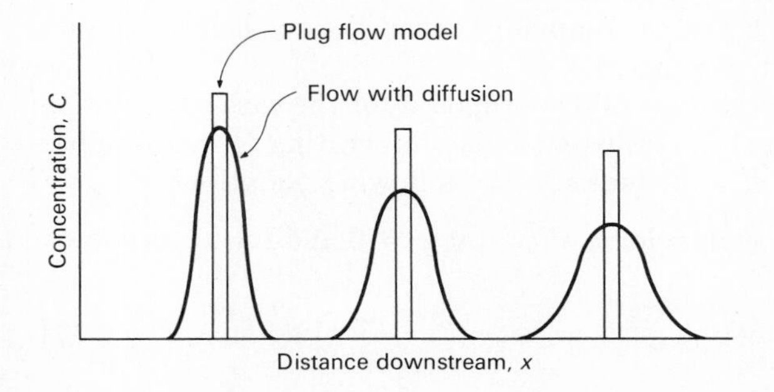

FIG. 15·7 Effect of diffusion on flow characteristics.

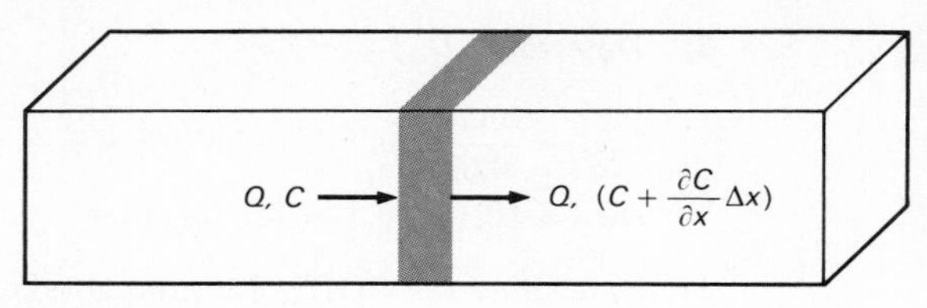

FIG. 15·8 Definition sketch for estuary diffusion model.

Inflow: $QC\,\Delta t - EA\frac{\partial C}{\partial x}\Delta t$

Outflow: $Q\,(C + \frac{\partial C}{\partial x}\Delta x)\,\Delta t - EA\,(\frac{\partial C}{\partial x} + \frac{\partial^2 C}{\partial x^2}\Delta x)\,\Delta t$

Sinks: $VK\bar{C}\,\Delta t$

were released instantaneously in a river in which there was no mixing in the lengthwise direction and only advection occurred, the concentration of the waste at various points downstream would be represented by the rectangles. If mixing is significant, however, the plug of dye will diffuse outward and mix with surrounding waters as it moves downstream. Then the dye concentration will assume the shape of the bell-shaped areas of Fig. 15·7.

The same type of analysis used for the case of flow with no mixing can be used in the case where dispersion is significant, provided a term is added that takes into account the effects of dispersion. The following equation is commonly used in describing these effects:

$$\frac{\partial M}{\partial t} = -EA\frac{\partial C}{\partial x} \tag{15·16}$$

where $\partial M/\partial t$ = mass flow
$\partial C/\partial x$ = concentration gradient
A = cross-sectional area
E = coefficient of turbulent mixing

Whenever a concentration gradient $\partial C/\partial x$ exists, a flow of mass $\partial M/\partial t$ occurs in such a way as to reduce the concentration gradient. It is assumed that the flowrate is proportional to the concentration gradient and the cross-sectional area over which this gradient acts. The proportionality constant is E and is commonly called the coefficient of eddy diffusion or turbulent mixing.

The model shown in Fig. 15·8 is proposed for the case of flow with diffusion. Performing the mass balance and integrating for the equilibrium case where $dC/dt = 0$ results in the following equations:

1. For the instantaneous release of dye at $x = 0$ and $t = 0$, as shown in Fig. 15·9,

$$C = C_0 e^{[-(x')^2/4Et]-Kt} \tag{15·17}$$

where $C_0 = \dfrac{M}{2A\sqrt{\pi E t}}$

M = mass discharged at $t = 0$

$x' = x - vt$

2. For the continuous discharge of waste at a rate W,

$$C = C_0 e^{jx} \tag{15·18}$$

where $C_0 = \dfrac{W}{Q\sqrt{1 + 4KE/v^2}}$

$$j = \frac{v}{2E}\left(1 \pm \sqrt{1 + \frac{4KE}{v^2}}\right)$$

The positive root for j refers to the upstream ($-x$) direction, and the negative root refers to the downstream ($+x$) direction.

The foregoing analysis is most often applied to estuarine systems where significant mixing occurs as a result of tidal action. The coefficient of eddy diffusion is used as the measure of this mixing. Several approaches have been taken in the determination of the magnitude of the coefficient of eddy diffusion, including (1) mathematical formulations and (2) field measurements using Eq. 15·17 with measured values of conservative tracers, such as dyes or salt. A mathematical formulation by Harleman is given in Eq. 15·19 [6]:

$$E = 77 n v R^{\frac{5}{6}} \tag{15·19}$$

where E = coefficient of eddy diffusion, sq ft/sec

n = Manning's roughness coefficient

v = velocity, fps

R = hydraulic radius, ft

Such an equation is useful in giving an order-of-magnitude answer. If E is significant, its magnitude should be determined from field measurements.

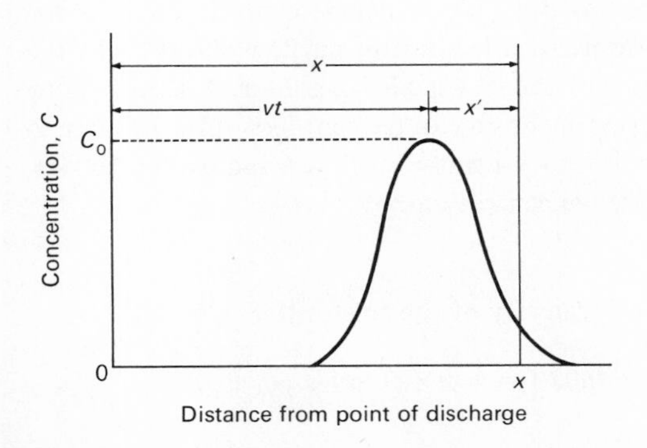

FIG. 15·9 Definition sketch for estuary diffusion model for instantaneous dye release.

One method of measuring E in the field is the instantaneous-dye-release method. A conservative dye ($K = 0$) is released from the point of waste discharge at slack tide, and the distribution of dye is measured at a later slack tide. Since t is known, E can be determined using Eq. 15·20.

$$\ln \frac{C}{C_0} = -\frac{x'^2}{4Et} \tag{15·20}$$

Values of $\ln C/C_0$ and x'^2 are plotted and the slope of the line is $-1/4Et$.

Another method of field measurement is the salinity intrusion method. The conservative substance is the salt concentration, and Eq. 15·18 is rewritten as

$$C = C_0 e^{(v/E)x} \tag{15·21}$$

The plus root is used because distances are measured upstream from the sea. Values of C, x, and v can be measured and E can be determined as in the dye-release method using Eq. 15·22.

$$\ln \frac{C}{C_0} = \frac{v}{E} x \tag{15·22}$$

Measurements should be made preferably at slack tide but in any case at the same point in the tidal cycle.

The use of the estuarine model with diffusion is illustrated in Example 15·3.

EXAMPLE 15·3 *Salinity Intrusion in an Estuary*

A city is located on an estuary, which it uses as a water supply. The city is 20 miles upstream from the ocean. The estuary has a uniform cross-sectional area of 5,000 sq ft, and E has been measured to be 10.0 sq mi/day. The chloride concentration in the ocean is 18,000 mg/liter. At a freshwater outflow of 1,000 cfs, the chloride concentration at the city is 25.4 mg/liter. A dam is to be built 20 miles upstream of the city to prevent loss of fresh water to the ocean. If the resultant freshwater outflow is reduced to 100 cfs, what will the chloride concentration be?

Solution

1. Determine the net velocity of the freshwater outflow.

$$v = \frac{Q}{A} = \frac{100}{5{,}000} = 0.02 \text{ fps} = 0.328 \text{ miles per day}$$

2. Determine j.

$$j = \frac{v}{E} = \frac{0.328}{10} = 0.0328 \text{ per mile}$$

3. Determine the chloride concentration.

$$C = C_0 e^{jx} = 18{,}000\ e^{0.0328(-20)}$$
$$C = 9{,}340 \text{ mg/liter}$$

This does not meet the USPHS standard of 250 mg/liter; therefore, the city will have to build a pipeline to take its drinking water from above the proposed dam.

Ocean Disposal

Ocean disposal is typically accomplished by submarine outfalls that consist of a long section of pipe to transport the sewage some distance from shore and, in the best examples, a diffuser section to dilute the waste with seawater. Diffusers are one of the most efficient methods of providing initial dilution of a waste in any waterway; but, because most of the design parameters for diffusers originated from work done on ocean outfalls, their design and operation is discussed in this section.

At the end of the outfall, treated or untreated wastewater is released in a simple stream or jetted through a manifold or multiple-port diffuser. At this point the sewage mixes with surrounding seawater; and the mixture, which is called the sewage field, rises to the surface and drifts in accordance with the prevailing ocean currents. This drift or movement with the currents is termed advection. At the same time, the field is also diffusing outward into the surrounding water. Hence, the coefficient of eddy diffusion is important. If the ocean is sufficiently stratified at the point of discharge it may be possible to maintain a submerged sewage field.

The design of an outfall should meet applicable receiving-water standards. Because the initial dilution from an efficient diffuser is so large that the reduction in dissolved oxygen is usually of no significance, bacterial, floatable material, nutrient, and toxicity requirements will govern the design and location of most outfalls. Accurate estimation of the number of coliform bacteria requires taking into account their reduction due to die-off, flocculation, and settling.

In summary, where the sewage field rises to the surface, the phenomena of importance in the design of ocean outfall systems are the initial dilution of the waste, the waste dispersion into surrounding waters, and the waste decay rate.

Initial Dilution When a waste is discharged from a single or multiple-port diffuser, the velocity of the jet will cause turbulent mixing with the surrounding water. If the waste is of lower specific gravity than the dilution water, the mixing jet will bend upward and may eventually reach the surface. If, as previously noted, the ocean is vertically stratified with an upper layer of warm water riding over colder water, it is possible to dilute the waste sufficiently with cold water so that the specific gravity of the waste–cold water mixture is greater than that of the warm upper layer. In such a case, the waste plume will remain submerged under the upper layer. Such vertical stratification is most often observed during warm summer months.

Rawn, Bowerman, and Brooks [16] have developed curves from field data for determining the initial dilution D_1 in a waste jet issuing from a horizontal port. Abraham [1] extended these curves to show that the dilution is a function of the depth Y_0 of the discharge port, the diameter of the discharge orifice D, and the Froude number F, for a liquid-liquid system, as shown in Fig. 15·10.

The Froude number is defined as

$$F = \frac{V_j}{\sqrt{(\Delta S/S)gD}} \tag{15·23}$$

where V_j = jet velocity
ΔS = difference in specific gravity between the waste and the surrounding seawater
S = specific gravity of the waste
g = acceleration due to gravity
D = discharge-jet diameter

The specific gravity of seawater normally varies between 1.010 and 1.030, while that for sewage ranges from 0.990 to 1.000. Along the California coast, the specific gravity of seawater is typically taken as 1.025 and that of sewage is taken as 0.999; therefore, the value of $\Delta S/S$ typically used in California is 0.026.

When moderately strong currents are encountered, the initial dilution may also be estimated from the equation

$$D_1 = \frac{V_x bd}{Q} \tag{15·24}$$

where V_x = current velocity
b = effective width of diffuser system
d = average depth of the sewage field
Q = sewage flowrate

This equation is simply a continuity relation between the sewage flowrate and the flowrate of fresh seawater over the outfall diffuser.

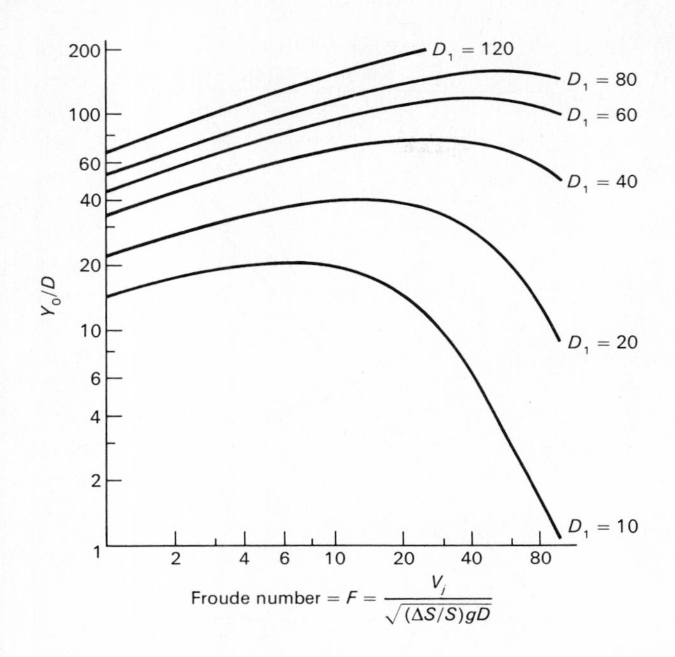

FIG. 15·10 Initial dilution by turbulent jet mixing [1].

Dispersion After initial dilution, a rather uniform sewage-seawater mixture is formed above the diffuser section. The sewage field then begins to move in response to the prevailing current. As it moves, the outer edges of the field entrain seawater as a result of turbulent mixing. The sewage field begins to diffuse outward and takes on the shape of a plume.

Float studies are frequently of value in determining both the strength and direction of prevailing currents as well as the extent of dispersion to be expected from a proposed outfall. For example, a description of floats and their use in studies made in the Great Lakes is given in Metcalf and Eddy [11].

Brooks has developed equations to describe the dispersion of a waste field by applying classical partial differential equations to the problem. The reader is referred to his excellent articles for a more detailed study [2, 16]. The result of his analysis for the case in which E is a function of the diffuser length raised to the four-thirds power is

$$D_2 = \frac{C_0}{C_t}$$

$$D_2 = \frac{1}{\operatorname{erf}\sqrt{\dfrac{\frac{3}{2}}{[1 + \frac{2}{3}\beta(x/b)]^3 - 1}}} \qquad (15\cdot25)$$

FIG. 15·11 Nomograph for the solution of Eq. 15·25 [3].

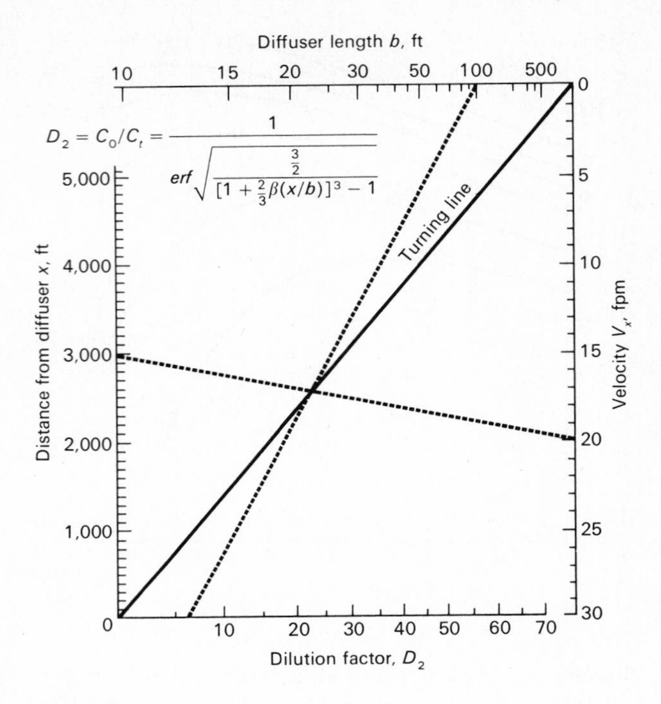

where D_2 = dilution due to eddy diffusion after initial dilution
C_t = maximum pollutant concentration at time t
C_0 = pollutant concentration after initial dilution
$\text{erf}(x)$ = error function (x)
$\beta = 12\ E/V_x b$
E = coefficient of eddy diffusion $= 0.001(b)^{\frac{4}{3}}$ (from Ref. [14])
V_x = current velocity, fps
x = distance along plume centerline, ft
b = effective diffuser system length, ft

A nomograph for the solution of this equation is presented in Fig. 15·11.

Waste Decay The third significant factor in waste dilution in the ocean is the decay rate of the waste. In the case of bacterial decay, this includes mortality as well as flocculation and sedimentation. Bacterial decay is most commonly assumed to follow a first-order relationship. In other words,

$$C_t = C_0 e^{-kt} \tag{15·26}$$

where C_t = bacterial concentration at time t
C_0 = bacterial concentration after initial dilution
k = bacterial decay constant
t = time

Much research has been conducted to determine the decay constant k. It has been shown that a 90 percent reduction in bacterial numbers can usually be obtained in 2 to 6 hr [7]. The variation in time is caused by differences in such characteristics of the seawater as temperature, salinity, and pH. The time in hours to achieve a 90 percent reduction in bacterial numbers is called the T_{90} time. If the decay constant k is changed to the equivalent T_{90} form, the dilution due to waste decay can be formulated as

$$D_3 = \frac{C_0}{C_t} = \exp \frac{2.3x}{T_{90}60(V_x)} \tag{15·27}$$

A nomograph for this equation is presented in Fig. 15·12. Use of the nomograph is illustrated in Example 15·4. The practical aspects considered in design include the outfall pipe, the diffuser section, and the overall hydraulics.

Outfall Design The outfall is used to convey the waste to the diffuser section. Its size is determined by the velocity, head loss, structural

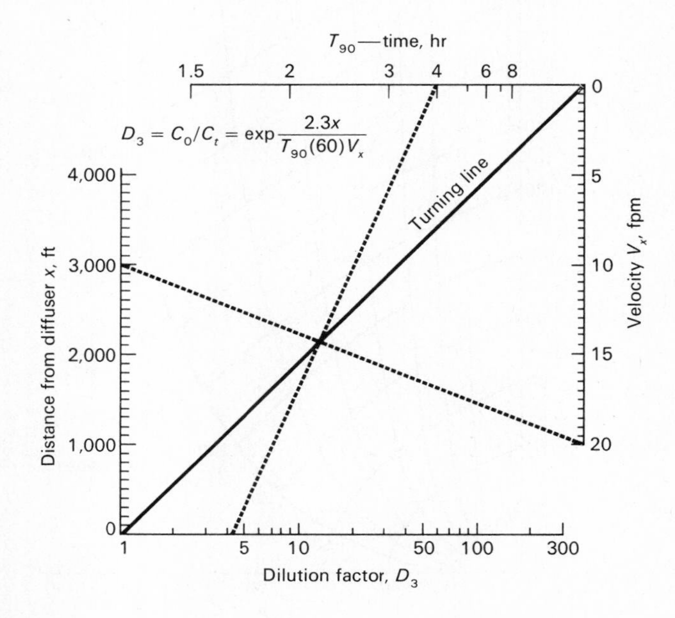

FIG. 15·12 Nomograph for the solution of Eq. 15·27 [3].

considerations, and economics of the situation. Velocities of 2 to 3 fps at average flow are normally recommended in pipeline design to avoid excessive head loss. Lower velocities will not be a problem provided the waste has received preliminary treatment to reduce the amount of settleable solids. On the other hand, velocities higher than 8 to 10 fps should be avoided because of excessive head loss.

The diffuser section should be oriented perpendicular to the prevailing ocean current. If, as in most cases, the currents are not predominant in any one direction, a Y- or V-shaped diffuser is commonly used. Several possible diffuser arrangements are given in Burchett, Tchobanoglous, and Burdoin [3].

Formerly, diffuser ports were large, between 6 and 9 in. The current trend, however, is to decrease the port diameter and increase the number of ports. Port diameters of less than 3 in. are being used on the fourth White's Point outfall of the Los Angeles County Sanitation Districts.

Diffuser ports are normally aligned horizontally, with ports alternating from side to side in order to avoid interference between adja-

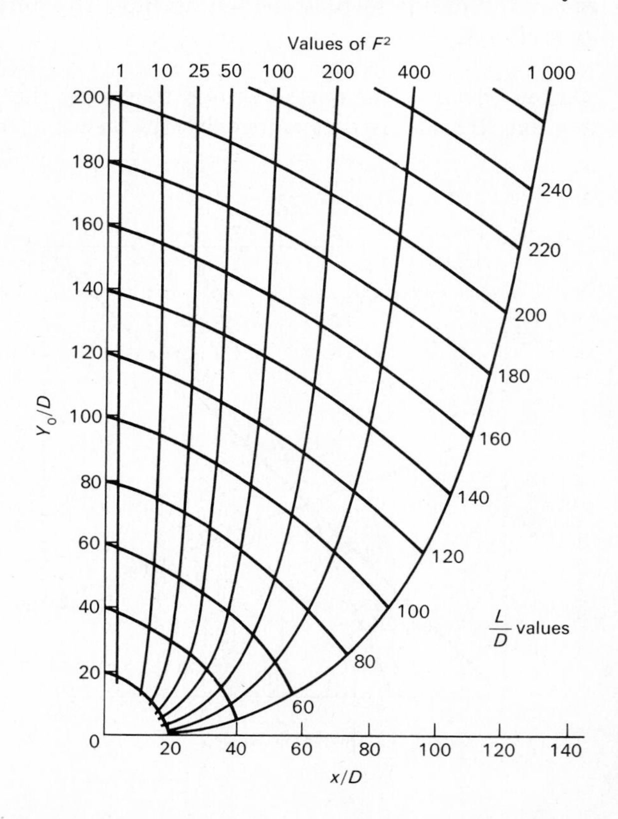

FIG. 15·13 Length of the turbulent jet path after discharge from a multiple-port diffuser [1].

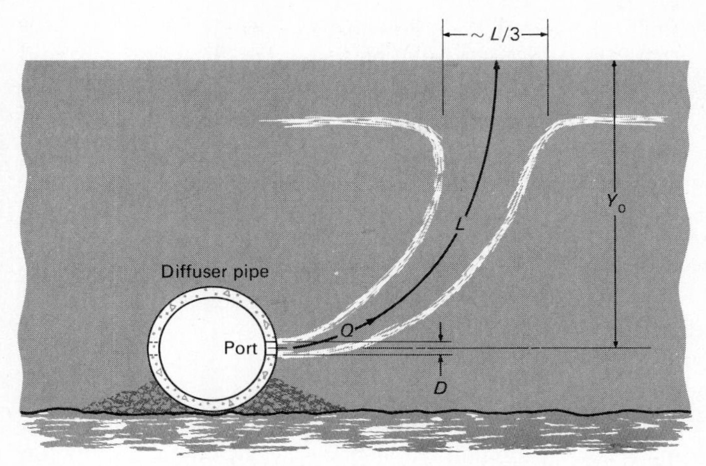

FIG. 15·14 Definition sketch for turbulent jet mixing analysis with a diffuser [16].

cent jet plumes as they rise toward the surface. Experimental data have indicated that the diameter of the plume is approximately equal to $L/3$, where L is the length of the plume trajectory. Values of L can be determined from Fig. 15·13. With ports discharging alternately, the individual ports can be as close as $L/6$. A rule of thumb, however, is that ports should be spaced 8 to 15 ft on centers. Diffuser ports of many different shapes, including specially designed mixing nozzles, have been used. A simple but effective diffuser port is illustrated in Fig. 15·14.

The hydraulics of a diffuser section are somewhat complex and beyond the scope of this presentation. The problem centers around equalizing the flow from each diffuser port. This may require a number of different sizes of ports along the diffuser section. The port diameters chosen during the preliminary design represent an average value. The final hydraulic analysis will determine the range of port sizes required to discharge the waste flow uniformly along the length of the diffuser section. An analysis of such hydraulics is presented in the article by Rawn, Bowerman, and Brooks [16]. The design of an ocean outfall is illustrated in Example 15·4, which is taken from Ref. [3].

EXAMPLE 15·4 *Ocean Outfall Design*

Design an ocean outfall for a wastewater flow of 10 mgd for the following conditions:

1. Peak flow = 20 mgd
2. Bottom slope = 12.5 ft/1,000 ft

3. Coliform concentration in wastewater = 10^6/ml
4. Critical onshore current = 20 fpm
5. Diffuser length = 14 ft/mgd = 140 ft
6. Diffuser spacing = 10 ft
7. Port discharge velocity at peak flow = 15 fps
8. Diffuser configuration is Y-shaped with 70 ft per leg
9. T_{90} = 4 hr

Solution

1. Determine the number of diffusers.

$$\text{Number of diffusers} = \frac{140 \text{ ft}}{10 \text{ ft}} = 14$$

2. Determine the port diameter.

$$\text{Total area of ports} = \frac{(20 \text{ mgd})(1.55)}{15 \text{ fps}} = 2.07 \text{ sq ft}$$

$$\text{Area per port} = \frac{2.07}{14} = 0.148 \text{ sq ft}$$

Use 14 ports with a $5\frac{1}{4}$-in. diameter.

3. Determine the outfall pipe size. Assume a velocity of 2.5 fps at average flow.

$$A = \frac{15.5 \text{ cfs}}{2.5 \text{ fps}} = 6.2 \text{ sq ft}$$

Use a 33-in. diameter pipe. The head loss at peak flow (f = 0.025) is 3.85 ft/1,000 ft.

4. Determine the total dilution factor for various lengths of outfall. At 3,000 ft from shore the depth is

$$3(12.5) = 37.5 \text{ ft}$$

(*a*) Compute the initial dilution.

$$\frac{Y_0}{D} = \frac{37.5}{5.25/12} = 86$$

$$F = \frac{15 \text{ fps}}{\sqrt{0.026(32.2)(5.25/12)}} = 24.8$$

From Fig. 15·10, D_1 = 43. (Note: The initial dilution should also be computed by Eq. 15·24, and the smaller of the two values should be used.)

(*b*) Determine the dilution due to dispersion. The effective length of the diffuser is 100 ft.

$$\frac{2}{\sqrt{2}}(70) = 100 \text{ ft}$$

From Fig. 15·11, D_2 = 6.

(*c*) Determine the dilution due to decay. From Fig. 15·12, $D_3 = 4.2$.

(*d*) Determine the total dilution.

$$D_t = D_1(D_2)(D_3) = 43(6)(4.2) = 1{,}084$$

5. Repeat the calculations in step 4 for several lengths of outfall. The dilutions obtained are shown in Table 15·2.

TABLE 15·2 DILUTION FACTORS VERSUS LENGTH OF OUTFALL FOR EXAMPLE 15·4 [3]

Distance from shore, ft	Dilution factor			
	D_1	D_2	D_3	D_T
1,000	17	2	1.8	61
2,000	30	4	2.8	336
3,000	43	6	4.2	1,080
4,000	60	8	6.9	3,310
5,000	80	11	11.0	9,680
6,000	1,000*		18.0	18,000
7,000	1,000		29.0	29,000
8,000	1,000		48.0	48,000
9,000	1,000		75.0	75,000

* Experience with existing outfall systems indicates that the maximum physical dilution achieved (initial dilution + transport dispersion) is in the neighborhood of 1,000.

6. Using the percent removal of coliforms for various processes given in Table 9·5, the total dilution factor and required length of outfall for several types of treatment to meet a standard of 10 coliforms per milliliter are given in Table 15·3.

TABLE 15·3 REQUIRED LENGTH OF OUTFALL FOR ALTERNATIVE TYPES OF TREATMENT IN EXAMPLE 15·4 [3]

Type of treatment	Required dilution to meet standard*	Length of outfall required, ft
Preliminary + Cl_2	1,500	3,300
Preliminary	75,000	8,800
Activated sludge + Cl_2	100	1,200
Activated sludge	10,000	5,100

* Assumed to be 10 coliforms per milliliter.

7. Using pertinent cost data for outfalls, treatment plants, and chlorine, combined with a suitable interest rate and estimated useful life, the most economical alternative can be selected.

The methods described in this section are applicable to outfalls extending into large lakes, such as the Great Lakes, and other large bodies of water that cannot be considered completely mixed. The following differences should be noted and allowed for:

1. In Eq. 15·23 ΔS is mainly due to temperature differences in fresh water.
2. A significant level of general or background pollution may exist.
3. A sustained flow, varying by season, may exist from inlet to outlet.
4. The lake may serve as a source of public water supply, as the Great Lakes do, and wastewater outfalls must terminate at locations that will not cause contamination of the water supply.

EFFLUENT DISPOSAL ON LAND

Agricultural use, recreational use, ground water recharge, spraying, and containment (ponding) are some of the methods that have been used to dispose of effluent directly on land. The first three of these are discussed in the following section dealing with direct and indirect reuse of wastewater. Land application is also discussed in Chap. 14.

Spraying

Spraying on irrigable land, wooded areas, and hillsides has been used primarily to dispose of industrial (generally canning) wastes. The amount of wastewater that can be disposed of depends on the climatic conditions, the infiltration capacity of the soil, the types of crops or grasses grown, and the quality standards imposed where runoff is allowed (see Table 15·4). A combination of evaporation, transpiration, percolation, and runoff is generally used for disposal.

Evaporation Ponds

In arid regions of the United States where evaporation losses are much greater than the amount of rainfall, evaporation ponds have been used for disposal of effluents. Evaporation losses are directly proportional to the area, so this method is applicable only if large areas of land are available.

TABLE 15·4 TYPICAL WASTEWATER LAND APPLICATION RATES [18]

Application method	Type of wastewater	Units	Application rate: Range	Application rate: Ave.	Remarks
Spreading basin	Clear water of good quality*	ft/day	1.6–10	4.2	Coarse soil
			1.7–3.6	2.2	Medium soil
			<1	0.16	Fine soil
	Primary effluent		0.05–0.50	0.15	Continuous inundation
	Treated effluent		0.2–6.0	1.5	Intermittent operation
Spray irrigation	Cannery wastes	gal/sq ft/day	0.2–2.0†	0.5	Applied at 0.2–0.6 in./hr
	Treated effluent		0.09–1.0†	0.2	
Leaching fields	Septic tank effluent	gal/sq ft/day	0.2–1.5	0.5	Continuous inundation
				1.5	Intermittent operation

* California data.
† Rate will vary with type of soil and type and degree of ground cover.

DIRECT AND INDIRECT REUSE OF WASTEWATER

It is generally impossible to reuse a wastewater completely or indefinitely. The reuse of treated effluent by direct or indirect means is a method of disposal that complements the other disposal methods. The amount of effluent that can be reused is affected by the availability and cost of fresh water, transportation and treatment costs, water-quality standards, and the reclamation potential of the wastewater. The material in this section briefly describes some of the possible reuses for waste effluent. The purpose is not to give a detailed description of all possibilities but to acquaint the student with the major areas of water reuse.

Water reuse may be classified according to use as (1) municipal, (2) industrial, (3) agricultural, (4) recreational, and (5) ground water recharge. Direct and indirect reuse applications for these uses are shown in Table 15·5.

Municipal Reuse

Direct reuse of treated wastewater as drinking water, after dilution in natural waters to the maximum possible extent and after coagulation,

TABLE 15·5 USES OF RENOVATED WATER

Use	Direct	Indirect
Municipal	Park or golf course watering. Lawn watering with separate distribution system. Potential source for municipal water supply	Ground water recharge to reduce aquifer overdrafts.
Industrial	Cooling tower water Boiler feed water Process water	Replenish ground water supply for industrial use
Agricultural	Irrigation of certain agricultural lands, crops, orchards, pastures, and forests Leaching of soils	Replenish ground water supply for agricultural overdrafts
Recreational	Forming artificial lakes for boating, swimming, etc. Swimming pools	Develop fish and waterfowl areas
Other	Ground water recharge to control saltwater intrusion Salt balance control in ground water Wetting agent-refuse compaction	Ground water recharge to control land subsidence problems Oil-well repressurizing Soil compaction

filtration, and heavy chlorination for disinfection, is practicable on an emergency basis. This practice varies only in degree from the situation existing on many rivers that are used for both water supply and waste disposal. One example is the Merrimack River in Massachusetts. Advanced methods of wastewater and water treatment, such as demineralization and desalination, are capable of almost complete removal of impurities, and water treated by such methods, after chlorination, is safe to drink. These methods are very expensive and, where they are found to be necessary due to inadequate water supplies, may be economically feasible only if a dual supply system is adopted. In such cases, adequately treated and disinfected sewage effluents could be reused for flushing toilets, yard watering, and other direct applications.

Industrial Reuse

Industry is probably the single greatest user of water in the world, and the largest of the industrial water demands is for process cooling water.

Waters with a high mineral content and those that do not meet bacterial drinking water standards have been used by industry in some cases. Public health dangers and aesthetic concerns are generally eliminated because of the use of closed-cycle processes.

Agricultural Reuse

The types of crops that can be irrigated with reclaimed wastewater depend on the quality of the effluent, the amount of effluent used, and the health regulations concerning the use of sewage or sewage effluent on crops. Health considerations in this country have dictated against the use of raw sewage. Furthermore, field crops that are normally consumed in a raw state cannot be irrigated with sewage of any kind. Preliminary treated or undisinfected sewage effluent is usually allowed for field crops, such as cotton, sugar beets, and vegetables for seed production. Additional data on wastewater disposal by irrigation may be found in Ref. [11].

Recreational Reuse

Golf course and park watering, establishment of ponds for boating and recreation, and maintenance of fish or wildlife ponds are methods for the recreational reuse of water. Today's technology allows the production of an excellent effluent that is well suited for the purposes described. The use of treated effluent for park watering has been practiced for many years in this country.

The Santee Project outside San Diego is an example of recreational reuse of wastewater in forming a series of lakes suitable for boating, fishing, and other recreational purposes [10].

Ground Water Recharge

Ground water recharge is one of the most common methods for combining water reuse and effluent disposal. Recharge has been used to replenish ground water supplies in many areas. The effluent from the Whittier Narrows Plant, operated by the Los Angeles County Sanitation Districts, is used for replenishment of the ground water in the Rio Hondo River Basin [9]. Typical application rates are given in Table 15·4. In New York, California, and other coastal areas, rapid development of industry and increase in population have caused a lowering of the ground water table, resulting in saltwater intrusion into the freshwater aquifers. Treated effluent has been used to replenish the ground water and to stop this intrusion. Another possible effluent use is in the recharging of oil-bearing strata. Oil companies have conducted much research on flooding techniques to increase the yield of oil-bearing strata.

PROBLEMS

15·1 Treated sewage containing 25 mg/liter of BOD_5 is discharged continuously at a rate of 1 mgd into a small lake. The lake has a surface area of 50 acres, a drainage area of 10 sq mi, and an average depth of 10 ft; its contents can be considered to be completely mixed. Aerobic conditions prevail throughout its depth. The runoff from the drainage area, containing 1 mg/liter of BOD_5, varies from 1 cfs/sq mi in the spring to 0.1 cfs/sq mi in the summer. The temperature of the lake contents is 5°C in the spring and 25°C in the summer. Determine the BOD_5 of the outlet stream in the spring and the summer. $K' = 0.3$ per day, $\theta_{spring} = 1.135$, $\theta_{summer} = 1.056$.

15·2 A wastewater containing 130 mg/liter of BOD_5 after preliminary treatment is discharged to a river at a rate of 20 mgd. The river has a minimum flowrate of 200 cfs, a BOD_5 of 2 mg/liter, and a velocity of 1.5 mph. After the wastewater is mixed with the river contents, the temperature is 20°C and the dissolved oxygen is 75 percent of saturation. Determine the oxygen sag at the critical point and at distances of $x_c/2$ above and below the critical point, and plot the curve. $K' = 0.25$ per day, $K_2' = 0.40$ per day.

15·3 The freshwater outflow in an estuary is 100 cfs and the average cross-sectional area is 1,000 sq ft. Assuming that seawater has a chloride concentration of 18,000 mg/liter, determine E from the following data:

x, miles	*C, mg/liter*
1	16,000
2	11,500
3	8,350
4	6,000
5	4,350

15·4 The freshwater runoff to an estuary has a chloride concentration of 30 mg/liter and amounts to 1,000 cfs. Assume that 100 cfs of wastewater with an average chloride concentration of 50 mg/liter is discharged to the estuary and that the chloride concentration at that point is 9,000 mg/liter. Determine the dilution available and the chloride concentration after mixing.

15·5 Derive Eq. 15·27 from Eq. 15·26.

15·6 Design an ocean outfall for an average sewage flow of 30 mgd and a peak flow of 45 mgd. The bottom slope is 20 ft/1,000 ft along the route of the outfall. The diffusers are to be located in 50 ft of water. The prevailing current is 15 fpm parallel to shore. Determine the dilution and coliform content at distances of 2,500 and 5,000 ft from the diffuser, assuming (1) the sewage has had preliminary treatment and (2) the sewage has had preliminary treatment plus chlorination. Use 10 ft of diffuser per mgd. $T_{90} = 3$ hr.

REFERENCES

1. Abraham, G.: *Jet Diffusion in Stagnant, Ambient Fluid,* Delft Hydraulics Laboratory Pub. 29, 1963.
2. Brooks, N. H.: Diffusion of Sewage Effluent in an Ocean Current, *Proceedings First International Conference on Waste Disposal in the Marine Environment,* University of California, Berkeley, Pergamon, New York, 1960.
3. Burchett, M. E., G. Tchobanoglous, and A. J. Burdoin: A Practical Approach to Submarine Outfall Calculations, *Public Works,* vol. 98, no. 5, 1967.
4. Committee on Quality Tolerances of Water for Industrial Uses: Progress Report, *J. New England Water Works Association,* vol. 54, p. 261, 1940.

5. Fair, G. M., E. W. Moore, and H. A. Thomas, Jr.: The Natural Purification of River Muds and Pollutional Sediments, *Sewage Works Journal*, vol. 13, pp. 270, 756, 1941.

6. Harleman, D. R. F.: The Significance of Longitudinal Dispersion in the Analysis of Pollution in Estuaries, *Proceedings 2nd International Conference on Water Pollution Research*, Tokyo, Pergamon, New York, 1964.

7. Hyperion Engineers: *Ocean Outfall Design*, Final Report to the City of Los Angeles, 1957.

8. Krenkel, P. A., and F. L. Parker (eds.): *Biological Aspects of Thermal Pollution*, Vanderbilt University Press, Nashville, Tenn., 1969.

9. McMichael, F. C., and J. E. McKee: *Wastewater Reclamation at Whittier Narrows*, California SWQCB Publication 33, 1965.

10. Merrell, J. E., *et al.*: *The Santee Recreation Project, Santee, California*, final report, U.S. Department of the Interior, Federal Water Pollution Control Administration, Cincinnati, Ohio, 1967.

11. Metcalf, L., and H. P. Eddy: *American Sewerage Practice*, vol. 3, 3d ed., McGraw-Hill, New York, 1935.

12. National Technical Advisory Committee: *Water Quality Criteria*, Federal Water Pollution Control Administration, Washington, D.C., 1968.

13. Parker, F. L., and P. A. Krenkel (eds.): *Engineering Aspects of Thermal Pollution*, Vanderbilt University Press, Nashville, Tenn., 1969.

14. Pearson, E. A.: *An Investigation of the Efficacy of Submarine Outfall Disposal of Sewage and Sludge*, California Water Pollution Control Board, Pub. 14, 1956.

15. Phelps, E. B.: *Stream Sanitation*, Wiley, New York, 1944.

16. Rawn, A. M., F. R. Bowerman, and N. H. Brooks: Diffusers for Disposal of Sewage in Sea Water, *Proceedings ASCE*, vol. 86, no. SA2, 1960.

17. *Report of the Engineering Board of Review*, Sanitary District of Chicago, part III, appendix I, 1925.

18. Tchobanoglous, G., and R. Eliassen: The Indirect Cycle of Water Reuse, *Water and Wastes Engineering*, vol. 6, no. 2, 1969.

19. Velz, C. J.: *Applied Stream Sanitation*, Wiley-Interscience, New York, 1970.

wastewater treatment studies

16

Industrial and other wastes that cannot be sucessfully treated by municipal wastewater treatment plants present a great diversity of treatment problems. Industrial operations and processes create wastes that are peculiar to the product manufactured. This variety of waste types is often quite visible after discharge to the environment. Floating and emulsified oils from refinery operations, white water from pulp and paper mills, colored wastes from cleaning and dyeing establishments, and silt and culm from mining operations are graphic examples. To further complicate the picture, new and unknown wastes are produced every year as a result of technological advances in the synthesis of organic compounds, in the development of new products, and in the application of new manufacturing techniques.

Because the number of difficult treatment problems with which the sanitary engineer must deal is increasing, the purpose of this chapter is to present and discuss a general approach that can be followed in solving wastewater treatment problems and to describe laboratory methods for evaluating the effectiveness of selected treatment operations and processes in treating a given wastewater. Although the procedures outlined are directed primarily toward studies of industrial wastewaters, they can be applied equally well to studies of municipal wastewaters.

PRELIMINARY STEPS

Identification of the problem and establishment of the requirements are the first two steps involved in the solution of industrial wastewater treatment problems. The importance of these two steps cannot be overstressed, for unless the problem is clearly defined and understood, the current requirements controlling the discharge of wastes are established, and the future requirements are projected, subsequent work will generally lack direction and will not often lead to creative solutions.

Problem Identification

Problem identification is perhaps the key step in the conduct of wastewater treatment studies. This involves much more than being told what the problem is by management. The engineer must be prepared to assess and define the problem on the basis of a review of the plant operations and physical facilities, discussions with plant operators and engineering personnel, research into the current literature, and his past experience. On the basis of his findings, he should advise management if he is at variance with their views concerning the nature and definition of the problem. Time spent on this phase of the study will, in most cases, lead to cost savings in the conduct of the study and will reduce the time required by minimizing the number of false starts.

Analysis of Requirements

In this step, a decision must be made as to the degree of treatment that the wastewater is to receive. This involves selecting target values for effluent-quality parameters to serve as a basis for future design. In many cases, effluent-quality requirements have already been established by local regulatory agencies in the form of discharge requirements. In reviewing such requirements, it is extremely important to consider what their status might be in the future. As a result of the current national emphasis being placed on upgrading the quality of receiving waters, discharge requirements are continually being made more stringent. A treatment facility designed to produce an effluent that meets today's requirements may not be a satisfactory solution in the near future because of the establishment of higher quality standards. Wherever possible, projected future quality requirements should be developed in cooperation with local regulatory agencies.

Many industries are required to pretreat their wastewater to render it amenable to treatment before discharging into municipal sewers.

If the waste study is of this type, the requirements for such pretreatment must be determined.

Where the potential for reusing treated wastewater exists, reuse requirements should be determined and compared with discharge requirements. It is anticipated that quality requirements for reuse will eventually be the major basis for design once discharge quality standards become so stringent that treatment costs become prohibitive.

WASTE SURVEY

After defining the problem and establishing the discharge requirements, a waste survey is conducted (1) to obtain the data necessary to characterize the various wastestreams and individual wastes that are being discharged and (2) to determine if the characteristics of the waste can be altered and if the volume of waste can be reduced by appropriate modifications in the manufacturing operations or processes or in the methods used to collect the wastes. The procedures involved in the conduct of a waste survey for an industrial waste treatment problem are outlined in this section.

Wastestream Characterization

Industrial plants with significant waste flows typically have waste flows from several different sources composing the total wastestream. The first and most important step is to analyze all significant sources by determining flowrates, obtaining wastewater samples, and characterizing samples through laboratory analysis. For large plants, when limitations prevent an exhaustive survey, a brief screening effort can be made to identify the important sources in terms of volume and composition. These sources can then be subjected to a more detailed survey.

Flow Measurement Flow measurements and the determination of the degree of variation in flowrate are necessary to establish a representative sampling program, to establish the volumes of wastes to be treated, and to determine the type of facilities required to handle fluctuating waste flows. If the waste conveyance lines are equipped with flow-metering devices at the proper points, the task of flow measurements is relatively simple. These are not often found, however, especially in older plants. In such cases, the investigator must choose the method, the equipment, and the location for obtaining flow data. The type of flow device, its location, cost of installation, quality of flow data, and suitability for the service must be considered. A discussion covering the kinds of flow-measuring devices, fundamentals of flow measurement, description and design of flow devices, and important points to consider in their selection is presented in Chap. 3, in the API Manual [6], in the ASTM Manual on Water [7], and in Spink [11].

Sampling The sampling techniques used in a waste survey must assure representative samples, because the data obtained from sample analysis will ultimately serve as a basis for designing treatment facilities. Because no universal procedure for sampling exists, sampling must be tailored to fit the operation of each manufacturing plant and the characteristics of the waste produced. Special procedures are necessary to handle problems in sampling wastes that vary considerably in composition. Thus suitable sampling locations must be selected and the frequency and type of sample to be collected must be determined.

Examination of drawings showing sewers and manholes in manufacturing areas will help locate appropriate points for sampling wastewaters. Such points should be located where flow conditions encourage a homogeneous mixture. In sewers and in deep, narrow channels, samples should be taken from a point one-third the water depth from the bottom. The collection point in wide channels should be rotated across the channel. The velocity of flow at the sample point should, at all times, be sufficient to prevent deposition of solids. When collecting samples, care should be taken to avoid creating excessive turbulence which may liberate dissolved gases and yield an unrepresentative sample.

The degree of flowrate variation will dictate the time interval for sampling. This period must be short enough to provide a true representation of the flow. Even when flowrates vary only slightly, the concentration of waste products may vary widely. Frequent sampling (10- or 15-min uniform intervals) allows estimation of the average concentration during the sampling period.

Selection of sampling equipment is an important consideration if continuous or automatic sampling is appropriate. A simple and inexpensive continuous sampler is shown in Fig. 16·1 [7]. An automatic sampling device is shown in Fig. 16·2*a*, and a diagrammatic arrangement of the installation is shown in Fig. 16·2*b* [7]. The scope of this chapter does not permit a complete description of the many automatic devices suitable for sampling industrial wastewaters. More detailed information may be found in Refs. [1, 4, and 9]. A discussion of precautions to be observed in taking samples and using sampling equipment is presented in Ref. [7].

A carefully performed sampling program will be for naught if the physical, chemical, and biological integrity of samples is not main-

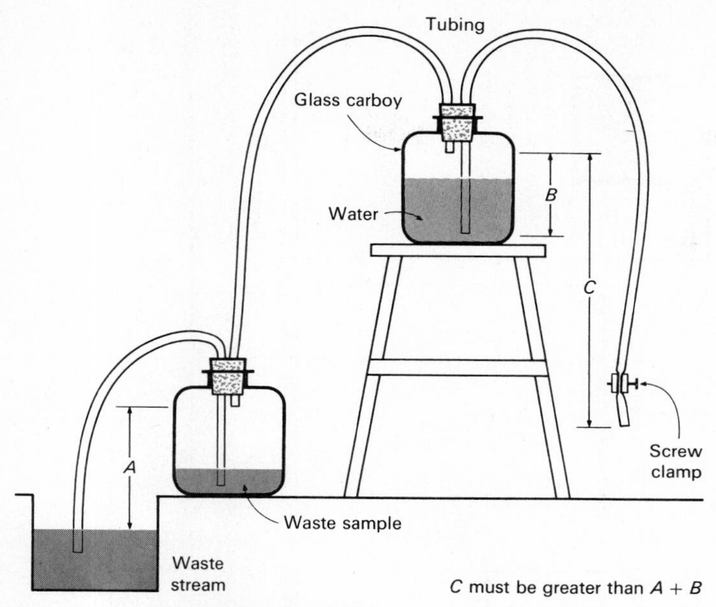

FIG. 16·1 Continuous-flow sampler [7].

tained during interim periods between sample collection and sample analysis. Considerable research on the problem of sample preservation has failed to perfect a universal treatment or method or to formulate a set of fixed rules applicable to samples of all types. Prompt analysis is undoubtedly the most positive assurance against error due to sample deterioration. When analytical and testing conditions dictate a lag between collection and analysis, such as when a 24-hr composite sample is collected, provisions must be made for preserving samples. A tabulation of preservative techniques for some selected parameters has been compiled recently and is shown in Table 16·1 [4]. Recommended methods of sample preservation for the analysis of properties subject to deterioration are covered in Ref. [12]. Probable errors due to deterioration of the sample should be noted in reporting analytical data.

Sample Analysis Analysis of the wastewater samples is the next step in the waste survey. The specific analyses to be performed will depend on the type of industrial activity and the purpose for which the survey is being conducted. The data on domestic wastewater given in Tables 7·1 and 7·3 in Chap. 7 can be used as a guide in determining the types of analyses that are performed routinely.

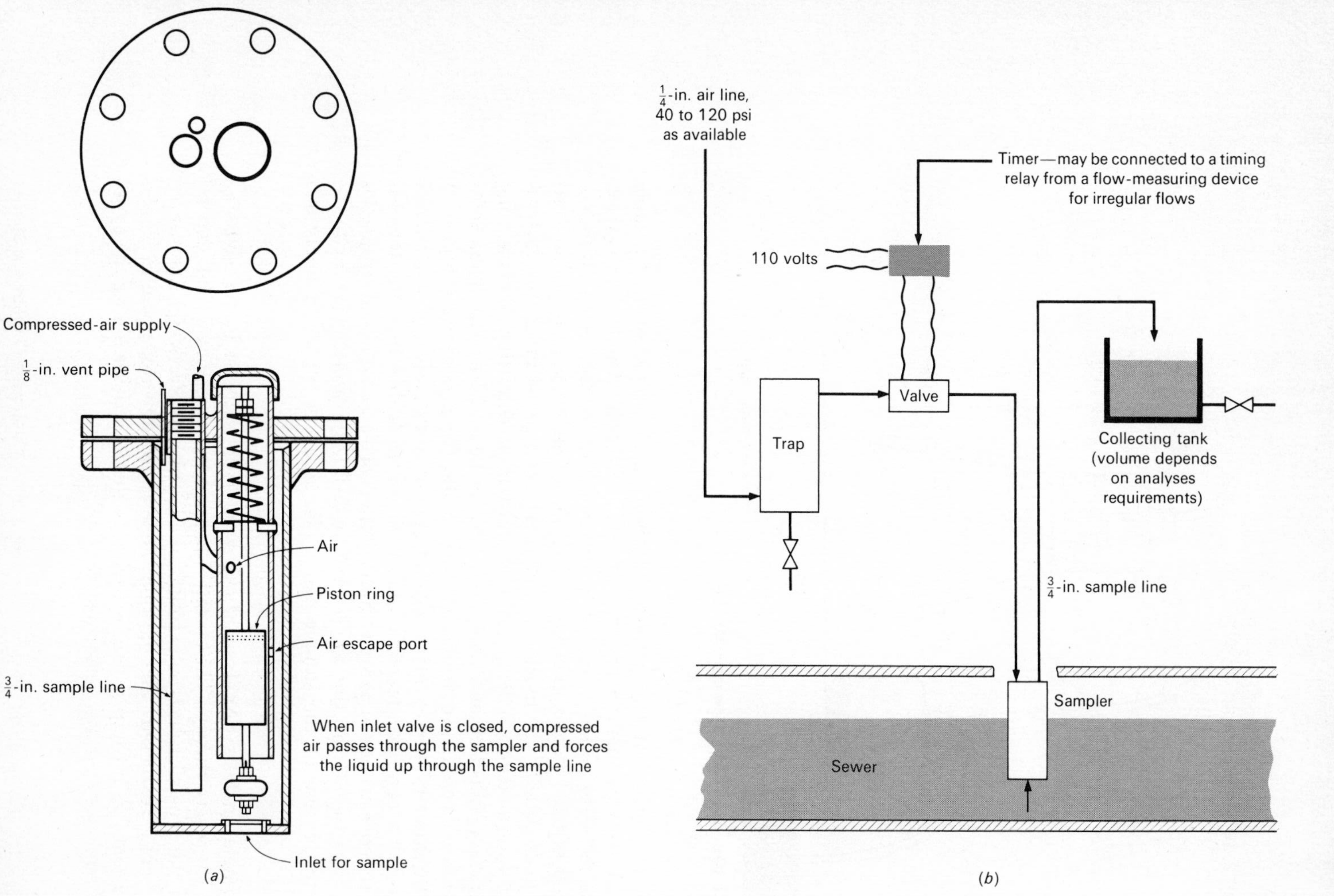

FIG. 16·2 *(a)* Automatic sampler. *(b)* Schematic arrangement of automatic sampler [7].

TABLE 16·1 SAMPLE PRESERVATION [4]

Parameter	Preservative	Maximum holding period
Acidity-alkalinity	Refrigeration at 4°C	24 hr
BOD	Refrigeration at 4°C*	6 hr
Calcium	None required	
COD	2 ml/liter H_2SO_4	7 days
Chloride	None required	
Color	Refrigeration at 4°C	24 hr
Cyanide	NaOH to pH 10	24 hr
Dissolved oxygen	Determine on-site†	No holding
Fluoride	None required	
Hardness	None required	
Metals, total	5 ml/liter HNO_3	6 months
Metals, dissolved	Filtrate: 3 ml/liter 1:1 HNO_3	6 months
Nitrogen, ammonia	40 mg/liter $HgCl_2$-4°C	7 days
Nitrogen, Kjeldahl	40 mg/liter $HgCl_2$-4°C	Unstable
Nitrogen, nitrate-nitrite	40 mg/liter $HgCl_2$-4°C	7 days
Oil and grease	2 ml/liter H_2SO_4-4°C	24–hr
Organic carbon	2 ml/liter H_2SO_4 (pH 2)	7 days
pH	None available	
Phenolics	1.0 g $CuSO_4$ + H_3PO_4 to pH 4.0-4°C	24 hr
Phosphorus	40 mg/liter $HgCl_2$-4°C	7 days
Solids	None available	
Specific conductance	None required	
Sulfate	Refrigeration at 4°C	7 days
Sulfide	2 ml/liter Zn acetate	7 days
Threshold odor	Refrigeration at 4°C	24 hr
Turbidity	None available	

* Slow-freezing techniques (to —25°C) can be used for preserving samples to be analyzed for organic content.
† For some methods of determination, 4- to 8-hr preservation can be accomplished with 0.7 ml conc H_2SO_4 and 20 mg NaN_3. Refer to *Standard Methods* [12] for prescribed applications. (Footnote not in original reference.)

Field Observations

In addition to gathering information on what is being discharged from various process units, it is important to obtain information on the operation of the units. Such information should include feed rates, feed composition and characteristics, and operating conditions, such as temperature, that might affect effluent characteristics. These data should then be correlated with the data obtained on wastewater quality and volume to establish the relationship between process operation and load factor and wastewater characteristics.

Process Analysis and Modification

Having generated the required data in the survey, the next step is to put it to use. The survey information should make it possible to identify the sources of wastewater that might be subject to segregated treatment. It is often the case that the treatment of a particular wastestream would be much less complicated and costly if it could be isolated and treated separately rather than combined with all other wastestreams and treated at one facility. An example of this situation is the combination of a very dilute waste, such as cooling wastewater, with a concentrated waste containing large amounts of settleable and floatable matter. The result is that the capacity of units designed to remove the settleable and floatable matter must be increased substantially without any particular benefit being derived. A similar situation results from combining a small volume of concentrated toxic wastes with a large volume of a waste containing soluble and colloidal organic matter that is amenable to biological treatment. Flowrate data will aid in identifying those sources with varying waste flows that might require the use of equalization basins to maintain constant flow necessary for consistent treatment.

A major objective in the process modification step is to reduce wastewater volume, especially if it can be accomplished at relatively little cost compared to that for waste treatment. The principal means of reducing waste volumes are improved housekeeping and maintenance to reduce leakage and spills, reuse of process wastewater, and process modification to reduce water requirements and waste concentration. The data collected in the survey will be valuable in identifying those waste sources for which volume reduction can be accomplished by the means suggested. The survey results will also be useful in predicting the possible effects of modified operation, as well as increased plant production, on wastewater quality. If there is any doubt whether the costs of making a process modification to reduce wastes will be offset by savings in treatment costs, then the decision to make the modification should be delayed until the alternative methods of treatment have been substantially screened and an economic comparison can be made.

ANALYSIS OF TREATMENT ALTERNATIVES

Having determined the nature of the problem, the treatment requirements, the wastewater characteristics, and, where possible, having made modifications in unit operations and processes, the engineer is now ready to proceed with selection of the treatment methods best suited to his particular project.

Preliminary Screening

The objective of this preliminary selection step is to screen all possible treatment processes applicable to the problem and to select for detailed investigation those that appear to be the most promising. The waste characterization data should indicate, in most cases, the general method of treatment that might best be suited to the particular problem. A summary of the more important treatment methods for industrial wastewaters is given in Table 16·2.

The screening process should involve a preliminary evaluation of the operational and economic feasibility of the various alternative processes. For this purpose, the engineer should refer to the literature to obtain information on the ways in which similar wastes have been handled by others and on new processes that have not yet had full-scale application. It is especially important that this investigation not be limited in the types of processes considered because of preconceived ideas or personal preferences on the part of the investigating engineer. Reasons for excluding a process from further consideration should be noted and be based on substantiated facts.

Detailed Screening

For those treatment alternatives selected in the preliminary screening step, a detailed screening should be conducted to evaluate their effectiveness in producing an effluent of the desired quality. Such an investigation may entail visiting waste treatment facilities with similar unit processes to gather detailed information on construction costs, maintenance cost, and unit performance. Contacts should be made with manufacturers of equipment and apparatus to ascertain the capabilities of available hardware. To distinguish between propaganda and fact when considering manufacturers' claims, all performance data should be documented.

By far the most important task to be performed in this step is conducting laboratory bench-scale studies to evaluate the applicability of the alternative processes using the actual wastewater in question. Because of the capital expenditures that are required for construction of treatment facilities and because process reliability is important, laboratory studies are considered to be a wise investment. This is especially true for the treatment of new wastes for which few or no

TABLE 16·2 INDUSTRIAL WASTEWATER TREATMENT OPERATIONS AND PROCESSES

Description	Application
	Physical unit operations
Dialysis-osmosis	Recovery of process materials, removal of dissolved solids
Distillation	Removal of dissolved solids, separation of liquid wastes for recovery or disposal
Evaporation:	Sludge dewatering, removal of volatile materials, concentration of liquid wastes
Dryers	
Multiple-effect	
Thermal-compression	
Filtration:	Removal of suspended solids, sludge dewatering
Filter-press	
Gravity	
Pressure	
Vacuum	
Flotation:	Removal of floating or suspended liquid and solid particles, concentration of sludges
Centrifugal	
Dispersed-air	
Dissolved-air	
Gravity	
Gas transfer:	Addition and removal of gases, removal of volatile oils
Adsorption	
Aeration	
Stripping	
Screening:	Removal of coarse floating and suspended solid material
Rotary	
Vibratory	
Sedimentation:	Removal of particulate matter, biological floc, chemical floc, concentration of sludges
Centrifugal	
Gravity	
Skimming	Removal of grease, oil, and floating material
Solvent extraction	Removal of particular soluble waste constituents, recovery of soluble-process material
	Chemical unit processes
Adsorption	Removal of soluble organic compounds
Coagulation-flocculation	Removal of colloidal material
Combustion	Conversion of sludge to ash, reduction of organic content, volume reduction
Ion exchange	Recovery of specific ions and compounds, removal of ionized organic and inorganic compounds
Neutralization:	pH control
Acid	
Bases	

TABLE 16·2 *(Continued)*

	Chemical unit processes
Oxidation: Air Cl_2 Ozone	Disinfection, reduction of COD, precipitation of soluble nutrients
Reduction	Conversion of soluble compound to volatile or precipitate form for removal
Sorption	Removal of soluble organics, selected inorganic compounds
	Biological unit processes
Aerobic: Activated-sludge Lagoons Trickling-filter	Removal of soluble or colloidal organic material
Anaerobic: Contact process Digestion Filters Imhoff tanks Lagoons	Stabilization of organic sludges and organic wastes
Spray irrigation	Removal of soluble or colloidal organic material
Stabilization ponds	Removal of soluble or colloidal organic material

data are available. In view of their importance, laboratory studies are discussed separately in the following section.

LABORATORY EVALUATION

An integral part of most wastewater treatment investigations is the conduct of bench-scale laboratory studies. The objectives of such studies are usually twofold. The first is to determine whether or not the waste in question is amenable to treatment with the proposed operations or processes. The second is to obtain data that may be used to design and operate pilot or full-scale facilities.

Of the many operations and processes listed in Table 16·2, only activated sludge, anaerobic digestion, carbon adsorption, and dissolved-air flotation are discussed in this section. These processes and operations were selected because of their wide application for the treatment of wastewaters from a wide variety of industrial activities. Details on the conduct of laboratory studies for many of the other treatment methods listed in Table 16·2 may be found in Refs. [2, 3, and 10].

Activated Sludge

Biological oxidation of wastewaters using some form of the activated-sludge process has proven to be one of the most effective methods for the treatment of both municipal wastes and organic industrial wastes. Bench-scale activated-sludge treatment units are used commonly to assess the treatability of wastewater and to determine the growth yield Y and decay coefficient k_d necessary for process design. If the scope of the investigation permits, the kinetic coefficients k, K_s can also be determined.

A bench-scale reactor, shown in Fig. 16·3, can be used for the conduct of both continuous-flow and batch studies, although it is recommended that continuous-flow studies be employed when possible because they better simulate actual conditions and yield more accurate results. The procedures involved in starting up and conducting a treatability study using a bench-scale continuous-flow reactor are outlined in Example 16·1.

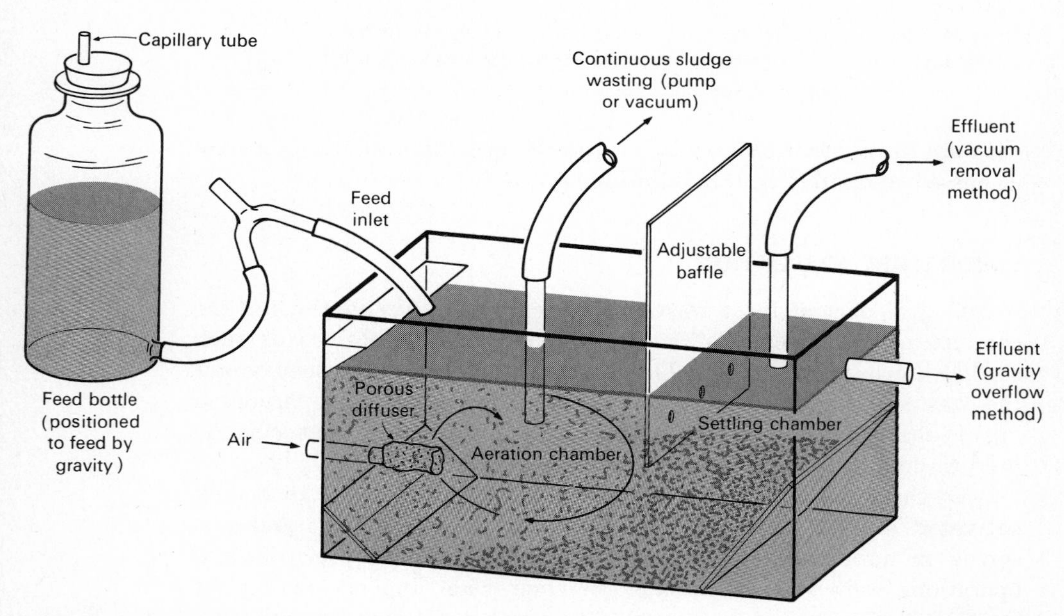

FIG. 16·3 Laboratory reactor used for the conduct of complete-mix continuous-flow activated-sludge treatment studies.

EXAMPLE 16·1 *Activated-Sludge Treatability Study*

Outline the steps involved in the conduct of an activated-sludge treatability study.

Solution

1. Select the operating conditions. As explained in Chaps. 10 and 12, either the mean cell residence time θ_c or the food-to-microorganism ratio U can be used as the controlling operating parameter. The θ_c method is suggested for use because of the ease and accuracy with which it can be applied. Units should be operated at several different θ_c's ranging from 3 to 20 days to evaluate the effect of θ_c on treatment efficiency.

   The temperature should remain constant during the test, which requires that the testing be conducted in a constant-temperature room. It is suggested that tests be conducted at two different temperatures representing average summer and winter conditions if they differ significantly for the particular project.

2. Prepare the feed and the reactors. If the wastewater contains a high concentration of suspended solids or floating oil or grease, some form of pretreatment must be used to remove it before feeding the waste to the units.

   The feed mechanism shown in Fig. 16·3 is a capillary-controlled gravity-feed system. Broken thermometers with the mercury removed serve as excellent capillary tubes, with flowrate adjustment provided by using tubes of different length. Other feed mechanisms that can be used are peristaltic and diaphragm pumps and an electrolysis pump (see Fig. 16·4), all of which provide a more constant flowrate than the capillary mechanism. It is recommended that all influent and effluent tubing be Tygon or similar transparent tubing, to allow observation of any clogging that might develop in the lines.

3. List the start-up procedure.
   (*a*) Fill units with seed sludge from a well-operating system. The initial volatile-solids concentration should be approximately 1,500 mg/liter.
   (*b*) Turn on the air supply and adjust the flowrate so that complete mixing is provided in the aeration zone, without preventing satisfactory sludge settling in the settling chamber. The air flow required for mixing far exceeds that required for biological synthesis in such small reactors. Experimentation with different baffle settings will

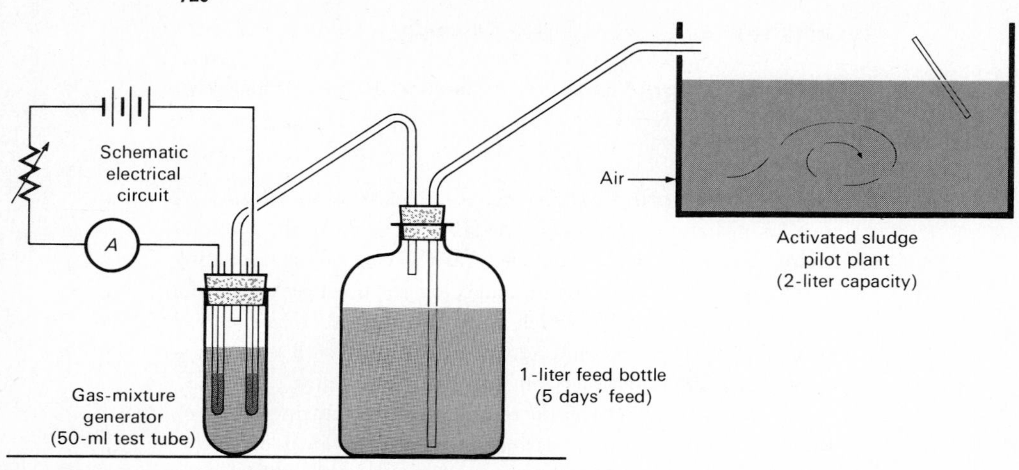

FIG. 16·4 Continuous-flow reactor system using electrolysis pump [14].

be required to achieve the optimum combination of mixing and settling.

(*c*) Begin feeding units at a flowrate necessary to achieve the desired hydraulic detention time. As shown in Fig. 16·3, effluent may be withdrawn from the reactor by means of a gravity overflow or by use of a vacuum suction tube set to maintain the desired volume in the reactor.

(*d*) Maintain the operation until steady-state conditions are reached. This is indicated when the BOD or COD of the effluent is stabilized. Another method of determining steady-state conditions is to monitor the oxygen-uptake rate of the mixed liquor using the Warburg respirometer described in Chap. 7. A constant uptake rate indicates a steady-state condition. The pH of mixed liquor should be measured daily, and neutralization procedures should be adjusted accordingly. The MLVSS concentration should be allowed to build up to a predetermined operating level computed using the equations given in Chap. 10; then solids wasting should be initiated. Continuous sludge wasting can be accomplished by wasting directly from the aeration chamber using laboratory peristaltic or diaphragm pumps or by using vacuum withdrawal. The amount to be wasted depends on the volume of the reactor and the value of θ_c. For example, if the volume of the reactor is 1.5 liters and θ_c is 10 days, then 150 ml of the mixed liquor would be removed per day. Alternatively, solids may be wasted on a daily basis. To waste solids the settling baffle is removed, the effluent port clamped shut, the air flow increased to assure complete mixing, and a portion of the mixed liquor is withdrawn. The liquid volume removed is then replaced with tap water.

If the settling characteristics of the sludge are poor and considerable solids carryover occurs, the mass of solids lost in the effluent should be taken into account with either of the aforementioned wasting systems to assure proper control. Solids carryover will always result in lower wasting rates. At steady state the total solids in the reactor should remain constant.

4. List the steady-state procedure. When steady-state conditions have been reached, a sampling and analysis schedule should be established. A suggested schedule is summarized in Table 16·3.

TABLE 16·3 ANALYSIS SCHEDULE FOR ACTIVATED–SLUDGE STUDY*

Analysis	Frequency	Raw waste	Mixed liquor	Effluent
BOD, mg/liter (filtered and unfiltered)	3/week	X		X
COD, mg/liter (filtered and unfiltered)	3/week	X		X
TOC (if desired), mg/liter (filtered and unfiltered)	3/week	X		X
Suspended solids, mg/liter	Daily		X	X
Volatile suspended solids, mg/liter	3/week		X	X
pH	Daily		X	
Color, turbidity	3/week			X
Nitrogen and phosphorus	1/week	X		X
Microscopic examination	1/week		X	
Specific compounds or ions	3/week	X		X
Settling characteristics	1/week		X	

*Adapted from Ref. [3].

The table is self-explanatory for the most part; however, some of the parameters are explained as follows:

(*a*) BOD, COD, TOC. It is suggested that these determinations be made on both filtered and unfiltered samples if the waste contains a significant amount of insoluble organic material.

(b) Nitrogen and phosphorus. The purpose of these tests is to determine if there are sufficient nutrients available for synthesis. If forms of these nutrients appear in the effluent, it can be assumed that there are sufficient nutrients. If no forms of nitrogen or phosphorus appear in the effluent, nutrient supplements, such as ammonium choride and sodium phosphate, must be added to the waste feed. The amount of nutrient added should be no greater than the calculated theoretical requirements, because excess nutrients may inhibit cell production.

(c) Microscopic examination. The type of microorganisms present in the mixed liquor in the largest numbers gives an indication of how "healthy" the sludge is. The presence of filamentous bacteria will manifest itself as bulking or floating sludge. Protozoa play an important role in polishing the effluent by consuming suspended bacteria. The presence of large numbers (thousands per milliliter) of active attached and crawling ciliates normally indicates a healthy sludge, while a preponderance of flagellates and free-swimming ciliates tentatively suggests that the sludge is not in its optimum condition. A detailed key for nonspecialists to identify different species of microorganisms commonly found in activated sludge is presented in Hawkes [5].

(d) Settling characteristics. It is important to have some measure of the settling characteristics of the sludge. This may be accomplished qualitatively simply by observing the sludge settle in the tank under quiescent conditions or by conducting a more quantitative test such as described in Chap. 8 and in "Return Sludge Requirements" in Chap. 12.

The sampling and analysis schedule should be continued for at least 2 weeks or until such time as consistent results are obtained.

5. Perform the data analysis.
   (a) Determine removal percentages based on influent and effluent BOD, COD, TOC, color, turbidity, and specific ions or compounds.
   (b) Plot percentage removals versus θ_c and temperature and feed rate if varied.
   (c) Correlate percentage removal and θ_c with sludge characteristics (microorganism population, settling characteristics).

This information can then be used to assess whether the wastewater is treatable by means of the activated-sludge process and will allow the determination of the optimum θ_c for best treatment efficiency and sludge-settling characteristics.

Determination of Kinetic Coefficients It will be recalled from the discussion in Chap. 10 on complete-mix systems with solids recycle that, for given values of flowrate Q, influent substrate concentration S_0, and the kinetic coefficients, k, K_s, Y, and k_d, unique values of efficiency E, effluent substrate concentration S, and total microbial mass in the system VX are determined for each value of θ_c. It is desired to operate the units over a range of effluent substrate concentrations; therefore, several different θ_c's (at least five) should be selected for operation ranging from 1 to 10 days. Using the data collected at steady-state conditions, mean values should be determined for Q, S_0, S, X, and ΔX. Recalling Eq. 10·5 and applying it on a finite basis yields

$$\frac{\Delta F}{\Delta t} = \frac{kXS}{K_s + S} = \frac{Q(S_0 - S)}{V} \tag{16·1}$$

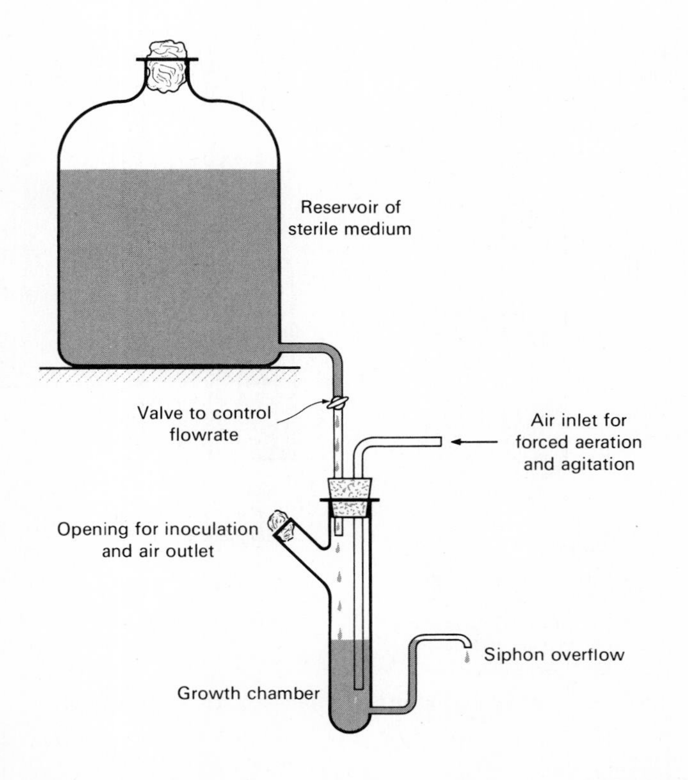

FIG. 16·5 A simplified diagram of a chemostat [13].

Dividing by X yields

$$\frac{\Delta F/\Delta t}{X} = \frac{kS}{K_s + S} \tag{16·2}$$

From the values determined, plot $(\Delta F/\Delta t)/X$ versus S and determine k and K_s (see Fig. 10·10). The values of Y and k_d may be determined by plotting $(\Delta X/\Delta t)/X$ versus $(\Delta F/\Delta t)/X$. The slope of the straight line passed through the plotted experimental datum points is equal to Y, and the intercept is equal to k_d.

Chemostat Analysis A chemostat apparatus, such as the one depicted in Fig. 16·5, can also be used for the determination of kinetic coefficients. Growth rates are determined by effluent turbidity measurements. Such units are normally used for pure culture determinations.

Anaerobic Treatment

If anaerobic treatment appears to be feasible on the basis of the wastewater characterization results but practical experience is limited in

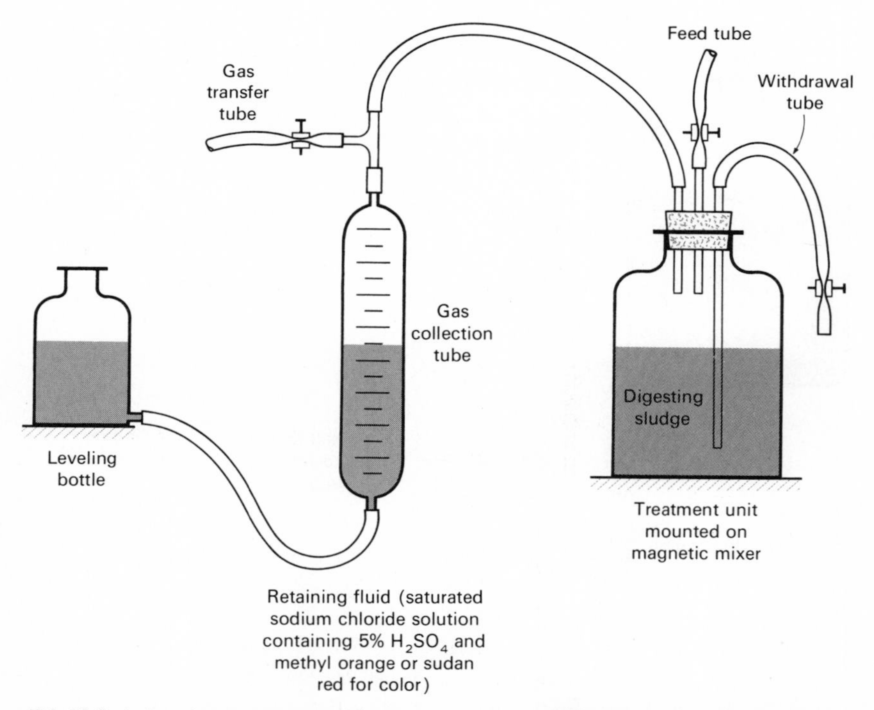

FIG. 16·6 Laboratory reactor used for the conduct of complete-mix batch-type anaerobic treatment studies.

applying this method to the particular waste, then bench-scale experiments should be conducted to assess anaerobic treatability. Either batch or continuous-flow reactors can be employed. For the purpose of assessing treatability, the batch system allows for a more convenient and less complicated operation. A schematic diagram of a batch reactor is shown in Fig. 16·6. The procedure is described in Example 16·2.

EXAMPLE 16·2 *Anaerobic Treatability Study*

Outline the steps involved in the conduct of an anaerobic treatability study using a batch system.

Solution

1. Select the operating conditions. The units in this system are essentially complete-mix–no-recycle reactors; therefore, θ_c is the same as the hydraulic retention time θ. Control of θ_c is thus maintained by wasting a constant volume from the reactor each day. At least three units should be operated at θ_c's in the range of 15 to 30 days.

   The feed rate will be limited by the amount of gas that can be measured in the collection tube (2-liter tubes are recommended, but 1-liter tubes will suffice). If 2-liter tubes are used, the feed rate may be determined by assuming that the gas produced is 60 percent methane so that 1.2 liters of methane are produced per day. At standard conditions, this corresponds to 3.43g of BOD_L per day.

   The waste should be diluted to a concentration that will allow the volume being fed to be equal to the volume wasted each day. The temperature should be maintained at 35°C throughout the test run.

2. List the start-up procedure.
   (*a*) Obtain digesting sludge from a well-operating municipal plant, taking precautions to reduce exposure to the air as much as possible.
   (*b*) Pass the sludge through a coarse screen to remove large particles and dilute it 1:1 with warm tap water.
   (*c*) Add the sludge to the digesting unit and seal the top. Wait 1 day to allow the facultative bacteria to consume the oxygen trapped in the digester.
   (*d*) Raise the aspirator bottle to apply a positive pressure on the system and withdraw the proper amount of mixed liquor dictated by the chosen θ_c.
   (*e*) Expel the gas in the digester so that the liquid surface

in the measuring tube reads zero when level with the liquid surface in the reservoir.

(f) Lower the aspirator bottle to apply a negative pressure on the unit and add waste, being careful not to allow air to enter the digester.

(g) Shake the digester well and set the aspirator bottle so that a slight positive pressure is applied to the system.

3. List the daily operating procedure.
   (a) Record the gas production with the aspirator bottle and measuring tube at equal levels.
   (b) Raise the aspirator bottle.
   (c) Shake the digesters well and waste the proper amount of mixed liquor.
   (d) Expel the gas to zero, leveling as before and lower the aspirator bottle.
   (e) Add the waste and raise the aspirator bottle. Steady-state conditions should be reached after approximately 1 week of operation and can be assumed when daily gas production remains constant.

4. Collect and analyze the samples. A suggested sampling and analysis schedule is shown in Table 16·4. An excellent discussion of digester control is given by McCarty [8]. Steady-state operation should be maintained for at least 4 weeks.

TABLE 16·4 ANALYSIS SCHEDULE FOR ANAEROBIC TREATMENT STUDY

Analysis	Frequency	Raw waste	Mixed liquor	Centrifuged effluent
COD	3/week	X		X
BOD	3/week	X		X
Volatile acids	3/week			X
pH	Daily		X	
Alkalinity	Daily		X	
Ammonia nitrogen	1/week			X
Gas analysis (% CH_4, % CO_2)	3/week			

5. Perform the data analysis.
   (a) Determine the removal efficiencies for each θ_c based on COD and BOD_5.
   (b) Determine the average daily gas production for each unit.
   (c) Determine the gas production and methane production in cu ft/lb BOD_L added for each θ_c.

(*d*) Determine the percent of COD stabilized based on CH_4 production (395 ml CH_4 = 1 g COD stabilized at 35°C).

(*e*) Plot the BOD removal, COD removal, volatile acids, pH, alkalinity, and gas production versus time for each unit.

Carbon Adsorption

Granular and powdered activated-carbon adsorption is finding expanded application as an advanced treatment process for the removal of low concentrations of soluble organic material. Industrial application of carbon adsorption is also increasing. In oil refineries it is being used as a means of removing low concentrations of phenol and trace organic molecules that might have toxic effects on receiving waters.

A test program to evaluate the economic and technical feasibility of granular activated-carbon wastewater treatment would consist of two parts. Batch isotherm tests would first be made to determine if the desired degree of treatment could be attained. Continuous-flow column tests would then be conducted to establish operating capacities, as well as to obtain column design data. The procedures involved are outlined in Examples 16·3 and 16·4.

EXAMPLE 16·3 *Development of Liquid Adsorption Isotherm*

Outline the steps involved in the conduct of a study to develop the Freundlich adsorption isotherm for the removal of a particular contaminant (BOD, TOC, color, or phenol) with powdered activated carbon.

Solution

1. Develop data for plotting the linear form of the Freundlich isotherm (see Chap. 9).

   Data for plotting isotherms are obtained by treating fixed volumes of wastewater with a series of known carbon dosages. For preliminary tests, dosages of 0, 5, 10, 20 and 50 mg/liter may be used. Subsequent adjustments in dosages can be made to obtain an adequate range of experimental points. It is recommended that the carbon be pulverized so that 95 percent will pass a 325-mesh screen. This will minimize the effect of particle-size variation on adsorption rate. The carbon-wastewater mixture is then agitated for a fixed time (minimum of 1 hr is recommended) at a constant temperature. The carbon is removed by filtration and the residual concentration of the

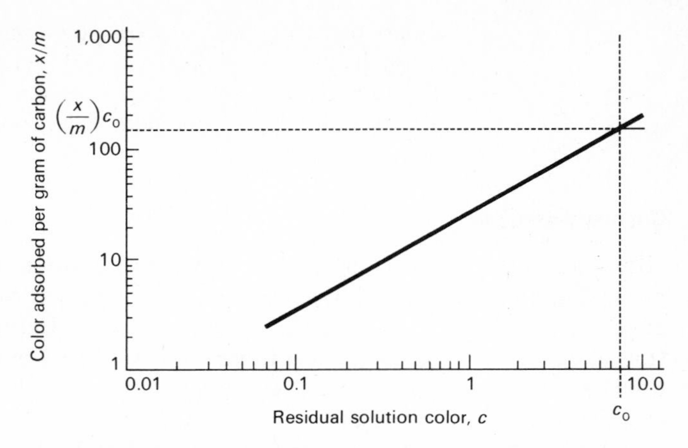

FIG. 16·7 Typical decolorization isotherm [15].

impurity in the filtrate is determined. All the values necessary to plot an isotherm can be calculated from these measurements.

2. Perform the data analysis. A tabulation of data from a color-removal isotherm test is shown in Table 16·5 and the resulting isotherm is illustrated in Fig. 16·7.

TABLE 16·5 TREATMENT OF ISOTHERM DATA [15]

M Weight of carbon, g/100 ml solution	C Residual-solution color	X Color adsorbed	X/M Color adsorbed per unit weight
0	7.70		
0.05	3.67	4.03	80.6
0.1	2.20	5.50	55.0
0.3	0.87	6.83	22.8
1.0	0.25	7.45	7.5

The adsorptive capacity of the carbon in a column application can be estimated by extending a vertical line from the point on the horizontal scale corresponding to the initial concentration, C_0, and extrapolating the isotherm to intersect this line. The X/M value at the point of intersection can be read from the vertical scale. This value, $(X/M)_{C_0}$ represents the amount of contaminant adsorbed per unit weight of carbon where the carbon is at equilibrium with the initial concentration of contaminant. This condition should exist in the upper section of a carbon bed during column treatment, and it therefore represents the ultimate capacity of the carbon for the particular waste. From the $(X/M)_{C_0}$ value, the ultimate capacity in terms of volume of waste treated per gram of carbon can be calculated simply by dividing $(X/M)_{C_0}$ by the amount of impurity adsorbed per unit volume.

Additional information is required before a full-scale or pilot system can be designed. The optimum flowrate and bed depth, as well as the operating capacity of the carbon, must be established to determine the dimensions and the number of columns necessary for continuous treatment. These parameters can be determined only by dynamic column tests.

EXAMPLE 16·4 *Activated-Carbon-Column Adsorption Study*

Outline the steps involved in the conduct of an activated-carbon-column adsorption study.

Solution

1. Select the operating conditions. Full-scale granular columns generally have bed depths ranging from 10 to 30 ft and diameters from 1 to 10 ft. Diameters can be scaled down in the laboratory to 2 or 4 in. with the flowrate correspondingly reduced. Data obtained from a 1- or 2-ft bed depth, however, cannot be extrapolated to determine performance in a 10-ft bed. Therefore, it is necessary to conduct tests in 10-ft or longer columns. This may be accomplished by using several 5-ft sections of plastic, metal, or Pyrex pipe in series operation.
2. List the operating procedures.
   (*a*) Waste should be delivered to the columns by a positive displacement pump at a predetermined rate. A good starting rate is 0.5 gpm/sq ft of cross-sectional column area. The rates should be gradually increased until the effluent quality is no longer acceptable.
   (*b*) Effluent samples should be withdrawn from each sampling port at hourly intervals or specific throughput volume intervals, and the concentration of the contaminant should be determined. Feed concentrations should also be determined and equalized if required.
3. Perform the data analysis.
   (*a*) A breakthrough curve should be constructed, showing effluent impurity concentration as a function of throughput volume for each bed depth and each flowrate. Having

selected the required degree of treatment, the volumes of acceptable effluent collected at the various bed depths can be determined from the curves.

(*b*) From the weight of carbon in each column, the carbon dosage resulting from each condition of bed depth and flowrate should be calculated.

(*c*) Investigate the advantages of using deeper beds by preparing a plot of acceptable throughput volume versus bed depth for each flowrate. Extrapolation of the straight-line portion of the curve to a greater depth will provide the information necessary to calculate carbon dosage at a given bed depth.

(*d*) Make design calculations for various adsorber configurations. The optimum system will provide adequate treatment at lowest total cost, considering capital, carbon, and reactivation costs.

Dissolved-Air Flotation

Dissolved-air flotation is finding increased application in municipal wastewater treatment plants for thickening of waste activated sludge

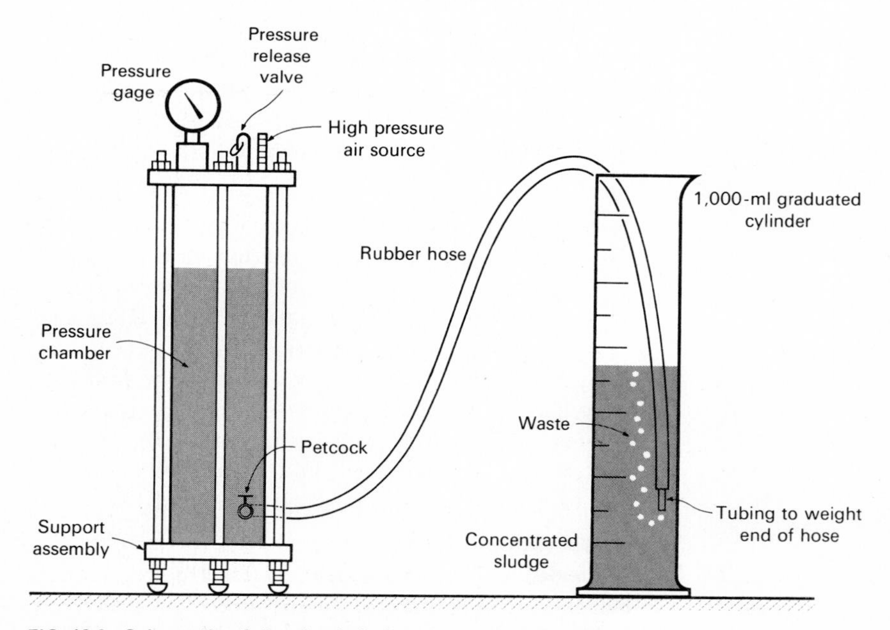

FIG. 16·8 Schematic of dissolved-air flotation test apparatus.

taken from the aeration tanks and flocculated chemical sludges and in industrial facilities for the separation of greases, oils, fibers, and other low-density contaminants from wastewater. The process can be used alone or in combination with flocculation. Three types of operation are used currently in practice: full-flow, split-flow, and recycle-flow.

The theoretical aspects of the design of the various types of flotation systems are discussed in Chap. 8. A reasonably good indication of what can be accomplished with dissolved-air flotation can be determined in the laboratory using the apparatus shown in Fig. 16·8. Tests may be performed for a range of conditions simulating all three types of operation, including adjustment of the proportion recycled in combination with various chemical aids. The test procedure is outlined in Example 16·5.

EXAMPLE 16·5 *Dissolved-Air Flotation Study*

Outline the procedures involved in the conduct of a dissolved-air flotation study.

Solution

1. List the operating procedure.
   (*a*) Fill the pressure chamber three-fourths full with wastewater and apply compressed air to the chamber to attain the desired pressure (30 to 50 psi).
   (*b*) Maintaining pressure in the chamber, shake the liquid-air mixture for 1 min and allow to stand for 3 min to achieve saturation.
   (*c*) Clear transfer tubing of air and unpressurized wastewater by dispelling fluid until it appears milky.
   (*d*) Place a specified volume of wastewater or sludge in the graduated cylinder and release a volume of pressurized liquid into the cylinder. If a recycle or split-flow process is being tested, the volume released is dictated by the desired flow ratio. The recycle ratio equals the volume of released liquid divided by the sample volume. If full-flow operation is being tested, a minimum of 500 ml should be released. The velocity of release should provide adequate mixing without shearing suspended solids in the feed mixture.
   (*e*) After flotation is complete (10 to 20 min), draw off the clarified effluent and the floated sludge separately and analyze the concentration of each.

(*f*) Repeat the test for several different pressures.

(*g*) If polyelectrolyte addition is to be studied, add polymer to sample and mix well prior to release of pressurized wastewater.

2. Perform the data analysis.

(*a*) Compute the air-solids ratio as explained in Chap. 8 from Eq. 8·24 or 8·25.

(*b*) Plot the relationship between effluent suspended solids and air-solids ratio if clarification is being investigated. If thickening is being studied, a graph of float solids versus air-solids ratio should be prepared. The optimum air-solids ratio can be determined from these plots and used in preliminary design calculations.

SELECTION AND DESIGN OF TREATMENT UNITS

Having eliminated those alternatives that are operationally infeasible on the basis of laboratory test results, a cost comparison is required to select among the remaining alternatives. Using the operating parameters developed from the laboratory studies and appropriate scale-up factors, preliminary design calculations can be made for full-scale facilities. In turn, cost estimates can be made on the basis of preliminary designs. For industrial plants, the cost of major process modifications identified earlier must also be considered at this time. For some applications, it may be necessary to construct pilot plant facilities to provide a more accurate simulation of the proposed full-scale operation. This is essential when experience is limited in the practical application of a process to the treatment of a particular type of wastewater. Final design will then be based on the parameters developed from the pilot plant studies.

PROBLEMS

16·1 Given the following data for four bench-scale continuous-flow activated-sludge units operated with a food-processing waste substrate, determine k_d and Y. Assume that 90 percent of the cells are active.

	Unit 1	Unit 2	Unit 3	Unit 4
Mean MLVSS, g	18.81	7.35	7.65	2.89
Mean sludge growth, g/day	0.88	1.19	1.42	1.56
Mean BOD applied/day/g MLVSS	0.17	0.41	0.40	1.09

16·2 Given the following data from five bench-scale activated-sludge units operating at steady state, determine the kinetic coefficients k, K_s, k_d, and Y.

Unit no.	S, mg/liter	$(dF/dt)/X$, day^{-1}	θ, hr	S_0, mg/liter	X, mg/liter	dX/dt, g/day	V, liters
1	7	0.55	4.6	300	2,700	0.91	1.5
2	13	1.15	4.0	300	1,500	1.18	1.5
3	18	1.56	4.0	300	1,090	1.20	1.5
4	30	1.91	4.0	300	850	1.17	1.5
5	41	1.94	4.0	300	800	1.10	1.5

16·3 Given the following results from a batch carbon-adsorption test for odor removal, plot the Freundlich adsorption isotherm and estimate the ultimate adsorptive capacity (odor removed per gram of carbon) of the carbon in a column application.

M Weight of carbon, g	C Residual odor
0	20
0.9	13
1.7	10
4	6
7	4
10	3

16·4 Review the literature and recommend appropriate scale-up factors for the application of laboratory aeration, vacuum filtration, flotation, and settling test data to full-scale design.

REFERENCES

1. Black, H. H.: Procedures for Sampling and Measuring Industrial Wastes, *Sewage and Industrial Wastes*, vol. 25, p. 45, 1952.
2. Besselievre, E. B.: *The Treatment of Industrial Wastes*, McGraw-Hill, New York, 1969.
3. Eckenfelder, W. W., Jr., and D. L. Ford: *Water Pollution Control*, Jenkins Publishing Co., Austin, Tex., 1970.
4. *FWPCA Methods for Chemical Analysis of Water and Wastes*, U.S. Department of Interior, Federal Water Pollution Control Administration, 1969.
5. Hawkes, H. A.: *The Ecology of Waste Water Treatment*, Pergamon, London, 1963.
6. *Manual on Disposal of Refinery Wastes: Volume on Liquid Wastes*, American Petroleum Institute, 1969.
7. *Manual on Water*, American Society for Testing and Materials, 1969.
8. McCarty, P. L.: Anaerobic Treatment Fundamentals, Parts 1–4, *Public Works*, vol. 95, nos. 9–12, 1964.
9. Ohio River Valley Water Sanitation Commission: *Planning and Making Industrial Waste Surveys*, Cincinnati, Ohio, 1952.

10. Ross, R. D.: *Industrial Waste Disposal,* Reinhold, New York, 1968.

11. Spink, L. K.: *Principles and Practice of Flow Meter Engineering,* The Foxboro Company, Foxboro, Mass., 1967.

12. *Standard Methods for the Examination of Water and Waste Water,* 13th ed., American Public Health Association, 1971.

13. Stanier, R. Y., M. Doudoroff, and E. A. Adelberg: *The Microbial World,* 3d ed., Prentice-Hall, Englewood Cliffs, N.J., 1970.

14. Symons, J. M.: Simple, Continuous-Flow, Low and Variable Rate Pump, *J. WPCF,* vol. 35, no. 11, 1963.

15. *The Laboratory Evaluation of Granular Activated Carbon for Liquid Phase Applications,* Pittsburgh Activated Carbon Co., 1966.

appendixes

A WEIGHTS, MEASURES, AND CONVERSION FACTORS

AREA

sq m	sq in.	sq ft	sq mi	acres
1	1,550	10.76	—	—
0.0929	144	1	—	—
—	—	—	1	640
4,047	—	43,560	—	1

DISCHARGE

cfs	mgd	gpm	acre-ft/day
1	0.646	449	1.98
1.547	1	694	3.06

HEAT, ENERGY, AND WORK

ft-lb	kw-hr	hp-hr	Btu	cal
1	—	—	—	0.324
2.655×10^6	1	1.341	3,473	860,565
1.98×10^6	0.746	1	2,545	641,615
778	—	—	1	252
3.086	—	—	0.00397	1

LENGTH

cm	in.	ft	yd	μ	miles
1	0.394	—	—	10^4	—
2.54	1	—	—	25,400	—
30.48	12	1	0.333	—	—
91.44	36	3	1	—	—
—	—	5,280	1,760	—	1

DENSITY

g/cc	lb/cu ft	lb/U.S. gal
1	62.4	8.34
0.016	1	0.134
0.1198	7.48	1

MASS

g	lb	ton
1	0.0022	—
454	1	—
—	2,000	1

PRESSURE

kg/sq cm	psi	atm	Water, ft	Mercury, in.
1	14.22	0.968	32.84	28.96
0.0703	1	0.068	2.31	2.04
1.033	14.7	1	33.93	29.92
0.0345	0.491	0.0334	1.13	1
—	0.433	—	1	0.883

TEMPERATURE

Degrees Fahrenheit $= 32 + \frac{9}{5} \times$ Degrees Centigrade

0	5	10	15	20	25	30	35	40	45	50	55	60
32	41	50	59	68	77	86	95	104	113	122	131	140

VELOCITY

mph	fps	in./min	cm/sec
1	1.47	1,054	44.7
—	1	720	30.5
—	—	1	0.043

VOLUME

cu ft	gal	liters
1	7.48	28.3
—	1	3.785

B PHYSICAL PROPERTIES OF WATER

Temperature, °F	Specific Weight, γ, lb/cu ft	Density, ρ, slug/cu ft	Viscosity, $\mu \times 10^5$, lb-sec/sq ft	Kinematic Viscosity, $\nu \times 10^5$, sq ft/sec	Surface Tension, σ, lb/ft	Vapor Pressure, pv, psia
32	62.42	1.940	3.746	1.931	0.00518	0.09
40	62.43	1.940	3.229	1.664	0.00514	0.12
50	62.41	1.940	2.735	1.410	0.00509	0.18
60	62.37	1.938	2.359	1.217	0.00504	0.26
70	62.30	1.936	2.050	1.059	0.00498	0.36
80	62.22	1.934	1.799	0.930	0.00492	0.51
90	62.11	1.931	1.595	0.826	0.00486	0.70
100	62.00	1.927	1.424	0.739	0.00480	0.95

Source: Adapted from J. K. Vennard: *Elementary Fluid Mechanics*, 4th ed., Wiley, New York, 1961, app. II.

C DISSOLVED–OXYGEN SOLUBILITY DATA

DISSOLVED OXYGEN,* mg/liter

Temperature, °C	Chloride Concentration, mg/liter				
	0	5,000	10,000	15,000	20,000
0	14.62	13.79	12.97	12.14	11.32
1	14.23	13.41	12.61	11.82	11.03
2	13.84	13.05	12.28	11.52	10.76
3	13.48	12.72	11.98	11.24	10.50
4	13.13	12.41	11.69	10.97	10.25
5	12.80	12.09	11.39	10.70	10.01
6	12.48	11.79	11.12	10.45	9.78
7	12.17	11.51	10.85	10.21	9.57
8	11.87	11.24	10.61	9.98	9.36
9	11.59	10.97	10.36	9.76	9.17
10	11.33	10.73	10.13	9.55	8.98
11	11.08	10.49	9.92	9.35	8.80
12	10.83	10.28	9.72	9.17	8.62
13	10.60	10.05	9.52	8.98	8.46
14	10.37	9.85	9.32	8.80	8.30
15	10.15	9.65	9.14	8.63	8.14
16	9.95	9.46	8.96	8.47	7.99
17	9.74	9.26	8.78	8.30	7.84
18	9.54	9.07	8.62	8.15	7.70
19	9.35	8.89	8.45	8.00	7.56
20	9.17	8.73	8.30	7.86	7.42
21	8.99	8.57	8.14	7.71	7.28
22	8.83	8.42	7.99	7.57	7.14
23	8.68	8.27	7.85	7.43	7.00
24	8.53	8.12	7.71	7.30	6.87
25	8.38	7.96	7.56	7.15	6.74
26	8.22	7.81	7.42	7.02	6.61
27	8.07	7.67	7.28	6.88	6.49
28	7.92	7.53	7.14	6.75	6.37
29	7.77	7.39	7.00	6.62	6.25
30	7.63	7.25	6.86	6.49	6.13

* Saturation values of dissolved oxygen in fresh and sea water exposed to dry air containing 20.90 percent oxygen under a total pressure of 760 mm of mercury.

Source: G. C. Whipple and M. C. Whipple: Solubility of Oxygen in Sea Water, *J. Am. Chem. Soc.*, vol. 33, p. 362, 1911. Calculated using data developed by C. J. J. Fox: On the Coefficients of Absorption of Nitrogen and Oxygen in Distilled Water and Sea Water and Atmospheric Carbonic Acid in Sea Water, *Trans. Faraday Soc.*, vol. 5, p. 68, 1909.

D MOST PROBABLE NUMBER OF COLIFORMS PER 100 ML OF SAMPLE

Number of Positive Tubes			MPN	Number of Positive Tubes			MPN	Number of Positive Tubes			MPN	Number of Positive Tubes			MPN	Number of Positive Tubes			MPN	Number of Positive Tubes			MPN
10 ml	1 ml	0.1 ml		10 ml	1 ml	0.1 ml		10 ml	1 ml	0.1 ml		10 ml	1 ml	0.1 ml		10 ml	1 ml	0.1 ml		10 ml	1 ml	0.1 ml	
0	0	0		1	0	0	2.0	2	0	0	4.5	3	0	0	7.8	4	0	0	13	5	0	0	23
0	0	1	1.8	1	0	1	4.0	2	0	1	6.8	3	0	1	11	4	0	1	17	5	0	1	31
0	0	2	3.6	1	0	2	6.0	2	0	2	9.1	3	0	2	13	4	0	2	21	5	0	2	43
0	0	3	5.4	1	0	3	8.0	2	0	3	12	3	0	3	16	4	0	3	25	5	0	3	58
0	0	4	7.2	1	0	4	10	2	0	4	14	3	0	4	20	4	0	4	30	5	0	4	76
0	0	5	9.0	1	0	5	12	2	0	5	16	3	0	5	23	4	0	5	36	5	0	5	95
0	1	0	1.8	1	1	0	4.0	2	1	0	6.8	3	1	0	11	4	1	0	17	5	1	0	33
0	1	1	3.6	1	1	1	6.1	2	1	1	9.2	3	1	1	14	4	1	1	21	5	1	1	46
0	1	2	5.5	1	1	2	8.1	2	1	2	12	3	1	2	17	4	1	2	26	5	1	2	64
0	1	3	7.3	1	1	3	10	2	1	3	14	3	1	3	20	4	1	3	31	5	1	3	84
0	1	4	9.1	1	1	4	12	2	1	4	17	3	1	4	23	4	1	4	36	5	1	4	110
0	1	5	11	1	1	5	14	2	1	5	19	3	1	5	27	4	1	5	42	5	1	5	130

Number of Positive Tubes			MPN	Number of Positive Tubes			MPN	Number of Positive Tubes			MPN	Number of Positive Tubes			MPN	Number of Positive Tubes			MPN	Number of Positive Tubes			MPN
10 ml	1 ml	0.1 ml		10 ml	1 ml	0.1 ml		10 ml	1 ml	0.1 ml		10 ml	1 ml	0.1 ml		10 ml	1 ml	0.1 ml		10 ml	1 ml	0.1 ml	
0	2	0	3.7	1	2	0	6.1	2	2	0	9.3	3	2	0	14	4	2	0	22	5	2	0	49
0	2	1	5.5	1	2	1	8.2	2	2	1	12	3	2	1	17	4	2	1	26	5	2	1	70
0	2	2	7.4	1	2	2	10	2	2	2	14	3	2	2	20	4	2	2	32	5	2	2	95
0	2	3	9.2	1	2	3	12	2	2	3	17	3	2	3	24	4	2	3	38	5	2	3	120
0	2	4	11	1	2	4	15	2	2	4	19	3	2	4	27	4	2	4	44	5	2	4	150
0	2	5	13	1	2	5	17	2	2	5	22	3	2	5	31	4	2	5	50	5	2	5	180
0	3	0	5.6	1	3	0	8.3	2	3	0	12	3	3	0	17	4	3	0	27	5	3	0	79
0	3	1	7.4	1	3	1	10	2	3	1	14	3	3	1	21	4	3	1	33	5	3	1	110
0	3	2	9.3	1	3	2	13	2	3	2	17	3	3	2	24	4	3	2	39	5	3	2	140
0	3	3	11	1	3	3	15	2	3	3	20	3	3	3	28	4	3	3	45	5	3	3	180
0	3	4	13	1	3	4	17	2	3	4	22	3	3	4	31	4	3	4	52	5	3	4	210
0	3	5	15	1	3	5	19	2	3	5	25	3	3	5	35	4	3	5	59	5	3	5	250

Number of Positive Tubes 10 ml	1 ml	0.1 ml	MPN	Number of Positive Tubes 10 ml	1 ml	0.1 ml	MPN	Number of Positive Tubes 10 ml	1 ml	0.1 ml	MPN	Number of Positive Tubes 10 ml	1 ml	0.1 ml	MPN	Number of Positive Tubes 10 ml	1 ml	0.1 ml	MPN	Number of Positive Tubes 10 ml	1 ml	0.1 ml	MPN
0	4	0	7.5	1	4	0	11	2	4	0	15	3	4	0	21	4	4	0	34	5	4	0	130
0	4	1	9.4	1	4	1	13	2	4	1	17	3	4	1	24	4	4	1	40	5	4	1	170
0	4	2	11	1	4	2	15	2	4	2	20	3	4	2	28	4	4	2	47	5	4	2	220
0	4	3	13	1	4	3	17	2	4	3	23	3	4	3	32	4	4	3	54	5	4	3	280
0	4	4	15	1	4	4	19	2	4	4	25	3	4	4	36	4	4	4	62	5	4	4	350
0	4	5	17	1	4	5	22	2	4	5	28	3	4	5	40	4	4	5	69	5	4	5	430
0	5	0	9.4	1	5	0	13	2	5	0	17	3	5	0	25	4	5	0	41	5	5	0	240
0	5	1	11	1	5	1	15	2	5	1	20	3	5	1	29	4	5	1	48	5	5	1	350
0	5	2	13	1	5	2	17	2	5	2	23	3	5	2	32	4	5	2	56	5	5	2	540
0	5	3	15	1	5	3	19	2	5	3	26	3	5	3	37	4	5	3	64	5	5	3	920
0	5	4	17	1	5	4	22	2	5	4	29	3	5	4	41	4	5	4	72	5	5	4	1,600
0	5	5	19	1	5	5	24	2	5	5	32	3	5	5	45	4	5	5	81				

E WATER QUALITY CRITERIA OF THE OHIO RIVER VALLEY WATER SANITATION COMMISSION

Minimum Conditions Applicable to All Waters at All Places and at All Times

1. Free from substances attributable to municipal, industrial or other discharges that will settle to form putrescent or otherwise objectionable sludge deposits.
2. Free from floating debris, oil, scum and other floating materials attributable to municipal, industrial or other discharges in amounts sufficient to be unsightly or deleterious.
3. Free from materials attributable to municipal, industrial or other discharges producing color, odor or other conditions in such degree as to create a nuisance.
4. Free from substances attributable to municipal, industrial or other discharges in concentrations or combinations which are toxic or harmful to human, animal, plant or aquatic life.

Stream-quality Criteria

For Public Water Supply The following criteria are for evaluation of stream quality at the point at which water is withdrawn for treatment and distribution as a potable supply:

1. *Bacteria* Coliform group not to exceed 5,000 per 100 ml as a monthly average value (either MPN or MF count); nor exceed this number in more than 20 percent of the samples examined during any month; nor exceed 20,000 per 100 ml in more than 5 percent of such samples.
2. *Threshold-odor number* Not to exceed 24 (at 60°C) as a daily average.
3. *Dissolved solids* Not to exceed 500 mg/liter as a monthly average value, nor exceed 750 mg/liter at any time. For Ohio River water, values of specific conductance of 800 and 1,200 micromhos/cm (at 25°C) may be considered equivalent to dissolved-solids concentrations of 500 and 750 mg/liter.
4. *Radioactive substances* Gross beta activity (in the known absence of Strontium-90 and alpha emitters) not to exceed 1,000 micro-microcuries per liter at any time.
5. *Chemical constituents* Not to exceed the following specified concentrations at any time:

Constituent	*Concentration (mg/liter)*
Arsenic	0.05
Barium	1.0
Cadmium	0.01
Chromium (hexavalent)	0.05
Cyanide	0.2
Fluoride	2.0
Lead	0.05
Selenium	0.01
Silver	0.05

For Industrial Water Supply The following criteria are applicable to stream water at the point where the water is withdrawn for use (with or without treatment) for industrial cooling and processing:

1. *Dissolved oxygen* Not less than 2.0 mg/liter as a daily-average value, nor less than 1.0 mg/liter at any time.
2. *pH* Not less than 5.0 nor greater than 9.0 at any time.
3. *Temperature* Not to exceed 95°F at any time.
4. *Dissolved solids* Not to exceed 750 mg/liter as a monthly average value, nor exceed 1,000 mg/liter at any time. For Ohio River water, values of specific conductance of 1,200 and 1,600 micromhos/cm (at 25°C) may be considered equivalent to dissolved-solids concentrations of 750 and 1,000 mg/liter.

For Aquatic Life The following criteria are for evaluation of conditions for the maintenance of a well-balanced, warm-water fish population. They are applicable at any point in the stream except for areas immediately adjacent to outfalls. In such areas cognizance will be given to opportunities for the admixture of waste effluents with river water.

1. *Dissolved oxygen* Not less than 5.0 mg/liter during at least 16 hr of any 24-hr period, nor less than 3.0 mg/liter at any time.
2. *pH* No values below 5.0 nor above 9.0, and daily average (or median) values preferably between 6.5 and 8.5.
3. *Temperature* Not to exceed 93°F at any time during the months of May through November, and not to exceed 73°F at any time during the months of December through April.
4. *Toxic substances* Not to exceed one-tenth of the 48-hr median tolerance limit, except that other limiting concentrations may be used in specific cases when justified on the basis of available evidence and approved by the appropriate regulatory agency.

For Recreation The following criterion is for evaluation of conditions at any point in waters designated to be used for recreational purposes, including such water-contact activities as swimming and water skiing:

Bacteria Coliform group not to exceed 1,000 per 100 ml as a monthly average value (either MPN or MF count); nor exceed this number in more than 20 percent of the samples examined during any month; nor exceed 2,400 per 100 ml (MPN or MF count) on any day.

For Agricultural or Stock Watering Criteria are the same as those shown for minimum conditions applicable to all waters at all places and at all times.

F SUGGESTED GRADUATE COURSE OUTLINE

Lecture No.	Topic	Assignment	
		Reading	Problems
1	Introduction	1–10	
2	Wastewater characteristics	227–241	7·3, 7·4, 7·9
3		254–269	7·9
4	BOD, COD, TOC, TOD, ThOD	241–254	7·5, 7·6
5	Physical unit operations	273–295	8·2, 8·5
6		296–321	8·8
7	Chemical unit processes	325–353	9·1, 9·6
8		353–369	9·8, 9·9
9	Design of facilities for physical and	423–446	11·2
10	chemical treatment of wastewater	446–470	11·6, 11·7
11		470–478	11·9
12	Biological unit processes	373–386	10·1
13		386–408	
14		408–419	10·5
15	Design of facilities for biological	481–496	12·2, 12·3
16	treatment of wastewater	496–533	
17		533–542	12·8
18		542–551	12·10
19		551–569	12·11
20	Design of facilities for treatment and	575–591	13·3
21	disposal of sludge	591–613	13·5
22		613–624	13·7
23		624–629	
24	Advanced wastewater treatment	633–653	14·2
25		653–670	14·6
26	Effluent disposal	673–687	15·1
27		687–700	15·2
28		700–703	15·6
29	Advanced wastewater treatment studies	707–717	
30		717–732	16·2

G SUGGESTED UNDERGRADUATE COURSE OUTLINE

Lecture No.	Topic	Assignment	
		Reading	Problems
1	Introduction	1–10	
2	Determination of sewage flowrate	13–33	2·1
3		33–44	2·8
4	Hydraulics of sewers	47–70	3·1, 3·2
5		70–84	3·5
6		84–100	3·8
7	Design of sewers	103–112	4·2
8		112–119	4·3
9		119–128	
10	Sewer appurtenances	131–150	5·1
11	Pumps and pumping stations	183–196	6·1
12		196–202	6·2, 6·3
13	Wastewater characteristics	227–241	7·1
14		241–262	7·3, 7·4
15		262–269	7·6
16	Physical unit operations	273–281	8·1
17		283–295	8·5
18	Chemical unit processes	325–330, 340–343	9·2
19		353–363	9·7
20	Biological unit processes	373–386	10·1
21		386–408	
22		408–419	
23	Design of wastewater treatment facilities	423–424, 441–455, 465–470	11·2, 11·6
24		481–496, 524–539	12·5, 12·6
25		542–567	
26		575–613	13·3
27	Advanced wastewater treatment	633–637	
28	Effluent disposal	673–687	15·2
29		687–703	
30	Review		

H SOME TERMS AND ABBREVIATIONS

Terms

Wastewater is defined by the Joint Editorial Board* as "a combination of the liquid and water-carried wastes from residences, commercial buildings, industrial plants, and institutions, together with any groundwater, surface water, and storm water that may be present." In recent years, wastewater has been used more commonly than the older term *sewage*.

Wastewater collection involves the removal and conveyance of these wastes through underground conduits called *sewers*. Where the sewers carry only the household and industrial wastes, they are called *sanitary sewers*. Those carrying storm water from the roof and street surfaces are known as *storm sewers*, while those carrying both household and industrial wastes and storm water are called *combined sewers*.

Wastewater treatment involves the removal of contaminants and may be accomplished by unit operations or processes, or combinations of operations and processes.

Wastewater disposal involves the ultimate disposal of the liquid portion of the wastewater and of the solids and other materials removed during treatment.

Sewerage, an older term, was often used to describe the facilities used for collection, treatment, and disposal of sewage. The term is declining in use and is now used to describe the piping system that collects and conveys wastewater from source to discharge.

Abbreviations

atm	atmosphere
av	average
bhp	brake horsepower
Btu	British thermal unit
cal	calorie
cfs	cubic feet per second

* Joint Editorial Board representing the American Public Health Association, the American Society of Civil Engineers, the American Water Works Association, and the Water Pollution Control Federation: *Glossary, Water and Wastewater Control Engineering*, 1969.

cm	centimeter
cu ft	cubic foot
fpm	feet per minute
fps	feet per second
ft	foot
ft-lb	foot-pound
g	gram
gal	gallon
gpad	gallons per acre per day
gpcd	gallons per capita per day
gpd	gallons per day
gpm	gallons per minute
hp	horsepower
hr	hour
in.	inch
lb	pound
min	minute
mgad	million gallons per acre per day
mgd	million gallons per day
mg	milligram
mg/liter	milligrams per liter
ml	milliliter
mm	millimeter
ppm	parts per million
psi	pounds per square inch
psia	pounds per square inch absolute
psig	pounds per square inch gage
rpm	revolutions per minute
rps	revolutions per second
scfm	standard cubic feet per minute
sec	second
sq ft	square foot
vol	volume

name index

subject index